Robust Statistical Procedures

WILEY SERIES IN PROBABILITY AND STATISTICS

Established by WALTER A. SHEWHART and SAMUEL S. WILKS

A complete list of the titles in this series appears at the end of this volume

Robust Statistical Procedures

Asymptotics and Interrelations

JANA JUREČKOVÁ
Charles University
Prague, Czech Republic

PRANAB KUMAR SEN
University of North Carolina
Chapel Hill, NC

A Wiley-Interscience Publication
JOHN WILEY & SONS, INC.
New York • Chichester • Brisbane • Toronto • Singapore

This text is printed on acid-free paper.

Library of Congress Cataloging in Publication Data:

Jurečková, Jana, 1940–
Robust statistical procedures : asymptotics and interrelations /
Jana Jurečková, Pranab Kumar Sen.
p. cm. — (Wiley series in probability and mathematical
statistics. Applied probability and statistics section)
"A Wiley-Interscience publication."
Includes bibliographical references.
ISBN 0-471-82221-3 (cloth : alk. paper)
1. Robust statistics. I. Sen, Pranab Kumar, 1937– .
II. Title. III. Series: Wiley series in probability and
mathematical statistics. Applied probability and statistics.
QA276.J87 1995 95-17912
519.5′4—dc20

Printed in the United States of America

10 9 8 7 6 5 4 3 2 1

Dedicated to the memory of Jaroslav Hájek,
our revered teacher and friend

Contents

PART II ROBUST STATISTICAL INFERENCE

Preface

The burgeoning growth of robust statistical procedures over the past thirty years has made this area one of the fastest growing fields in contemporary statistics. Basic concepts of robustness have permeated both statistical theory and practice, creating a genuine need for a full study of their interactions.

Linearity of regression, stochastic independence, homoscedasticity, and normality of errors, typically assumed in a parametric formulation, provide access to standard statistical tools for drawing conclusions about parameters of interest, but there may not be any guarantee that such regularity assumptions are tenable in a given context. Therefore, natural interest turned toward probing into the effects of plausible departures from model-assumptions on the performance characteristics of classical statistical procedure, and this was the genesis of robust statistics. In the assessment of such departures and/or interpretations of robustness, theoretical researchers and applied statisticians have differed, often sharply, resulting in a diversity of concepts and measures of influence functions, among these, breakdown points, qualitative versus quantitative robustness, and local versus global robustness. All of these related ideas can be presented in a logical and unified manner, integrated by a sound theoretical basis, and this is the motivation for the theoretical focus of this book.

Nonparametric procedures are among the precursors of modern robust statistics. In nonparametrics, robustness is generally interpreted in a global sense, while in robust statistics, a relatively local up to an infinitesimal interpretation is more commonly adopted. The classical M-, L-, and R-procedures are the fundamental considerations in this respect; other developments include differential statistical functions, regression quantiles, and regression rank score statistics. Huber's innovation of M-procedures, followed by the contributions of Hampel, revolutionized the field, their excellent treatments of this aspect of robustness are to be found in Huber (1981) and Hampel et al. (1986). Some other books that are not entirely devoted to robustness but contain enlightening applications of specialized viewpoints are those by Shorack and Wellner (1986) and Koul (1992), which place more emphasis on weighted empiricals, and by Serfling (1980), which particularly emphasizes von Mises's statistical functions. The more recent book of Rieder (1994) is a valuable source of mathematical abstractions relating to infinitesimal concepts of robustness and nonparametric optimality based on a least-favorable local alternative approach. Some other contemporary books on robustness have excellent interpretations and sound motivations for real data analysis, and are more applications oriented. Integration of mathematical concepts with statistical interpretations often precludes the exact (finite) treatment of robustness. The incorporation of asymptotics in robustness has eliminated this impasse to a great extent, and sometimes has yielded finite sample interpretations. Yet, while the different types of robust procedures have evolved in individual patterns their ultimate goals are often the same. This fact is reflected by asymptotic relations and equivalences of various (class of) procedures, and there is a need to focus on their interrelations so as to unify the general methodology and channel the basic concepts into a coherent system. This perspective has not fully been explored in contemporary works. Thus it is our primary objective. In this book we attempt to harmoniously blend asymptotics and its interrelations to depict robustness in a proper and broader setup.

A discussion of the organization of this book is deferred to Chapter 1. In general, Part I (Chapters 2 through 7) forms the core of theoretical foundations and deals with asymptotics and interrelations, and Part II (Chapter 8, 9, and 10) is devoted to robust statistical inference, given the general theory and methodology of Part I. Admittedly, Part II has greater statistical depth and is therefore of greater interest to applied statisticians, whereas Part I provides handy resource material for the requisite theoretical background. Still both theoretical statisticians and graduate students in mathematical statistics may desire this integrated view. Although our approach is largely theoretical, mathematical abstractions have been carefully kept at bay. Section 2.5 provides a pool of basic mathematical tools that have been used in later chapters in derivations of main results. To many readers, these tools are familiar in some form or another, while for most applied statisticians, this collection may preclude unnecessary probe into relevant source materials which are often presented at higher levels of abstractions. Our treatment of robustness is by no means complete. For example, there has been little emphasis on linear models with "mixed effects," multivariate linear models, and more general nonlinear models. However, we sincerely hope that the methodology presented here with due emphasis on asymptotics and interrelations will pave the way for further developments on robust statistical procedures in more complex models that are yet to be developed or could not be included in this book for various reasons.

Jana Jurečková
Pranab K. Sen

Prague, Czech Republic
Chapel Hill, NC
October 8, 1995

Acknowledgments

Our work on this book has consumed a period of about ten years.Even before that and continually during the past fifteen years, we have been collaborating with each other and also with a number of colleagues and advisers at Charles Univeristy, Prague, University of North Carolina, Chapel Hill, University of Illinois at Urbana-Champaign, and at Chicago, Hamburg Universität, Mathematical Centre, Amsterdam, Limburgs Universitair Centrum, Belgium, and other academic centers in Australia, Brazil, Denmark, France, Germany, the Czech and Slovak Republics, India, Poland, Russia, and the United States. The impact of such collaboration on this work is overwhelming, and to all our colleagues, we owe a deep sense of gratitude. A good part of this task was completed during the first author's visits to the United States and Canada, and the second author's visits to Prague. These visits were supported by Research Grants from Natural Sciences and Engineering Research Council of Canada, the United States Science Foundation, Office of Naval Research, Cary C. Bosamer Foundation at the University of North Carolina, Charles University Grant #GAUK 365, and Czech Republic Grant #GAČR 201/93/2168. The support of these institutions is gratefully acknowledged.

Ms. Beatrice Shube, the past editor of Wiley-Interscience, has been an inspiration behind this project, and we have had continual support and cooperation from our editors, Margaret Irwin, Kate Roach, and Angela Volan at John Wiley and Sons. Our copyeditor, Dana Andrus, and production manager, Rosalyn Farkas, have been very helpful in bringing this project to a successful completion.

We are grateful to the three reviewers: Rudy Beran, University of California, Berkeley, Robert G. Staudte, and Elena Kulinskaya, both at LaTrobe University, Australia, for their penetrating reading of the manuscript and the penultimate stage and for their useful and constructive suggestions, which we have tried to incorporate to the extent possible in this final version. Finally, we are indebted to Dr. Shubhabrata Das for his commendable task in preparing the LaTex version of this work as well as making numerous suggestions and corrections on the manuscript. Both Pranab Kumar Mandal and Antonio Carlos Pedroso de Lima have helped us in the electronic processing of the LaTex version when Dr. Das and the two authors were separated by geographical barriers; their assistance is greatly appreciated.

J.J.

P.K.S.

Robust Statistical Procedures

Chapter 1

Introduction and Synopsis

1.1 INTRODUCTION

The past three decades have witnessed a phenomenal growth of research literature on *robust statistical inference*, and its implications in *statistical methodology* and *data analysis* pertaining to a broad field of applications. Researchers in social, engineering, physical, clinical, biomedical, public health and environmental, and a variety of other experimental sciences are not only becoming increasingly aware of the power of statistical reasoning (i.e., planning, modeling, and analysis) for their experimental findings, but also they are themselves contributing significantly to further useful developments in statistical sciences (including theory and methodology) by raising pertinent queries on the rationality and appropriateness of the usual statistical methods (and their underlying regularity assumptions) in their experimental setups, where there is ample room for plausible departures from model assumptions in various possible directions. Standard (and mostly parametric) statistical models rest on certain structures (generally referred to as the *deterministic* components) that are camouflaged by *chaos* (chance variations) which are usually referred to as *errors* or the *stochastic* component. For this reason certain regularity assumptions are usually introduced with a dual view toward asserting the degree of uncertainty due to the stochastic component and probing the deterministic component in a valid yet efficient manner. In this respect statistical reasoning is indispensable. Both the *validity* and *efficiency* criteria refer to the general *performance characteristics* of usual statistical reasoning. These characteristics depend as much on the underlying model as on the regularity assumptions surrounding it. From a practical applicational point of view, there is a genuine need to look critically at the stipulated regularity assumptions in order to ensure that they are appropriately used. On the other hand, *model flexibility* generally calls for an increased number of parameters. More parameters generally mean more complexity in the statistical analysis. Ideally the more flexible the basic model is and the less stringent are the regularity assumptions, the more easily it can be adapted. However, one has to pay a price for flexibility: An increased number of model parameters or constraints generally demands a much larger sample size to match a satisfactory level of precision/validity for the statistical conclusions from the experimental outcome.

These conflicting undercurrents have somewhat confounded the strategies of statisticians who generally take a middle-line approach that ensures a satisfactory resolution. Through proper planning (i.e., *designing*) of a study, appropriate *sampling schemes*, and, if necessary, appropriate *transformations* on response variable, often, it is possible to design suitable models that closely match the unknown circumstances without of course ruling out possible departures from the assumed model which may have low significance. Therefore, in model selection, one needs to look critically at the assumptions, and examine the effects of plausible departures from these model based assumptions on the performance characteristics of statistical procedures (i.e., *tests* and *estimators*). A good statistical procedure that is not sensibly affected by "small" departures from the model based assumptions is termed *(locally) robust*; procedures that do not require this "smallness" part of the departures from the model, are termed *globally robust.* This is a very broad interpretation of *robustness.* In effect a robust statistical procedure does not experience a serious change in its performance characteristics if there is a small departure from the model based regularity assumptions. The interpretations of "serious" change as well as "departure" have been subjected to indepth statistical discussion over the past 25 years, often pervaded by mathematical ingenuities, and by statistical and practical foresight. Therefore, it may be quite appropriate for us to launch a thorough (theoretical-cum-methodological) study at the current stage of this continual and profound development of robustness.

To motivate this study of the robustness of statistical inference, let us consider the classical problem of estimating the *location parameter* interpreted as the mean, median or center of a symmetric distribution. Usually one assumes that the distribution associated with the population is normal, with mean θ and variance σ^2 (which are generally both unknown). Then the *maximum likelihood estimator* (MLE) of the population mean θ is the sample mean (e.g., $\bar{X}$), and when $\sigma^2 < \infty$, $\bar{X}$ is an optimal estimator of θ in some well defined manner (as will be established in later chapters). In reality, the distribution (F) may not be strictly normal, and departures from the assumed normality may occur along various avenues. For example, F may be a mixture of two normal distributions - one would be predominant with given mean θ and variance σ^2, and the other one, though insignificant in proportional representation, would have such a profound difference in the mean and/or variance that the performance of the sample mean $\bar{X}$ would become different from that of a single normal distribution. In such a case $\bar{X}$ may lose its optimality, and its efficiency would deteriorate quickly as the divergence of the two mixing normal distributions increases. A similar situation may arise when the actual F has *heavier tails* than a normal distribution. For example, if F is a Laplace, logistic, or even Student's t-distribution, $\bar{X}$ may not retain its optimality properties. An extreme case is the Cauchy distribution for which $\bar{X}$ becomes totally inefficient and inconsistent. It may be quite relevant to allow such departures from the assumed normality (to a smaller or greater extent depending on the context) and to look for alternative estimators that are less sensitive to such departures. In biomedical studies the response variable Y is typically nonnegative and with a positively skewed distribution. Therefore often one can make a transformation $X = g(Y)$, where $g(.)$ is a monotone and continuous function [on $\mathbb{R}_1 = (-\infty, \infty)$]. Typical examples of $g(.)$ are

(1) $g(y) = \log y$ (*log-transformation*);
(2) $g(y) = y^{1/2}$ (*square-root transformation*);
(3) $g(y) = y^{1/3}$ (*cube-root transformation*, appropriate when Y refers to the weight/volume of some organ that behaves with relation to the third power of the diameter X;
(4) $g(y) = y^{\lambda}$ for some $\lambda > 0$.

It may be noted that (2) and (3) are particular cases of (4) when λ is taken as 1/2 and 1/3 respectively. Even (1) is a special case of (4) as may be easily verified by allowing λ to go to 0 and using the L´ Hŏpital rule. It may be noted further that due to the unspecified value of λ there are some additional complications in a practical adoption of (4). A data-adaptive choice of λ entails some complications in a valid use of standard statistical analysis schemes. For this reason, the first three transformations are more commonly used in practice. In the first case Y has a lognormal distribution when X has a normal one. In the other cases X remain a positive random variable, and hence its normality is reasonable only when the ratio of its mean and standard deviation is large. Thus, from theoretical standpoint, assuming X to have a normal law is not always well-founded. Even so, if the underlying d.f. of Y is a mixture of two distributions (both positively skewed), such a transformation may fail to induce normality of X to a satisfactory extent. Is it not therefore reasonable to allow for possible departures from the assumed normal law? The extent of such a departure may depend on the transformation adopted as well as on the original distribution. If instead of the sample mean $\bar{X}$, we consider the sample median $\tilde{X}$, its performance will be fairly insensitive to heavy tails and be more *robust* than the mean $\bar{X}$. On the other hand, for normal F, $\tilde{X}$ is about one-third less efficient than $\bar{X}$. Therefore one may have to compromise between robustness and efficiency of estimators. Fortunately, there are alternative estimators that blend efficiency and robustness more coherently, and we will have the occasions to discuss these estimators in detail in the later chapters. For test statistics this robustness (or sensitivity) may apply to the *size (level of significance)* as well as *power* properties.

Statisticians (and mathematicians) have been receptive to the basic notion of robustness and its effective implementations in statistical methodology. A significant contributor to the development of robust statistical procedures is Peter J. Huber, and his graduate level textbook (1981) provides a good coverage of the basic theory developed during the 1960s and 1970s. At least two other contemporary advanced graduate level monographs, namely, by Serfling (1980) and Sen (1981a) contain some detailed discussions on the asymptotic properties of robust estimators and test statistics for the location as well as simple linear models. More recently the book by Hampel, Ronchetti, Rousseeuw, and Stahel (1985) addresses some of the robustness issues from (mostly) a finite sample point of view. In these developments, the primary focus has been on the location, location-scale, simple regression, and general linear models. In this perspective robustness against plausible departures from the assumed form of the error distribution, *heteroscadasticity* of the errors and lack of *independence* of these error components (*serial dependence, intraclass correlations* etc.), and presence of *outliers* (or *error-contamination*) have all been identified as principal issues. Two other monographs, by Shorack and Wellner (1986) and Koul (1992) on asymptotics of the *(weighted) empirical process*, have touched on robustness primarily from an asymptotic point of view. The recent book by Rieder (1994) is devoted solely to robust asymptotic statistics, but it emphasizes mostly the infinitesimal concept of robustness and nonparametric optimality based on the notion of least favorable local alternatives. There are yet other works dealing with more specialized aspects of robustness; for more on the literature, we refer to Staudte and Sheather (1990) and Rey (1980) where the primary emphasis has been on motivation and practical utility at an introductory to intermediate level of presentation.

A significant development in this general area relates to *Monte Carlo* simulation and *resampling* plan-based studies. Although numerical evidences gathered from such studies are overwhelming, the theoretical and methodological justifications have been acquired solely from the relevant asymptotic theory developed in the past 20 years. Since these developments took place in different models and levels of mathematical sophistication, there is a need for a thoroughly unified treatment of this asymptotic theory. With this motivation, our primary objective is to focus on such theoretical and methodological aspects of asymptotics of robust statistics in a unified and logically integrated manner. In that vein a network of concepts of robustness has been developed here,with due emphasis on *local*, *global*, *qualitative* and *quantitative* robustness aspects. The theory (and methodology) of robust statistics (covering inference and data analysis) has evolved not only by the mathematical ingenuity but also by statistical and data-analytical insights. The primary tools are based on *M-*, *L-*, and *R-statistics* (and estimators), and to them *differentiable sta-*

tistical functions and *minimum-distance* procedures have found their rightful way. Discussions of most of these procedures appear in Huber (1981), Serfling (1980), Sen (1981a), and Shorack and Wellner (1986), among others, although the motivations and general objectives of these monographs differ. *Interrelations* of these diverse statistics (and estimators) deserve a systematic and indepth study even for the simple location or regression models. Nevertheless, the treatment of this important topic has been scant in the monographs, and there has not been much indepth study in any unified form. Consequently we shall focus on these interrelationships beyond the usual first order asymptotics and incorporate such affine relations to explore the robustness properties to a finer extent. In our discussion, asymptotics will play a unifying and vital role.

For sample quantiles the celebrated *Bahadur* (1966) *representation*, provided a novel approach to the study of the asymptotic theory of order statistics, quantile functions, and a broad class of statistical functionals. Some of these developments are reported in Serfling (1980, ch. 2), Sen (1981a, ch. 7) and other contemporary advanced monographs. However, the past two decades have witnessed a phenomenal growth of research literature on Bahadur representations for various statistics (which are typically nonlinear in structure). Bahadur's (1966) own results, as further extended by Kiefer (1967) and supplemented by Ghosh (1971) in a weaker and yet elegant form, led the way to various types of representations for statistics and estimators. Along this direction, we have observed that *first-order asymptotic distributional representations* (FOADR) and *second order asymptotic distributional representations* (SOADR), having their roots in the Bahadur representations, have emerged as very useful tools for studying the general asymptotic theory for a broad class of robust estimators and statistics. Such FOADR and SOADR results also provide a convenient way of assessing the interrelations of various competing estimators or statistics in a meaningful asymptotic setup. In a conventional asymptotic setup generally there are many competing estimators or test statistics which share the same asymptotic properties, and in order to discriminate among such rival estimators or tests, more refined tools such as the SOADR are generally needed. A major objective of this book is to focus on such FOADR and SOADR results for various robust procedures, to exploit their role thoroughly and to incorporate them to present the relevant asymptotics in a unified manner.

It is worth mentioning here that higher order asymptotics have an important role in optimal statistical inference problems even in a broader perspective. Generally the minimization of mean squared errors (or quadratic risks) criteria or other conventional risk measures employed to derive an optimal estimator within a class requires some moment conditions. Such conditions may not be needed for some other alternative criteria, so alternative classes of estimators can be compared by such measures. Among such possibilities, the Pitman (1937) attention in the recent past, and it deserves a special mention. It is known [see for example, Keating, Mason, and Sen (1993) for the single parameter case, and Sen (1994a) for a general multiparameter case] that, in a conventional first-order optimality perspective, PMC and quadratic risk criteria are isomorphic for a broad class of estimators that admit a first-order asymptotic normal representation. The interesting fact is that in a higher-order optimality perspective this asymptotic normality or the isomorphism may not totally serve the purpose. The asymptotic behavior of the residual terms properly normalized provides further clues toward such higher-order comparisons. Therefore the SOADR results which may not necessarily require stringent moment conditions may cast light on the appropriateness and motivation of various risk measures in the study of higher-order efficiency properties. A survey of such higher-order asymptotics in a conventional risk formulation has recently been made by Ghosh (1994). Further the more recent work of Ghosh et al. (1994, 1995), though pursued mostly in a parametric mold, may be regarded as the precursors for deeper studies of higher-order PMC asymptotics that are well anticipated in the near future in the robust inference case as well.

Among other noteworthy developments over the past 15 years, *regression quantiles*, have steadily reshaped the domain and scope of robust statistics. It started with a humble aim of regression L-estimators by Koenker and Bassett (1978) and traversed the court of robustness onto the domain of *regression rank scores*, see Gutenbrunner and Jurečková (1992). In this way it provides a natural link to various classes of robust estimators and strenghtens their interrelationships as well. Moreover *adaptive estimation theory* has also emerged as a vital component, and in this book we will give an adequate treatment of robust adaptive estimation of location and regression parameters. In passing, we refer to the recent monograph of Bickel et al. (1993) which contains a systematic account of adaptive estimation theory, albeit largely in a semi-parametric mold. We will also consider recursive and sequential estimation based on robust statistics. In our coverage of topics and display of (asymptotic) theory and interrelationships, we have attempted to be more comprehensive than in any other source. By making this book mathematically rigorous, theoretically sound, and statistically motivating, we have primarily aimed at the advanced graduate level, for coursework on asymptotic theory of robust statistical inference with special emphasis on the interrelationships among families of (competing) statistics (or estimators). The asymptotic theory is further streamlined to match the needs of practical applications. In this respect it would be helpful for readers to have familiarity with the basic theory of robustness e.g., the introductory discussion in Huber (1981), although we do provide a survey of robust statistics. The reader is of course expected to be familiar with the basic theory of statistical inference and decision theory including estimation theory, hypothesis testing, and classical linear models, at least, at an intermediate level, though a measure-theoretic orientation is not that essential. Advanced calculus, real analysis, linear algebra, and some matrix theory are also important mathematical prerequisites.

1.2 SYNOPSIS

To set the robust statistical inference themes in proper perspective, we present our general treatment of asymptotics and interrelations in Part I, consisting of Chapters 2 through 7. In Part 2 which comprises the rest of the monograph, we deal with statistical inference.

The basic theory and methodology is initiated in Section 2.2. Section 2.3 discusses qualitative, quantitative, local, and global aspects of robustness. *Minimax theory, scale-equivariance,* and ***studentization*** from the robustness perspective are considered in Section 2.4. Results in Sections 2.3, 2.3, and 2.4 are presented without derivations. Likewise some basic results in probability theory and large sample statistical theory are presented without derivations in Section 2.5. The inclusion of this final section of Chapter 2 provides a pool of basic mathematical and (asymptotic) statistical tools. They will be useful in later chapters in the derivations of main results. These results will be familiar to many mathematically oriented readers, but applied statisticians may be releived of probing into the relevant source materials which are often presened at high levels of abstraction.

Chapter 3 provides a basic survey of robust estimators of location and regression parameters, special emphasis on certain families of estimators:

1. Maximum likelihood type or M-estimators
2. Linear functions of order statistics or L-estimators, including Regression quantiles
3. Rank statistics based or R-estimators, with access to Regression rank scores estimators
4 Minimum distance estimators
5. Pitman-type estimators, or P-estimators
6. Bayes-type or B-estimators
7. Statistical functionals

Basic motivations and formulations of these estimators, and their first-order asymptotic representations, are presented in a systematic order. These findings pave way for the study of asymptotic normality, consistency, and other related properties. Many of the results of Chapter 3 are strengthened and presented in a comparatively more general mold in subsequent chapters dealing with individual classes. However, the implications of these first order results are lucidly presented in this chapter. The representation of a robust statistic (which is typically nonlinear) in terms of an average (or sum) of independent, centered r.v.'s, plus a remainder term converging to 0, in probability, as the sample size increases, not only provides a visible access to the related asymptotic normality results but also to other properties, such as breakdown points, influence functions, and robust variance estimation. All of these constitute an important aspect of the contemplated, integrated study of robustness. The genesis of higher-order representations lies in such first-order ones. In this respect the basic insight is due to Bahadur (1966), who formulated a very precise representation for a sample quantile, under very simple regularity conditions; he cleverly revealed the limitations of higher-order asymptotics of robust statistics, and called for the need of "smoothing" to eliminate some of these drawbacks. In our presentation we will examine various robust statistics in the light of Bahadur representations to facilitate a comprehensive study of their intricate interrelationships.

Chapter 4 is devoted to the study of L-estimators (of location and regression parameters). It demonstrates appropriate asymptotic representations for such a (possibly, nonlinear) statistic in terms of a linear statistic and a remainder term, and incorporates this representation in more refined statistical analysis as far as possible. Bahadur representations occupy a focal point in this perspective. For sample quantiles, such results are presented in Section 4.2; extensions to finite mixtures of sample quantiles are embraced in this mold. Some generalizations to possibly nonlinear functions of a (fixed) number of sample quantiles are also discussed in the section. L-statistics with *smooth scores* occupy a prominent place in statistical inference, and a detailed account of these statistics is given in Section 4.3, and some further refinements for general L-estimators are presented in Section 4.4. Because of the affinity of L-statistics to statistical functionals introduced in Section 3.7), we integrate out the findings in Sections 4.2, 4.3, and 4.4 with those of differentiable statistical functionals, and this is presented in Section 4.5. Allied second-order representation results are considered in Section 4.6. L-estimation in linear models constitutes an important topic in robust estimation. In this respect the advent of regression quantiles (with access to trimmed least squares estimation theory) marked the initiation of a novel area of research work, and in section 4.7 we consider an elaborate treatment of this important topic. The relevance of regression quantiles in R-estimation theory is relegated to the last section of Chapter 6.

M-estimators of location, regression, and general parameters are studied in Chapter 5. Unlike the case of L-estimators, these M-estimators are not generally scale equivariant. Moreover M-estimators may be expressed as implicit statistical functionals, based on estimating equations. This in turn may limit the associate score functions to be bounded and piecewise continuous. Although such a condition may often be justified on the ground of bounded influence functions, some alternative theoretical methods are more appealing, and these are presented in a unified manner. *One-step, studentized*, and *adaptive scale equivariant* M-estimation procedures are also given due considerations. The major emphasis on M-estimation theory is on local robustness, and FOADR and SOADR results have important impacts in this context too.

In Chapter 6 results parallel to Chapter 5 are developed for R-estimators of location and regression parameters. Such R-estimators are scale equivariant and enjoy robustness on a global basis. For the location/regression model, such R-estimators can be conveniently computed by some simple iterative method, and in some cases, exact algebraic forms are also available. However, often for general linear models such iterative procedures involve extensive computations. The computational task may of course be simplified by using a preliminary estimator and incorporating a basic *uniform asymptotic linearity* result due to Jurečková (1969, 1971 a,b). This fundamental result is the unifying power in the whole development, and it is treated here with utmost care. The last few years have witnessed some significant developments on rank-based estimation theory, where *regression quantiles* based *regression rank scores* estimators occupy a focal point. Also based on a common score function the regression rank scores estimators and the classical R-estimators of parameters in a linear model are asymptotically stochastically equivalent, see Jurečková and Sen (1993). Robust estimation of scale parameters based on regression rank scores is also treated in this chapter.

For location as well as linear models, R-, L-, and M-estimators enjoy some interrelations. The findings of the earlier chapters enable us to explore these interrelations more thoroughly and systematicall in Chapter 7. There are certain other problems that require some careful examination. Among these are the k-step versions of M-estimators of location and regression parameters; they are treated in the chapter as well. Similar results hold for other types of k-step estimators.

Chapters 8, 9, and 10 comprise Part II. They primarily deal with statistical inference based on robust statistics, covering both (point as well as interval) estimation theory and testing of hypotheses. Asymptotic theory of point estimation in a sequential setup constitute the main theme of Chapter 8. For a large (but nonstochastic) sample size, it is of natural interest to find an optimal robust estimator. Since the underlying distribution F may not be that precisely known, the choice of an optimal score function for L-, M-, or R-estimators (of location/regression) may not be available. There are several options to eliminate this indeterminacy. First, one may attempt to estimate such optimal score functions (which may depend on the unspecified

F in a rather involved way) from the data set itself, and to incorporate such an estimated optimal score function in the formulation of *adaptive estimation*, and in a robust setup, such adaptive estimators are considered in Chapter 8. Alternatively, for a chosen score function and a particular type of estimators (namely, L-, M-, or R-estimators), incorporating the cost of sampling, one may formulate a suitable risk function (which also generally depends on the unknown F). Minimization of this risk (over the sample size n) leads to the *minimum risk estimation* theory. *Sequential procedures* to attain such minimum risks at least in some meaningful asymptotic sense exist, and their robust counterparts are considered in Chapter 8. One of the problems cropping up in such sequential estimation problems is the computational convenience of such nonlinear estimators often requiring iterative solutions. For this reason *robust recursive estimators* are presented in the last section of Chapter 8.

The dual problem of robust confidence set estimation theory is considered in Chapter 9. Confidence sets are used to interpret the margin of fluctuations of the point estimators from their population counterparts. First, fixed-sample-size procedures based on robust estimators and test statistics are considered. In this context to estimate the (asymptotic) variance function in a robust and consistent manner. Suitable *resampling plans*, such as the *jackknife* and the *bootstrap*, can be adopted. But such adaptations are to be judged not only by their consistency but also by their robustness properties. Both jackknifing and bootstrapping are considered in a systematic manner, and their relative merits and faults are studied. It appears that for R- and M-estimators of location/regression, the uniform asymptotic linearity results developed in Part I provide a more convenient alternative to such variance estimation problems. One other problem in confidence set estimation is referred to as the *bounded-width* (or *bounded-diameter*) *confidence interval* (or set) problem. Here,not only we like to have a desired coverage probability for the confidence set but also to limit the width (or diameter) of the confidence interval (or region) to some pre-assigned positive number. If the underlying distribution F is not adequately specified, no fixed sample size procedure exists for this problem, and suitable sequential procedures are sought to achieve the dual goals. Robust sequential confidence sets of this type are considered in Chapter 9, covering both uniparameter and multiparameter problems. Asymptotic interrelations of such robust confidence sets (intervals) are studied along with the asymptotic normality and related properties of *stopping times*. Some of the findings are updated versions of parallel representations in Sen (1981a, ch. 10).

Chapter 10 deals with *robust statistical tests* both in the fixed sample size and sequential setups. With the concepts and tools developed in Part I as well as in the preceding two chapters of Part II, general interpretations of robustness of testing procedures (for statistical hypotheses) are made, with due emphasis on the location/scale and linear models treated in earlier chapters. Tests based on robust estimators as well as robust statistics rank order and M-statistics are considered in detail, along with the classical likelihood ratio type tests (and their variants). *Repeated significance tests* based on such statistics are discussed, with emphasis on their role in applied statistics. Also included are the *tests with power one* for which the general asymptotics presented in Part I play a basic role. Analogues of classical *sequential probability* (and *likelihood*) ratio type tests based on robust statistics are considered for location/scale and regression models, and their performance characteristics are studied in the light of general robustness criteria developed earlier. Our coverage of sequential tests goes beyond Chapter 9 of Sen (1981a). It incorporates additional relatively recent results with extensive discussion of robustness.

In statistical interpretations and dissemination often some mathematical manipulations are unavoidable; one cannot be presented without the incorporation of the other. To make the presentation of the statistical theory smoother, we have relegated proofs of some of the more complex results to the Appendix. These derivations have genuine interest of their own.

A full though by no means exhaustive bibliography is provided at the end of the book with the hope that the references will facilitate additional studies in related problems.

PART I

Asymptotics and Interrelations

There has to be a finite limit to an expectation,
and asymptotics for every statistical resolution.
Can parallel lines retain identity at infinity?
In asymptotics, why then succumb to compatibility?
Why indulge in robustness when a sample is large?
But, outliers seek asylum in an infinitesimal barge!
Why strive to trace the breakdown points' sovereignty,
while error contaminations acquire asymptotic nullity?
With a robust influence function, scan the interrelations,
and let uniform linearity eliminate all the dissolutions.

Chapter 2

Preliminaries

2.1 INTRODUCTION

In this chapter we focus on robust statistical inference for the classical linear models. Section 2.2 presents some basic aspects of the classical theory of statistical inference. The concept of robustness arising in this context is then outlined in Section 2.3. Quantitative as well as qualitative aspects of robustness and local as well as global aspects of robustness are discussed. The minimax theory, scale-equivariance, and studentization are all important features related to robust statistical procedures. Their basic formulation is considered in Section 2.4. In the dissemination of the theory of robust statistical inference, we need to make use of basic results in probability theory as well as large sample theory, and a systematic account of these results (mostly without proofs) is given in Section 2.5. Throughout the chapter, the main emphasis is on motivations (rather than derivations). More technical derivations will be given in subsequent chapters.

2.2 INFERENCE IN LINEAR MODELS

We start with the classical normal theory of linear model in an univariate setup. Let $X_1, \ldots, X_n$ be n (≥ 1) independent random variables (r.v.) with

$$\mathbb{E}X_i = \theta_i, \quad i = 1, \ldots, n; \quad \boldsymbol{\theta}_n = (\theta_1, \ldots, \theta_n)'. \tag{2.2.1}$$

Assume that (1) each X_i has a normal distribution, (2) these normal distributions all have a common (and finite) variance σ^2 (usually unknown), and (3) $\boldsymbol{\theta}_n$ belongs to a p-dimensional subspace $\mathbb{R}_p^\star$ of the n-dimensional Euclidean space $\mathbb{R}_n$, where $n > p$. In the classical *linear model* we have

$$\boldsymbol{\theta}_n = \mathbf{C}_n\boldsymbol{\beta}, \quad \boldsymbol{\beta}_n = (\beta_1, \dots, \beta_p)', \tag{2.2.2}$$

where $\boldsymbol{\beta}$ is a vector of unknown regression parameters and $\mathbf{C}_n$ is a known design matrix (of order $n \times p$). In the particular case where all the θ_i are the same (e.g., equal to μ), we have $\mathbf{C}_n = (1, ..., 1)'$ and $\beta = \mu$ so that $p = 1$. Thus the location model is a particular case of the linear model in (2.2.2). The linear model in (2.2.2) is often expressed as

$$\mathbf{X}_n = (X_1, \dots, X_n)' = \mathbf{C}_n\boldsymbol{\beta} + \mathbf{e}_n; \quad \mathbf{e}_n = (e_1, \dots, e_n)', \tag{2.2.3}$$

where the e_i are independent and identically distributed (i.i.d.) r.v.'s with a normal distribution $\mathcal{N}(0, \sigma^2)$ and $0 < \sigma^2 < \infty$.

For the model in (2.2.1), consider the estimation of $\boldsymbol{\theta}_n$ (or of $\boldsymbol{\beta}$ in the model (2.2.2)). Note that the joint density of the X_i is given by

$$(2\pi)^{-n/2}\sigma^{-n} \exp\left\{ -(2\sigma^2)^{-1}\sum_{i=1}^{n}(X_i - \theta_i)^2\right\}. \tag{2.2.4}$$

The *maximum likelihood estimator* (MLE) of $\boldsymbol{\theta}_n$ may be obtained by maximizing (2.2.4) with respect to $\boldsymbol{\theta}_n$ subject to the constraint that $\boldsymbol{\theta}_n \in \mathbb{R}_p^\star$. This reduces to minimizing (with respect to $\boldsymbol{\theta}_n$)

$$||\mathbf{X}_n - \boldsymbol{\theta}_n||^2 = (\mathbf{X}_n - \boldsymbol{\theta}_n)'(\mathbf{X}_n - \boldsymbol{\theta}_n) \quad \text{when} \quad \boldsymbol{\theta}_n \in \mathbb{R}_p^\star. \tag{2.2.5}$$

The solution to the minimization problem in (2.2.5), denoted by $\hat{\boldsymbol{\theta}}_n$, is the classical *least squares estimator* (LSE) of $\boldsymbol{\theta}_n$. In this setup the X_i are given and the minimization is carried out with respect to the $\boldsymbol{\theta}_i$. For the model in (2.2.3), if we denote the ith row of $\mathbf{C}_n$ by $\mathbf{c}_i'$, for $i = 1, \dots, n$, then $\hat{\boldsymbol{\beta}}_n$, the LSE of $\boldsymbol{\beta}$, is obtained by minimizing

$$||\mathbf{X}_n - \mathbf{C}_n\boldsymbol{\beta}||^2 = (\mathbf{X}_n - \mathbf{C}_n\boldsymbol{\beta})'(\mathbf{X}_n - \mathbf{C}_n\boldsymbol{\beta}) \tag{2.2.6}$$

(with respect to $\boldsymbol{\beta}$), and it is explicitly given by

$$\hat{\boldsymbol{\beta}}_n = (\mathbf{C}_n'\mathbf{C}_n)^{-1}(\mathbf{C}_n'\mathbf{X}_n). \tag{2.2.7}$$

We assume that $\mathbf{C}_n'\mathbf{C}_n$ is of full rank p ($\leq n$); otherwise, a generalized inverse has to be used in (2.2.7). In passing, we may remark that this equivalence of the LSE and the MLE holds for the entire class of *spherically symmetric distributions* (for which the joint density function of $X_1, \dots, X_n$ is a sole function of the norm $||\mathbf{X}_n - \boldsymbol{\theta}_n||^2$). Though homoscedastic normal densities belong to this class, most of the nonnormal densities do not. In general, the MLE and LSE are not the same, and the algebraic form of the MLE is highly dependent on the form of the underlining error distributions.

The *optimality* of the LSE $\hat{\boldsymbol{\theta}}_n$ may be interpreted with respect to various criteria. First, for an arbitrary vector $\mathbf{a}_n = (a_1, \ldots, a_n)'$ of real constants, $\hat{\xi}_n = \sum_{i=1}^n a_i\hat{\theta}_i$ $(= \mathbf{a}_n'\hat{\boldsymbol{\theta}}_n)$ uniformly minimizes the variance among the unbiased estimators of $\xi = \mathbf{a}_n'\boldsymbol{\theta}_n$. Second, an estimator $T_n = T(\mathbf{X}_n)$ is an *equivariant* estimator of ξ , if for every $\mathbf{b}_n = (b_1, \ldots, b_n)' \in I\!R_p^\star$ and all $\mathbf{X}_n$,

$$T(\mathbf{X}_n + \mathbf{b}_n) = T(\mathbf{X}_n) + \mathbf{a}_n'\mathbf{b}_n. \tag{2.2.8}$$

Among the class of all such equivariant estimators, $\hat{\xi}_n$ minimizes the *quadratic risk* $I\!E_\xi(T - \xi)^2$. Note that for the model in (2.2.3), $\hat{\boldsymbol{\beta}}_n$ is a linear function of $\hat{\boldsymbol{\theta}}_n$, and hence these optimality properties are also shared by $\hat{\boldsymbol{\beta}}_n$. There are other optimality properties too.

If we drop the assumptions of normality and independence of $X_1, \ldots,$ X_n and only assume that

$$I\!EX_i = \theta_i, \;\; i = 1, \ldots, n, \boldsymbol{\theta}_n = (\theta_1, \ldots, \theta_n)' \in I\!R_p^\star,$$

$$Cov(X_i, X_j) = \delta_{ij}\sigma^2 \quad \text{for } i, j = 1, \ldots, n; \; \sigma^2 < \infty, \tag{2.2.9}$$

where δ_{ij} the usual Kronecker delta (i.e., $\delta_{ij} = 1$ if $i = j$, and 0, otherwise), then an optimality property of $\hat{\boldsymbol{\theta}}_n$, can still be proved but in a more restricted sense. Among all linear unbiased estimators of $\mathbf{a}_n'\boldsymbol{\theta}_n$, $\mathbf{a}_n'\hat{\boldsymbol{\theta}}_n$ uniformly minimizes the variance. This is provided by the classical Gauss-Markov theorem, and is referred to as the BLUE (*best linear unbiased estimator*) property. In this setup the form of the distribution of the X_i is not that important, but the X_i need to have finite variances. If we want to remove the restriction on linear unbiased estimators and establish the optimality of the LSE in a wider class of estimators, we may need more specific conditions on these distributions. This may be illustrated by the following results of Rao (1959, 1967, 1973) and Kagan, Linnik and Rao (1967, 1973):
Consider the linear model in (2.2.3) sans the normality of the errors. Assume that the e_i are i.i.d. r.v.'s with the distribution function d.f. F, defined on $I\!R$, such that for a positive integer s,

$$\int_{\boldsymbol{R}} x\,dF(x) = 0 \text{ and } \int_{\boldsymbol{R}} x^{2s}\,dF(x) < \infty. \tag{2.2.10}$$

Assume that $\mathbf{C}_n$ is of rank p, and consider p linearly independent linear functions of $\boldsymbol{\beta}$, denoted by $l_1, \ldots, l_p$, respectively. Let $\hat{l}_{n1}, \ldots, \hat{l}_{np}$ be the LSE of $l_1, \ldots, l_p$, respectively. If $\hat{l}_{n1}, \ldots, \hat{l}_{np}$ are optimal estimators of their respective expectations in the class of all polynomial unbiased estimators of order less than or equal to s, then under suitable regularity conditions on $\mathbf{C}_n$, the first $(s + 1)$ moments of F should coincide with the corresponding moments of a normal distribution. Specifically F should coincide with a normal distribution if $\hat{l}_{n1}, \ldots, \hat{l}_{np}$ are optimal in the class of all unbiased estimators. Furthermore Kagan (1970) proved that in the location model, the latter characterization

of the normal law is continuous in the following sense:
Let $X_1, \ldots, X_n$ $(n \geq 3)$ be i.i.d. r.v.'s with the d.f. $F(x-\theta)$ such that $\int_{\mathbf{R}} x dF(x) = 0$ and $\int_{\mathbf{R}} x^2 dF(x) = \sigma^2 < \infty$, and let $\bar{X}_n = n^{-1}\sum_{i=1}^n X_i$, the LSE of θ, be *ϵ-admissible* in the sense that there exists no unbiased estimator $\tilde{\theta}$ of θ satisfying

$$\frac{n\mathbb{E}_\theta(\tilde{\theta}-\theta)^2}{\sigma^2} < \frac{n\mathbb{E}_\theta(\bar{X}_n-\theta)^2}{\sigma^2} - \epsilon, \quad \epsilon > 0. \tag{2.2.11}$$

Then

$$\sup_{x \in \mathbf{R}} |F(x) - \Phi(\frac{x}{\sigma})| \leq K(-\log \epsilon)^{-1/2}, \quad \epsilon > 0, \tag{2.2.12}$$

where $\Phi(.)$ stands for the standard normal d.f. and K $(0 < K < \infty)$ is some constant, independent of ϵ.
For additional characterization results, we may again refer to Kagan et al. (1973). Let us look at the LSE $\hat{\boldsymbol{\beta}}_n$ in (2.2.7). Using (2.2.3), we have

$$\hat{\boldsymbol{\beta}}_n = \boldsymbol{\beta} + (\mathbf{C}_n'\mathbf{C}_n)^{-1}(\mathbf{C}_n'\mathbf{e}_n), \tag{2.2.13}$$

Therefore, when the elements of $\mathbf{e}_n$ are i.i.d. r.v.'s with the normal distribution with 0 mean and a finite positive variance σ^2, we obtain that $(\hat{\boldsymbol{\beta}}_n - \boldsymbol{\beta})$ has the p-variate normal distribution with mean vector $\mathbf{0}$ and dispersion matrix $\sigma^2(\mathbf{C}_n'\mathbf{C}_n)^{-1}$. Further $(\mathbf{X}_n - \mathbf{C}_n\hat{\boldsymbol{\beta}}_n)'(\mathbf{X}_n - \mathbf{C}_n\hat{\boldsymbol{\beta}}_n)/\sigma^2$ has the central chi-square distribution with $n-p$ *degrees of freedom* (DF), independently of $\hat{\boldsymbol{\beta}}_n$. Therefore, using the fact that $(\hat{\boldsymbol{\beta}}_n - \hat{\boldsymbol{\beta}})'(\mathbf{C}_n'\mathbf{C}_n)(\hat{\boldsymbol{\beta}}_n - \hat{\boldsymbol{\beta}})/\sigma^2$ (by the Cochran theorem) has the central chi square distribution with p DF, we obtain from above that

$$\frac{(\hat{\boldsymbol{\beta}}_n - \boldsymbol{\beta})'(\mathbf{C}_n'\mathbf{C}_n)(\hat{\boldsymbol{\beta}}_n - \boldsymbol{\beta})/p}{||\mathbf{X}_n - \mathbf{C}_n\hat{\boldsymbol{\beta}}_n||^2/(n-p)} \tag{2.2.14}$$

has the central variance-ratio $(F-)$ distribution with $(p, n-p)$ DF. If we denote the upper $100\alpha\$ < 1)$ of this F-distribution by $F_{p,n-p,\alpha}$, we obtain from (2.2.14) that

$$\mathbb{P}\Big\{\frac{(\hat{\boldsymbol{\beta}}_n - \boldsymbol{\beta})'(\mathbf{C}_n'\mathbf{C}_n)(\hat{\boldsymbol{\beta}}_n - \boldsymbol{\beta})}{||\mathbf{X}_n - \mathbf{C}_n\hat{\boldsymbol{\beta}}_n||^2} \leq p(n-p)^{-1}F_{p,n-p:\alpha}\Big\} = 1-\alpha. \tag{2.2.15}$$

Now, (2.2.15) provides a *confidence ellipsoid* for $\boldsymbol{\beta}$ with the *confidence coefficient* (or *coverage probability*) $1-\alpha$. Note that (by the Courant theorem)

$$\sup\Big\{\frac{|\mathbf{a}'(\hat{\boldsymbol{\beta}}_n - \boldsymbol{\beta})|}{(\mathbf{a}'(\mathbf{C}_n'\mathbf{C}_n)^{-1}\mathbf{a})^{\frac{1}{2}}} : \mathbf{a} \in \mathbb{R}_p\Big\} = \Big\{(\hat{\boldsymbol{\beta}}_n - \boldsymbol{\beta})'(\mathbf{C}_n'\mathbf{C}_n)(\hat{\boldsymbol{\beta}}_n - \boldsymbol{\beta})\Big\}^{1/2}.$$

Thus (2.2.15) may equivalently be written as

$$\mathbb{P}\Big\{|\mathbf{a}'(\hat{\boldsymbol{\beta}}_n - \boldsymbol{\beta})| \leq \{\frac{p}{(n-p)}F_{p,n-p:\alpha}(||\mathbf{X}_n - \mathbf{C}_n\hat{\boldsymbol{\beta}}_n||^2)(\mathbf{a}'(\mathbf{C}_n'\mathbf{C}_n)^{-1}\mathbf{a})\}^{1/2}$$
$$\text{for every } \mathbf{l} \in \mathbb{R}_p\Big\} = 1-\alpha. \tag{2.2.16}$$

This provides a *simultaneous confidence interval* for all possible linear combinations of the $\boldsymbol{\beta}_j$. If our interest centers only on a subset of q of the p parameters $(\boldsymbol{\beta}_1, \ldots, \boldsymbol{\beta}_p)$ in (2.2.14), in the numerator, we may choose an appropriate quadratic form in these q estimators. It will lead to a parallel form of (2.2.15) or (2.2.16), where $F_{p,n-p:,\alpha}$ has to be replaced by $F_{q,n-p:\alpha}$. Since, for $q < p$, $F_{q,n-p:,\alpha} < F_{p,n-p:,\alpha}$, this would lead to a shorter confidence region (for the given subset of parameters). Like the point estimators, these confidence regions have also some optimal properties under the basic assumption that the e_i are i.i.d. r.v.'s with a normal distribution. The exact distribution theory in (2.2.14) may not hold when the e_i may not have the normal distribution. As a result the coverage probability in (2.2.15) or (2.2.16) may cease to be equal to $1 - \alpha$. The confidence intervals are more vulnerable to any departure from the assumed normality of the errors. For large values of n we may of course have some justifications, as will be discussed later on.

The optimal tests of various (linear) hypotheses in the model (2.2.1) when the X_i are normally distributed are typically based on the MLE or, equivalently, the LSE. Consider the general linear hypothesis :

$$H_0 : \boldsymbol{\theta}_n \in \mathbb{R}_s^{\star\star}, \quad \text{a linear subspace of } \mathbb{R}_p^{\star} \text{ with } s < p. \tag{2.2.17}$$

This formulation covers various hypotheses in the linear model (2.2.3), and it is usually referred to in the literature as the *analysis of variance* (ANOVA) model. Rather than the unrestricted LSE in (2.2.5), we will consider the restricted LSE $\hat{\boldsymbol{\theta}}_n^{\star}$, which leads to a minimization of $||\mathbf{X}_n - \boldsymbol{\theta}_n||^2$ under the restriction that $\boldsymbol{\theta}_n \in \mathbb{R}_s^{\star\star}$. If we define $F_{p-s,n-p:\alpha}$ as in (2.2.14), for the case where the X_i are normally distributed with a common variance $\sigma^2 : 0 < \sigma < \infty$, we have the classical test for H_0 with the *critical region* :

$$\frac{||\hat{\boldsymbol{\theta}}_n^{\star} - \hat{\boldsymbol{\theta}}_n||^2}{||\mathbf{X}_n - \hat{\boldsymbol{\theta}}_n||^2} > \frac{(p-s)}{(n-p)} F_{p-s,n-p:\alpha}. \tag{2.2.18}$$

Since under H_0, $(n-p)(p-s)^{-1}||\hat{\boldsymbol{\theta}}_n^{\star} - \hat{\boldsymbol{\theta}}_n||^2/||\mathbf{X}_n - \hat{\boldsymbol{\theta}}_n||^2$ has the central F-distribution with $(p-s, n-p)$ DF, the *level of significance* (or *size*) of this test is equal to α : $(0 < \alpha < 1)$. Under any departure from the null hypothesis, this statistic has a noncentral F-distribution with DF (p - s,n - p) and a nonnegative noncentrality parameter. Then the probability of the critical region in (2.2.18) is always greater than or equal to α. That is, the test is *unbiased.* Moreover the test is invariant under linear (orthogonal) transformations : $\mathbf{X}_n \to \mathbf{Y}_n = \mathbf{D}_n\mathbf{X}_n$. In fact, it is the most powerful among *invariant* and *unbiased size* α tests of of H_0. While, for the above-mentioned invariance, the normality of the X_i is not that crucial, negation of this normality may nevertheless distort the distribution of *maximal invariants* and take away the most powerful property of the test in (2.2.18). Even the size of the test may be different from α when the X_i are not necessarily homoscadastic and normal. Thus the optimality of the classical ANOVA tests based on the LSE is confined to homoscadastic normal distributions of the X_i, and they may not be robust against any departure from this basic assumption.

To focus on this latter aspect, we may consider the usual linear model in (2.2.3) under the Gauss-Markov setup in (2.2.9). By (2.2.13), whenever the e_i are i.i.d. r.v.'s with a finite positive variance σ^2, under quite general conditions on $\mathbf{C}_n$, the asymptotic (as $n \to \infty$) normality of $n^{1/2}(\hat{\boldsymbol{\beta}}_n - \boldsymbol{\beta})$ may be proved. For example, if we have a positive definite (p.d.) matrix $\mathbf{Q}$, such that

$$\lim_{n\to\infty} n^{-1}(\mathbf{C}_n'\mathbf{C}_n) = \mathbf{Q}, \tag{2.2.19}$$

then the multivariate central limit theorem in Section 2.5 may be directly adapted to establish this asymptotic normality. The condition in (2.2.19) may be replaced by a Noether-type condition that as $n \to \infty$,

$$\max\{\mathbf{c}_i'(\mathbf{C}_n'\mathbf{C}_n)^{-1}\mathbf{c}_i : 1 \leq i \leq n\} \to 0, \tag{2.2.20}$$

along with a milder growth condition on $tr(\mathbf{C}_n'\mathbf{C}_n)$. We leave some of these as exercises (Problems 2.2.1 and 2.2.2). Thus we have under (2.2.19),

$$n^{1/2}(\hat{\boldsymbol{\beta}}_n - \boldsymbol{\beta}) \xrightarrow{\mathcal{D}} \mathcal{N}_p(\mathbf{0}, \sigma^2\mathbf{Q}^{-1}) \quad \text{as } n \to \infty. \tag{2.2.21}$$

For the d.f. F of the e_i, we may define the *Fisher information* $I(f)$ as

$$I(f) = \int_{\boldsymbol{R}} \Big(\frac{f'(x)}{f(x)}\Big)^2 dF(x) \quad \text{(assumed to be finite)}\ , \tag{2.2.22}$$

and use the classical *Cramér-Rao inequality* (see Problem 2.2.3):

$$\sigma^2 \geq [I(f)]^{-1}, \tag{2.2.23}$$

where the strict equality sign holds only when $f'(x)/f(x) = kx$ a.e., for some nonzero finite constant k. This is the case when F is itself a normal d.f. Instead of the LSE, if we would have used the MLE based on the true density f then we would have (2.2.21) with σ^2 replaced by $I(f)^{-1}$ (see Problem 2.2.4). Clearly the true model MLE is asymptotically efficient, and the LSE is asymptotically efficient if and only if (*iff*) the errors e_i are i.i.d. and normally distributed. Under the same setup it can be shown (see Problem 2.2.5) that

$$\frac{||\mathbf{X}_n - \mathbf{C}_n\hat{\boldsymbol{\beta}}_n||^2}{(n-p)} \xrightarrow{\mathcal{P}} \sigma^2 \quad \text{as } n \to \infty, \tag{2.2.24}$$

so that (2.2.14) holds with the F-distribution being replaced by the central chi square d.f. with p DF. Thus (2.2.15) and (2.2.16) hold with the only change that $F_{p,n-p:\alpha}$ is replaced by $F_{p,\infty:\alpha}$ $(= p^{-1}\chi^2_{p,\alpha})$. By (2.2.23) and the shortness-criterion of the diameters of the confidence ellipsoids, we conclude that the LSE-based confidence region is asymptotically efficient iff the errors e_i are i.i.d. and normally distributed. Similarly, using (2.2.21) and (2.2.24), it can be shown (see Problem 2.2.6) that for the restricted LSE $\hat{\beta}_n^\star$ (under the model in H_0),

$$\frac{\|\hat{\boldsymbol{\theta}}_n^\star - \hat{\boldsymbol{\theta}}_n\|^2 (n-p)}{\|\mathbf{X}_n - \mathbf{C}_n \hat{\boldsymbol{\beta}}_n\|^2} \xrightarrow{\mathcal{D}} \chi^2_{p-s}, \tag{2.2.25}$$

and that its *Pitman efficacy* (to be defined more precisely in a later section) with respect to local alternatives is proportional to σ^{-2}. For the true model MLE-based *likelihood ratio test*, we have the same asymptotic chi square d.f. under H_0 and the corresponding Pitman efficacy is proportional to $I(f)$ (see Problem 2.2.7). Hence, in the light of the conventional Pitman efficiency , the test in (2.2.18) is asymptotically optimal iff the underlying d.f. is normal.

In the above discussion, some of the derivations are left out in the form of problems, set at the end of this chapter. These derivations require some of the probabilistic tools discussed in Section 2.5, and hence the reader is suggested to look into that section before attempting to solve these problems.

2.3 ROBUSTNESS CONCEPTS

In Section 2.2 we have noticed that the performance characteristics of the LSE-based inference procedures are rather sensitive to possible variations from assumed normal distributions of the errors, so the LSE is *distributionally non-robust*, i.e. *non-robust* with respect to deviations from the assumed (normal) distribution. The same criticism may be made against the MLE based statistical inference procedures when the assumed and the true distributions of the errors are not the same. This brings us to feel intuitively that if we are not sure of the underlying model (e.g., the form of the error d.f.), we should rather use procedures that do not show this kind of sensitivity (even if they are not quite optimal under the model) or, otherwise speaking, that are (distributionally) *robust*.

These ideas are not new. The historic studies by Stigler (1973, 1980) reveal that even the earlier statisticians were aware of the consequences of using an incorrect model in a statistical inference procedure. Pearson (1931) observed the sensitivity of the classical ANOVA procedures to possible departures from the assumed normal model (mostly, in the form of skewness and kurtosis), and scores of papers appeared over the next 20 years (mostly in *Biometrika*) examining the effect of such deviations on the size and power of the classical ANOVA tests. The term *robustness* was first used by Box (1953). We should also mention the work of the Princeton Statistical Research Group incorporated in the late 1940s (under the leadership of J. W. Tukey) which tried to find robust alternatives of the classical methods then in practice.

Though it is intuitively clear what robustness should be, its definition did not evolve in a universal way. Even today, there exist several (related) mathematical concepts of robustness. This should not be considered a shortcoming; rather, it illustrates the basic fact that there are diverse aspects of robustness and, hence, diverse criteria as well. Box and Anderson (1955) argued that a good statistical procedure should be insensitive to changes not involving the parameters (to be estimated) or the hypothesis (to be tested), but should be *effective* in being sensitive to the changes of parameters to be estimated or hypotheses being tested. This idea is generally accepted. It has the natural implications of insensitivity to model departures and sensitivity to good performance characteristics under the model. At this point it may be useful to elucidate this diverse general framework.

1. *Local and global robustness.* In a local model we allow only small changes. For example, we may assume that the errors are normally distributed according to a given variance, but this distribution is locally *contaminated* by a heavier tailed distribution, a case that may arise in the *outlier* or *gross error model.* In a global sense we admit that the real distribution is not precisely known and can be a member of a large set of distributions. For example, the error d.f. may be assumed to be a member of the class of all symmetric d.f.'s (where the assumption of finite variance or Fisher information may or may not be imposed). In a local model we are naturally interested in retaining the performance optimality of a procedure (for the given model) to a maximum possible extent, allowing its insensitivity only to small departures from the assumed model. In the global case *validity-robustness* of a procedure (over the entire class of models contemplated) dominates the picture, and there are also issues relating to typical measures of optimality or desirability of a procedure for a broader class of models.

2. *Specific property-based robustness.* We try to determine which special property of the procedure should be given priority in the robustness picture. For example, we could try to limit the sensitivity of the bias or the mean square error of the estimator (as well as its asymptotic versions) or some other characteristic of the distribution of this estimator. For testing procedure we may similarly emphasize the size, power or some other performance characteristic and seek robustness specifically for it.

3. *Type of model departures.* We must specify what kinds of deviations we have in mind. One possibility is the deviations in the shape of the error distribution, discussed so far. It is equally possible to use some other measures to depict such departures. Other assumptions, such as the independence of the errors, should also be given consideration in determining the scope of robustness.

We now turn to some basic concepts of robustness in estimation for a simple model. Other illustrations can be found in Huber (1981) and Hampel et al. (1985). Consider the simple model where $X_1, \ldots, X_n$ are i.i.d.r.v.'s with a d.f. G generally unknown. We have a *parametric model* $\{F_\theta : \theta \in \Theta\}$ formed by a dominated system of distributions and wish to estimate θ for which F_θ

is as close to G as possible. Let $T_n = T_n(X_1, \ldots, X_n)$ be an estimator of θ which we express as a functional $T(G_n)$ of the *empirical* (sample) d.f. G_n of $X_1, \ldots, X_n$. In this context, we naturally let T_n be a *Fisher-consistent estimator* of θ ; that is $T(F_\theta) = \theta$, $\theta \in \Theta$. We assume that T_n has high efficiency of some kind (such as, its risk or (asymptotic) variance if it is reasonably small). First consider the *local* (infinitesimal) concept of robustness, introduced by Hampel (1968, 1971). Denote by $\mathcal{L}_F(T_n)$ the probability distribution (law) of T_n where the true d.f. of X_1 is F. The sequence $\{T_n\}$ of estimators or test statistics is called *qualitatively robust* at $F = F_0$ if the sequence $\{\mathcal{L}_F(T_n)\}$ is equicontinuous at F_0 with respect to the Prokhorov metric Π on a probability space as will be be considered in Section 2.5. In other words, the sequence $\{T_n\}$ is called qualitatively robust if for each $\epsilon > 0$, there exist an $\delta > 0$ and an integer n_0 such that for any d.f. F and all $n \geq n_0$,

$$\Pi(F, F_0) < \delta \Rightarrow \Pi(\mathcal{L}_F(T_n), \mathcal{L}_{F_0}(T_n)) < \epsilon. \tag{2.3.1}$$

If $T_n = T(G_n)$, where (as before) G_n is the empirical d.f. of $X_1, \ldots, X_n$, then the definition in (2.3.1) conveys the continuity of $T(.)$ with respect to the weak topology metricized by the Prokhorov metric, or some other one (e.g., the Levi metric). The weak continuity of T_n at G (here G is the true d.f. of X_1) and its consistency at G in the sense that $T_n \to T(G)$ almost surely (a.s.) as $n \to \infty$ characterize the robustness of T_n in a neighborhood of G. Hampel (1968, 1974) introduced the concept of *influence curve* (IC) $IC(x; G, T)$ whose value at the point x ($\in \mathbb{R}_1$) is equal to the directional derivative of $T(G)$ at G in the direction of the one-point distribution function $\delta_x(t)$:

$$IC(x; G, T) = \lim_{\epsilon \downarrow 0} \{\epsilon^{-1}[T((1-\epsilon)G + \epsilon\delta_x) - T(G)]\}, \quad x \in \mathbb{R}_1 \tag{2.3.2}$$

where for every $t, x \in \mathbb{R}_1$,

$$\delta_x(t) = 0 \text{ or } 1 \quad \text{according as } t \text{ is} \leq x \text{ or not.} \tag{2.3.3}$$

By taking $IC(x)$ as a measure of sensitivity of T to the single point x, we can restrict ourselves to functionals with bounded influence curves. However, not every qualitatively robust functional has a bounded IC. A simple example is the R-estimator of location (i.e., the center of symmetry θ of the d.f. $G(x) = F(X - \theta)$) based on the classical normal-scores signed-rank statistic:

$$T_n = \frac{1}{2}(T_n^- + T_n^+), \tag{2.3.4}$$

where

$$T_n^- = \sup\{t : S_n(\mathbf{X}_n - t\mathbf{1}) > 0\}, \; T_n^+ = \inf\{t : S_n(\mathbf{X}_n - t\mathbf{1}) < 0\}, \tag{2.3.5}$$

and

$$S_n(\mathbf{X}_n - t\mathbf{1}) = \sum_{i=1}^{n} \text{sign}(X_i - t) a_n(R_{ni}^+(t)), \tag{2.3.6}$$

where $R_{ni}^{+}(t)$ is the rank of $|X_i - t|$ among $|X_1 - t|, \ldots, |X_n - t|$, for $i = 1, \ldots, n$, and $a_n(k)$ is the expected value of the kth smallest order statistic in a sample of size n drawn from the chi distribution with 1 DF, for $k = 1, \ldots, n$. We denote by $\Phi(.)$ the standard normal d.f. Then, if F possesses an absolutely continuous symmetric density function f with a finite Fisher information $I(f)$, defined by (2.2.22), we have (see Problem 2.3.1)

$$IC(x; F, T) = \Phi^{-1}(F(x))[\int \{-f'(y)\}\Phi^{-1}(F(y))dy]^{-1}, \quad x \in \mathbb{R}_1 \qquad (2.3.7)$$

so that as $x \to \pm\infty$, $IC(x)$ goes to $\pm\infty$. We will deal with such R-estimators in detail in Chapter 6.

Hampel (1986, 1974) introduced the terminology *gross error sensitivity* of T_n at F by defining it as

$$\gamma^{\star} = \gamma^{\star}(T, F) = \sup\{|IC(x; F, T)| : \ x \in \mathbb{R}_1\}. \qquad (2.3.8)$$

Note that for unbounded influence functions, (2.3.8) is not usable, and that also for either bounded or unbounded influence functions, it fails to take into account an average picture but rather depicts the extreme case. Perhaps because of these shortcomings, he also proposed to look for an estimator that minimizes the asymptotic variance at the assumed model subject to a prescribed bound on the gross error sensitivity (and to the Fisher consistency as well). This way the gross error sensitivity is given some importance in the choice of an optimal robust estimator, though the main emphasis is still on the model-based optimality of the estimator. The solution of such a minimization problem is an estimator asymptotically efficient among the qualitatively robust (or infinitesimally robust) estimators with influence curve bounded by a given constant. One may find it more realistic to look for an estimator that is reasonably efficient not only in an infinitesimal neighborhood but also in a compact neighborhood of the given model. However, such a solution can only rarely be attempted, except for some symmetric and invariant models. The relevant estimators are optimal among the equivariant estimators as are the tests among the invariant tests.

Let us illustrate this point with a simple example of the classical location model : $X_1, \ldots, X_n$ are i.i.d. r.v.'s with the d.f. $G(x) = F_0(x - \theta)$, defined on $\mathbb{R}_1$. Since our risk functions must be invariant to the translation, we can restrict our considerations to estimators that are *translation-equivariant*; namely they should satisfy

$$T_n(X_1 + c, \ldots, X_n + c) = T_n(X_1, \ldots, X_n) + c, \ \ \forall c \in \mathbb{R}_1. \qquad (2.3.9)$$

The neighborhood of the model corresponding to the given model with the d.f. F_0 can be represented by one of the following cases:

(1) *Model of ϵ-contaminacy.* The neighborhood is defined by

$$\mathcal{P}_\epsilon = \{F : \ F = (1 - \epsilon)F_0 + \epsilon H, \ H \in \mathcal{H}\}, \quad \epsilon > 0, \qquad (2.3.10)$$

where $\mathcal{H}$ is some set of distribution functions , defined on $\mathbb{R}_1$.

(2) The *Kolmogorov-neighborhood* of F_0 is defined by

$$\mathcal{P}_\epsilon = \{F : \sup_{x\in\boldsymbol{R}} |F(x) - F_0(x)| < \epsilon\}, \quad \epsilon > 0. \tag{2.3.11}$$

For the finite sample case Huber (1968) suggested the following criterion of performance of an estimator T_n : For some fixed $a > 0$, set

$$S(T;a) = \sup_{\mathcal{P}_\epsilon}\{\max[\mathbb{P}\{T_n < \theta - a\}, \mathbb{P}\{T_n > \theta + a\}]\}. \tag{2.3.12}$$

The estimator T_n is called *minimax robust* if it minimizes $S(T;a)$ over the class of translation-equivariant estimators.

Another possible criterion is the *minimax asymptotic bias* or the *minimax asymptotic variance of the estimator* over the family $\mathcal{P}_\epsilon$. This approach, first considered by Huber (1964), depends on the symmetry of the model and requires regularity conditions that ensure the asymptotic normality of the estimators under consideration. In admitting that $100\epsilon\%$ of the observations come from a contaminating population, this approach is more realistic than that of qualitative robustness. These two performance characteristics can be reconciled by (2.3.12) under quite general conditions when n is large if we allow a to depend on n (e.g., $a = n^{-1/2}c$, for some fixed $c > 0$). Some further results along this line having relevance to our subsequent studies are briefly presented in the following section.

Some authors, e.g., Huber-Carol (1970); Jaeckel (1971); Beran (1977); Rieder (1978, 1979, 1980, 1981, 1994), have considered similar criteria but looked for a *minimax solution* over a *shrinking neighborhood* of the model [in the sense that in (2.3.10) or (2.3.11), ϵ is made to depend on n, and it converges to 0 as $n \to \infty$, e.g., $\epsilon = \epsilon_n = O(n^{-1/2})$]. In addition to the possibility of using models in (2.3.10) and (2.3.11), they have considered the possibility of representing such shrinking neighborhoods by means of the *Hellinger* or the *total variation* metrics. Although motivations for such an approach may be obtained from *Pitman efficiency* (as will be shown later), a return to the infinitesimal approach may yield more robust results as Hampel (1968, 1974) has demonstrated.

Hampel (1971), aware of the shortcomings of the infinitesimal approach, introduced the concept of the *breakdown point* as a global measure of robustness; the breakdown point of T_n with respect to the d.f. F is defined as

$$\begin{aligned} \epsilon^\star \;=\; \sup_{0<\epsilon\leq 1} \Big\{ \epsilon : &\ \exists \text{ a compact subset } K(\epsilon) \text{ of the parameter space} \\ &\text{s. t. } \Pi(F,G) < \epsilon \Rightarrow \lim_{n\to\infty} 2\mathbb{P}_G\{T_n \in K(\epsilon)\} = 1 \Big\}, \end{aligned} \tag{2.3.13}$$

where $\Pi(\cdot)$ is defined in (2.3.1). The breakdown point is then the largest possible fraction of the population F, which is replaced by arbitrary data but still allows the estimator to be informative for the parameter. Hampel emphasized that a good estimator should be qualitatively robust, have a bounded influence function and have a high breakdown point (ideally equal to 1/2). With this in view we present the following concepts.

2.3.1 Finite-sample breakdown and tail-performance

Donoho and Huber (1983) introduced a finite-sample version of the breakdown point that is an extension of the original idea of Hodges (1967). Jurečková (1981) introduced a finite sample measure of tail behavior of an estimator of location, which was then extended to the linear regression in He et al. (1990). These two finite sample measures of robustness, though defined differently, appear to be very close to each other, and this is a point that we will illustrate in more detail.

Consider the linear regression model

$$Y_i = \mathbf{x}_i'\boldsymbol{\beta} + \mathbf{e}_i, \quad \mathbf{x}_i \in \mathbb{R}_p, \ i = 1, \ldots, n, \tag{2.3.14}$$

and denote by $\mathbf{Z}$ the set of n data points, i.e.

$$\mathbf{Z}' = \left(\mathbf{z}_1' \ldots, \mathbf{z}_n'\right), \quad \mathbf{z}_i = (\mathbf{x}_i', y_i) = (x_{i1}, \ldots, x_{ip}, y_i), \tag{2.3.15}$$

$i = 1, \ldots, n$. Let $\mathbf{T}_n = \mathbf{T}_n(\mathbf{Z})$ be an estimator of $\boldsymbol{\beta}$. Consider all possible contaminated samples $\mathbf{Z}'$ that are obtained from $\mathbf{Z}$ by replacing any m of the original points $\mathbf{z}_1, \ldots, \mathbf{z}_n$ by arbitrary values. Then the *finite sample breakdown point* $m_n^\star(\mathbf{T}, \mathbf{Z})$ of estimator $\mathbf{T}_n$ is defined as

$$m_n^\star(\mathbf{T}, \mathbf{Z}) = \min\{m : \sup_{\mathbf{Z}'} \|\mathbf{T}_n(\mathbf{Z}') - \mathbf{T}_n(\mathbf{Z})\| = \infty\}. \tag{2.3.16}$$

Somc authors instead call the ratio $\epsilon_n^\star(\mathbf{T}, \mathbf{Z}) = m_n^\star(\mathbf{T}, \mathbf{Z})/n$ a breakdown point. For instance, if $\mathbf{T}$ is the least squares estimator, we can easily see that $m_n^\star(\mathbf{T}, \mathbf{Z}) = 1, \forall \mathbf{Z}$. On the other hand, assuming that the errors $e_1, \ldots, e_n$ are i.i.d. with a joint d.f. F such that

$$0 < F(x) < 1, \quad \forall x \in \mathbb{R}_1, \tag{2.3.17}$$

we can define the $B(a, \mathbf{T}_n) : \mathbb{R}_1^+ \to \mathbb{R}_1^+$,

$$B(a, \mathbf{T}_n) = \frac{-\log \mathbb{P}_{\boldsymbol{\beta}}(\max_i |\mathbf{x}_i'(\mathbf{T}_n - \boldsymbol{\beta})| > a)}{-\log\left(1 - F(a)\right)}, \quad a \in \mathbb{R}_1^+. \tag{2.3.18}$$

Naturally, for a reasonable estimator $\mathbf{T}_n$,

$$\lim_{a \to \infty} \mathbb{P}_{\boldsymbol{\beta}}(\max_{1 \le i \le n} |\mathbf{x}_i'(\mathbf{T}_n - \boldsymbol{\beta})| > a) = 0. \tag{2.3.19}$$

We are interested in estimators for which this rate of convergence is as fast as possible. Since we are not able to control $B(a, \mathbf{T}_n)$ for all $a \in (0, \infty)$, we consider as a *measure of tail behavior* of $\mathbf{T}_n$: $\lim_{a\to\infty} B(a, \mathbf{T}_n)$ if it exists, and eventually $\limsup_{a\to\infty} B(a, \mathbf{T}_n)$ and $\liminf_{a\to\infty} B(a, \mathbf{T}_n)$. If the estimator $\mathbf{T}_n$ is such that there exist at least one positive and at least one negative residuals $y_i - \mathbf{x}_i'\mathbf{T}_n$, $i = 1, \ldots, n$, then we can easily verify that

$$\bar{B} = \limsup_{a\to\infty} B(a, \mathbf{T}_n) \leq n. \tag{2.3.20}$$

In other words, $\mathbb{P}_{\boldsymbol{\beta}}(\max_i |\mathbf{x}_i'(\mathbf{T}_n - \boldsymbol{\beta})| > a)$, will tend to zero at most n times faster than $\mathbb{P}(e_1 > a)$ as $a \to \infty$. We will refer to distributions satisfying

$$\lim_{a\to\infty} \Big[-(ca^r)^{-1} \log\{1 - F(a)\}\Big] = 1, \tag{2.3.21}$$

for some $c > 0$, $r > 0$, as to exponentially tailed or to distributions of type I, and we will refer to distributions satisfying

$$\lim_{a\to\infty} \Big[-(m \log a)^{-1} \log\{1 - F(a)\}\Big] = 1, \tag{2.3.22}$$

for some $m > 0$, as algebraically tailed or type II. It turns out that the finite-sample breakdown point is just in correspondence with the tail behavior of $\mathbf{T}_n$ for distributions of type II. Let us illustrate it on the location model. We will remark on the regression model in connection with the LSE and later in connection with some specific estimators.

In the location model

$$Y_i = \theta + e_i, \quad i = 1, \ldots, n, \tag{2.3.23}$$

we have $\mathbf{X} = \mathbf{1}_n$ and $\boldsymbol{\beta} = \theta$. Hence

$$B(a, T_n) = \frac{-\log \mathbb{P}_\theta(|T_n - \theta| > a)}{-\log\{1 - F(a)\}}, \quad a \in \mathbb{R}_1^+. \tag{2.3.24}$$

A close link between tail performance and breakdown point for large class of estimators is illustrated in the following theorem.

Theorem 2.3.1 *Suppose that $T_n(Y_1, \ldots, Y_n)$ is a location equivariant estimator of θ such that T_n is increasing in each argument Y_i. Then T_n has a universal breakdown point $m^\star = m_n^\star(T)$, and for any symmetric, absolutely continuous F such that $0 < F(x) < 1\ \forall x \in \mathbb{R}_1$ and such that*

$$\lim_{a\to\infty} \frac{-\log\big(1 - F(a + c)\big)}{-\log\big(1 - F(a)\big)} = 1 \quad \forall c \in \mathbb{R}_1, \tag{2.3.25}$$

it holds that

$$m^\star \leq \liminf_{a\to\infty} B(a, T_n) \leq \limsup_{a\to\infty} B(a, T_n) \leq n - m^\star + 1. \tag{2.3.26}$$

Theorem will be proved with the aid of the following lemma:

Lemma 2.3.1 *Let $Y_{(1)} \leq \ldots \leq Y_{(n)}$ be the order statistics corresponding to $Y_1, \ldots, Y_n$. Then, under the conditions of Theorem 2.3.1, (i) $m^\star$ is universal, and (ii) there exists a constant A such that*

$$Y_{(m^\star)} - A \leq T_n \leq Y_{(n-m^\star+1)} + A. \tag{2.3.27}$$

Proof. (i) It suffices to show that, if $m^\star(T_n, \mathbf{Y}) = m$ for an arbitrary vector $\mathbf{Y} = (Y_1, \ldots, Y_n)$, then $m^\star(T_n, \mathbf{Y}^0)$ when $\mathbf{Y}^0 = \mathbf{0}$. Set $\mathbf{C} = \max_{1 \leq i \leq n} |Y_i|$. If $m^\star(T_n, \mathbf{Y}) = m$ and T_n is translation equivariant then (a) $\forall B > 0$, there exists a contaminated sample $\mathbf{Z} = (\mathbf{V}, \mathbf{Y}')$, consisting of m new elements $\mathbf{V}$ and n-m old elements from $\mathbf{Y}$ such that $|T_n(\mathbf{Z})| > B + C$ and (b) there exists $B > 0$ such that for all contaminated samples $\mathbf{Z} = (\mathbf{V}, \mathbf{Y}')$ with $(m-1)$ new elements V and $n - m + 1$ old elements, $|T_n(\mathbf{Z})| < B - C$.
Let $\mathbf{Z}$ satisfy (a) for fixed B and $T_n(\mathbf{Z}) > 0$. Then, by equivariance, $T_n(\mathbf{Z} - C) > B$, and hence by monotonicity $T_n(\mathbf{V} - C, \mathbf{0}) > B$. Similarly

$$T_n(\mathbf{Z}) < 0 \Rightarrow T_n(\mathbf{Z} + C) < -B \Rightarrow T_n(\mathbf{V} - C, \mathbf{0}) < -B.$$

A similar argument shows that if for all perturbations of $(m-1)$ elements of $(m-1)$ elements of $\mathbf{Y}$ is $|T_n(\mathbf{Z})| < B - C$, then for all perturbations of $(m-1)$ elements of $\mathbf{Y}^0 = (0, \ldots, 0)$ is $|T_n(\mathbf{Z})| < B$.
(ii) Set $m = m^\star$ and let R_i be the rank of Y_i among $Y_1, \ldots, Y_n$; then, by equivariance,

$$\begin{aligned} T_n(Y_1, \ldots, Y_n) &= T_n(Y_1 - Y_{(m)}, \ldots, Y_n - Y_{(m)}) + Y_{(m)} \\ \geq Y_{(m)} &+ T_n((Y_1 - Y_{(m)})\mathbb{I}[R_1 \leq m], \ldots, (Y_n - Y_{(m)})\mathbb{I}[R_n \leq m]) \end{aligned}$$

where only $(m-1)$ arguments of T_n are nonzero in the last expression. By the definition of $m^\star$, $|T_n(0, \ldots, 0)|$ with only $(m^\star - 1)$ zeros by outliers is bounded, say by A; hence $T_n \geq Y_{(m)} - A$. Similarly we obtain the other inequality. □
Proof of Theorem 2.3.1. By Lemma 2.3.1,

$$\begin{aligned} \mathbb{P}_\theta(T_n - \theta > a) &= P_0(T_n > a) \geq P_0(Y_{(m)} > a + A) \\ &\geq P_0(Y_1 > a + A, \ldots, Y_{n-m+1} > a + A) \\ &= (1 - F(a + A))^{n-m+1}, \end{aligned}$$

hence

$$\begin{aligned} \limsup_{a\to\infty} B(a, T_n) &\leq \limsup_{a\to\infty} \frac{-\log 2\mathbb{P}_0(T_n > a)}{-\log(1 - F(a))} \\ &\leq \lim_{a\to\infty} \frac{\log 2 + (n - m + 1)\log(1 - F(a + A))}{\log(1 - F(a))} \\ &= n - m + 1 \end{aligned} \tag{2.3.28}$$

Conversely,

$$\begin{aligned}
\mathbb{P}_0(T_n > a) &\leq \mathbb{P}_0(Y_{(n-m+1)} > a - A) \\
&= n\binom{n-1}{m-1}\int_{a-A}^{\infty} F^{n-m}(x)\big(1-F(x)\big)^{m-1}dF(x) \\
&\leq n\binom{n-1}{m-1}\int_{F(a-A)}^{1}(1-u)^{m-1}du \\
&= \binom{n}{m}\big(1-F(a-A)\big)^m, \qquad (2.3.29)
\end{aligned}$$

and hence

$$\liminf_{a\to\infty} B(a, T_n) \geq \lim_{a\to\infty} \frac{\log\binom{n}{m^\star} + m^\star \log\big(1 - F(a-A)\big)}{\log\big(1 - F(a)\big)} = m^\star$$

□

Remark. Note that Theorem 2.3.1 holds for both types I and II distributions. If T_n has high breakdown points, like the sample median $m^\star = n/2$, then

$$\frac{n}{2} \leq \liminf_{a\to\infty} B(a, T_n) \leq \limsup_{a\to\infty} B(a, T_n) \leq \frac{n}{2} + 1,$$

for both types of distributions. Hence $\mathbb{P}_\theta(|T_n - \theta| > a)$ tends to zero as $a \to \infty$ at least $(n/2)$times faster than the tails of the underlying error distributions.

The estimators of location parameter can be affected only by outliers in outcomes $\mathbf{Y}$. For this reason there are many robust estimators of location with good breakdown and tail-behavior properties.

In the regression model (2.3.14), there may also be outliers among the rows of matrix $\mathbf{X}$. In fact, an influential (*leverage*) point $\mathbf{x}_i$ can completely change the direction of the regression hyperplane. There are various regression diagnostics methods to identify the leverage points, but it is generally more difficult to identify outliers in $\mathbf{X}$ than those in $\mathbf{Y}$. Mathematically the concept of leverage point is not precisely defined in the literature. While the most influential points in LS regression are the rows of $\mathbf{X}$ leading to great diagonal elements of the *hat matrix* $\mathbf{H} = \mathbf{X}(\mathbf{X}'\mathbf{X})^{-1}\mathbf{X}'$, that is, i_0 such that

$$\mathbf{x}_{i_0}'(\mathbf{X}'\mathbf{X})^{-1}\mathbf{x}_{i_0} \gg 0, \qquad (2.3.30)$$

(Huber 1981), the most influential points in L_1-regression are apparently determined by quantities

$$\sup_{\|\boldsymbol{b}\|=1}\Big\{|\mathbf{x}_i'\mathbf{b}|/(\sum_{j=1}^{n}\mathbf{x}_j'\mathbf{b})\Big\}, \qquad (2.3.31)$$

see Bloomfield and Steiger (1983).

Very illuminating is the effect of a leverage point on the tail performance of the least squares estimator. The LSE is optimal in the location model for the normal distribution. However, in a regression model, the leverage point causes the LSE to have very poor tail performance, even with Gaussian errors. Consider the linear regression model (2.3.14) and the measure of tail performance (2.3.18). Let $\bar{h}$ denote the maximal diagonal element of the hat-matrix $\mathbf{H} = \mathbf{X}(\mathbf{X}'\mathbf{X})^{-1}\mathbf{X}'$, i.e.

$$\bar{h} = \max_{1 \le i \le n} h_{ii}, \quad h_{ii} = \mathbf{x}_i'(\mathbf{X}'\mathbf{X})^{-1}\mathbf{x}_i, \quad i = 1, \ldots, n. \tag{2.3.32}$$

Remember that $\mathbf{H}$ is a projection matrix, $\mathbf{H}'\mathbf{H} = \mathbf{H}$, of rank p, with the trace p; $0 \le h_{ii} \le 1$, $1 \le i \le n$.

Theorem 2.3.2 *Let $\mathbf{T}_n$ be the LSE of $\boldsymbol{\beta}$ in the model (2.3.14) with i.i.d. errors e_i, $i = 1, \ldots, n$, distributed according to a symmetric d.f. F with non-degenerate tails (i.e., $0 < F(x) < 1 \; \forall x \in \mathbb{R}_1$). Denote*

$$\overline{B} = \limsup_{a \to \infty} B(a, \mathbf{T}_n), \quad \underline{B} = \liminf_{a \to \infty} B(a, \mathbf{T}_n). \tag{2.3.33}$$

1. *If F is of type I with $1 \le r \le 2$, then*

$$\bar{h}^{1-r} \le \underline{B} \le \overline{B} \le \bar{h}^{-r} \wedge n.$$

2. *If F is of type I with $r = 1$, then*

$$\bar{h}^{-1/2} \le \underline{B} \le \bar{B} \le \bar{h}^{-1}.$$

3. *If F is normal, then*

$$\underline{B} = \overline{B} = \bar{h}^{-1}.$$

4. *If F is of type II, then*

$$\underline{B} = \overline{B} = 1.$$

Proof. Without loss of generality, assume that $\bar{h} = h_{11}$. Let $\mathbf{h}_i'$ be the ith row of $\mathbf{H}$ and $\hat{Y}_i = \mathbf{x}_i'\mathbf{T}_n = \mathbf{h}_i'\mathbf{Y}$, $i = 1, \ldots, n$. Then

$$\begin{aligned}
&\mathbb{P}_{\boldsymbol{\beta}}(\max_i |\mathbf{x}_i'(\mathbf{T}_n - \boldsymbol{\beta})| > a) = \mathbb{P}_0(\max_i |\mathbf{h}_i'\mathbf{Y}| > a) \\
&\ge \mathbb{P}_0(\mathbf{h}_1'\mathbf{Y} > a) \ge \mathbb{P}_0(\bar{h}Y_1 > a, h_{12}Y_2 \ge 0, \ldots, h_{1n}Y_n \ge 0) \\
&\ge \mathbb{P}_0(Y_1 > a/\bar{h}) \cdot 2^{-(n-1)} \\
&= 2^{-(n-1)}\Big(1 - F(a/\bar{h})\Big).
\end{aligned}$$

Hence

$$\begin{aligned}\overline{B} &\leq \lim_{a\to\infty} \frac{-\log\left(1-F(a/\bar{h})\right)}{-\log\left(1-F(a)\right)} \\ &= \lim_{a\to\infty} \frac{c(a/\bar{h})^r}{ca^r} = \bar{h}^{-r}\end{aligned}$$

for F of type I, which gives the upper bounds in conditions (1) and (2), respectively. For F of type II,

$$\bar{B} \leq \lim_{a\to\infty} \frac{m\log(a/\bar{h})}{m\log a} = 1, \tag{2.3.34}$$

while $\underline{B} \geq 1$, because there are nonnegative as well as nonpositive residuals $Y_i - \hat{Y}_i,\ i = 1,\dots,n$.

If F is normal $N(0,\sigma^2)$, then $\hat{\mathbf{Y}} - \mathbf{X}\boldsymbol{\beta}$ has an n-dimensional normal distribution $N_n(\mathbf{0},\sigma^2\mathbf{H})$. Hence

$$I\!P_0(\max_i |\hat{Y}_i| > a) \geq I\!P_0(\mathbf{h}_1'\mathbf{Y} > a) = 1 - \Phi(a\sigma^{-1}\bar{h}^{-1/2})$$

and

$$\bar{B} \leq \bar{h}^{-1}. \tag{2.3.35}$$

On the other hand, if F is of type I with $1 < r \leq 2$, then applying the Markov inequality, we get for any $\epsilon \in (0,1)$,

$$\begin{aligned}&I\!P_{\boldsymbol{\beta}}(\max_i |\mathbf{x}_i'(\mathbf{T}_n - \boldsymbol{\beta})| > a) = I\!P_0(\max_i |\hat{Y}_i| > a) \\ \leq\ & I\!E_0[\exp\{(1-\epsilon)c\bar{h}^{1-r}(\max_i |\hat{Y}_i|)^r\}]/\exp\{(1-\epsilon)c\bar{h}^{1-r}a^r\} \qquad (2.3.36)\end{aligned}$$

and if the expectation in (2.3.36) is finite, say $\leq C_\epsilon < \infty$, then

$$-\log I\!P_0(\max_i |\hat{Y}_i| > a) \geq -\log C_\epsilon + (1-\epsilon)c\bar{h}^{1-r}a^r. \tag{2.3.37}$$

We thus obtain the lower bound in conditions (1) and (3).

We need to find the upper bound for the expectation in (2.3.36). Denoting $||\mathbf{x}||_s = \left(\sum_{i=1}^n x_i^s\right)^{1/s}$, by the Hölder inequality for $s = r/(r-1) \geq 2$, we get

$$\begin{aligned}(\max_i |\hat{Y}_i|)^r &= \max_i |\mathbf{h}_i'\mathbf{Y}|^r \leq\leq \max_i \left(||\mathbf{h}_i||_s ||\mathbf{Y}||_r\right)^r \\ &\leq \max_i (\sum_{j=1}^n h_{ij}^2)^{r/s} \sum_{k=1}^n |Y_k|^r \leq \bar{h}^{r-1} \sum_{k=1}^n |Y_k|^r. \qquad (2.3.38)\end{aligned}$$

Hence

$$\begin{aligned}&I\!E_0 \exp\left\{(1-\epsilon)c\bar{h}^{1-r}\left(\max_i |\hat{Y}_i|\right)^r\right\} \\ \leq\ & I\!E_0 \exp\left\{(1-\epsilon)c\sum_{k=1}^n |Y_k|^r\right\} = \left[I\!E_0 \exp\left\{(1-\epsilon)c|Y_1|^r\right\}\right]^n \qquad (2.3.39)\end{aligned}$$

Because of (2.3.21), there exists $K \geq 0$ such that

$$1 - F(x) \leq \exp\{-(1 - \frac{\epsilon}{2})bx^r\}, \quad \text{for } x > k. \tag{2.3.40}$$

Thus, dividing the integration domain of the integral

$$\mathbb{E}_0 \exp\{(1-\epsilon)c|Y-1|^r\} = -2\int_0^\infty \exp\{(1-\epsilon)bx^r\} d(1-F(x)) \tag{2.3.41}$$

in $(0, K)$ and (K, ∞) and using (2.3.40) and integration by parts in the second integral, we conclude that the expectation in (2.3.41) is finite and that (2.3.37) applies. Analogously, if $r = 1$, then

$$\begin{aligned} &\mathbb{P}_0\big(\max_i |\hat{Y}_i| > a\big) \\ &\leq \mathbb{E}_0 \exp\Big\{(1-\epsilon)c\bar{h}^{-\frac{1}{2}} \max_i |\hat{Y}_i|\Big\} \exp\Big\{-(1-\epsilon)c\bar{h}^{-\frac{1}{2}}a\Big\} \end{aligned}$$

and

$$\max_i |\hat{Y}_i| = \max_i |\mathbf{h}_i' \mathbf{Y}| \leq \bar{h}^{\frac{1}{2}} \sum_{k=1}^n |Y_k|,$$

which gives the lower bound in condition (2). □

2.4 ROBUST AND MINIMAX ESTIMATION OF LOCATION

Let $X_1, \ldots, X_n$ be n i.i.d. r.v.'s with a d.f. F , where F is generally unknown and is usually assumed to be a member of a broad class of d.f.'s. As a first step toward a plausible characterization of F, we try to establish some descriptive statistics (e.g., location and dispersion measure of F). Let us illustrate the location case. There are quite a few characteristics of the location or the center of the population, but none of these is universal.

Following Bickel and Lehmann (1975b), we will begin by summarizing some desirable properties of a measure of location. Let $\mathcal{F}$ be a family of distribution functions (defined on $\mathbb{R}_1$) such that if X is distributed according to the d.f. $F \in \mathcal{F}$ (denoted by $X \sim F$), then the d.f. of $aX + b$ also belongs to $\mathcal{F}$ for all $a, b \in \mathbb{R}_1$. A measure of location is then a functional $\theta(F)$ on $\mathcal{F}$. Writing $\theta(F)$ as $T(X)$ (when $X \sim F$), we have

1. $T(X+b) = T(X)+b$, $b \in \mathbb{R}_1$ and $F \in \mathcal{F}$ (translation-equivariance).
2. $T(aX) = aT(x)$, $a \in \mathbb{R}_1$ and $F \in \mathcal{F}$ (scale equivariance).

3. If G is stochastically larger than F [i.e., with respect to some partial ordering $\prec$, $F \prec G$, such as $F(x) \geq G(x)$ for all $x \in \mathbb{R}_1$], then for all such F and G belonging to $\mathcal{F}$, $\theta(F) \leq \theta(G)$.

4. T is qualitatively robust at every $F \in \mathcal{F}$, in the Hampel sense, as described in the preceding section.

5. It may also be natural to expect $T(X)$ to change sign under reflection with respect to the origin, $T(-X) = -T(X)$, and this is contained in 2 (by letting $a = -1$). Thus, the *sign equivariance* is contained in scale equivariance in an extended sense.

There are many functionals that satisfy the first four properties. A very notable exception is the mean $\theta(F) = \int x dF(x)$, which satisfies properties 1 to 3 but not property 4. Some of these functionals will be characterized in Chapter 3. Here we mainly intend to illustrate two broad classes of functionals and some criteria of choosing an appropriate one among them.

L-Functionals Let $Q(t) = F^{-1}(t)$ [$= \inf\{t : F(x) \geq t\}$], $t \in (0,1)$ be the usual quantile function associated with the d.f. F , and let $K(t)$ be a distribution function on [0,1]. Then, an L-functional is typically of the form

$$\theta(F) = \int_0^1 Q(t)dK(t). \tag{2.4.1}$$

It is easy to verify that $\theta(F)$ in (2.4.1) satisfy properties 1,2 and 3. However, $\theta(F)$ is not qualitatively robust (in the Hampel sense) if the support of the d.f. $K(.)$ is the entire unit interval [0,1]. If, on the other hand, for some a_0, a^0, such that $0 < a_0 < a^0 < 1$, $[a_0, a^0]$ is the shortest support of $K(.)$, then $\theta(F)$ is qualitatively robust at F_0, provided K and F_0 have no common points of discontinuity; see Problem 2.4.1. This provides an intuitive justification for the use of trimmed or Winsorized means in actual practice; we will make more comments on it in a later place.

M-functionals Let $\psi : \mathbb{R}_1 \mapsto \mathbb{R}_1$ be a nondecreasing function that takes on negative as well as positive values such that $\int \psi(x)dF(x) = 0$. For simplicity of presentation, we may assume that ψ is monotone, and define $\theta(F)$ as

$$\theta(F) = \frac{1}{2}[\theta^-(F) + \theta^+(F)], \tag{2.4.2}$$

where

$$\theta^-(F) = \sup\{t : \int \psi(x-t)dF(x) > 0\}, \tag{2.4.3}$$

$$\theta^+(F) = \inf\{t : \int \psi(x-t)dF(x) < 0\}. \tag{2.4.4}$$

Then $\theta(F)$ is termed an M-functional. It satisfies properties 1 and 3, but generally it is not scale-equivariant. If ψ is a bounded function, then $\theta(F)$

satisfies property 4. The particular case of $\psi(x) \equiv x$ leads us to the mean (which satisfies 2), but when ψ is not necessarily linear, property 2 may not hold. To make $\theta(F)$ in (2.4.2) scale equivariant, in (2.4.3) one replaces $\psi(x)$ by $\psi(x/s(F))$, $x \in \mathbb{R}^1$, where the functional $s(F)$ (written as $s(X)$ equivalently) satisfies a basic equivariance condition that $s(aX+b) = |a|s(X)$, for every $a, b \in \mathbb{R}_1$. However, with this amendment, $\theta(F)$ may not satisfy property 3 any more.

Now recall that $X_1, \ldots, X_n$ are i.i.d. r.v.'s with the d.f. F, defined on $\mathbb{R}_1$, and F_n, the corresponding empirical (sample) d.f. is a step functional, also defined on $\mathbb{R}_1$. If in (2.4.1) or (2.4.2), we replace the d.f. F by F_n we get a natural estimator $T_n = \theta(F_n)$. Under quite general regularity conditions (to be discussed in several later chapters), $n^{-1/2}(T_n - \theta(F))$ is asymptotically normally distributed with 0 mean and a finite, positive variance $\sigma^2(T, F)$. This *asymptotic variance* (i.e., the variance of the asymptotic distribution) plays a very basic role in the study of the performance of an estimator. As before, we let Φ stand for the normal d.f. Also, we let

$$e(T, \bar{X}; F) = \sigma^2(\bar{X}, F)/\sigma^2(T, F), \tag{2.4.5}$$

when F is the true d.f. Then, we have the following natural requirement for a suitable robust estimator T_n.

$$e(T, \bar{X}; F) \text{ is very close to 1 when } F \equiv \Phi,$$

$$\sup\{e(T, \bar{X}; F) : F \in \mathcal{F}_1\} = \infty, \inf\{e(T, \bar{X}; F) : F \in \mathcal{F}_1\} \geq c > 0,$$

$$\text{for a large subfamily} \mathcal{F}_1 \text{ of } \mathcal{F}, \text{ whereas for a nonempty subset}$$

$$\mathcal{F}_1^0 \text{ of } \mathcal{F},\ e(T, \bar{X}; F) \geq 1 \ \forall F \in \mathcal{F}_1^0. \tag{2.4.6}$$

Among the class of L-estimators and M-estimators of location, there exists no universally optimal measure of location satisfying all the requirements in (2.4.5). The R-estimators defined in (2.3.4) through (2.3.6) may have some distinct advantages in this respect. For example, if we let $\mathcal{F}_1$ stand for the class of symmetric (absolutely continuous) d.f.'s on $\mathbb{R}_1$, then, for the classical *normal-scores* estimator, (2.4.6) holds with $c = 1$. However, it does not have a bounded influence function.

Confined to the class of L- and M-estimators, Bickel and Lehmann (1975b) recommended the α-trimmed mean of the form

$$\theta(F) = (1-2\alpha)^{-1} \int_{\alpha}^{1-\alpha} Q(t)dt,\ 0 < \alpha < 1/2, \tag{2.4.7}$$

with $\alpha \in [0.05, 0.10]$ and they showed that

$$\inf_{F \in \mathcal{F}_1} e(T, \bar{X}; F) = (1-2\alpha)^2,\ \sup_{F \in \mathcal{F}_1} e(T, \bar{X}; F) = \infty, \tag{2.4.8}$$

where $T_n = \theta(F_n)$ [corresponding to $\theta(F)$ in (2.4.6)] and $\mathcal{F}_1$ is the family of d.f.'s F, which are absolutely continuous in an interval $(Q(\alpha)-\eta, Q(1-\alpha)+\eta)$, for some $\eta > 0$ and have positive and finite derivatives in a neighborhood of $Q(\alpha)$ and $Q(1-\alpha)$, respectively.

Intuitively it is clear that the more restrictive is the class $\mathcal{F}_1$ of d.f.'s (we may choose), the easier it is to find a satisfactory measure of location over $\mathcal{F}_1$. To illustrate this point, we consider the Huber model of *ϵ-contaminacy* in (2.3.10) with F_0 as the standard normal d.f. (Φ) and with a symmetric contamination. Thus in this location problem our model is of the form

$$\epsilon = \{F : \; F = (1-\epsilon)\Phi + \epsilon H, \; H \in \mathcal{H}\}, \; 0 < \epsilon < 1, \tag{2.4.9}$$

where $\mathcal{H}$ is the set of substochastic distribution functions symmetric about zero. Let us restrict our attention to the class of functionals T that are translation equivariant [see property 1] and such that $n^{1/2}(T(F_n) - \theta(F))$ is asymptotically normally distributed. With a *squared error loss*, the risk function will be the variance $\sigma^2(T, F)$ computed from this asymptotic normal distribution (and may as well be termed the *asymptotic distributional risk* (ADR). We are looking for the *minimax functional* that minimizes

$$\sup\{\sigma^2(T, F) : \; F \in \mathcal{F}_\epsilon \}. \tag{2.4.10}$$

First, we look for a solution of the above problem in the subclass of M-functionals in (2.4.2) and(2.4.3). If the d.f. F has an absolutely continuous density function f with a finite Fisher information $I(f)$, defined by (2.2.22) and if the score function ψ in (2.4.3) is square integrable (with respect to F),

$$\sigma_\psi^2 = \int \psi^2(x) dF(x) < \infty, \tag{2.4.11}$$

then, under quite mild regularity conditions (as we will see in Chapter 3), the asymptotic variance of $n^{1/2}(T(F_n) - \theta(F))$ is

$$\sigma^2(T, F) = \sigma_\psi^2 \{\int \psi(x)\{-f'(x)/f(x)\} dF(x)\}^{-2}. \tag{2.4.12}$$

Let us denote by

$$\mathcal{F}_\epsilon^\star = \{F \in \mathcal{F}_\epsilon : \; I(f) < \infty\}. \tag{2.4.13}$$

Huber (1964) proved that the *least favorable distribution* F_0 for which $I(f_0) \leq I(f)$ for all $F \in \mathcal{F}_\epsilon^\star$ exists and satisfies $\int dF_0 = 1$, and that the MLE of the shift θ in $F_0(x-\theta)$ minimizes the risk in (2.4.9). This estimator is an M-estimator (M-functional) generated by the score function

$$\psi_0(x) = \begin{cases} x & \text{for } |x| \leq k, \\ k\text{sign}(x) & \text{for } |x| > k, \end{cases} \tag{2.4.14}$$

where k (> 0) depends on ϵ through

$$\int_{-k}^{+k} \phi(x)dx + k^{-1}2\phi(k); \; \phi(x) = (d/dx)\Phi(x). \qquad (2.4.15)$$

Hence, for any M-functional T_{ψ}, for every $F \in \mathcal{F}_1$,

$$\sigma^2(T_{\psi_0}, F) \le \sigma^2(T_{\psi_0}, F_0) = 1/I(f_0) \le \sigma^2(T_{\psi}, F) \le \sup_{F \in \mathcal{F}_1} \sigma^2(T, F). \quad (2.4.16)$$

Noting that T_{ψ_0} is translation-equivariant and that the asymptotic risk of the Pitman estimator of θ in $F_0(x - \theta)$ is equal to the reciprocal of $I(f_0)$, we conclude that T minimizes (2.4.9) also in the class of translation-equivariant estimators. This estimator (functional) satisfies properties 3 and 4 but not 2. That is, it is not scale-equivariant. As we will see in Chapter 4, another minimax solution to this problem is the trimmed mean in (2.4.6) with

$$\alpha = (1 - \epsilon)(1 - \phi(k)) + \epsilon/2 \text{ and } k \text{ satisfying (2.4.14).} \qquad (2.4.17)$$

It may further be noted that the trimmed mean is scale-equivariant, but not the estimator T_{ψ_0}. Another (scale-equivariant) solution lies in the class of R-estimators, and this will be considered in Chapter 6. A scale-equivariant version of the functional T may be constructed by using the score function $\psi_0(x/s(F_0))$ instead of $\psi_0(x)$ (as has already been mentioned after (2.4.3)). However, this amended M-estimator loses the minimax property. This is mainly due to the fact that the asymptotic variance of this latter estimator at the model F_0 (as we will see in Chapter 5) is given by

$$\sigma^2(T_{\psi_0}, F_0, s) = s^2(F_0)\{\psi_0^2(x)(\int \psi_0'(\frac{x}{s(F_0)})dF_0(x))^2\}, \qquad (2.4.18)$$

and, in view of the segmented form of ψ_0 in (2.4.14), it can be easily shown that (2.4.18) fails to attain the information limit $I(f_0)$ for any finite k, (see Problem 2.4.2).

2.5 CLIPPINGS FROM PROBABILITY AND ASYMPTOTIC THEORY

Consider a probability space $(\Omega, \mathcal{A}, \mathbb{P})$ where $\mathcal{A}$ is the σ-field of subsets of a nonempty set Ω and $\mathbb{P}(.)$ is a normed measure. A set of observations $\mathbf{X} = (X_1, \ldots, X_m)$ along with a distribution function $F(\mathbf{x}) = \mathbb{P}\{X < x\}$ generate a probability space $(\Omega, \mathcal{A}, \mathbb{P})$ where the range of Ω is $\mathbb{R}_m$ and $\mathbb{P}(.)$ is uniquely determined by $F(.)$. Thus a random vector (r.v.) $\mathbf{X}$ is defined as a measurable function from $(\Omega, \mathcal{A}, \mathbb{P})$ into $\mathbb{R}_m$, for some $m \geq 1$; for $m = 1$, it is termed a random variable. This definition extends to more general (random) functions as well. For example, a *stochastic process* $X = \{X(t),\ t \in T\}$ (or more generally, a sequence $X_{\gamma} = \{X_{\gamma}(t),\ t \in T\},\ \gamma \in \Gamma$) is a collection of r.v.'s defined over some probability space $(\Omega, \mathcal{A}, \mathbb{P})$, where T is an *index set* such that, for each $t \in T$, $X(t)$ (or $X_{\gamma}(t)$) is $\mathcal{A}$-measurable. In asymptotic theory of statistical inference, such a stochastic process can often be represented as *random elements* in an appropriate *function space*. Thus we may consider a *topological space* (T) endowed with a *(semi-)metric* $\mathbf{d} : T \times T \mapsto \mathbb{R}_1^+$, and we denote the class of Borel subsets of T by $\mathcal{B}(T)$. Then a function $U : \Omega \mapsto T$ is said to be a random element of T if $\{\omega \in \Omega :\ U(\omega) \in B\} \in \mathcal{A},\ \forall B \in \mathcal{B}(T)$. Notable examples of T are the following:

1. $C = C[0,1]$, the space of continuous functions on the unit interval $[0,1]$ with $d(x,y) = ||x(t) - y(t)|| = \sup\{|x(t) - y(t)| : t \in [0,1]\}$.

2. $D = D[0,1]$, the space of functions on [0,1] having only discontinuities of the first kind (i.e., they are right continuous and have left-hand limits). Now letting $\Lambda = \{\lambda_t,\ t \in [0,1]\}$ be the class of strictly increasing, continuous mapping of [0,1] onto itself, $d(.)$ is the *Skorokhod* J_1*-metric (topology)* defined for $x, y \in D[0,1]$ by

$$\rho_S(x,y) = \inf\{\epsilon > 0 : ||\lambda_t - t|| \leq \epsilon,\ ||x(t) - y(\lambda_t)|| \leq \epsilon\}. \tag{2.5.1}$$

Clearly $C[0,1]$ is a subspace of $D[0,1]$, and their extensions to $[0,K]$ for some $K \in (0,\infty)$, are straightforward. With a modified metric, the domains may also be extended to $\mathbb{R}_1^+$ (see, Whitt 1970; Lindvall 1973). Further, extensions to $[0,1]^p$ for some $p \geq 1$, have also been worked out; we may refer to Sen (1981a,ch.2) for a survey of these results.
Consider next a stochastic process $X_{\gamma};\ \gamma \in \Gamma$, where Γ is a linearly ordered index set (and in the discrete case, Γ can be taken equivalently as $\mathbb{N}$,the set of nonnegative integers or any subset of it). Let $\{F_{\gamma},\ \gamma \in \Gamma\}$ be a system of subsigma fields of $\mathcal{A}$, such that X_{γ} is F_{γ}-measurable, for each $\gamma \in \Gamma$. Then $\{X_{\gamma},\ \gamma \in \Gamma\}$ is said to be adapted to $\mathcal{F} = \{F_{\gamma},\ \gamma \in \Gamma\}$.
Further $\{F_{\gamma},\ \gamma \in \Gamma\}$ is nondecreasing or nonincreasing if for $\gamma, \gamma' \in \Gamma$ and $\gamma < \gamma'$, $F_{\gamma} \subset F_{\gamma'}$ or $F_{\gamma} \supset F_{\gamma'}$.

(a) If F_{γ} is nondecreasing (or nonincreasing) in $\gamma \in \Gamma$, $\mathbb{E}|X_{\gamma}| < \infty$, $\gamma \in \Gamma$, and if for every $\gamma, \gamma' \in \Gamma$, and $\gamma' >$ (or $<$) γ,

$$\mathbb{E}\{X_{\gamma'}|F_{\gamma}\} = X_{\gamma} \text{ almost everywhere (a.e.),} \tag{2.5.2}$$

then $\{X_{\gamma}, \gamma \in \Gamma\}$ is termed a *martingale* (or *reverse martingale*). If in (2.5.2) the " = " sign is replaced by " $\geq$ "(or "$\leq$ ")sign, then $\{X_{\gamma},\ \gamma \in \Gamma\}$ is a *sub*-(or *super-) martingale* or reversed ones according as F_{γ} is nondecreasing or nonincreasing.

(b) If $\{X_\gamma, \mathcal{F}_\gamma;\ \gamma \in \Gamma\}$ is a (forward or reverse) martingale and $g(.)$ is an integrable, *convex* function, then $\{g(X_\gamma), \mathcal{F}_\gamma;\ \gamma \in \Gamma\}$ is a (forward or reverse) submartingale; if $g(.)$ is *concave*, then the submartingale structure changes to a supermartingale structure.

(c) If the X_i are (centered) independent r.v.'s, then the $S_n (= \sum_{j=1}^n X_j)$ form a martingale, and $|S_n|^p$, $p > 0$, form a submartingale or supermartingale sequence according as p is $\geq$ or < 1. If the X_i are i.i.d. r.v.'s, then the $\bar{X}_n$ (= $n^{-1}S_n$) form a reverse martingale and the $|\bar{X}_n|^p$ form a reverse submartingale or supermartingale according as p is $\geq$ or < 1. In this setup the X_i does notneed to be centered.

(d) *The Doob moment inequality* (Doob 1967, p. 318). Let $\{X_n, \mathcal{F}_n;\ n \geq 1\}$ be a nonnegative submartingale. Then

$$\mathbb{E}\Big\{\max_{1 \leq k \leq n} X_k^\alpha\Big\} \leq \begin{array}{ll} e(e-1)^{-1}\{1 + \mathbb{E}(X_n \log^+ X_n)\}, & \alpha = 1, \\ \{\alpha/(\alpha-1)\}^\alpha \mathbb{E}\{X_n^\alpha\}, & \alpha > 1, \end{array} \qquad (2.5.3)$$

where $\log^+ x = \max\{1, \log x\}$, $x \geq 0$.

(e) *Submartinagle convergence theorem* (Doob, 1967). Let $\{X_n, \mathcal{F}_n;\ n \geq 1\}$ be a submartingale, and suppose that $\sup_n \mathbb{E}|X_n| < \infty$. Then, there exists a r.v. X, such that

$$X_n \to X \text{ a.s.}, \quad \text{as } n \to \infty, \quad \mathbb{E}|X| \leq \sup_n \mathbb{E}|X_n|. \qquad (2.5.4)$$

Now let $\{X_n, \mathcal{F}_n;\ n \geq 1\}$ be a submartingale or a reverse submartingale, and assume that the X_n are uniformly integrable (i.e., $\mathbb{E}\{|X_n| I(|X_n| > c)\} \to 0$ as $c \to \infty$, uniformly in n). Then there exists a r.v. X, such that $X_n \to X$ a.s., and in the first mean, as $n \to \infty$. In particular, if $\{X_n, \mathcal{F}_n;\ n \geq 1\}$ is a reversed martingale, then the X_n are uniformly integrable. Hence

$$X_n \to X \text{ a.s. and in the first mean as } n \to \infty. \qquad (2.5.5)$$

Some other inequalities will be introduced later. As will be seen in the subsequent chapters, asymptotic theory of statistical inference rests on deeper adaptations of various basic concepts and tools in probability theory. While a complete coverage of this topics is beyond the scope of this limited introduction, we present the most pertinent ones, along with their sources, so that a further indepth reading may be facilitated when necessary. Chapters 2 to 4 of Sen and Singer (1993) provide a survey of most of these results.

2.5.1 Modes of Convergence of Stochastic Elements

Unlike the elements in real analysis whose convergence is well defined, those in stochastic analysis which are in fact themselves stochastic have different convergence properties that are formulated depending on the specific model of interest. Critical in their formulation are the definitions of the probability space and of random elements on functional spaces, which were introduced earlier.

Definition 2.5.1 (*Stochastic convergence* or *convergence in probability*) Let $\{X_n\}$ and X be random elements defined on a probability space $(\Omega, \mathcal{A}, \mathbb{P})$. If the X_n and X have realizations in a space T endowed with a distance or norm $d : T \times T \mapsto \mathbb{R}_1^+$, then we say that X_n converges in probability to X [relative to the metric $d(.)$] if for every $\epsilon > 0$,

$$\lim_{n \to \infty} \mathbb{P}\{d(X_n, X) > \epsilon\} = 0. \tag{2.5.6}$$

In notation, we express this as

$$d(X_n, X) \xrightarrow{P} 0 \text{ or } X_n \xrightarrow{P} X, \quad \text{as } n \to \infty. \tag{2.5.7}$$

If the X_n and X are real valued, then $d(x, y) = |x - y|$. So $d(X_n, X) \to 0 \Rightarrow |X_n - X| \to 0$. If these elements belong to $\mathbb{R}_p$, for some $p \geq 1$, we can take $d(\mathbf{x}, \mathbf{y}) = ||\mathbf{x} - \mathbf{y}||$, the Euclidean norm. In this case, it is also possible to choose some other norms, such as $d(\mathbf{x}, \mathbf{y}) = ||\mathbf{x} - \mathbf{y}||_\infty = \max\{|x_j - y_j| : 1 \leq j \leq p\}$. Since $||\mathbf{x} - \mathbf{y}||_\infty \leq ||\mathbf{x} - \mathbf{y}|| \leq p||\mathbf{x} - \mathbf{y}||_\infty$, either definition of the metric $d(.)$ suffices. Consider next the case where X_n and X are random elements in an appropriate function space T, endowed with a (semi-) metric $d : T \times T \mapsto \mathbb{R}^+$. We have already introduced such function spaces and discussed their appropriate metrics $d(.)$. For example, for the $C[0, 1]$ space, we may choose the uniform metric $d(x, y) = \sup\{|x(t) - y(t)| : 0 \leq t \leq 1\}$, and the definition in (2.5.6) remains valid.

Definition 2.5.2 (*Convergence with probability 1* or *almost sure convergence* or *strong convergence* or *convergence almost everywhere*) In the same setup as in the preceding definition, suppose that for every $\epsilon > 0$,

$$\lim_{n \to \infty} \mathbb{P}\{\bigcup_{m \geq n} [d(X_m, X) > \epsilon]\} = 0. \tag{2.5.8}$$

Then we say that X_n converges to X with probability 1 as $n \to \infty$. The other terminologies are synonymous. In notation, we write (2.5.8) as [relative to the metric $d(.)$]

$$d(X_n, X) \xrightarrow{\text{a.s.}} 0 \text{ or } X_n \to X, \text{ a.s.,} \quad \text{as } n \to \infty. \tag{2.5.9}$$

Note that (2.5.8) implies (2.5.6), and hence,

$$X_n \to X, \text{a.s.} \Rightarrow X_n \xrightarrow{P} X, \tag{2.5.10}$$

but the converse may not be true.

Definition 2.5.3 (*Convergence in the rth mean*, $r > 0$) Suppose that the X_n and X are real or vector valued r.v.'s. If

$$\lim_{n\to\infty} \mathbb{E}\|X_n - X\|^r = 0 \text{ for some } r > 0, \tag{2.5.11}$$

we say that X_n converges in the rth mean to X; in our earlier notation

$$X_n \overset{rth}{\to} X, \text{ or } X_n \to X \text{ in } L_r \text{ norm as } n \to \infty. \tag{2.5.12}$$

The definition extends to function spaces wherein $\|x - y\|$ is to be replaced by $d(x, y)$. It is easy to verify that

$$X_n \overset{rth}{\to} X \text{ for some } r > 0 \Rightarrow X_n \overset{P}{\to} X. \tag{2.5.13}$$

But (2.5.11), unless accompanied by suitable rate of convergence or some other intrinsic properties of the X_n, may not necessarily imply (2.5.8). We will present some related results later.

Definition 2.5.4 (*Complete convergence*) In the setup of Definition 2.5.2, suppose that

$$\lim_{n\to\infty} \sum_{N\geq n} \mathbb{P}\{d(X_N, X) > \epsilon\} = 0 \ \forall \epsilon > 0. \tag{2.5.14}$$

Then we say that X_n converges completely to X as $n \to \infty$, or that $X_n \overset{C}{\to} X$, as $n \to \infty$. Note that

$$X_n \overset{C}{\to} X \Rightarrow X_n \to X, \quad \text{a.s., as} \quad n \to \infty. \tag{2.5.15}$$

The converse may not be universally true.

We have tacitly assumed that the X_n and X are defined on a common probability space. In many statistical adaptations, this coherence condition is not needed. A much weaker mode of convergence can be formulated (under weaker regularity conditions) that eliminates this coherence, and encompasses a wider domain of adaptations. Before we look at the formulation of "weak convergence", let us consider a very simple example.

Example 2.5.1 (*Binomial distribution*) Consider a sequence $\{Y_i;\ i \geq 1\}$ of i.i.d. r.v.'s such that

$$\mathbb{P}(Y_i = 0) = 1 - \mathbb{P}(Y_i = 1) = 1 - \pi,\ 0 < \pi < 1.$$

Let $S_n = Y_1 + \ldots + Y_n$, $n \geq 1$. Note that $\mathbb{E}S_n = n\pi$ and $V(S_n) = n\pi(1-\pi)$. Consider the r.v.'s :

$$X_n = \frac{S_n - n\pi}{\sqrt{n\pi(1-\pi)}}, \quad n \geq 1. \tag{2.5.16}$$

It is well known that S_n has the $Bin(n, \pi)$ distribution and that S_n can only assume values in $\{0, 1, \ldots, n\}$. Thus in (2.5.16), X_n is a discrete r.v., although the mesh becomes dense as n increases. If we denote by $G_n(y) = \mathbb{P}\{X_n \leq y\}$, $y \in \mathbb{R}$, the d.f. of X_n, and if $G(Y)$ stands for the d.f. of a standard normal r.v. (which we denote by X), then we will see that as n increases, G_n converges to G:

$$\lim_{n\to\infty} \sup_{y\in \mathbb{R}} |G_n(y) - G(y)| = 0. \tag{2.5.17}$$

We can rewrite this as

$$G_n \xrightarrow{w} G \quad \text{as } n \to \infty. \tag{2.5.18}$$

We need to keep two things in mind here: (1) X may not be defined on the same probability space as the X_n are defined, and (2) if G were not continuous everywhere, then in (2.5.17) we might have to exclude the points of discontinuity (or jump-points) of G (where the left- and right-hand limits may not agree). Given these technicalities, we may rewrite (2.5.18) as $X_n \to X$, in law/distribution, or by definition

$$X_n \xrightarrow{\mathcal{D}} X \quad \text{or} \quad \mathcal{L}(X_n) \to \mathcal{L}(X), \quad \text{as } n \to \infty. \tag{2.5.19}$$

This leads to the following

Definition 2.5.5 (*Weak convergence* or *convergence in law/distribution*) If for a sequence $\{G_n\}$ of d.f.'s, defined on $\mathbb{R}_k$, for some $k \geq 1$, there exists a d.f. G, such that (1) G is the d.f. of a random element X not necessarily defined on the same probability space and (2) at all points of continuity of G, $G_n \to G$ as $n \to \infty$, then we say that X_n, which has the d.f. G_n, converges in law (or distribution) to X. We denote it by the notation of (2.5.18) or (2.5.19).

It is evident that if $X_n \xrightarrow{P} X$, then $X_n \xrightarrow{\mathcal{D}} X$ as well, but the converse is not necessarily true. The concept of weak convergence plays a most fundamental role in the asymptotic theory of statistical inference. To see the full impact of this concept, we need to extend the notion of convergence to encompass function spaces with appropriate topologies that eliminate the basic difficulties caused by discontinuities of various types and and other topological metrization problems. We will deal with these issues more adequately in a later section.

2.5.2 Basic Probability Inequalities

The following elemetary inequality provides the genesis of all subsequent inequalities.

1. *Chebychev's Inequality.* Let U be a nonnegative r.v. such that $\mu = \mathbb{E}U$ exists. Then for every positive t,

$$\mathbb{P}\{U \geq \mu t\} \leq t^{-1}. \tag{2.5.20}$$

An immediate consequence of (2.5.20) is the Markov inequality. Let T_n be a statistic such that for some $r > 0$, for some T (possibly a r.v.), $\nu_{nr} = \mathbb{E}|T_n - T|^r$ exists. Then letting $U = U_n = |T_n - T|^r$, we have for For every $t > 0$,

$$\begin{aligned} \mathbb{P}\{|T_n - T| \geq t\} &= \mathbb{P}\{U_n \geq t^r\} \\ &= \mathbb{P}\{U_n \geq \nu_{nr}(t^r/\nu_{nr})\} \\ &\leq \nu_{nr}/t^r. \end{aligned} \tag{2.5.21}$$

Thus, if $\nu_{nr} \to 0$ as $n \to \infty$, we obtain from (2.5.21) that $T_n - T \xrightarrow{P} 0$. This establishes the implification relation in (2.5.13). Moreover, if $\sum_{n\geq 1} \nu_{nr}$ converges, then (2.5.21) implies that $T_n \to T$a.s., as $n \to \infty$. In most statistical applications this simple prescription works well. However, for (2.5.8) or (2.5.14) to hold, we do not need this series convergence criterion. Often, we may obtain some of these results by more powerful inequalities (and under more specific structures on the r.v.'s).

Another variant of (2.5.20), under more stringent conditions, is the following:

2. *Bernstein Inequality.* Let U be a r.v. such that $M_U(t) = \mathbb{E}(e^{tU})$ exists for all $t \in [0, K]$, for some $K > 0$. Then, for every real u, we have

$$\mathbb{P}\{U \geq u\} \leq \inf_{t\in[0,K]} \left\{e^{-tU} M_U(t)\right\}. \tag{2.5.22}$$

We may observe here that if U is the sum of n (centered) r.v.'s, independent and having finite moment generating functions, then $M_U(t)$ is factorizable into n terms. So the right-hand side of (2.5.22) reduces to a term that exponentially (in n) converges to 0 as $n \to \infty$. This generally gives a much sharper bound than that obtainable using (2.5.20) or (2.5.21). Further, if the components of U are all bounded r.v., the moment generating function exists, and hence such sharper bounds can easily be obtained.

Example 2.5.1 (Revisited) Let $T_n = n^{-1}S_n$ and $U_n = S_n - n\pi$. Then $M_{U_n}(t) = [M_Y(t)]^n = \{1 + \pi(e^t - 1)\}^n$. Substituting this in (2.5.22) and following a few routine steps, we have for every $0 < \pi < 1$ and $\epsilon > 0$,

$$\mathbb{P}\{|T_n - \pi| > \epsilon\} \leq [\rho(\epsilon, \pi)]^n + [\rho(\epsilon, 1 - \pi)]^n \tag{2.5.23}$$

where

$$\rho(\epsilon, \pi) = \{\frac{\pi}{\pi + \epsilon}\}^{\pi+\epsilon}\{\frac{1-\pi}{1-\pi-\epsilon}\}^{1-\pi-\epsilon}, \tag{2.5.24}$$

and it is easy to verify that $0 \leq \rho(\epsilon, \pi) < 1$, $\forall \epsilon > 0$, $\pi \in (0, 1)$.

Hoeffding (1963) incorporated this simple probability inequality in the derivation of a probability inequality for a general class of bounded r.v.'s. His ingenious tool is based on two other inequalities: that the geometric mean of a set of positive numbers can not be larger than their arithmetic mean, and that for $x \in [0,1]$, $t \geq 0$, $e^{tx} \leq 1 + x(e^t - 1)$.

3. *Hoeffding inequality.* Let $\{X_k; k \geq 1\}$ be independent (but not necessarily identically distributed) r.v.'s such that $\mathbb{P}\{0 \leq X_k \leq 1\} = 1 \, \forall k \geq 1$. Set $\mu_k = \mathbb{E}X_k$, $k \geq 1$, and $\bar{\mu}_n = n^{-1}\sum_{k=1}^n \mu_k$. Also let $\bar{X}_n = n^{-1}\sum_{k=1}^n X_k$, $n \geq 1$. Then for every $\epsilon > 0$,

$$\mathbb{P}\{|\bar{X}_n - \bar{\mu}_n| \geq \epsilon\} \leq [\rho(\epsilon, \bar{\mu}_n)]^n + [\rho(\epsilon, 1 - \bar{\mu}_n)]^n, \tag{2.5.25}$$

where $\rho(.)$ is defined by (2.5.24).

The above inequality extends directly to the case where for some (a, b) : $-\infty < a < b < +\infty$, $\mathbb{P}\{a \leq X_k \leq b\} = 1 \, \forall k$. In this case, we define $Y_k = (X_k - a)/(b - a)$, and apply (2.5.25), replacing ϵ by $\epsilon' = \epsilon/(b-a)$ and the μ_k by the $\mathbb{E}Y_k$, $k \geq 1$. As in the earlier cases, (2.5.25) holds for every $n \geq 1$ and $\epsilon > 0$. It is also possible to allow that ϵ to depend on n, for example, $\epsilon = tn^{-1/2}$, $t > 0$. Looking at (2.5.24), we may verify that $(\partial/\partial\epsilon)\log\rho(\epsilon, \pi)|_{\epsilon=0} = 0$ and

$$(\partial^2/\partial\epsilon^2)\log\rho(\epsilon, \pi) = -\{(\pi + \epsilon)(1 - \pi - \epsilon)\}^{-1} \leq -4. \tag{2.5.26}$$

As such, $\log\rho(\epsilon, \pi) \leq -2\epsilon^2$, $\forall \epsilon > 0$, while for $\epsilon = tn^{-1/2}$, $\log\rho(tn^{-1/2}, \pi) = -(2n)^{-1}t^2\{\pi(1-\pi)\}^{-1}\{1 + O(n^{-1/2})\}$, so (2.5.25) yields a rate $e^{-t^2/2\pi(1-\pi)}$, that is similar to the rate obtained using the asymptotic normal distribution of $\sqrt{n}(\bar{X}_n - \bar{\mu}_n)$.

There are some other probability inequalities more intimately associated with the laws of large numbers, which will be reviewed later on.

2.5.3 Some Useful Inequalities and Lemmas

Some moment and other inequalities basic results in probability theory. We start with those mostoften used.

1. *Jensen inequality.* Let X be a r.v., and let $g(x)$, $x \in \mathbb{R}$, be a convex function such that $\mathbb{E}g(X)$ exists. Then

$$g(\mathbb{E}X) \leq \mathbb{E}g(X), \tag{2.5.27}$$

where the equality sign holds only when $g(.)$ is linear a.e.

Recall that for $x \in \mathbb{R}$, $p \geq 1$, $|x|^p$ is convex, and hence (2.5.27) when applied to a sequence $\{T_n - T\}$ yields the following:

$$T_n \overset{rth}{\to} T, \text{ for some } r > 0 \Rightarrow T_n \overset{sth}{\to} T, \quad \forall s \leq r. \tag{2.5.28}$$

2. *Hölder inequality.* Let X and Y be two not necessary independent r.v.'s, such that for $p > 0$, $q > 0$ with $p^{-1} + q^{-1} = 1$, $\mathbb{E}|X|^p$ and $\mathbb{E}|Y|^q$ both exist. Then

$$\mathbb{E}|XY| \leq (\mathbb{E}|X|^p)^{1/p}(\mathbb{E}|Y|^q)^{1/q} \tag{2.5.29}$$

The special case of $p = q = 2$ is known as the *Cauchy-Schwarz inequality.*

3. *Minkowski Inequality.* For r.v.'s $X_1, \ldots, X_n$, not necessary independent, and $p \geq 1$, whenever $\mathbb{E}|X_i|^p < \infty$, $\forall i$,

$$(\mathbb{E}\{|\sum_{i=1}^{n} X_i|^p\})^{1/p} \leq \sum_{i=1}^{n}[\mathbb{E}|X_i|^p]^{1/p}. \tag{2.5.30}$$

A variant form of this inequality is

$$[\{|\sum_{i=1}^{n}(X_i + Y_i)|^p\}]^{1/p} \leq (\sum_{i=1}^{n}\{|X_i|^p)^{1/p} + (\sum_{i=1}^{n}\{|Y_i|^p)^{1/p}. \tag{2.5.31}$$

4. *C_r-inequality.* Consider

$$|a + b|^r \leq C_r\{|a|^r + |b|^r\}, \quad r \geq 0, \tag{2.5.32}$$

where

$$C_r = \begin{cases} 1, & 0 \leq r \leq 1 \\ 2^{r-1}, & r > 1 \end{cases} \tag{2.5.33}$$

For $m \geq 2$ items, $C_r = 1$ or m^{r-1} according as $0 \leq r \leq 1$ or $r \geq 1$.

5.*Arithmetic-geometric-harmonic mean (A.M.-G.M.-H.M.) inequality.* Let $a_1, \ldots, a_n$ be nonnegative members, $n \geq 1$. Then

$$\begin{aligned} \text{A.M.} = \bar{a}_n = \frac{1}{n}\sum_{i=1}^{n} a_i \quad &\geq \quad (\prod_{i=1}^{n} a_i)^{1/n} = G.M. \\ &\geq \quad (\frac{1}{n}\sum_{i=1}^{n}\frac{1}{a_i})^{-1} = H.M., \end{aligned} \tag{2.5.34}$$

where equality holds only when $a_1 = \ldots = a_n$. Weightewd and integral versions of (2.5.34) are also available in the literature.

6. *Entropy inequality.* Let $\{a_i\}$, $\{b_i\}$ be convergent sequence of positive numbers such that $\sum_{i=1}^{n} a_i \geq \sum_{i=1}^{n} b_i$. Then

$$\sum_{i=1}^{n} a_i \log(b_i/a_i) \leq 0, \tag{2.5.35}$$

where the equality sign holds only when $a_i = b_i$, $\forall i \geq 1$.

7. *Monotone convergence lemma.* If the events $\{A_n\}$ are monotone, that is, either $A_1 \subset A_2 \subset A_3 \subset \ldots$ or $A_1 \supset A_2 \supset A_3 \supset \ldots$, with limit A, then

$$\lim_{n\to\infty} \mathbb{P}(A_n) = \mathbb{P}(A). \tag{2.5.36}$$

(Compare the result with the limit of a monotone sequence of numbers.)

8. *Borel-Cantelli lemma.* Let $\{A_n\}$ be a sequence of events, and denote by $\mathbb{P}(A_n)$ the probability that A_n occurs, $n \geq 1$. Also let A denote the event that A_n occur infinitely often (i.o.). Then

$$\sum_{n\geq 1} \mathbb{P}(A_n) \leq +\infty \Rightarrow \mathbb{P}(A) = 0, \tag{2.5.37}$$

whether or not the A_n are independent.If the A_n are independent, then

$$\sum_{n\geq 1} \mathbb{P}(A_n) = +\infty \Rightarrow \mathbb{P}(A) = 1, \tag{2.5.38}$$

9. *Khintchine equivalence lemma, I.* Let $\{X_n\}$ and $\{Y_n\}$ be two arbitrary sequence of r.v.'s. Then if $\sum_{n\geq 1} \mathbb{P}(X_n \neq Y_n) \leq +\infty$, then the Strong Law of Large Numbers holds for both sequences or none.

10. *Khintchine equivalence lemma, II.* If X is a r.v., then

$$\mathbb{E}|X| < \infty \Longleftrightarrow \sum_{k\geq 1} k\mathbb{P}\{k \leq |X| < k+1\} < \infty. \tag{2.5.39}$$

11. *Fatou lemma.* If the X_n are nonnegative r.v.'s, then

$$\mathbb{E}\{\liminf_{n\to\infty} X_n\} \leq \liminf_{n\to\infty} \mathbb{E}(X_n). \tag{2.5.40}$$

The Fatou lemma holds under a conditional setup too. That is,for $X_n|\mathcal{B}$, where $\mathcal{B}$ is a sub-sigma field of $\mathcal{A}$ and $(\Omega, \mathcal{A}, \mathbb{P})$ is the probability space.

12. *Uniform integrability.* Let $\{G_n,\ n \geq n_0\}$ be a sequence of d.f.'s, defined on $\mathbb{R}_q$, for some $q \geq 1$, and let $h(y) : \mathbb{R}_q \mapsto \mathbb{R}_p$, $p \geq 1$, be a continuous function such that

$$\sup_{n\geq n_0} \int_{\{||y||>a\}} ||h(y)|| dG_n(y) \to 0. \tag{2.5.41}$$

Then $h(.)$ is uniformly integrable (relatively to $\{G_n\}$).

13. *Lebesgue dominated convergence theorem.* For a sequence $\{X_n\}$ of measurable functions, suppose that there exists an Y, on the same probability space $(\Omega, \mathcal{A}, \mathbb{P})$,

$$|X_n| \leq Y \text{ a.e., where } \mathbb{E}|Y| < \infty \tag{2.5.42}$$

and either $X_n \to X$a.e., or $X_n \xrightarrow{\mathcal{D}} X$, for a suitable r.v. X. Then

$$\mathbb{E}|X_n - X| \to 0, \quad \text{as } n \to \infty. \tag{2.5.43}$$

A related version deemphasizing the dominating r.v. Y is the following: Let $\{X,\ X_n,\ n \geq n_0\}$ be a sequence of r.v.'s, such that $X_n - X \xrightarrow{\mathbb{P}} 0$ and

$$\mathbb{E}\{\sup_{n \geq n_0} |X_n|\} < \infty. \tag{2.5.44}$$

Then (2.5.43) holds, and hence $\mathbb{E}X_n \to \mathbb{E}X$ as $n \to \infty$. The Lebesgue dominated convergence theorem also holds under a conditional setup, namely sub-sigma fields of $\mathcal{A}$ given $\mathcal{B}$.

14. *Kolmogorov three series criterion.* The series $\sum_{n \geq 1} X_n$ of independent summands converges a.s. to a r.v. if and only if, for a fixed $c > 0$, the three series

$$\sum_{n \geq 1} \mathbb{P}\{|X_n| > c\}, \ \sum_{n \geq 1} \text{Var}(X_n^c) \ \text{ and } \ \sum_{n \geq 1} \mathbb{E}(X_n^c) \tag{2.5.45}$$

all converge, where

$$X_n^c = X_n I(|X_n| \leq c). \tag{2.5.46}$$

15. *Hewit-Savage zero-one Law:* Let $\{X_i;\ i \geq 1\}$ be i.i.d. r.v.'s. Then every exchangeable event (that remains invariant under any permutation of the indices $1, \ldots, n$ of $X_1, \ldots, X_n$) has probability either equal to 0 or 1.

16. *Fubini theorem.* Let $(\Omega, \mathcal{A}, \mathbb{P})$ be a product probability space with $\Omega = \Omega_1 \times \Omega_2$, $\mathcal{A} = \mathcal{A}_1 \times \mathcal{A}_2$ and $\mathbb{P} = \mathbb{P}_1 \times \mathbb{P}_2$ and let $(\bar{\mathcal{A}}_1, \bar{\mathbb{P}}_1)$, $(\bar{\mathcal{A}}_2, \bar{\mathbb{P}}_2)$, and $(\bar{\mathcal{A}}_1 \times \bar{\mathcal{A}}_2, \bar{\mathbb{P}}_1 \times \bar{\mathbb{P}}_2)$ be their completions. Let $f(w_1, w_2)$ be a measurable function with respect to $\bar{\mathcal{A}}_1 \times \bar{\mathcal{A}}_2$ and integrable with respect to $\bar{\mathbb{P}}_1 \times \bar{\mathbb{P}}_2$. Then
(a) $f(w_1, .)$ is $\bar{\mathcal{A}}_2$-measurable and $\bar{\mathbb{P}}_2$- integrable.
(b) The integral $\int f(., w_2) d\bar{\mathbb{P}}_2(w_2)$ is $\bar{\mathcal{A}}_1$-measurable and $\bar{\mathbb{P}}_1$-integrable.
(c) For $f(.,.)$ nonnegative or $\mathbb{P}_1 \times \mathbb{P}_2$ integrable,

$$\int\int f(w_1, w_2) d\bar{\mathbb{P}}_1(w_1) d\bar{\mathbb{P}}_2(w_2)$$

$$= \int [\int f(w_1, w_2) d\bar{\mathbb{P}}_2(w_2)] d\bar{\mathbb{P}}_1(w_1)$$
$$= \int [\int f(w_1, w_2) d\bar{\mathbb{P}}_1(w_1)] d\bar{\mathbb{P}}_2(w_2) \qquad (2.5.47)$$

17. *Faddeev lemma.* Let the sequence $\{f_n(t,u); (t,u) \in (0,1)^2, n \geq 1\}$ be densities in t for fixed $u \in (0,1)$, such that for every $\epsilon > 0$,

$$\lim_{n\to\infty} \int_{u-\epsilon}^{u+\epsilon} f_n(t,u)dt = 1.$$

Further assume that there exists another sequence $\{g_n(t,u), (t,u) \in (0,1)^2, n \geq 1\}$, such that $g_n(t,u)$ is $\nearrow$ or $\searrow$ in t according as t is in $(0,u)$ or $(u,1)$, when u, n are fixed, and

$$f_n(t,u) \leq g_n(t,u), \ \forall (t,u) \in (0,1)^2, \ \ n \geq 1;$$

$$\sup_n \int_0^1 g_n(t,u)dt < \infty.$$

Then for every integrable $\varphi(.) : [0,1] \mapsto \mathbb{R}$,

$$\lim_{n\to\infty} \int_0^1 \varphi(t) f_n(t,u)dt = \varphi(u) \ \text{ a.e. } \ (u \in [0,1]). \qquad (2.5.48)$$

18. *Uniform continuity and equicontinuity.* A function $f(.)$ defined on a metric space S (equipped with a metric $\rho : S \times S \mapsto \mathbb{R}^+$), and taking values in a metric space T (equipped with a metric $\delta(.)$), is said to be uniformly continuous on S, if for every $\epsilon > 0$, there exists an $\eta > 0$, such that

$$\rho(x_1, x_2) < \eta \Rightarrow \delta(f(x_1), f(x_2)) < \epsilon, \ \ \forall x_1, x_2 \in S. \qquad (2.5.49)$$

Let us now extend this definition to the case where $f(.)$ is stochastic. We introduce $\mathcal{B}(T)$ as the class of Borel subsets of T, and consider the probability space $(T, \mathcal{B}, \mathbb{P})$ where T is equipped with the metric $\delta : T \times T \mapsto \mathbb{R}^+$. In this way, we introduce $\mathcal{P}$ as the space of all probability measures on $(T, \mathcal{B})$, and we also interpret uniform continuity as *equicontinuity* with respect to the space $\mathcal{P}$.

19. The *weak***-topology* in $\mathcal{P}$ is the weakest topology such that, for every bounded continuous function φ and $P \in \mathcal{P}$, the mapping

$$P \to \int \varphi dP : \mathcal{P} \mapsto \mathbb{R} \ \text{ is continuous.} \qquad (2.5.50)$$

The space $\mathcal{P}$ of probability measures, topologized by the *weak**-topology, is itself a complete separable and metrizable space.

20. The *Prokhorov metric* (distance) between two members P_1 and $P_2 \in \mathcal{P}$ is defined by

$$\rho_P(P_1, P_2) = \inf\{\epsilon > 0 | P_1(A) \leq P_2\{A^\epsilon\} + \epsilon \; \forall A \in \mathcal{B}\} \tag{2.5.51}$$

where A^ϵ is the closed ϵ-neighborhood of A $(= \{x \in T : \inf_{y \in A} d(x, y) < \epsilon\})$, $\epsilon > 0$. The *Levi distance* between two d.f.'s F_1 and F_2 defined on $\mathbb{R}$ is given by

$$\rho_L(F_1, F_2) = \inf\{\epsilon : F_1(x - \epsilon) - \epsilon \leq F_2(x) \leq F_1(x + \epsilon) + \epsilon, \; \forall x \in \mathbb{R}\}. \tag{2.5.52}$$

Thus, for a fixed F and ϵ, we have the *Levi neighborhood* $\mathbf{n}_\epsilon^L(F)$ of d.f. F_2's that satisfy (2.5.52) for $F_1 = F$. Similarly, for a fixed $P \in \mathcal{P}$, (2.5.51) defines for every $\epsilon : 0 < \epsilon < 1$, a neighborhood of $\mathbf{n}_\epsilon^P(P)$ of probability measures ($P_2 \in \mathcal{P}$) termed the *Prokhorov neighborhood* of P [see (2.3.11)]. [Note that on $\mathbb{R}$, the *Kolmogorov distance* $d_K(F, G) = ||F - G|| = \sup\{|F(x) - G(x)| : x \in \mathbb{R}\}$ or for general $\mathcal{P}$, the *total variation* distance $d_{TV}(P_1, P_2) = \sup\{|P_1(A) - P_2(A)| : A \in \mathcal{B}\}$ do not generate the *weak**-topology mentioned before. Similarly, a *contamination neighborhood* $\mathbf{n}_\epsilon^c(F) = \{G : \; G = (1 - \epsilon)F + \epsilon H, \; H \in \mathcal{P}\}$ (see (2.3.10)) is not a neighborhood in a topological sense. In Section 2.3 we have referred to this as the *gross-error model.*]

2.5.4 Laws of Large Numbers and Related Inequalities

The probability inequality in (2.5.21), for $r = 2$, known as the Chebyshev inequality, is the precursor of the so-called laws of large numbers (LLN) which were originally developed for sums of independent r.v.'s, but have been extended to cover other dependent cases as well.

1. *Khintchine weak LLN.* Let $\{X_k;\; k \geq 1\}$ be a sequence of i.i.d. r.v.'s with a finite $\theta = \mathbb{E}X$. Let $\bar{X}_n = n^{-1}\sum_{i=1}^{n} X_i,\; n \geq 1$. Then

$$\bar{X}_n \to \theta, \;\text{ in probability, as }\; n \to \infty. \tag{2.5.53}$$

In particular, if the X_k can only assume values 0 and 1 with respective probabilities $1 - \pi$ and π $(0 < \pi < 1)$, then $\theta = \pi$ and (2.5.53) is referred to as the *Bernoulli LLN.*

2. *Khintchine strong LLN.* If the X_i are i.i.d. r.v's, then $\bar{X}_n \to c$ a.s., as $n \to \infty$, if and only if $\mathbb{E}X_i$ exists and $c = \mathbb{E}X_i$. Again, if the X_i are Bernoulli r.v.'s, $\bar{X}_n \to \pi$ a.s., as $n \to \infty$, and this is known as the *Borel strong LLN.*

3. *Markov weak LLN.* Let $\{X_k,\ k \geq 1\}$ be independent r.v.'s with $\mu_k = \mathbb{E}X_k$, and $\nu_{k,\delta} = \mathbb{E}|X_k - \mu_k|^{1+\delta} < \infty$, for some $\delta :\ 0 < \delta \leq 1,\ k \geq 1$. If

$$n^{-1-\delta} \sum_{k=1}^{n} \nu_{k,\delta} \to 0 \text{ as } \quad n \to \infty, \tag{2.5.54}$$

then

$$\bar{X}_n - \mathbb{E}\bar{X}_n \to 0, \text{ in probability, as } n \to \infty. \tag{2.5.55}$$

Note that (2.5.54), known as the *Markov condition*, compensates for the possible non-i.d. nature of the X_k.

4. *Kolmogorov strong LLN.* Let $X_k,\ k \geq 1$, be independent r.v.'s such that $\mu_k = \mathbb{E}X_k$ and $\sigma_k^2 = \text{Var}(X_k)$ exist for every $k \geq 1$, and further assume that

$$\sum_{k \geq 1} k^{-2}\sigma_k^2 < \infty. \tag{2.5.56}$$

Then

$$\bar{X}_n - \mathbb{E}\bar{X}_n \to 0 \text{ a.s.,} \quad \text{as } n \to \infty. \tag{2.5.57}$$

5. *Kolmogorov maximal inequality.* Let $\{T_n,\ n \geq 1\}$ be a zero mean martingale, such that $\mathbb{E}T_n^2$ exists for every $n \geq 1$. Then for every $t > 0$,

$$\mathbb{P}\{\max_{1 \leq k \leq n} |T_k| > t\} \leq t^{-2}\mathbb{E}(T_n^2). \tag{2.5.58}$$

If $\{T_n,\ n \geq m\}$ is a (zero mean) reversed martingale, then for every $N \geq n \geq m$ and $t > 0$,

$$\mathbb{P}\{\max_{n \leq k \leq N} |T_k| > t\} \leq t^{-2}\mathbb{E}(T_n^2). \tag{2.5.59}$$

The inequality (2.5.58) was originally established by Kolmogorov for T_n as the sum of (centered) independent r.v.'s; (2.5.59) holds for $\bar{X}_n$ as well as U-statistics, which are known to be reversed martingales. In this way, the Bernstein inequality (2.5.22) extends readily to the next case

6. *Bernstein inequality for submartingales.* Let $\{T_n,\ n \geq 1\}$ be a submartingale, such that $M_n(\theta) = \mathbb{E}\{e^{\theta T_n}\}$ exists for all $\theta \in (0, \theta_0)$. Then $\forall t > 0$,

$$\mathbb{P}\{\max_{1 \leq k \leq n} |T_k| > t\} \leq \inf_{\theta > 0}\{e^{-\theta t}M_n(\theta)\}. \tag{2.5.60}$$

If $\{T_n,\ n \geq m\}$ is a reversed submartingale, and $M_n(\theta)$ exists, then

$$\mathbb{P}\{\sup_{N \leq n} |T_N| > t\} \leq \inf_{\theta > 0}\{e^{-\theta t}M_n(\theta)\}, \quad \forall t > 0. \tag{2.5.61}$$

7. *Hájek-Rényi-Chow inequality:* Let $\{T_n, n \geq 1\}$ be a submartingale, and let $\{c_n, n \geq 1\}$ be a nonincreasing sequence of positive numbers. Assume that $T_n^+ = (T_n \vee 0)$ has a finite expectation for $n \geq 1$. Then for every $t > 0$,

$$\mathbb{P}\{\max_{1 \leq k \leq n} c_k T_k > t\} \leq t^{-1}\{c_1 \mathbb{E} T_1^+ + \sum_{k=2}^{n} c_k \mathbb{E}(T_k^+ - T_{k-1}^+)\}. \tag{2.5.62}$$

If $\{T_n, n \geq m\}$ is a reversed martingale, and the c_k are $\nearrow$, then

$$\mathbb{P}\{\max_{n \leq k \leq N} c_k T_k > t\} \leq t^{-1}\{c_n \mathbb{E} T_n^+ + \sum_{k=n+1}^{N} (c_k - c_{k-1}) \mathbb{E} T_k^+\}. \tag{2.5.63}$$

Note that (2.5.58) [or (2.5.59)] is a particular case of (2.5.62) or (2.5.63) when the c_k are all equal to 1.

8. *Kolmogorov strong LLN for martingales.* Let $\{T_n = \sum_{k \leq n} Y_k, n \geq 1\}$ be a martingale such that $\mathbb{E}|Y_k|^p$ exists for some $p: 1 \leq p \leq 2\ \forall k \geq 1$. Thus $\{T_n, n \geq 1\}$ is a zero mean L_p-martingale. Further let $\{b_n, n \geq 1\}$ be an increasing sequence of positive numbers, such that $b_n \nearrow \infty$ as $n \to \infty$, and

$$\sum_{n \geq 2} b_n^{-p} \mathbb{E}\{|Y_n|^p | Y_j, j < n\} < \infty \text{ a.s.} \tag{2.5.64}$$

Then

$$b_n^{-1} T_n \to 0 \text{ a.s.}, \quad \text{as } n \to \infty. \tag{2.5.65}$$

Note that (2.5.57) is a special case of (2.5.65) when $p = 2$, $b_n \equiv n$.

2.5.5 Central Limit Theorems

Test-statistics or estimators based on sample observations, termed *statistics*, are the basic tools for drawing statistical conclusions from the sample observations. Here the main objective of a thorough study relates to the distribution theory of statistics, termed the *sampling distribution.* Unfortunately, except in some specialized cases, the exact sampling distribution can not be easily derived and the task becomes prohibitively laborious as the sample size becomes large. On the other hand, a statistic T_n can be so normalized [say, $(T_n - a_n)/b_n$, with suitable $\{a_n\}$ and $\{b_n\}$] that as n becomes large, the distribution of the normalized sequence can well be approximated by some simple ones (such as the Poisson, normal, chi square, beta, gamma etc). In other words, the notion of *weak convergence* introduced in Definition 2.5.5 becomes applicable. In this book we will incorporate the basic concepts of weak convergence (at various levels of generality) in various contexts. For this reason we intend to present the basic results in a logical order, with basic lemmas and theorems preceding the theory they underlie.

Consider a sequence $\{\mathbf{T}_n,\, n \geq n_0\}$ of (normalized) stochastic vectors, and let $F_n(\mathbf{t}) = \mathbb{P}\{\mathbf{T}_n \leq \mathbf{t}\}$, $\mathbf{t} \in \mathbb{R}_p$, and for some $p \geq 1$. We conceive of a d.f. F also defined on $\mathbb{R}_p$, and we let $\mathbf{J}$ ($\subset \mathbb{R}_p$) be the set of points of continuity of F, and let $\mathbf{T}$ be a stochastic vector whose d.f. is F. Then, if

$$F_n(\mathbf{t}) \to F(\mathbf{t}),\ \forall \mathbf{t} \in \mathbf{J} \ \text{ as } n \to \infty, \tag{2.5.66}$$

we say that $F_n \xrightarrow{w} F$, or $\mathbf{T}_n \xrightarrow{\mathcal{D}} \mathbf{T}$, or $\mathcal{L}(\mathbf{T}_n \to \mathcal{L}(\mathbf{T})$. This definition will be extended later to more general stochastic elements (defined on appropriate function spaces).

1. *Helly-Bray lemma*: Let $\{F_n\}$ satisfy (2.5.66), and let $g(.)$ be a continuous function on a compact $\mathbf{C} \subset \mathbb{R}_p$ whose boundary lies in the set $\mathbf{J}$. Then

$$\int_{\mathbf{C}} g(.)dF_n(.) \to \int_{\mathbf{C}} g(.)dF(.) \ \text{ as } n \to \infty. \tag{2.5.67}$$

If $g(.)$ is bounded and continuous on $\mathbb{R}_p$, then (2.5.67) also holds when $\mathbf{C}$ is replaced by $\mathbb{R}_p$.

2. *Sverdrup Lemma*: Let $\{\mathbf{T}_n\}$ be a sequence of random vectors such that $\mathbf{T}_n \xrightarrow{\mathcal{D}} \mathbf{T}$, and let $\mathbf{g}(.) : \mathbb{R}_p \mapsto \mathbb{R}_q$, $q \geq 1$, be a continuous mapping. Then $\mathbf{g}(\mathbf{T}_n) \xrightarrow{\mathcal{D}} \mathbf{g}(\mathbf{T})$.

3. *Cramér-Wold theorem*. Let $\{\mathbf{T}_n\}$ be a sequence of random vectors. Then $\mathbf{T}_n \xrightarrow{\mathcal{D}} \mathbf{T}$, if and only if, for every fixed $\boldsymbol{\lambda} \in \mathbb{R}_p$, $\boldsymbol{\lambda}'\mathbf{T}_n \xrightarrow{\mathcal{D}} \boldsymbol{\lambda}'\mathbf{T}$.

4. *Lévy-Cramér theorem*. Let $\{F_n,\, n \geq n_0\}$ be a sequence of d.f.'s, defined on $\mathbb{R}_p$, for some $p \geq 1$, with the corresponding sequence of characteristic functions $\{\phi_n(\mathbf{t}),\, n \geq n_0\}, \mathbf{t} \in \mathbb{R}_p$. Then a necessary and sufficient condition for $F_n \xrightarrow{w} F$ is that $\phi_n(\mathbf{t}) \to \phi(\mathbf{t})\ \forall \mathbf{t} \in \mathbb{R}_p$, where $\phi(\mathbf{t})$ is continuous at $\mathbf{t} = \mathbf{0}$ and is the characteristic function of F.
First, we consider the case of $p = 1$ and T_n as sum of independent r.v.'s. This constitutes the main domain of the CLT (central limit theorem).

5. *Classical CLT*. Let $\{X_k,\, k \geq 1\}$ be i.i.d. r.v.'s with $\mathbb{E}X = \mu$ and $\text{Var}(X) = \sigma^2 < \infty$. Let $T_n = \sum_{k=1}^n X_k$, $n \geq 1$, and let

$$Z_n = (T_n - n\mu)/(\sigma\sqrt{n}). \tag{2.5.68}$$

Then, $Z_n \xrightarrow{\mathcal{D}} Z$, where Z has the standard normal d.f. If the $\mathbf{X}_k$ are p-vectors with $\boldsymbol{\mu} = \mathbb{E}\mathbf{X}$ and $\boldsymbol{\Sigma} = \text{Var}(\mathbf{X})$, then by incorporating the Cramér-Wold theorem, we have $\mathbf{Z}_n = n^{-1/2}(\mathbf{T}_n - n\boldsymbol{\mu}) \xrightarrow{\mathcal{D}} Z \sim \mathcal{N}_p(\mathbf{0}, \boldsymbol{\Sigma})$.

Specialized to the Bernoulli case where the X_k can only assume the values 0 and 1 with probability $1-\pi$ and π, respectively, we have

$$\big(n\pi(1-\pi)\big)^{-\frac{1}{2}}(T_n - n\pi) \xrightarrow{\mathcal{D}} \mathcal{N}(0,1).$$

This result is known as the *de Moivre-Laplace theorem.* The notation $\mathcal{N}_p(\mathbf{a},\mathbf{B})$ stands for a random p-vector that has a multi-normal d.f. with the mean vector $\mathbf{a}$ and dispersion matrix $\mathbf{B}$; for $p=1$, we drop the subscript p (or 1).

6. *Liapounov CLT.* Let X_k, $k \geq 1$, be independent r.v.'s with $\mu_k = \mathbb{E}X_k$, $\sigma_k^2 = \mathrm{Var}(X_k)$, and for some $\delta \in (0,1]$,

$$\nu_{2+\delta}^{(k)} = \mathbb{E}\{|X_k - \mu_k|^{2+\delta}\} < \infty \ \forall k \geq 1. \tag{2.5.69}$$

Now let $T_n = \sum_{k=1}^{n} X_k$, $\xi_n = \mathbb{E}T_n = \sum_{k=1}^{n} \mu_k$, $s_n^2 = \mathrm{Var}(T_n) = \sum_{k=1}^{n} \sigma_k^2$, $Z_n = (T_n - \xi_n)/s_n$, and let

$$\rho_n = s_n^{-2-\delta}\Big(\sum_{k=1}^{n} \nu_{2+\delta}^{(k)}\Big), \quad n \geq 1. \tag{2.5.70}$$

Then

$$\rho_n \to 0 \Rightarrow Z_n \xrightarrow{\mathcal{D}} \mathcal{N}(0,1) \ \text{ as } n \to \infty. \tag{2.5.71}$$

7. *Lindeberg-Feller theorem.* Define T_n, ξ_n, s_n^2, Z_n and so on, as in Liapounov's theorem. [There is no need of (2.5.69) and (2.5.70).] Consider the following three conditions.

A. *Uniform asymptotic negligibility (UAN)* condition,

$$\max\{\sigma_k^2/s_n^2 : 1 \leq k \leq n\} \to 0, \quad \text{as } n \to \infty. \tag{2.5.72}$$

B. *Asymptotic normality (AN)* condition,

$$\mathbb{P}\{Z_n \leq x\} \to \Phi(x) = (2\pi)^{-1/2} \int_{-\infty}^{x} e^{-\frac{1}{2}t^2}\,dt, \quad x \in \mathbb{R}. \tag{2.5.73}$$

C. *Lindeberg-Feller (uniform integrability)* condition, for every $\epsilon > 0$

$$\frac{1}{s_n^2}\sum_{k=1}^{n} \mathbb{E}\{(X_k-\mu_k)^2 I(|X_k-\mu_k| > \epsilon s_n)\} \to 0 \quad \text{as } n \to \infty. \tag{2.5.74}$$

Then **A** and **B** hold simultaneously if and only if **C** holds.
Observe that if we let $Y_{nk} = s_n^{-1}(X_k - \mu_k)$, $1 \leq k \leq n$, then (2.5.72) implies that $\max_{1\leq k\leq n} \mathbb{P}\{|Y_{nk}| > \epsilon\} \to 0$ as $n \to \infty$, for every $\epsilon > 0$, that is, the Y_{nk} are *infinitesimal.* The "if" part (i.e., **C**$\Rightarrow$ **A**,**B**) of the theorem is due to Lindeberg, while the "only if" part is due to Feller.

8. *Bounded CLT.* If the X_k are all bounded r.v.'s, it follows readily that $Z_n \xrightarrow{\mathcal{D}} \mathcal{N}(0,1)$ if and only if $s_n^2 \to \infty$ as $n \to \infty$.

9. *Hájek-Šidák CLT.* Let $\{Y_k,\ k \geq 1\}$ be a sequence of i.i.d. r.v.'s such that $\mathbb{E}Y = \mu$ and $\mathrm{Var}(Y) = \sigma^2$ exist, and let $\mathbf{c}_n = (c_{n1}, \ldots, c_{nn}),\ n \geq 1$, be a triangular scheme of real numbers such that as $n \to \infty$,

$$\max\{c_{nk}^2 / \sum_{k=1}^{n} c_{nk}^2 :\ 1 \leq k \leq n\} \to 0. \tag{2.5.75}$$

Then

$$Z_n = \frac{\sum_{k=1}^n c_{nk}(Y_k - \mu)}{\sigma(\sum_{k=1}^n c_{nk}^2)^{1/2}} \xrightarrow{\mathcal{D}} \mathcal{N}(0,1) \quad \text{as } n \to \infty.$$

By incorporating the Y_{nk} in the Lindeberg-Feller theorem or in the above theorem, we may conceive of a more general *triangular array* of row-wise independent r.v.'s:

10. *Triangular array CLT.* Consider a triangular array (of row-wise independent r.v.'s) $X_{nj},\ j \leq k_n,\ n \geq 1$, where $k_n \to \infty$ as $n \to \infty$. Then the X_{nk} form an infinitesimal system of r.v.'s and

$$Z_n = \sum_{k=1}^{k_n} X_{nk} \xrightarrow{\mathcal{D}} \mathcal{N}(0,1) \tag{2.5.76}$$

if and only if, for every $\epsilon > 0$, as $n \to \infty$,

$$\sum_{k=1}^{k_n} \mathbb{P}\{|X_{nk}| > \epsilon\} \to 0, \tag{2.5.77}$$

$$\sum_{k=1}^{k_n} \{\int_{\{|x|<\epsilon\}} x^2 dP_{nk}(x) - (\int_{\{|x|<\epsilon\}} x dP_{nk}(x))^2\} \to 1, \tag{2.5.78}$$

where $P_{nk}(x) = \mathbb{P}\{X_{nk} < x\},\ x \in \mathbb{R},\ k \leq k_n,\ n \geq 1$.

11. *Dependent CLT* (Dvoretzky 1972). For a triangular sequence $\{X_{nk},\ k \leq k_n;\ n \geq 1\}$, let $\mathcal{B}_{nk} = \mathcal{B}(X_{nj},\ j \leq k),\ k \geq 1$, and

$$\mu_{nk} = \mathbb{E}(X_{nk}|\mathcal{B}_{nk-1}) \text{ and } \sigma_{nk}^2 = \mathbb{E}(X_{nk}^2|\mathcal{B}_{nk-1}) - \mu_{nk}^2, \tag{2.5.79}$$

for $k \geq 1$, and let $Z_n = \sum_{k \leq k_n} X_{nk}$. Suppose that as $n \to \infty$,

$$\sum_{k=1}^{k_n} \mu_{nk} \to 0, \quad \sum_{k=1}^{k_n} \sigma_{nk}^2 \to 1, \tag{2.5.80}$$

and the (conditional) Lindeberg condition holds, that is, $\forall \epsilon > 0$,

$$\sum_{k=1}^{k_n} \mathbb{E}(X_{nk}^2 I(|X_{nk}| > \epsilon)|\mathcal{B}_{nk-1}) \to 0, \quad \text{as } n \to \infty. \tag{2.5.81}$$

Then $Z_n \xrightarrow{\mathcal{D}} \mathcal{N}(0,1)$.

12. *Martingale (array) CLT.* If for each n, the X_{nk} are martingale differences, then the μ_{nk} are 0 (a.e.), while $\mathbb{E}Z_n^2 = s_n^2 = \mathbb{E}(\sum_{k \leq k_n} \sigma_{nk}^2)$. Therefore, if

$$\{\sum_{k \leq k_n} \sigma_{nk}^2\}/s_n^2 \xrightarrow{\mathcal{P}} 1, \tag{2.5.82}$$

and the Lindeberg condition in (2.5.74) holds for the X_{nk}, then by the Chebyshev inequality (2.5.81) holds too. Both (2.5.82) and (2.5.74) imply that $Z_n/s_n \xrightarrow{\mathcal{D}} \mathcal{N}(0,1)$ as $n \to \infty$. This result applies to a martingale sequence as well (a result due to Brown 1971).

13. *Reverse martingale CLT.* Consider a sequence $\{T_k,\ k \geq 1\}$ that form a reverse martingale, and set $\mathbb{E}T_n = 0$, $\forall n \geq 1$. Let $Y_k = T_k - T_{k+1}$, $k \geq 1$ and $v_k^2 = \mathbb{E}(Y_k^2|T_{k+1}, T_{k+2}, \ldots)$ and let

$$w_n^2 = \sum_{k \geq n} v_k^2 \text{ and } s_n^2 = \mathbb{E}w_n^2 = \mathbb{E}T_n^2. \tag{2.5.83}$$

If (a) $w_n^2/s_n^2 \xrightarrow{\mathcal{P}} 1$ as $n \to \infty$, and (b) $\forall \epsilon > 0$,

$$w_n^{-2} \sum_{k \geq n} \mathbb{E}(Y_k^2 I(|Y_k| > \epsilon w_n | T_{k+1}, T_{k+2}, \ldots) \xrightarrow{\mathcal{P}} 0 \tag{2.5.84}$$

or (c)

$$w_n^{-2} \sum_{k \geq n} Y_k^2 \to 1, \text{ a.s.,} \quad \text{as } n \to \infty, \tag{2.5.85}$$

then $T_n/s_n \xrightarrow{\mathcal{D}} \mathcal{N}(0,1)$.

14. *Multivariate CLT.* Take any version of the preceding CLT's and use the Cramér-Wold theorem (3), to conclude that the CLT extends to the multivariate case under parallel regularity conditions.

15. *Renewal (CL) theorem.* Let $\{X_k,\ k \geq 1\}$ be a sequence of identically distributed nonnegative r.v.'s with mean $\mu > 0$ and variance σ^2 $(< \infty)$. Further assume that either the X_k are independent or $\{X_k - \mu,\ k \geq 1\}$ form a martingale-difference sequence. Let

$$T_n = X_1 + \ldots + X_n,\ n \geq 1,\ T_0 = 0. \tag{2.5.86}$$

For every $t > 0$, define

$$N_t = \max\{k : T_k \le t\},\ N_0 = 0. \tag{2.5.87}$$

Then, by definition, $T_{N_t} \le t < T_{N_t+1}$, $\forall t \ge 0$. Therefore, as $t \to \infty$,

$$t^{-1/2}(N_t - t/\mu) \xrightarrow{\mathcal{D}} \mathcal{N}(0, \sigma^2/\mu^3). \tag{2.5.88}$$

16. *Berry-Esséen theorem.* Consider the setup of the classical CLT in (2.5.68) and assume that for some $\delta :\ 0 < \delta \le 1,\ \mathbb{E}|X-\mu|^{2+\delta} = \nu_{2+\delta} < \infty$. Let $\Phi_n(x) = \mathbb{P}\{Z_n \le x\},\ x \in \mathbb{R}$ and $\Phi(x)$ be the standard normal d.f. Then there exists a constant C such that

$$\Delta_n = \sup_{x\in\mathbb{R}} |\Phi_n(x) - \Phi(x)| \le Cn^{-\delta/2}(\nu_{2+\delta}/\sigma^{2+\delta}). \tag{2.5.89}$$

Various modifications of this uniform bound are available in the literature.

17. *Edgeworth expansions.* For the statistic Z_n in the preceding theorem, assume the existence of higher-order (central) moments up to a certain order. Then

$$\begin{aligned}\Phi_n(x) &= \Phi(x) - \frac{1}{\sqrt{n}}(\mu_3/6\sigma^3)\Phi^{(3)}(x)\\ &+ \frac{1}{n}\{(\frac{\mu_4 - 3\sigma^4}{\sigma^4})\Phi^{(4)}(x) + \frac{1}{72}(\frac{\mu_3^2}{\sigma^6})\Phi^{(6)}(x)\}\\ &+O(n^{-3/2}),\end{aligned} \tag{2.5.90}$$

where $\Phi^{(k)}(x) = (d^k/dx^k)\Phi(x),\ k \ge 0$, and $\mu_k = \mathbb{E}(X-\mu)^k,\ k \ge 1$, are the central moments of the r.v. X, $\mu = \mathbb{E}X$. Instead of $O(n^{-3/2})$, we could have gone up to $O(n^{-(s-2)/2})$ for some $s \ge 3$ by imposing appropriate moment conditions on X. For a smooth function $H(.) : \mathbb{R}_p \mapsto \mathbb{R}_q,\ p, q \ge 1$, a more general version of this expansion is given by Bhattacharya and Ghosh (1978).

18. *Law of the iterated logarithms.* If the $X_i,\ i \ge 1$ are i.i.d. r.v.'s with finite mean μ and variance σ^2, then

$$\limsup_{n\to\infty} \frac{\sum_{i=1}^n (X_i - \mu)}{(2\sigma^2 n \log\log n)^{1/2}} = 1, \text{ with probability 1.} \tag{2.5.91}$$

The clause that the X_i are identically distributed can be eliminated by either a boundedness condition or a moment condition in the following manner:

Let $X_i, i \ge 1$ be independent r.v.'s with means μ_i and finite variance $\sigma_i^2,\ i \ge 1$, such that

$$B_n^2 = \sum_{i=1}^n \sigma_i^2 \to \infty \text{ as } n \to \infty.$$

Moreover assume that either of the following two conditions hold:

Kolmogorov. There exists a suitable sequence $\{m_n\}$ of positive numbers, such that with probability 1,

$$|X_n - \mu_n| \leq m_n \text{ where } m_n = o(B_n(\log\log B_n)^{-1/2}), \text{ as } n \to \infty.$$

Chung. For some $\epsilon > 0$,

$$B_n^{-3}\{\sum_{i=1}^{n} \mathbb{E}|X_i - \mu_i|^3\} = O((\log B_n)^{-1-\epsilon}).$$

Then

$$\limsup_{n\to\infty} \frac{\sum_{i=1}^{n}(X_i - \mu_i)}{(2B_n^2 \log\log B_n)^{1/2}} = 1, \text{ with probability } 1. \tag{2.5.92}$$

For extensions to some dependent sequences of r.v.'s, we may refer to Stout (1974). The derivations of these laws of iterated logarithms (LIL) are based on the CLT with a suitable rate of convergence.

2.5.6 Limit Theorems Allied to CLTs

The results presented in the preceding subsection are mostly for linear statistics, and may not cover other important cases arising in practice (particularly in robust and nonparametric statistics). There are are some allied results that provide the access to the CLT by suitable approximations, and we present some of these in this subsection.

1. *Slutzky theorem.* Let $\{X_n\}$ and $\{Y_n\}$ be two sequences of r.v.'s (not necessarily independent), such that

$$X_n \xrightarrow{\mathcal{D}} X \text{ and } Y_n \xrightarrow{\mathcal{P}} c, \text{ a constant, as } n \to \infty, \tag{2.5.93}$$

where X has a d.f. F (which may be degenerate at a point). Then

$$\begin{aligned} &X_n \pm Y_n \xrightarrow{\mathcal{D}} X \pm c, \ X_n Y_n \xrightarrow{\mathcal{D}} cX, \\ &X_n/Y_n \xrightarrow{\mathcal{D}} X/c, \text{ if } c \neq 0. \end{aligned} \tag{2.5.94}$$

The result extends directly to the case where the X_n or Y_n are vector valued random elements. A couple of examples can be cited that illustrate the utility of the Slutzky theorem.

Example 2.5.2. Let X_i, $i \geq 1$ be i.i.d. r.v.'s with mean μ and variance σ^2. An unbiased estimator of σ^2 is

$$S_n^2 = \frac{1}{n-1}\sum_{i=1}^{n}(X_i - \bar{X}_n)^2$$

$$= \frac{n}{n-1}\{\frac{1}{n}\sum_{i=1}^{n}(X_i-\mu)^2 - (\bar{X}_n-\mu)^2\}, \tag{2.5.95}$$

where $\sqrt{n}(\bar{X}_n-\mu)^2 \xrightarrow{\mathcal{P}} 0$ by the Chebyshev inequality, while by the classical CLT, whenever $\mathbb{E}X^4 < \infty$,

$$\sqrt{n}\{\frac{1}{n}\sum_{i=1}^{n}(X_i-\mu)^2 - \sigma^2\} \xrightarrow{\mathcal{D}} \mathcal{N}(0, \mu_4-\sigma^4), \tag{2.5.96}$$

where $\mu_4 = \mathbb{E}(X-\mu)^4$. Finally $n/(n-1) \to 1$ as $n \to \infty$. As a result, by the Slutzky theorem, we claim that under $\mu_4 < \infty$,

$$\sqrt{n}(S_n^2 - \sigma^2) \xrightarrow{\mathcal{D}} \mathcal{N}(0, \mu_4 - \sigma^4), \tag{2.5.97}$$

although S_n^2 does not have independent summands.

Example 2.5.3. In the same setup as in the preceding example, let

$$t_n = \frac{\sqrt{n}(\bar{X}_n-\mu)}{S_n} = \frac{\sqrt{n}(\bar{X}_n-\mu)/\sigma}{(S_n/\sigma)}, \tag{2.5.98}$$

and note that by (2.5.95), $S_n^2/\sigma^2 \xrightarrow{\mathcal{P}} 1$ as $n \to \infty$, while the CLT applies to $\sqrt{n}(\bar{X}_n-\mu)/\sigma$. Therefore by the Slutzky theorem, we conclude that $t_n \xrightarrow{\mathcal{D}} \mathcal{N}(0,1)$ as $n \to \infty$.

2. *Transformation on statistics.* Suppose that $\{T_n\}$ is a sequence of statistics and $g(T_n)$ is a transformation on T_n such that

$$n^{1/2}(T_n - \theta)/\sigma \xrightarrow{\mathcal{D}} \mathcal{N}(0,1), \tag{2.5.99}$$

and suppose also that $g : \mathbb{R} \mapsto \mathbb{R}$ is a continuous function such that $g'(\theta)$ exists and is different from 0. Then

$$n^{1/2}[g(T_n) - g(\theta)]/\{\sigma g'(\theta)\} \xrightarrow{\mathcal{D}} \mathcal{N}(0,1). \tag{2.5.100}$$

As an example, consider S_n^2 in (2.5.95), and take $g(S_n^2) = S_n = \sqrt{S_n^2}$. It is easy to verify that whenever $\mu_4 < \infty$,

$$\sqrt{n}(S_n - \sigma) \xrightarrow{\mathcal{D}} \mathcal{N}(0, (\mu_4-\sigma^4)/4\sigma^2). \tag{2.5.101}$$

We proceed similarly in the case where $g(.)$ and/or T_n are vector-valued functions or statistics.

3. *Variance stabilizing transformations.* Suppose that in (2.5.99), $\sigma^2 = h^2(\theta)$ is a nonnegative function of θ. If we choose $g(.)$ such that $g'(\theta)h(\theta) = c$, so that

$$g(\theta) = c\int_0^\theta [h(y)]^{-1}dy, \tag{2.5.102}$$

then, we have

$$n^{1/2}[g(T_n) - g(\theta)] \xrightarrow{\mathcal{D}} \mathcal{N}(0, c^2). \tag{2.5.103}$$

Important examples are
(a) $g(T_n) = \sin^{-1}\sqrt{T_n}$ (binomial proportion),
(b) $g(T_n) = \sqrt{T_n}$ (Poisson r.v.)
(c) $g(T_n) = \log T_n$ ($T_n \equiv S_n^2$ in (2.5.95)), and
(d) $g(T_n) = \tanh^{-1} T_n$ ($T_n \equiv$ sample correlation coefficient).

4. *Projection theorem* (Hoeffding-Hájek). Let $T_n = T(X_1, \ldots, X_n)$ be a symmetric function of $X_1, \ldots, X_n$. For each i $(= 1, \ldots, n)$, let

$$T_{ni} = \mathbb{E}[T_n|X_i] - \mathbb{E}T_n; \;\; T_n^\star = \sum_{i=1}^{n} T_{ni}, \tag{2.5.104}$$

and note that the T_{ni} are independent r.v.'s. Moreover

$$\mathbb{E}T_n^\star = 0 \;\text{ and }\; \mathrm{Var}(T_n^\star) = \sum_{i=1}^{n} \mathbb{E}(T_{ni}^2). \tag{2.5.105}$$

Further

$$\begin{aligned} &\mathbb{E}\{T_n^\star(T_n - \mathbb{E}T_n)\} \\ &= \sum_{i=1}^{n} \mathbb{E}\{T_{ni}(T_n - \mathbb{E}T_n)\} \\ &= \sum_{i=1}^{n} \mathbb{E}\{[\mathbb{E}(T_n - \mathbb{E}T_n|X_i)](T_n - \mathbb{E}T_n)\} \\ &= \sum_{i=1}^{n} \mathbb{E}\{[\mathbb{E}(T_n - \mathbb{E}T_n|X_i)]\mathbb{E}\{(T_n - \mathbb{E}T_n)|X_i\}\} \\ &= \sum_{i=1}^{n} \mathbb{E}(T_{ni}^2) = \mathrm{Var}(T_n^\star). \end{aligned} \tag{2.5.106}$$

Therefore

$$\begin{aligned} &\mathbb{E}\Big\{[(T_n - \mathbb{E}T_n) - T_n^\star]^2\Big\} \\ &= \mathbb{E}(T_n - \mathbb{E}T_n)^2 + \mathrm{Var}(T_n^\star) - 2\mathrm{Var}(T_n^\star) \\ &= \mathrm{Var}(T_n) - \mathrm{Var}(T_n^\star) \end{aligned} \tag{2.5.107}$$

Thus, whenever

$$[\mathrm{Var}(T_n) - \mathrm{Var}(T_n^\star)]/\mathrm{Var}(T_n^\star) \to 0 \;\text{ as } n \to \infty, \tag{2.5.108}$$

$(T_n - \mathbb{E}T_n - T_n^\star)/\sqrt{\text{Var}(T_n^\star)} \xrightarrow{\mathcal{P}} 0$, while a suitable version of the CLT can be adopted to show that

$$T_n^\star/\sqrt{\text{Var}(T_n^\star)} \xrightarrow{\mathcal{D}} \mathcal{N}(0,1). \qquad (2.5.109)$$

Consequently, by the Slutzky theorem,

$$(T_n - \mathbb{E}T_n)/\sqrt{\text{Var}(T_n^\star)} \xrightarrow{\mathcal{D}} \mathcal{N}(0,1), \qquad (2.5.110)$$

and it is also possible to replace $\text{Var}(T_n^\star)$ by $\text{Var}(T_n)$. Hoeffding (1948) incorporated this projection technique for U-statistics which are symmetric and unbiased estimators but may not have generally independent summands; Hájek (1968) popularized this technique for a general class of rank statistics where also the independence of the summands is vitiated. Van Zwet (1984) used this projection (termed the *Hoeffding-decomposition)* for general symmetric statistics for deriving Berry-Esséen-type bounds.

2.5.7 CLTs for Quadratic Forms

In asymptotic statistical inference, quadratic function(al)s play a basic role, especially in multiparameter problems. First, we present two basic (finite sample size) results that provide the access to the desired asymptotics.

1. *Courant theorem.* Let $\mathbf{x} \in \mathbb{R}_p$ for some $p \geq 1$, and let $\mathbf{A}$ and $\mathbf{B}$ be two positive semidefinite (p.s.d.) matrices, $\mathbf{B}$ being nonsingular. Then

$$\begin{aligned} \lambda_p &= ch_{\min}(\mathbf{A}\mathbf{B}^{-1}) \\ &= \inf\{\frac{\mathbf{x}'\mathbf{A}\mathbf{x}}{\mathbf{x}'\mathbf{B}\mathbf{x}} : \mathbf{x} \in \mathbb{R}_p\} \\ &\leq \sup\{\frac{\mathbf{x}'\mathbf{A}\mathbf{x}}{\mathbf{x}'\mathbf{B}\mathbf{x}} : \mathbf{x} \in \mathbb{R}_p\} \\ &= ch_{\max}(\mathbf{A}\mathbf{B}^{-1}) = \lambda_1 \end{aligned} \qquad (2.5.111)$$

2. *Cochran theorem.* Let $\mathbf{X} \sim \mathcal{N}_p(\boldsymbol{\mu}, \boldsymbol{\Sigma})$ where $rank(\boldsymbol{\Sigma}) = q \leq p$. Let $\mathbf{A}$ be a p.s.d. matrix such that $\mathbf{A}\boldsymbol{\Sigma}\mathbf{A} = \mathbf{A}$. Then

$$\mathbf{X}'\mathbf{A}\mathbf{X} \sim \chi^2_{q,\Delta}; \quad \Delta = \boldsymbol{\mu}'\mathbf{A}\boldsymbol{\mu}, \qquad (2.5.112)$$

where $\chi^2_{a,\delta}$ stands for a r.v. having the noncentral chi-square distribution with a DF and noncentrality parameter δ. For $\boldsymbol{\mu} = \mathbf{0}$, $\mathbf{X}'\mathbf{A}\mathbf{X}$ has the central chi square distribution with q DF.

3. *Slutzky-Cochran theorem:* Let $\{\mathbf{T}_n\}$ be a sequence of stochatic p-vectors such that $\sqrt{n}(\mathbf{T}_n - \boldsymbol{\theta}) \xrightarrow{\mathcal{D}} \mathcal{N}_p(\mathbf{0}, \boldsymbol{\Sigma})$, where $R(\boldsymbol{\Sigma}) = q : 1 \leq q \leq p$. Let $\{\mathbf{A}_n\}$

be a sequence of (random) matrices such that $\mathbf{A}_n \xrightarrow{\mathcal{P}} \mathbf{A}$ and $\mathbf{A}\boldsymbol{\Sigma}\mathbf{A} = \mathbf{A}$. Then

$$Q_n = n(\mathbf{T}_n - \boldsymbol{\theta})'\mathbf{A}_n(\mathbf{T}_n - \boldsymbol{\theta}) \xrightarrow{\mathcal{D}} \chi^2_{q,0}. \tag{2.5.113}$$

This result extends directly to local alternative setups, and will be considered later. For a class of statistical models, $\boldsymbol{\Sigma}$ itself is an idempotent matrix of rank q $(\leq p)$, so that

$$\boldsymbol{\Sigma} = \boldsymbol{\Sigma}^2 \text{ and } Trace(\boldsymbol{\Sigma}) = R(\boldsymbol{\Sigma}) = q\ (\leq p). \tag{2.5.114}$$

In that case $\mathbf{A} = \boldsymbol{\Sigma}$ satisfies the condition that $\mathbf{A}\boldsymbol{\Sigma}\mathbf{A} = \boldsymbol{\Sigma}^3 = \boldsymbol{\Sigma}^2 = \boldsymbol{\Sigma} = \mathbf{A}$, and hence (2.5.13) holds.

4. *Cochran partition theorem:* Suppose that Q_n is defined as in (2.5.113) and that there exist k (≥ 1) p.s.d. matrices $\mathbf{A}_{n1}, \ldots, \mathbf{A}_{nk}$, such that $\mathbf{A}_{n1} + \ldots + \mathbf{A}_{nk} = \mathbf{A}_n$ and $\mathbf{A}_{ni} \to \mathbf{A}_i : \mathbf{A}_i\boldsymbol{\Sigma}\mathbf{A}_i = \mathbf{A}_i$, $R(\mathbf{A}_i) = q_i$, $i = 1, \ldots, k$. Let Qni be defined as in (2.5.113) with $\mathbf{A}_n$ replaced by $\mathbf{A}_{ni}$, $i = 1, \ldots, k$. Then a necessary and sufficient condition that (a) $Q_{ni} \xrightarrow{\mathcal{D}} \chi^2_{q_i,0}$, $i = 1, \ldots, k$, and (b) $Q_{n1}, \ldots, Q_{nk}$ are asymptotically independent, is that $\mathbf{A}_i\mathbf{A}_j = \mathbf{0}$, $\forall i \neq j = 1, \ldots, k$.

2.5.8 Contiguity of Probability Measures

An excellent account of this concept and its impact on asymptotics is available in Hájek and Šidák (1967, ch. 6).

Definition 2.5.6 Let $\{P_n\}$ and $\{Q_n\}$ be two sequence of (absolutely continuous) probability measures on measure spaces $\{(\Omega_n, \mathcal{A}_n, \mu_n)\}$. Let $p_n = dP_n/d\mu_n$ and $q_n = dQ_n/d\mu_n$. Then, if for any sequence of events $\{A_n\}$: $A_n \in \mathcal{A}_n$,

$$[P_n(A_n) \to 0] \Rightarrow [Q_n(A_n) \to 0], \tag{2.5.115}$$

the sequence of measures $\{Q_n\}$ is said to be contiguous to $\{P_n\}$.
Note that if P_n and Q_n are L_1-norm equivalent,

$$||P_n - Q_n|| = \sup\{|P_n(A_n) - Q_n(A_n)| : A_n \in \mathcal{A}_n\} \to 0, \tag{2.5.116}$$

then (2.5.115) holds, but the converse may not be true. If T_n is $\mathcal{A}_n$- measurable, then the contiguity ensures that

$$T_n \to 0, \text{ in } P_n\text{-probability } \Rightarrow T_n \to 0, \text{ in } Q_n\text{-probability}. \tag{2.5.117}$$

In this context, we may define the *likelihood ratio* statistic as

$$L_n = \begin{cases} q_n/p_n, & \text{if } p_n > 0, \\ 1, & \text{if } p_n = q_n = 0, \\ \infty, & \text{if } 0 = p_n < q_n. \end{cases} \tag{2.5.118}$$

LeCam (1960) characterized the contiguity of $\{Q_n\}$ to $\{P_n\}$ by two basic lemmas: First, if

$$\log L_n \xrightarrow{\mathcal{D}} \mathcal{N}(-\frac{1}{2}\sigma^2, \sigma^2) \text{ (under } \{P_n\}), \tag{2.5.119}$$

then $\{Q_n\}$ is contiguous to $\{P_n\}$.

Second, if

$$(T_n, \log L_n) \xrightarrow{\mathcal{D}} \mathcal{N}_2(\boldsymbol{\mu}, \boldsymbol{\Sigma}) \text{ (under } \{P_n\}), \tag{2.5.120}$$

where $\boldsymbol{\mu} = (\mu_1, \mu_2)'$ and $\boldsymbol{\Sigma} = ((\sigma_{ij}))_{i,j=1,2}$ with $\mu_2 = -\frac{1}{2}\sigma_{22}$, then

$$T_n \xrightarrow{\mathcal{D}} \mathcal{N}(\mu_1 + \sigma_{12}, \sigma_{11}) \text{ (under } \{Q_n\}). \tag{2.5.121}$$

An important feature of this result is that the limit distribution under $\{P_n\}$ lends itself directly to the one under $\{Q_n\}$ by a simple adjustment of the mean. In nonparametrics simpler ways of proving (2.5.120) exist, but the distribution theory under $\{Q_n\}$ can be quite involved. The beauty of this contiguity-based proof lies in the simplicity of the proof for such contiguous alternatives. For an excellent treatise of this subject matter, we refer to the classical textbook of Hájek and Šidák (1967, ch. 6). The concept of contiguity has much broader scope in statistical asymptotics, and we shall refer to some of this later. A significant development in this direction is the Hájek-Inagaki-LeCam theorem.

2.5.9 Hájek-Inagaki-LeCam Theorem and the LAN Condition

Following LeCam and Yang (1990), consider for each n (≥ 1), a family of probability measures $\{P_{\boldsymbol{\theta},n} : \boldsymbol{\theta} \in \boldsymbol{\Theta}\}$ on some measure space $(\Omega_n, \mathcal{A}_n)$, where $\boldsymbol{\Theta} \in \mathbb{R}_k$, for some $k \geq 1$. Let δ_n (typically, $n^{-1/2}$) be a positive number, and assume that (1) the true value of $\boldsymbol{\theta}$ is interior to $\boldsymbol{\Theta}$, and (2) for $\mathbf{t}_n \in$ bounded set B, $\{P_{\boldsymbol{\theta}+\delta_n\mathbf{t}_n,n}\}$ and $\{P_{\boldsymbol{\theta},n}\}$ are contiguous. Denote the log-likelihood ratio by

$$\Lambda_n(\boldsymbol{\theta} + \delta_n\mathbf{t}_n, \boldsymbol{\theta}) = \log\{dP_{\boldsymbol{\theta}+\delta_n\mathbf{t}_n,n}/dP_{\boldsymbol{\theta},n}\}. \tag{2.5.122}$$

Then the family $\varepsilon_n = \{P_{\boldsymbol{\eta},n}, \boldsymbol{\eta} \in \boldsymbol{\Theta}\}$ is called *locally asymptotically quadratic* (LAQ) at $\boldsymbol{\theta}$ if there exist a stochastic vector $\mathbf{S}_n$ and a stochastic matrix $\mathbf{K}_n$, such that

$$\Lambda_n(\boldsymbol{\theta} + \delta_n\mathbf{t}_n, \boldsymbol{\theta}) - \mathbf{t}_n'\mathbf{S}_n + \frac{1}{2}\mathbf{t}_n'\mathbf{K}_n\mathbf{t}_n \to 0, \tag{2.5.123}$$

in $P_{\boldsymbol{\theta},n}$-probability, for every $\mathbf{t}_n \in B$, where $\mathbf{K}_n$ is a.s. p.d. The LAN *(locally asymptotically normal)* conditions refer to the particular case of LAQ where $\mathbf{K}_n$ can be taken as nonstochastic, while *locally asymptotically mixed normal* (LAMN) differ from the LAQ conditions in that the limiting distribution of the matrices $\mathbf{K}_n$, if it exists, does not depend on $\mathbf{T}_n$ $(\in B)$. Also, note that a density $f(x, \boldsymbol{\eta})$ is DQM *(differentiable in quadratic mean)* at $\boldsymbol{\theta}$ if there exist vectors $\mathbf{V}(x)$, such that

$$\int \{|\boldsymbol{\theta\eta}|^{-1}|\sqrt{f(x,\boldsymbol{\eta})} - \sqrt{f(x,\boldsymbol{\theta})} - (\boldsymbol{\eta}-\boldsymbol{\theta})'\mathbf{V}(x)|\}^2 d\mu(x) \to 0 \text{ as } \boldsymbol{\eta} \to \boldsymbol{\theta}. \tag{2.5.124}$$

Therefore DQM $\Rightarrow$ LAQ.

Convolution Theorem. For the model $\mathcal{F}_{\boldsymbol{\theta},n}\{P_{\boldsymbol{\theta}+\boldsymbol{\tau}\delta_n,n}; \boldsymbol{\tau} \in I\!\!R_k\}$, satisfying the LAMN condition, let $\mathcal{L}(\mathbf{K}_n|P_{\boldsymbol{\theta},n}) \to \mathcal{L}(\mathbf{K})$ with nonrandom $\mathbf{K}$. If $\mathbf{T}_n$ is an estimator of $\mathbf{A}\boldsymbol{\tau}$ for a given nonrandom matrix $\mathbf{A}$ such that $\mathcal{L}(\mathbf{T}_n - \mathbf{A}\boldsymbol{\tau}|\boldsymbol{\theta} + \boldsymbol{\tau}\delta_n)$ tends to a limit H independent of $\boldsymbol{\tau}$, then $H(.)$ is the distribution of $\mathbf{A}\mathbf{K}^{-1/2}\mathbf{Z} + \mathbf{U}$, where $\mathbf{Z}$ and $\mathbf{U}$ are independent, and $\mathbf{Z} \sim \mathcal{N}(\mathbf{0}, \mathbf{I})$. If $\hat{\boldsymbol{\xi}}_n$ is a BAN estimator of $\mathbf{A}\boldsymbol{\tau}$, $\hat{\boldsymbol{\xi}}_n \xrightarrow{\mathcal{D}} \mathbf{A}\mathbf{K}^{-1/2}\mathbf{Z}$, we may write

$$\mathbf{T}_n \stackrel{\mathcal{D}}{=} \hat{\boldsymbol{\xi}}_n + \mathbf{U} \quad \text{as } n \to \infty, \tag{2.5.125}$$

where $\hat{\boldsymbol{\xi}}_n$ and $\mathbf{U}$ are asymptotically independent.

2.5.10 Weak Convergence of Probability Measures

We refer to the function space setups (including the $C[0,1]$ and $D[0,1]$ spaces) formulated at the beginning of this section. During the past 30 years, weak convergence of probability measures has been studied under increasing generality for such function spaces, and an up-to-date treatment of this is beyond the scope of the present book. Rather, we provide here only those parts of this development having relevance to the theory outlined in subsequent chapters. Dudley (1985) provides an excellent and extensive coverage of this topic.
Let S be a *metric space* and let $\mathcal{S}$ be the class of Borel sets in S. Let $\{(\Omega_n, \mathcal{B}_n, \mu_n);\ n \geq n_0\}$ be a sequence of probability spaces, let T_n be a measurable mapping : $(\Omega_n, \mathcal{B}_n) \mapsto (S, \mathcal{S})$, and let P_n be the probability measure induced in $(S, \mathcal{S})$ by T_n, for $n \geq n_0$. Further let $(\Omega, \mathcal{B}, \mu)$be a probability space, $T: \ (\Omega, \mathcal{B}) \mapsto (S, \mathcal{S})$, be a measurable map, and let P be the probability measure induced by T in $(S, \mathcal{S})$. The space spanned by Ω_n may not necessarily contain Ω.

Definition 2.5.7 The sequence $\{P_n,\ n \geq n_0\}$ of probability measures on $(S, \mathcal{S})$ is said to converge weakly to P, if

$$\int g dP_n \to \int g dP, \ \ \forall g \in C(S) \tag{2.5.126}$$

where $C(S)$ is the class of bounded, continuous, real functions on S. In notation, we denote it by $P_n \Rightarrow P$.

It follows that (2.5.126) implies that for every continuous functional h assuming values in $\mathbb{R}_k$, for $k \geq 1$ as $n \to \infty$,

$$h(T_n) \xrightarrow{\mathcal{D}} h(T) \quad \text{when} \quad P_n \Rightarrow P. \tag{2.5.127}$$

The continuity assumption on h may be replaced by assuming that the set of discontinuities of h has P-measure 0. There is a basic difference between Definitions 2.5.5 and 2.5.7. Definition 2.5.5 entails the *convergence of finite-dimensional* laws while (2.5.126) entails additionally the *tightness* or *relative compactness* of $\{P_n\}$.

Definition 2.5.8 (LeCam-Prokhorov). A family Π of probability measures on $(S, \mathcal{S})$ is said to be *tight* if for every $\epsilon > 0$, there exists a compact set K_ϵ such that

$$P(K_\epsilon) > 1 - \epsilon, \quad \forall P \in \Pi. \tag{2.5.128}$$

In subsequent chapters various stochastic processes, some of which are multiparamter, will be encountered. Mostly one would then have a functional space that can be reduced by a proper transformation to the $D[0,1]^p$ space - a multi-parameter extension of the $D[0,1]$ space, as outlined at the beginning of this section. In the same way the $C[0,1]$ space lends itself to the $C[0,1]^p$ space, where $C[0,1]^p$ is a subspace of $D[0,1]^p$. With the $C[0,1]^p$ space we associate the *uniform topology* specified by the metric

$$\rho(x, y) = \sup\{|x(\mathbf{t}) - y(\mathbf{t})| : \mathbf{t} \in [0,1]^p\}. \tag{2.5.129}$$

Under the metric ρ, $C[0,1]^p$ is a complete and separable metric space. But the space $D[0,1]^p$ is the space of all real-valued functions $f : [0,1]^p \mapsto \mathbb{R}$, where f has only discontinuities of the first kind. For this we extend the metric ρ_S in (2.5.1) as follows: Let Λ denote the class of all strictly increasing, continuous mappings $\lambda : [0,1] \mapsto [0,1]$. Also let $\boldsymbol{\lambda}(\mathbf{t}) = (\lambda_1(t_1), \ldots, \lambda_p(t_p))$, $\boldsymbol{\lambda} \in \Lambda^p$ and $\mathbf{t} \in [0,1]^p$. Then the Skorokhod distance between two elements x and y in $D[0,1]^p$ is defined by

$$\rho_S(x, y) = \inf\Big\{\epsilon > 0 : \exists \boldsymbol{\lambda} \in \Lambda^p : \text{ s.t. } \sup_{\mathbf{t} \in [0,1]^p} ||\boldsymbol{\lambda}(\mathbf{t}) - \mathbf{t}|| < \epsilon$$
$$\text{and } \sup_{\mathbf{t} \in [0,1]^p} |x(\mathbf{t}) - y(\boldsymbol{\lambda}(\mathbf{t}))| < \epsilon\Big\}. \tag{2.5.130}$$

Consider now a partition of $[0,1]^p$ formed by finitely many hyperplanes parallel to the p principal axes such that each element of this partition is a left-closed right-open rectangle of diameter at least δ (> 0). A typical rectangle is denoted by R, and the partition generated by the hyperplanes is denoted by $\mathcal{R}$. Let

$$w_f^0(R) = \sup\{|f(\mathbf{t}) - f(\mathbf{s})| : \mathbf{s}, \mathbf{t} \in R\} \tag{2.5.131}$$

and

$$w'_f(\delta) = \inf_{\mathcal{R}}\{\max_{R \in \mathcal{R}}[w_f^0(R)]\}. \qquad (2.5.132)$$

Then a function $f : [0,1]^p \mapsto \mathbb{R}$, belongs to $D[0,1]^p$, iff,

$$\lim_{\delta \to 0} w'_f(\delta) = 0. \qquad (2.5.133)$$

1. *Compactness theorem:* A sequence $\{P_n\}$ of probability measures on $(D[0,1]^p, \mathcal{D}^p)$ is tight (relatively compact) iff
(a) for every $\epsilon > 0$, there exists a $M_\epsilon < \infty$ such that

$$P_n(\{f \in D[0,1]^p : ||f|| > M_\epsilon\}) \leq \epsilon, \ \ \forall n \qquad (2.5.134)$$

and (b) for every $\epsilon > 0$,

$$\lim_{\delta \downarrow 0} \limsup_{n \to \infty} P_n(\{f \in D[0,1]^p : w'_f(\delta) \geq \epsilon\}) = 0. \qquad (2.5.135)$$

In practice, by letting

$$w_f(\delta) = \sup\{|f(\mathbf{t}) - f(\mathbf{s})| : \ \mathbf{s}, \mathbf{t} \in [0,1]^p, \ ||\mathbf{s} - \mathbf{t}|| \leq \delta\}, \qquad (2.5.136)$$

and noting that

$$w'_f(\delta) \leq w_f(2\delta), \ \forall\, 0 < \delta \leq \frac{1}{2} \ \text{and} \ f \in D[0,1]^p, \qquad (2.5.137)$$

we may replace $w'_f(\delta)$ in (2.5.135) by $w_f(2\delta)$. Recall that (2.5.134) shows that f is uniformly bounded in probability (on $D[0,1]^p$), so some "maximal inequality" may be used to establish this condition. On the other hand, verification of (2.5.135) is generally more delicate; this is usually accomplished by using certain tricky inequalities known as *Billingsley-type inequalities* after Billingsley (1968) who gives an elaborate treatment for $p = 1$. A multiparameter extension, due to Bickel and Wichura (1971), is the following.

2. *Theorem (Billingsley inequality).* The compactness theorem holds if for every pair of neighboring blocks in $[0,1]^p$, $\lambda > 0$, and n,

$$\begin{aligned} &P_n(\{m(B(\mathbf{s},\mathbf{t}], B(\mathbf{s}',\mathbf{t}']) \geq \lambda\}) \\ &\quad \leq \ \lambda^{-\gamma}\{\mu(B(\mathbf{s},\mathbf{t}] \cup B(\mathbf{s}',\mathbf{t}'])\}^{1+\beta}, \end{aligned} \qquad (2.5.138)$$

for some $\gamma > 0$, $\beta > 0$, where $\mathbf{s}$, $\mathbf{t}$ $\mathbf{s}'$, $\mathbf{t}' \in [0,1]^p$,

$$m(B(\mathbf{s},\mathbf{t}], B(\mathbf{s}',\mathbf{t}']) = \min\{|f(B(\mathbf{s},\mathbf{t}])|, |f(B(\mathbf{s}',\mathbf{t}'])|\}, \qquad (2.5.139)$$

$f(B(\mathbf{s},\mathbf{t}])$ is the increment of f around the block $B(\mathbf{s},\mathbf{t}]$, and μ is a measure on the Borel sets of $[0,1]^p$.

A more stringent but more easily verifiable condition for (2.5.138) is the following:

$$\begin{aligned} & \mathbb{E}(|f_n(B(\mathbf{s},\mathbf{t}])|^{\alpha_1}|f_n(B(\mathbf{s}',\mathbf{t}'])|^{\alpha_2} \\ \leq & \ M\{\mu(B(\mathbf{s},\mathbf{t}])\mu(B(\mathbf{s}',\mathbf{t}'])\}^{(1+\beta)/2}, \end{aligned} \tag{2.5.140}$$

where M, α_1, α_2, β are positive constants and $\mu(.)$ is a sigma-finite measure on $[0,1]^p$.

For the convergence of *finite-dimensional distributions* (f.d.d.) (of $\{P_n\}$ to P), usually some simpler techniques based on the Cramér-Wold theorem and CLT's work out well, but generally the compactness part of the weak convergence in (2.5.116) is difficult to establish. This task can be made simpler by using the following generalization of the Slutzky theorem:
If $\{X_n\}$ and $\{Y_n\}$ both belong to a common separable space S, equipped with a topology $\rho_S(.)$, then

$$X_n \xrightarrow{\mathcal{D}} X \text{ and } \rho(X_n, Y_n) \xrightarrow{\mathcal{P}} 0 \Rightarrow Y_n \xrightarrow{\mathcal{D}} X. \tag{2.5.141}$$

Verification of $\rho(X_n, Y_n) \xrightarrow{\mathcal{P}} 0$ can proceed by incorporating the following.

3. *Convexity lemma.* Let $\{Y_n(\mathbf{t}),\ \mathbf{t} \in \mathbf{T}\}$ be a sequence of random convex functions defined on a convex, open subset $\mathbf{T} \subset \mathbb{R}_p$, for some $p \geq 1$. Suppose that $\{\xi(\mathbf{t}),\ \mathbf{t} \in \mathbf{T}\}$ is a real-valued function on $\mathbf{T}$, such that

$$Y_n(\mathbf{t}) \to \xi(\mathbf{t}) \text{ in probability/a.s., for each } \mathbf{t} \in \mathbf{T}. \tag{2.5.142}$$

Then for each compact set $\mathcal{C}$ of $\mathbf{T}$,

$$\sup\{|Y_n(\mathbf{t}) - \xi(\mathbf{t})| : \mathbf{t} \in \mathcal{C}\} \to 0, \text{ in probability/a.s.,} \tag{2.5.143}$$

and $\xi(\mathbf{t})$ is necessarily convex on $\mathbf{T}$.

For nonstochastic Y_n, this is essentially theorem 10.8 of Rockafellar (1970), and its adaptation to stochastic analysis is due to Anderson and Gill (1982), Heiler and Willers (1988) and Pollard (1991), among others. If $Y_n(\mathbf{t})$ can be written as $Y_n^{(1)}(\mathbf{t}) \pm Y_n^{(2)}(\mathbf{t})$ (or a finite mixture, not necessarily convex, of several $Y_n^{(j)}(\mathbf{t})$), where the individual $Y_n^{(j)}(\mathbf{t})$ satisfy the hypothesis of the lemma, then (2.5.143) also holds for Y_n. Two important results follow as corollaries to the above lemma.

1. If $\xi(\mathbf{t})$, $\mathbf{t} \in \mathcal{C}$ attains an extrema, say, ξ_0, at an interior point $\mathbf{t}_0$ of $\mathcal{C}$, then $Y_n(\mathbf{t})$ has also an extrema that converges to ξ_0 in the same mode as (2.5.143).

2. If $\xi = 0$, then $Y_n(\mathbf{t}) \to 0$ uniformly in $\mathcal{C}$, in the same mode as in (2.5.143).

2.5.11 Some Important Gaussian Functions

Let $\mathbb{R}_p^+$ be the nonnegative orthant of $\mathbb{R}_p, p \geq 1$, and consider the probability space $(\Omega, \mathcal{C}^p, W)$, where $\mathcal{C}^p$ is the sigmafield generated by the open subsets of $C[\mathbb{R}_p^+]$ and W is the probability measure induced in $C[\mathbb{R}_p^+]$ by a real-valued stochastic process $X(.) = X(\mathbf{t}), \mathbf{t} \in \mathbb{R}_p^+\}$. Then W is defined to be a Wiener measure on $(C[\mathbb{R}_p^+], \mathcal{C}^p)$, if the following hold:

1. $W\{[X(\mathbf{t}), \mathbf{t} \in \mathbb{R}_p^+] \in C[\mathbb{R}_p^+]\} = 1.$ (2.5.144)

2. Let B and C be any two rectangles in $\mathbb{R}_p^+$, such that $B \cap C = 0$ (i.e., they are disjoint, though they might be adjacent). Let $X(B)$ and $X(C)$ be the *increment* of the process $X(.)$ over the rectangles B and C respectively. Then $X(B)$ and $X(C)$ are stochastically independent.

3. For every $B \subset \mathbb{R}_p^+$ and real x

$$W\{X(B) \leq x\} = \frac{1}{\sqrt{2\pi\sigma^2\lambda(B)}} \int_{-\infty}^{x} \exp\Big\{ -u^2/2\sigma^2\lambda(B)\Big\} du \qquad (2.5.144)$$

where $0 < \sigma^2 < \infty$ and $\lambda(B)$ is the Lebesgue measure of B. In particular, if $\sigma = 1$, W is termed a *standard Wiener measure* on $C[\mathbb{R}_p^+]$.

Note that result 1 implies that the sample paths of $X(.)$ are continuous a.e., result 2 implies that $X(.)$ has independent increments, while result 3 specifies the Gaussian nature. The increment of $X(.)$ over a block $B = B(\mathbf{s}, \mathbf{t}]$ is defined as

$$X(B) = \sum_{j=1}^{p} \sum_{l_j=0,1} (-1)^{p-\sum_{j=1}^{p} l_j} X(l_j t_j + (1-l_j)s_j, \ 1 \leq j \leq p). \qquad (2.5.145)$$

The above definition simplifies for $p = 1$ where $X(t)$ has sample paths continuous a.e. on $\mathbb{R}^+$. Moreover for any partition: $(0, t_1)$, $[t_1, t_2)$, $[t_2, t_3), \ldots$ $\mathbb{R}^+, X(0) = 0$, $X(t_1)$, $X(t_2) - X(t_1)$, $X(t_3) - X(t_2), \ldots$ are all independently normally distributed with 0 means and variances $\sigma^2(t_j - t_{j-1})$, $j \geq 1$, where $t_0 = 0$. We say that $X(.)$ is a *Brownian motion* or *Wiener process* (standard if $\sigma = 1$) on $\mathbb{R}^+$. By construction, $\{X(t), \ t \in [0, T]\}$ remains a Wiener process for every $T: \ 0 < T < \infty$. In this context, consider a related process $X^0(.)$ defined by

$$X^0(t) = X(t) - tX(1), \ \ 0 \leq t \leq 1, \qquad (2.5.146)$$

so that $X^0(.)$ is also Gaussian and has continuous (a.e.) sample paths. Furthermore $\mathbb{E}X^0(t) = 0 \ \forall t \in [0, 1]$, and for every $s, t \in [0, 1]$,

$$\mathbb{E}[X^0(s)X^0(t)] = \sigma^2(s \wedge t - st), \qquad (2.5.147)$$

so that the process no longer has independent increments, but $X^0(0) = X^0(1) = 0$, with probability 1. In other words, $X^0(.)$ is tied-down at the two ends $\{0, 1\}$. X^0 is termed a *Brownian bridge* or a *tied down Wiener process* (standard when $\sigma^2 = 1$). For general p (≥ 1), we define $X^0(.) = \{X^0(\mathbf{t}),\ \mathbf{t} \in [0,1]^p\}$, by letting

$$X^0(\mathbf{t}) = X(\mathbf{t}) - |\mathbf{t}|X(\mathbf{1}),\ \mathbf{t} \in [0,1]^p. \tag{2.5.148}$$

It is easy to verify that $\mathbb{E}X^0(\mathbf{t}) = 0\ \forall \mathbf{t} \in [0,1]^p$, and that

$$\mathbb{E}[X^0(B)X^0(C)] = \sigma^2[\lambda(B \cap C) - \lambda(B)\lambda(C)], \tag{2.5.149}$$

for every $B, C \subset [0,1]^p$, where the increments are defined as in (2.5.145). $X^0(.)$ is Gaussian and is tied-down (to 0, w.p. 1) at each edge of $[0,1]^p$ (standard one when $\sigma = 1$).

Consider now the case of $p = 2$, and let $X^\star(.) = \{X^\star(t_1, t_2),\ 0 \leq t_1 \leq \infty,\ 0 \leq t_2 \leq 1\}$ be defined by

$$X^\star(t_1, t_2) = X(t_1, t_2) - t_2 X(t_1, 1),\ \ \mathbf{t} \in \mathbf{T} \tag{2.5.150}$$

where $\mathbf{T} = \mathbb{R}_1^+ \times [0,1]$, and $\{X(\mathbf{t}),\ \mathbf{t} \in \mathbb{R}_2^+\}$ is a standard Wiener process on $\mathbb{R}_2^+$ (also termed a *Brownian sheet*). Then $\mathbb{E}X^\star(\mathbf{t}) = 0$, $\mathbf{t} \in \mathbf{T}$, and

$$\mathbb{E}[X^\star(s)X^\star(t)] = (s_1 \wedge t_1)(s_2 \wedge t_2 - s_2 t_2) \tag{2.5.151}$$

for every $\mathbf{s}, \mathbf{t} \in \mathbf{T}$. $X^\star(.)$ is termed a *Kiefer process*. This Gaussian function is also tied down at $t_1 = 0$ and $t_2 = 0, 1$. The definition extends directly to $\mathbf{T} = \mathbb{R}_{p_1}^+ \times [0,1]^{p_2}$, $p_1 \geq 1$, $p_2 \geq 1$, $p_1 + p_2 = p \geq 2$.

Let us consider next the case where $X(.)$ is itself a vector (stochastic) process. We write

$$\mathbf{X}(\mathbf{t}) = [X_1(\mathbf{t}), \ldots, X_k(\mathbf{t})]',\ \ \mathbf{t} \in \mathbf{T}, \tag{2.5.152}$$

where $k \geq 1$ and $\mathbf{T} \subset \mathbb{R}_p^+$ for some $p \geq 1$. Suppose that $\{X_j(\mathbf{t}),\ \mathbf{t} \in \mathbb{R}_p^+\}$ are independent copies of a Wiener process. Then a multivariate Gaussian process $\mathbf{M}(.) = \{\mathbf{M}(\mathbf{t}),\ \mathbf{t} \in \mathbf{T}\}$ can be defined by introducing a nonstochastic matrix-valued $\mathbf{A}(.) = \{\mathbf{A}(\mathbf{t}), \mathbf{t} \in \mathbf{T}\}$ and letting

$$\mathbf{M}(\mathbf{t}) = \mathbf{A}(\mathbf{t})\mathbf{X}(\mathbf{t}),\ \ \mathbf{t} \in \mathbf{T}. \tag{2.5.153}$$

Because of this representation, distributional properties of $\mathbf{X}(.)$ can be incorporated for the study of parallel results for $\mathbf{M}(.)$. Let us confine ourselves to a special case where $p = 1$ and $T = \mathbb{R}^+$. Then, we can define a *k-parameter Bessel process* $B = \{B(t),\ t \in \mathbb{R}^+\}$ by letting

$$B^2(t) = [\mathbf{X}(t)]'[\mathbf{X}(t)] = ||\mathbf{X}(t)||^2,\ \ t \in \mathbb{R}^+. \tag{2.5.154}$$

If we define

$$\mathbf{Q}(t) = [\mathbf{A}(t)][\mathbf{A}(t)]',\ \ t \in \mathbb{R}^+, \tag{2.5.155}$$

then we may also define

$$B^2(t) = [\mathbf{M}(t)]'[\mathbf{Q}(t)]^-[\mathbf{M}(t)],\ \ t \in \mathbb{R}^+, \tag{2.5.156}$$

where $[\,]^-$ stands for a *generalized inverse*. Thus (2.5.154) is a canonical representation for Bessel processes.

2.5.12 Weak Invariance Principles

Functional CLT's relate to the extensions of the basic results in Section 2.5.5 along the lines of the main themes in Section 2.5.10 and 2.5.11. Such functional CLT's are also known as *weak invariance principles.* Since the seminal work on this subject in classic texts of Parthasarathy (1967) and Billingsley (1968), there has been a steady flow of research. We only present some important results having direct relevance to subsequent chapters.

1. *Dvoretzky-McLeish theorem.* Let $\{Z_{n,k},\ 0 \le k \le k_n; n \ge 1\}$ be a triangular array of r.v.'s (not necessarily independent), and we set $Z_{n,0} = 0\, w.p.1 \forall n \ge 1$. Let

$$S_{n,k} = Z_{n,0} + \ldots, Z_{n,k},\quad k \le k_n,\ n \ge 1, \tag{2.5.157}$$

and let $\mathcal{F}_{n,k}$ be the sigma-field generated by $S_{n,k}$, for $k \ge 1$ and $n \ge 1$. Assume that $\mathbb{E}Z^2_{n,k} < \infty$ for every k, n. Let

$$\mu_{n,k} = \mathbb{E}(Z_{n,k}|\mathcal{F}_{n,k-1}) \quad \text{and} \quad \sigma^2_{n,k} = \mathrm{Var}(Z_{n,k}|\mathcal{F}_{n,k-1}), \tag{2.5.158}$$

for $k \ge 1$, $n \ge 1$, and assume that $k_n \to \infty$ as $n \to \infty$. Define a stochastic process $W_n = \{W_n(t),\ t \in [0,1]\}$ by letting

$$W_n(t) = S_{n,k_n(t)},\ t \in [0,1], \tag{2.5.159}$$

where $k_n(t)$ is an integer-valued, nondecreasing, and right-continuous function of t on [0,1], with $k_n(0) = 0$, $n \ge 1$. Suppose that the (conditional) Lindeberg condition holds, that is, for every $\epsilon > 0$,

$$\sum_{k \le k_n} \mathbb{E}\{Z^2_{n,k} I(|Z_{n,k}| > \epsilon)|\mathcal{F}_{n,k-1}\} \xrightarrow{P} 0, \tag{2.5.160}$$

as $n \to \infty$. Further assume that for each $t \in [0,1]$,

$$\sum_{k \le k_n(t)} \mu_{n,k} \xrightarrow{P} 0 \text{ and } \sum_{k \le k_n(t)} \sigma^2_{n,k} \xrightarrow{P} t \tag{2.5.161}$$

as $n \to \infty$. Then

$$W_n \xrightarrow{\mathcal{D}} W, \text{ in the Skorohod topology on } D[0,1], \tag{2.5.162}$$

where W is a standard Wiener process on [0,1].

Scott (1973) studied elaborately (2.5.162) when the $S_{n,k}$ form a martingale array and formulated alternative (but equivalent) regularity conditions pertaining to this weak invariance principle. Earlier Brown (1971) considered a martingale sequence $\{S_k,\ k \geq 0\}$ and established the same resultunder (i) (2.5.160) and (ii)

$$(\sum_{k\leq n} \sigma_k^2)/\mathbb{E}(\sum_{k\leq n} \sigma_k^2) \xrightarrow{P} 1, \text{ as } n \to \infty. \tag{2.5.163}$$

An important inequality developed in this context is the one listed below.

2. *Brown submartingale inequality:* Let $\{X_n, \mathcal{F}_n;\ n \geq 1\}$ be a submartingale. Then for every $\epsilon > 0,\ n \geq 1$,

$$\begin{aligned} \mathbb{P}\{\max_{1\leq k\leq n} |X_k| > 2\epsilon\} &\leq \mathbb{P}\{|X_n| > \epsilon\} + \mathbb{E}\{(\frac{1}{\epsilon}|X_n| - 2)I(|X_n| \geq 2\epsilon)\} \\ &\leq \epsilon^{-1}\mathbb{E}\{|X_n|I(|X_n| \geq \epsilon)\}. \end{aligned} \tag{2.5.164}$$

This inequality leads to the following theorem.

3. *Tightness theorem.* For a martingale sequence $\{S_k,\ k \geq 1\}$, if W_n is defined as in (2.5.157), then the convergence of f.d.d.'s to those of a Gaussian process W ensures the tightness of W_n (and hence the weak convergence to W).

For a reversed martingale sequence $\{X_n, \mathcal{F}_n;\ n \geq 1\}$, define $Z_{n,i} = X_{n+i-1} - X_{n+i},\ i \geq 1$, without any loss of generality, we set $\mathbb{E}X_n = 0$ and let

$$V_n = \sum_{i\geq 1} \mathbb{E}(Z_{n,i}^2|\mathcal{F}_{n+1}) \text{ and } s_n^2 = \mathbb{E}V_n = \mathbb{E}(X_n^2). \tag{2.5.165}$$

Then we have the following.

4. *Loynes functional CLT.* Let $n(t) = \min\{k :\ s_k^2/s_n^2 \leq t\}$, and $W_n(t) = s_n^{-1}X_{n(t)},\ t \in [0,1]$. Then, if $V_n/s_n^2 \xrightarrow{P} 1$ and the $Z_{n,i}$ satisfy the (conditional) Lindeberg condition, W_n weakly converges to a standard Wiener process.

It is possible to strengthen the above-mentioned results. Let $q(.) = \{q(t),\ 0 < t < 1\}$ be a continuous, nonnegative function on $[0,1]$ such that for some $a > 0$, $q(t)$ is $\nearrow$ in $t \in (0,a]$, and $q(t)$ is bounded away from 0 on $(a,1]$ and $\int_0^1 q^{-2}(t)dt < \infty$. Let

$$d_q(x,y) = \sup\{q^{-1}(t)|x(t) - y(t)| :\ t \in [0,1]\}. \tag{2.5.166}$$

Then the weak convergence results in the preceding three theorems hold even when the J_1-topology is replaced by the $d_q(.)$-metric (Sen 1981a, th. 2.4.8). Such functional CLT's also hold for another important class of empirical processes, which is considered next.

2.5.13 Empirical Distributional Processes

Let $\{\mathbf{X}_i,\ i \geq 1\}$ be a sequence of i.i.d. r.v.'s with a d.f. F defined on $\mathbb{R}_p$, for some $p \geq 1$. For every $n \geq 1$, the *empirical d.f.* (EDF) F_n is defined by

$$F_n(\mathbf{x}) = n^{-1}\sum_{i \leq n} I(\mathbf{X}_i \leq \mathbf{x}),\ \mathbf{x} \in \mathbb{R}_p. \tag{2.5.167}$$

Note that $\mathbb{E}F_n(\mathbf{x}) = F(\mathbf{x}),\ \mathbf{x} \in \mathbb{R}_p$, and that

$$Cov[F_n(\mathbf{x}), F_n(\mathbf{y})] = n^{-1}\{F(\mathbf{x} \wedge \mathbf{y}) - F(\mathbf{x})F(\mathbf{y})\} \tag{2.5.168}$$

where $\mathbf{x} \wedge \mathbf{y} = (x_1 \wedge y_1, \ldots, x_p \wedge y_p)'$. The usual *empirical process* $V_n = n^{1/2}(F_n - F)$ is defined by

$$V_n(\mathbf{x}) = n^{1/2}(F_n(\mathbf{x}) - F(\mathbf{x})),\ \mathbf{x} \in \mathbb{R}_p. \tag{2.5.169}$$

If $F_{[j]}$ is the jth marginal d.f. corresponding to the d.f. F, and we assume that $F_{[j]}$ is continuous a.e. for $1 \leq j \leq p$, then we may consider the transformation $\mathbf{X} \to \mathbf{Y} = (Y^{(1)}, \ldots, Y^{(p)})'$ where $Y^{(j)} = F_{[j]}(X^{(j)}),\ 1 \leq j \leq p$. The $Y^{(j)}$ will have marginally uniform (0,1) d.f. and their joint d.f. is denoted by $G(\mathbf{y}) = \mathbb{P}\{\mathbf{Y} \leq \mathbf{y}\},\ \mathbf{y} \in [0,1]^p$. The reduced empirical d.f. G_n is then defined by

$$G_n(\mathbf{y}) = n^{-1}\sum_{i \leq n} I(\mathbf{Y}_i \leq \mathbf{y}),\ \mathbf{y} \in [0,1]^p. \tag{2.5.170}$$

We consider then a p-parameter *empirical process* $U_n = \{U_n(\mathbf{t}),\ \mathbf{t} \in [0,1]^p\}$ by letting

$$U_n(\mathbf{t}) = n^{1/2}[G_n(\mathbf{t}) - G(\mathbf{t})],\ \mathbf{t} \in [0,1]^p. \tag{2.5.171}$$

Note that by (2.5.169) and (2.5.171),

$$U_n(\mathbf{t}) = V\Big(((F_{[1]}(x^{(1)}), \ldots, F_{[p]}(x^{(p)}))\Big) \tag{2.5.172}$$

whenever $\mathbf{t} = ((F_{[1]}(x^{(1)}), \ldots, F_{[p]}(x^{(p)}))$. Clearly it suffices to study U_n alone. By (2.5.170), for finitely many $\mathbf{t}$'s, say, $\mathbf{t}_1, \ldots, \mathbf{t}_m$, all belonging to $[0,1]^p$, a linear combination of $U_n(\mathbf{t}_1), \ldots, U_n(\mathbf{t}_m)$ is expressible as an average of i.i.d. r.v.'s so that the classical CLT apply. We conclude that

$$[U_n(\mathbf{t}_1), \ldots, U_n(\mathbf{t}_m)] \xrightarrow{\mathcal{D}} [U(\mathbf{t}_1), \ldots, U(\mathbf{t}_m)], \tag{2.5.173}$$

where $U(.)$ is Gaussian with null mean and

$$\mathbb{E}[U(\mathbf{s})U(\mathbf{t})] = G(\mathbf{s} \wedge \mathbf{t}) - G(\mathbf{s})G(\mathbf{t}), \tag{2.5.174}$$

for every $\mathbf{s}, \mathbf{t} \in [0,1]^p$. Moreover, since the summands in (2.5.171) are i.i.d. zero-one-valued r.v.'s, Billingsley-type inequalities are easy to verify. This leads to the following two theorems.

1. *Weak convergence theorem for ED process.* If F is continuous, then U_n converges in law (on $D[0,1]^p$ to U).
In particular, for $p = 1$, U reduces to a standard Brownian bridge on $[0, 1]$. Also, if $F(\mathbf{x}) \equiv \prod_{j=1}^p F_{[j]}(x^{(j)})$, then $G(\mathbf{t}) = |\mathbf{t}| = t_1 \cdot \ldots \cdot t_p$ so that U reduces to a standard tied-down Brownian sheet.
We consider another $(p+1)$-parameter stochastic process, $U_n^\star(s, \mathbf{t})$, $s \in \mathbb{R}^+$, $\mathbf{t} \in [0,1]^p$, by letting

$$U_n^\star(s,\mathbf{t}) = n^{-1/2}(n_s U_{n_s}(\mathbf{t})),\ n_s = [ns],\ \mathbf{t} \in [0,1]^p, \tag{2.5.175}$$

where $U_m(.)$ is defined as in (2.5.171) for $m \geq 1$. As in (2.5.151), we consider a *Kiefer-process* $U^\star = \{U^\star(s,\mathbf{t}),\ s \in \mathbb{R}^+,\ \mathbf{t} \in [0,1]^p\}$. Then

$$U_n^\star \xrightarrow{\mathcal{D}} U^\star,\ \text{ in the } J_1\text{-topology on } D[0,\infty) \times [0,1]^p. \tag{2.5.176}$$

We may also define $U_n^0 = \{U_n^0(s,\mathbf{t}) = n^{1/2} U_{n_s}(\mathbf{t}),\ s \in [0,1],\ \mathbf{t} \in [0,1]^p\}$ by letting $n_s = \min\{k : n/k \leq s\}$. Then, as in Neuhaus and Sen (1977), we have

$$U_n^0 \xrightarrow{\mathcal{D}} U^\star,\ \text{ in the } J_1\text{-topology on } D[0,1]^{p+1}. \tag{2.5.177}$$

The special case of $p = 1$ leads $U^\star$ to a Kiefer process as defined in (2.5.150)-(2.5.151). For the case of $p = 1$, the weak convergence results for U_n and $U_n^\star$ or U_n^0 can be strengthened to suitable $d_q(.)$ metrics. We have the following.

2. *Weak convergence theorem 2.* Let $q_1 = \{q_1(t),\ t \in \mathbb{R}^+\}$ satisfy the conditions stated in (2.5.166). Also let $q_2 = \{q_2(t),\ t \in [0,1]\}$ be a continuous, nonnegative function on [0,1], bounded away from 0 on $[\gamma, 1-\gamma]$ for some $0 < \gamma \leq 1/2$, and nondecreasing (nonincreasing) on $[0,\gamma)$ $((1-\gamma, 1])$. Consider the metric

$$\rho_q(x,y) = \sup\{|x(s,t) - y(s,t)|/q_1(s)q_2(t) :\ (s,t) \in [0,1]^2\} \tag{2.5.178}$$

Then $U_n \xrightarrow{\mathcal{D}} U$ in ρ_q-metric on $D[0,1]^2$, whenever $\int_0^1 q_1^{-2}(t)dt < \infty$ and $\int_0^{1/2} t^{-1} \exp\{-\epsilon t^{-1} q_2^{02}(t)\}dt < \infty$, $\epsilon > 0$ and $q_2^0(t) = q_2(t) \wedge q_2(1-t)$, $t \in [0,1]$.

A more easily verifiable condition (Wellner 1974) is that

$$\int_0^1 \int_0^1 [q(t,s)]^{-2} ds dt < \infty, \tag{2.5.179}$$

where $q(t,s) = q_1(t)q_2(s)$, $(t,s) \in [0,1]^2$. A direct consequence of the above two theorems is the following: For $p = 1$ for every $\lambda > 0$,

$$\mathbb{P}\{\sup_{x \in \mathbf{R}} |V_n(x)| \geq \lambda\} \quad = \quad \mathbb{P}\{\sup_{0 \leq t \leq 1} |U_n(t)| \geq \lambda\}$$

$$\begin{aligned} &\to \quad \mathbb{P}\{\sup_{0\le t\le 1}|U(t)|\ge\lambda\} \\ &= \quad 2\sum_{k\ge 1}(-1)^{k-1}\exp(-2k^2\lambda^2), \end{aligned} \tag{2.5.180}$$

where the last step follows from the classical boundary crossing probability results for Brownian motions and bridges due to Doob (1949). Similarly

$$\begin{aligned} \mathbb{P}\{\sup_{x\in\mathbf{R}} V_n(x)\ge\lambda\} &= \mathbb{P}\{U_n(t)\ge\lambda, \quad \text{for some } t\in[0,1]\} \\ &\to \mathbb{P}\{U(t)\ge\lambda, \quad \text{for some } t\in[0,1]\} \\ &= \exp(-2\lambda^2), \end{aligned} \tag{2.5.181}$$

Thus, $\sup\{|V_n(x)|: x\in\mathbb{R}\} = \sup\{|U_n(t)|: t\in[0,1]\} = \mathrm{O}_p(1)$. Further, Dvoretzky, Kiefer and Wolfowitz (1956) have shown that for every finite n, the left-hand sides in (2.5.180) or (2.5.181) are actually $\le$ the right-hand side limits. Although for $p>1$ such a dominating bound is not precisely known, Kiefer (1961) managed to show that for every $\epsilon>0$ ($\epsilon<2$) there exists a positive $c(\epsilon)<\infty$ such that for every $n\ge 1$, $p\ge 1$ and $\lambda>0$,

$$\mathbb{P}\{\sup_{\mathbf{t}\in[0,1]^p}|U_n(\mathbf{t})|\ge\lambda\}\le c(\epsilon)\exp(-(2-\epsilon)\lambda^2). \tag{2.5.182}$$

The last inequality in turn yields as $n\to\infty$,

$$\sup\{|U_n(\mathbf{t})|: \mathbf{t}\in[0,1]^p\} = \mathrm{O}((\log n)^{1/2}) \text{ a.s.} \tag{2.5.183}$$

Of course more precise a.s. convergence rates are available for the case of $p=1$, and some of these are reported in a later subsection.

Closely related to the empirical d.f. F_n (or G_n) is the *sample quantile function* Q_n, $\{Q_n(t), 0<t<1\}$, where

$$Q_n(t) = F_n^{-1}(t) = \inf\{x: F_n(x)\ge t\}. \tag{2.5.184}$$

For the uniform d.f., induced by the mapping $X\to Y=F(X)$, we have the *uniform quantile function* $Q_n^0(t)$, $t\in(0,1)$, where

$$Q_n^0(t) = G_n^{-1}(t) = \inf\{y: G_n(y)\ge t\}. \tag{2.5.185}$$

As such, it is quite intuitive to formulate a *(reduced) sample quantile process* $U_n^Q(.)$, by letting

$$U_n^Q(t) = n^{1/2}[Q_n^0(t)-t], \; 0\le t\le 1. \tag{2.5.186}$$

A similar formulation can be made of $Q_n(t)$ in (2.5.184). By definition (2.5.185), $Q_n^0(G_n(y))\ge y$, $\forall y\in[0,1]$, and the equality sign holds at the n points $Y_{n:k}$, $1\le k\le n$, which are the order statistics of the n i.i.d. r.v.'s

$Y_1, \ldots, Y_n$ from the uniform (0,1) d.f. It follows by simple argument that as $n \to \infty$

$$\sup\{|U_n(t) + U_n^Q(t)| : t \in (0,1)\} \to 0 \text{ a.s.}, \tag{2.5.187}$$

so that the weak convergence principles studied for $U_n(.)$ remain applicable for $U_n^Q(.)$ as well [use (2.5.140] in this context]. With suitable rate of convergence for (2.5.184), similar a.s. convergence results for $U_n^Q(.)$.

2.5.14 Weak Invariance Principle: Random Change of Time

Using the notations of Section 2.5.10, consider the $D[0,1]$ space endowed with the Skorohod J_1-topology induced by the metric $\rho_S(.)$ in (2.5.1), where $\{\lambda : [0,1] \mapsto [0,1]$ onto itself. Let Λ_0 be an (extended) class of λ belong to the $D[0,1]$ space such that λ_t is nondecreasing and $0 \leq \lambda_t \leq 1$ $\forall t \in [0,1]$ (note that the strict monotonicity and continuity condition for Λ have been weakened for Λ_0). For $x \in D[0,1]$ and $y \in \Lambda_0$, let $x \circ y$ denote the composition: $(x \circ y)(t) = x(y(t))$, $t \in [0,1]$. Let X be a random element of $D[0,1]$ and Y a random element of Λ_0 so that (X,Y) is a random element of $D[0,1] \times \Lambda_0$. Further consider a sequence $\{(X_n, Y_n);\ n \geq 1\}$ of random elements of $D[0,1] \times \Lambda_0$ such that

$$(X_n, Y_n) \xrightarrow{\mathcal{D}} (X, Y) \tag{2.5.188}$$

and $\mathbb{P}\{X \in C[0,1]\} = 1 = \mathbb{P}\{Y \in C[0,1]\}$. Then

$$X_n \circ Y_n \xrightarrow{\mathcal{D}} X \circ Y. \tag{2.5.189}$$

Recall that for the uniform metric $\rho(.)$, $\rho(x_n \circ y_n, x \circ y) \leq \rho(x_n \circ y, x \circ y) + \rho(x \circ y_n, x \circ y)$ so that the uniformity guaranteed by the $C[0,1]$ space (for X, Y) and the fact that for each n, X_n and Y_n have a common domain (which may vary with n) extends (2.5.188) to (2.5.189). An important application of (2.5.189) relates to the case of stochastic sizes for the partial sum process or the empirical distributional process, treated in earlier subsections.

2.5.15 Embedding Theorems and Strong Invariance Principles

The basic idea of *embedding of Wiener processes* is due to Skorohod (1956) and it was consolidated by Strassen (1964, 1967). Let $\{X_n, \mathcal{F}_n;\ n \geq 1\}$ be a martingale so that $\{Y_n = X_n - X_{n-1}, \mathcal{F}_n; n \geq 1\}$ form a martingale difference sequence. We take $X_0 = 0$ with probability 1 so that $\mathbb{E}X_n = 0\ \forall n \geq 1$. Also, $\mathbb{E}(Y_n|\mathcal{F}_{n-1}) = 0$ a.e. $\forall n \geq 1$. We assume that

$$v_n = \mathbb{E}(Y_n^2|\mathcal{F}_{n-1}) \text{ exists a.e. } \forall n \geq 1. \tag{2.5.190}$$

We write

$$V_n = \sum_{k \leq n} v_k, \quad n \geq 1, \tag{2.5.191}$$

and assume that $V_n \xrightarrow{\text{a.s.}} \infty$, as $n \to \infty$.

1. *Skorohod-Strassen theorem.* For a martingale $\{X_n, \mathcal{F}_n;\ n \geq 1\}$ satisfying (2.5.190) – (2.5.191), there exists a probability space $(\Omega^\star, \mathcal{A}^\star, P^\star)$ with a standard Wiener process W $(= \{W(t),\ t \in \mathbb{R}^+\})$ and a sequence $\{T_n;\ n \geq 1\}$ of nonnegative r.v.'s such that the sequence $\{X_n\}$ has the same distribution as $\{W(\sum_{k \leq n} T_k)\}$. Strassen (1967) prescribed a construction based on the following mapping. Let

$$S_{V_n} = X_n,\ n \geq 1, \quad S_0 = X_0 = 0 \ \ w.p.1; \tag{2.5.192}$$

$$S_t = S_{V_n} + Y_{n+1}(t - V_n)/v_{n+1}, \quad V_n \leq t \leq V_{n+1}, \quad n \geq 1. \tag{2.5.193}$$

Thus $S = \{S_t,\ t \in \mathbb{R}^+\}$ has continuous sample paths (on $\mathbb{R}^+$). We may set without loss of generality that v_n are all strictly positive, with probability 1.

2. *Strassen Theorem.* Let $S = \{S_t,\ t \in \mathbb{R}^+\}$ be defined as above. Also let $f(.) = \{f(t),\ t \in \mathbb{R}^+\}$ be a nonnegative and nondecreasing function such that

$$f(t) \text{ is } \uparrow \ \text{ but } t^{-1}f(t) \text{ is } \downarrow \text{ in } t (\in \mathbb{R}^+). \tag{2.5.194}$$

Suppose now that V_n goes a.s. to ∞ as $n \to \infty$, and that

$$\sum_{n \geq 1} \frac{1}{f(V_n)} \mathbb{E}\Big\{Y_n^2 I(|Y_n| > \sqrt{f(V_n)})|\mathcal{F}_{n-1}\Big\} < \infty \text{ a.e.} \tag{2.5.195}$$

Then there exists a standard Wiener process W on $\mathbb{R}^+$ such that

$$S_t = W(t) + o((\log t)(tf(t))^{1/4}) \text{ a.s.,} \quad \text{as} \quad t \to \infty. \tag{2.5.196}$$

Both these theorems rest on the basic embedding result:

$$X_n = W(T_1 + \ldots + T_n) \text{ a.s.,} \quad \text{for all } n \geq 1, \tag{2.5.197}$$

To comprehend this basic reprtesentation, we need to introduce a new probability space $(\Omega^\star, \mathcal{A}^\star, P^\star)$ on which W and $\{T_k;\ k \geq 1\}$ are defined and also to consider a sequence $\{X_n^\star,\ n \geq 0\}$ of r.v.'s on this space such that $\{X_n^\star,\ n \geq 1\} \stackrel{\mathcal{D}}{=} \{X_n,\ n \geq 1\}$; (2.5.197) actually holds for the $X_n^\star$.

By the law of iterated logarithms for $\{X_n^\star\}$, it suffices to choose $f(t) = t(\log\log t)^2(\log t)^{-4}$ as $t \to \infty$, while sharper choices of $f(t)$ can be made under higher-order moment conditions on the X_n. In fact, for $\{Y_n^\star\}$ i.i.d. r.v.'s with unit varince and a finite fourth moment, (2.5.195) can be strengthened to

$$S(n) = W(n) + \mathrm{O}((n \log\log n)^{1/4}(\log n)^{1/2}) \text{ a.s., } \quad \text{as } n \to \infty. \tag{2.5.198}$$

For independent r.v.'s, a major contribution to the Wiener process embedding is due to the Hungarian school, and reported in Csörgö and Révész (1981). For Y_i that are i.i.d. r.v.'s with 0 mean and unit variance, and a finite moment-generating function $\mathbb{E}\{\exp(tY)\}$ $\forall t \leq t_0$, for some $t_0 > 0$, they were able to show that

$$S(n) = W(n) + \mathrm{O}(\log n) \text{ a.s., } \quad \text{as } n \to \infty. \tag{2.5.199}$$

and that (2.5.197) or (2.5.198) holds uniformly in a class of d.f.'s for which the fourth moment or moment generating function is bounded uniformly. Brillinger (1969) was the first person to derive a parallel embedding theorem for the empirical process U_n in (2.5.171) for $p = 1$. He showed that there exits a probability space with sequence of Brownian bridges $\{B_n(t), 0 \leq t \leq 1\}$ and processes $\{\tilde{U}_n(t),\ 0 \leq t \leq 1\}$ such that

$$\{\tilde{U}_n(t),\ 0 \leq t \leq 1\} \stackrel{\mathcal{D}}{=} \{U_n(t),\ 0 \leq t \leq 1\}, \tag{2.5.200}$$

For each $n (\geq 1)$,

$$\sup_{0 \leq t \leq 1} |\tilde{U}_n(t) - B_n(t)| = \mathrm{O}(n^{-1/4}(\log n)^{1/2}(\log\log n)^{1/4}) \text{ a.s.} \tag{2.5.201}$$

Kiefer (1972) extension relates to the two-parameter processes $U_n^\star$ [(2.5.175)] and $U^\star$ [(2.5.176)], with a rate $O(n^{1/3}(\log n)^{2/3})$ a.s . A variety of related in depth results for U_n, as well as U_n^Q, are discussed in Csörgö and Révész (1981). We conclude this subsection with two a.s. results.

1.[Csörgö and Révész result]. Let $\epsilon_n = n^{-1}(\log n)^4$. Then

$$(\log\log n)^{-1/2} \sup_{\epsilon_n \leq t \leq 1-\epsilon_n} \frac{|U_n(t)|}{\sqrt{t(1-t)}} \leq 2 \text{ a.s., } \quad \text{as } n \to \infty. \tag{2.5.202}$$

2.[Csáki result]. for every $\epsilon > 0$,

$$(\log\log n)^{-1/2} \sup_{0<t<1} \frac{|U_n(t)|}{\{t(1-t)}\}^{1/2-\epsilon} = \mathrm{O}(1) \text{ a.s., } \quad \text{as } n \to \infty. \tag{2.5.203}$$

2.5.16 Asymptotic Relative Efficiency: Concept and Measures

In Sections 2.2, 2.3, and 2.4 the basic concepts in statistical inference were laid down, with special emphasis on robustness and efficiency considerations. For estimators, (asymptotic) variance or mean square error can be used as a yardstick for their asymptotic efficiency. As such the *asymptotic efficacy* of a real valued estimator T_n is defined as the reciprocal of its asymptotic mean square error(AMSE). In a parameter setup the classical Fréchet-Cramér-Rao bound and the asymptotic normality of $\sqrt{n}(T_n - \theta)$ provide a good justification for this adoption. Actually the convolution theorem in (2.5.125) ensures that a BAN estimator $\hat{\xi}_n$ is asymptotically optimal, even in multiparameter setup. The FOADR results explored in the subsequent chapters provide a similar justification for the class of robust estimators treated in this book. But there is a basic question regarding the AMSE in the multiparameter case, where it would be a matrix of order, say, $p \times p$, and the multivariate Cramér-Rao information inequality,

$$\mathbf{V} - \mathbf{I}_{\boldsymbol{\theta}}^{-1} = p.s.d. \tag{2.5.204}$$

specifies that $\mathbf{V}$, the AMSE of $\sqrt{n}(\mathbf{T}_n - \boldsymbol{\theta})$ dominates the reciprocal of the Information matrix $\mathbf{I}_{\boldsymbol{\theta}}$ in a matrix sense. Since $\hat{\boldsymbol{\xi}}_n$ would have AMSE $\mathbf{I}_{\boldsymbol{\theta}}^{-1}$, comparison of $\mathbf{T}_n$ with $\hat{\boldsymbol{\xi}}_n$ will not create any problem, and one may consider various measures of *asymptotic relative efficiency (ARE)* of $\mathbf{T}_n$ with respect to $\hat{\boldsymbol{\xi}}_n$. For example,

1. *D-ARE:* $|\mathbf{I}_{\boldsymbol{\theta}}\mathbf{V}|^{-1/p}$ (*generalized variance*)
2. *A-ARE:* $p^{-1}Trace(\mathbf{I}_{\boldsymbol{\theta}}\mathbf{V})^{-1}$ (*trace criterion*), and
3. *E-ARE:* $Ch_{\max/\min}((\mathbf{I}_{\boldsymbol{\theta}}\mathbf{V})^{-1})$ (*extreme root*).

All of these measures will be ≤ 1 for all $\mathbf{T}_n$, but they may not be the same. However, if we have two competing estimators, say, $\mathbf{T}_1$ and $\mathbf{T}_2$, with AMSE matrices $\mathbf{V}_1$ and $\mathbf{V}_2$, respectively, then $\mathbf{V}_1^{-1}\mathbf{V}_2$ may lead to different values for the D-, A-, and E- criteria. For univariate linear models we have an intermediate setup where $\mathbf{V} = \sigma_V^2\mathbf{A}$, and hence all three criteria reduce to $(I_{\boldsymbol{\theta}}\sigma_V^2)^{-1}$. Another way of comparing various estimators is to consider their *large deviation probabilities*, i.e., $-n^{-1}\log P_{\boldsymbol{\theta}}\{\|\mathbf{T}_n - \boldsymbol{\theta}\| \geq \epsilon\}$ (see Hoeffding 1965). There is opportunity for further development in this direction, although in this book we do not place much emphasis on such large deviation probability results.

Let us consider briefly the case with statistical tests. In a single parameter case, if the test statistic T_n is asymptoticallly normal under the null hypothesis, and the alternative hypothesis is contiguous to the null one, by (2.5.120)-(2.5.121) it has also asymptotically normal law under the alternative with only a shift in the mean. If we consider a one-sided test at α-level of significance $(0 < \alpha < 1)$, and if $\Phi(\tau_\alpha) = 1 - \alpha$, then the limiting power of the test based on T_n is $1 - \Phi(\tau_\alpha - kc)$, where the number k comes from the alternative hypothesis specification $(k > 0)$, while the constant c (≥ 0) depends on the asymptotic mean and variance of T_n and varies from one test statistic to another. Thus, if we equate the limiting powers of two competing tests (e.g., based on $T_n^{(1)}$ and $T_n^{(2)}$,) and if c_1 and c_2 are respectively the constants appearing in the limiting power functions (for the common contiguous alternative), then the *asymptotic relative efficiency* (ARE) of $T_n^{(2)}$ with respect to $T_n^{(1)}$ is given by

$$e(T^{(2)}|T^{(1)}) = (\ c_2^2\ )/(\ c_1^2\ ). \tag{2.5.205}$$

In the literature this is termed the *Pitman ARE*. It can be justified similarly for two-sided alternatives. In the multiparameter case, if the limiting distribution of the test statistics are both central chi squared d.f. with p DF, and noncentral ones under contiguous alternatives, then PARE can be computed from their respective noncentrality parameters, both of which are suitable quadratic forms. In general, their ratio may depend on the direction of the alternatives relative to the null hypothesis. Based on the large deviation probabilities, the concept of Bahadur (1960) ARE has gained popularity, despite the fact that it is difficult to derive its exact form. The approximate form is generally insensitive to departures in the asymptotic power functions in various directions.

A few other results from asymptotic and probability theory will be presented in the Appendix.

2.6 PROBLEMS

2.2.1. Show that (2.2.19) implies (2.2.20) and that the converse may not hold. Verify this further with the example where $c_{2j-1} = -j$, $c_{2j} = +j$ for $j \geq 1$.

2.2.2. Let $\mathbf{D}_n = \mathbf{C}_n'\mathbf{C}_n$ and $\boldsymbol{\Delta}_n = diag(d_{n11}^{1/2}, \ldots, d_{npp}^{1/2})$. Assume that (2.2.20) holds and that the $n^{-1}d_{njj}$, $1 \leq j \leq p$ are all bounded away from 0. Then show that as $n \to \infty$,

$$\mathbf{D}_n^{1/2}(\hat{\boldsymbol{\beta}}_n - \boldsymbol{\beta}) \xrightarrow{\mathcal{D}} \mathcal{N}_p(\mathbf{0}, \sigma^2\mathbf{I}_p).$$

2.2.3. Define the e_i as in (2.2.3), and let $f(e)$ be the density function of e. Then note that for every real θ,

$$\theta = \mathbb{E}(e+\theta) = \int_{-\infty}^{\infty} (e+\theta)f(e)de = \int_{-\infty}^{\infty} ef(e-\theta)de,$$

and that differentiating w.r.t. θ (state the regularity conditions),

$$1 = \int_{-\infty}^{\infty} e\{-f'(e-\theta)/f(e-\theta)\}f(e-\theta)de, \ \forall\theta.$$

Use $\theta = 0$ and the Cauchy-Schwartz inequality to show that

$$\begin{aligned} 1^2 = 1 \ &\leq \ \Big(\int_{-\infty}^{\infty} e^2 dF(e)\Big)\Big(\int_{-\infty}^{\infty} \{-f'(e)/f(e)\}^2 dF(e) \\ &= \ \sigma^2 I(f). \end{aligned}$$

Hence, or otherwise, verify (2.2.23).

2.2.4. For the MLE of $\boldsymbol{\beta}$, use the estimating equations

$$\sum_{i \leq n} \mathbf{c}_i \{f'(X_i - \mathbf{c}_i'\hat{\boldsymbol{\beta}}_n)/f(X_i - \mathbf{c}_i'\hat{\boldsymbol{\beta}}_n)\} = \mathbf{0},$$

and show that [state the regularity conditions]

$$\lim_{n\to\infty} \mathbf{D}_n \mathbb{E}[(\hat{\boldsymbol{\beta}}_n - \boldsymbol{\beta})(\hat{\boldsymbol{\beta}}_n - \boldsymbol{\beta})'] = I(f)\mathbf{I}_p.$$

2.2.5. Write $\mathbf{X}_n - \mathbf{C}_n\hat{\boldsymbol{\beta}}_n = \mathbf{e}_n - \mathbf{C}_n(\hat{\boldsymbol{\beta}}_n - \boldsymbol{\beta})$ and express

$$||\mathbf{X}_n - \mathbf{C}_n\hat{\boldsymbol{\beta}}_n||^2 = ||\mathbf{e}_n||^2 + (\hat{\boldsymbol{\beta}}_n - \boldsymbol{\beta})'\mathbf{C}_n'\mathbf{C}_n(\hat{\boldsymbol{\beta}}_n - \boldsymbol{\beta}) - 2\mathbf{e}_n'\mathbf{C}_n(\hat{\boldsymbol{\beta}}_n - \boldsymbol{\beta}).$$

Show that $n^{-1}||\mathbf{e}_n||^2 \to \sigma^2$ a.s.as $n \to \infty$. Use Problem 2.2.2 to verify that $(\hat{\boldsymbol{\beta}}_n - \boldsymbol{\beta})'\mathbf{C}_n'\mathbf{C}_n(\hat{\boldsymbol{\beta}}_n - \boldsymbol{\beta}) = \mathrm{O}_p(1)$ so that the third term on the r.h.s. is $\mathrm{O}_p(n^{1/2})$. Hence, or otherwise, verify (2.2.24).

2.2.6. Show that for the likelihood ratio test under the normal error model [based on (2.2.25)] involves the LSE of $\boldsymbol{\beta}$, and if the underlying d.f. F is not necessary normal (but has a finite variance σ^2), its Pitman efficacy is proportional to σ^{-2}.

2.2.7. For the likelihood ratio test based on the density f $(= F')$, show that the Pitman efficacy is proportional to $I(f)$. You may use the quadratic form (in the MLE) approximation for the likelihood ratio test statistic and Problems 2.2.2 and 2.2.4.

2.3.1. Verify (2.3.7). [You may use the representations in Chapter 6.]

2.4.1. Assess the qualitative robustness of $\theta(F)$ in (2.4.1) when the support of K is a subset of [0,1].

2.4.2. Show that for segmented Ψ_0, $\sigma^2(T_{\Psi_0}, F_0, s)$ cannot attain the lower bound $[I(f_0)]^{-1}$ for any finite k.

Chapter 3

Robust Estimation of Location and Regression

3.1 INTRODUCTION

In this chapter we turn to a systematic account of robust estimators of location and regression parameters. We place special emphasis on the motivations of several important classes of robust estimators and on their basic properties. We will consider two general situations.

1. *Location models.* Let $X_1, \ldots, X_n$ be $n(\geq 1)$ independent and identically distributed (i.i.d.) random variables (r.v.) with an unknown distribution function G, defined on the real line $\mathbb{R}_1$. We conceive of a parametric family of d.f.'s $\{F_\theta : \theta \in \Theta\}$ (usually a dominated system of distributions) with $\boldsymbol{\Theta} \subseteq \mathbb{R}_p$, for some $p \geq 1$, and we wish to estimate θ for which F_θ provides the closest approximation of G. For the location model, we assume that

$$F_\theta(x) \quad = \quad F_0(x - \theta), \tag{3.1.1}$$

where θ is real and F_0 belongs to a class $\mathcal{F}_0$. This model extends immediately to the multivariate case, where X_i's, as well as x and θ in (3.1.1), are q-vectors for some $q \geq 1$.

2. *Regression models.* Suppose that $X_1, \ldots, X_n$ are independent r.v.'s where X_i has d.f. $F(x - \theta_i)$, for $i = 1, \ldots, n$, and the vector $\boldsymbol{\theta} = (\theta_1, \ldots, \theta_n)'$ of parameters satisfy the condition that for some p $(1 \leq p \leq n)$,

$$\boldsymbol{\theta} \in \Pi_p, \tag{3.1.2}$$

where Π_p a linear p-dimensional subspace of $\mathbb{R}_n$. As in the location model, the d.f. F is of unknown form. We conceive of a class $\mathcal{F}$ of distributions, and we assume that $F \in \mathcal{F}; \mathcal{F}$ can either be a compact neighborhood of a fixed d.f. F_0 or it may even be a broad class of (absolutely continuous) d.f.'s. The choice of this $\mathcal{F}$ has an important bearing on the choice of robust estimators for the corresponding models. Note that the location model is a special case of the regression model for which $\theta = \theta_1$, for $\theta \in \boldsymbol{\Theta} \subset \mathbb{R}_1$.

Among various robust estimators, three broad classes, namely the M-, L- and R- estimators, have turned out to be the most interesting, and they have been studied extensively in the literature. The preceding chapter gave a brief introduction to these estimators. In this as well as later chapters we focus on various properties of these estimators in this as well as later chapters. For detailed studies of some other properties of these estimators not discussed thoroughly in this book, we refer the reader to Azencott et al. (1977), Serfling (1980), Huber (1981), Sen (1981a), Bickel (1981), Lehmann (1983), Koul (1992), and Rieder (1994), among others.

In Section 3.2, we consider the basic formulation of the M-estimators. These apear to be the most flexible among the three classes mentioned above. They are well defined for a variety of models for which maximum likelihood estimators (MLE) are also defined. M-estimators cover both the MLE and least square estimators (LSE) as subclasses.

In Section 3.3, L-estimators are considered. These L-estimators were originally conceived as linear combinations of (functions of) order statistics for efficient estimation of location or scale parameters, and they are generally computationally appealing and possess various desirable properties. L-estimators have also been considered for linear models, and we will refer to that later on.

Section 3.4 is devoted to the study of R-estimators. These estimators are generally based on the ranks of observations (or signed ranks), and generally they correspond to suitable rank tests for symmetry or randomness against shift or regression alternatives. We will mainly consider the R-estimators of location and regression parameters.

Besides these principal classes of robust estimators, some other notable classes have been considered in this chapter. *Minimum distance estimators* (MDE) and *Pitman-type estimators* (PE) are briefly treated in Section 3.5. The concluding section presents a general introduction to *differentiable statistical functionals* with due emphasis on their robustness properties.

3.2 M-ESTIMATORS

As an illustration we start with the case of i.i.d. r.v.'s and consider a general single parameter model. Suppose that $X_1, \ldots, X_n$ have common d.f. F_θ, with the real (unknown) parameter θ, and assume that there exists a real valued function $\rho(x,t) : \mathbb{R}_2 \to \mathbb{R}_1$ such that for the model $\{F_\theta : \theta \in \Theta\}, \Theta \subseteq \mathbb{R}_1$,

$$\int \rho(x,t)dF_\theta(x) \text{ has a unique minimum at } t = \theta, \forall \theta \in \boldsymbol{\Theta}. \tag{3.2.1}$$

Then, an M-estimator of θ is defined as a statistic $M_n = M_n(X_1, \ldots, X_n)$, which is a solution (with respect to t) of the minimization of $\sum_{i=1}^n \rho(X_i,t)$. We may therefore write

$$M_n = Arg \cdot \min\{\sum_{i=1}^n \rho(X_i,t) : t \in \mathbb{R}_1\}. \tag{3.2.2}$$

In many cases, the minimization problem in (3.2.2) leads to the equation

$$\sum_{i=1}^n \psi(X_i,t) = 0 \tag{3.2.3}$$

where $\psi(x,t) = k\{(\partial/\partial t)\rho(x,t)\}$, for all x,t, and k is a nonzero real number. For example, for the location model pertaining to the normal family of d.f.'s, we have for the MLE, $\rho(x,t) = (x-t)^2$, and hence $\psi(x,t) = (x-t)$, for every real x,t. For the same problem, instead of the MLE, we may consider the L_1-norm estimator for which $\rho(x,t) = |x-t|$ for $x,t \in \mathbb{R}_1$, so that we have $\psi(x,t) = \text{sign}(x-t)$, and the solution in (3.2.2) reduces to the median of $X_1, \ldots, X_n$. Other forms of $\rho(x,t)$ for location or other models may be chosen from relevant considerations. If $F_\theta(x)$ has the density $f(x,\theta)$ (with respect to a sigmafinite measure μ) and the density is differentiable in θ, then on choosing $\rho(x,\theta) = -\log f(x,\theta)$, we obtain that the corresponding $\psi(x,t) = -f'(x,t)/f(x,t)$, and hence the class of M-estimators in (3.2.2) include the MLE as a subclass. For the simple location model in (3.1.1), if we restrict ourselves to the class $\mathcal{F}_0$ of d.f.'s symmetric around 0, then we usually consider $\rho(x,t) = h(x-t)$ where $h(-x) = h(x)$ for any x. But the picture may change if the underlying d.f. is not a member of this class $\mathcal{F}_0$. Looking from a wider perspective, the following questions immediately arise:

1. What are we really estimating in the case where G, the true d.f. of the X_i does not belong to the class $\{F_\theta : \theta \in \Theta\}$?

2. Under what regularity conditions does there exist a consistent sequence of solutions of (3.2.2)?

3. What is the compensation for efficiency in the robust estimation derived from (3.2.2)?

4. What can we say in general about the distributional and other related properties of the M-estimators?

The answer to the first question is relatively simple. If we denote the empirical d.f. F_n by $F_n(x) = n^{-1}\sum_{i=1}^n I(X_i \le x), x \in \mathbb{R}_1$, then by the classical Glivenko-Cantelli lemma, when G is the d.f., we have

$$||F_n - G|| = \sup_{x \in \boldsymbol{R}_1} |F_n(x) - G(x)| \to 0 \quad \text{a.s., as } n \to \infty \tag{3.2.4}$$

Thus parallel to (3.2.1), we may consider the minimization

$$\int \rho(x,t)dG(x) = \min, \quad t \in \mathbb{R}_1 \tag{3.2.5}$$

and if (3.2.5) admits a solution $M(G)$, then the M-estimator in (3.2.2) can be described generally as an estimator of $M(G)$. However, the existence of a unique solution of (3.2.5) is not guaranteed by (3.2.4) alone. Additional conditions on the function $\rho(.,.)$ and/or the d.f. G are generally needed in this context. Characterization of $M(G)$ plays an important role for M-estimation.

In addressing the second question, we should keep in mind the classical problem of the existence of a consistent solution of the likelihood equation providing the usual MLE. Some thoughts on this question have been provided in the literature by Huber (1981, ch. 6), Serfling (1980, ch. 7), and Lehmann (1983, ch. 6), among others. If $\psi(x,t)$ is monotone in t, then the existence of M-estimators can be established under very general regularity conditions, and the consistent sequence of solutions of (3.2.3) can easily be identified under parallel regularity conditions. However, the monotonicity of $\psi(x,t)$ in t is only a sufficient condition, and the existence of a consistent M-estimator can also be established for some nonmonotone $\psi(x,t)$, which are sufficiently smooth in t. Typically, in such a case, there are multiple solutions of (3.2.3), and it may be difficult to identify the one among them that will be consistent. Various approaches have been suggested to eliminate this basic problem. We will find convenient to use a *one-step version* of M-estimators that can be made arbitrarily close to the consistent solution of (3.2.2) when n is large. We deter the details of this theory to Chapter 7.

The last two questions relate to some specific problems that arise in the study of asymptotic normality of M_n (when suitably normalized). For monotone $\psi(.,t)$ (in t), this asymptotic normality result is generally obtained by inversion, and we will discuss this more later on. Our main emphasis here is on asymptotic representation results of a consistent sequence of M-estimators by a sum of independent r.v.'s that not only yields the asymptotic normality result as a by-product, but also provides deeper insight into the asymptotic behavior of M_n and some other important asymptotic results. Basically we aim to show that if there exists a consistent sequence $\{M_n\}$, then under some regularity conditions on $F(\equiv G)$ and ψ

$$n^{1/2}(M_n - M(F)) = -n^{-1/2}(\gamma(F))^{-1}\sum_{i=1}^{n}\psi(X_i, M(F)) + o_p(1), \tag{3.2.6}$$

where $f(x) = (d/dx)F(x)$,

$$\gamma(F) = -\int \psi(x, M(F))df(x) \quad \text{or} \quad \int f(x)d\psi(x, M(F)), \tag{3.2.7}$$

and $\psi(x,t) = (\partial/\partial t)\rho(x,t)$. Such representations, supplemented by more precise orders for the remainder term, will be considered in Chapter 5. This representation among others implies [via the central limit theorem on $\psi(X_i, M(F))$]

that $n^{1/2}(M_n - M(F))$ is asymptotically normal with zero mean and variance $(\gamma(F))^{-2}\text{Var}[\psi(X_1, M(F))]$, and this, in turn, provides an answer to question 3. We may remark that for the symmetric location model, Huber (1964) studied first the asymptotic normality of M_n, and later (1967), he gave generalizations to other problems. This topic has been a subject of many studies over the past 30 years, and diverse techniques have been employed by a host of researchers toward the same goal. Our approach based on asymptotic representations of the type (3.2.6) has some additional advantages, and we will explore them systematically in this and subsequent chapters. Let us concentrate first on the location model, briefly introduced in Section 3.1. For this model we have $F_\theta(x) = F(x-\theta)$, and F is an unknown d.f., symmetric around 0, and we assume that $F \in \mathcal{F}_0$, the class of absolutely continuous (symmetric) d.f.'s. The M-estimator is then defined as a solution of the minimization

$$\sum_{i=1}^{n} \rho(X_i - t) = \min, \tag{3.2.8}$$

where, due to the assumed symmetry of F, it is recommended that a symmetric ρ be employed in (3.2.8). The *influence curve* of M_n is then given by

$$IC(x; F, M) = \psi(x)/\gamma(F), \quad \gamma(F) = \int f(x)d\psi(x). \tag{3.2.9}$$

Note that IC(x;ψ, M) is proportional to ψ, where the proportionality constant $(\gamma(F))^{-1}$ depends on both ψ and the density $f(.)$. Also, IC(x;ψ, M) is bounded whenever ψ is bounded and $\gamma(F) \neq 0$. Moreover, whenever f admits of a derivative f' for almost all x, then we obtain by partial integration that

$$\gamma(F) = \int \{-f'(x)/f(x)\}\psi(x)dF(x), \tag{3.2.10}$$

a form that is more suitable for manipulation in a number of situations. In passing, we may recall that the Huber (1964) estimator is asymptotically minimax over a contaminated neighborhood of a fixed symmetric distribution F_0 (see Section 5.2 for some details). In particular when F_0 is the standard normal d.f., we are led to the Huber M-estimator generated by $\psi = \psi_0$ in (2.4.13).

The M-estimator M_n [the solution of (3.2.8)] is clearly *translation-equivariant* [since $\rho(X_i - t) = \rho((X_i + a) - (t + a))$, for every $a \in \mathbb{R}_1$], but it may not be *scale-equivariant*, since in general, $\psi(cx) \neq c\psi(x)\ \forall x \in \mathbb{R}_1, c > 0$. The M-estimator can be made scale-equivariant by some modifications. Huber (1964) suggested that either the *simultaneous M-estimation* of location and scale parameters be employed or the M-estimator of location be *studentized* by a statistic S_n of the form $S(X_1, \ldots, X_n)$ such that $S(aX_1, \ldots, aX_n) = |a|S(X_1, \ldots, X_n)$ for every real a. In (3.2.8) we use $\rho((X_i - t)/S_n)$ instead of $\rho(X_i - t)$, for $i = 1, \ldots, n$. Unlike in the case of the location parameter (which may be interpreted as the center of symmetry of a distribution), there is perhaps no natural interpretation of a scale parameter, and hence in a real situation *studentization* leads to a better choice of the function $\rho(.)$. Thus, despite the modest aim to make M-estimators scale-equivariant, studentization serves a better purpose in practical applications. In addition it has been observed in some situations to be very reasonable in estimating the common location and the ratio of scale parameters in the two-sample model (Jurečková and Sen 1982a).

Let us return to the regression model in (3.1.2). As a natural extension of (3.2.3), we now consider a distance function

$$\sum_{i=1}^{n} \rho(X_i, t_i) : \mathbf{t} = (t_1, \ldots, t_n)' \in \Pi_p, \tag{3.2.11}$$

where Π_p is defined in (3.2.3). Our task is to minimize (3.2.11) with respect to t. The M-estimator of $\boldsymbol{\theta} = (\theta_1, \ldots, \theta_n)'$ is a solution of

$$\sum_{i=1}^{n} \rho(X_i, t_i) := \min, \quad \mathbf{t} \in \Pi_p \tag{3.2.12}$$

For the general regression model we choose a $n \times p$ matrix $\mathbf{C}= ((c_{ij}))$ of known regression constants and p-vector $\boldsymbol{\beta} = (\beta_1, \ldots, \beta_p)'$ of unknown regression parameters, and set

$$\boldsymbol{\theta} = \mathbf{C}\boldsymbol{\beta} \quad \text{and} \quad \rho(a, b) = \rho(a - b), \tag{3.2.13}$$

so that, identifying $\psi(y)$ as $(d/dy)\rho(y)$ in the case of a continuous ψ we can equivalently write (3.2.12) as

$$\sum_{i=1}^{n} c_{ij}\psi\Big(X_i - \textstyle\sum_{k=1}^{p} c_{ik} b_k\Big) = 0, \quad j = 1, \ldots, p. \tag{3.2.14}$$

This equation may be regarded as a natural generalization of the location model to the regression case, and, as we will see later, the M-estimators in both the models (3.2.12) and (3.2.14) have similar properties. We will focus rather on a small difference (in terms of computational simplicity) in the two situations.

Usually the function ρ in (3.2.8) is taken to be convex, and often to induce more robustness, ψ is taken to be of bounded variation. For monotone (nondecreasing) ψ, the solution in (3.2.8) can be expressed as

$$M_n = \frac{1}{2}(M_{n,1} + M_{n,2}); \tag{3.2.15}$$

$$M_{n,1} = \sup\{t : \sum_{i=1}^{n} \psi(X_i - t) > 0\}, M_{n,2} = \inf\{t : \sum_{i=1}^{n} \psi(X_i - t) < 0\}. \tag{3.2.16}$$

Note that for nondecreasing $\psi, \sum_{i=1}^{n} \psi(X_i - t)$ is nonincreasing in $t \in \mathbb{R}_1$. Hence M_n in (3.2.15) represents the centroid of the set of solutions of (3.2.14), and it eliminates the possible arbitrariness of such a solution. If $p = 1$, the situation in regression model is very similar to the location model. If $p \geq 2$, $\mathbf{c}_i$ satisfy some concordance-discordance conditions, then the left-hand side of (3.2.14) may be decomposed as a finite linear combination of terms that are monotone in each of the p arguments $b_1, \ldots, b_p$. In any case, whether or not ψ is monotone, we may use an initial (square-root n-) consistent estimator and adopt an iterative procedure. Generally, the consistency of the solution of (3.3.11) follows when $\rho(x, t)$ is convex in t. If this is not the case but ψ is continuous in t, existence of a $\sqrt{n}$-consistent estimator of $\boldsymbol{\beta}$ can be established, although it may not coincide with the global minimum of (3.2.12). In this context, as well as in other asymptotic results, asymptotic linearity M-statistics in the location or regression parameter plays a central role, and we will explore this systematically in this book.

To establish the asymptotic normality of $n^{1/2}(M_n - \theta)$ in the location model (with monotone ψ), we must show by (3.2.15) and (3.2.16) that under $\theta = 0$ and for every fixed t,

$$n^{-1/2} \sum_{i=1}^{n} \psi(X_i - n^{-1/2}t)$$

has asymptotically a normal distribution with mean $-t\gamma(F)$ and variance σ_ψ^2, defined by (2.4.10) ; see Problem 3.2.1. Though this inversion technique is applicable for the regression model when $p = 1$, we generally need a different approach for $p \geq 2$, and the linearity approach provides an easy access. Actually, for the regression model, we have parallel to (3.2.6) for $n \to \infty$,

$$n^{1/2}(\mathbf{M}_n - \boldsymbol{\beta}) = n^{-1/2}(\gamma(F))^{-1} \sum_{i=1}^{n} \mathbf{c}_i \psi(X_i - \mathbf{c}_i'\boldsymbol{\beta}) + o_p(1) \qquad (3.2.17)$$

where $\gamma(F)$ is defined in (3.2.7). Then under the Noether condition,

$$\max\left\{\mathbf{c}_i'\left(\sum_{j=1}^{n} \mathbf{c}_j \mathbf{c}_j'\right)^{-1} \mathbf{c}_i : 1 \leq i \leq n\right\} \to 0 \quad \text{as } n \to \infty, \qquad (3.2.18)$$

on the $\mathbf{c}_i$, by the usual multivariate central limit theorem,

$$n^{1/2}(\mathbf{M}_n - \boldsymbol{\beta}) \xrightarrow{\mathcal{D}} \mathcal{N}_p(\mathbf{0}, \nu^2 \mathbf{C}^{\star -1}), \qquad (3.2.19)$$

where

$$\mathbf{C}^{\star} = \lim_{n \to \infty} n^{-1}\left(\sum_{j=1}^{n} \mathbf{c}_j \mathbf{c}_j'\right) \text{ and } \nu^2 = (\gamma(F))^{-2} \sigma_\psi^2, \qquad (3.2.20)$$

see Problem 3.2.2 in this connection. This asymptotic linear representation-based proof of asymptotic normality applies to the simple location model as well, where $c_i = 1\ \forall i, p = 1$ and hence $C^\star = 1$. Thus for the location model

$$n^{1/2}(M_n - \theta) \xrightarrow{\mathcal{D}} \mathcal{N}(0, \nu^2). \tag{3.2.21}$$

Recalling that

$$\nu^2 = \{\int_{\mathbb{R}_1} \psi^2(x)dF(x)\}/\{\int_{\mathbb{R}_1} f(x)d\psi(x)\}^2 = \nu^2(\psi, F) \tag{3.2.22}$$

and that $\mathbf{C}^\star$ is a given positive definite (p.d.) matrix, we may conclude that for both the location and regression models, the choice of the score function ψ is governed by the same considerations. As such, various considerations laid down in Sections 2.3 and 2.4 pertain to both the location and regression models. Further, as in the location model, generally, $\psi(cx) \neq c\psi(x)$ for every $x \in \mathbb{R}_1$ and $c > 0$, and this lack of scale-equivariance of ψ implies that the M-estimators of the regression parameters are also not generally scale-equivariant. Parallel to the location model, we may also be interested in considering scale-equivariant M-estimators of regression parameters, and these will be considered in a general setup in Chapter 5. The one-step version of M-estimators of regression parameters will be considered in Chapter 7.

We may note that for every $\mathbf{b} \in \mathbb{R}_p, \mathbf{t} \in \mathbb{R}_p$, and $x \in \mathbb{R}_1$ and $\mathbf{c}_i$,

$$\psi(x - \mathbf{c}_i'\mathbf{t}) = \psi\big((x + \mathbf{c}_i'\mathbf{b}) - \mathbf{c}_i'(\mathbf{b} + \mathbf{t})\big). \tag{3.2.23}$$

By looking at (3.2.14) and (3.2.23), we can immediately see that the M-estimators of location or regression are *regression-equivariant*:

$$\mathbf{M}_n(X_1 + \mathbf{b}'\mathbf{c}_1, \ldots, X_n + \mathbf{b}'\mathbf{c}_n) = \mathbf{M}_n(X_1, \ldots, X_n) + \mathbf{b}, \quad \forall \mathbf{b} \in \mathbb{R}_p. \tag{3.2.24}$$

Because of this regression-equivariance, we set without loss of generality $\boldsymbol{\beta} = \mathbf{0}$. Then, for ψ skew-symmetric about 0, by virtue of (3.2.14),

$$\mathbf{M}_n(X_1, \ldots, X_n) = -\mathbf{M}_n((-1)X_1, \ldots, (-1)X_n), \quad \text{w. p. 1.} \tag{3.2.25}$$

On the other hand, under $\mathbf{b} = \mathbf{0}$, whenever the d.f. F is symmetric about 0 (as is generally assumed), we note that X_i and $(-1)X_i$ have the same distribution (F). Hence by (3.2.25) we conclude that $\mathbf{M}_n$ has a distribution (diagonally) symmetric about $\mathbf{0}$ (when $\boldsymbol{\beta}$ is $\mathbf{0}$). Incorporating the regression-equivariance in (3.2.24), we conclude that for the general regression model whenever the ψ-function is skew-symmetric and the d.f. F is symmetric, $\mathbf{M}_n$, the M-estimator of the regression parameter (vector) $\boldsymbol{\beta}$ is distributed (diagonally) symmetrically, in the sense that

$$(\mathbf{M}_n - \boldsymbol{\beta}) \quad \text{and} \quad (\boldsymbol{\beta} - \mathbf{M}_n) \quad \text{both have the same d.f.} \tag{3.2.26}$$

This implies that marginally each element of $\mathbf{M}_n$ has a distribution symmetric about the corresponding element of $\boldsymbol{\beta}$ so that

$$\mathbf{M}_n \text{ is a median-unbiased estimator of } \boldsymbol{\beta}. \tag{3.2.27}$$

We may remark that both the assumptions of skewsymmetry of ψ and symmetry of F are crucial for the median-unbiasedness of $\mathbf{M}_n$. The median-unbiasedness may not imply unbiasedness in the classical sense. To see this, we need to verify the uniform integrability condition of the $\mathbf{M}_n$. This places extra regularity conditions on ψ and F (the location model case has been studied in Jurečková and Sen 1982b). If we use a smooth and bounded function ψ, this integrability condition does not entail a restriction on d.f. F (we refer to Problems 3.2.4 – 3.2.5). Note that (3.2.24) - (3.2.27) may hold, even for the nonmonotone score functions, like for some (Hampel-type) functions, mentioned in Section 2.4. However, we generally recommend monotone score functions for their computational ease, and in this case the M-estimators of location and regression retain their validity and robustness for a broad class of (symmetric) distributions. Within this broad class of d.f.'s, optimal score functions for specific subclasses (in the sense of local robustness) can be chosen as in Sections 2.3 and 2.4.

Finally, let us consider the breakdown and tail-behavior properties of M-estimator of location parameter with bounded influence function. Assume that ρ in (3.2.8) is convex and symmetric and that $\psi = \rho'$ is continuous and bounded. Then M_n is translation equivariant and monotone in each $X_i,\ i = 1, \ldots, n$; hence Theorem 2.3.1 applies. Assume first that n is odd, and denote $r = (n+1)/2$. Let

$$K = \sup\{\psi(x) : x \geq 0\} \Big[= -\inf\{\psi(x) : x \leq 0\}\Big]. \tag{3.2.28}$$

Then

$$\begin{aligned}
&I\!P_\theta(|M_n - \theta| > a)\\
&= I\!P_0(M_n > a) + I\!P_0(M_n < -a)\\
&\geq I\!P_0(\sum_{i=1}^{n} \psi(X_i - a) > 0) + I\!P_0(\sum_{i=1}^{n} \psi(X_i + a) < 0)\\
&\geq I\!P_0(\psi(X_{n:r} - a) > \frac{n-1}{n+1}K)\\
&+ I\!P_0(\psi(X_{n:r} + a) < -\frac{n-1}{n+1}K)\\
&\geq I\!P_0(X_{n:r} > a + c_1) + I\!P_0(X_{n:r} < -a - c_1)\\
&\geq 2\binom{n}{r}\big(F(a+c_1)\big)^{r-1}\big(1 - F(a+c_1)\big)^{r},
\end{aligned} \tag{3.2.29}$$

where $c_1 = \inf\{x : \psi(x) \geq \frac{n-1}{n+1}K\} \in [0, \infty)$. Similarly,

$$\begin{aligned}
&\mathbb{P}_0(|M_n| > a) \\
&\leq \mathbb{P}_0(\sum_{i=1}^{n} \psi(X_i - a) \geq 0) + \mathbb{P}_0(\sum_{i=1}^{n} \psi(X_i + a) \leq 0) \\
&\leq \mathbb{P}_0(\psi(X_{n:r} - a) \geq -\frac{n-1}{n+1}K) \\
&+ \mathbb{P}_0(\psi(X_{n:r} + a) \leq \frac{n-1}{n+1}K) \\
&\leq \mathbb{P}_0(X_{n:r} \geq a - c_2) + \mathbb{P}_0(X_{n:r} \leq -a + c_2) \\
&\leq 2r \binom{n}{r} (1 - F(a - c_2))^r, \qquad (3.2.30)
\end{aligned}$$

where $c_2 = \sup\{x : \psi(x) \leq \frac{n-1}{n+1}K\} \in [0, \infty)$. Hence by (3.2.29),

$$\limsup_{a\to\infty} B(a, M_n) \leq \frac{n+1}{2} \lim_{a\to\infty} \frac{\log(1 - F(a + c_1))}{\log(1 - F(a))} = \frac{n+1}{2}, \qquad (3.2.31)$$

provided F satisfies (2.3.25), and similarly (3.2.11) implies

$$\liminf_{a\to\infty} B(a, M_n) \geq \frac{n+1}{2}. \qquad (3.2.32)$$

Thus (3.2.33), (3.2.34), and Theorem 2.3.1 imply that

$$\lim_{a\to\infty} B(a, T_n) = \frac{n+1}{2} = m^\star. \qquad (3.2.33)$$

If n is even, we proceed analogously and obtain

$$m^\star = \frac{n}{2}, \quad \frac{n}{2} \leq \liminf_{a\to\infty} B(a, M_n) \leq \limsup_{a\to\infty} B(a, M_n) \leq \frac{n}{2} + 1. \qquad (3.2.34)$$

We come to the conclusion that the M-estimator of the location parameter in a symmetric model, generated by a nondecreasing bounded function, has the largest possible breakdown point and its tail performance is not affected by the tails of the distribution. Hence the M-estimator is robust.

The above considerations could be extended in a straightforward manner to the sample median. Hence (3.2.35) and (3.2.36) also apply to the median.

3.3 L-ESTIMATORS

L-estimators are termed after linear combinations of (functions of) order statistics. In the location-scale family of distributions, it was observed that the sample order statistics may generally provide simple and efficient estimators; recall the completeness of sample order statistics (i.e., Problem 3.3.1). To motivate the L-estimators, let us first look at the simple location model. Let $X_1, \ldots, X_n$ be n i.i.d. r.v.'s with a d.f. F_θ, defined on the real line $\mathbb{R}_1$, and let

$$X_{n:1} \leq \ldots \leq X_{n:n} \tag{3.3.1}$$

be the sample order statistics. Then an L-estimator of a parameter θ is defined by

$$L_n = \sum_{i=1}^{n} c_{ni} X_{n:i}, \tag{3.3.2}$$

where the coefficients $c_{n1}, \ldots, c_{nn}$ are known. The coefficients are chosen so that L_n in (3.3.2) is a desirable estimator of θ in the sense of unbiasedness, consistency and other efficiency properties. We may remark that if the d.f. F_θ admits a density f_θ (on its support), then the ties in (3.3.1) may be neglected w. p. 1, so the order statistics $X_{n:1}, \ldots, X_{n:n}$ are all distinct w. p. 1. As in Section 2.4 we introduce the *quantile function* $Q(t) = inf\{x : F_\theta(x) \geq t\}, 0 < t < 1$ and its sample counterpart

$$Q_n(t) = \inf\{x : F_n(x) \geq t\}, \quad 0 < t < 1, \tag{3.3.3}$$

where the sample d.f. F_n is defined before (3.2.4). We further conceive of a sequence $\{J_n = [J_n(u) : u \in (0,1)]\}$ of functions defined on the unit interval (0,1) such that $J_n(u)$ assumes a constant value $J_n(i/(n+1))$ on $((i-1)/n, i/n]$, for $i = 1, \ldots, n$. On letting

$$c_{ni} = n^{-1} J_n(i/(n+1)), \quad i = 1, \ldots, n, \tag{3.3.4}$$

and noting that F_n has the jumps of magnitude n^{-1} at the n-order statistics (and 0 elsewhere), we obtain from (3.3.2) -(3.3.4) that

$$\begin{aligned} L_n &= n^{-1} \sum_{i=1}^{n} Q_n(i/(n+1)) J_n(i/(n+1)) \\ &= \int_{\mathbb{R}_1} Q_n(F_n(x)) J_n(F_n(x)) dF_n(x). \end{aligned} \tag{3.3.5}$$

In this form, L_n is identifiable as the sample counterpart corresponding to the L-functional in (2.4.1). Due to their simple and explicit forms [(3.3.2) or (3.3.5)], the L-estimators are computationally appealing (relative to the other types of estimators). It is clear that the c_{ni} or equivalently the sequence $\{J_n(.)\}$ are major considerations in the study of properties of the L-estimators. To motivate this study, we consider the location-scale model: X_i's are i.i.d. r.v.'s with a density function $f_{\mu,\delta}(x), x \in \mathbb{R}_1$, given by

$$f_{\mu,\delta}(x) = \delta^{-1} f_0((x-\mu)/\delta), \quad x \in \mathbb{R}_1, \mu \in \mathbb{R}_1, \delta \in \mathbb{R}_1^+, \tag{3.3.6}$$

where $f_0(.)$ has a specified form (independent of μ and δ). For the time being, we assume that f_0 has a finite second moment, and without any loss of generality, we assume that the variance of f_0 is equal to 1. Then we write

$$X_{n:i} = \mu + \delta Y_{n:i}, \quad i = 1, \ldots, n, \tag{3.3.7}$$

where $Y_{n:i}$'s are the sample order statistics of a sample of size n from a d.f. F_0 corresponding to the density f_0. If we let

$$\mathbb{E}Y_{n:i} = a_{n:i} \text{ and } \mathrm{Cov}(Y_{n:i}, Y_{n:j}) = b_{nij} \quad \text{for } i, j = 1, \ldots, n, \tag{3.3.8}$$

then both $\mathbf{a}_n = (a_{n:1}, \ldots, a_{n:n})'$ and $\mathbf{B}_n = ((b_{nij}))_{i,j=1,\ldots,n}$ are independent of (μ, δ) and can be computed with the density f_0. Let $\mathbf{B}_n^{-1} = ((b_n^{ij}))$. For the sake of simplicity, let us assume that $\mathbf{B}_n^{-1}$ exists. We are naturally tempted to use the weighted least squares method (see, Lloyd 1952) to obtain the best (i.e., minimum variance unbiased) estimators of (μ, δ) based on the $X_{n:i}$ in (3.3.7). This amounts to the minimization of

$$W(\mu, \delta) = \sum_{i=1}^{n} \sum_{j=1}^{n} b_n^{ij} (X_{n:i} - \mu - \delta a_{n:i})(X_{n:j} - \mu - \delta a_{n:j}) \tag{3.3.9}$$

with respect to the two unknown parameters (μ, δ). The resulting estimators derived from the estimating equations:

$$(\partial/\partial s)W(s,t) = 0, (\partial/\partial t)W(s,t) = 0 \quad \text{at } (s,t) = (\hat{\mu}_n, \hat{\delta}_n), \tag{3.3.10}$$

are clearly linear in $X_{n:i}$ and conform to the form in (3.3.2) with the c_{ni} depending on $\mathbf{a}_n$ and $\mathbf{B}_n^{-1}$. In the literature, this is known as BLUE (best linear unbiased estimator) of (μ, δ). The theory sketched above can easily be extended to cover the case of censored data where some of the order statistics at the lower and/or upper extremity are omitted from consideration (on the ground of robustness or for other reasons). In this case we need to confine ourselves to a subset of the $X_{n:i}$'s (i.e., $X_{n:j}, j \in J \subset \{1, \ldots, n\}$). Using the weighted least squares method on this subset, we again obtain an optimal BLUE (within this subset). In this case some of the c_{ni}'s in (3.3.2) (at the beginning or end) are equal to 0. A very similar situation arises when we have a subset of selected order statistics, $\{X_{n:[np_j]}, j = 1, \ldots, k\}$ where k ($\leq n$) is a fixed positive integer and $0 < p_1 < \ldots < p_k < 1$ are some given numbers. Again the weighted least squares theory, as applied to this specific subset, yields a BLUE that conforms to (3.3.2), but only k of the c_{ni}'s are different from 0. In the extreme case we may use a single quantile $X_{n:[np]}$ or a pair $(X_{n:[np_1]}, X_{n:[np_2]})$, and these may be regarded as L-estimators too. During the 1950s and 1960s, BLUE theory for various densities and under diverse censoring patterns emerged as a very popular area of research. We may refer to Sarhan and Greenberg (1962) and David (1981) for useful accounts of the developments. We may remark that the $a_{n:i}$ and b_{nij}

in (3.3.8) depend on the underlying density f_0 in a rather involved manner, and their exact computations (for any given n) require extensive numerical integration, and hence, for large n, the task may become prohibitively laborious (even with the advent of modern computers). Fortunately the quantiles $F_0^{-1}(i/(n+1))$ often provide good approximations for the $a_{n:i}$, and similar approximations are available for the b_{nij} (see, Problems 3.3.5 and 3.3.6). Use of these approximations has greatly simplified the computational aspects leading to estimators that are termed nearly or asymptotically BLUE. A very useful account of this theory in various specific models is given in Sarhan and Greenberg (1962). From this theory we might get a clear justification for (3.3.4), with the further insight that there exists a function $J = \{J(u), 0 < u < 1\}$ defined on (0,1) such that

$$J_n(u) \to J(u), \text{ as } n \to \infty, \text{ for every (fixed) } u \in (0,1). \tag{3.3.11}$$

This expression also suggests that the optimal J generally depends on the density f_0. We may even write

$$J(u) = J(u; f_0) \quad \text{for } u \in (0,1). \tag{3.3.12}$$

One important implication of (3.3.12) is the vulnerability of the optimal score J to possible departures from the assumed density. Even when the actual density (e.g., g_0) and the assumed one (f_0) are fairly close to each other, the b_n^{ij}'s are very sensitive to small departures, and hence the asymptotically optimal score $J(., f_0)$ may not be that close to the true score $J(., g_0)$. This feature can easily be verified with the "error-contamination" model (2.3.10). This suggests that for BLUE, it may be wiser to look into possible departures from the assumed model and to choose the score function $J(.)$ in a more robust manner. While such a robust choice of J may not lead to the asymptotic best property (under the model f_0), it may have other minimaxity or similar properties for a class of distributions "close" to the assumed one (F_0).

With this introduction we are in a position to generalize the pure parametric BLUE theory to robust BLUE theory, and this is in line with the robust M-estimators discussed in the previous section. We start with usual location model where $F_\theta(x) = F(x-\theta), x \in I\!R_1$, and $\theta \in I\!R_1$. We assume that F is symmetric about the origin. We consider a general class of L-estimators of location (3.3.2) satisfying

$$\sum c_{ni} = 1 \text{ and } c_{ni} = c_{n,n-i+1}, \quad \text{for every } i. \tag{3.3.13}$$

The first condition in (3.3.13) ensures the translation-equivariance of L_n, while the second condition is associated with the assumed symmetry of the d.f. F_0. It ensures that L_n has a symmetric distribution around θ so that L_n is median-unbiased. We refer to Problem 3.3.7 in this context. An important class of L_n belonging to this family of L-estimators is the so-called kth order rank-weighted mean (see, Sen 1964) which can be expressed as

$$T_{nk} = \binom{n}{2k+1}^{-1} \sum_{i=k+1}^{n-k} \binom{i-1}{k}\binom{n-i}{k} X_{n:i}, \tag{3.3.14}$$

for $k = 0, \ldots, [(n+1)/2]$. Note that for $k = 0$, $T_{n0} = \bar{X}_n = n^{-1}\sum_{i=1}^{n} X_i$, the sample mean, while for $k = [(n+1)/2]$, (3.3.14) reduces to the sample median. For $k \geq 1$, T_{nk} does not need to be fully efficient for the basic model where F is a normal d.f. Thus the class $\{T_{nk}; k \geq 0\}$ covers a scale from the sample median to the sample mean, namely from highly robust to highly nonrobust estimators of location. Since this is a subclass of L-estimators satisfying (3.3.13), we conclude that the class in (3.3.13) has the same characterization too. Though the sample median is a robust estimator of θ, it is generally not very efficient for nearly normal d.f. Among the L-estimators of location that are less sensitive to deviations from the normal distribution and yet are more efficient than the sample median at the normal distribution (model), we may mention the following:

1. *Trimmed mean.* For an $\alpha : 0 < \alpha < 1/2$, let

$$L_n(\alpha) = \Big\{ \sum_{i=[n\alpha]+1}^{n-[n\alpha]} X_{n:i} \Big\} / (n - 2[n\alpha]), \tag{3.3.15}$$

where $[x]$ denotes the largest integer $\leq x$.

2. *Winsorized mean.* For an $\alpha : 0 < \alpha < 1/2$, let

$$L_n^\star(\alpha) = \Big\{ [n\alpha] X_{n:[n\alpha]} + \sum_{i=[n\alpha]+1}^{n-[n\alpha]} X_{n:i} + [n\alpha] X_{n:n-[n\alpha]+1} \Big\} / n. \tag{3.3.16}$$

Note that (3.3.13) holds for either of these estimators. Thus both the trimmed and Winsorized means are translation-equivariant estimators of location, and for symmetric F, they are both median-unbiased for θ. The idea to trim off some of the (extreme) observations and calculate the average of the remaining (central) observations is so old that we can hardly say by whom it was originally proposed. Looking at (3.3.14), (3.3.15) and (3.3.16), we observe that by letting k such that $k/n \approx \alpha$, all of these estimators basically aim to eliminate the effect of the extreme values, although we have a relatively constant weight in (3.3.15) for the central values, along with the jumps in (3.3.16) and a bell-shaped weight function in (3.3.14). For all of them

$$c_{ni} = 0 \quad \text{for } i \leq k, \;\; i \geq n-k+1,$$

and the c_{ni}'s are all nonnegative. Each is a convex combination of the set of order statistics. Though generally the L-estimators of location are convex combinations of order statistics, there may be situations where some of the

c_{ni}'s may turn out to be negative. To stress this point, we may consider a trimmed L-estimator of the form (3.3.2), where

$$c_{ni} = 0 \quad \text{for } i \leq [n\alpha] \text{ and } i \geq n - [n\alpha] \text{ for some } \alpha \in (0, \frac{1}{2}). \tag{3.3.17}$$

From the asymptotic distribution of L_n (as will be studied in Chapter 4), we see that under (3.3.17), L_n can be asymptotically efficient for some d.f. F but that such an F should then have typically an exponentially decreasing tail (on either side). If, however, F has tails heavier than an exponential tail, then an L-estimator to be asymptotically efficient must have negative c_{ni}'s in the two extremes. Thus we can not impose the nonnegativity on the c_{ni} as a blanket rule for L-estimators of location.

Let us now study the consistency and asymptotic efficiency properties of L-estimators of location and, in general, of other L-estimators as well. From what has been discussed after (3.3.10) we gather that in a relatively more general situation, we may decompose an L-estimator in the form

$$\begin{aligned} L_n &= n^{-1}\sum_{i=1}^{n} J_{n,1}(i/(n+1))X_{n:i} + \sum_{j=1}^{k} a_{nj}X_{n:[np_j]} \\ &= L_{n1} + L_{n2}, \text{say}, \end{aligned} \tag{3.3.18}$$

where $J_{n,1} : (0,1) \rightarrow \mathbb{R}_1, 0 < p_1 < \ldots < p_k < 1$, $a_{ni}, \ldots, a_{nk}$ are given numbers, and we may assume that (3.3.11) holds for $J_{n,1}(.)$, while the a_{nj}'s have suitable limits. Note that Winsorized mean in (3.3.16) is a particular case of (3.3.18), where $k = 2$, $a_{n1} = a_{n2} = [n\alpha]/n$, $p_1 = 1-p_2 = \alpha$, and $J_{n,1}(u) = 0$ outside the interval $[\alpha, 1-\alpha]$ and it is equal to 1 otherwise. Similarly (3.3.14) and (3.3.15) are also special cases of (3.3.18) where $L_{n2} = 0$. The sample median corresponds to the case of $k = 1, p_1 = 1/2$, and $L_{n1} = 0$, while the sample mean to the case of $L_{n2} = 0$ and $J_{n,1}(u) = 1 \forall u \in (0,1)$. Often the *interquartile range*

$$X_{n:[3n/4]} - X_{n:[n/4]} \tag{3.3.19}$$

is used as a measure of the scatter of a distribution, and this also corresponds to (3.3.18) where $L_{n1} = 0$. We will study the consistency of L_n under the setup in (3.3.18). Defining the quantile function $Q(t)$ as in earlier, we assume that

$$F_\theta^{-1}(= Q(p)) \quad \text{is uniquely defined for} \quad p = p_j, j = 1, \ldots, k. \tag{3.3.20}$$

[Note that the strict monotonicity of F_θ^{-1} at $t = p$ ensures (3.3.20).] Then, for every $\epsilon > 0$,

$$\begin{aligned} &P\Big\{X_{n:k} > F_\theta^{-1}(p) + \epsilon\Big\} \\ &\quad = P\Big\{(k-1 \text{ or less number of } X_1, \ldots, X_n \text{ are less than } F_\theta^{-1}(p) + \epsilon\Big\} \\ &\quad = P\Big\{\sum_{i=1}^{n} I(F_\theta^{-1}(p) + \epsilon > X_i) \leq k-1\Big\}. \end{aligned} \tag{3.3.21}$$

The Hoeffding inequality [see (2.5.25)] may be used to obtain an exponential rate of convergence (to 0) whenever $k/n \to p$; a similar inequality holds for the lower tail. Using this inequality (and relegating the proof as an exercise, Problem 3.3.8,) we easily obtain that under (3.3.20) as $n \to \infty$

$$X_{n:[np_j]} \to F_\theta^{-1}(p_j) \quad \text{a.s., for each} \quad j = 1, \dots k. \tag{3.3.22}$$

Thus, if we assume that

$$\lim_{n\to\infty} a_{nj} = a_j \quad \text{exists and } |a_j| < \infty \text{ for } \quad j = 1, \dots, k, \tag{3.3.23}$$

we obtain from (3.3.22) and (3.3.23) that under (3.3.20) and (3.3.23),

$$L_{n2} \overset{w.p.1}{\to} \sum_{j=1}^{k} a_j F_\theta^{-1}(p_j), \quad \text{as } n \to \infty. \tag{3.3.24}$$

The treatment for L_{n1} is a bit more delicate. Further regularity conditions on $J_{n,1}$, such as in (3.3.11), and on the d.f. F are necessary to establish the desired result. First, we note that under some smoothness conditions on the score function, $\{L_{n1}, n \geq n_0\}$ may form a reverse martingale sequence, and in a less stringent setup it can be approximated by a reverse martingale (in the sense that the remainder term converges a.s., or in some norm to 0 as $n \to \infty$). In such a case the reverse martingale convergence theorem can easily be called on to show that L_{n1} converges a.s. to its centering constant, as $n \to \infty$). For the particular case of L_n in (3.3.14) this reverse martingale property has been exploited in Sen (1981a, ch. 7) to cover a general class of L_{n1} for which $L_{n1} = L_n^0 + \xi_n$, where

$$\{L_n^0\} \text{ is a reverse martingale and } \quad \xi_n \overset{a.s.}{\to} 0 \text{ as } n \to \infty. \tag{3.3.25}$$

[See Lemma 7.4.1 and Theorem 7.5.1 of Sen (1981a) for some generalizations of (3.3.25).] Actually, looking at (3.3.5), we may also rewrite L_{n1} as

$$L_{n1} = \int_{\boldsymbol{R}_1} J_{n,1}(F_n(x)) dF_n(x), \tag{3.3.26}$$

where by the Glivenko-Cantelli theorem, $||F_n - F_\theta|| \to 0$ a.s. as $n \to \infty$. If we assume that $J_{n,1}(.)$ converges to some $J_1(.)$ (as $n \to \infty$), then, under suitable conditions on $J_1(.)$ and F_θ, the consistency of L_{n1} can be studied. If $J_1(.)$ can be expressed as difference of two monotone and continuous functions [on (0,1)], then the following holds (see Sen 1981a, th. 7.6.1):
Let r and s be two positive numbers such that $r^{-1}+s^{-1} = 1$ and $\int_{\boldsymbol{R}_1} |x|^r dF_\theta(x) < \infty$ and $\limsup_n ||J_{n,1}||_s < \infty$. Then writing $\lambda_1(\theta) = \int_{\boldsymbol{R}_1} J_1(F_\theta(x)) dF_\theta(x)$,

$$L_{n1} - \lambda_1(\theta) \to 0 \quad \text{a.s., as} \quad n \to \infty. \tag{3.3.27}$$

Van Zwet (1980) relaxed the assumption of pointwise convergence of $J_{n,1}(u)$ to some $J_1(u), u \in (0,1)$ [under $\|J_{n,1}\|_s < \infty$, uniformly in n] and also obtained a parallel result for $r = \infty$ (i.e., for bounded r.v.'s); we leave these as exercises (see Problems 3.3.9 and 3.3.10). In the context of robust estimation of location (and regression), we usually have a score function J of bounded variation, and also, as in the case of the trimmed or Winsorized mean, (3.3.17) may hold. In such a case the moment condition on F may be avoided, and we have the following result (see Huber 1969, ch. 3):

If $J(.)$ and $F_\theta^{-1}(.)$ have no common points of discontinuity, and for some $\alpha \in (0, 1/2)$, $(a) J(u) = 0$ whenever u does not lie in $[\alpha, 1-\alpha]$ and (b) both J and F_θ^{-1} have bounded variation on $[\alpha, 1-\alpha]$, then for $J_{n,1}(u) = J(u), u \in (0,1)$, as $n \to \infty$,

$$L_{n1} \xrightarrow{p} \int_0^1 J(t) F_\theta^{-1}(t) dt, \text{ under } F_\theta. \tag{3.3.28}$$

The proof follows directly by using (3.3.26) with $J_{n,1}(t) = J(t)$ and bounded variation conditions on $J(.)$ and $F_\theta^{-1}(.)$, along with the Glivenko-Cantelli Theorem. We omit it (see Problem 3.3.11).

By virtue of (3.3.18), (3.3.24), and (3.3.28), writing

$$\mu(F_\theta) = \int_0^1 J(t) F_\theta^{-1}(t) dt + \sum_{j=1}^{k} a_j F_\theta^{-1}(p_j), \tag{3.3.29}$$

we conclude that under the regularity conditions assumed before,

$$L_n \to \mu(F_\theta), \text{ in probability (actually, a.s.), as } n \to \infty \tag{3.3.30}$$

So far we have assumed that X_i's have the d.f. F_θ. If X_i's have some other d.f. G on $\mathbb{R}_1$, we may replace F by G in (3.3.20) through (3.3.30) and parallel to (3.3.30), we may conclude that $L_n \to \mu(G)$, in probability, as $n \to \infty$. For the model F_θ relating to (3.3.1), we have that a general L_n in (3.3.18) is a consistent estimator of θ provided that

$$\mu(F_\theta) = \theta, \quad \text{for } \theta \in \Theta. \tag{3.3.31}$$

This is the case, for instance, in the location model: $F_\theta(x) = F(x-\theta), \theta \in \mathbb{R}_1$ with F symmetric about 0, provided

$$J(1-t) = J(t), \quad t \in (0,1),$$

$$a_j = a_{k-j+1}, \; p_j = p_{k-j+1}, \text{ for } j = 1, \ldots, k,$$

and

$$\int_0^1 J(t) dt + (a_1 + \ldots + a_k) = 1. \tag{3.3.32}$$

Of course, (3.3.32) is a sufficient condition for (3.3.31) (to hold for all F belonging to the class of symmetric d.f's), while for a special form of F, all we need to verify is that (3.3.29) equals to 0 at $\theta = 0$. (3.3.32) holds for the trimmed and Winsorized means.

From the point of view of robustness, generally trimmed L-estimators are preferred. Toward this end, (3.3.28) - (3.3.30) provide a useful picture for the consistency, although various authors have tried to establish the consistency and asymptotic normality of L_n under less restrictive conditions on the score function, purely for academic interest. Some of these results are displayed in detail by Serfling (1980), Sen (1981a), and others. Keeping equation (3.3.32) in mind, we consider here the following asymptotic normality result (see Boos 1979) for a trimmed L-estimator.

Let $J(.)$ be bounded and continuous on [0,1] up to a set of Lebesgue measure 0 and G^{-1}-measure 0, and let $J(u)$ be equal to 0 for $u \notin [\alpha, 1-\alpha]$, for some $\alpha : \ 0 < \alpha < 1/2$. Also let the d.f. G (of the X_i) have finite and positive derivative (g) at $G^{-1}(p_j)$ for $j = 1, \ldots, k$. Then

$$n^{1/2}(L_n - \mu(G)) \xrightarrow{\mathcal{D}} \mathcal{N}(0, \sigma^2(G)), \tag{3.3.33}$$

provided that $0 < \sigma^2(G) < \infty$, where

$$\begin{aligned} \sigma^2(G) &= \operatorname{Var}\Big\{ - \int [I(X_1 \leq y) - G(y)] J(G(y) dy \\ &+ \sum_{j=1}^{k} \frac{a_j}{g(G^{-1}(p_j))} [p_j - I(X_1 \leq G^{-1}(p_j))] \Big\}. \end{aligned} \tag{3.3.34}$$

If the second component in (3.3.18) vanishes, that is, $L_n = L_{n1}$, then $\sigma^2(G)$ in (3.3.34) reduces to

$$\int\int J(G(x))J(G(y))[G(x \wedge y) - G(x)G(y)]dxdy = \sigma_0^2(G), \text{ say.} \tag{3.3.35}$$

Further, if $G(x) = F_\theta(x) = F(x - \theta)$ where F is symmetric about 0 and if $J(u) = J(1-u)$, $\forall u \in (0,1)$, then [see (3.3.32)] $\mu(G) = \theta$ and $\sigma_0^2(G) = \sigma_0^2(F)$. Thus, if F admits first- and second-order derivatives (f and f', respectively) a.e. such that $I(f) = \int [f'(x)/f(x)]^2 dF(x) < \infty$, then on letting

$$J(u) = J_F(u) = \psi'(F^{-1}(u))/I(f); \ \psi(x) = -f'(x)/f(x), \ x \in \mathbb{R}_1, \tag{3.3.36}$$

we immediately obtain that the corresponding L_n has the asymptotic variance $\sigma_0^2(F)$, where

$$\sigma_0^2(F) = \{I(f)\}^{-1}. \tag{3.3.37}$$

Therefore, based on the score function $J_F(.)$, L_n is asymptotically efficient. Also note that in order that $J_F(u) = 0$ for $u < \alpha$ or $u > 1 - \alpha$, for some $\alpha : \ 0 < \alpha < 1/2$, we must have $\psi(x) =$ constant for $x \leq$

$F^{-1}(\alpha)$, or $x \geq F^{-1}(1-\alpha)$. In other words, in the two tails, $f(x)$ should be exponentially decreasing. The exponentially decreasing tail behavior of f characterizes the asymptotic optimality of trimmed L-estimators. (3.3.36) also tells us that if $\psi(x)$ is not nondecreasing (i.e., f is not log-convex) then $\psi'(.)$ may be negative somewhere, so that an asymptotically efficient L_n may not have all nonnegative c_{ni}. This may be verified by a simple example of a model where $F(x) = \pi F_1(x) + (1-\pi)F_2(x)$, both F_1 and F_2 are symmetric (but they are not the same) and $0 < \pi < 1$. The proof of (3.3.33) is not presented here, but in the next chapter we will develop a deeper linearity result for L-estimators that will contain the proof of this asymptotic normality as a by-product (and under more general regularity conditions). For other approaches to the asymptotic normality of L-estimators, we refer to Serfling (1980), Sen (1981a), and Shorack and Wellner (1986), among others. However, we will only elaborate the "linearity" approach.

Based on what has been discussed so far, we may summarize that for the location problem the class of L-estimators discussed before contains asymptotically efficient elements. Whenever the tails of a distribution are of exponential form (i.e., they are heavier than the normal ones), the trimmed L-estimators appear to be very desirable from robustness and asymptotic efficiency considerations. On the other hand, for even heavier type of tails (of a symmetric d.f.), such as in the Cauchy type d.f., whenever the density possesses a finite Fisher information, one may obtain an optimal score function leading to an asymptotically optimal L-estimator. However, in such a case the density is not generally log-convex, and hence this optimal score function may even lead to some negative scores outside an interval $[\alpha, 1-\alpha]$, where $0 < \alpha < 1/2$, although it may still be bounded and continuous (a.e.). Finally, if the true d.f. G does not belong to the model $\{F_\theta : \theta \in \Theta\}$, then the L-estimator is a consistent estimator of $\mu(G)$, defined by (3.3.29) with F_θ replaced G. Therefore, (3.3.31), namely, $\mu(G) = \theta$, provides a characterization for the consistency of an L-estimator of location when the true d.f. G may not belong to the assumed model. In particular, if G is symmetric about its location (θ), then $\mu(G) = \theta$ for all symmetric score functions (for which the integral in (3.3.29) exists, with F replaced by G) and hence, L-estimators may generally be considered in a more general setup. Within this broad framework, particular score functions may be chosen in such a way that they retain asymptotic efficiency for suitable subclasses of symmetric d.f.'s. This also points out that for symmetric score functions a departure from the assumed symmetry of the d.f. G may have some effect, and the L-estimator may cease to be a consistent estimator of the location parameter. This aspect of the vulnerability of L-estimators to departures from the assumed symmetry of the underlying d.f. G has prompted many workers to study the robustness of L-estimators to infinitesimal asymmetry, and compare the same with other types of estimators. While there seems to be no problem in studying this in an asymptotic setup, the small sample study requires extensive numerical computations, and so will not be considered here. We refer to Problems 3.3.12 and 3.3.13 in this context.

In (3.3.19), we have considered an L-estimator of dispersion of a distribution. Unlike in the case of location parameters, here, in general, a suitably posed measure of dispersion may not be identified with the scale parameter of a d.f.,even if the d.f. may have a location-scale form. For example, the population counterpart of (3.3.19), namely the population interquartile range $G^{-1}(3/4) - G^{-1}(1/4)$, is a multiple of the scale parameter, where this multiplicity factor is itself a functional of the underlying d.f. (Problem 3.3.14 is set to workout this factor for some well known forms of location-scale type d.f.'s and to stress the variability of this factor). Thus whereas in the location model a measure of location may be identified with the point of symmetry of the d.f. for the entire class of symmetric d.f.'s, in the scale model we need to proceed in a different manner. A very natural setup is to allow G to belong to a suitable family $\mathcal{F}$ of d.f. and to consider a *dispersion-functional* $T(G)$ as a suitable measure of dispersion. In this context we may naturally impose the two basic conditions on $T(.)$:

1.The functional is *translation-invariant*; that is, if the original r.v.'s $X_1, \ldots, X_n$ are simultaneously translated by a constant a ($\in \mathbb{R}_1$), then for every $a \in \mathbb{R}_1$, the measure computed from the $X_i - a$ $(1 \leq i \leq n)$ is the same as that computed from $X_1, \ldots, X_n$. (The same conclusion applies to the population counterparts too, for every $a \in \mathbb{R}_1$.)

2.The functional is *scale-equivariant*; that is, if the X_i's are replaced by bX_i, $i = 1, \ldots, n$, for some $b \in \mathbb{R}_1$, then the functional is also $|b|$ times the original one (in the sample as well as population case) for every $b \in \mathbb{R}_1$.

Quantitatively we also expect $T(G)$ should be greater than $T(F)$ if G is more spread out than F. When F and G both have the same functional form and differ possibly in location/scale parameters, then of course the above requirement can easily be verified for various specific measures. However, two unrelated d.f.'s may not always have this inequality. One approach to this problem is to consider the set of paired differences $X_i - X_j, 1 \leq i < j \leq n$, and to formulate a measure of dispersion based on $M = n(n-1)/2$ observations $\{|X_i - X_j| = Z_{ij}$, say, $1 \leq i < j \leq n\}$. The Z_{ij}'s are clearly translation-invariant (on the X_i) and scale-equivariant, and a very natural way may be to consider an L_p-norm:

$$U_n^{(p)} = \left\{ \binom{n}{2}^{-1} \sum_{1 \leq i < j \leq n} Z_{ij}^p \right\}^{1/p}, \quad \text{for some } p > 0. \tag{3.3.38}$$

For the particular case of $p = 2$, (3.3.38) is a scalar multiple of the usual standard deviation (and is known to be a nonrobust estimator), while for $p = 1$, it is related to the so-called Gini's mean difference. In any case $(U_n^{(p)})^p$ is a simple U-statistic, and hence the basic theory developed by Hoeffding (1948), and others, may easily be incorporated to study the optimality and other properties of this estimator; we leave some of these attempts in the form of exercises (see Problems 3.3.15 - 3.3.16). Another possibility is to order the

$M = n(n-1)/2$ values of the Z_{ij}'s, denote them by $Z_{M:1}, \ldots, Z_{M:M}$, and consider a linear combination of these order statistics as a suitable measure of the dispersion of the original X_i's. In the above formulation, it is also possible to use the order statistics $g(Z_{M:k})$, $k = 1, \ldots, M$, where $g(z)$ is monotone on $\mathbb{R}^+$ [e.g., $g(z) = z^2$]. Bickel and Lehmann (1976, 1979) have advocated the use of the trimmed mean of $g(Z_{M:k})$ as a suitable measure of dispersion. The basic difference between the two sets of order statistics $\{X_{n:i}\}$ and $\{Z_{M:k}\}$ is that whereas $X_{n:i}$'s are generated by n independent r.v.'s, the $Z_{M:k}$'s correspond to the Z_{ij}'s that are not all independent. This difference has been taken into account in the formulation of the *generalized L-estimators* (see Serfling 1980), and we will briefly comment on them later. Concerning measures of dispersion for symmetric G, Bickel and Lehmann (1979) suggested a broad class of measures of the form

$$\triangle(G) = \left[\int_{1/2}^{1} \left\{G^{-1}(t) - G^{-1}(1-t)\right\}^p d\Lambda(t)\right]^{1/p}, \quad p > 0, \tag{3.3.39}$$

where $\Lambda(.)$ is any finite measure on [1/2,1]. Replacing $G^{-1}(t)$ by the sample quantile function $Q_n(t)$ [see (3.3.3)], we arrive at an estimator of $\triangle(G)$. For $p = 1$ this is a linear function of the order statistic $\{X_{n:k}\}$, although for $p \neq 1$ it is no longer the case. Nevertheless, such estimators can well be approximated by a linear combination of functions of order statistics, and hence they will also be termed L-estimators.

Rivest (1982) combined M- and L-estimators of location into a form (allowing for studentization) that can be called L-M-estimators. Such an estimator (T_n) is defined as a solution of the equation

$$\sum_{i=1}^{n} J_n(i/(n+1))\psi((X_{n:i} - t)/s_n) = 0 \quad \text{(with respect to } t\text{)}, \tag{3.3.40}$$

where s_n is a suitable estimator of the scale parameter, and $J_n(.)$ and $\psi(.)$ are defined as in Section 3.2 and in this section. For $J(.) \equiv 1$, T_n reduces to an M-estimator (considered in Section 3.2), while for $\psi(x) \equiv x$, T_n is an L-estimator. Like the generalized L-estimators, these L-M-estimators may also be characterized in terms of some empirical d.f.'s, and the theory to be developed in Section 6.3 will be applicable to these estimators as well.

Before we pass on to the regression model, we may remark that there is an extensive literature on L-estimators and allied order statistics. For some early developments on order statistics, we may refer to Daniell (1920), Lloyd (1952), as well as the monograph edited by Sarhan and Greenberg (1962) (where other references are cited). The monograph of David (1981) contains a very thorough treatment of important statistical uses of order statistics. In the literature L-estimators arise very frequently, and the early need for the treatment of asymptotic theory for L-estimators was stressed by Bennett (1952), Jung (1955), Chernoff, Gastwirth and Johns (1965), Birnbaum and

Laska (1967), among others. More general (and diverse) asymptotic theory appeared in later works by Stigler (1969,1973, 1974), Sen (1977, 1980, 1981a) Helmers (1977, 1980, 1981), Boos (1979), Boos and Serfling (1979), and others. Some detailed (and unified) studies of the asymptotic theory of L-estimators are made by Shorack and Wellner (1986), while various Monte Carlo studies were reported by the Princeton group, Andrews et al. (1972). We will discuss some aspects of this asymptotic theory in subsequent text.

In a regression model the observations are not all identically distributed, but the errors are so. Bickel's (1973) idea sparked the activity incorporating the residuals based on some preliminary estimators in the formulation of suitable quantile processes. Bickel's estimators have good efficiency properties, but they are generally computationally complex and are not invariant under a reparametrization of the vector space spanned by the columns of $\mathbf{C}$, defined after (3.2.12). However, Bickel's ingenuity lies in the demonstration of regression counterparts of location L-estimators. Ruppert and Carroll (1980) proposed another L-estimator of regression parameters along the same vein, but its asymptotic properties depend on the initial estimator.

A more promising direction is due to Koenker and Bassett (1978). They extended the concept of quantiles to linear models (termed *regression quantiles*). To illustrate their methodology, let us consider the usual linear model:

$$X_i = c_{i1}\beta_1 + \cdots + c_{ip}\beta_p + e_i, \quad i = 1, \ldots, n \ (> p),$$

where $\beta_1, \ldots, \beta_p$ are the unknown regression parameters, the c_{ij}'s are known regression constant, and the error components e_i's are i.i.d. r.v.'s with a d.f. G. For a fixed $\alpha : \ 0 < \alpha < 1$, we define

$$\psi_\alpha(x) = \alpha - I(x < 0) \quad \text{and} \quad \rho_\alpha(x) = x\psi_\alpha(x), \quad \text{for } x \in \mathbb{R}_1. \tag{3.3.41}$$

Then the α-regression quantile is defined as a solution $[\hat{\boldsymbol{\beta}}(\alpha)]$ of the minimization [with respect to $\mathbf{t} = (t_1, \ldots, t_p)'$] :

$$\sum_{i=1}^{n} \rho_\alpha\Big(X_i - \sum_{j=1}^{n} c_{ij}t_j\Big) := \min. \tag{3.3.42}$$

Notice that $\hat{\boldsymbol{\beta}}(\alpha)$ is in fact an M-estimator: Koenker and Bassett (1978) proved that $\hat{\boldsymbol{\beta}}(\alpha)$ is a consistent estimator of $(\beta_1 + G^{-1}(\alpha), \beta_2, \ldots, \beta_p)'$ whenever $c_{i1} = 1$, for every $i = 1, \ldots, n$, and they also showed that the asymptotic distribution of $\hat{\boldsymbol{\beta}}(\alpha)$ is quite analogous to that of the sample α-quantile in the location case. This regression quantile seems to provide a reasonable basis for L-estimation in linear models. Koenker and Bassett (1978) also suggested the use of trimmed least squares estimators (TLSE) in linear models. This idea may be posed as follows: First, consider two numbers α_1, α_2, such that $0 < \alpha_1 < \alpha_2 < 1$ (typically$\alpha_1 = 1 - \alpha_2$), and calculate the regression quantiles $\hat{\boldsymbol{\beta}}(\alpha_1)$ and $\hat{\boldsymbol{\beta}}(\alpha_2)$ along the lines of (3.3.41) and (3.3.42)].

Second, trim off all observations such that

$$X_i \leq \sum_{j=1}^{P} c_{ij}\hat{\beta}_j(\alpha_1) \text{ or } X_i \geq \sum_{j=1}^{P} c_{ij}\hat{\beta}_j(\alpha_2), \quad i = 1, \ldots, n. \tag{3.3.43}$$

Finally, compute the ordinary LSE from the remaining (untrimmed) observations. Koenker and Bassett (1978) made a conjecture that the trimmed LSE may be regarded as a natural extension of the classical trimmed mean (for the location model) to the regression model. Later on this conjecture was supported by Ruppert and Carroll (1980) who derived the asymptotic distribution of the trimmed LSE and illustrated the close resemblance between the two trimmed estimators. Based on a Monte Carlo study, Antoch, Collomb and Hassani (1984) demonstrated some other good properties of the trimmed LSE. Besides the trimmed LSE, other estimators and tests may also be effectively based on the regression quantiles in (3.3.42). We will have occasion to discuss some of these estimators and tests in greater depths in Chapters 4 and 6. For completeness, we may mention the work of Antoch (1984), Bassett and Koenker (1978, 1982, 1984), Jurečková (1983 a,b,c, 1984c), Jurečková and Sen (1984), and Bassett (1984), Portnoy (1983) among others. Incorporating some of these developments in a broader setup, we will study the first- and second-order asymptotic relations of L-estimators to other estimators (in general linear models) and utilize these results in the study of asymptotic properties of L-estimators and L-tests in linear models.

We see easily that the finite sample breakdown point of the L-estimator (3.3.2) of location is $k+1$ when k extremes are trimmed off on either side, that is,

$$c_i = 0, \quad i = 1, \ldots, k, n-k+1, \ldots, n, \quad k \leq \frac{n}{2}, \tag{3.3.44}$$

On the other hand, L_n is translation equivariant and monotone in each argument provided that $c_i \geq 0$, $i = 1, \ldots, n$, and hence Theorem 2.3.1 applies. We can show that for a distribution of type II,

$$\lim_{a\to\infty} B(a, L_n) = k+1. \tag{3.3.45}$$

Hence we say that $\mathbb{P}_\theta(|L_n - \theta| > a)$ converges to 0 as $a \to \infty$, $(k+1)$ times faster than $\mathbb{P}(|X_1| > a)$. The sample median is certainly the most robust in this context. To prove (3.3.45), let us assume that exactly k points are trimmed off on each side while $c_{n,k+1} = c_{n,n-k} > 0$. Then we can write

$$\begin{aligned}
&\mathbb{P}_0(L_n > a) \\
&\quad \geq \mathbb{P}_0(X_{n:i} > -a, \ i = k+1, \ldots, n-k-1) \\
&\quad \geq \mathbb{P}_0\Big(X_{n:i} > -a, \ i = k+1, \ldots, n-k-1, \ X_{n:n-k} > a(\frac{2}{c_{n-k}} - 1)\Big)
\end{aligned}$$

$$\begin{aligned}
&\geq \mathbb{P}_0\Big(X_1 > -a, \ldots, X_{n-k-1} > -a, X_{n-k} > a(\frac{2}{c_{n-k}} - 1), \ldots \\
&\qquad \ldots, X_n > a(\frac{2}{c_{n-k}} - 1)\Big) \\
&= \big(F(a)\big)^{n-k-1}\Big[1 - F\big(a(\frac{2}{c_{n-k}} - 1)\big)\Big]^{k+1}.
\end{aligned}$$

We obtain the same inequality for $\mathbb{P}_0(L_n < -a)$. Thus

$$\begin{aligned}
&\limsup_{a\to\infty} B(a, L_n) \\
&\quad \leq \limsup_{a\to\infty} \frac{-(k+1)\log\big[1 - F\{a(2/c_{n-k} - 1)\}\big]}{-\log\big(1 - F(a)\big)} \\
&\quad \leq \lim_{a\to\infty}(k+1)\frac{m\log a(2/c_{n-k} - 1)}{m\log a} = k+1, \qquad (3.3.46)
\end{aligned}$$

while $\liminf B(a, L_n) \geq k+1$ by Theorem 2.3.1, and we arrive at (3.3.45). For breakdown properties of regression L-estimators we refer to Chapter 4.

3.4 R-ESTIMATORS

We may note that the ranks $(R_1, \ldots, R_n)$ of the observations $(X_1, \ldots, X_n)$ are invariant under a large class of monotone transformations. This invariance property yields robustness of rank-based tests against outliers and other distributional departures, and hence estimators of location based on rank tests are expected to enjoy parallel robustness properties. This basic observation lays the foundation of rank (R-) estimators of location, and the same idea extends to the linear models as well. Recall that the sample median is essentially tied down to the classical sign test, and a very similar phenomenon holds for general rank statistics and their derived estimators. In the literature this is referred to as the *alignment principle.* That is, the observations may be so aligned that a suitable rank statistic based on these aligned observations is least significant, and from such aligned observations the R-estimators can easily be derived. Dealing with rank statistics, we have generally much less restrictive assumptions on the underlying distributions, and hence from robustness perspective, rank tests are *globally robust* (compared to the M- and L-estimators, which are generally *locally robust*). The picture remains the same in the location as well as regression models, although the computational aspects for the R-estimators in linear models may be far more complicated than in the location model. In general, though an algebraic expression for an R-estimator in terms of the observations may not exist, there are already fast computational algorithms for ranking. The robustness and efficiency considerations sometimes make the R-estimators preferable to other competing ones.

We will consider the various properties of these R-estimators in a more unified manner. For example, we will begin with R-estimators of location based on the sign statistics and then move on to more general case. Let $X_1, \ldots, X_n$ be n i.i.d. r.v.'s with a continuous d.f. F_θ, where θ is the median of the distribution and is assumed to be unique (so that F is strictly monotone at θ). Then we write

$$\theta = F_\theta^{-1}(0.5). \tag{3.4.1}$$

For testing the null hypothesis

$$H_0 : \theta = \theta_0 \text{ against } H : \theta \neq \theta_0, \tag{3.4.2}$$

the simple sign test statistic is given by

$$S_n^0 = \sum_{i=1}^{n} \text{sign}(X_i - \theta_0). \tag{3.4.3}$$

Note that under H_0, S_n^0 has a completely specified distribution (see Problem 3.4.1) with mean 0 and variance n. For notational simplicity, we write

$$S_n(t) = S(X_1 - t, \ldots, X_n - t) = \sum_{i=1}^{n} \text{sign}(X_i - t), \quad t \in \mathbb{R}_1. \tag{3.4.4}$$

Note that $S_n(t)$ is nonincreasing in $t \in \mathbb{R}_1$, while $S_n(0)$ has a distribution centered at 0. Hence we define an estimator $\hat{\theta}_n$ of θ by equating $S_n(t)$ to 0. We may note that $S_n(t) = 0$ may not have an exact root or the root may not be unique. To illustrate this point, let $X_{n:1}, \ldots, X_{n:n}$ be the order statistics corresponding to $X_1, \ldots, X_n$. Consider then a typical t lying between $X_{n:k}$ and $X_{n:k+1}$, for $k = 0, \ldots, n$, where $X_{n:0} = -\infty$, and $X_{n:n+1} = +\infty$. Then it is easy to show (see Problem 3.4.2) that

$$S_n(t) = (n - 2k) \quad \text{for } X_{n:k} < t < X_{n:k+1}, \ k = 0, \ldots, n, \tag{3.4.5}$$

and, if we let sign(0) = 0, then $S_n(X_{n:k}) = (n-2k+1)$ for $k = 1, \ldots .n$. Thus, if n is odd ($= 2m+1$, say), we obtain that $\hat{\theta}_n = X_{n:m+1}$, while if n is even ($=2m$, e.g.), we have by (3.4.5) that $S_n(t) = 0$ for every $t \in (X_{n:m}, X_{n:m+1})$. In the later case we may take the centroid of the set of admissible solutions as the estimator $\hat{\theta}_n = (X_{n:m} + X_{n:m+1})/2$, and this agrees with the conventional definition of the sample median. Note that we do not need to assume that F_θ is symmetric about θ in this setup. If, however, we are able to make this additional assumption on the d.f., then we can use a general class of signed-rank statistics for the testing problem in (3.4.2), and these in turn provide us a general class of R-estimators of location. Thus we assume that

$$F_\theta(x) = F(x - \theta), \quad \text{where } F(y) + F(-y) = 1 \ \forall y \in \mathbb{R}_1, \tag{3.4.6}$$

so that (3.4.1) holds and θ is median as well as the center of symmetry of the d.f. F_θ. For the hypothesis testing problem in (3.4.2), we may consider a general signed rank statistic

$$S_n(\mathbf{X}_n - \theta_0 \mathbf{1}_n) = \sum_{i=1}^{n} \text{sign}(X_i - \theta_0) a_n(R_{ni}^+(\theta_0)), \tag{3.4.7}$$

where $R_{ni}^+(\theta_0)$ is the rank of $|X_i - \theta_0|$ among $|X_1 - \theta_0|, \ldots, |X_n - \theta_0|$ for $i = 1, \ldots, n$, and $a_n(1) \leq a_n(2) \leq \ldots a_n(n)$ are given scores. The particular case of $a_n(k) = k/(n+1)$, $k = 1, \ldots, n$, leads to the *Wilcoxon signed-rank statistic*. Further, for $a_n(k) =$ expected value of the k-th smallest observation in a sample of size n from the chi distribution with 1 degree of freedom, for $k = 1, \ldots, n$, we have the *one-sample normal score statistic*. Note that under H_0, $X_i - \theta_0$ has a d.f. F symmetric about 0, and hence, (see Problem 3.4.3) $\text{sign}(X_i - \theta_0)$ and $R_{ni}^+(\theta_0)$ are stochastically independent. The vector of signs assumes all possible 2^n realisations with the common probability $1/2^n$, and the vector of ranks assumes all possible $n!$ realisations [the permutations of $(1, \ldots, n)$] with the common probability $1/(n!)$. Hence under H_0, the statistic in (3.4.7) has a distribution with mean 0, independently of the underlying continuous F. As in (3.4.4), we may write $S_n(t)$ for $S_n(\mathbf{X}_n - t\mathbf{1}_n), t \in \mathbb{R}_1$. Then (see Problem 3.4.4) it is easy to see that whenever the $a_n(k)$'s are monotone (in k), $S_n(t)$ is nonincreasing in $t (\in \mathbb{R}_1)$. Thus, as in the case of the sign statistic, we may equate $S_n(t)$ to 0 to obtain an estimator of θ. In general, $S_n(t) = 0$ does not have a unique root. We can set

$$R_n^- = \sup\{t : S_n(t) > 0\}, \; R_n^+ = \inf\{t : S_n(t) < 0\}, \tag{3.4.8}$$

$$R_n = \frac{1}{2}\{R_n^- + R_n^+\}. \tag{3.4.9}$$

Then R_n is an R-estimator of θ. Note that for $a_n(k) = 1$, $k = 1, \ldots, n$, (3.4.7) reduces to the sign statistic in (3.4.3), and hence R_n reduces to the sample median. Thus the sample median may be interpreted as an L-estimator as well as an R-estimator. It may also be interpreted as an M-estimator when in (3.2.15)-(3.2.16) we take $\psi(t) = \text{sign}(t)$, $t \in \mathbb{R}_1$. Similarly, for the case of the Wilcoxon signed rank statistic [in (3.4.7)], we have by arguments similar to the case of the sign statistic (see Problem 3.4.5)

$$R_n(W) = \text{median}\{\frac{1}{2}(X_i + X_j) : 1 \leq i \leq j \leq n\}. \tag{3.4.10}$$

However, for general $a_n(k)$ (which are nonlinear in k/n, $k = 1, \ldots, n$), such an algebraic expression for the R-estimator may not be available. But, using the sample median or the median of the mid-averages [i.e., (3.4.10)] as an initial estimator, an iterative solution for (3.4.8)-(3.4.9) can be obtained in a finite number of steps. For large samples, a one-step version may also be considered.

Note that for every real c and positive d, $\text{sign}(d(X_i + c - t - c)) = \text{sign}(X_i - t)$, for $t \in \mathbb{R}_1$, while the rank of $d|X_i + c - t - c|$ among $d|X_1 + c - t - c|, \ldots, d|X_n + c - t - c|$ is the same as the rank of $|X_i + c - t - c|$ among $|X_1 + c - t - c|, \ldots, |X_n + c - t - c|$ for $i = 1, \ldots, n$, $t \in \mathbb{R}_1$. Hence it is easy to verify that (see Problem 3.4.6)

$$R_n(\mathbf{X}_n + c\mathbf{1}_n) = R_n(\mathbf{X}_n) + c \quad \text{for every real } c, \tag{3.4.11}$$

that is, R_n is a translation-equivariant estimator of θ;

$$R_n(d\mathbf{X}_n) = dR_n(\mathbf{X}_n) \quad \text{for every } d, \in \mathbb{R}_1^+ \tag{3.4.12}$$

that is, R_n is scale-equivariant estimator of θ. Further, by the monotonicity of $S_n(t)$ in t and by (3.4.8) - (3.4.9), we obtain that for $t, \theta \in \mathbb{R}_1$,

$$\begin{aligned} P_\theta\{S_n(\mathbf{X}_n - t\mathbf{1}_n) < 0\} &\leq P_\theta\{R_n \leq t\} \\ &\leq P_\theta\{S_n(\mathbf{X}_n - t\mathbf{1}_n) \leq 0\}. \end{aligned} \tag{3.4.13}$$

Note that the statistic $S_n(\mathbf{X}_n - t\mathbf{1}_n)$ does not have a continuous distribution even if the d.f. F is continuous (Problem 3.4.7). However, when θ holds, $S_n(\mathbf{X}_n - \theta\mathbf{1}_n)$ has a d.f. independent of F such that

$$P_\theta\{S_n(\mathbf{X}_n - \theta\mathbf{1}_n) = 0\} = P_0\{S_n(\mathbf{X}_n) = 0\} = \eta_n, \tag{3.4.14}$$

where independently of F, $0 \leq \eta_n < 1, \forall n$ and $\lim_{n\to\infty} \eta_n = 0$. It follows that

$$\frac{1}{2}(1 - \eta_n) \leq P_\theta\{R_n < \theta\} \leq P_\theta\{R_n \leq \theta\} \leq \frac{1}{2}(1 + \eta_n), \quad \forall\theta \in \mathbb{R}_1. \tag{3.4.15}$$

Actually, for symmetric F, when θ holds, $X_i - \theta$ and $\theta - X_i$ both have the same d.f. F. On the other hand, by (3.4.7) and the fact that $\text{sign}(X_i - t) = -\text{sign}(t - X_i)$, we obtain that $R_n((-1)\mathbf{X}_n) = -R_n(\mathbf{X}_n)$ so that $R_n - \theta$ has a d.f. symmetric about 0 (although this d.f. may not be continuous everywhere); see Problem 3.4.8. We may conclude that for a symmetric F,

$$R_n \text{ is a median-unbiased estimator of } \theta. \tag{3.4.16}$$

Since, generally, R_n is not a linear estimator, evaluation of its moments requires more refined analysis. However, it will be shown later that R_n possesses finite mean and second moment under quite general conditions on the scores and the d.f. F to have finite variance for R_n to have so. For some details, we may refer to Sen (1980); see Problem 3.4.9.

For location and shift parameters, the idea of using rank tests to derive R-estimators was elaborated in Hodges and Lehmann (1963) (see also Sen 1963). However, we should mention that in Walker and Lev (1953, ch. 18), Lincoln Moses coined the same principle for the derivation of distribution-free confidence intervals for location and shift parameters; Moses (1965) also contains an useful graphical device to obtain such estimators. The asymptotic normality and other related properties of R-estimators of location can be studied using (3.4.13) and the asymptotic normality of signed rank statistic for local (contiguous) alternatives (on letting $t = \theta + n^{-1/2}u, \ u \in \mathbb{R}_1$); we pose this as Problem 3.4.10. However, this direct technique may not work out that well for the general linear model. As in the earlier sections, it will be convenient to develop an asymptotic representation of the form

$$n^{1/2}(R_n - \theta) = n^{-1/2}(\gamma(F))^{-1} \sum_{i=1}^{n} \phi(F(X_i - \theta)) + o_p(1), \qquad (3.4.17)$$

where the scores $a_n(k)$ are generated by the score function $\phi = \{\phi(t), \ 0 < t < 1\}$ and $\gamma(F)$ is a functional of the d.f. F and ϕ. This asymptotic representation extends directly to the regression model, and will be studied in Chapter 6. Generally, ϕ is a skew-symmetric function on (0,1), and hence, using the central limit theorem on the principal term on the right hand side of (3.4.17) we get the asymptotic normality of R_n as an immediate corollary. This representation also justifies the adaptation of the Pitman efficiency of rank tests for the study of the asymptotic efficiency of R-estimators. We will discuss this in Chapter 6.

For two distributions differing only in location parameters, the usual two-sample rank tests can similarly be used to derive R-estimators of the shift parameter. This has been done by Hodges and Lehmann (1963), Sen (1963), and others. However, we may characterize the two sample problem as a particular case of the simple regression model where the regression constants c_i can only assume two values 0 and 1 with respective frequencies n_1 and n_2, and $n = n_1 + n_2$. As such, the estimator of regression slope based on Kendall (1938) tau statistic (see Sen 1968a) contains the two-sample R-estimator based on Wilcoxon rank sum statistic as a particular case. Similarly, R-estimators of the regression parameter based on general linear rank statistics, initiated by Adichie (1967), pertain to the two-sample model as well. We will not enter into the detailed discussion on the two-sample models. Rather, we shall first present a general regression model and then append some simplifications in a simple regression model and in a two-sample model. Notable works in this direction are due to Jurečková (1969,1971a,b), Sen (1969), Koul (1971), and Jaeckel (1972), and some good accounts are given in Humak (1983; ch. 2) and Puri and Sen (1985, ch. 6).

Consider the usual linear model:

$$\mathbf{X}_n = (X_1, \ldots, X_n)' = \mathbf{C}\boldsymbol{\beta} + \mathbf{e}_n : \ \mathbf{e}_n = (e_1, \ldots, e_n)' \qquad (3.4.18)$$

where $\mathbf{C} = (\mathbf{c}_1', \ldots, \mathbf{c}_n')'$ is a known design matrix of order $n \times p$, $\boldsymbol{\beta} = (\beta_1, \ldots, \beta_p)'$ is the vector of unknown (regression) parameters, and the e_i's are i.i.d. r.v.'s with the d.f. F. For the time being, we are assuming that none of the n rows of $\mathbf{C}$ has equal elements. Later we will take the first column of $\mathbf{C}$

as $(1,\ldots,1)'$ and introduce the necessary modifications. In the linear model in (3.4.18), for testing the null hypothesis $H_0 : \boldsymbol{\beta} = \boldsymbol{\beta}_0$ (specified) against $H : \boldsymbol{\beta} \neq \boldsymbol{\beta}_0$, one considers a vector of linear rank statistics

$$\mathbf{S}_n(\boldsymbol{\beta}_0) = (S_{n1}(\boldsymbol{\beta}_0), \ldots, S_{np}(\boldsymbol{\beta}_0))', \tag{3.4.19}$$

where

$$S_{nj}(\boldsymbol{\beta}_0) = \sum_{i=1}^{n}(c_{ij} - \bar{c}_{nj})a_n(R_{ni}(\boldsymbol{\beta}_0)), \quad j = 1, \ldots, p; \tag{3.4.20}$$

$R_{ni}(\mathbf{t})$ is the rank of the residual

$$\delta_i(\mathbf{t}) = X_i - \mathbf{c}_i'\mathbf{t} \tag{3.4.21}$$

among $\delta_1(\mathbf{t}), \ldots, \delta_n(\mathbf{t})$, $\mathbf{t} \in \mathbb{R}_p$, $\bar{c}_{nj} = n^{-1}\sum_{i=1}^{n} c_{ij}$, $j = 1, \ldots, p$, and $a_n(1) \leq \ldots \leq a_n(n)$ are given scores. Note that $R_{ni}(\boldsymbol{\beta}_0)$ are exchangeable r.v.'s, $i = 1, \ldots, n$ under H_0, and the vector of these ranks takes on each of the $n!$ permutations of $(1, \ldots, n)$ with the common probability $(n!)^{-1}$. With respect to the uniform permutation distribution, we may enumerate the actual null hypothesis distribution (and moments) of the statistics in (3.4.20), and verify that (see Problem 3.4.12)

$$\mathbb{E}[\mathbf{S}_n(\boldsymbol{\beta}_0)|H_0] = \mathbf{0} \quad \text{and} \quad \text{Var}[\mathbf{S}_n(\boldsymbol{\beta}_0)|H_0] = \mathbf{C}_n^{\star} \cdot A_n^2, \tag{3.4.22}$$

where

$$\mathbf{C}_n^{\star} = \sum_{i=1}^{n}(\mathbf{c}_i - \bar{\mathbf{c}}_n)(\mathbf{c}_i - \bar{\mathbf{c}}_n)', \quad A_n^2 = \frac{1}{n-1}\sum_{i=1}^{n}[a_n(i) - \bar{a}_n]^2, \tag{3.4.23}$$

with $\bar{a}_n = n^{-1}\sum_{k=1}^{n} a_n(k)$ and $\bar{\mathbf{c}}_n = n^{-1}\sum_{i=1}^{n}\mathbf{c}_i$. As in the location model we may define here $\mathbf{S}_n(\mathbf{t})$, $\mathbf{t} \in \mathbb{R}_p$ by replacing $\boldsymbol{\beta}_0$ by $\mathbf{t}$, and are naturally tempted in equating $\mathbf{S}_n(\mathbf{t})$ to $\mathbf{0}$ to solve for an R-estimator of the regression parameter (vector) $\boldsymbol{\beta}$. For the case of the simple regression model (i.e., p=1), we can show that $S_{n1}(t_1)$ is nonincreasing in t_1 (see Problem 3.4.13), and hence, the justification for equating $S_{n1}(t)$ to 0 to obtain an estimator of β_1 follows the same alignment principle as in the location model. However, for $p \geq 2$, though $S_{nj}(\mathbf{t})$ is nonincreasing in t_j for each $j = 1, \ldots, p$, its behavior with the variation in t_s, $s \neq j$, may depend a lot on the design matrix $\mathbf{C}$. For example, if the p columns of $\mathbf{C}$ are pairwise concordant, then $S_{nj}(\mathbf{t})$ is nonincreasing in each of the elements of $\mathbf{t}$, and if the sth column of $\mathbf{C}$ is discordant to the jth column, then $S_{nj}(\mathbf{t})$ is nonincreasing in t_s. In the usual (fixed-effects) linear models it is possible to parametrize in such a manner that each c_{ij} can be decomposed into $c_{ij}^{(1)} + c_{ij}^{(2)}$, such that the $\mathbf{c}_j^{(1)}$, $j = 1, \ldots, p$ are concordant, $\mathbf{c}_j^{(2)}$, $j = 1, \ldots, p$ are concordant, while $\mathbf{c}_j^{(1)}$ and $\mathbf{c}_j^{(2)}$ are discordant, for $j \neq s = 1, \ldots, p$. Under this setup, it is possible to express

$S_{nj}(\mathbf{t})$ as a finite mixture of terms, each one monotone in the arguments $(\mathbf{t})$, and with this the alignment principle can be invoked to solve for $S_n(\mathbf{t}) = \mathbf{0}$ to derive the R-estimators in a meaningful way. This was essentially the idea of Jurečková (1971) for generating R-estimators in the multiple regression model. Again, since a unique solution may not generally be available, she defined an R-estimator $\mathbf{R}_n$ of $\boldsymbol{\beta}$ as a solution of the minimization problem

$$\sum_{j=1}^{p} |S_{nj}(\mathbf{t})| = (\text{ minimum with respect to } \mathbf{t} \in \mathbb{R}_p). \tag{3.4.24}$$

For the simple regression model, in view of the monotonicity of $S_{n1}(t_1)$ in t_1, the situation is parallel to that in (3.4.8) - (3.4.9). In this context the asymptotic linearity of $S_{ni}(t)$ (uniformly in t in a small neighborhood of 0), in probability, established by Jurečková (1969), provides an easy access to the study of the asymptotic properties of R-estimators of regression. Parallel results for the general regression model (under additional regularity conditions on the design matrix $\mathbf{C}$) were also obtained by Jurečková (1971a, b). Later on, the regularity conditions on $\mathbf{C}$ were slightly relaxed by Heiler and Willers (1988). Jaeckel (1972) defined an R-estimator of regression as a solution of the minimization:

$$\sum_{i=1}^{n} \{a_n(R_{ni}(\mathbf{t}) - \bar{a}_n\}\delta_i(\mathbf{t}) = \text{ minimum (with respect to } \mathbf{t} \in \mathbb{R}_p). \tag{3.4.25}$$

Jaeckel interpreted (3.4.25) as a measure of dispersion of the residuals $\delta_i(t)$, $i = 1, \ldots, n$ and advocated the use of this measure instead of the usual residual variance which is generally used in the method of the least squares. He pointed out that both the solutions in (3.4.24) and (3.4.25) are asymptotically equivalent in probability, and hence they share the same asymptotic properties. Another version of R_n was proposed by Koul (1971). Instead of the L_1-norm in (3.4.24), he suggested that the use of an L_2-norm involving a quadratic form in $\mathbf{S}_n(\mathbf{t})$ with $(\mathbf{C}_n^{\star})^{-1}$ (or some other p.d. matrix) as the discriminant. All these three versions are asymptotically equivalent, and we will discuss the first version more thoroughly than others. It is difficult to say which of the three minimizations is computationally the simplest one. However, in each case an iterative procedure is generally needed. Using some initial estimator of $\boldsymbol{\beta}$ in defining the residuals as well as the aligned rank statistics, a Newton-Raphson method may then be used in the desired minimization problem. For large samples this idea leads us to the adaptation of the so-called one-step R-estimators of regression, which will be considered in detail in Chapter 7.

It is assumed that the scores $a_n(k)$ are generated by a score function $\phi : (0,1) \to \mathbb{R}_1$ [i.e., $a_n(k) = \phi(k/(n+1))$, $k = 1, \ldots, n$], where usually ϕ is taken to be nondecreasing and square integrable. We set

$$\sigma^2(\phi, F) = \Big\{ \int_0^1 \phi^2(u)du - \Big(\int_0^1 \phi(u)du \Big)^2 \Big\} \Big\{ \int_{\mathbb{R}_1} \phi(F(x))f'(x)dx \Big\}^{-2}, \tag{3.4.26}$$

where we assume that the d.f. F admits an absolute continuous density function f (with derivative f') almost everywhere, such that the Fisher information $I(f)$ is finite. Then, parallel to (3.4.17), we would have an asymptotic representation of R-estimators of regression in the form:

$$\mathbf{C}_n^\star(\mathbf{R}_n - \boldsymbol{\beta}) = \mathbf{C}_n^\star(\gamma(F,\phi))^{-1}\sum_{i=1}^{n}(\mathbf{c}_i - \bar{c}_n)\phi(F(X_i - \mathbf{c}_i'\boldsymbol{\beta})) + o_p(1), \quad (3.4.27)$$

where $\mathbf{C}_n^\star$ is defined by (3.4.23) and $\gamma(F,\phi) = \int_{R_1}\phi(F(x))f'(x)dx$; a more detailed discussion of this representation will be considered in Chapter 6. From (3.4.27) and the multivariate central limit theorem, we immediately obtain that

$$(\mathbf{C}_n^\star)^{1/2}(\mathbf{R}_n - \boldsymbol{\beta}) \xrightarrow{\mathcal{D}} \mathcal{N}(\mathbf{0}, \sigma^2(\phi, F)\mathbf{I}), \quad \text{when } \boldsymbol{\beta} \text{ holds.} \quad (3.4.28)$$

Note that for the location model, (3.4.28) holds with $\mathbf{C}_n^\star = n$, $\beta = \theta$ and $I = 1$ (all scalar). As a result $\sigma^2(\phi, F)$ emerges as the key factor in the study of the asymptotic efficiency and other related properties of R-estimator. In this setup we put

$$\psi(u) = -f'(F^{-1}(u))/f(F^{-1}(u)), \quad u \in (0,1). \quad (3.4.29)$$

Then, noting that $\int \psi(u)du = 0$ and setting (without any loss of generality) that $\int_0^1 \phi(u)du = 0$, we obtain from (3.4.26) and (3.4.29) that

$$\sigma^2(\phi, F)I(f) = \frac{\langle\psi,\psi\rangle\,\langle\phi,\phi\rangle}{\langle\psi,\phi\rangle^2} \geq 1 \;\; \forall\phi \quad (3.4.30)$$

where $\langle\ ,\ \rangle$ stands for the inner product (in the Hilbert space) and the equality sign in (3.4.30) holds only when $\psi(u)$ and $\phi(u)$ agrees up to any multiplicative constant. Thus, for $\phi \equiv \psi$, we have $\sigma^2(\phi, F) = (I(f))^{-1}$ (the Cramér-Rao information limit), and hence the class of R-estimators contains an asymptotically efficient member. Drawing a parallel to the results discussed in Section 2.4, we also see that the class of R-estimators contains an asymptotically minimax member over a contaminated family of distributions (an analogue of the Huber minimax M-estimator). This also brings us to the need to study the interrelationships of M-, L-, and R-estimators for the common (location or regression) model, and this will be taken up in Chapter 7. We may remark here that the very fact that R-estimators are based on the ranks rather than on the observations themselves guarantees that they are less sensitive to the gross errors as well as to most heavy-tailed distributions.

Let us now consider a natural extension of the model in (3.4.18), namely

$$\mathbf{X}_n = \theta\mathbf{1}_n + \mathbf{C}\boldsymbol{\beta} + \mathbf{E}_n, \quad \theta \in \mathbb{R}_1, \quad (3.4.31)$$

where all the other symbols have the same interpretation as in (3.4.18). Our task is to provide R-estimators of θ and $\boldsymbol{\beta}$. Note that the ranks $R_{ni}(\mathbf{t})$ remain invariant under any translation of the X_i (by a real $c \in \mathbb{R}_1$), and hence $\mathbf{S}_n(\mathbf{t})$ in (3.4.19) fails to provide an estimator of θ. On the other hand, the R-estimators considered earlier for the estimation of $\boldsymbol{\beta}$ does not require F to be symmetric. Under the additional assumption that F is symmetric, we may use R-estimators of $\boldsymbol{\beta}$ to construct residuals, on which we may use the signed rank statistics to obtain R-estimators of θ. For the simple regression model, this was proposed by Adichie (1967), while for the general linear model, a more thorough treatment is due to Jurečková (1971b). For estimating the regression parameter vector $\boldsymbol{\beta}$ in the model (3.4.31), we use the linear rank statistic in (3.4.19) - (3.4.20) and denote the derived R-estimator by $\hat{\boldsymbol{\beta}}_{n(R)}$. Consider then the residuals

$$\hat{X}_i(t) = X_i - t - \mathbf{c}_i'\hat{\boldsymbol{\beta}}_{n(R)}, \quad i = 1, \ldots, n, \quad t \in \mathbb{R}_1. \tag{3.4.32}$$

On these residuals, we construct the signed rank statistic $S_n(t)$, as defined in (3.4.7) [with θ_0 replaced by t and $\mathbf{X}_n$ by $\mathbf{X}_n - \mathbf{C}\hat{\boldsymbol{\beta}}_{n(R)}$]. Then the R-estimator of θ is given by (3.4.8) - (3.4.9), as adapted to these residuals. Note that the residuals in (3.4.32) are not generally independent of each other, nor are they marginally identically distributed. Hence some of the exact properties of the R-estimators of location discussed earlier may not be generally true. (3.4.11) and (3.4.12) hold for this aligned estimator of θ, while in (3.4.13) the inequalities continue to hold if we replace $\mathbf{X}_n$ by $\mathbf{X}_n - \mathbf{C}\hat{\boldsymbol{\beta}}_{n(R)}$. However, under true $(\theta, \boldsymbol{\beta})$ the distributions of $S_n(\mathbf{X}_n - \mathbf{C}\hat{\boldsymbol{\beta}}_{n(R)} - \theta\mathbf{1}_n)$ may no longer be independent of the underlying F, so (3.4.15) and the median unbiasedness in (3.4.16) may no longer hold. This may also call for a more refined proof of the asymptotic normality of R-estimators of θ (taking into account the possible dependence of the residuals and their nonhomogeneity), and the proof set in Problem 3.4.7 may not work out here. Again, in this context, an asymptotic representation of R-estimators of the form of a mixture of (3.4.17) and (3.4.27) is available in the literature (viz., Jurečková 1971a, b), and this may be used with advantage to study the asymptotic properties of such aligned R-estimators. We will discuss such representations in more detail in Chapter 6. We may add here that for the simultaneous estimation of $(\theta, \boldsymbol{\beta})$, we need to assume that F is symmetric. Under this assumption it is possible to use more general signed rank statistics

$$S_{nj}(t_0, \mathbf{t}) = \sum_{i=1}^{n} \operatorname{sign}(X_i - t_o - \mathbf{c}_i'\mathbf{t})a_n(R_{ni}^+(t_0, \mathbf{t})), \ t_0 \in \mathbb{R}_1, \mathbf{t} \in \mathbb{R}_p, \tag{3.4.33}$$

for $j = 0, 1, \ldots, p$, where $c_{i0} = 1$, for $i = 1, \ldots, n$, and $R_{ni}^+(t_0, \mathbf{t})$ is the rank of $|X_i - t_0 - \mathbf{c}_i'\mathbf{t}|$ among $|X_1 - t_0 - \mathbf{c}_i'\mathbf{t}|, \ldots, |X_n - t_0 - \mathbf{c}_i'\mathbf{t}|$, for $i = 1, \ldots, n$; the rest of the symbols have the same meaning as in before. At the true $(\theta, \boldsymbol{\beta})$ the vector $(S_{n0}(t_0, \mathbf{t}), \ldots, S_{np}(t_0, \mathbf{t}))'$ has a joint distribution independent of F and has the mean vector $\mathbf{0}$. Thus, by an adaptation of the same alignment principle as in the case of linear rank statistics, we have an R-estimator of

$(\theta, \boldsymbol{\beta})$ as a solution of the minimization problem:

$$\sum_{j=0}^{p} |S_{nj}(t_0, \mathbf{t})| = \text{ minimum (with respect to } t_0 \in \mathbb{R}_1,\ \mathbf{t} \in \mathbb{R}_p). \quad (3.4.34)$$

Computational algorithm of (3.3.34) is naturally more involved than (3.4.24), although some simple approximations (based on iterations) are available for large samples. For the case of Wilcoxon signed rank statistics, this method was suggested by Koul (1969), although a more general situation is covered in Jurečková (1971a, b). In this context the idea of using one-step R-estimators of location and regression parameters is appealing. For such R-estimators based on general signed rank statistics, results parallel to (3.4.27) and (3.4.28) hold, wherein we need to replace $\mathbf{C}_n^\star$ by

$$\mathbf{C}_n^{0\star} = \sum_{i=1}^{n} (1,\ \mathbf{c}_i')'(1,\ \mathbf{c}_i) \quad [\text{of order } (p+1) \times (p+1)]. \quad (3.4.35)$$

The discussions on asymptotically efficient scores and asymptotically minimax R-estimators made after (3.4.50) also pertain to these estimators.

In the case of the simple regression model,

$$X_i = \theta + \beta c_i + e_i, \quad i = 1, \ldots, n,$$

the linear rank statistic $S_{n1}(t_1)$ is invariant under translation, and monotone nonincreasing in t_1, and hence the R-estimator of slope β may be obtained in a convenient form (i.e., as the centroid of a closed interval; see Problem 3.4.14). In this case the R-estimator of intercept parameter θ may also be obtained in a convenient form by using (3.4.8)-(3.4.9) based on the residuals $X_i - \hat{\beta}_n c_i, i = 1, \ldots, n$. On the other hand, a simultaneous solution for the intercept and slope based on (3.3.34) will be computationally heavier. This feature is generally true for the multiple regression model as well. For the R-estimator of regression based on linear rank statistics, the symmetry of F is not that necessary, while for the other R-estimators based on the signed rank statistics, symmetry of F constitutes an essential part of the basic assumptions. Any departure from this assumed symmetry of F may therefore have some impact on the R-estimators based on signed statistics. Hence, for the estimation of the slope alone, the linear rank statistics based R-estimators seem to have more robustness (against plausible asymmetry of the error d.f.) than the ones based on signed rank statistics (and the M-estimators considered in Section 3.2), where F is also assumed to be symmetric. A similar feature will be observed on L-estimators in the linear model (see Chapter 4), where the asymmetry leads to the bias only in estimation of the intercept and not of the slopes. Unlike the case of least square estimators, the M- and R-estimators of location or regression parameters are generally nonlinear, and hence a re-parametrization of the model in (3.4.31), such as (for a nonsingular $\mathbf{B}$)

$$\mathbf{X}_n = \theta \mathbf{1}_n + \mathbf{D}\boldsymbol{\gamma} + \mathbf{E}_n, \quad \text{where } \mathbf{D} = \mathbf{CB} \quad \text{and } \boldsymbol{\gamma} = \mathbf{B}^{-1}\boldsymbol{\beta}, \quad (3.4.36)$$

may lead to R-estimators of $\boldsymbol{\gamma}$ that do not agree (exactly) with the corresponding ones by premultiplying by $\mathbf{B}^{-1}$ the R-estimator of $\boldsymbol{\beta}$ when $p > 1$. For $p = 1$ the scale-equivariance property holds (see Problem 3.4.15). This lack of invariance property is also shared by the M-estimators. This drawback of R- and M-estimators is particularly serious when the basic model is not be written in a unique way, and one may therefore want to preserve this invariance (under re-parametrization). However, as we will see later, this problem ceases to be of any significance in large samples. The asymptotic representation of the form (3.4.27), among other things, also ensures this re-parametrization invariance up to the principal order. The equivariance property holds in the simple regression model where the re-parametrization reduces to the translation and scale changes (see (3.4.11)-(3.4.12) and Problem 3.4.14). For a general linear model (not necessarily of full rank) there may not be a unique way of defining a reduced model of full rank, and in such a case the particular basis should be chosen with care (unless n is large). The choice of canonical form for (3.4.31) eliminates this problem. Both R_n and M_n are affine-equivariant: If $\mathbf{d}_i' = \mathbf{c}_i'\mathbf{B}$, $i = 1, \ldots, n$, then $\sum_i \mathbf{d}_i\psi(X_i - \mathbf{d}_i'\mathbf{t}) = \mathbf{0} \Rightarrow \mathbf{B}\sum_i \mathbf{d}_i\psi(X_i - \mathbf{d}_i'\mathbf{t}) = \mathbf{0}$, and hence $\mathbf{BM}_n^{\star} = \mathbf{M}_n$, similar for $\mathbf{R}_n$.

For the two-sample case a linear rank statistic may equivalently be written in terms of a two-sample rank test statistics. Often this permits an easier way of solving for the minimization problem involved in the computation of the R-estimators. Based on the Wilcoxon two-sample rank-sum statistic, the R-estimator of the difference of locations is given by

$$R_n = \text{ median}\left\{Y_j - X_i : 1 \leq j \leq n_2,\ 1 \leq i \leq n_1\right\}, \qquad (3.4.37)$$

where $Y_j = X_{n_1+j}, j = 1, \ldots, n_2$ (see Problem 3.4.11). However, for general rank statistic, the derived R-estimator may not have such a closed form even in the two-sample problem.

The multisample model of the shift in location is also a special case of the general regression model in (3.4.18). Here, for some fixed k (≥ 2), we have positive integers $n_1, \ldots, n_k$ such that $n = n_1 + \ldots + n_k$, and the vectors $\mathbf{c}_i$ can have only k possible realizations $(1, 0, \ldots, 0)'$, $(0, 1, \ldots, 0)', \ldots, (0, \ldots, 0, 1)'$ (with respective frequencies $n_1, \ldots, n_k$). Our main interest lies in the estimation of a *contrast* $\lambda = \mathbf{l}'\boldsymbol{\beta}$ where $\mathbf{l}'\mathbf{1} = 0$. Such a contrast may also be equivalently expressed as a linear combination of all possible paired differences $\beta_j - \beta_{j'}$, $1 \leq j < j' \leq k$. For each $\beta_j - \beta_{j'}$ an R-estimator can be obtained from the relevant pair of samples, and hence the corresponding linear combinations of these R-estimators may be taken as an estimator. However, for different combinations of paired differences, we may have different R-estimators (although these will be asymptotically equivalent, in probability). This nonuniqueness is mainly due to the nonlinear structure of the R-estimators. Lehmann (1963) suggested some *compatible* R-estimators in one way analysis of variance models. We pose some of these in the form of exercises (see Problems 3.4.16- 3.4.17). Such compatible estimators are also relevant to the M-estimation theory in the usual analysis of variance models. An alternative solution to this compatibility criterion is to work with the individual sample (R- or M-)estimators of locations and to form a natural contrast in these estimators. In the context of biological assays e.g., parallel line or slope ratio assays and other practical application, it is often possible to partition the set of observations $\mathbf{X}_n$ into subsets where within each subset we have a simple regression model. In such a case one may develop the theory of compatible R- (or M-)estimators by using the usual R- (or M-) estimation theory for each subset and then pooling these subset estimators by the usual least squares or other methods. We pose this exercise as Problem 3.4.18.

Summarizing, we see that the R-estimators in linear models may not be compatible under re-parametrization but can be made so with some additional manipulations. Generally the R-estimators have good global robustness and efficiency properties, and they are translation and scale equivariant. On the other hand, they are usually calculated by some iterative procedure, though for large samples they may be approximated by one-step versions which require only one ranking. Further properties of such good R-estimators of location and regression parameters will be studied in Chapter 6.

Gutenbrunner (1986) and Gutenbrunner and Jurečková (1992) developed the concept of regression rank scores (which are extensions of the ordinary rank scores to linear models) and incorporated them in estimation in linear models, invariant to additional nuisance regression. Such estimators are based on appropriate linear programming algorithms, and usually computationally simpler than the usual R-estimators. Further developments due to Jurečková (1991, 1992) and Jurečková and Sen (1993) will be incorporated in a unified treatment of regression rank scores estimators in Chapter 6.

3.5 MINIMUM DISTANCE AND PITMAN ESTIMATORS

In this section we briefly consider the basic features of the minimum distance estimation theory (in a parametric setup but retaining robustness considerations) and the usual Pitman-type estimators of location. The results of this section are of rather special interest, but they will not be followed at length in subsequent chapters.

3.5.1 Minimum Distance Estimation

The motivation is basically due to Wolfowitz (1957). The estimation procedure is set in a parametrized framework, although it can easily be modified to suit a broader class of models. For any given data set and a model to be fitted to the data set, a natural way to estimate the parameters may be to minimize the suitable distance between the data and the fitted model. With a proper choice of this distance norm, the resulting estimators can be rendered robust, consistent as well as efficient. A full bibliography of the minimum distance method was prepared by Parr (1981), and an excellent review appears in Beran (1984). Koul (1992) gives an account of asymptotic theory of MDE in the linear regression model based on weighted empirical processes.

The MDE method can briefly be described as follows. Suppose that $X_1, \ldots, X_n$ are i.i.d. r.v.'s with a d.f. G defined on $\mathbb{R}_1$, and let

$$G_n(x) = n^{-1} \sum_{i=1}^{n} I(X_i \leq x), \ x \in \mathbb{R}_1,$$

be the corresponding empirical (sample) d.f. We conceive of a parametric family $\mathcal{F} = \{F_\theta : \ \theta \in \Theta\}$ of d.f.'s (called the *projection family*) as an appropriate model, and for every pair of d.f.'s G, F, we conceive of an appropriate measure of distance

$$\delta(G, F). \tag{3.5.1}$$

Then the minimum distance estimator (MDE) $\tilde{\theta}_n$ of θ is defined as the value of t for which $\delta(G_n, F_t)$ is a minimum over $t \in \Theta$:

$$\tilde{\theta}_n : \ \delta(G_n, F_{\tilde{\theta}_n}) = \inf\Big\{\delta(G_n, F_t) : \ t \in \Theta\Big\}. \tag{3.5.2}$$

For example, for a location family $F_\theta(x) = F(x - \theta), \ x, \theta \in \mathbb{R}_1$, we may take $\delta(G, F_t) = \int[x - \int y dF_t(y)]^2 dG(x)$, and the resulting estimator in (3.5.2) is the sample mean $\bar{X}_n = n^{-1}\sum_{i=1}^{n} X_i$ (the least square estimator of θ). A very popular measure of the distance is the so-called weighted Cramér-von Mises distance, which may be defined as

$$\delta(G_n, F) = \int [G_n(x) - F(x)]^2 w(F(x)) dF(x), \tag{3.5.3}$$

where $w(t), \ t \in (0, 1)$ is a suitable (nonnegative) weight function. Again, for the location family, (3.5.3) relates to the minimization of the distance $\int[G_n(x) - F(x)]^2 w(F(x)) dF(x)$ (with respect to $t \in \mathbb{R}$). Other possibilities for this measure of distance are the well-known chi squared distance, *Kolmogorov-Smirnov distance* (i.e., $\sup\{|G_n(x) - F_\theta(x)|, \ x \in \mathbb{R}_1\}$), and the *Hellinger distance* (i.e.,$\int[g_n^{1/2}(x) - f_\theta^{1/2}(x)]^2 dx$ where g_n is a suitable (nonparametric) estimator of the density f_θ), among others. Even the maximum likelihood estimator can be characterized as a MDE with a suitable distance function (essentially related to *Kullback-Leibler information*). With specific distance functions relating to the parametric estimators, the MDE are generally non-robust. The MDE procedure is not restricted to the location model and/or to the univariate d.f.'s. In a parametric setup it can as well be considered for more general forms. With the Cramér-von Mises or Kolmogorov-Smirnov distance, the resulting MDE's are generally more robust, while the Hellinger distance makes it even more natural and leads to a class of more general minimum divergence estimators, studied by Vajda (1984a-e). In this context, the work of Parr and Schucany (1980) on minimum distance and robust estimation is illuminating. Millar (1981) has studied the robustness of MDE (in the minimax sense over a shrinking neighborhood of a distribution). Robustness of Cramér-von Mises and Hellinger MDE was also demonstrated by Donoho and Liu (1988 a,b). Related asymptotics are discussed by Koul (1992).

Computationally the MDE may not be particularly convenient. No closed expression for the MDE is generally available (apart from some well-known parametric cases), and iterative solutions are usually required. A unique solution does not exit in all cases, and additional regularity conditions on the model are therefore required to ensure this. In a general framework (see Beran 1984), the MDE may not be asymptotically normal. An example of asymptotically normal MDE is that of Blackman (1955) whose asymptotic normality can be proved by incorporating the weak convergence of the empirical distributional process. This technique may be applicable in a variety of other models. In a general setup the minimization of the distance may result in a highly nonlinear form (even in an asymptotic representation) for the MDE, and hence asymptotic normality results may not be tenable. From the robustness considerations, the class of MDE's is quite flexible: With a proper choice of the weight function $w(.)$ in the Cramér-von Mises distance function, one can practically get estimators with arbitrary influence functions. A similar feature is also observed in some other measures of distance. The most appealing feature of the MDE is its clear statistical motivation and easy interpretation. Even if the true d.f. G may not belong to the class $\mathcal{F}$, we still know what we are estimating (i.e., the value of θ corresponding to $F_\theta \in \mathcal{F}$ nearest to G). Parallel to the MDE, minimum distance tests based upon the shortest distance between the data set and the fitted model (which are governed by appropriate null hypotheses) were studied. These tests share some of the robustness aspects of the MDE.

3.5.2 Pitman Estimators

For the location parameter model $F_\theta(x) = F(x-\theta)$, $x, \theta \in \mathbb{R}_1$ with F having density f and finite first-order moment, the Pitman (1939) estimator (PE) of location (θ) is defined by

$$T_n = \frac{\left\{ \int_{\mathbb{R}_1} t \prod_{i=1}^n f(X_i - t)dt \right\}}{\left\{ \int_{\mathbb{R}_1} \prod_{i=1}^n f(X_i - t)dt \right\}}. \tag{3.5.4}$$

If the true density function $f(.)$ is known, the Pitman estimator is translation-equivariant [i.e., $T_n(\mathbf{X}_n + c\mathbf{1}_n) = T_n(\mathbf{X}_n) + c \ \forall c \in \mathbb{R}_1$, and T_n has the minimal risk with respect to quadratic loss among equivariant estimators]. The PE are unbiased for θ, although they need not be UMV (as UMV estimators may not generally exist for all f). The PE are consistent under essentially the same regularity conditions as pertaining to the maximum likelihood estimators of θ, and in fact, they are asymptotically equivalent, in probability. By virtue of this stochastic equivalence, from robustness considerations, the PE shares the nonrobustness properties of the usual MLE, and hence they may not be that desirable. For various large sample properties of PE, we may refer to Strasser (1981a-c).

3.5.3 Pitman-Type Estimators of Location

A new class of robust PE, termed Pitman-type estimators (PTE), was suggested by Johns (1979). Suppose that $X_1, \ldots, X_n$ are i.i.d. r.v.'s with a d.f. $G_\theta(x) = G(x - \theta)$, $x, \theta \in \mathbb{R}_1$, where the form of G is not necessarily known. For an appropriate nonnegative $h(.)$, [i.e., $h : \mathbb{R}_1 \to \mathbb{R}_1^+$], the PTE of θ is defined as

$$T_n^\star = \frac{\left\{ \int_{\mathbb{R}_1} t \prod_{i=1}^n h(X_i - t)dt \right\}}{\left\{ \int_{\mathbb{R}_1} \prod_{i=1}^n h(X_i - t)dt \right\}}. \tag{3.5.5}$$

Whereas the PE are "close to" the MLE, the PTE are "close to" M-estimators generated by $\rho(x) = -\log h(x)$. Asymptotic equivalence of $T_n^\star$ with an appropriate M-estimator (with the pertaining order) was proved by Hanousek (1988). The analogy between (3.5.4) and (3.5.5) is quite clear. Whereas (3.5.4) uses the assumed form of the density f, the function $h(.)$ is of more arbitrary form, and in this way a class of robust competitors can be generated. The influence curve of $T_n^\star$ naturally depends on this choice of $h(.)$. It is given by

$$IC(x; G, T^\star) = \frac{\psi(x)}{\int_{\mathbb{R}_1} \psi'(x)dG(x)}, \tag{3.5.6}$$

where

$$\psi(x) = (d/dx)\log h(x). \tag{3.5.7}$$

The idea behind this type of estimators (i.e., the PTE) is that for the PTE in (3.5.4), we have $h(x) = (d/dx)G(x)$. This PE minimizes the quadratic risk among the translation-equivariant estimators of θ (when $F = G$). Thus the

PTE class contains an efficient member and is robust by appropriate choice of the function $h(.)$ in (3.5.5). Huber (1984) studied the finite sample breakdown points of $T_n^\star$ (which may be in some cases). We refer to Section 2.3 for the computation of the influence curve in (3.4.6) and the breakdown points. Note that $\prod_{i=1}^n f(X_i - t)$ is the likelihood function at the point $\theta = t$. Under the local asymptotic normality (LAN) condition, the log-likelihood function is quadratic in $t - \hat{\theta}_n$ in an $n^{1/2}$-neighborhood of the MLE $\hat{\theta}_n$, and this leads to the IC (3.5.6) of T_n with $h(.) = f(.)$. A similar expansion may be used for the product $\prod_{i=1}^n h(X_i - t)$ in (3.5.5) to verify (3.5.6) - (3.5.7); we leave the details as an exercise (see Problem 3.5.4).

We may remark that $T_n^\star$ in (3.5.5) can be expressed as a functional of the empirical d.f. F_n, and hence the von Mises functional (to be studied in Section 3.6) can be used for the study of the relevant properties of these PTE. In fact the general class of M-, L-, and R-estimators, as well as the estimators considered in this section, can be expressed in a general form of *differentiable statistical functionals.* Thus with a view to unify some of the diverse results presented in this and earlier sections, we will present an introduction to such functionals and some of their basic properties.

3.5.4 Bayes-Type Estimators of general parameter

Following the idea of Pitman-type estimators, a new class of robust Bayesian-type (or B-)estimators was suggested by Hanousek (1990). Suppose that $X_1, \ldots, X_n$ are i.i.d. r.v.'s with a d.f. $G_{\theta_0}(x)$, $x, \theta_0 \in \Theta \subseteq \mathbb{R}_1$. The B-estimator of θ_0 corresponding to the prior density $\pi(\theta)$ on Θ and generated by the function $\rho(x, \theta) : \mathbb{R}_1 \times \Theta \to \mathbb{R}_1$ is defined as

$$T_n^\star = \frac{\int_{\mathbb{R}_1} \theta \exp\{-\sum_{i=1}^n \rho(X_i, \theta)\}\pi(\theta)d\theta}{\int_{\mathbb{R}_1} \exp\{-\sum_{i=1}^n \rho(X_i, \theta)\}\pi(\theta)d\theta}. \tag{3.5.8}$$

As in the case of PTE, Hanousek (1990) showed that $T_n^\star$ is asymptotically equivalent to the M-estimator generated by ρ for a large class of priors and under some conditions on ρ and G_θ. This is in correspondence with the asymptotic equivalence of Bayes and ML estimators. For a study of this relation, we refer the reader to LeCam (1953), Stone (1974), Fu and Gleser (1975), Ibragimov and Hasminskii (1981), among others.

3.6 DIFFERENTIABLE STATISTICAL FUNCTIONS

As in the earlier sections, we conceive of a statistic $T_n = T(X_1, \ldots, X_n)$ based on n i.i.d. r.v.'s $X_1, \ldots, X_n$ where X_i has the d.f. G. Also we denote the sample (empirical) d.f. by F_n. Often T_n can be expressed as a functional of the empirical d.f. F_n. The simplest example of this type is the sample mean

$$\bar{X}_n = n^{-1} \sum_{i=1}^{n} X_i = \int_{\mathbb{R}_1} x dF_n(x). \tag{3.6.1}$$

A functional of a d.f. G of the form

$$T(G) = \int_{\mathbb{R}_1} \phi(x) dG(x) \tag{3.6.2}$$

is called a *linear statistical functional*; the range space for x can be $\mathbb{R}_p$ for any $p \geq 1$. Similarly, a functional of the form

$$T(G) = \int \int \phi(x, y) dG(x) dG(y) \tag{3.6.3}$$

is called a *second degree functional.* In general, we can introduce a regular functional of degree m (≥ 1) by incorporating a *kernel* $\phi(.)$ of degree $m(\geq 1)$ and letting

$$T(G) = \int \ldots \int \phi(x_i, \ldots, x_m) dG(x_1) \ldots dG(x_m). \tag{3.6.4}$$

Von Mises (1936, 1947) recognized the basic importance of such statistical functional and considered the natural estimators

$$T(F_n) = \int \ldots \int \phi(x_i, \ldots, x_m) dF_n(x_1) \ldots dF_n(x_m). \tag{3.6.5}$$

For $m = 1$ (i.e., for linear functionals), $T(F_n)$ is an unbiased estimator of $T(G)$, but for $m \geq 2$, it may not be so. Halmos (1946) and Hoeffding (1948) considered unbiased and optimal (symmetric) estimators of $T(G)$; in the literature these are known as U-statistics. Typically a U-statistic can be expressed as a linear combination of some $T(F_n)$ and vice versa. In fact these two statistics are asymptotically equivalent in the sense that their difference is $O(n^{-1})$ a.s., as $n \to \infty$. We pose some of these results on $T(F_n)$ and the related form of U_n, the U-statistic, in the form of exercises (see Problems 3.6.1 - 3.6.3). Detailed treatment of U-statistics and von Mises functionals are given in Serfling (1980) and Sen (1981a), among other places, and we will not enter into the details here.

The ingenuity of von Mises (1947) lies in an expansion of $T(F_n)$ around $T(G)$ in such a form that the first-order term is a linear functional (and provides the access to the first-order asymptotic theory) and the remainder term is negligible in an appropriate sense. Naturally the possibility of such an expansion rests on some *differentiability* properties of statistical functionals. Granted such differentiability properties of a statistical functional, we will later see that the influence function, asymptotic normality, and other related properties can all be studied in a unified manner. Before we enter into that discussion, we will consider briefly the connection of M-, L-, and R-estimators, treated in earlier sections, with such differentiable statistical functionals.

We start with the location model, for which the M-estimator of θ, defined by (3.2.2) with $\psi(x,t) = \psi(x-t)$, $x, t \in \mathbb{R}_1$, can be written in the form

$$\int_{\mathbb{R}_1} \psi(x - M_n)dF_n(x) = 0. \tag{3.6.6}$$

Thus the functional corresponding to (3.6.6) is defined to be a root $T(G) = \theta$ of the implicit equation

$$\int_{\mathbb{R}_1} \psi(x - \theta)dG(x) = 0. \tag{3.6.7}$$

Next, as in Section 3.3, we consider a typical L-estimator of location. We take such an estimator in the following (smooth) version:

$$L_n = \int_{\mathbb{R}_1} xJ(F_n(x))dF_n(x) \tag{3.6.8}$$

so that the functional corresponding to (3.6.8) is of the form

$$\int_{\mathbb{R}_1} xJ(G(x))dG(x) = \int_0^1 G^{-1}(t)J(t)dt. \tag{3.6.9}$$

[In Section 3.3, we have considered more general L_n for which the score function $J(.)$ could depend on n as well, and for such a representation we need more involved functionals (of G) depending on n.] Unlike the case of M-estimators, the L-estimators are explicitly defined. The R-estimators of location considered in Section 3.4, like the M-estimators, are also defined as roots of some implicit equations. Let us look back at (3.4.7), and for simplicity, we let $a_n(k) = \phi^+(k/(n+1))$, for $k = 1, \ldots, n$, where $\phi^+ : [0,1] \to \mathbb{R}^+$. We also let $\phi^\star(u) = \phi^+(u)\text{sign}(u)$, $u \in (-1,1)$. We may write (3.4.7) equivalently as

$$n\int_{\mathbb{R}_1} \phi^\star\Big(F_n(x) - F_n(2\theta_0 - x)\Big)dF_n(x),\ \theta_0 \in \mathbb{R}_1. \tag{3.6.10}$$

Thus the functional corresponding to (3.6.10) is defined as a root $T(G) = \theta$ of the implicit equation

$$\int_{\mathbb{R}_1} \phi^\star\Big(G(x) - G(2\theta - x)\Big)dG(x) = 0. \tag{3.6.11}$$

Here, for scores defined in a more general manner, some further uniformity conditions may also be needed in the definition of $T(G)$ (which may then depend on n as well). Likewise the minimum distance estimators considered in Section 3.5 may be written in terms of some (implicit) functionals, depending on the norm $\delta(.)$ in (3.5.1); the very form in (3.5.2) suggests this. Using

$$\begin{aligned}\prod_{i=1}^{n} h(X_i - t) &= \exp\Big\{\sum_{i=1}^{n} \log h(X_i - t)\Big\} \\ &= \exp\Big\{n \int_{\boldsymbol{R}_1} \log h(x - t) dF_n(x)\Big\}, \end{aligned} \tag{3.6.12}$$

we rewrite (3.5.5) as

$$\frac{\int_{\boldsymbol{R}_1} t \exp\Big\{n \int \log h(x - t) dF_n(x)\Big\} dt}{\int \exp\Big\{n \int \log h(x - t) dF_n(x)\Big\} dt}. \tag{3.6.13}$$

Notice that the functional representation of $T_n^\star$ and its population counterpart $T(G) = T(G; n)$ depends on the sample size n as well. These examples suggest that for a broad class of estimators, one can consider a general statistical functional of the form

$$T_n = T(F_n; n), \ n \geq n_0 \tag{3.6.14}$$

which may be defined implicitly or explicitly, and then use appropriate expansion of $T(F_n; n)$ to approximate it by a convenient linear functional. Toward this end, we introduce the basic differentiability condition on $T(G; n)$ which permits the desired operation. For simplicity of presentation, we first consider the case of $T(G; n) = T(G)$ independent of n.

Taking the lead from von Mises (1947), we are tempted to write

$$T(F_n) = T(G) + T_G^{(1)}(F_n - G) + R_n(F_n, G), \tag{3.6.15}$$

where $T_G^{(1)}$ is the *derivative* of the functional at G and $R_n(F_n, G)$ is the *remainder term*. To make this expansion meaningful, first we consider two topological vector spaces A and B and denote by $L(A, B)$ the set of (continuous) linear transformations from A to B. We identify $T(F_n)$ as a function $T : O \to B$, so that F_n and G both belong to the set O, an open subset of A. We consider a class C of compact subsets of A. A function $T : O \to B$ is *Hadamard (or compact) differentiable* at $G \in O$ if there exists a $T_G^{(1)} \in L(A, B)$, such that for any $K \subset C$, uniformly in $H \in K$,

$$\lim_{t \to 0} \Big\{t^{-1}[T(F + tH) - T(F) - T_G^{(1)}(tH)]\Big\} = 0. \tag{3.6.16}$$

The linear functional $T_G^{(1)}$ is called the *Hadamard* (or *compact*) *derivative* of T at G.

If instead of the class C of compact subsets of A, we consider the class of all bounded subsets of A, then (3.6.16) corresponds to the *Fréchet differentiability* of $T(.)$ at G. Similarly, if the class C is the class of all single point subsets of A, we have *Gateaux differentiability.* From the above definitions, it follows that the mode of differentiability is dependent on the topology of the space involved. Fernholz (1983) gives an excellent account of such differentiability properties of various statistical functionals. The Fréchet mode is more stringent than the Hadamard mode which in turm implies the Gateaux mode. It seems that the Hadamard differentiability appears to serve more purposes in statistical inference; there have been problems using the Gateaux differentiability method, while the Fréchet mode is often too stringent. For this reason we will confine ourselves to the mode of differentiability in (3.6.16) for our statistical analysis.

Looking at (3.6.15) and (3.6.16), we observe that

$$R_n(F_n, G) = o\Big(||F_n - G||\Big) \text{ on the set } ||F_n - G|| \to 0. \tag{3.6.17}$$

Here $||.||$ stands for the Kolmogorov supnorm (i.e., $\sup\{|F_n(x) - F(x)| : x \in \mathbb{R}_1\}$), although other measures introduced in Section 2.4 could also be used. Note that by the classical results on the Kolmogorov-Smirnov statistics, we have

$$n^{1/2}||F_n - G|| = O_p(1) \tag{3.6.18}$$

so that (3.6.17) is $o_p(n^{-1/2})$ as $n \to \infty$. On the other hand, the linear functional $T_G^{(1)}(F_n - G)$ may be written equivalently as

$$\int T^{(1)}(G; x)d[F_n(x) - G(x)], \tag{3.6.19}$$

where, without any loss of generality, we may set $\int T^{(1)}(G; x)dG(x) = 0$. Thus (3.6.19) can be written as $n^{-1}\sum_{i=1}^{n} T^{(1)}(G; X_i)$ so that the classical central limit theorem and related probability tools can be used for this average of i.i.d. r.v.'s. This yields a direct proof of the asymptotic normality of

$$\frac{\sqrt{n}(T_n - T(G))}{\sigma_{(G)}}, \text{ where } \sigma^2_{(G)} = \mathrm{Var}[T^{(1)}(G; x)]. \tag{3.6.20}$$

If we only want to prove the consistency of $T(F_n)$, [as an estimator of $T(G)$], we do not need even the Hadamard differentiability of $T(\cdot)$ at G. It suffices to assume the *Hadamard continuity,* $T(F) \to T(G)$ as $||F - G|| \to 0$. Further, if we define the influence function $IC(x; G, T)$ as in (2.3.2) and (2.3.3), then it is easy to see that

$$IC(x : G, t) = T^{(1)}(G; x), \quad x \in \mathbb{R}_1. \tag{3.6.21}$$

Thus the first-order Hadamard differentiability of $T(.)$ at G also provides the desired information on the influence curve of $T(.)$. For the more general case where $T_n = T(F_n; n)$, we need to consider a sequence $\{T(F; n); n \geq n_0\}$ of differentiable statistical functionals. We may assume that (3.6.15) and (3.6.16) hold uniformly in $n \geq n_0$. In other words, we need to replace $T(.)$ and $T^{(1)}(.)$ in (3.6.16) by $T(.; n)$ and $T^{(1)}(.; n)$, respectively, and assume that the limits hold uniformly in n.

The above treatment can easily be extended to a more general case where the basic r.v.'s are not identically distributed or are not independent. For the remainder term $o_p(n^{-1/2})$, all we need is that (3.6.18) holds for such a general scheme, and we know that this may be true in a broad spectrum of situations. The d.f. G may be replaced by some other d.f. (e.g., average d.f.), while to apply the central limit theorem to the first order term, we need to assume appropriate regularity conditions on the $T^{(1)}(G; X_i)$, $i = 1, \ldots, n$ (which may not be i.i.d. r.v.'s). From the theoretical point of view, the Hadamard differentiability permits of an expansion that can be used with advantage in a variety of situations where the i.i.d. structure of the basic random variables may not be true. A very typical example of this type is the one-step version of $T(G)$. Suppose that $\hat{F}_n$ is the empirical d.f. for the residuals (fitted from some specified (e.g., regression) models involving some unknown parameters), and we consider a statistic $T(\hat{F}_n)$. So long as $T(F)$ is Hadamard continuous at some G and $\hat{F}_n$ converges to G in some sense (i.e., weakly or a.s.), $T(\hat{F}_n)$ converges in the same sense to $T(G)$. Similarly, whenever (3.6.18) holds for F_n being replaced by $\hat{F}_n$, and a version of the central limit theorem holds for the $T^{(1)}(G; \hat{X}_i)$, $\hat{X}_i$ being the residuals, the asymptotic normality result in (3.6.20) holds for T_n being replaced by $T(\hat{F}_n)$ although, in such a case $\sigma^2_{(G)}$, should be replaced, for example, by

$$\lim_{n\to\infty} n^{-1}\mathrm{Var}\Big[\sum_{i=1}^{n} T^{(1)}(G; \hat{X}_i)\Big] = \sigma^{\star 2}_{(G)}. \tag{3.6.22}$$

We leave some of these details in the form of exercises (see Problems 3.6.4 - 3.6.5). For further illustrations of this point, we refer the reader to generalized L-estimators, discussed before (3.3.39). If $X_1, \ldots, X_i$ are i.i.d. r.v.'s and $\phi(X_1, \ldots, X_m)$ is a kernel of degree m (≥ 1), we could consider the pseudovariables

$$Z_{i_1 \ldots i_m} = \phi\Big(X_{i_1}, \ldots, X_{i_m}\Big) \quad \text{for } 1 \leq i_1 < i_2 < \ldots < i_m \leq n. \tag{3.6.23}$$

Based on these $N = \binom{n}{m}$ pseudovariables, we construct the empirical d.f. H_N and, side by side, denote the actual d.f. of $Z_{1 \ldots m}$ by H. We assume that H is continuous almost everywhere. Then a generalized L-statistic can be defined as

$$L_n^\star = \int_{\mathbb{R}_1} m(x) J_N(H_N(x)) dH_N(x), \tag{3.6.24}$$

where $m(.) : \mathbb{R}_1 \to \mathbb{R}_1$ is a given function and $J_N(t)$, $t \in (0, 1)$, is the adapted score function. The functional $T(H)$ corresponding to (3.6.24) can be written as

$$\int_0^1 m(H^{-1}(t))J(t)dt. \tag{3.6.25}$$

thus, whenever $n^{1/2}||H_N - H|| = O_p(1)$, we can proceed as in (3.6.16) through (3.6.19) and obtain the principal term as

$$\int T^{(1)}(H;x)d[H_N(x) - H(x)], \tag{3.6.26}$$

The weak convergence of $n^{1/2}[H_N - H]$ (to a Gaussian function) ensures the asymptotic normality of (3.6.26), and the asymptotic normality of $n^{1/2}(L_n^\star - T(H))$ immediately follows. Fortunately weak convergence results for such U-processes have been studied in detail (e.g., see Silverman 1983, Serfling 1984, and Sen 1983, among others), and hence for such generalized L-estimators, the asymptotic theory can be studied without much difficulty. Incidently, for the L-M-estimator in (3.3.40), a representation similar to (3.6.7)-(3.6.8) works out with an additional factor $J_n(F_n(x))$ or $J(G(x))$ in the nonstudentized case. However, the studentized r.v.'s $(X_i - t)/s_n$ are not independent, but if $n^{1/2}(s_n - \gamma)$ is $O_p(1)$ and $\psi(.)$ is skewsymmetric a very similar proof works out (see Problem 3.6.7).

From what has been discussed so far, it is clear that this Hadamard differentiability approach provides a convenient way of studying the asymptotic properties of a large class of estimators; (3.6.15) is analogous to the (first-order) representation result for M-, L-, and R-estimators, considered in the earlier sections. This naturally raises a question whether to treat the M-, L-, and R-estimators separately and to put more emphasis on these estimators instead of on the so-called Hadamard differentiable functionals. We now proceed to settle this basic question.

Though we have characterized some of the M-, L-, and R-estimators as members of the class of differentiable statistical functionals, verification of the (first-order) Hadamard differentiability condition for such functionals may generally require some regularity conditions that are more stringent than the ones treated in the earlier sections. As an illustration, consider the set of sufficient conditions for the differentiability of a functional corresponding to an M-estimator. Let $G(x) = F_\theta(x) = F(x - \theta)$ for some fixed F (i.e., the location model), and consider the score function $\psi(t)$, $t \in \mathbb{R}_1$. The following lemma gives sufficient conditions for the desired Hadamard differentiability.

Lemma 3.6.1 *Suppose that $\psi(t)$ is continuous and piecewise differentiable with bounded derivative $\psi'(t)$, such that $\psi'(t) = 0$ for all $t \notin J$, where J is a compact interval containing* 0 *as an inner point. Also suppose that F is continuous with a positive and piecewise continuous density f. Then the statistical (implicit) functional corresponding to the M-estimator is Hadamard differentiable.*

We leave the proof as an exercise (see Problem 3.6.8). Comparing the results in Section 3.2 and Lemma 3.6.1, we conclude that the differentiable statistical functional approach for M-estimators may demand comparatively more stringent regularity conditions. In fact the MLE with an unbounded likelihood score function (as well as the classical least square estimators) do not always satisfy the regularity conditions of Lemma 3.6.1. Of course this lemma provides only sufficient conditions.It may be possible to verify the first-order differentiability condition under even less stringent regularity conditions on $\psi(.)$, but such a proof could depend very much on the particular form of $\psi(.)$ or F, and the treatment in Section 3.2 may then appear to be much less involved. Let us next move on to the case of L-estimators, which are of course explicit functionals. We have the following:

Lemma 3.6.2 *Suppose that for some* $\alpha: \ 0 < \alpha < 1/2$, $J(t)$ *is 0 for* $t \leq \alpha$ *and* $\geq 1 - \alpha$, $J(t)$ *is continuous and piecewise differentiable with bounded derivative for* $t \in [\alpha, 1-\alpha]$. *Also assume that* $G^{-1}(t)$ *is square integrable inside* $[\alpha, 1-\alpha]$, *and that* G *is absolutely continuous. Then* $T(G)$ *in (3.6.9) is first-order Hadamard differentiable.*

We leave the proof as an exercise (see Problem 3.6.9). Again, we can compare the results of Section 3.3 with Lemma 3.6.2 and come to the same conclusion that the differentiable statistical functional approach may restrict the class of L-estimators to the trimmed members. Finally, we comment on R-estimators. In this case, for the first order Hadamard differentiability of the implicit functional in (3.6.11), a set of sufficient conditions includes that $\phi^\star(u)$, $u \in (-1, 1)$, is continuous and piecewise differentiable with bounded and piecewise continuous derivatives $\phi^{\star\prime}(t)$. Problem 3.6.10 is setup to verify this. This appears to be rather restrictive. For example, for the location model the Wilcoxon signed rank statistic based R-estimator is a member of this class, but the one based on the normal scores statistic (for which $\phi^{\star\prime}(t) = \Phi^{-1}(t)$, Φ being the standard normal d.f.) is not. In fact most of the unbounded score functions are excluded from this set. As such, we would prefer to treat the R-estimators in a more general framework without appealing to the differentiable statistical function approach. In later chapters we will include further discussions on such differentiable functionals for their plausible adaptation in a general framework of robust estimation. Nevertheless, we will continue to put emphasis on the three basic classes of robust estimators, namely M-, L-, and R-estimators (along with their inter-relationships). Although the Hadamard differentiability provides a good amount of flexibility in the treatment of the asymptotic theory of differentiable statistical functions, we should not overemphasize its utility. For functionals of the type (3.6.4)-(3.6.5), this Hadamard differentiability demands quite stringent regularity conditions on the kernel $\phi(.)$, yet, by making use of the simple identity that $dF_n(X_i) = dG(X_i) + d[F_n(X_i) - G(X_i)]$, $i = 1, \ldots, m$, we obtain the classical Hoeffding (1961) decomposition:

$$T(F_n) = T(G) + \sum_{h=1}^{m} \binom{m}{h} T^{(h)}(F_n), \tag{3.6.27}$$

where

$$T^{(h)}(F_n) = \int \cdots \int \phi_h(x_1, \ldots, x_h) d[F_n(x_1) - G(x_1)] \ldots d[F_n(x_h) - G(x_h)], \tag{3.6.28}$$

and ϕ_h a kernel of degree h, is given by

$$\phi_h((x_1, \ldots, x_h) = \int \cdots \int \phi(x_1, \ldots, x_h, y_{h+1}, \ldots, y_m) dG(y_{h+1}) \ldots dG(y_m), \tag{3.6.29}$$

for $h = 1, \ldots, m$ (see Problem 3.6.12). As such the asymptotic normality result can be deduced under the usual second moment conditions on the ϕ_h, $h = 1, \ldots, m$, while the influence function comes out as

$$T^{(1)}(F_n) = \frac{1}{n} \sum_{i=1}^{n} [\phi_1(X_i) - \int_{\boldsymbol{R}_1} \phi_1(x) dG(x)],$$

which is a linear statistics. Thus, whenever the statistical functionals are more structured (in the sense of von Mises 1947), the usual von Mises expansion may come out to be much more adaptable than the Hadamard differentiability approach. For some related asymptotic theory of such von Mises functionals, we refer the reader to von Mises (1947), Hoeffding (1948), Miller and Sen (1972), and Sen (1974a,b,c), among others. Hoeffding (1948, 1961) introduced (for U-statistics) a novel idea of L_2-projection (into sum of independent r.v.'s), and this technique can be used for a general statistical functional as well. Let us denote by

$$T_{ni} = \mathbb{E}[T(F_n)|X_i], i = 1, \ldots, n; \; \hat{T}_n = \sum_{i=1}^{n} T_{ni}. \tag{3.6.30}$$

Then, under quite general regularity conditions, it can be shown that $\hat{T}_n$ and $T_n - \hat{T}_n$ are uncorrelated and $V(T_n)/V(\hat{T}_n) \to 1$ as $n \to \infty$. On the other hand, $\hat{T}_n$ involves independent summands on which the central limit theorem and other standard asymptotic theory can easily be adapted. Thus the asymptotic properties of $T(F_n)$ can be studied with the help of the quadratic mean approximation by $\hat{T}(F_n)$. For rank statistics, this approach has been popularized by the pioneering work of Hájek (1968). For a general symmetric statistic, this approach has been elegantly pursued by van Zwet (1984) in the context of Berry-Esseen type bounds for the asymptotic distribution theory of T_n. It is easy to see that for a Hadamard differentiable statistical functional $T(F_n)$, the L_2-projection method leads to $\hat{T}_n$, the same principal term [i.e., $n^{-1}\sum_{i=1}^{n} T^{(1)}(G; X_i)$] so that the study of the influence function, for instance, can be carried out in the usual manner. We advocate the use of this L_2-projection method whenever feasible.

3.7 PROBLEMS

3.2.1. Apply a suitable version of the central limit theorem, and show that under $\theta = 0$, $n^{-1/2}\sum_{i=1}^{n}\psi(X_i - n^{-1/2}t)$ is asymptotically normal with mean $-\gamma t$ and variance σ_ψ^2, where γ and σ_ψ^2 are defined by (3.2.10) and (2.4.10) respectively. Hence, or otherwise, show that for every real t,

$$\begin{aligned} &\lim_{n\to\infty} P_\theta\{n^{1/2}(M_n - \theta) \le t\} \\ &= \lim_{n\to\infty} P_0\{n^{-1/2}\sum_{i=1}^{n}\psi(X_i - n^{-1/2}t) \le 0\} \\ &= \Phi(\gamma t/\sigma_\psi), \end{aligned}$$

where $\Phi(.)$ is the standard normal d.f.

3.2.2. Assume that (3.2.17) holds (a formal proof is given in Chapter 5). Identifying that the $\psi(X_i - \mathbf{c}_i'\boldsymbol{\beta})$ are i.i.d. r.v.'s with 0 mean and variance σ_ψ^2 (when $\boldsymbol{\beta}$ holds), show that the classical central limit theorem holds for the right hand side of (3.2.17) when the Noether condition in (3.2.18) is satisfied. Hence, or otherwise, verify (3.2.19). Consider the particular case of $\psi(x) \equiv \text{sign}(x)$ so that $\mathbf{M}_n$ reduces to the L_1-norm estimator of $\boldsymbol{\beta}$. Work out the expression for γ^2 in (3.2.19) in this case. For $\psi(x) \equiv x$, show that (3.2.17) is an identity for the least square estimator of $\boldsymbol{\beta}$.

3.2.3. If $\mathbf{X}$ and $(-1)\mathbf{X}$ both have the same d.f. F, defined on $\mathbb{R}_p$, then F is said to be diagonally symmetric about $\mathbf{0}$. Show that if F is diagonally symmetric about $\mathbf{0}$, then for every $\mathbf{l} \in \mathbb{R}_p$, $\mathbf{l}'\mathbf{X}$ has a d.f. symmetric about 0. This property is related to the multivariate median unbiasedness: If for every $\mathbf{l} \in \mathbb{R}_p$, $\mathbf{l}'(\mathbf{T}_n - \theta)$ has 0 median, then $\mathbf{T}_n$ is multivariate median unbiased for θ. Verify (3.2.17). [Sen 1990]

3.2.4. Show that corresponding to the score function $\psi(x) = \text{sign}(x)$, $x \in \mathbb{R}$, for the location model, the M-estimator is the sample median. Suppose now that for some $\alpha > 0$, $\mathbb{E}_F|X|^\alpha < \infty$, where α need not be greater than or equal to 1. Then verify that $\mathbb{E}M_n = \theta$ whenever $(n+1)\alpha \ge 2$. Thus for large n, the unbiasedness of M_n may not require the existence of the first moment of F. [Sen 1959].

3.2.5. Consider the Huber (1964) estimator, for which

$$\psi(x) = xI(|x| \le k) + k \,\text{sign}(x) I(|x| > k), \quad \text{for some finite } k \ (> 0).$$

Suppose that $\mathbb{E}_F|X|^\alpha < \infty$, for some $\alpha > 0$ (not necessarily ≥ 1). What is the minimum sample size (n), such that the Huber estimator has a finite rth moment, for some $r > 0$? Compare the situation with that of the sample mean.

3.2.6. Consider the Cauchy density $f(x) = \pi^{-1}(1+x^2)^{-1}$, $x \in \mathbb{R}$. For a sample size 3 from the Cauchy density $f(x-\theta)$, find the MLE of θ, and discuss whether it has a finite first moment or not. Is there any unbiased estimator of θ for $n = 3$?

3.2.7. Consider a mixture model:

$$f(x) = \frac{1}{2}[\phi(x) + l(x)], \ x \in \mathbb{R},$$

where ϕ is the standard normal density and $l(.)$ is the laplace density ($\frac{1}{2}e^{-|x|}$). For a sample of n observations from $f(x-\theta)$, which of the estimators: mean or median, would you prefer on the ground of robustness and/or asymptotic variance?

3.2.8. For the uniform $[-\theta, \theta]$ d.f., $\theta > 0$, which ψ-function would you recommend for the M-estimation of θ ? What about a mixture of two uniform distributions ?

3.3.1. Show that $t(\mathbf{X}_n) = \{X_{n:1}, \ldots, X_{n:n}\}$ is a symmetric function of $\mathbf{X}_n = (X_1, \ldots, X_n)$ and, conversely, that any symmetric function of $\mathbf{X}_n$ is a function of $t(\mathbf{X}_n)$. Also show that $t(\mathbf{X}_n)$ is complete for the class of d.f.'s on $\mathbb{R}_n$ when F is (a) any discrete distribution or (b) any absolutely continuous d.f. As $t(\mathbf{X}_n)$ a sufficient statistics? Is it minimal sufficient in general?

3.3.2.Define the $a_{n:i}$ and b_{nij} as in (3.3.8). Show that

$$a_{n:i} = i\binom{n}{i}\int_{\mathbb{R}_1} x[F_0(x)]^{i-1}[1-F_0(x)]^{n-i} dF_0(x),$$

for every $i = 1, \ldots, n, n \ge 1$. Use the results in Problems 3.2.4 and 3.2.6 to show that the $a_{n:i}$'s all exist whenever $\mathbb{E}_{F_0}(X)$ exists. What happens when $\mathbb{E}_{F_0}|X|^\alpha < \infty$ for some $\alpha :\ 0 < \alpha < 1$? Similarly show that the b_{nij}'s all exist whenever $\mathbb{E}_{F_0}(X^2) < \infty$. What happens when $\mathbb{E}_{F_0}|X|^\alpha < \infty$ for some $\alpha :\ 0 < \alpha < 2$?

3.3.3. Write down the estimating equations in (3.3.10) in terms of the $a_{n:i}$, b_{nij}, and $X_{n:i}$. Under what conditions (on the $a_{n:i}$ and b_{nij}), are the estimators $\hat{\mu}_n$ and $\hat{\delta}_n$ uncorrelated?

3.3.4. Show that whenever $\mathbb{E}_{F_0}|X|^\alpha < \infty$ for some $\alpha > 0$, there exists a sample size n_0 such that for all $n \geq n_0$, treating the set J as $\{k, \ldots, n-k+1\}$ with $k : k\alpha \geq 2$, one can obtain the BLUE of (μ, δ) as in (3.3.10) with the set J replacing $\{1, \ldots, \text{n}\}$.

3.3.5. Rewrite $a_{n:i}$ as $i\binom{n}{i}\int_0^1 (F_0^{-1}(t))t^{i-1}(1-t)^{n-i}dt$, and make use of the beta distribution converges to a normal one to claim that as $n \to \infty$ with $i/n \to p : 0 < p < 1$, $a_{n:i} \to F_0^{-1}(p)$. Show that in this respect the continuity and the strict monotonicity of $F_0^{-1}(t)$ at $t = p$ suffice.

3.3.6. Show that if $f(F^{-1}(p_1))$ and $f(F^{-1}(p_2))$ are both positive and $i/n \to p_1$, $j/n \to p_2 : 0 < p_1 < p_2 < 1$, as $n \to \infty$, then

$$nb_{nij} \to \frac{p_1(1-p_2)}{f(F^{-1}(p_1))f(F^{-1}(p_2))}.$$

What happens when $p_1 = 0$ or $p_2 = 1$?

3.3.7. Under (3.3.13), take $\theta = 0$ (WLOG). Then show that

$$L_n(X_1, \ldots, X_n) = -L_n(-X_1, \ldots, -X_n) \stackrel{\mathcal{D}}{=} -L_n(X_1, \ldots, X_n),$$

since the d.f. F_0, symmetric about 0. Hence L_n has a d.f. symmetric about 0. Use the result in Problem 3.2.3 to claim that L_n is median unbiased for θ.

3.3.8. Verify (3.3.22) and (3.3.24) using the Hoeffding inequality and (3.3.23).

3.3.9. Verify the a.s. convergence result on $L_{n_1} - \lambda_1(\theta)$ [after (3.3.27)] under the Hölder-type condition on F and $\{J_{n,1}\}$ and the pointwise convergence of $J_{n,1}$ to some J_1.

3.3.10. (Continuation). Eliminate the pointwise convergence of $J_{n,1}$ to J_1 by the uniformity $||J_{n,1}||_s$ in n. Also show that if F has a compact support, then the needed condition on $\{J_{n,1}\}$ can be reduced further.

3.3.11. Provide a proof of (3.3.28).

3.3.12. Let $X_1, \ldots, X_n$ be a random sample from the distribution with d.f.

$F_{(n)}(x-\theta)$, where

$$F_{(n)} \equiv (1-cn^{-1/2})G + cn^{-1/2}H,$$

G and H are fixed d.f.'s, $G(x)+G(-x) = 1$, $\forall x \in \mathbb{R}_1$. Let M_n be an M-estimator defined as a solution of $\sum_{i=1}^{n} \psi(X_i - M) = 0$, where ψ is a nondecreasing odd function such that ψ' and ψ'' are continuous and bounded up to a finite number of points. Then

$$n^{\frac{1}{2}}(M_n - \theta) \xrightarrow{\mathcal{D}} \mathcal{N}(b, \nu^2(\psi, G)),$$

where $b = cA/\mathbb{E}_G\psi'$, $A = \int \psi(x)dH(x)$, and $\nu^2(\psi, G)$ is defined in (3.2.22). Hence M_n has asymptotic bias b and the asymptotic mean square error $\nu^2(\psi, G) + b^2$. [Jaeckel 1971]

3.3.13. Consider the model of Problem 3.3.12, and assume that, for a fixed $0 < \alpha < 1/2$, H puts all of its mass to the right of $G^{-1}(1-\alpha)$. Then the α-trimmed mean has the asymptotic bias $b(\alpha) = cG^{-1}(1-\alpha)/(1-2\alpha)$, and the asymptotic mean square error

$$\sigma^2(\alpha) + b^2(\alpha) = \frac{1}{(1-2\alpha)^2}\{2\int_0^{G^{-1}1-\alpha} x^2 dG(x) + (2\alpha + c^2)(G^{-1}(\alpha))^2\}.$$

[Jaeckel 1971]

3.3.14. Let $G(x) = G_0((x-\theta)/\delta)$, $\theta \in \mathbb{R}_1$, $\delta \in \mathbb{R}_1^+$ and G_0 does not depend on (θ, δ). Then show that

$$G^{-1}(3/4) - G^{-1}(1/4) = \delta\{G_0^{-1}(3/4) - G_0^{-1}(1/4)\} = \delta T(G_0),$$

where the functional $T(G_0)$ does not depend on (θ, δ). Study the nature of $T(G_0)$ for G_0 normal, Laplace, and Cauchy, and comment on the errors in the definition of the interquartile range if the true F_0 and assumed G_0 are not the same.

3.3.15. Show that $(U_n^{(2)})^2/2$ defined in (3.3.38) is equal to the sample variance (unbiased form) and is an optimal nonparametric estimator of the functional $T_2(F) = \int_{\mathbb{R}_1}\int_{\mathbb{R}_1}(x_1 - x_2)^2/2dF(x_1)dF(x_2) = \sigma_F^2$. Verify that $(U_n^{(2)})^2/2$ is a U-statistic with a kernel of degree 2.

3.3.16. Show that $U_n^{(1)}$ in (3.3.38) is a U-statistic with a kernel of degree 2 and is an optimal nonparametric estimator of the functional

$$T_1(F) = \int_{\mathbb{R}_2} |x_1 - x_2| dF(x_1)dF(x_2).$$

3.3.17. For both $U_n^{(1)}$ and $U_n^{(2)}$, work out the asymptotic normality results through U-statistics theory. Compare the two functionals $T_1(F)$ and $T_2(F)$ when F is normal and Laplace, and also comment on their robustness properties.

3.3.18. Define the Z_{ij} as in before (3.3.38). Consider an L-statistic based on these Z_{ij}'s, and verify its stochastic convergence and asymptotic normality properties through empirical (U-) processes convergence results. Here the empirical U-process $F_{Un}(x)$ is defined by $\binom{n}{2}^{-1}\sum_{1\le i<j\le n} I(Z_{ij}\le x)$, $x\in \mathbb{R}_1^+$,. Verify that if $F_U(x) = P\{Z_{ij}\le x\}$, $x\in\mathbb{R}_1^+$, then $\|F_{Un}-F_U\|\to 0$ *a.s.*, as $n\to\infty$ and $\sqrt{n}(F_{Un}-F_U)$ converges weakly to a tied-down Gaussian process.

3.4.1. For S_n^0, defined in (3.4.3), let $T_n = (n+S_n^0)/2$. Show that under H_0, $T_n \sim \text{Bin}(n, 1/2)$, and hence verify that S_n^0 is distributionfree under H_0, and as n increases $n^{-1/2}S_n^0 \xrightarrow{\mathcal{D}} \mathcal{N}(0,1)$.

3.4.2. Verify (3.4.5).

3.4.3. Suppose that $X_1,\ldots,X_n$ are i.i.d. r.v.'s with a continuous d.f. F, symmetric about 0. Let R_{ni}^+ = rank of $|X_i|$ among $|X_1|,\ldots,|X_n|$, and $S_i = \text{sign}(X_i)$, for $i=1,\ldots,n$. Show that the two vectors $\mathbf{R}_n^+ = (R_{n1}^+,\ldots,R_{nn}^+)'$ and $\mathbf{S}_n = (S_1,\ldots,S_n)'$ are stochastically independent, where $\mathbf{R}_n^+$ can assume all possible $(n!)$ realizations (the permutations of $1,\ldots,n$), with the common probability $(n!)^{-1}$ and $\mathbf{S}_n$ takes on all possible 2^n sign-inversions (i.e., $(\pm1,\ldots,\pm1)'$) with the common probability 2^{-n}. What happens if F is not symmetric about 0?

3.4.4. Define $S_n(t) = S_n(\mathbf{X}_n - t\mathbf{1}_n)$ as in after (3.4.7). Show that (a) $\text{sign}(X_i - t)$ is nonincreasing in $t\in\mathbb{R}_1$, (b) $R_{ni}^+(t)$ is nonincreasing in t when $X_i > t$ and is nondecreasing in t when $X_i < t$, and hence, $S_n(t)$ is nonincreasing in $t\in\mathbb{R}_1$. For the particular case of $a_n(k) = 1\ \forall 1\le k\le n$, show that (3.4.5) leads to the same result for the sign statistic.

3.4.5. The Wilcoxon signed-rank statistic W_n is given by $\sum_{i=1}^n \text{sign}(X_i)R_{ni}^+$. Verify that

$$\begin{aligned} W_n^+ &= \sum_{i=1}^n \{\frac{1}{2}\{1+\text{sign}(X_i)\}R_{ni}^+ = \frac{n(n+1)}{4} + \frac{1}{2}W_n \\ &= \sum_{1\le i\le j\le n} I(X_i + X_j > 0). \end{aligned}$$

As such, $W_n(t) = 2\sum_{1\le i\le j\le n} I(X_i + X_j - 2t > 0) - \binom{n+1}{r}$, $t \in \mathbb{R}_1$. Hence verify (3.4.10).

3.4.6. Verify the equivariance relations in (3.4.11) and (3.4.12).

3.4.7. Show that for S_n defined by (3.4.7), under $\theta_0 = 0$, $P_0(S_n = 0) = \eta_n > 0$, for every finite n (≥ 1). However, $\eta_n \to 0$ as $n \to \infty$.

3.4.8. Decompose the d.f. F as $F_1 + F_2$, where F_1 is absolutely continuous and F_2 is a step function, and assume that both of them are symmetric about 0 and that F_2 is nondegenerate. Show that $R_n - \theta$ has a symmetric d.f. around 0 but the d.f. is not continuous everywhere.

3.4.9. Suppose that F admits an absolutely continuous, symmetric, and strongly unimodal density f, and that for some $a > 0$, $\mathbb{E}_F|X|^a < \infty$. Then there exists a positive integer n_0, such that $\mathrm{Var}(R_n)$ exists for every $n \ge n_0$. Further, if the score generating function $\phi(.)$ satisfies the conditions that for $0 < u < 1$,

$$|(\partial^r/\partial u^r)\phi(u)| \le K(1-u)^{-\delta-r}, \ r = 0, 1, 2,$$

then $\lim_{n\to\infty} n\,\mathrm{Var}(R_n) = \gamma^2$, where γ^2 is the variance of the asymptotic distribution of $\sqrt{n}(R_n - \theta)$.

3.4.10. In (3.4.13), let $t = t_n = \theta + n^{-1/2}u$. Then express both the left- and right-hand sides as probabilities, under a contiguous alternative, for the signed rank statistic S_n. Hence, or otherwise, derive the asymptotic normality of $\sqrt{n}(R_n - \theta)$.

3.4.11. For the simple regression model

$$X_i = \theta + \beta c_i + e_i, \quad i = 1, \ldots, n,$$

consider the Kendall tau statistic

$$K_n = \sum_{1\le i<j\le n} \mathrm{sign}(c_i - c_j)\mathrm{sign}(X_i - X_j),$$

and let $K_n(t)$ be the aligned statistic when the X_i's are replaced by $X_i - tc_i$, $1 \le i \le n$. Show that $K_n(\beta)$ is distributionfree with location 0, and that $K_n(t)$ is nonincreasing in t. Hence, or otherwise, show that the derived R-estimator of β is

$$R_n = \mathrm{median}\{(X_i - X_j)/(c_i - c_j) : 1 \le i < j \le n, \ c_i \ne c_j\}.$$

Further show that if the c_i's are binary then R_n reduces to the two-sample Wilcoxon scores estimator of the difference of location parameters. Extend the result for a general linear rank statistic.

3.4.12. Show that under H_0, for every $k \geq 1$, and distinct $i_1, \ldots, i_k$,

$$P\{R_{ni_1}(\beta_0) = j_1, \ldots, R_{ni_k}(\beta_0) = j_k\} = (n^{[k]})^{-1},$$

for every $j_1 \neq \ldots \neq j_k = 1, \ldots, n$, where $n^{[k]} = n \ldots (n-k+1)$. Hence, or otherwise, verify the two formulae in (3.4.22).

3.4.13. Define the $R_{ni}(t)$ as in (3.4.20)-(3.4.21) (when $p = 1$), and show that when $a_n(k)$ is nonincreasing in $k : 1 \leq k \leq n$, $(c_i - \bar{c}_n)a_n(R_{ni}(t))$ is nonincreasing in t. Hence show that there exists a half-open interval (or a singleton point) for which $S_n(t)$ is equal to 0 (or > 0 or < 0), and that an R-estimator of β can be defined as in the location model.

3.4.14. Show that for the simple regression model in Problem 3.4.11, the R-estimator of β is translation-invariant, and hence that working with the residuals $X_i - R_n c_i$, $1 \leq i \leq n$, one can get a closed expression for the R-estimator of θ whenever such an expression exists for the simple location model.

3.4.15. Consider the R-estimator of β in Problem 3.4.11. Reparametrize the model by letting $c_i = dc_i^\star + g$, where $d > 0$ and $g \in I\!R_1$, and $\beta^\star = d\beta$, $\theta^\star = \theta + \beta g$. Then $R_n(\beta^\star = dR_n(\beta)$. What can you say for $p = 2$?

3.4.16. For the one-way ANOVA model $X_{ij} \sim F(x - \theta_i)$, $1 \leq j \leq n_i$, $1 \leq i \leq c$, let $\Delta_{ii'} = \theta_i - \theta_{i'}$, $i, i' = 1, \ldots, c$, a contrast $\mathbf{l}'\boldsymbol{\theta}$, $\mathbf{l}'\mathbf{1} = 0$, can also be written as $\mathbf{l}'\boldsymbol{\Delta}_0$, where $\boldsymbol{\Delta}_0 = (\Delta_{10}, \ldots, \Delta_{c0})'$ and $\Delta_{i0} = c^{-1}\sum_{j=1}^{c} \Delta_{ij}$, $1 \leq i \leq c$. Thus one can use the two-sample R-estimators of the $\Delta_{ii'}$ and, minimizing $\sum_{i \neq i'}^{c}[\hat{\Delta}_{ii'} - (\theta_i - \theta_{i'})]^2$, can obtain the compatible estimators of the $\Delta_{ii'}$ (and $\mathbf{l}'\boldsymbol{\Delta}_0$). Alternatively, one may take $l_1\Delta_{11} + \ldots + l_c\Delta_{c1}$ and use $\hat{\Delta}_{11}, \ldots, \hat{\Delta}_{c1}$ to estimate the contrast. Show that they are asymptotically equivalent, in probability.

3.4.17. (Continuation). Under an additional assumption that F is symmetric, use the R-estimators of the θ_i and construct a contrast estimator. Show that under conjugate score functions (for one- and two-sample rank statistics), this estimator is also asymptotically equivalent to the compatible one in Problem 3.4.16.

3.4.18. In an indirect bioassay problem, consider the model

$$X_{Sij} = \alpha_S + \beta_S d_i + e_{ij}, \; X_{Tij} = \alpha_T + \beta_T d_i + e_{ij},$$

$1 \leq j \leq m$, $1 \leq i \leq k$, where the d_i's are given dose variables, X_S and X_T are the response variables of the standard and test preparations, α_S, β_S, and

α_T, β_T are associated parameters, and the errors e_{ij}'s are i.i.d. r.v.'s with a continuous d.f. F. Consider a *parallel line assay* wherein it is assumed that $\beta_S = \beta_T = \beta$ (unknown), and the parameter of interest is $\theta = (\alpha_S - \alpha_T)/\beta$. For each of the 4 parameters, use R- (or M-) estimators, and then use the least square method (on $\hat{\alpha}_S$, $\hat{\beta}_S$, $\hat{\alpha}_T$, $\hat{\beta}_T$,) to combine them into pooled estimates for $\beta(= \beta_S = \beta_T)$ and α_S, α_T. Use the later one to estimate θ. [Sen 1971]

3.5.1. Show that the Pitman estimator in (3.5.4) is translation-equivariant for the location model. Moreover, if there exists a sufficient statistic for θ, show that T_n is a function of that statistic.

3.5.2. Show that T_n and the MLE are equivalent up to the order $o_p(n^{-1/2})$. Hence, or otherwise, comment on its robustness.

3.5.3. Examine the role of $h(.)$ in (3.5.5) in the context of robustness and asymptotic efficiency of $T_n^\star$.

3.5.4. Work out the IC for $T_n^\star$ in (3.5.5), and verify that when $h(.)$ agrees with the true density $f(.)$, the IC agrees with that of the MLE of θ.

3.5.5. Examine the relation between $h(.)$ in (3.5.5) and $\rho(.)$ in (3.5.8) such that both estimators have a common IC.

3.6.1. For a symmetric kernel $\phi(.)$ of degree m (≥ 1), show that the U-statistic $U_n = \binom{n}{m}^{-1} \sum_{1 \leq < i_1 < \ldots < i_m \leq n} \phi(X_{i_1}, \ldots, X_{i_m})$ is a symmetric, unbiased, and optimal (nonparametric) estimator of $T(G)$. Define $T(F_n)$ as in (3.6.5), and verify that (a) for $m = 1$, $T(F_n) = U_n$, and (b) for $m \geq 2$, in general, $T(F_n) \neq U_n$, but $|T(F_n) - U_n| = O(n^{-1})$ a.s., as $n \to \infty$.

3.6.2. For $m = 2$, show that if $\mathbb{E}_G \phi(X_1, X_1) \neq \mathbb{E}_G \phi(X_1, X_2)$, then $T(F_n)$ is not unbiased for $T(G)$, although U_n is so. Extend this result for an arbitrary $m \geq 2$.

3.6.3. Use the completeness of sample order statistics (see Problem 3.3.1), and verify that U_n is minimum risk (nonparametric) estimator in the class of unbiased ones. Can you claim the same property for $T(F_n)$? How about asymptotically?

3.6.4. For the residuals $\hat{X}_i$, define the empirical d.f. $\hat{F}_n$ as in (3.6.22). Study the weak convergence of $\sqrt{n}(F_n - G)$ under appropriate model regularity conditions, and hence, or otherwise, verify (3.6.22).

3.6.5. In the case of models involving nuisance parameters, show that, in general, $\sqrt{n}(\hat{F}_n - F_n)$ does not converge in law to a degenerate random function, and this in turn induces additional complications in the expansion for $\sigma_G^{\star 2}$ in (3.6.22). Verify this with the location-scale model when $T(G)$ refers to the locations parameter.

3.6.6. For Z_s in (3.6.23)-(3.6.24), use the Hoeffding decomposition in (3.6.27), and verify that $\sqrt{n}||H_N - H|| = O_p(1)$ and the $\sqrt{n}(H_N - H)$ converges weakly to a Gaussian process.

3.6.7. For the studentized case $\hat{X}_i(t) = (X_i - t)/s_n$ where $s_n = \gamma + n^{-1/2}U_n + o_p(n^{-1/2})$ and $U_n \sim \mathcal{N}(0, \beta^2)$, show that for a skew-symmetric ψ, the results of Problem 3.6.5 hold.

3.6.8. Provide a formal proof of Lemma 3.6.1.

3.6.9. Provide a formal proof of Lemma 3.6.2.

3.6.10. For R-estimators of location, show that Hadamard differentiability holds when the score function is continuous and piecewise differentiable with a bounded and piecewise continuous derivative. [Fernholz 1983]

3.6.11. Provide a parallel result for M-estimators of location.

3.6.12. Provide a formal proof of (3.6.27) for U-statistics. Show that if the kernel is unbounded, then the U-statistic is not necessarily Hadamard differentiable.

Chapter 4

Asymptotic Representations for L-Estimators

4.1 INTRODUCTION

In the preceding two chapters we observed that robustification, by its very formulation, induces certain structural changes. The resulting robust estimators (or test statistics) are typically nonlinear functions of the sample observations. For this reason, the standard probabilistic and asymptotic tools presented in Section 2.5 are not capable of providing us with a unified approach toward the study of (asymptotic) distribution theory and related properties of general robust estimators and test statistics. Indeed, inspired by the effectiveness of the Hoeffding-Hájek projection theorems for nonlinear statistics (Section 2.5), we may be tempted to extend the projection technique to general robust statistics. But while the projection does yield a linear approximation of a nonlinear statistics, it can be difficult to find its explicit form and estimate the remainder term. We would be better off to accept another, perhaps less optimal linear approximation of a nonlinear statistic that is of reasonably simpler form and whose remainder term is negligible and can be handled. The past 25 years have witnessed significant developments along this line, and our emphasis will therefore be primarily on such asymptotic representations.

Let us recall the motivation on statistical functionals treated in Section 3.6 and take a closer at (3.6.9). The representation in (3.6.15) is a first step in this direction, while the simplification in (3.6.19) provides the access to the desired asymptotic normality of the estimator. If, however, we want to proceed beyond the asymptotic normality (e.g., the law of iterated logarithms, weak and strong invariance principles, etc.), we need to study the order of the remainder term in a relatively more refined manner. Such a study will demand extra regularity conditions on the functional. These regularity conditions on the functional will preclude some important members of the class of robust estimators. For example, R-estimators of location/regression based on unbounded score functions do not belong to such regular functionals' class, and the same criticism may be voiced for M-estimators. On the other hand, there might no longer be available some alternative formulations that do not require such stringent regularity conditions. For a special class of L-estimators, namely the sample quantiles, such linear representations were elegantly formulated by Bahadur (1966) under very mild regularity conditions, and are known as *Bahadur representations* for sample quantiles. The Bahadur representations were a precursor to the general asymptotic representations to be considered in this and subsequent chapters. As such, we motivate our presentation as follows: If $X_1, \ldots, X_n$ are i.i.d. r.v's with a d.f. F, defined on $\mathbb{R}_1$, and if for a given $p: 0 < p < 1$, $F^{-1}(p) = \xi$ is defined uniquely, then, denoting the sample order statistics by $X_{n:1} \leq \ldots \leq X_{n:n}$, we may consider the sample quantile $X_{n:k_n}$ as an estimator of ξ, when $n^{-1}k_n$ is close to p. Bahadur (1966) showed that if F is twice differentiable at ξ and the pdf $f(\xi)$ is positive, then

$$X_{n:k_n} - \xi = \{nf(\xi)\}^{-1} \sum_{i=1}^{n} \{p - I(X_i \leq \xi)\} + r_n, \tag{4.1.1}$$

where for any $k_n : |n^{-1}k_n - p| = o(n^{-1/2})$, we have

$$r_n = O\Big(n^{-3/4}(\log n)^{1/2}(\log\log n)^{1/4}\Big) \quad \text{a.s.,} \quad \text{as } n \to \infty. \tag{4.1.2}$$

The representation in (4.1.1) expresses $X_{n:k_n} - \xi$ as an average of n i.i.d. r.v.'s, up to an asymptotically negligible remainder term r_n; (4.1.1) and (4.1.2) imply the asymptotic normality of $n^{1/2}\{X_{n:k_n} - \xi\}$. As we will illustrate in the next two chapters, the utility of this linear representation goes far beyond the asymptotic normality. Kiefer (1967) obtained an exact order for $r_n = r_n(p)$ in (4.1.1) and derived the asymptotic (nondegenerate) distribution of $n^{3/4}r_n(p) : \forall x \in \mathbb{R}$, and every $p \in (0,1)$,

$$\begin{aligned}\lim_{n\to\infty} \mathbb{P}\Big\{n^{3/4}f(\xi)r_n(p) \leq x\Big\} \\ = 2\Big\{p(1-p)\Big\}^{-\frac{1}{2}} \int_0^\infty \Phi(t^{-1/2}x)\phi(t\{p(1-p)\}^{-1/2})dt,\end{aligned} \tag{4.1.3}$$

where Φ and ϕ stand for the d.f. and density function of a standard normal variable. Duttweiler (1973) studied the moment convergence results and showed that for any $\epsilon > 0$ as $n \to \infty$

$$\mathbb{E}\Big\{n^{3/4}f(\xi)r_n(p)\Big\}^2 = \Big\{2p(1-p)/\pi\Big\}^{1/2} + o(n^{-1/4+\epsilon}). \tag{4.1.4}$$

Representations similar to (4.1.1) hold for a general class of estimators and related statistics, and that will be systematically presented in this and the next two chapters. For a statistic $T_n = T(X_1, \ldots, X_n)$ based on n i.i.d. r.v.'s $X_1, \ldots, X_n$, having a d.f. $F_\theta(x)$, $\theta \in \Theta \subset \mathbb{R}_1$, we would like to show that in estimating θ, under suitable regularity conditions, there exists a function $\xi : \mathbb{R}_1 \times \Theta \mapsto \mathbb{R}_1$, such that

$$T_n - \theta = n^{-1} \sum_{i=1}^{n} \xi(X_i, \theta) + r_n(\theta), \text{ where } r_n(\theta) = o_p(n^{-1/2}). \tag{4.1.5}$$

We will term the representation of the type (4.1.5) as the *first-order* or *weak Bahadur-type representation.* Equation (4.1.5) can often be strengthened to

$$r_n(\theta) = o_p(n^{-1/2}), \text{ uniformly in } \theta \in K, \text{ for every compact } K \subseteq \Theta. \tag{4.1.6}$$

T_n satisfying (4.1.5) and (4.1.6), such that $n^{1/2}(T_n - \theta)$ is asymptotic normal, is termed the *CLUAN (consistent, linear, uniformly asymptotically normal)* estimator of θ (Bickel 1981). Thus, for a CLUAN estimator T_n, (4.1.5) holds, and

$$\mathbb{P}_\theta\left\{n^{1/2}|r_n(\theta)| > \epsilon\right\} \to 0 \quad \text{as } n \to \infty, \tag{4.1.7}$$

uniformly on every compact $K \subseteq \Theta$, and

$$\mathbb{E}_\theta\{\xi(X_1, \theta)\} = 0 \quad \text{and} \quad \mathbb{E}_\theta\{\xi(X_1, \theta)\}^2 < \infty, \ \theta \in \Theta; \tag{4.1.8}$$

$$n^{-1/2} \sum_{i=1}^{n} \xi(X_i, \theta) \xrightarrow{\mathcal{D}} \mathcal{N}\left(0, \mathbb{E}_\theta\{\xi(X_1, \theta)\}^2\right), \quad \theta \in \Theta. \tag{4.1.9}$$

Clearly, if (4.1.5) holds uniformly in θ over compact subspaces of Θ, then the CLUAN estimators can most conveniently be studied with the help of the scores $\xi(X_i, \theta)$, $i \geq 1$. Such representations have other useful applications too. For example, if we have representations of the type (4.1.5) for two sequences $\{T_n\}$ and $\{T_n^\star\}$ of estimators (for a common parameter θ), then not only can we easily derive the joint asymptotic distribution of $(T_n, T_n^\star)$ but also looking at the remainder terms, we can gather useful information on the *rate of stochastic equivalence* of $\{T_n\}$ and $\{T_n^\star\}$ (as n becomes large). This representation also enables us to study the quality of a *one-step version* $(T_n^\star)$ of T_n which is roughly of the form

$$T_n^\star = \hat{\theta}_n + n^{-1} \sum_{i=1}^{n} \hat{\xi}(X_i, \hat{\theta}_n), \tag{4.1.10}$$

where $\hat{\theta}_n$ is a consistent initial estimator of θ and $\hat{\xi}$ is either equal to ξ (when ξ does not involve any nuisance parameter) or an estimator of ξ based on the incorporation of suitable (consistent) estimators of the nuisance parameters.

In the present chapter, we wiall study systematically the Bahadur-type representations of L-estimators; the cases of M- and R-estimators will be taken up in the next two chapters. The higher is the order of r_n in (4.1.5), the more stringent are the regularity conditions on the weight functions. An interesting characteristic to observe is that the highest possible rate in (4.1.5) is $r_n = O_p(n^{-1})$, although for nonsmooth scores we may have to be satisfied with a rate of $n^{-3/4}$. Later in the chapter, we will find it convenient to deal separately with the two cases of nonsmooth and smooth weights and append a general result on a combined case.

4.2 BAHADUR'S REPRESENTATIONS FOR SAMPLE QUANTILES

For the preliminary notations, we refer to Section 3.3. For a sample quantile $X_{n:k_n}$ corresponding to the population quantile $\xi_p : F(\xi_p) = p \ (0 < p < 1)$, where $k_n = np + o(n^{1/2})$, Ghosh (1971) obtained the weaker version (4.1.5) for (4.1.1) under a weaker condition that F is once differentiable at ξ_p with $f(\xi_p) > 0$ (compare this with the original Bahadur 1966 a.s. representation where F was assumed to be twice differentiable at ξ_p). We find it convenient to present first the following result on sample quantiles where the proof is adapted to the sample d.f. F_n.

Theorem 4.2.1 *If F is differentiable at ξ_p with $f(\xi_p) > 0$ and $k_n = np + o(n^{1/2})$, then the representation in (4.1.1) holds with $r_n = o_p(n^{-1/2})$.*

Proof: Note that for a real t,

$$\mathbb{P}\Big\{X_{n:k_n} \le \xi_p - n^{-1/2}t\Big\} = \mathbb{P}\Big\{F_n(\xi_p - n^{-1/2}t) \ge n^{-1}k_n\Big\}$$

$$= \mathbb{P}\{n^{1/2}[F_n(\xi_p - n^{-1/2}t) - F(\xi_p - n^{-1/2}t)] \ge n^{1/2}[n^{-1}k_n - F(\xi_p - n^{-1/2}t)]\Big\}. \tag{4.2.1}$$

Since $n^{1/2}[n^{-1}k_n - F(\xi_p - n^{-1/2}t)] = tf(\xi_p) + o(1)$ while $n^{1/2}[F_n(y) - F(y)]$ is asymptotically normal with zero mean and variance $F(y)[1 - F(y)]$ ($\le 1/4$), we can choose t so large (but finite) that the right-hand side of (4.2.1) can be made arbitrarily small for all n sufficiently large. A similar result holds for the right hand side tail. Hence, we conclude that whenever $f(\xi_p)$ is positive,

$$n^{1/2}|X_{n:k_n} - \xi_p| = O(1), \quad \text{in probability.} \tag{4.2.2}$$

Next we note that if F were continuous a.e., $\{n^{1/2}[F_n(F^{-1}(t)) - t],\ t \in [0, 1]\}$ would have the weak convergence to a standard Brownian bridge on

$D[0,1]$. This weak convergence in turn ensures the tightness of the same sequence of empirical processes. By virtue of the assumed differentiability of F at ξ_p, we may consider a small neighborhood of ξ_p (where F is continuous), and using the weak convergence of the above process only on this small neighborhood, we obtain that as $n \to \infty$,

$$\sup\left\{n^{1/2}|F_n(x) - F(x) - F_n(\xi_p) + F(\xi_p)| : |x - \xi_p| \le \eta\right\} = o_p(1), \quad (4.2.3)$$

for every (small) $\eta > 0$. Since $F_n(X_{n:k_n}) = n^{-1}k_n = p + o(n^{-1/2})$, from (4.2.2) and (4.2.3), the desired result follows (by the first order Taylor expansion of $F(X_{n:k_n})$ around $F(\xi_p) = p$). □

The first-order asymptotic representation in Theorem 4.2.1 has been strengthened to a second-order representation by Bahadur (1966) under an additional differentiability condition on F. We have the following:

Theorem 4.2.2 *If F is twice differentiable at ξ_p with $f(\xi_p) > 0$ then (4.1.1) and (4.1.2) hold.*

Outline of the proof. First, in (4.2.1), we replace t by $t(\log n)^{1/2}$, and for $[F_n(y) - F(y)]$ (bounded r.v.) we use the Hoeffding inequality (see, Section 2.5). Then, parallel to (4.2.2), we have

$$|X_{n:k_n} - \xi_p| = O\left((n^{-1}\log n)^{1/2}\right) \quad \text{a.s.,} \quad \text{as } n \to \infty. \quad (4.2.4)$$

Next we consider the interval $[\xi_p - Cn^{-1/2}(\log n)^{1/2}, \xi_p + Cn^{-1/2}(\log n)^{1/2}]$, for some arbitrary C ($< \infty$), and divide this interval into $2b_n$ subintervals of width $Cn^{-1/2}(\log n)^{1/2}b_n^{-1}$ each. In this context we choose $b_n \sim n^{1/4}$. Note that for any $a \le x \le b$,

$$F_n(a) - F(b) \le F_n(x) - F(x) \le F_n(b) - F(a). \quad (4.2.5)$$

Hence $\sup\left\{|F_n(x) - F(x) - F_n(\xi_p) + F(\xi_p)| : |x - \xi_p| \le C(n^{-1}\log n)^{1/2}\right\}$ can be bounded by $\max\left\{|F_n(x_j) - F(x_j) - F_n(\xi_p) + F(\xi_p)| : j = 1, \dots, 2b_n\right\} + O(n^{-3/4}(\log n))$, where the x_j stand for the grid points $\xi_p - C(n^{-1}\log n)^{1/2} + jC(n^{-1}\log n)^{1/2}b_n^{-1}$, $j = 1, \dots, 2b_n$. For each j, we use the Hoeffding inequality to $\mathbb{P}\{|F_n(X_j) - F(X_j) - F_n(\xi_p) + F(\xi_p)| > Cn^{-3/4}\log n(\log\log n)^{1/4}\}$, $j = 1, 2, \dots, 2b_n$, and complete the proof by showing that the sum of these $2b_n$ probabilities is $o(n^{-m})$, for some $m > 1$. Hence, by the Borel-Cantelli lemma, we have

$$\sup\left\{n^{1/2}|F_n(x) - F(x) - F_n(\xi_p) + F(\xi_p)| : |x - \xi_p| \le (n^{-1}\log n)^{1/2}C\right\}$$
$$= O\left(n^{-3/4}(\log n)^{1/2}(\log\log n)^{1/4}\right) \ a.s., \ \text{as } n \to \infty. \quad (4.2.6)$$

The desired result follows from (4.2.4) and (4.2.6). □

We are also interested in an intermediate result where in (4.1.2) instead of an *a.s.* order of $n^{-3/4}(\log n)^{1/2}(\log\log n)^{1/4}$, we like to have an $O_p(n^{-3/4})$ Keeping (3.3.19) in mind, consider now a relatively more general form

$$L_{n2} = \sum_{j=1}^{k} a_j X_{n:[np_j]+1}, \tag{4.2.7}$$

where

$$0 < p_1 < \ldots < p_k < 1 \text{ and } k \text{ is a positive integer (fixed);} \tag{4.2.8}$$

$$a_1, \ldots, a_k \text{ are given real numbers.} \tag{4.2.9}$$

We wish to establish a representation of the form

$$L_{n2} - \mu = n^{-1} \sum_{i=1}^{n} \psi_2(X_i) + R_n, \tag{4.2.10}$$

supplemented by the stochastic order of R_n, with some μ depending on F and the a_j and $\psi_2 : \mathbb{R}_1 \mapsto \mathbb{R}_1$, such that $\int_{-\infty}^{\infty} \psi_2(x) dF(x) = 0$.

Theorem 4.2.3 *Suppose that F is twice differentiable at $F^{-1}(p_j)$ and $F'(F^{-1}(p_j)) > 0$, for each $j(= 1, \ldots, k)$. Then*

$$L_{n2} - \sum_{j=1}^{k} a_j F^{-1}(p_j) = n^{-1} \sum_{i=1}^{n} \psi_2(X_i) + r_n \tag{4.2.11}$$

where r_n satisfy (4.1.2) and

$$\psi_2(x) = \sum_{j=1}^{k} a_j \Big[f(F^{-1}(p_j))\Big]^{-1} \Big\{p_j - I(x \le F^{-1}(p_j))\Big\}, \; x \in \mathbb{R}_1. \tag{4.2.12}$$

Outline of the proof. For each j $(= 1, \ldots, k)$ we make use of (4.2.2). In (4.2.5) and (4.2.6) we therefore concentrate on the interval $[F^{-1}(p_j) - C(n^{-1}\log n)^{1/2}, F^{-1}(p_j) + C(n^{-1}\log n)^{1/2}]$, for some arbitrary C, and use b_n $(\sim n^{1/4})$ subintervals, as in the proof of Theorem 4.2.2. The rest of the proof follows by using the Hoeffding inequality again. The details are therefore left as an exercise (Problem 4.2.1). □

For the case of a single sample quantile, (4.1.3) depicts the asymptotic distribution of the remainder term in (4.2.11). We will consider a generalization of this result for M-estimators with discontinuous score functions in the next chapter. However, in the case of more than one quantiles, such an asymptotic distribution may not come in a very handy form. In fact it may depend heavily on k as well as the a_j. For the quantile process (related to the uniform distribution) Kiefer (1970) considered the distance between the empirical and quantile processes [i.e., $\sup\{|n^{3/4}(\log n)^{-1/2}f(F^{-1}(p))r_n(p)| : 0 < p < 1\}$] and demonstrated its close alliance with that of the Kolmogorov-Smirnov statistic. He actually showed that the limiting distribution of this distance measure is

$$1 + 2\sum_{k=1}^{\infty}(-1)^k e^{-2k^2t^4},\ t \geq 0. \tag{4.2.13}$$

For certain related results, we may refer to the impressive monograph of Csörgő and Révész (1981) (see also Csörgő and Horváth 1993). In the case of M-estimators with discontinuous score functions (as we will see in Chapter 5), we are able to find the asymptotic distribution of the remainder term (as shown in Chapter 5). However, the remainder term in (4.2.11) with arbitrary weights $a_1, \ldots, a_k$ may not have simple (the second-order) asymptotic distribution.

Remarks. For the case of a single quantile $X_{n:k_n}$, $k_n \sim np : 0 < p < 1$, the proof of Theorem 4.2.2 can be adopted directly (see Problem 4.2.2) to show that under the same regularity conditions, $r_n = \mathrm{O}_p(n^{-3/4})$. The same argument may be used for each of the $X_{n:[np_j]+1}$, $1 \leq j \leq k$, to show that under the hypothesis of Theorem 4.2.3 the remainder term is $\mathrm{O}_p(n^{-3/4})$ (see Problem 4.2.3). When we do not require the second derivative, we can at least claim the representation with $r_n = \mathrm{o}_p(n^{-1/2})$. On the other hand, as we will see later, for smooth functions we can obtain the order $r_n = \mathrm{O}_p(n^{-1})$, faster than for a combination of single quantiles. This is the idea behind "smoothing" sample quantiles; see (4.3.34)-(4.3.36) for more details. In applications we often have a statistic $T_n = h(X_{n:k_n})$ with a smooth function $h(.)$:

$$\frac{h(X_{n:k_n}) - h(\xi_p)}{X_{n:k_n} - \xi_p} \to h'(\xi_p), \text{ in probability}, \tag{4.2.14}$$

whenever $h'(.)$ exists at ξ_p, and $X_{n:k_n} \to \xi_p$, in probability. Thus, the FOADR result extends directly for T_n under this additional differentiability condition on $h(.)$. Suppose now that $h'(x)$ exists in a neighborhood of ξ_p, and it satisfies a Lipschitz condition of order α, for some $\alpha > 0$:

$$|h'(x) - h'(\xi_p)| \leq K|x - \xi_p|^\alpha \ \forall x : |x - \xi_p| \leq \delta, \tag{4.2.15}$$

where $0 < K < \infty$ and $\delta > 0$. Without loss of generality, we may set $0 < \alpha \leq 1$ and δ small. Then, whenever (4.2.15) holds for some $\alpha > 1/2$, Theorem 4.2.2 holds for T_n as well (see Problem 4.2.4). On the other hand, if (4.2.15) holds for some $\alpha : 0 < \alpha \leq 1/2$, then Theorem 4.2.2 holds for T_n but for the remainder term r_n, we will have (see Problem 4.2.5)

$$r_n = \mathrm{O}(n^{-(1+\alpha)/2}(\log n)^{1/2}(\log\log n)^{1/4+\alpha/2}) \text{ a.s.}, \tag{4.2.16}$$

as $n \to \infty$. Along the same lines let us consider the case of k (≥ 1) quantiles $X_{n:[np_j]+1}$, $j = 1, \ldots, k$ as treated in (4.2.7), but in a more general form:

$T_n = h(X_{n:[np_1]+1}, \ldots, X_{n:[np_k]+1})$ for some smooth $h : \mathbb{R}_k^> \to \mathbb{R}$, where $\mathbb{R}_k^> = \{\mathbf{x} \in \mathbb{R}_k : x_1 \leq \ldots \leq x_k\}$. We assume that $\boldsymbol{\xi} = (\xi_{p_1}, \ldots, \xi_{p_k})'$ lies in the interior of $\mathbb{R}_k^>$ and that $(\partial/\partial\mathbf{x})h(\mathbf{x}) = \mathbf{h}'_{\mathbf{x}}$ exists in a neighborhood of $\boldsymbol{\xi}$. Then Theorem 4.2.1 extends directly to T_n and (4.2.10) holds with $\mathbf{a} = (a_1, \ldots, a_k)' = \mathbf{h}'_{\boldsymbol{\xi}}$ (see Problem 4.2.6). Finally, if we consider a natural extension of (4.2.15) and assume that

$$||\mathbf{h}_{\mathbf{x}} - \mathbf{h}'_{\boldsymbol{\xi}}|| \leq K||\mathbf{x} - \boldsymbol{\xi}||^{\alpha}, \tag{4.2.17}$$

for some $\alpha : 0 < \alpha \leq 1$ and $0 < K < \infty$, whenever $||\mathbf{x} - \boldsymbol{\xi}|| \leq \delta$ for some $\delta > 0$, then Theorem 4.2.2 holds for T_n with R_n in (4.2.10) being (4.1.2) or (4.2.16) according as α is $> 1/2$ or $0 < \alpha \leq 1/2$. Again, we relegate the proof to an exercise (Problem 4.2.7). We may conclude that the degree of smoothness of $h(.)$, as reflected by the order of convergence of the *gradient* of $h(.)$, dictates the rate of convergence of the remainder term as well as its mode of convergence. This remainder term is not generally $O_p(n^{-1})$. In Section 4.3 we will derive such better SOADR results under alternative smoothness conditions.

4.3 REPRESENTATIONS FOR SMOOTH SCORES L-STATISTICS

Let us consider a linear combination of order statistics of the form

$$L_{n1} = \sum_{i=1}^{n} c_{ni} X_{n:i}, \tag{4.3.1}$$

where the scores c_{ni} are generated by a smooth function $J : (0,1) \mapsto \mathbb{R}_1$ in either of the following two ways:

$$c_{ni} = n^{-1} J(i/(n+1)), \quad i = 1, \ldots, n; \tag{4.3.2}$$

$$c_{ni} = \int_{(i-1)/n}^{i/n} J(u)du, \quad i = 1, \ldots, n. \tag{4.3.3}$$

For the genesis of L_{n1}, we may again refer to Section 3.3. Our goal is to consider a representation for L_{n1} in the form of (4.2.10) with specific order for R_n. In fact we will see that $R_n = O_p(n^{-r})$ for some $r \in (\frac{1}{2}, 1]$ and that the higher the value of r that we achieve, the greater is the stringency of the smoothness conditions on J.

The representation for L_{ni} will be considered under the following smoothness conditions on J and F:

A. J is continuous on (0,1) up to a finite number of points $s_1, \ldots, s_m$ where m is a nonnegative integer and $0 < s_1 < \ldots < s_m < 1$.

B. F is continuous *a.e.*, and $F^{-1}(s) = \inf\{x : F(x) \geq s\}$ satisfies the Lipschitz condition of order 1 in a neighborhood of $s_1, \ldots, s_m$.

Moreover either of the following is assumed to hold:

C1. *Trimmed J.* $J(u) = 0$ for $u \in [0, \alpha) \cup (1 - \alpha, 1]$ for some positive α, such that $\alpha < s_1$, $s_m < 1 - \alpha$, and J satisfies the Lipschitz condition of order ν (≤ 1) in $(\alpha, s_1), (s_1, s_2), \ldots, (s_m, 1 - \alpha)$. Moreover, for some $\tau > 0$, $F^{-1}(1 - \alpha + \tau) - F^{-1}(\alpha - \tau) < \infty$.

C2. *Untrimmed J.* For some $\beta > 0$, $\sup\{|x|^\beta F(x)[1 - F(x)] : x \in \mathbb{R}_1\} < \infty$, and J satisfies the Lipschitz condition of order $\nu \geq (2/\beta) + \triangle - 1$ (for some $\triangle \in (0,1)$) in each of the intervals $(0, s_1), (s_1, s_2), \ldots, (s_m, 1)$.

A variation of **C2** will also be considered later. For the moment we have the following:

Theorem 4.3.1 *Define L_{n1} with the scores c_{ni} in (4.3.3). Then, under* **A**, **B**, *and* **C1** *or* **C2**,

$$L_{n1} - \mu = n^{-1} \sum_{i=1}^{n} \psi_1(X_i) + R_n \text{ with } R_n = O_p(n^{-r}), \tag{4.3.4}$$

where

$$r = ((\nu + 1)/2) \wedge 1, \tag{4.3.5}$$

$$\mu = \mu(J, F) = \int_0^1 J(u) F^{-1}(u) du, \tag{4.3.6}$$

$$\psi_1(x) = -\int_{\mathbb{R}_1} \{I(y \geq x) - F(y)\} J(F(y)) dy, \ x \in \mathbb{R}_1. \tag{4.3.7}$$

Proof. Let $U_i = F(X_i)$ and $U_{n:i} = F(X_{n:i})$, $i = 1, \ldots, n$. Also let

$$F_n(x) = n^{-1} \sum_{i=1}^{n} I(X_i \leq x), \ x \in \mathbb{R}_1,$$

$$U_n(t) = n^{-1} \sum_{i=1}^{n} I(U_i \leq t), \ t \in [0, 1]. \tag{4.3.8}$$

Then, the function

$$\phi(s) = -\int_0^1 \{I(u \geq s) - u\} J(u) du, \ s \in [0, 1], \tag{4.3.9}$$

is bounded, absolutely continuous with $\phi'(s) = J(s)$, $s \in (0,1)$. By virtue of (4.3.3), (4.3.6), and (4.3.9), we have

$$\begin{aligned} L_{n1} - \mu &= \int_{\boldsymbol{R}_1} x d[\phi(F_n(x)) - \phi(F(x))] \\ &= -\int_{\boldsymbol{R}_1} [\phi(F_n(x)) - \phi(F(x))] dx \\ &= -\int_0^1 [\phi(U_n(s)) - \phi(s)] dF^{-1}(s). \end{aligned} \tag{4.3.10}$$

On the other hand, by (4.3.7) and (4.3.9)

$$\begin{aligned} n^{-1}\sum_{i=1}^{n} \psi_1(X_i) &= \int_{\boldsymbol{R}_1} \psi_1(x) d[(F_n(x) - F(x)] \\ &= -\int_{\boldsymbol{R}_1} [F_n(x) - F(x)]\phi'(F(x)) dx \\ &= -\int_0^1 \{U_n(s) - s\}\phi'(s) dF^{-1}(s). \end{aligned} \tag{4.3.11}$$

Hence,

$$R_n = L_{n1} - \mu - n^{-1}\sum_{i=1}^{n} \psi_1(X_i) = \int_0^1 V_n(s) dF^{-1}(s), \tag{4.3.12}$$

where

$$V_n(s) = \phi'(s)[U_n(s) - s] - [\phi(U_n(s)) - \phi(s)], \; s \in (0,1). \tag{4.3.13}$$

Let us then fix an η (> 0), where

$$0 < \eta < \triangle/(2(\gamma+1)) \text{ under } \mathbf{C2}, \text{ and } 0 < \eta < \frac{1}{2}, \text{ under } \mathbf{C1}. \tag{4.3.14}$$

Recall that for each n, $\{(U_n(s) - s)/(1-s);\; s \in (0,1)\}$ is a martingale, (see Section 2.5), so that using the Hájek-Rényi-Chow inequality (for submartingales), we readily obtain

$$D_n = \sup\left\{ n^{\frac{1}{2}} |U_n(s) - s| / \{s(1-s)\}^{\frac{1}{2}-\eta} :\; 0 < s < 1 \right\} = \mathrm{O}_p(1). \tag{4.3.15}$$

[For an alternative derivation of (4.3.15), we may refer to (3.48) of Csáki (1984)]. Thus, given $\epsilon > 0$, there exist finite C (> 0) and n_0 such that

$$\mathbb{P}\{D_n > C\} \le \epsilon/2, \text{ for every } n \ge n_0. \tag{4.3.16}$$

Without any loss of generality we may simplify the presentation by assuming that there is a single point s_0 $(0 < s_0 < 1)$ of discontinuity of J.

Then, using (4.3.12), we may decompose, for example,

$$\begin{aligned} R_n &= \int_0^{s_0-\delta_n} V_n(s)dF^{-1}(s) + \int_{s_0-\delta_n}^{s_0+\delta_n} V_n(s)dF^{-1}(s) \\ &\quad + \int_{s_0+\delta_n}^{1} V_n(s)dF^{-1}(s) \\ &= R_{n1} + R_{n2} + R_{n3}, \end{aligned} \tag{4.3.17}$$

where we take

$$\delta_n = 2Cn^{-\frac{1}{2}} \text{ with } C \text{ defined in (4.3.16).} \tag{4.3.18}$$

We will study the order of each component R_{nj}, $j = 1, 2, 3$ separately. First, we may note that by virtue of (4.3.16),

$$\begin{aligned} &\mathbb{P}\Big\{|R_{n1}| \geq Kn^{-r}\Big\} \\ &= \mathbb{P}\Big\{|R_{n1}| \geq Kn^{-r},\ D_n \leq C\Big\} + \mathbb{P}\Big\{|R_{n1}| \geq Kn^{-r},\ D_n > C\Big\} \\ &\leq \mathbb{P}\Big\{|R_{n1}| \geq Kn^{-r},\ D_n \leq C\Big\} + (\epsilon/2)\ \forall n \geq n_0. \end{aligned} \tag{4.3.19}$$

Since, under **C1**, $J(u) = 0$, for $u \in (0, \alpha)$ and $s_0 > \alpha$, the first term on the right-hand side of (4.3.19) is bounded from above by

$$\begin{aligned} &\mathbb{P}\Big\{\int_{s_0-\delta_n}^{s_0+\delta_n} \Big(I[U_n(s) > s] \int_0^{U_n(s)-s} |J(u+s) - J(s)|du \\ &+ \ I[U_n(s) \leq s] \int_{U_n(s)-s}^{0} |J(u+s) - J(s)|du\Big)dF^{-1}(s) \geq \frac{K}{n^r}, D_n \leq C\Big\} \\ &\leq \ \mathbb{P}\Big\{MC^2n^{-r}(F^{-1}(s_0 - \delta_n) - F^{-1}(\alpha - \tau)) \geq Kn^{-r}\Big\} \\ &\leq \ \epsilon/2, \end{aligned} \tag{4.3.20}$$

for sufficiently large K (> 0) and all $n \geq n_1$ $(\geq n_0)$; here M is a positive constant. Hence, under **A**, **B**, and **C1**, $R_{n1} = O_p(n^{-r})$. Let us next consider the case where trimmed J is replaced by umtrimmed one (i.e., **C1** by **C2**). Notice that under **C2**,

$$|F^{-1}(s)| \leq M_1[s(1-s)]^{-1/\beta},\ 0 < s < 1,\ 0 < M_1 < \infty, \tag{4.3.21}$$

and hence, using the definition in (4.3.14), we can bound the first term on the right hand side of (4.3.19) by

$$\begin{aligned} &\mathbb{P}\Big\{M \int_0^1 |U_n(s) - s|^{1+\nu} dF^{-1}(s) \geq Kn^{-r},\ D_n \leq C\Big\} \\ &\leq \ \mathbb{P}\Big\{M^{1+\nu} n^{-(1+\nu)/2} \int_0^1 \{s(1-s)\}^{-1+\Delta/2} ds \geq Kn^{-r}\Big\} \\ &\leq \ \epsilon/2, \end{aligned} \tag{4.3.22}$$

for sufficiently large K (> 0) and for $n \geq n_2$ ($\geq n_0$). This implies that under **A**, **B**, and **C2**, $R_{n1} = \mathrm{O}_p(n^{-r})$. The same treatment holds for R_{n3}. Thus, it remains only to prove that $R_{n2} = \mathrm{O}_p(n^{-r})$. Toward this end, we may note that for $s \in (s_0 - \delta_n, s_0 + \delta_n)$, both $J(s)$ and $F^{-1}(s)$ are bounded, and hence,

$$
\begin{aligned}
& \mathbb{P}\Big\{|R_{n2}| \geq Kn^{-r}\Big\} \\
\leq \; & \mathbb{P}\Big\{|R_{n2}| \geq Kn^{-r},\, D_n \leq C\Big\} + \mathbb{P}\Big\{D_n > C\Big\} \\
\leq \; & \mathbb{P}\Big\{\sup\{|V_n(s)| :\; s_0 - \delta_n \leq s \leq s_0 + \delta_n\} \\
& [F^{-1}(s_0 + \delta_n) - F^{-1}(s_0 - \delta_n)] \geq Kn^{-r}\Big\} + \epsilon/2 \\
\leq \; & \mathbb{P}\Big\{MCn^{-1/2}[F^{-1}(s_0 + \delta_n) - F^{-1}(s_0 - \delta_n)] \geq Kn^{-r}\Big\} + \epsilon/2 \\
\leq \; & \epsilon
\end{aligned}
\tag{4.3.23}
$$

for sufficiently large K (> 0) and $n \geq n_3$ (≥ 0). This completes the proof of the theorem. □

Let us now consider the case where the scores are defined by (4.3.2). When J is trimmed (see **C1**), the results in Theorem 4.3.1 hold for the scores in (4.3.2) as well. If, however, J is not trimmed, then for the scores in (4.3.2), a representation of the type (4.3.4) with $r = 1$ may demand more stringent conditions on J as well as F. Toward this end, we may have to assume that for some positive a,

$$
\sup\Big\{|x|^{1+a}F(x)[1 - F(x)] :\; x \in \mathbb{R}_1\Big\} < \infty. \tag{4.3.24}
$$

Note that [e.g., Sen(1959)] (4.3.24) implies that $\mathbb{E}|X|^b < \infty$, for some $b \geq 1$, so that (4.3.24) entails the existence of the first order moment of X. In fact, (4.3.24) is equivalent to the existence of $\mathbb{E}|X|^c$ for some $c > 1$. As a variant of **C2**, we then formulate the following:

C3. J satisfies the Lipschitz condition of order 1 in $(0, s_1), \ldots, (s_k, 1)$ and (4.3.24) holds for some $a > 0$.

Theorem 4.3.2 *Assume that either* **A**, **B**, *and* **C1** *hold with* $\nu = 1$, *or* **A**, **B**, **C3** *hold. Then, for the scores defined by (4.3.2), the representation in (4.3.4) holds with* $r = 1$ *for* R_n.

Proof. Let us denote L_{n1} by $L^{\star}_{n1}$ when the scores are given (4.3.2), and let

$$
J_n(s) = J(i/(n+1)) \quad \text{for } (i-1)/n < s \leq i/n,\; i = 1, \ldots, n, \tag{4.3.25}
$$

$$
\mu_n = \int_0^1 J_n(s)dF^{-1}(s), \tag{4.3.26}
$$

$$
\phi_n(s) = -\int_0^1 \{I(u \geq s) - u\}J_n(u)du, \;\; s \in (0,1). \tag{4.3.27}
$$

Then, proceeding as in (4.3.10), we obtain that for L_{n1} defined by the scores in (4.3.3),

$$\begin{aligned}
&|L^\star_{n1} - \mu_n - L_{n1} + \mu| \\
&= \left| \int_0^1 [\phi_n(U_n(s)) - \phi_n(s) - \phi(U_n(s)) + \phi(s)] dF^{-1}(s) \right| \\
&\le \int_0^1 \Big\{ I(U_n(s) \ge s) \int_0^{U_n(s)-s} |J_n(s+t) - J(s+t)| dt \\
&+ I(U_n(s) < s) \int_{U_n(s)-s}^0 |J_n(s+t) - J(s+t)| dt \Big\} dF^{-1}(s). \qquad (4.3.28)
\end{aligned}$$

Next, note that by (4.3.25) and the assumed first-order Lipschitz condition on J,

$$|J_n(s) - J(s)| \le Mn^{-1}, \text{ for every } s \in (0,1),\ 0 < M < \infty. \qquad (4.3.29)$$

Further, by (4.3.15) and (4.3.16), we have $|U_n(s) - s| = O_p(n^{-1/2})$, uniformly in $s \in (0,1)$. As such, in the trimmed case, e.g., **C1**, the right-hand side of (4.3.28) is $O_p(n^{-1})$. Under **C3**, the right-hand side of (4.3.28) is bounded by

$$\begin{aligned}
&2Mn^{-1} \int_0^1 |U_n(s) - s| dF^{-1}(s) \\
&\le 2Mn^{-1} \int_{\boldsymbol{R}_1} |x| d[F_n(x) - F(x)] \\
&= O_p(n^{-1}), \qquad (4.3.30)
\end{aligned}$$

where the last step follows by using the Khintchine law of large numbers (under **C3**). Thus, under the hypothesis of the theorem, (4.3.28) is $O_p(n^{-1})$. On the other hand, for J being Lipschitz of order one (under **C1** or in **C2** allowing $\nu = 1$), we see that in (4.3.5) r reduces to 1, so (4.3.4) holds for L_{n1} with $R_n = O_p(n^{-1})$. Hence we obtain

$$\begin{aligned}
L^\star_{n1} &= L_{n1} - \mu + \mu_n + O_p(n^{-1}) \\
&= (\mu_n - \mu) + n^{-1} \sum_{i=1}^n \psi_1(X_i) + O_p(n^{-1}). \qquad (4.3.31)
\end{aligned}$$

Finally, by (4.3.29) and **C1** with $\nu = 1$ or **C3**, we obtain under **C3**,

$$|\mu_n - \mu| \le Mn^{-1} \int_0^1 |F^{-1}(u)| du = Mn^{-1} \mathbb{E}|X| = O(n^{-1}). \qquad (4.3.32a)$$

Under **C1**, the left-hand side is

$$\le Mn^{-1} \int_\alpha^{1-\alpha} |F^{-1}(u)| du = O(n^{-1}). \qquad (4.3.32b)$$

Hence the proof of the theorem is complete. □

Remark. If X_i's are i.i.d. r.v.'s with the d.f. $F_\theta(x) = F(x-\theta)$, $x \in \mathbb{R}_1$, such that

1. $F(x) + F(-x) = 1 \forall x \in \mathbb{R}_1$ (i.e., F symmetric about 0),
2. $J(u) = J(1-u)$ $\forall u \in (0,1)$, and $\int_0^1 J(u)du = 1$,

then under the hypothesis of Theorem 4.3.1,

$$L_{n1} - \theta = n^{-1} \sum_{i=1}^{n} \psi_1(X_i - \theta) + \mathrm{O}_p(n^{-r}), \tag{4.3.33}$$

where r and ψ_1 are defined by (4.3.5) and (4.3.7), and under the hypothesis of Theorem 4.3.2, we have

$$L_{n1}^\star(\text{or } L_{n1}) = \theta + n^{-1} \sum_{i=1}^{n} \psi_1(X_i - \theta) + \mathrm{O}_p(n^{-1}). \tag{4.3.34}$$

This provides a general representation for L-estimators of location parameters when the score function is smooth. As notable examples, we may consider the following:

1. *Sample mean* $\bar{X}_n = n^{-1}\sum_{i=1}^n X_i$. This corresponds to (4.3.2) or (4.3.3) with $J(u) = 1$, for every $u \in (0,1)$.

2. *Trimmed mean* $T_n = (n-2k)^{-1}\sum_{j=k+1}^{n-k} X_{n:j}$ where k is a positive integer with $k/n \to \alpha$, for some $\alpha \in (0, \frac{1}{2})$. Here (C1) holds with $J(u) = (1-2\alpha)^{-1}$, $\alpha \le u \le 1-\alpha$, and $J(u) = 0$ for $u \in (0,\alpha) \cup (1-\alpha, 1)$.

3. *Rank-weighted mean* (Sen 1964)

$$T_{n,k} = \binom{n}{2k+1}^{-1} \sum_{i=k+1}^{n-k} \binom{i}{k}\binom{n-i-1}{k} X_{n:i+1}, \quad \text{for } k \in (0, n/2).$$

Note that for $T_{n,0} = \bar{X}_n$, while for $k = [(n+1)/2]$, it reduces to the sample median. For any fixed k (≥ 1), although $\binom{n}{2k+1}^{-1}\binom{i}{k}\binom{n-i-1}{k}$ $(i = 1, \ldots, n,)$ may not correspond to (4.3.2) or (4.3.3), they may be approximated by a smooth J (up to the order n^{-1}) where $J(u) = \{u(1-u)\}^k$, $u \in (0,1)$.

4. Harrell and Davis (1982) *quantile estimator* defined by

$$T_n = \sum_{i=1}^{n} c_{ni} X_{n:i} \tag{4.3.35}$$

$$c_{ni} = \frac{\Gamma(n+1)}{\Gamma(k)\Gamma(n-k+1)} \int_{(i-1)/n}^{i/n} u^{k-1}(1-u)^{n-k} du, \quad i = 1, \ldots, n, \tag{4.3.36}$$

where $k \sim np$, and p $(0 < p < 1)$ relates to the order of the quantile to be estimated. An alternative estimator, due to Kaigh and Lachenbruch (1982), also belongs to this class (with more affinity to the rank weighted mean). For (4.3.36) the scores in (4.3.3) are relevant with a bounded and continuous J.

5. *BLUE estimator* of location (Jung 1955,1962; Blom 1956). For a d.f. $F_\theta(x) = F(x-\theta)$, $x \in \mathbb{R}_1$, when F admits an absolutely continuous density function f (with derivative f'), an asymptotically best unbiased linear estimator of the location parameter θ is given by L_{n1} with the scores defined by (4.3.2) and where

$$J(F(x)) = \gamma'(x) \quad \text{and} \quad \gamma(x) = -f'(x)/f(x),\ x \in \mathbb{R}_1. \tag{4.3.37}$$

For a large class of location-scale family of distributions, (4.3.37) leads to bounded J which satisfies **C1**, **C2** or **C3**, and **A**, **B**, and hence, the theorems in this section pertain to these BLUE estimators of locations. A similar case holds for the BLUE of the scale parameters.

4.4 ASYMPTOTIC REPRESENTATIONS FOR GENERAL L-ESTIMATORS

Let us now consider the general case of an L-estimator of the form

$$L_n = L_{n1} + L_{n2} \text{ (or } L_{n1}^\star + L_{n2}) \tag{4.4.1}$$

where L_{n2} is defined by (4.2.7) and L_{n1} or $L_{n1}^\star$ by (4.3.1)-(4.3.3). A very notable example of this type is the *Winsorized mean* defined by

$$W_n = n^{-1}\Big\{\sum_{i=k+1}^{n-k} X_{n:i} + k(X_{n:k} + X_{n:n-k+1})\Big\}, \tag{4.4.2}$$

for some positive integer k $(\leq n/2)$. As in the case of the trimmed mean, we may let $k/n \to \alpha$; $0 < \alpha < 1/2$. For $L_{n1} = n^{-1}\sum_{i=k+1}^{n-k} X_{n:i}$, we have the scores defined by (4.3.2) or (4.3.3) with $J(u) = 1$, for $u \in (\alpha, 1-\alpha)$ and 0, otherwise. L_{n2} corresponds to the midrange $kn^{-1}(X_{n:k} + X_{n:n-k+1})$. Thus Theorem 4.2.3 applies to L_{n2} and Theorem 4.3.2 to L_{n1}. However, a closer look at these two theorems reveals that the remainder term for L_n $(= L_{n1} + L_{n2})$ can at best be of the order $n^{-3/4}$. This observation provides a key to the asymptotic representation for L_n in the general setup in (4.4.1). Keeping in mind the representations in Theorems 4.2.3, 4.3.1, and 4.3.2, we define

$$\psi(x) = \psi_1(x) + \psi_2(x),\ x \in \mathbb{R}_1, \tag{4.4.3}$$

where ψ_1 and ψ_2 are defined by (4.3.7) and (4.2.12), respectively. Then we have the following theorem [where

$$\mu^0 = \sum_{j=1}^{k} a_j F^{-1}(p_j) + \int_0^1 J(u)du \tag{4.4.4}$$

and all the notations are borrowed from (4.2.11) and (4.3.6)]:

Theorem 4.4.1 *Under the hypotheses of Theorem 4.2.3 and 4.3.1,*

$$L_n - \mu^0 = n^{-1}\sum_{i=1}^{n} \psi(X_i) + \mathrm{O}_p(n^{-r}); \;\; r = (\frac{\gamma+1}{2}) \wedge (\frac{3}{4}). \tag{4.4.5}$$

Under the hypothesis of Theorem 4.2.3, (4.4.5) holds with an order of the remainder term as $\mathrm{O}_p(n^{-3/4})$.

The proof is direct adaptation of the relevant theorems under the hypotheses and hence is omitted.

It is clear that as far as the stochastic order of the remainder terms in the asymptotic representation of an L-estimator is concerned, the use of discrete quantiles (i.e., L_{n2}) leads to a slower order, even when the smooth part (i.e., L_{n1}) may have an order $\mathrm{O}_p(n^{-1})$ for the remainder term. If we are interested in the study of the second-order asymptotic distributional representation (SOADR) of L-estimators, the choice of the normalizing factor will depend on the component L_{n2}. Only for smooth scores with some further regularity conditions, we are able to use the desired normalizing factor n. Then it may be of interest to specify the asymptotic distribution of $n[L_{n1}-\mu-n^{-1}\sum_{i=1}^n \psi_1(X_i)]$. This study is greatly facilitated by the use of statistical functionals that are "differentiable" in a certain sense, and we will consider this important topic in the next section. Generally, for L-estimators of location, smooth and bounded scores lead to the satisfaction of such differentiability conditions.

4.5 REPRESENTATIONS FOR STATISTICAL FUNCTIONALS

In Section 3.6 we have introduced a class of statistical functionals and discussed their relevance to other conventional robust estimation procedures. Particularly, the L-estimators with smooth scores have close affinity to such functions, and hence we find it convenient to consider here some asymptotic (Bahadur-type) representations for such functionals and to incorporate these results in the study of parallel results for L-estimators.

Let $X_1, \ldots, X_n$ be i.i.d. r.v.'s with the d.f. F and let F_n be the sample d.f. Then for a parameter $\theta = T(F)$, expressible as a functional of the d.f.

F, a natural estimator is $T_n = T(F_n)$. In (3.6.15) through (3.6.21) we have considered certain differentiability aspects of $T(.)$ leading to the asymptotic normality result in (3.6.20). In the current section we consider more general results in this direction. We refer to (3.6.16) and (3.6.17) for the definition of (first-order) compact or Hadamard differentiability of $T(.)$ at F. As such, if $T(.)$ is first-order Hadamard differentiable at F, we have by virtue of (3.6.15), (3.6.16), and (3.6.17),

$$\begin{aligned} T_n &= T(F_n) + \int T^{(1)}(F;x)d[F_n(x) - F(x)] + R_n(F_n, F) \\ &= T(F) + n^{-1}\sum_{i=1}^{n} T^{(1)}(F;X_i) + \mathrm{o}(\|F_n - F\|), \end{aligned} \tag{4.5.1}$$

where $T^{(1)}(F;x)$ is the compact derivative of $T(.)$ at F (also is more popularly known as the *influence function*). Note that by construction

$$\mathbb{E}T^{(1)}(F;X) = \int T^{(1)}(F;x)dF(x) = 0. \tag{4.5.2}$$

The second moment σ_F^2 of $T^{(1)}(F;X_i)$ is defined by (3.6.20), and we assume that it is finite and positive. Recall that for every real t (≥ 0),

$$\mathbb{P}\Big\{\|F_n - F\| > tn^{-\frac{1}{2}}\Big\} \leq 2\exp\{-2t^2\} \quad \text{for every } n \geq 1, \tag{4.5.3}$$

so that

$$\|F_n - F\| = \mathrm{O}_p(n^{-\frac{1}{2}}). \tag{4.5.4}$$

From (4.5.1) and (4.5.4), we conclude that for all first-order Hadamard differentiable $T(.)$, the first-order asymptotic representation in (4.1.5) holds with $\xi(X_i, \theta) = T^{(1)}(F;X_i)$. Note that θ is itself a functional of F, and hence, in this representation, we have a more general formulation of the influence function scores. In the same spirit as in Sections 4.2 and 4.3, we will study more refined orders for the remainder term in (4.5.1) (under additional regularity conditions on the functionals). For this reason we introduce the following.

Let V and W be topological vector spaces, let $L_1(V,W)$ be the set of continuous linear transformations from V to W, and let

$$L_2(V,W) = \{f : f \in \mathcal{C}(V,W),\ f(tH) = t^2 f(H)\ \forall H \in V,\ t \in \mathbb{R}_1\},$$

where $\mathcal{C}(V,W)$ is the set of continuous functionals from V to W. Let $\mathcal{A}$ be an open set of V. A functional $T : \mathcal{A} \to W$ is second-order Hadamard (or compact) differentiable at $F \in \mathcal{A}$, if there exist $T_F{}' \in L_1(V,W)$ and $T_F{}'' \in L_2(V,W)$ such that for any compact set Γ of V,

$$\lim_{t\to 0} t^{-2}\{T(F + tH) - T(F) - T_F'(tH) - \frac{1}{2}T_F''(tH)\} = 0, \tag{4.5.5}$$

uniformly for $H \in \Gamma$. Then T'_F and T''_F are called the *first-* and *second-order Hadamard* (or *compact) derivative* of T at F, respectively.

We denote the remainder term of the second order expansion in (4.5.5) as $Rem_2(tH)$, so that $Rem_2(tH) = T(F+tH) - T(F) - T'_F(tH) - \frac{1}{2}T''_F(tH)$. Then (4.5.5) may also be represented as

$$\lim_{t \to 0} t^{-2} Rem_2(F + tH) = 0, \tag{4.5.6}$$

for any sequence $\{H_n\}$ with $H_n \to H \in V$. This definition of the second-order Hadamard-differentiability is adapted from Ren and Sen (1995b) and is consistent with the one given by Sen (1988a). Further note that $T'_F(\delta_x - F) = IC(x; F, T) = \frac{d}{dt}T(F + t(\delta_x - F))|_{t=0}$, and $T''_F(\delta_x - F) = \frac{d^2}{dt^2}T(F + t(\delta_x - F))|_{t=0}$, where δ_x is the d.f. with the unit point mass at x, and $IC(.)$ is the influence function, introduced in Chapter 3. Further the existence of the second-order Hadamard derivative implies the existence of the first-order one. Finally, we may always normalize T'_F and T''_F in such a way that (4.5.2) holds, with

$$T^{(2)}(F; x, y) = T^{(2)}(F; y, x) \quad \text{a.e.}(F) \tag{4.5.7}$$

$$\int T^{(2)}(F; x, y) dF(y) = 0 = \int T^{(2)}(F; y, x) dF(x) \quad \text{a.e.}(F) \tag{4.5.8}$$

Using (4.5.5)-(4.5.8), we obtain that for a second-order Hadamard differentiable functional $T(.)$ (at F),

$$\begin{aligned} T_n = T(F) \quad &+ \quad n^{-1}\sum_{i=1}^{n} T^{(1)}(F; X_i) + (2n^2)^{-1}\sum_{i=1}^{n}\sum_{j=1}^{n} T^{(2)}(F; X_i, X_j) \\ &+ \quad o(n^{-1}), \ \text{ in probability.} \end{aligned} \tag{4.5.9}$$

This yields a second order asymptotic expansion of $T(F_n)$. The third term on the right-hand side of (4.5.9) corresponds to the leading term in the remainder term for the first-order expansion for T_n, and hence the asymptotic distribution of the normalized version of this term yields the desired result.

Let us denote by

$$\begin{aligned} R_n^{\star\star} &= n^{-2}\sum_{i=1}^{n}\sum_{j=1}^{n} T^{(2)}(F; X_i, X_j) \\ &= \int\int T^{(2)}(F; x, y) dF_n(x) dF_n(y). \end{aligned} \tag{4.5.10}$$

Then note that $R_n^{\star\star}$ is a von Mises functional of degree 2. However, in this case, by virtue of (4.5.8), the underlying parameter is stationary of order 1 (in the sense of Hoeffding 1948), so $T_1^{(2)}(F; x) = \mathbb{E}T^{(2)}(F; x, X_i) = 0$ a.e. (F). As such, the asymptotic distribution of $R_n^{\star\star}$ is not generally normal, although the asymptotic distribution theory of Hoeffding's U-statistics in the degenerate case may be used to derive the form of this asymptotic distribution.

Let us denote by

$$\bar{T}_n^{(2)} = n^{-1} \sum_{i=1}^{n} T^{(2)}(F; X_i, X_i), \tag{4.5.11}$$

$$U_n^{(2)} = \binom{n}{2}^{-1} \sum_{\{1 \le i < j \le n\}} T^{(2)}(F; X_i, X_j), \tag{4.5.12}$$

$$\tau^{(2)}(F) = \int T^{(2)}(F; x, x) dF(x) \tag{4.5.13}$$

and assume that $\tau^{(2)}(F)$ is well defined at F (and is finite). Moreover we can assume that

$$\mathbb{E}_F \left[\{T^{(2)}(F; X_1, X_2)\}^2 \right] < \infty. \tag{4.5.14}$$

Then we note that

$$nR_n^{\star\star} = \bar{T}_n^{(2)} + nU_n^{(2)}. \tag{4.5.15}$$

Now, using the Khintchine strong law of large numbers, we obtain from (4.5.11) and (4.5.13) that

$$\bar{T}_n^{(2)} \to \tau^{(2)}(F) \quad \text{a.s., as} \quad n \to \infty. \tag{4.5.16}$$

From (4.5.15) and (4.5.16), we conclude that

$$nR_n^{\star\star} - \tau^{(2)}(F) \text{ has the same asymptotic distribution as } nU_n^{(2)}, \tag{4.5.17}$$

if they have any at all. Recall that by (4.5.8) and (4.5.14),

$$\mathbb{E}_F \left[T^{(2)}(F; X_1, X_2) T^{(2)}(F; X_1, X_3) \right] = 0, \tag{4.5.18}$$

so the U-statistic $U_n^{(2)}$ corresponds to a parameter that is stationary of order 1. In this case the asymptotic distribution of $nU_n^{(2)}$ is nondegenerate, and hence, by (4.5.17), $nR_n^{\star\star} - \tau^{(2)}(F)$ has a nondegenerate distribution (in the asymptotic case). Thus we find it more convenient to study first the asymptotic distribution theory of $nU_n^{(2)}$ and then to move on to the case of $nR_n^{\star\star}$.

Note that by the assumption made in (4.5.14), $T^{(2)}(.) \in L_2(F)$, and hence there exists a set (finite or infinite) of eigenvalues $\{\lambda_k\}$ of $T^{(2)}(.)$ corresponding to orthonormal functions $\{\tau_k(.);\ k \ge 0\}$ such that

$$\int T^{(2)}(F; x, y) \tau_k(x) dF(x) = \lambda_k \tau_k(y) \quad \text{a.e. } (F),\ \forall k \ge 0 \tag{4.5.19}$$

$$\int \tau_k(x) \tau_q(x) dF(x) = \delta_{kq}, \tag{4.5.20}$$

where δ_{kq}, the Kronecker delta, is 1 or 0 according as $k = q$ or not, $k, q \ge 0$.

Theorem 4.5.1 *Under (4.5.14) and the convention in (4.5.8), as $n \to \infty$,*

$$nU_n^{(2)} \xrightarrow{\mathcal{D}} \sum_{\{k \geq 0\}} \lambda_k \{Z_k^2 - 1\}, \tag{4.5.21}$$

where the Z_k's are i.i.d. r.v.'s with the standard normal distribution, and the λ_k's are defined by (4.5.19).

Proof. For every positive integer M, let us denote by

$$Q_{(M)}(x, y) = \sum_{k \leq M} \lambda_k \tau_k(x) \tau_k(y), \quad x, y \in \mathbb{R}_1, \tag{4.5.22}$$

and let

$$Q(x, y) = Q_{(\infty)}(x, y) = \sum_{k \geq 0} \lambda_k \tau_k(x) \tau_k(y), \quad x, y \in \mathbb{R}_1. \tag{4.5.23}$$

Also let

$$\begin{aligned} U^\star_{n(M)} &= n^{-1} \sum_{\{1 \leq i \neq j \leq n\}} Q_{(M)}(X_i, X_j) \\ &= \sum_{k \leq M} \lambda_k \Big[n^{-\frac{1}{2}} (\sum_{i=1}^n \tau_k(X_i))^2 - n^{-1} \sum_{i=1}^n \tau_k^2(X_i) \Big]. \end{aligned} \tag{4.5.24}$$

Now, the $\tau_k(.)$ form an orthonormal system, so by the central limit theorem,

$$n^{-\frac{1}{2}} \sum_{i=1}^n \tau_k(X_i) \xrightarrow{\mathcal{D}} \mathcal{N}(0,1) \quad \text{as } n \to \infty, \text{ for every (fixed) } k. \tag{4.5.25}$$

Also, by the Khintchine strong law of large numbers, as $n \to \infty$,

$$n^{-1} \sum_{i=1}^n \tau_k^2(X_i) \to 1 \text{ a.s., for every (fixed) } k. \tag{4.5.26}$$

Therefore, by (4.5.24), (4.5.25), and (4.5.26), we conclude that for every (fixed) M,

$$U^\star_{n(M)} \xrightarrow{\mathcal{D}} \sum_{k \leq M} \lambda_k \{Z_k^2 - 1\}, \quad \text{as} \quad n \to \infty. \tag{4.5.27}$$

On the other hand,

$$\mathbb{E}\Big[\{U^\star_{n(M)} - U^\star_{n(\infty)}\}^2 \Big] \to 0 \quad \text{as } M \to \infty, \tag{4.5.28}$$

and hence, for every $\epsilon > 0$, there exists an M $(= M_\epsilon)$ such that

$$\mathbb{P}\Big\{ |U^\star_{n(M_\epsilon)} - U^\star_{n(\infty)}| > \epsilon \Big\} \leq \epsilon. \tag{4.5.29}$$

Therefore, identifying $nU_n^{(2)}$ as equivalent to $U^{\star}_{n(\infty)}$, the desired result follows by using (4.5.2), (4.5.29) and choosing M appropriately large. □

By virtue of (4.5.17) and (4.5.21), we immediately arrive at the following

Theorem 4.5.2 *For a second-order Hadamard differentiable functional $T(.)$, under the hypothesis of Theorem 4.5.1, as $n \to \infty$,*

$$nR_n^{\star\star} - \tau^{(2)}(F) \xrightarrow{\mathcal{D}} \sum_{k\geq 0} \lambda_k\{Z_k^2 - 1\}, \tag{4.5.30}$$

where the λ_k and Z_k are defined as in Theorem 4.5.1.

In the literature $\tau^{(2)}(F)$ is generally referred to as the *first-order bias* term [it stands for the location of the asymptotic distribution in (4.5.30)]. The asymptotic distribution in (4.5.30) is generally referred to as the *second-order asymptotic distribution.* Thus the representation in (4.5.9) with (4.5.30) may be termed a *second-order asymptotic distributional representation* (SOADR) for T_n.

From (4.5.9), (4.5.10), and Theorem 4.5.2 we have the following:

COROLLARY 4.5.2.1. If the functional $T(F)$ is stationary of order 1 at F [i.e., $T^{(1)}(F;x) = 0$ *a.e.* (F)], then as $n \to \infty$,

$$n\Big[T_n - T(F) - n^{-1}\tau^{(2)}(F)\Big] \xrightarrow{\mathcal{D}} \sum_{k\geq 0} \lambda_k\{Z_k^2 - 1\}, \tag{4.5.31}$$

where the λ_k and Z_k are defined as in Theorem 4.5.1.

It is clear from the above results that for a second order Hadamard differentiable functional $T(.)$, the second order compact derivative $T^{(2)}(.)$ and its Fourier series representation relative to an orthonormal system provide all the necessary information on the form of the remainder term in (4.5.1) and its asymptotic distribution. This procedure also enables us to consider the SOADR results in a unified manner. However, in this context we may mention that not all functionals are necessarily second-order Hadamard differentiable, and hence the theorems in this section may not apply to a very wide class of functionals. This point will be made more clear in the next chapter dealing with M-estimators. Even in the context of L-estimators, an estimator of the type L_{n2} may notbe in general Hadamard differentiable (of the second order), and hence the theorems in this section may not be applicable for the L_{n2} in Section 4.2. Actually for L-statistics of the type L_{n2}, the normalizing factor for the remainder term r_n is $n^{3/4}$ [and not n^1 as in (4.5.30)], and hence the end product will be a degenerate r.v. even if we apply a similar expansion. It seems appropriate to incorporate these theorems only to the case where in Section 4.3., the rate r is equal to 1 [comparable to (4.5.30)]. We will explain these applications with some smooth L-functionals in the next section. We may, however, note that even in this smooth case, not all L-functionals are covered in this formulation. As a very glaring example we consider the case of the population mean $T(F) = \int x dF(x)$, where for simplicity we assume that F is defined on the real line $\mathbb{R}_1$. Recall that

$$|T(F) - T(F_n)|/\|F_n - F\| = |\int_{\mathbb{R}_1} x d[F(x) - F_n(x)]|/\|F_n - F\|. \quad (4.5.32)$$

When the support of the d.f. F is infinite [i.e., $0 < F(x) < 1$, for all real and finite x], (4.5.2) cannot be bounded a.e.(F), even for n very large. Thus $T(F)$ is not generally Hadamard differentiable, although written in terms of an L-functional, this corresponds to the smooth case where $J(u) = 1$ for every $u \in (0, 1)$.

4.6 SOADR FOR CERTAIN SMOOTH L-FUNCTIONALS

Suppose that we have a general L-statistic of the form L_n in (4.4.1) where for L_{n1} the scores are defined by (4.3.3). Then we can write $L_n = T(F_n)$, where

$$T(F) = \int_0^1 F^{-1}(u)J(u)du + \sum_{j=1}^{k} a_j F^{-1}(p_j). \quad (4.6.1)$$

We now introduce a function $K = K_1 + K_2 = \{K_1(t) + K_2(t),\ t \in [0, 1]\}$, by letting

$$K_1(t) = \int_0^t J(s)ds,\ t \in (0, 1); \quad (4.6.2)$$

$$K_2(t),\ \text{a pure step function with jumps } a_j \text{ at } t = p_j,\ 1 \le j \le k. \quad (4.6.3)$$

We can write, for example,

$$\begin{aligned} T(F) &= \int_0^1 F^{-1}(t)dK(t) = \int_0^1 F^{-1}(t)dK_1(t) + \int_0^1 F^{-1}(t)dK_2(t) \\ &= T_1(K) + T_2(K). \end{aligned} \quad (4.6.4)$$

The influence function ψ_1 for $T_1(.)$ is given by (4.3.7), and similarly, for $T_2(.)$ the corresponding influence function is given by ψ_2 in (4.2.12). We rewrite these in the following form:

$$\psi_1(F^{-1}(t)) = \int_0^1 \Big[\{s - I(t \le s)\}/f(F^{-1}(s))\Big] dK_1(s), \quad (4.6.5)$$

$$\psi_2(F^{-1}(t)) = \int_0^1 \Big[\{s - I(t \le s)\}/f(F^{-1}(s))\Big] dK_2(s). \tag{4.6.6}$$

Since $K_2(.)$ is a step function, (4.6.6) can be written as a finite sum of k terms. Thus the influence function for $T(F)$ in (4.6.4) can be written as

$$\psi(F^{-1}(t)) = \int_0^1 \Big[\{s - I(t \le s)\}/f(F^{-1}(s))\Big] dK(s). \tag{4.6.7}$$

The second order Hadamard differentiability of $T(.)$ at F naturally depends on the positivity of $f(F^{-1}(t))$, its differentiability and the differentiability of $K(.)$. If the component $K_2(.)$ is nondegenerate, then there is at least one jump point for $J(.)$. Hence it can be shown that the second order Hadamard derivative of $T_2(.)$ at F does not exist. In order to make use of the results in preceding section, we may have to confine ourselves to the class of L-functionals of the type $T_1(F)$. Even so, we need further regularity conditions. Toward this end, assuming that $K_1(t)$ is twice differentiable, we set

$$\begin{aligned} K_1(t+a) &= K_1(t) + aK_1'(t) + (\frac{1}{2}a^2)K_1''(t+ha) \\ &= K_1(t) + aJ(t) + (\frac{1}{2}a^2)J'(t+ha), \end{aligned} \tag{4.6.8}$$

$0 < h < 1$; $t, t+a \in (0,1)$, where $J'(.)$ stands for the derivative of the score function $J(.)$. If we assume that the condition (C1) in Section 4.3 holds with $\nu = 1$ and further that $J'(u)$ is continuous in $u \in (\alpha, 1-\alpha)$, then we can set

$$\begin{aligned} &T_1(F_n) - T_1(F) - \int \psi_1(F;x)d[F_n(x) - F(x)] \\ &= \int xdK_1(F_n(x)) - \int xdK_1(F(x)) + \int [F_n(x) - F(x)][f(x)]^{-1}dK_1(F(x)) \\ &= \int [K_1(F(x)) - K_1(F_n(x))]dx + \int [F_n(x) - F(x)]J(F(x))dx \\ &= \int [K_1(F(x)) - K_1(F_n(x)) + J(F(x))[F_n(x) - F(x)]]dx \\ &= \int \Big\{\frac{[K_1(F(x)) - K_1(F_n(x))]}{[F(x) - F_n(x)]} - J(F(x))\Big\}[F(x)) - F_n(x)]dx \\ &= \frac{1}{2}\int [F_n(x) - F(x)]^2 J'(F_n^\star(x))dx, \end{aligned} \tag{4.6.9}$$

where $F_n^\star \in (F_n, F)$. Let us define

$$T_{2n}^{0^\star} = \frac{1}{2}\int [F_n(x) - F(x)]^2 J'(F(x))dx \tag{4.6.10}$$

so that, by (4.6.9) and (4.6.10), we have

$$T_1(F_n) = T_1(F) + \int \psi_1(F;x)d[F_n(x) - F(x)] + T_{2n}^{0^\star}$$

$$+ \quad \frac{1}{2}\int [F_n(x) - F(x)]^2 \Big\{ J'(F_n^\star(x)) - J'(F(x)) \Big\} dx. \quad (4.6.11)$$

Since $J(u)$ [and $J'(u)$] vanish outside $[\alpha, 1-\alpha]$ $(0 < \alpha < \frac{1}{2})$, and $F^{-1}(1-\alpha) - F^{-1}(\alpha)$ is bounded, $J'(F_n^\star) - J'(F)$ is uniformly continuous in the norm $||F_n - F||$, and hence the last term on the right-hand side of (4.6.11) is $o(||F_n - F||^2)$. This shows that the second-order Hadamard differentiability of $T_1(.)$ at F holds under **C1** with $\nu = 1$ and an additional continuity condition on $J'(.)$ (which exists under the first order Lipschitz condition on J). Note that in the above derivation, we may have to take into account that $J(.)$ may have jump at $u = \alpha$ and $u = 1 - \alpha$. However, the above proof is quite capable of handling any finite number of jump points for $J'(.)$ on $[\alpha, 1-\alpha]$. For simplicity of presentation, we will sacrifice this minor refinement.

Let us now incorporate this characterization of smooth L-functionals in the study of their SOADR results in some specific cases. Consider, first, the case of the trimmed mean described in Section 4.3 [following (4.3.34)]. Here we have

$$J(u) = (1-2\alpha)^{-1} I(\alpha \le u \le 1-\alpha), \;\; u \in (0,1), \;\; 0 < \alpha < \frac{1}{2}. \quad (4.6.12)$$

Some direct manipulations lead to the following explicit formulation of the associated influence function:

$$T^{(1)}(P;x) = \begin{cases} (1-2\alpha)^{-1}[F^{-1}(\alpha) - F^{-1}(\frac{1}{2})], & x < F^{-1}(\alpha), \\ (1-2\alpha)^{-1}[F^{-1}(1-\alpha) - F^{-1}(\frac{1}{2})], & x > F^{-1}(1-\alpha), \\ (1-2\alpha)^{-1}[x - F^{-1}(\frac{1}{2})], & \text{otherwise.} \end{cases} \quad (4.6.13)$$

For $\lambda > 0$,

$$(F + \lambda(G-F))^{-1}(u) = t, \;\; \text{that is, } u = F(t) + \lambda[G(t) - F(t)], \quad (4.6.14)$$

we obtain that

$$F^{-1}(u) = F^{-1}(F(t) + \lambda[G(t) - F(t)]) = t + \lambda[G(t) - F(t)]/f(F^{-1}(t))$$

$$- \frac{1}{2}\lambda^2[G(t) - F(t)]^2 f'(F^{-1}(t))/f^3(F^{-1}(t)) + o(\lambda^2), \text{ as } \lambda \to 0. \quad (4.6.15)$$

Hence

$$\lim_{\lambda \to 0} \Big\{ \lambda^{-2}[T(F) - 2T(F + (G-F)) + T(F + 2(G-F))] \Big\}$$

$$= -\frac{\Big\{ \int_{\boldsymbol{R}_1} [G(x) - F(x)]^2 f^{-3}(x) f'(x) dK_1(F(x)) \Big\}}{(1-2\alpha)}$$

$$= -\frac{\Big\{ \int_{\boldsymbol{R}_1} [G(x) - F(x)]^2 f^{-2}(x) f'(x) J(F(x)) \Big\}}{(1-2\alpha)}. \quad (4.6.16)$$

We have for $G = F_n$,

$$
\begin{aligned}
&T^{(2)}(F; X_i, X_j) \\
&= \Big\{ \int_{\boldsymbol{R}_1} \{[I(X_i \le x) - F(x)][I(X_j \le x) - F(x)]/f(x)f(x)\} \\
&\quad \{-f'(x)/f(x)\} J(F(x)) dF(x) \Big\} / (1 - 2a). \qquad (4.6.17)
\end{aligned}
$$

With this formulation we are in a position to use Theorem 4.5.1 in finding out the SOADR for the trimmed mean. Note that here we need $f(x)$ to be strictly positive over $F^{-1}(\alpha) \le x \le F^{-1}(1-\alpha)$, $f'(x)/f(x)$ to be well behaved in the same domain, and the second moment of the right-hand side of (4.6.17) to be finite. In the truncated case (where $0 < \alpha < 1/2$), this entails no further restriction of f or f'.

Let us next consider the case of the Harrell-Davis quantile estimator defined by (4.3.35)-(4.3.36). For simplicity of presentation, we consider the specific value of $p = 1/2$ so that T_n corresponds to a version of the median. Here the c_{ni} do not vanish at the two tails. However, the c_{ni} converge to 0 at an exponential rate when i does not lie in the range $[np_1, np_2]$ where $0 < p_1 < 1/2 < p_2 < 1$. Thus, if the underlying d.f. F admits a finite νth order (absolute) moment for some $\nu > 0$ (not necessarily greater than or equal to 1), then in (4.3.35) we may concentrate on the range $np_1 \le i \le np_2$, and the remainder term can be made $o_p(n^{-1})$ for n adequately large. On this truncated version, the conditions **A**, **B**, **C1** in Section 4.3 hold with a smooth $J_n(u)$ which is, however, dependent on n. In fact, for $k = n - k$, we have

$$J_n(u) = n^{\frac{1}{2}}(2\pi)^{-\frac{1}{2}}[4u(1-u)]^{\frac{n}{2}}u^{-1}\{1 + \mathrm{O}(n^{-1})\}, \quad u \in (0,1), \qquad (4.6.18)$$

where, we may note that on letting $u = 1/2 + n^{-1/2}t$, t real,

$$[4u(1-u)]^{\frac{n}{2}} = [1 - \frac{4}{n}t^2]^{\frac{n}{2}} = \exp(-2t)^2[1 + \mathrm{O}(n^{-1})]. \qquad (4.6.19)$$

This strong concentration of $J_n(u)$ around $u = 1/2$ [in a neighborhood of the order $n^{-1/2}$] leads us to the following:

$$|T_n - X_{n:[n/2+1]}| = \mathrm{O}(n^{-3/4}(\log n)^{3/4}) \quad \text{a.s., as} \quad n \to \infty. \qquad (4.6.20)$$

We may refer to Yoshizawa et al. (1985) for the details of this proof and leave the same as an exercise. Since for the sample median $X_{n:[n/2+1]}$, the SOADR result in (4.5.30) does not hold [the rate of convergence being $\mathrm{O}(n^{-3/4})$], (4.6.20) may suggest that the same SOADR result does not hold for the T_n. However, with a bit more intricate analysis, we can show that because of (4.6.18) and (4.6.19), T_n has enough of smoothness to possess a SOADR result.

Let us define

$$K_n(t) = \int_0^t J_n(u)du, \quad t \in [0,1], \qquad (4.6.21)$$

where $J_n(u)$ is based on the scores in (4.3.36) and is equivalently expressed in (4.6.18) when n is large. Also let

$$\theta_n = \int_{\mathbb{R}_1} x dK_n(F(x)). \tag{4.6.22}$$

Note that if F is symmetric about the median θ, then $\theta_n = \theta$, for every $n \geq 1$. Even otherwise, an asymptotic expansion of θ_n around θ can be made using the density function f and its first derivative f'. Thus whenever we assume that f' exists at θ and is continuous in a neighborhood of θ we may write $\theta_n = \theta + n^{-1}\delta + \mathrm{o}(n^{-1})$, where δ depends on the density f and f'. Then we have

$$\begin{aligned} T_n - \theta_n &= \int_{\mathbb{R}_1} x d[K_n(F_n(x)) - K_n(F(x))] \\ &= -\int_{\mathbb{R}_1} [K_n(F_n(x)) - K_n(F(x))]dx \\ &= -\int [F_n(x) - F(x)]J_n(F(x))dx \\ &\quad -\frac{1}{2}\int [F_n(x) - F(x)]^2 J_n'(F_n^\star)dx, \end{aligned} \tag{4.6.23}$$

where $F_n^\star[= hF + (1-h)F_n$ for some $h \in (0,1)]$ lies between F_n and F. We define $W_n = \{W_n(x),\ x \in \mathbb{R}_1\}$ by letting

$$W_n(x) = n^{\frac{1}{2}}[F_n(x) - F(x)], \quad x \in \mathbb{R}_1. \tag{4.6.24}$$

We can write the first term on the right hand side of (4.6.23) as

$$\Big[(2\pi)^{-\frac{1}{2}} \int_{\mathbb{R}_1} W_n(\theta + n^{-\frac{1}{2}}y)[F(\theta + n^{-\frac{1}{2}}y)]^{-1} \\ \exp\{-2n[F(\theta + n^{-\frac{1}{2}}y) - F(\theta)]^2\} n^{-\frac{1}{2}} dy\Big][1 + \mathrm{O}(n^{-1})]. \tag{4.6.25}$$

Let us write then

$$\begin{aligned} c_n &= c_n(F) = (2\pi)^{-\frac{1}{2}} \int_{\mathbb{R}_1} [F(\theta + n^{-\frac{1}{2}}y)]^{-1} \\ &\quad \exp\{-2n[F(\theta + n^{-\frac{1}{2}}y) - F(\theta)]^2\}dy \end{aligned} \tag{4.6.26}$$

and we note that $c_n \to 1/f(\theta)$, as $n \to \infty$. In fact we may set

$$c_n = \{f(\theta)\}^{-1} + n^{-\frac{1}{2}} c_1^\star(F) + \mathrm{o}(n^{-\frac{1}{2}}), \text{ as } n \to \infty, \tag{4.6.27}$$

where $c_1^\star(F)$ depends on the density f and its first derivative f' (at θ). As such, we have from (4.6.23), (4.6.25), and (4.6.26) that

$$\begin{aligned}
&T_n - \theta_n + c_n[n^{-\frac{1}{2}}W_n(\theta)] \\
&= n^{-\frac{1}{2}}(2\pi)^{-\frac{1}{2}} \int_{\mathbb{R}_1} [W_n(\theta) - W_n(\theta n^{-\frac{1}{2}})][F(\theta + n^{-\frac{1}{2}}y)]^{-1} \\
&\quad [\exp\{-2n[F(\theta + n^{-\frac{1}{2}}y) - F(\theta)]^2\}]dy + \mathrm{O}_p(n^{-\frac{3}{2}}) \\
&\quad -(2n)^{-1} \int_{\mathbb{R}_1} W_n^2(x) J_n'(F_n^\star(x))dx. \qquad (4.6.28)
\end{aligned}$$

In view of the symmetry of $J(u)$ around $u = 1/2$ and the uniform boundedness of $W_n(.)$, in probability, the last term on the right hand side of (4.6.28) is $\mathrm{o}_p(n^{-1})$. On the other hand, $\{n^{1/4}[W_n(\theta) - W_n(\theta + n^{-1/2}y)],\ y \in \mathbb{R}_1\}$ is asymptotically Gaussian, so the first term on the right-hand side of (4.6.28), when adjusted by the normalizing factor $n^{3/4}$, has asymptotically a normal distribution with zero mean and a finite variance. Hence we obtain from the above that as $n \to \infty$,

$$n^{\frac{3}{4}}\Big\{T_n - \theta_n + c_n[n^{-\frac{1}{2}}W_n(\theta)]\Big\} \xrightarrow{\mathcal{D}} Z \sim \mathcal{N}(0, \gamma^2), \qquad (4.6.29)$$

where γ^2 is a finite positive number. This provides a simple SOADR result for the estimator T_n. However, in this context we may note that (1) the rate is still $n^{3/4}$ (and not n), and (2) unlike in (4.1.3), here the limiting law is normal (not a mixed normal). Thus the smoothness in the score function serves an effective purpose in simplifying the SOADR result, although it does not lead to the usual rate of convergence (i.e., n) for smooth L-functionals. Finally looking at (4.6.27) and (4.6.29), we may conclude that in (4.6.29), c_n may be replaced by $\{f(\theta)\}^{-1}$; however, the replacement of θ_n by θ may demand the symmetry of f around θ (at least in a neighborhood).

Let us next consider the case of the rank-weighted means $\{T_{n,k}\}$, defined after (4.3.34). For any fixed k (≥ 0), let $\phi_k(X_1, \ldots, X_{2k+1})$ be the median of $X_1, \ldots, X_{2k+1}$. Then we may write equivalently

$$T_{n,k} = \binom{n}{2k+1}^{-1} \sum_{\{1 \leq i_1 < \ldots < i_{2k+1} \leq n\}} \phi_k(X_{i_1}, \ldots, X_{i_{2k+1}}). \qquad (4.6.30)$$

See Problem 4.6.1 (Sen 1964). For each h ($0 \leq h \leq 2k+1$), let

$$\phi_{k,h}(x_1, \ldots, x_h) = \mathbb{E}\Big\{\phi_k(x_1, \ldots, x_h, X_{h+1}, \ldots, X_{2k+1})\Big\}, \qquad (4.6.31)$$

$$\begin{aligned}
\phi_{k,h}^\star(x_1, \ldots, x_h) &= \phi_{k,h}(x_1, \ldots, x_h) \\
&- \sum_{j=1}^{h} \phi_{k,h-1}(x_1, \ldots, x_{j-1}, x_{j+1}, \ldots, x_h) \\
&+ \ldots + (-1)^h \phi_{k,0}(\cdot), \qquad (4.6.32)
\end{aligned}$$

for $h = 0, 1, \ldots, 2k+1$. Corresponding to the kernel in (4.6.32), the U-statistic based on $X_1, \ldots, X_n$ is denoted by $U^{\star}_{n,k}(h)$, for $h = 0, 1, \ldots, 2k+1$. Then, by (4.6.30), (4.6.31), and (4.6.32), we obtain that

$$T_{n,k} = \sum_{h=0}^{2k+1} \binom{2k+1}{h} U^{\star}_{n.k}(h). \tag{4.6.33}$$

Note that $U^{\star}_{n,k}(0)$ = expected value of the median of $(X_1, \ldots, X_{2k+1}) = \theta_{(k)}$, say, and hence is nonstochastic. $U^{\star}_{n,k}(1)$ is an average over n i.i.d. r.v.'s with mean 0, $U^{\star}_{n,k}(2)$ is a U-statistic of degree 2 with expectation 0, and it corresponds to the case of a parameter which is stationary of order 1 [in the sense of (4.5.18)]. Similarly, for every $h \geq 3, U^{\star}_{n,k}(h)$ has a variance of the order n^{-h}. Thus we have from the above decomposition,

$$n\Big[T_{n,k} - \theta_{(k)} - (2k+1)U^{\star}_{n,k}(1)\Big] = \binom{2k+1}{2} nU^{\star}_{n.k}(2) + \mathrm{O}_p(n^{-\frac{1}{2}}). \tag{4.6.34}$$

In this representation we assume that the kernel $\phi_k(.)$ has a finite second moment (i.e., the median of a sample of size $2k+1$ has a finite variance). A sufficient condition for the finiteness of this second moment is that the underlying d.f. F (of X) has a finite absolute moment of order r (> 0) : $r(k+1) \geq 2$ (see Problem 4.6.2; Sen, 1959). For $nU^{\star}_{n.k}(2)$ we are in a position to use directly Theorem 4.5.1, and this provides the desired SOADR result (for a fixed K). The situation is somewhat different when k $(= n_k)$ is made to depend on n in such a way that $k_n \to \infty$ as $n \to \infty$. The picture depends very much on the particular setup for the order of k_n. For example, if we let $k_n \cong (n+1)/2$, T_{n,k_n} reduces to the sample median, and hence, it will have the SOADR result given by (4.1.3) (with an order $n^{3/4}$ instead of n). Similarly, if we let $k_n = (n+1)/2 \pm q$, for some fixed positive integer q, then the weights $\binom{n}{2k+1}^{-1}\binom{i-1}{k}\binom{n-i}{k}$ are concentrated only in the range $(n+1)/2 - q \leq i \leq (n+1)/2 + q$, so the SOADR result will be close to (4.1.3) with the order $n^{3/4}$. For this reason we consider the case of k_n increasing with n in such a way that $n^{-1}k_n$ is bounded from above by a number $t < 1/2$. Then we have the following SOADR result:

For simplicity of presentation, we take $n = 2m+1$. Then for $i = m+1 \pm r$, we have

$$\begin{aligned}
&\binom{2m+1}{2k+1}^{-1}\binom{i-1}{k}\binom{2m+1-i}{k} \\
&= \binom{2m+1}{2k+1}^{-1}\binom{m+r}{k}\binom{m-r}{k} \\
&= \Big[\binom{2m+1}{2k+1}^{-1}\binom{m}{k}^2\Big]\cdot\Big[\binom{m+r}{k}\binom{m-r}{k}/\binom{m}{k}^2\Big] \\
&= \binom{2m+1}{2k+1}^{-1}\binom{m}{k}^2 \frac{(m^2-r^2)\ldots\{(m-k+1)^2-r^2\}}{m^2\ldots(m-k+1)^2}.
\end{aligned} \tag{4.6.35}$$

Next we note that

$$\begin{aligned}
&\frac{(m^2 - r^2)\ldots\{(m-k+1)^2 - r^2\}}{m^2\ldots(m-k+1)^2} \\
&= \exp\Big\{\sum_{i=1}^{k}\log(1 - \frac{r^2}{(m-i+1)^2})\Big\} \\
&= \Big\{\exp(-\frac{r^2 k}{m(m-k)})\Big\}\Big\{1 + \mathrm{O}((m-k+1)^{-1})\Big\}. \qquad (4.6.36)
\end{aligned}$$

If we let

$$r = r_n = \Big\{m(m-k_n)k_n^{-1}\Big\}^{\frac{1}{2}}y, \quad \psi(n) = \Big\{n(n-2k_n)/2k_n\Big\}^{\frac{1}{2}}, \qquad (4.6.37)$$

where $y \in \mathbb{R}_1$, then we have a representation very similar to that in (4.6.28) with $W_n(\theta+n^{-1/2}y)$ and $F(\theta+n^{-1/2}y)$ replaced by $W_n(\theta+y/\psi(n))$ and $F(\theta+y/\psi(n))$ respectively. Therefore, proceeding as in (4.6.28) through (4.6.29), we obtain that as $n \to \infty$,

$$\begin{aligned}
&(n\psi(n))^{\frac{1}{2}}\Big\{T_{n,k_n} - \theta_{(k_n)} + \{f(\theta_{(k_n)})\}^{-1}[n^{-\frac{1}{2}}W_n(\theta_{(k_n)})]\Big\} \\
&\xrightarrow{\mathcal{D}} Z^\star \sim \mathcal{N}(0,\gamma^{\star 2}) \quad \text{for some } \gamma^\star : 0 < \gamma^\star < \infty. \qquad (4.6.38)
\end{aligned}$$

It is interesting to note that if $k_n = \mathrm{O}(n)$, then $n\psi(n) = \mathrm{O}(n^{3/2})$, so the rate of convergence is comparable to (4.6.29). On the other hand, if $k_n = \mathrm{o}(n)$, then $n\psi(n) = \mathrm{O}(n^2/k_n^{1/2})$, so in (4.6.38) we have $(n\psi(n))^{1/2} = \mathrm{O}(n^1 k_n^{-1/4})$, and a better rate of convergence is achieved. In practice, if we are able to assume that the pdf $f(.)$ is symmetric about the median θ, then we are able to replace $\theta_{(k_n)}$ by θ for all n, and hence with a slowly increasing $\{k_n\}$, we are able to achieve a SOADR almost of the order n. This explains the relative merits of the rank-weighted average over the sample median or the other quantile estimator.

4.7 L-ESTIMATION IN LINEAR MODEL

Consider the linear regression model

$$\mathbf{Y} = \mathbf{X}\boldsymbol{\beta} + \mathbf{Z} \qquad (4.7.1)$$

where $\mathbf{Y} = (Y_1, \ldots, Y_n)'$ is the vector of observations, $\mathbf{X} = \mathbf{X}_n$ is an $n \times p$ matrix of known constants with the rows $\mathbf{x}_i'$ such that $x_{1i} = 1,\ i = 1, \ldots, n$

and $\mathbf{Z} = (Z_1, \ldots, Z_n)'$ is the vector of errors which are i.i.d. random variables with distribution function F.

While M- and R-estimators in the linear model have been extensively studied, until recently there have been no straightforward extension of L-estimators from the location to the linear regression model. The main problem was the lack of a successful extension of the sample quantile to the regression case. Bickel's (1973) analogues of L-estimators, which were apparently the first ones, had good efficiency properties but were computationally complex and were not equivariant to reparametrization.

A suitable definition of regression quantile is due to Koenker and Bassett (1978). They defined the α-regression quantile $\hat{\boldsymbol{\beta}}_n(\alpha)$, $0 < \alpha < 1$, as the solution of the minimization

$$\hat{\boldsymbol{\beta}} = \text{ arg min}\{\sum_{i=1}^{n} \rho_\alpha(Y_i - \mathbf{x}_i'\mathbf{t}) : \ \mathbf{t} \in I\!R_p\} \tag{4.7.2}$$

where

$$\rho_\alpha(x) = |x|\Big\{(1-\alpha)I[x<0] + \alpha I[x>0]\Big\}. \tag{4.7.3}$$

Koenker and Bassett (1978) pointed out that $\hat{\boldsymbol{\beta}}(\alpha)$ could be found as an optimal solution $\hat{\boldsymbol{\beta}}$ of the linear program

$$\begin{aligned} &\alpha \sum_{i=1}^{n} r_i^+ + (1-\alpha)\sum_{i=1}^{n} r_i^- = \min \quad \text{subject to} \\ &\sum_{j=1}^{p} x_{ij}\beta_j + r_i^+ - r_i^- = Y_i, \ i = 1, \ldots, n; \\ &\beta_j \in I\!R_1, \ j = 1, \ldots, p; \ r_i^+ \geq 0, \ r_i^- \geq 0, \ \ i = 1, \ldots, n; \end{aligned} \tag{4.7.4}$$

where r_i^+ and r_i^- are positive and negative parts of the residuals $Y_i - \mathbf{x}_i'\boldsymbol{\beta}$, $i = 1, \ldots, n$, respectively. The fact that the regression quantiles could be found as a solution of (4.7.4) is not only important for its computational aspect, it reveals the structure of the concept. It implies that the set $B_n(\alpha)$ of solutions of (4.7.4) [and hence also that of (4.7.2)] is non-empty, convex, compact, and polyhedral. If there are no additional restrictions, we may fix a version $\hat{\boldsymbol{\beta}}(\alpha)$ of α-regression quantile as the lexicographically maximal element of $B_n(\alpha)$. Moreover the dual program to (4.7.4) is of interest; it takes on the form

$$\begin{aligned} &\sum_{i=1}^{n} Y_i \Delta_i = \max \quad \text{subject to} \\ &\sum_{i=1}^{n} x_{ij}\Delta_i = 0, \ \ j = 1, \ldots p; \ \ \alpha - 1 \leq \Delta_i \leq \alpha, \ \ i = 1, \ldots n. \end{aligned} \tag{4.7.5}$$

Transforming $\triangle_i + (1-\alpha) = a_i^\star$, $i = 1, \ldots, n$, we obtain an equivalent version (in terms of $a_i^\star \in (0,1)$, $i = 1, \ldots, n$) of the dual program (4.7.5):

$$\sum_{i=1}^{n} Y_i a_i^\star = \max \quad \text{subject to}$$
$$\sum_{i=1}^{n} x_{ij} a_i^\star = (1-\alpha) \sum_{i=1}^{n} x_{ij}, \quad j = 1, \ldots p; \tag{4.7.6}$$

The optimal solution $\mathbf{a}_n^\star(\alpha) = (a_{n1}^\star(\alpha), \ldots, a_{nn}^\star(\alpha))'$ of (4.7.6), dual regression quantiles, has been termed the *regression rank scores* by Gutenbrunner (1986). If $\mathbf{X}_n$ is of rank p and $0 < \alpha < 1$, then the vector $\Big((1-\alpha)\sum_{i=1}^{n} x_{i1}, \ldots, (1-\alpha)\sum_{i=1}^{n} x_{ip}\Big)'$ is an interior point of the set $\{\mathbf{X}_n'\mathbf{z} : \mathbf{z} \in [0,1]^n\}$. Hence the optimal solution of (4.7.6) satisfies the following inequalities:

$$\hat{a}_{ni}^\star(\alpha) = \begin{cases} 1, & \text{if } Y_i > \mathbf{x}_i'\hat{\boldsymbol{\beta}}_n(\alpha), \\ 0, & \text{if } Y_i < \mathbf{x}_i'\hat{\boldsymbol{\beta}}_n(\alpha), \ i = 1, \ldots, n. \end{cases} \tag{4.7.7}$$

We will refer to these relations in the subsequent text.

We will show that the asymptotic behavior of $\hat{\boldsymbol{\beta}}_n(\alpha)$ is analogous to that of the sample quantile in the location model. Among the asymptotic properties of $\hat{\boldsymbol{\beta}}_n(\alpha)$, the asymptotic distribution and the Bahadur-type representation are of primary interest. Besides that it could be shown that the regression quantile process

$$\mathbf{L}_n(\alpha) = n^{\frac{1}{2}}\Big(\hat{\boldsymbol{\beta}}_n(\alpha) - \tilde{\boldsymbol{\beta}}(\alpha)\Big), \ 0 < \alpha < 1 \tag{4.7.8}$$

with $\tilde{\boldsymbol{\beta}}(\alpha) = (\beta_1 + F^{-1}(\alpha), \beta_2, \ldots, \beta_p)'$, considered as a random element of $D^p[0,1]$, has weak convergence properties analogous to the one-dimensional quantile process. This enables one to consider a broad class of estimators of $\boldsymbol{\beta}$ in the form of linear combinations of regression quantiles:

$$\mathbf{T}_n^\nu = \int_0^1 \hat{\boldsymbol{\beta}}_n(\alpha) d\nu(\alpha) \tag{4.7.9}$$

where ν is an appropriate signed measure on (0,1) (finite with a compact support $\subset (0,1)$). Specific choices of ν lead to extensions of various location L-estimators. Thus an atomic ν leads to a combination of single regression quantiles. We obtain an extension of α_1, α_2-trimmed mean if we take ν absolutely continuous with respect to the Lebesgue measure with the density

$$J = (\alpha_2 - \alpha_1)^{-1} I_{[\alpha_1, \alpha_2]}, \ 0 < \alpha_1 < \alpha_2 < 1. \tag{4.7.10}$$

However, regarding the computational aspect, we will rather define the trimmed least square estimator $\mathbf{T}_n(\alpha_1, \alpha_2)$ in the Koenker and Bassett (1978) manner: Let

$$a_i = a_{ni} = I\Big[\mathbf{x}_i'\hat{\boldsymbol{\beta}}_n(\alpha_1) < Y_i < \mathbf{x}_i'\hat{\boldsymbol{\beta}}_n(\alpha_2)\Big] \tag{4.7.11}$$

and calculate the weighted least squares estimator with the weights a_i, $i = 1, \ldots, n$. Denoting $\mathbf{A}_n = diag(a_i)$, we get the following explicit expression for the trimmed LSE:

$$\mathbf{T}_n(\alpha_1, \alpha_2) = (\mathbf{X}_n' \mathbf{A}_n \mathbf{X}_n)^{-1} \mathbf{X}_n' \mathbf{A}_n \mathbf{Y}_n \tag{4.7.12}$$

In the case that the inverse in (4.7.12) does not exist, it should be replaced by a generalized inverse. However (Problem 4.7.1), under mild regularity conditions, $n^{-1}(\mathbf{X}_n' \mathbf{A}_n \mathbf{X}_n)^{-1}$ converges to a positive definite matrix, in probability as $n \to \infty$. The estimator (4.7.12) differs from the one defined in (4.7.9) and (4.7.10) in the weights of observations for which $Y_i = \mathbf{x}_i' \hat{\boldsymbol{\beta}}(\alpha_1)$, $Y_i = \mathbf{x}_i' \hat{\boldsymbol{\beta}}(\alpha_2)$, or $\mathbf{x}_i' \hat{\boldsymbol{\beta}}(\alpha_1) \geq \mathbf{x}_i' \hat{\boldsymbol{\beta}}(\alpha_2)$, $i = 1, \ldots, n$. However, the probability of the last inequality tends to 0 as $n \to \infty$ (Problem 4.7.2).

In the present section, we shall concentrate on two asymptotic propositions on L-estimators in linear model; namely we will fully derive the asymptotic representations of the sample quantiles and of the trimmed LSE of (4.7.12). The technique used for regression quantiles will be later applied to M-estimators generated by a discontinuous ψ-function.

We will impose the following regularity conditions on the matrix $\mathbf{X}_n$:

A1. $\lim_{n\to\infty} \mathbf{Q}_n = \mathbf{Q}$, where $\mathbf{Q}_n = n^{-1} \mathbf{X}_n' \mathbf{X}_n$ and $\mathbf{Q}$ is a positive definite matrix.

A2. $n^{-1} \sum_{i=1}^n x_{ij}^4 = O(1)$, as $n \to \infty$, for $j = 1, \ldots, p$.

A3. $x_{i1} = 1$, $i = 1, \ldots n$.

Theorem 4.7.1 *Suppose that the distribution function F of Z_i in (4.7.1) $(i = 1, \ldots, n)$ is continuous and is twice differentiable in a neighborhood of $F^{-1}(\alpha)$ and $F'(F^{-1}(\alpha)) = f(F^{-1}(\alpha)) > 0$, $0 < \alpha < 1$. Then, under the conditions* **A1** - **A3**,

$$\begin{aligned} &\hat{\boldsymbol{\beta}}_n(\alpha) - \tilde{\boldsymbol{\beta}} \\ &= n^{-1}[f(F^{-1}(\alpha))]^{-1} \mathbf{Q}^{-1} \sum_{i=1}^n \mathbf{x}_i \phi_\alpha(Z_i - F^{-1}(\alpha)) + \mathbf{R}_n(\alpha), \end{aligned} \tag{4.7.13}$$

where $||\mathbf{R}_n(\alpha)|| = O_p(n^{-3/4})$ *as* $n \to \infty$ *and*

$$\phi_\alpha(x) = \alpha - I[x < 0], \quad x \in I\!R_1. \tag{4.7.14}$$

Proof. Denote $\mathbf{S}(\mathbf{t}) = (S_1(\mathbf{t}), \ldots, S_p(\mathbf{t}))'$ by

$$\begin{aligned} S_j(\mathbf{t}) \;=\; & n^{-\frac{1}{2}} \sum_{i=1}^n x_{ij} \Big[\phi_\alpha(Z_i - F^{-1}(\alpha) - n^{-\frac{1}{2}} \mathbf{x}_i' \mathbf{t}) \\ & - \phi_\alpha(Z_i - F^{-1}(\alpha)) \Big], \; \mathbf{t} \in I\!R_p, \; j = 1, \ldots, p. \end{aligned} \tag{4.7.15}$$

By (A.2.1)-(A.2.8) (see the Appendix),

$$\sup_{||\mathbf{t}|| \leq C} ||\mathbf{S}(\mathbf{t}) + f(F^{-1}(\alpha)) \mathbf{Q} \mathbf{t}|| = O_p(n^{-\frac{1}{4}}) \tag{4.7.16}$$

for every $C > 0$ as $n \to \infty$.

Let $\hat{\boldsymbol{\beta}}(\alpha)$ be a solution of the minimization (4.7.2). Then

$$\|n^{-\frac{1}{2}} \sum_{i=1}^{n} \mathbf{x}_i \phi_\alpha(Y_i - \mathbf{x}_i' \hat{\boldsymbol{\beta}}(\alpha))\| = \mathrm{O}(n^{-\frac{1}{4}}) \quad \text{a.s.} \tag{4.7.17}$$

Further let $G_j^+(\epsilon)$ be the right derivative of the function

$$n^{-\frac{1}{2}} \sum_{i=1}^{n} \mathbf{x}_i \rho_\alpha \Big(Y_i - \mathbf{x}_i' \hat{\boldsymbol{\beta}}(\alpha) + \epsilon \mathbf{e}_j \Big) \tag{4.7.18}$$

for a fixed j, $1 \le j \le p$, where $\mathbf{e}_j \in I\!R_p$, $e_{jk} = \delta_{jk}, j, k = 1, \dots, p$. Then $G_j^+(\epsilon) = -n^{-1/2} \sum_{i=1}^n x_{ij} \phi_\alpha(Y_i - \mathbf{x}_i'\hat{\boldsymbol{\beta}}(\alpha) + \epsilon \mathbf{e}_j))$ is nondecreasing in ϵ. Hence, for $\epsilon > 0$, $G_j^+(-\epsilon) \le G_j^+(0) \le G_j^+(\epsilon)$ and further $G_j^+(-\epsilon) \le 0$ and $G_j^+(\epsilon) \ge 0$ because (4.7.18) attains the minimum at $\epsilon = 0$. Thus $|G_j^+(0)| \le G_j^+(\epsilon) - G_j^+(-\epsilon)$ and letting $\epsilon \downarrow 0$, we obtain

$$|G_j^+(0)| \le n^{-\frac{1}{2}} \sum_{i=1}^{n} |x_{ij}| I \Big[Y_i - \mathbf{x}_i' \hat{\boldsymbol{\beta}}(\alpha) = 0 \Big]. \tag{4.7.19}$$

Because the matrix $\mathbf{X}_n$ is of the rank p for every $n \ge n_0$ (see condition **A1**) and the distribution function F is continuous, at most p indicators on the right-hand side of (4.7.19) are nonzero a.s.; this together with condition **A2** leads to (4.7.17).

The representation (4.7.13) will follow from (4.7.16), by letting $\mathbf{t} \to n^{1/2}(\hat{\boldsymbol{\beta}}(\alpha) - \tilde{\boldsymbol{\beta}}(\alpha))$, provided that we show that

$$\|n^{\frac{1}{2}}(\hat{\boldsymbol{\beta}}(\alpha) - \tilde{\boldsymbol{\beta}}(\alpha))\| = \mathrm{O}_p(1), \text{ as } n \to \infty. \tag{4.7.20}$$

Regarding (4.7.17), it is sufficient to prove that, given $\epsilon > 0$, there exist $C > 0$, $\eta > 0$ and a positive integer n_0 so that for $n > n_0$,

$$I\!P\Big\{ \inf_{\|\mathbf{t}\| \ge C} \|n^{-\frac{1}{2}} \sum_{i=1}^{n} \mathbf{x}_i \phi_\alpha(Z_i - F^{-1}(\alpha) - n^{-\frac{1}{2}} \mathbf{x}_i' \mathbf{t})\| < \eta \Big\} < \epsilon. \tag{4.7.21}$$

To prove (4.7.21), let us first note that there exist $K > 0$ and n_1 such that

$$I\!P\Big\{ \|n^{-\frac{1}{2}} \sum_{i=1}^{n} \mathbf{x}_i \phi_\alpha(Z_i - F^{-1}(\alpha))\| > K \Big\} < \epsilon/2. \tag{4.7.22}$$

for $n > n_1$. Take C and η such that

$$C > 2K/(f(F^{-1}(\alpha)\lambda_0) \quad \text{and} \quad \eta < K/2, \tag{4.7.23}$$

where λ_0 is the minimum eigenvalue of $\mathbf{Q}$. Then there exists n_2 so that for $n \geq n_2$,

$$\mathbb{P}\Big\{\inf_{\|\mathbf{t}\|=C}\Big[-n^{-\frac{1}{2}}\sum_{i=1}^{n}\mathbf{t}'\mathbf{x}_i\phi_\alpha(Z_i - F^{-1}(\alpha) - n^{-\frac{1}{2}}\mathbf{x}_i'\mathbf{t})\Big] < C\eta\Big\} < \epsilon. \quad (4.7.24)$$

Actually the left-hand side of (4.7.24) is less than or equal to

$$\begin{aligned}
&\mathbb{P}\Big\{\inf_{\|\mathbf{t}\|=C}\Big[-n^{-\frac{1}{2}}\sum_{i=1}^{n}\mathbf{t}'\mathbf{x}_i\phi_\alpha(Z_i - F^{-1}(\alpha) - n^{-\frac{1}{2}}\mathbf{x}_i'\mathbf{t})\Big] < C\eta, \\
&\inf_{\|\mathbf{t}\|\geq C}\Big[-n^{-\frac{1}{2}}\sum_{i=1}^{n}\mathbf{t}'\mathbf{x}_i\phi_\alpha(Z_i - F^{-1}(\alpha)) + f(F^{-1}(\alpha))\mathbf{t}'\mathbf{Q}\mathbf{t}\Big] \geq 2C\eta\Big\} \\
&+\mathbb{P}\Big\{\inf_{\|\mathbf{t}\|=C}\Big[-n^{-\frac{1}{2}}\sum_{i=1}^{n}\mathbf{t}'\mathbf{x}_i\phi_\alpha(Z_i - F^{-1}(\alpha)) \\
&\qquad + \quad f(F^{-1}(\alpha))\mathbf{t}'\mathbf{Q}\mathbf{t}\Big] < 2C\eta\Big\}. \qquad (4.7.25)
\end{aligned}$$

The first term of (4.7.25) is less than or equal to

$$\begin{aligned}
&\mathbb{P}\Big\{\sup_{\|\mathbf{t}\|=C}\Big[\mathbf{t}'\mathbf{S}(\mathbf{t}) + f(F^{-1}(\alpha))\mathbf{t}'\mathbf{Q}\mathbf{t}\Big] \geq 2C\eta\Big\} \\
&\leq \mathbb{P}\Big\{\sup_{\|\mathbf{t}\|=C}\|\mathbf{S}(\mathbf{t}) + f(F^{-1}(\alpha))\mathbf{Q}\mathbf{t}\| \geq \eta\Big\} < \epsilon/2 \qquad (4.7.26)
\end{aligned}$$

for $n \geq n_2$. The second term of (4.7.25) is less than or equal to

$$\mathbb{P}\Big\{\inf_{\|\mathbf{t}\|=C}\Big[-n^{-\frac{1}{2}}\sum_{i=1}^{n}\mathbf{t}'\mathbf{x}_i\phi_\alpha(Z_i - F^{-1}(\alpha)) + f(F^{-1}(\alpha))C^2\lambda_0\Big] < 2C\eta\Big\}$$

$$\leq \mathbb{P}\Big\{-C\|n^{-\frac{1}{2}}\sum_{i=1}^{n}\mathbf{x}_i\phi_\alpha(Z_i - F^{-1}(\alpha))\| \leq -KC\Big\} < \epsilon/2 \qquad (4.7.27)$$

for $n > m_1$ by (4.7.22). Combining (4.7.25)-(4.7.27) with $n_0 = \max(n_1, n_2)$, we arrive at (4.7.24).

Now, take $\mathbf{s} \in \mathbb{R}_p$ such that $\|\mathbf{s}\| = C$, and put $x_i^\star = -n^{-\frac{1}{2}}\mathbf{x}_i'\mathbf{s}$, $i = 1, \ldots, n$. Then $M(\tau) = \sum_{i=1}^{n} x_i^\star\phi_\alpha(Z_i - F^{-1}(\alpha) + \tau x_i^\star)$ is nondecreasing in $\tau \in \mathbb{R}_1$ so that for $\tau \geq 1$,

$$\begin{aligned}
M(\tau) &= -n^{-\frac{1}{2}}\sum_{i=1}^{n}\mathbf{s}'\mathbf{x}_i\phi_\alpha(Z_i - F^{-1}(\alpha) - n^{-\frac{1}{2}}\tau\mathbf{x}_i'\mathbf{s}) \\
\geq M(1) &= -n^{-\frac{1}{2}}\sum_{i=1}^{n}\mathbf{s}'\mathbf{x}_i\phi_\alpha(Z_i - F^{-1}(\alpha) - n^{-\frac{1}{2}}\mathbf{x}_i'\mathbf{s}).
\end{aligned}$$

If $||\mathbf{t}|| \geq C$, then $\mathbf{t} = \tau \mathbf{s}$ where $\mathbf{s} = (C/||\mathbf{t}||)\mathbf{t}$ and $\tau = ||\mathbf{t}||/C \geq 1$. Hence

$$\begin{aligned} & \mathbb{P}\Big\{ \inf_{||\mathbf{t}|| \geq C} || - n^{-\frac{1}{2}} \sum_{i=1}^{n} \mathbf{x}_i \phi_\alpha (Z_i - F^{-1}(\alpha) - n^{-\frac{1}{2}} \mathbf{x}_i' \mathbf{t}) || < \eta \Big\} \\ \leq \ & \mathbb{P}\Big\{ \inf_{||\mathbf{t}|| \geq C} \Big[- n^{-\frac{1}{2}} \sum_{i=1}^{n} \mathbf{t}' \mathbf{x}_i \phi_\alpha (Z_i - F^{-1}(\alpha) - n^{-\frac{1}{2}} \mathbf{x}_i' \mathbf{t} \Big] (C/||\mathbf{t}) || < C\eta \Big\} \\ \leq \ & \mathbb{P}\Big\{ \inf_{||\mathbf{s}|| = C} \Big[- n^{-\frac{1}{2}} \sum_{i=1}^{n} \mathbf{s}' \mathbf{x}_i \phi_\alpha (Z_i - F^{-1}(\alpha) - n^{-\frac{1}{2}} \mathbf{x}_i' \mathbf{s}) \Big] < C\eta \Big\} < \epsilon, \end{aligned}$$

for $n > n_0$, and this completes the proof of (4.7.20). The proposition of the theorem will follow from (4.7.16) where we insert $\mathbf{t} \to n^{1/2}(\hat{\boldsymbol{\beta}}(\alpha) - \tilde{\boldsymbol{\beta}}(\alpha))$ and take (4.7.17) into account. □

Remark. The technique used in the proof of Theorem 4.7.1 leads to an asymptotic representation of a broad class of M-estimators. In view of the details provided here, we will often refer to it in a single step in the subsequent text.

COROLLARY 4.7.1.1. The sequence $\{n^{1/2}(\hat{\boldsymbol{\beta}}(\alpha) - \tilde{\boldsymbol{\beta}}(\alpha))\}$ has asymptotic p-dimensional normal distribution

$$\mathcal{N}_p\Big(\mathbf{0}, \alpha(1-\alpha)(f(F^{-1}(\alpha))^{-2}\mathbf{Q}^{-1}\Big). \tag{4.7.28}$$

We now turn to the trimmed least squares estimator in the form defined in (4.7.12). Fix α_1, α_2, $0 < \alpha_1, < \alpha_2 < 1$. We will try to derive an asymptotic representation of $\mathbf{T}_n$ with the remainder term of order $O_p(n^{-1})$, which is in correspondence with the rate of the trimmed mean in the location case. We should impose slightly stronger condition on $\mathbf{X}_n$ and on F. Namely the standardization of elements of $\mathbf{X}_n$ given in condition **A2** is strengthened to

A2'. $\text{Max}\{|x_{ij}| : 1 \leq i \leq n; 1 \leq j \leq p\} = O(1)$ as $n \to \infty$.

Moreover the following conditions will be imposed on the d.f. F:

B1. F is absolutely continuous with density f and $0 < f(x) < \infty$ for $F^{-1}(\alpha_1) - \epsilon < x < F^{-1}(\alpha_2) + \epsilon$, $\epsilon > 0$.

B2. The density f has a continuous and bounded derivative f' in neighborhoods of $F^{-1}(\alpha_1)$ and $F^{-1}(\alpha_2)$.

Let $\mathbf{T}_n = \mathbf{T}_n(\alpha_1, \alpha_2)$ be the trimmed LSE;

$$\mathbf{T}_n = (\mathbf{X}_n' \mathbf{A}_n \mathbf{X}_n)^{-1} \mathbf{X}_n' \mathbf{A}_n \mathbf{Y}_n,$$

where $\mathbf{A}_n$ is the $n \times n$ diagonal matrix with the diagonal $(a_1, \ldots, a_n)$ of (4.7.11). The asymptotic representation for $\mathbf{T}_n$ is given in the following theorem:

Theorem 4.7.2 *Suppose that the sequence $\{\mathbf{X}_n\}$ of matrices satisfies conditions* **A1**,**A2'** *and* **A3**, *and that the distribution function F of errors Z_i, $i = 1, \ldots, n$, satisfies conditions* **B1** *and* **B2**. *Then*

$$\begin{aligned}\mathbf{T}_n(\alpha_1, \alpha_2) - \boldsymbol{\beta} - \mathbf{e}_1 \eta (\alpha_2 - \alpha_1)^{-1} \\ = \quad (\alpha_2 - \alpha_1)^{-1} n^{-1} \mathbf{Q}_n^{-1} \sum_{i=1}^{n} \mathbf{x}_i \psi(Z_i) + \mathbf{R}_n \end{aligned} \qquad (4.7.29)$$

where $\|\mathbf{R}_n\| = O_p(n^{-1})$ *as* $n \to \infty$,

$$\eta = (1 - \alpha_2) F^{-1}(\alpha_2) + \alpha_1 F^{-1}(\alpha_1), \quad \mathbf{e}_1 = (1, 0, \ldots, 0)' \in \mathbb{R}_p, \qquad (4.7.30)$$

$$\psi(z) = \begin{cases} F^{-1}(\alpha_1) & \textit{if } z < F^{-1}(\alpha_1), \\ z & \textit{if } F^{-1}(\alpha_1) \le z \le F^{-1}(\alpha_2), \\ F^{-1}(\alpha_2) & \textit{if } z > F^{-1}(\alpha_2). \end{cases} \qquad (4.7.31)$$

Equivalently

$$\begin{aligned}\mathbf{T}_n(\alpha_1, \alpha_2) - \boldsymbol{\beta} - \mathbf{e}_1 \delta \\ = \quad n^{-1} (\alpha_2 - \alpha_1)^{-1} \mathbf{Q}_n^{-1} \sum_{i=1}^{n} \mathbf{x}_i (\psi(Z_i) - \mathbb{E}\psi(Z_i)) + \mathbf{R}_n, \end{aligned} \qquad (4.7.32)$$

where

$$\delta = (\alpha_2 - \alpha_1)^{-1} \int_{\alpha_1}^{\alpha_2} F^{-1}(u) du. \qquad (4.7.33)$$

Notice that the first component of $\mathbf{T}_n(\alpha_1, \alpha_2)$ is generally asymptotically biased while the other components are asymptotically unbiased. $\mathbf{T}_n$ is asymptotically unbiased in the symmetric case. Let us formulate this special case as a corollary.

COROLLARY 4.7.2.1. If F is symmetric around 0, and $\alpha_1 = \alpha$, $\alpha_2 = 1 - \alpha$, $0 < \alpha < 1/2$, then as $n \to \infty$,

$$\begin{aligned}\mathbf{T}_n(\alpha) - \boldsymbol{\beta} \\ = \quad n^{-1}(1 - 2\alpha)^{-1} \mathbf{Q}_n^{-1} \sum_{i=1}^{n} \mathbf{x}_i \psi(Z_i) + O_p(n^{-1}). \end{aligned} \qquad (4.7.34)$$

The asymptotic representation in (4.7.32) immediately implies that $n^{1/2}(\mathbf{T}_n(\alpha_1, \alpha_2) - \boldsymbol{\beta} - \mathbf{e}_1 \delta)$ is asymptotically normally distributed:

COROLLARY 4.7.2.2. Under the condition of Theorem 4.7.2, the sequence $n^{1/2}(\mathbf{T}_n(\alpha_1, \alpha_2) - \boldsymbol{\beta} - \mathbf{e}_1 \delta)$ has asymptotic p-dimensional normal distribution with expectation $\mathbf{0}$ and with the covariance matrix $\mathbf{Q}^{-1} \sigma^2(\alpha_1, \alpha_2)$

where

$$\begin{aligned}
&\sigma^2(\alpha_1, \alpha_2, F) \\
&= (\alpha_2 - \alpha_1)^{-1}\Big\{\int_{\alpha_1}^{\alpha_2} \alpha_2(F^{-1}(u) - \delta)^2 du \\
&+ \alpha_1(F^{-1}(\alpha_1) - \delta)^2 + (1-\alpha_2)(F^{-1}(\alpha_2) - \delta)^2 \\
&- \Big[\alpha_1(F^{-1}(\alpha_1 - \delta) + (1-\alpha_2)(F^{-1}(\alpha_2) - \delta)\Big]^2\Big\}.
\end{aligned} \tag{4.7.35}$$

If moreover $F(x)+F(-x) = 1,\ x \in I\!R_1$, and $\alpha_1 = \alpha,\ \alpha_2 = 1-\alpha,\ 0 < \alpha < 1/2$, then $n^{1/2}(\mathbf{T}_n(\alpha) - \boldsymbol{\beta})$ has p-dimensional asymptotic normal distribution with expectation $\mathbf{0}$ and with the covariance matrix $\sigma^2(\alpha, F)\mathbf{Q}^{-1}$ where

$$\sigma^2(\alpha, F) = (1-2\alpha)^{-2}\Big\{\int_{\alpha}^{1-\alpha} (F^{-1}(u))^2 du + 2\alpha(F^{-1}(\alpha))^2\Big\}. \tag{4.7.36}$$

Remark. Notice that $\sigma^2(\alpha_1, \alpha_2, F)$ coincides with the asymptotic variance of the (α_1, α_2)-trimmed mean (compare with Theorem 4.3.1.)

Proof of Theorem 4.7.2. First, we prove the following approximation which we will use in the sequel: For $\alpha = \alpha_1$ or α_2,

$$\mathbf{x}_i'(\hat{\boldsymbol{\beta}}(\alpha) - \boldsymbol{\beta}) = F^{-1}(\alpha) + O_p(n^{-\frac{1}{2}}) \tag{4.7.37}$$

uniformly in $i = 1, \ldots, n$, as $n \to \infty$. By Theorem 4.7.1, we then write $\mathbf{x}_i'(\hat{\boldsymbol{\beta}}(\alpha) - \boldsymbol{\beta}) = F^{-1}(\alpha) + A_{in} + B_{in}$, where

$A_{in} = n^{-1}\sum_{k=1}^{n} h_{ik}\phi_\alpha(Z_k - F^{-1}(\alpha)),\ \ h_{ik} = \mathbf{x}_i'\mathbf{Q}^{-1}\mathbf{x}_k$,
and $B_{in} = \mathbf{x}_i'\mathbf{R}_n(\alpha),\ i = 1, \ldots, n;\ k = 1, \ldots, n$.

Next $A_{in} = O_p(n^{-\frac{1}{2}})$, uniformly in $i = 1, \ldots, n$; this follows from the Chebyshev inequality because by **A1** and **A2'**, $I\!E A_{in} = 0$ and that

$\text{Var}A_{in} = \alpha(1-\alpha)n^{-2}\sum_{k=1}^{n} h_{ik}^2 = \alpha(1-\alpha)n^{-2}\mathbf{x}_i'\mathbf{Q}^{-1}(\mathbf{X}_n'\mathbf{X}_n)\mathbf{Q}^{-1}\mathbf{x}_i$
$= O(n^{-1})$, uniformly in $i = 1, \ldots, n$.

Hence we obtain (4.7.37) if we notice from (4.7.13) and **A2'** that $B_{in} = O_p(n^{-3/4})$ uniformly in $i = 1, \ldots, n$.

Now fix $0 < \alpha_1 < \alpha_2 < 1$, and denote

$$a_i^- = I\Big[Y_i > \mathbf{x}_i'\hat{\boldsymbol{\beta}}(\alpha_1)\Big],\quad a_i^+ = I\Big[Y_i \geq \mathbf{x}_i'\hat{\boldsymbol{\beta}}(\alpha_2)\Big] \tag{4.7.38}$$

and

$$b_i^- = I\Big[Z_i > F^{-1}(\alpha_1)\Big],\ b_i^+ = I\Big[Z_i \geq F^{-1}(\alpha_2)\Big],\ b_i = b_i^- - b_i^+, \tag{4.7.39}$$

$i = 1, \ldots, n$. Then $a_{ni} = a_i^- - a_i^+$ by (4.7.11) ($i = 1, \ldots, n,$) and hence, regarding (4.7.6) and (4.7.7), we have

$$|\sum_{i=1}^{n} a_i^+ x_{ij} - (1-\alpha_2)\sum_{i=1}^{n} x_{ij}|$$

$$\begin{aligned}
&= |\sum_{i=1}^{n}(a_i^+ - \hat{a}_i^\star(\alpha_2)x_{ij}| \\
&= |\sum_{i=1}^{n}(1 - \hat{a}_i^\star(\alpha_2))I[Y_i = \mathbf{x}_i'\hat{\boldsymbol{\beta}}(\alpha_2)]x_{ij}| \\
&\leq p \cdot \max\{|x_{ij}| : 1 \leq i \leq n;\ 1 \leq j \leq p\} = p \cdot \mathrm{O}(1). \qquad (4.7.40)
\end{aligned}$$

We obtain an analogous inequality for $|\sum_{i=1}^{n}(a_i^- - (1-\alpha_1))x_{ij}|$, $j = 1, \ldots, p$.

In the next step we prove the following approximation:

$$n^{-1}\sum_{i=1}^{n} x_{ij}[a_i Z_i - \psi(Z_i)] = \eta n^{-1}\sum_{i=1}^{n} x_{ij} + \mathrm{O}_p(n^{-1}), \qquad (4.7.41)$$

$j = 1, \ldots, p$. Actually

$$\begin{aligned}
& n^{-1}\sum_{i=1}^{n} x_{ij}[a_i Z_i - \psi(Z_i)] \\
&= n^{-1}\sum_{i=1}^{n} x_{ij}[(a_i^- - b_i^-) - (a_i^+ - b_i^+)]Z_i \\
&- n^{-1}\sum_{i=1}^{n} x_{ij}[b_i^+ F^{-1}(\alpha_2) + (1 - b_i^-)F^{-1}(\alpha_1)]. \qquad (4.7.42)
\end{aligned}$$

It follows from (4.7.37) that $Z_i = F^{-1}(\alpha_1) + \mathrm{O}_p(n^{-1/2})$ if $|a_i^- - b_i^-| = 1$, and that $Z_i = F^{-1}(\alpha_2) + \mathrm{O}_p(n^{-1/2})$ if $|a_i^+ - b_i^+| = 1$, respectively, uniformly in $i = 1, \ldots, n$. Hence, by (4.7.11),

$$\begin{aligned}
& n^{-1}\sum_{i=1}^{n} x_{ij}[a_i Z_i - \psi(Z_i)] \\
&= n^{-1}(F^{-1}(\alpha_1) + \mathrm{O}_p(n^{-\frac{1}{2}}))\Big[\sum_{i=1}^{n}(1 - \alpha_1 - b_i^-)x_{ij} + \mathrm{O}_p(1)\Big] \\
& \quad -n^{-1}F^{-1}(\alpha_1)\sum_{i=1}^{n}(1 - b_i^-)x_{ij} \\
&= -n^{-1}(F^{-1}(\alpha_2) + \mathrm{O}_p(n^{-\frac{1}{2}}))\Big[\sum_{i=1}^{n}(1 - \alpha_2 - b_i^+)x_{ij} + p\mathrm{O}(1)\Big] \\
& \quad -n^{-1}F^{-1}(\alpha_2)\sum_{i=1}^{n} b_i^+ x_{ij} \\
&= -\eta n^{-1}\sum_{i=1}^{n} x_{ij} + \Big\{n^{-1}\sum_{i=1}^{n}(1 - \alpha_1 - b_i^-)x_{ij} \\
&- n^{-1}\sum_{i=1}^{n}(1 - \alpha_2 - b_i^+)x_{ij}\Big\}\mathrm{O}_p(n^{-\frac{3}{2}}))
\end{aligned}$$

$$= \quad -\eta n^{-1}\sum_{i=1}^{n} x_{ij} + o_p(n^{-1})$$

by **A1**, (4.7.39) and by the Chebychev inequality; this gives (4.7.41).

Returning to definitions of a_i^+, a_i^-, b_i^+, and b_i^- and making use of **A1** and **A2'**, we may analogously prove that

$$n^{-1}\sum_{i=1}^{n} x_{ij}x_{ik}(a_{in} - (\alpha_2 - \alpha_1)) = O_p(n^{-\frac{1}{2}}),\ j,k = 1,\ldots,p. \tag{4.7.43}$$

Put differently,

$$n^{-1}\mathbf{X}_n'\mathbf{A}_n\mathbf{X}_n = n^{-1}(\alpha_2 - \alpha_1)\mathbf{X}_n'\mathbf{X}_n + O_p(n^{-\frac{1}{2}}).$$

Then it follows from (4.7.41), (4.7.43), and from **A1**, **A2'** and **A3** that

$$\mathbf{T}_n - \boldsymbol{\beta} - \mathbf{e}_1\delta = (\alpha_2 - \alpha_1)^{-1}(\mathbf{X}_n'\mathbf{X}_n)^{-1}\sum_{i=1}^{n}\mathbf{x}_i[\psi(Z_i) - \mathbb{E}\psi(Z_i)] + O_p(n^{-1}), \tag{4.7.44}$$

and this completes the proof of the theorem. □

We conclude this section with some alternative formulations of regression L-estimators due to Welsh (1987), Carroll and Welsh (1988), Jurečková and Welsh (1990), and Ren (1994). Consider the same linear model as in (4.7.1), and write

$$Y_i = \mathbf{x}_i'\boldsymbol{\beta} + Z_i,\quad i = 1,\ldots,n. \tag{4.7.45}$$

Let $\breve{\boldsymbol{\beta}}_n$ be a preliminary estimator of $\boldsymbol{\beta}$, and consider the residuals

$$\hat{Y}_i = Y_i - \mathbf{x}_i'\breve{\boldsymbol{\beta}}_n,\quad 1 \le i \le n. \tag{4.7.46}$$

Here also we take $x_{1i} = 1, \forall i \ge 1$ and $\sum_{i=1}^{n} x_{ji} = 0,\ 2 \le j \le p$. We stick to the assumptions **A1**, **A2** and **A3** made in Theorem 4.7.1. Let then

$$\hat{G}_n(y) = n^{-1}\sum_{i=1}^{n} I(\hat{Y}_i \le y),\quad y \in \mathbb{R}_1, \tag{4.7.47}$$

and let $\phi_n(t)$ be any pointwise consistent estimator of $\{f'(F^{-1}(t))\}^{-1}$, $t \in (0,1)$; we may refer to Welsh (1987) for such $\phi_n(.)$. Further let $h(.) = \{h(t),\ 0 \le t \le 1\}$ be a smooth weight function, and let $w_1, \ldots w_m$ be non-stochastic weights, $0 < q_1 < \ldots < q_m < 1$, $m < \infty$, and define

$$T(\hat{G}_n) = \int_0^1 \hat{G}_n^{-1}(t)h(t)dt + \int_0^1 \hat{G}_n^{-1}(t)dm(t), \tag{4.7.48}$$

where $m(t) = \sum_{i=1}^{m} w_i I(q_i \le t)$, $t \in (0,1)$. Welsh (1987) considered the matrix

$$\mathbf{Q}_w = \sum_{i=1}^{n}\mathbf{x}_i\mathbf{x}_i'\{h(\hat{G}_n(Y_i)) + \sum_{j=1}^{m} w_j I(Y_i \le q_j)\}, \tag{4.7.49}$$

and proposed the one-step estimator

$$\breve{\boldsymbol{\beta}}_n^{(1)} = \breve{\boldsymbol{\beta}}_n + \Big[T(\hat{G}_n) - \mathbf{Q}_w^{-1}\sum_{i=1}^{n}\mathbf{x}_i\{\int I(\hat{Y}_i \le y)h(\hat{G}_n(y))dy$$
$$+ \sum_{j=1}^{m} w_j\phi_n(q_j)I(\hat{Y}_i \le \hat{G}_n^{-1}(q_j))\}\Big]. \tag{4.7.50}$$

Asymptotic properties of $\breve{\boldsymbol{\beta}}_n^{(1)}$ were studied by Jurečková and Welsh (1990). Their regularity conditions are slightly more stringent than the ones by Ren (1994) who followed a Hadamard differentiability approach developed earlier by Ren and Sen (1991). Since this approach will be treated in greater detail in Chapter 5 (dealing with M-estimators), we will not repeat them here. Recall that by assumption $\mathbf{Q}_n = \begin{pmatrix} 1 & \mathbf{0} \\ \mathbf{0} & \mathbf{Q}_n^\star \end{pmatrix}$, where $\mathbf{Q}_n^\star = n^{-1}\sum_{i=1}^n \mathbf{x}_i^\star \mathbf{x}_i^{\star'}$ and $\mathbf{x}_i^\star = (x_{i2}, \ldots, x_{ip})'$, $i = 1, \ldots, n$. Ren (1994), considered the estimator

$$\breve{\boldsymbol{\beta}}_n^{(2)} = \breve{\boldsymbol{\beta}}_n + \Big[T(\hat{G}_n) - \mathbf{Q}_n^{\star -1} n^{-1}\sum_{i=1}^{n}\mathbf{x}_i^\star\{\int I(\hat{Y}_i \le y_i)h(\hat{G}_n(y))dy$$
$$+ \sum_{j=1}^{m} w_j\phi_n(q_j)I(\hat{Y}_i \le \hat{G}_n^{-1}(q_j))\}\Big], \tag{4.7.51}$$

where $\breve{\boldsymbol{\beta}}_n$ and other quantities are defined as before. Note that $\mathbf{Q}_w$ in (4.7.49) depends on the preliminary estimator $\breve{\boldsymbol{\beta}}_n$, while $\mathbf{Q}_n^\star$ does not. This introduces some simplifications in the computational aspects of the estimator. Moreover, for large n, $\hat{G}_n(\hat{G}_n^{-1}(t) \approx t$ for all $t \in [0,1]$ [actually the difference is $O(1/n)$], and hence (4.7.51) is also related closely to the one given by Carroll and Welsh (1988). If we let

$$\mathbf{Q}_n^0 = Diag(q_{njj}^{1/2},\ 1 \le j \le p) = Diag(1, \mathbf{Q}_n^{0\star}), \tag{4.7.52}$$

then $\mathbf{Q}_n^0(\breve{\boldsymbol{\beta}}_n^{(2)} - \breve{\boldsymbol{\beta}}_n)$ is a statistical functional of two weighted empirical processes $\mathbf{S}_n^\star(t, \mathbf{u})$ and $J_n^\star(t, \mathbf{u})$, $t \in [0,1]$, where $\mathbf{u} = n^{1/2}\mathbf{Q}_n^0(\breve{\boldsymbol{\beta}}_n - \boldsymbol{\beta}) \in \mathbb{R}_p$, and

$$\mathbf{S}_n^\star(t, \mathbf{u}) = \sum_{i=1}^{n} x_{ni}^\star I(Y_i^0 \le \mathbf{x}_{ni}'\mathbf{u} + F^{-1}(t)), \tag{4.7.53}$$

$$\mathbf{J}_n^\star(t, \mathbf{u}) = n^{-1}\sum_{i=1}^{n} I(Y_i^0 \le \mathbf{x}_{ni}'\mathbf{u} + F^{-1}(t)), \tag{4.7.54}$$

with $Y_i^0 = Y_i - \mathbf{x}_i'\boldsymbol{\beta}$ (i.i.d. r.v.'s with d.f. F) and

$$\mathbf{x}_{ni} = n^{-\frac{1}{2}}(\mathbf{Q}_n^0)^{-1}\mathbf{x}_i = (n^{-\frac{1}{2}}, \mathbf{x}_{ni}^{\star'})',\ i = 1, \ldots, n. \tag{4.7.55}$$

Since similar weighted empirical processes appear in more general forms in Chapter 5, we omit the details, and present the following result due to Ren (1994).

Suppose that $h(.)$ is the difference of two nonnegative functions and is bounded and continuous a.e. with $h(t) = 0$ for $t \notin [\alpha_1, \alpha_2]$ for some $0 < \alpha_1 < \alpha_2 < 1$. Also assume that **A1**, **A2** and **A3** hold and that $\int_0^1 h(t)dt + \sum_{j=1}^m w_j = 1$. Moreover the Y_i^0 are the true errors with a d.f. F having location 0, and hence we may set without any loss of generality that

$$T(F) = \int_0^1 F^{-1}(t)h(t)dt + \int_0^1 F^{-1}(t)dm(t) = 0 \qquad (4.7.56)$$

Finally, assume that

$$n^{1/2}\mathbf{Q}_n^0(\breve{\boldsymbol{\beta}}_n - \boldsymbol{\beta}) = \mathrm{O}_p(1), \qquad (4.7.57)$$

and

$$\phi_n(q_j) - \phi(q_j) \xrightarrow{P} 0, \text{ as } \ n \to \infty \ 1 \leq j \leq p. \qquad (4.7.58)$$

Then we have

$$n^{1/2}\mathbf{Q}_n^0(\breve{\boldsymbol{\beta}}_n^{(2)} - \boldsymbol{\beta}) + \mathbf{Q}_n^0\mathbf{Q}_n^{-1}\sum_{j=1}^n \mathbf{x}_{ni}\psi(Y_i^0) \xrightarrow{P} 0, \qquad (4.7.59)$$

where

$$\begin{aligned}\psi(x) \ &= \ \int [I(x < y) - F(y)]h(F(y))dy \\ &+ \ \sum_{j=1}^m w_j\phi(q_j)[I(x \leq F^{-1}(q_j)) - q_j] \ \text{ for } x \in \mathbb{R}_1. \qquad (4.7.60)\end{aligned}$$

From (4.7.59) and (4.7.60), we obtain that

$$n^{1/2}(\breve{\boldsymbol{\beta}}_n^{(2)} - \boldsymbol{\beta}) \xrightarrow{\mathcal{D}} \mathcal{N}_p(\mathbf{0}, \sigma_\psi^2\mathbf{Q}^{-1}), \qquad (4.7.61)$$

where

$$\sigma_\psi^2 = \int_{\boldsymbol{R}} \psi^2(x)dF(x) < \infty. \qquad (4.7.62)$$

We relegate the proof of the above main result as exercises (see Problems 4.7.3 and 4.7.4). Recall that σ_ψ^2 is comparable to the location model treated earlier, so such one-step versions provide the natural analogues of the location L-statistics for the regression model.

4.8 BREAKDOWN POINT OF SOME L- AND M-ESTIMATORS

Consider the model (4.7.1) with fixed design matrix $\mathbf{X}$. If we define the breakdown point of an estimator $\mathbf{T}_n$ of $\boldsymbol{\beta}$ by perturbing $(\mathbf{x}_i', Y_i)$ pairs, then the L_1-estimator as well as any other M-estimator has breakdown $1/n$ (e.g., see Bloomfield and Steiger 1983).

Alternatively, define the breakdown with respect to outliers in $\mathbf{Y}$ only: Let $m^\star = m^\star(\mathbf{Y})$ be the minimum number of replacements of components of $\mathbf{Y}$ by arbitrary values that can lead to the infinite norm $||\mathbf{T}_n||$ with a fixed $\mathbf{X}$. Even with a fixed design the breakdown point in the linear model is not in correspondence with that in the location model, as we will soon illustrate.

Let $m_\star$ denote the largest integer m such that for any subset M of $N = \{1, 2, \ldots, n\}$ of size m,

$$\inf_{||\mathbf{b}||=1} \left\{ \left[\sum_{i \in N \setminus M} |\mathbf{x}_i'\mathbf{b}| \right] / \left[\sum_{i \in N} |\mathbf{x}_i'\mathbf{b}| \right] \right\} > \frac{1}{2}. \tag{4.8.1}$$

There is a close relation between $m^\star$ and $m_\star$ in the case of L_1-estimator and some M-estimators including the Huber estimator, as is described in the following theorem:

Theorem 4.8.1 *(i) Let $\mathbf{T}_n$ be the L_1-estimator of $\boldsymbol{\beta}$ defined through the minimization*

$$\sum_{i=1}^{n} |Y_i - \mathbf{x}_i'\mathbf{t}| := \min, \quad \mathbf{t} \in I\!R_p. \tag{4.8.2}$$

Then

$$m_\star + 1 \le m^\star \le m_\star + 2. \tag{4.8.3}$$

(ii) The relations (4.8.3) are also true for the l_1-type estimators of $\boldsymbol{\beta}$ defined through the minimization

$$\sum_{i=1}^{n} \rho(Y_i - \mathbf{x}_i'\mathbf{t}) := \min, \tag{4.8.4}$$

with the continuous function ρ satisfying

$$\big|\rho(u) - |u|\big| \le K, \quad u \in I\!R_1, \ 0 < K < \infty. \tag{4.8.5}$$

Remark. We do not know the upper bound for $m_\star/n$. In the special case of regression line coming through the origin and equidistant values x_i's running over [0,1], $m_\star/n \to 1 - 2^{-1/2} \approx 0.29289$. Generally the breakdown point of

M-estimators in the linear model may be quite low, even when calculated with respect to outliers in $\mathbf{Y}$ only.

Proof. Without loss of generality, put $\boldsymbol{\beta} = \mathbf{0}$.
(i) If all but $m \leq m^{\star}$ of the Y's are bounded (e.g., by 1), then $||\mathbf{T}_n||$ will be uniformly bounded. Actually, by the triangle inequality,

$$\begin{aligned}\sum_{i \in N} |Y_i - \mathbf{x}_i'\mathbf{b}| &\geq \sum_{i \in N \setminus M} |\mathbf{x}_i'\mathbf{b}| - \sum_{i \in N \setminus M} |Y_I| - \Big(\sum_{i \in M} |\mathbf{x}_i'\mathbf{b}| - \sum_{i \in M} |Y_i| \Big) \\ &= \sum_{i \in N \setminus M} |\mathbf{x}_i'\mathbf{b}| - \sum_{i \in M} |\mathbf{x}_i'\mathbf{b}| + \sum_{i \in N} |Y_i| - 2 \sum_{i \in N \setminus M} |Y_i|.\end{aligned}$$

By definition of $m_{\star}$, there exists $c > 1/2$ such that

$$\sum_{i \in N \setminus M} |\mathbf{x}_i'\mathbf{b}| \geq c \sum_{i \in M} |\mathbf{x}_i'\mathbf{b}|, \quad \text{and} \quad \sum_{i \in M} |\mathbf{x}_i'\mathbf{b}| \leq (1-c) \sum_{i \in N} |\mathbf{x}_i'\mathbf{b}|$$

for all subsets M of size $m \leq m_{\star}$. Therefore

$$\sum_{i \in N} |Y_i - \mathbf{x}_i'\mathbf{b}| - \sum_{i \in N} |Y_i| \geq (2c-1) \sum_{i \in N} |\mathbf{x}_i'\mathbf{b}| - 2 \sum_{i \in N \setminus M} |Y_i|,$$

and if $|Y_i| \leq 1$ for $i \in N \setminus M$, there exists a constant C such that if $||\mathbf{b}|| > C$, then

$$\sum_{i \in N} |Y_i - \mathbf{x}_i'\mathbf{b}| - \sum_{i \in N} |Y_i| \geq 0.$$

Thus $m^{\star} \geq m_{\star} + 1$.

By the definition of $m_{\star}$, there exists a subset M of size $m_{\star} + 1$ and a vector $\mathbf{b}_0$, $||\mathbf{b}_0|| = 1$ such that

$$\sum_{i \in N \setminus M} |\mathbf{x}_i'\mathbf{b}_0| \leq \sum_{i \in M} |\mathbf{x}_i'\mathbf{b}_0|.$$

Thus, for $m = n_{\star} + 2$, there exists a subset $M^{\star} \subset N$ such that

$$\sum_{i \in N \setminus M^{\star}} |\mathbf{x}_i'\mathbf{b}_0| < \sum_{i \in M^{\star}} |\mathbf{x}_i'\mathbf{b}_0| \quad \text{with strict inequality.}$$

Then $\eta(\mathbf{b}_0) > 0$, where

$$\eta(\mathbf{b}) = \sum_{i \in M^{\star}} |\mathbf{x}_i'\mathbf{b}| - \sum_{i \in N \setminus M^{\star}} |\mathbf{x}_i'\mathbf{b}|.$$

Suppose that $Y_i = 0$ for $i \in N \setminus M^\star$, and $Y_i = c\mathbf{x}_i'\mathbf{b}_0$ for $i \in M^\star$, $c > 0$. Then

$$\sum_{i\in N} |Y_i - c\mathbf{x}_i'\mathbf{b}_0| = \sum_{i\in N\setminus M^\star} |c\mathbf{x}_i'\mathbf{b}_0| = \sum_{i\in M^\star} |c\mathbf{x}_i'\mathbf{b}_0| - c\eta(\mathbf{b}_0) = \sum_{i\in N} |Y_i| - c\eta(\mathbf{b}_0).$$

On the other hand, for a bounded $\boldsymbol{\beta}$ and sufficiently large c,

$$\begin{aligned} \sum_{i\in N} |Y_i - \mathbf{x}_i'\boldsymbol{\beta}| &\geq \sum_{i\in N} |Y_i| - \sum_{i\in N} |\mathbf{x}_i'\boldsymbol{\beta}| \\ &\geq \sum_{i\in N} |Y_i| - n||\boldsymbol{\beta}|||(\max_i ||\mathbf{x}_i||) > \sum_{i\in N} |Y_i - c\mathbf{x}_i'\mathbf{b}_0|. \end{aligned}$$

This means that $m_\star + 2$ outliers in $\mathbf{Y}$ may lead to a breakdown of $\mathbf{T}_n$.

(ii) Let $\mathbf{T}_1$ be an M-estimator minimizing (4.8.4) with ρ satisfying (4.8.5), and let $\mathbf{T}$ be the L_1-estimator. Then there exists a constant $c > 0$ such that

$$||\mathbf{T}_1 - \mathbf{T}|| \leq (2nK)/c. \tag{4.8.6}$$

Similarly, as in (4.7.18)-(4.7.19), we have for $||\mathbf{b}|| = 1$ (the directional derivative of the objective function a $\mathbf{T}_n$ in direction $\mathbf{b}$)

$$-\sum_{i\notin h} \text{sign}(Y_i - \mathbf{x}_i'\mathbf{T})\mathbf{x}_i'\mathbf{b} + \sum_{i\in h} |\mathbf{x}_i'\mathbf{b}| > 0,$$

where $h = \{i : Y_i = \mathbf{x}_i'\mathbf{t}_n\}$. Then

$$c = \inf_{||\mathbf{b}||=1} \left\{ \min \left[\epsilon_i = \pm 1, \; i \notin h : -\sum_{i\notin h} \epsilon_i \mathbf{x}_i'\mathbf{b} + \sum_{i\in h} |\mathbf{x}_i'\mathbf{b}| > 0 \right] \right\} > 0,$$

and for $||\boldsymbol{\beta} - \mathbf{T}|| \geq 2nK/c$,

$$\begin{aligned} \sum_{i=1}^n \rho(Y_i - \mathbf{x}_i'\boldsymbol{\beta}) &\geq \sum_{i=1}^n |Y_i - \mathbf{x}_i'\boldsymbol{\beta}| - nK \\ &\geq \sum_{i=1}^n |Y_i - \mathbf{x}_i'\mathbf{T}| + \frac{2nK}{c} \cdot c - nK \\ &\geq \sum_{i=1}^n \rho(Y_i - \mathbf{x}_i'\mathbf{T}) \geq \sum_{i=1}^n \rho(Y_i - \mathbf{x}_i'\mathbf{T}_1). \end{aligned}$$

Hence (4.8.6) follows. Together with proposition (i), this result implies proposition (ii). □

Remark. He et al. (1990) has studied the close connection between the breakdown point and the measure of tail performance (defined in Chapter 3) for a broad class of robust estimators in linear models.

4.9 PROBLEMS

4.2.1. Use the Outline of the proof of Theorem 4.2.2, and verify (4.2.11).

4.2.2. Use the technique in Theorem 4.2.2 to show that in Theorem 4.2.1, $r_n = \mathrm{O}_p(n^{-3/4})$ when $f'(x)$ exists in a neighborhood of ξ_p. Use this decomposition for each $X_{n:[np_j]+1}, 1 \leq j \leq k$, and verify that in Theorem 4.2.3, $r_n = \mathrm{O}_p(n^{-3/4})$.

4.2.3. Under (4.2.15), for $\alpha > 1/2$, show that Theorem 4.2.2 holds for $T_n = h(X_{n:k_n})$, with r_n satisfying (4.1.2).

4.2.4. Use the first order Taylor's expansion for $h(X_{n:k_n})$ and (4.2.15) to show that for $\alpha : 0 < \alpha \leq 1/2$, r_n satisfies (4.2.16).

4.2.5. Verify that Theorem 4.2.1 holds for $h(X_{n:[np_1]+1}, \ldots, X_{n:[np_k]+1})$ where its gradient $\mathbf{h}'_{\boldsymbol{\xi}}$ exists.

4.2.6. Extend the result in the previous problem to the order of r_n in (4.1.2) or (4.2.16), under (4.2.17) with $\alpha > 1/2$ or $0 < \alpha \leq 1/2$.

4.3.1. Provide a formal proof of (4.3.15).

4.3.2. For the rank weighted mean $T_{n,k}$, consider the asymptotic situation where $k \sim n\alpha$ for some $0 < \alpha < 1/2$. Use the Stirling approximation to the factorials, and obtain the expression for the limiting score function $J(u)$, $0 < u < 1$. Compare this score function with that arising when k is held fixed. Which, if any, conforms to (4.3.3)? Comment on the case where $k = n/2 - \mathrm{O}(1)$. Finally for $k \sim np$, compare the rank-weighted mean and (4.3.35)-(4.3.36).

4.3.3. Consider the logistic d.f. $F(x) = \{1 + e^{-(x-\theta)}\}^{-1}$, $x \in \mathbb{R}$. Show that $-f'(x)/f(x) = 2F(x) - 1$, $x \in \mathbb{R}$. Hence, or otherwise, find the ABLUE of θ.

4.3.4. Consider the Laplace d.f.: $f(x) = (1/2)e^{-|x-\theta|}$, $x \in \mathbb{R}$. What is the BLUE of θ ? Verify whether or not **C1**, **C2**,or **C3** holds.

4.3.5. Consider the density $f(x)$ given by

$$I(x < \theta)\sigma_1^{-1}\phi((x-\theta)/\sigma_1) + I(x > \theta)\phi((x-\theta)/\sigma_2), \quad x \in \mathbb{R},$$

where $\theta \in \Theta \subset I\!R$, $\sigma_1 > \sigma_2 > 0$. Note that f has a jump discontinuity at θ. Find out the ABLUE of θ when $\sigma_1/\sigma_2 = c$ is known.

4.4.1. Provide a formal proof of Theorem 4.4.1.

4.4.2. For the density $f(.)$ in Problem 4.3.5, verify whether or not the representation in (4.4.5) holds. For the same model, show that for the sample median, (4.4.5) does not hold.

4.4.3. In a life-testing model, let $f(x) = \theta^{-1}e^{-x/\theta}$, $x \geq 0$, $\theta > 0$, be the density, and suppose that corresponding to a given $r : r \leq n$ and $r/n \sim p$: $0 < p < 1$, the observed data set relates to the order statistics $X_{n:1}, \ldots, X_{n:r}$. Obtain the MLE $(\hat{\theta}_{nr})$ of θ, and show that it is of the form (4.4.1). Hence, or otherwise, verify that (4.4.5) holds.

4.4.4. For Problem 4.4.3, consider a type I censoring at a point T : $0 < T < \infty$, and let r_T be the number of failures during the time $(0, T)$. Obtain the MLE of θ in this case. What can you say about the representation in (4.4.5) for this MLE of θ?

4.5.1. Consider the sample median $\tilde{X}_n = T(F_n) = \inf\{x : \ F_n(x) \geq 1/2\} = F_n^{-1}(0.5)$. Thus $T(F) = F^{-1}(0.5)$ is the population counterpart of $\tilde{X}_n$. Use the definition (4.5.5) to verify whether or not $T(F_n)$ is second-order Hadamard differentiable. Is $T(F_n)$ first-order Hadamard differentiable?

4.5.2. As an estimator of the scale parameter, consider the following:

$$U_n = \binom{n}{2}^{-1} \sum_{\{1 \leq i < j \leq n\}} |X_i - X_j|.$$

Show that U_n may also be written as

$$\begin{aligned} &\binom{n}{2}^{-1} \sum_{i=1}^{n} (n - 2i + 1) X_{n:n-i+1} \\ &= \binom{n}{2}^{-1} \sum_{i \leq (n+1)/2} (n - 2i + 1)[X_{n:n-i+1} - X_{n:i}]. \end{aligned}$$

Hence or otherwise, verify whether or not U_n is first-/second-order Hadamard differentiable. What happens when the d.f. F (of X) has a compact support?

4.5.3. Consider a general von Mises' functional

$$T(F_n) = \int \cdots \int \phi(x_1, \ldots, x_m) dF_n(x_1) \ldots dF_n(x_m),$$

where $m \geq 1$ and $\phi(.)$ is a kernel of degree m. Show that when $\phi(.)$ is bounded a.e., Hadamard differentiability of $T(F)$ (at F) holds.

4.5.4. Let $f(.)$ be a normal density with mean θ and variance σ^2. Consider the ABLUE of σ based on n observations $X_1, \ldots, X_n$. Verify whether or not this estimator satisfies the Hadamard-differentiability condition in (4.5.5).

4.6.1. Verify the identity in (4.6.30).

4.6.2. If $\mathbb{E}_F|X|^r < \infty$ for some $r > 0$, show that for every $k \geq 1$ such that $(k+1)r \geq p$, $\mathbb{E}|\tilde{X}_{2k+1}|^p < \infty$. In particular, for $p = 2$, comment on the Cauchy d.f., and show that $k \geq 2$ is sufficient for the sample median $\tilde{X}_{2k+1}$ to have a finite variance.

4.7.1.Define $\mathbf{A}_n = Diag(a_i)$ as in (4.7.11), and verify that under appropriate regularity conditions, as $n \to \infty$,

$$n^{-1}\mathbf{X}_n'\mathbf{A}_n\mathbf{X}_n \xrightarrow{P} \mathbf{Q}^\star \quad \text{p.d.}$$

Obtain an expression for $\mathbf{Q}^\star$.

4.7.2. Show that as $n \to \infty$, for $\alpha_1 < 1/2 < \alpha_2$,

$$\mathbb{P}\{\mathbf{x}_i'\hat{\boldsymbol{\beta}}(\alpha_1) \geq \mathbf{x}_i'\hat{\boldsymbol{\beta}}(\alpha_2)\} \to 0.$$

4.7.3. Verify (4.7.59) with the aid of the Hadamard differentiability approach. (Ren 1994)

4.7.4. Hence, or otherwise, verify (4.7.61).

Chapter 5

Asymptotic Representations for M-Estimators

5.1 INTRODUCTION

In Chapter 4 we thoroughly studied the intricate interrelationships of L-estimators, L-functionals, and differentiable statistical functionals and, from some basic resuilts, in a rather simple manner established asymptotic representations. M-estimators of location and/or regression parameters are also expressible as statistical functionals but in an implicit manner. This in turn requirea a more elaborate treatment. Verification of Hadamard (or compact) differentiability of M-functionals (yielding M-estimators) requires bounded score functions as well as other regularity conditions. Although such a bounded condition is often justified on the ground of robustness, the formulation excludes the MLE, the precursors of M-estimators. From theoretical standpoint some alternative methods can be more conveniently applied. In fact the "uniform asymptotic linearity of M-statistics" (in the associated parameter(s)) can play a fundamental role, and this context constitutes the main theme of the current chapter. A variety of one-step and studentized versions of M-estimators of location/regression parameters, will be considered along with their related FOADR and SOADR results are presented systematically.

Section 5.2 is devoted to the study of M-estimators of general parameters. Section 5.3 probes into deeper results for the location model, under somewhat simpler regularity conditions, when the scale parameter is treated as fixed (i.e., not as nuisance). Studentized M-estimators of location are presented in Section 5.4. As in the preceding chapter, the linear (regression) model is treated fully, and discussion on M-estimators for the fixed scale and their studentized versions is deferred to Section 5.5. Section 5.6 examines the basic notion of scale statistics in this development. For completeness, a broad review of the Hadamard differentiability of M-functionals is provided in Section 5.7. Some comparisons of the alternative methods are also made in this section.

5.2 M-ESTIMATION OF GENERAL PARAMETERS

Let $\{X_i;\ i \geq 1\}$ be a sequence of i.i.d. r.v.'s with a d.f. $F(x, \boldsymbol{\theta})$ where $\boldsymbol{\theta} \in \boldsymbol{\Theta}$, an open set in $\mathbb{R}_p$. The true value of $\boldsymbol{\theta}$ is denoted by $\boldsymbol{\theta}_0$. Let $\rho(x, \mathbf{t}) : \mathbb{R} \times \boldsymbol{\Theta} \mapsto \mathbb{R}$ be a function, absolutely continuous in the elements of $\mathbf{t}$ (i.e., $t_1, \ldots, t_p$), and such that the function

$$h(\mathbf{t}) = \mathbb{E}_{\boldsymbol{\theta}_0} \rho(X_1, \mathbf{t}) \tag{5.2.1}$$

exists for all $\mathbf{t} \in \boldsymbol{\Theta}$ and has a unique minimum over $\boldsymbol{\Theta}$ at $\mathbf{t} = \boldsymbol{\theta}_0$. Then the M-estimator $\mathbf{M}_n$ of θ_0 is defined as the point of global minimum of $\sum_{i=1}^n \rho(X_i, \mathbf{t})$ with respect to $\mathbf{t} \in \boldsymbol{\Theta}$, i.e.,

$$\mathbf{M}_n = \text{Arg. min}\{\sum_{i=1}^{n} \rho(X_i, \mathbf{t}) \,:\, \mathbf{t} \in \boldsymbol{\Theta}\}. \tag{5.2.2}$$

It is of natural interest to study the following:

1. Under what regularity conditions (on $F, \boldsymbol{\theta}, \rho$), does there exist a solution of (5.2.2) that is a $\sqrt{n}$-consistent estimator of $\boldsymbol{\theta}_0$?

2. Under what regularity conditions, do FOADR (and in some cases SOADR) results hold?

In this generality some asymptotic results are considered, albeit under comparatively more stringent regularity conditions on $\rho(.)$. The situation is quite different for $p = 1$ (i.e., for a single parameter), where a SOADR result holds under quite general regularity conditions. This will be followed by some further simplifications for the location model, to be treated in Section 5.3. For simplicity we will denote $\mathbb{E}_{\boldsymbol{\theta}_0}(.)$ and $P_{\boldsymbol{\theta}_0}(.)$ by $\mathbb{E}(.)$ and $P(.)$, respectively.

Let us consider first the general vector parameter. Here we will impose the following regularity conditions on ρ and on F:

A1. *First-order derivatives.* The functions $\psi_j(x, \mathbf{t}) = (\partial/\partial t_j)\rho(x, \mathbf{t})$ are assumed to be absolutely continuous in t_k with the derivatives $\dot{\psi}_{jk}(x, \mathbf{t}) = (\partial/\partial t_k)\psi_j(x, \mathbf{t})$, such that $\mathbb{E}[\dot{\psi}_{jk}(X_1, \boldsymbol{\theta}_0)]^2 < \infty$, $(j, k = 1, \ldots, p)$. Further we assume that the matrices $\boldsymbol{\Gamma}(\boldsymbol{\theta}_0) = [\gamma_{jk}(\boldsymbol{\theta}_0)]_{j,k=1}^p$ and $\mathbf{B}(\boldsymbol{\theta}_0) = [b_{jk}(\boldsymbol{\theta}_0)]_{j,k=1}^p$ are positive definite, where

$$\gamma_{jk}(\boldsymbol{\theta}) = \mathbb{E}_{\boldsymbol{\theta}} \dot{\psi}_{jk}(\boldsymbol{\theta}), \tag{5.2.3}$$

and

$$b_{jk}(\boldsymbol{\theta}) = \text{Cov}_{\boldsymbol{\theta}}(\psi_j(X_1, \boldsymbol{\theta}), \psi_k(X_1, \boldsymbol{\theta})), \quad j, k = 1, \ldots, p. \tag{5.2.4}$$

A2. *Second- and third-order derivatives.* $\dot{\psi}_{jk}(x, \mathbf{t})$ are absolutely continuous in the components of $\mathbf{t}$ and there exist functions $M_{jkl}(x, \boldsymbol{\theta}_0)$ such that $m_{jkl} = \mathbb{E} M_{jkl}(X_1, \boldsymbol{\theta}_0) < \infty$ and

$$|\ddot{\psi}_{jkl}(x, \boldsymbol{\theta}_0 + \mathbf{t})| \leq M_{jkl}(x, \boldsymbol{\theta}_0), \quad x \in \mathbb{R}_1, \ ||\mathbf{t}|| \leq \delta, \ \delta > 0,$$

where

$$\ddot{\psi}_{jkl}(x, \mathbf{t}) = \frac{\partial^2 \psi_j(x, \mathbf{t})}{\partial t_k \partial t_l}, \quad j, k, l = 1, \ldots, p.$$

Under the conditions **A1** and **A2**, the point of global minimum in (5.2.2) is one of the roots of the system of equations:

$$\sum_{i=1}^{n} \psi_j(X_i, \mathbf{t}) = 0, \quad j = 1, \ldots, p. \tag{5.2.5}$$

The following theorem states an existence of a solution of (5.2.5) that is $\sqrt{n}$-consistent estimator of $\boldsymbol{\theta}_0$ and admits an asymptotic representation.

Theorem 5.2.1 *Let $X_1, X_2, \ldots$ be i.i.d. r.v.'s with d.f. $F(x, \boldsymbol{\theta}_0)$, $\boldsymbol{\theta}_0 \in \boldsymbol{\Theta}$, $\boldsymbol{\Theta}$ being an open set of $\mathbb{R}_p$. Let $\rho(x, \mathbf{t}) : \mathbb{R}_1 \times \boldsymbol{\Theta} \mapsto \mathbb{R}_1$ be a function absolutely continuous in the components of $\mathbf{t}$ and such that the function $h(\mathbf{t})$ of (5.2.1) has a unique minimum at $\mathbf{t} = \boldsymbol{\theta}_0$. Then, under the conditions* **A1** *and* **A2**, *there exists a sequence $\{M_n\}$ of solutions of (5.2.5) such that*

$$n^{\frac{1}{2}} ||\mathbf{M}_n - \boldsymbol{\theta}_0|| = O_p(1) \ \textit{as} \ n \to \infty, \tag{5.2.6}$$

and

$$\mathbf{M}_n = \boldsymbol{\theta}_0 - n^{-1} (\boldsymbol{\Gamma}(\boldsymbol{\theta}_0))^{-1} \sum_{i=1}^{n} \boldsymbol{\psi}(X_i, \boldsymbol{\theta}_0) + \mathbf{O}_p(n^{-1}), \tag{5.2.7}$$

where $\boldsymbol{\psi}(x, \boldsymbol{\theta}) = (\psi_1(x, \boldsymbol{\theta}), \ldots, \psi_p(x, \boldsymbol{\theta}))'$.

The representation (5.2.7) implies the asymptotic normality of $\mathbf{M}_n$ which we will state as a corollary:

COROLLARY 5.2.1. *Under the conditions of Theorem 5.2.1, $n^{1/2}(\mathbf{M}_n - \boldsymbol{\theta}_0)$ has asymptotically p-dimensional normal distribution $\mathcal{N}_P(\mathbf{0}, \mathbf{A}(\boldsymbol{\theta}_0))$ with $\mathbf{A}(\boldsymbol{\theta}_o) = (\boldsymbol{\Gamma}(\boldsymbol{\theta}_0))^{-1} \mathbf{B}(\boldsymbol{\theta}_0)$.*

Proof of Theorem 5.2.1. First we prove the uniform second order asymptotic linearity for the vector of partial derivatives of ρ: for fixed C, $0 < C < \infty$,

$$\sup_{||\mathbf{t}|| \leq C} ||n^{-\frac{1}{2}} \sum_{i=1}^{n} \left[\boldsymbol{\psi}(X_i, \boldsymbol{\theta}_0 + n^{-\frac{1}{2}} \mathbf{t}) - \boldsymbol{\psi}(X_i, \boldsymbol{\theta}_0) \right] - \boldsymbol{\Gamma}(\boldsymbol{\theta}_0) \mathbf{t}|| = O_p(n^{-\frac{1}{2}}), \tag{5.2.8}$$

as $n \to \infty$. When proved, (5.2.8) implies the consistency and asymptotic representation in the following way: first, given an $\epsilon > 0$, there exists $K > 0$ and an integer n_0 so that, for $n \geq n_0$,

$$P(||n^{-\frac{1}{2}} \sum_{i=1}^{n} \boldsymbol{\psi}(X_i, \boldsymbol{\theta}_0)|| > K) < \epsilon/2. \tag{5.2.9}$$

Take $C > (\epsilon + K)/\lambda_1(\boldsymbol{\theta}_0)$ where $\lambda_1(\boldsymbol{\theta}_0)$ is the minimum eigenvalue of $\boldsymbol{\Gamma}(\boldsymbol{\theta}_0)$. Then for $n \geq n_0$,

$$P(\sup_{||\mathbf{t}||=C} ||\mathbf{S}_n(\mathbf{t})|| > \epsilon) < \epsilon/2, \tag{5.2.10}$$

where $\mathbf{S}_n(\mathbf{t}) = n^{-\frac{1}{2}} \sum_{i=1}^{n} \left[\boldsymbol{\psi}(X_i, \boldsymbol{\theta}_0 + n^{-\frac{1}{2}}\mathbf{t}) - \boldsymbol{\psi}(X_i, \boldsymbol{\theta}_0)\right] + \boldsymbol{\Gamma}(\boldsymbol{\theta}_0)\mathbf{t}$. By (5.2.9) and (5.2.10),

$$\begin{aligned}
&\inf_{||\mathbf{t}||=C} \left\{n^{-\frac{1}{2}}\mathbf{t}' \sum_{i=1}^{n} \boldsymbol{\psi}(X_i, \boldsymbol{\theta}_0 + n^{-\frac{1}{2}}\mathbf{t})\right\} \\
&\geq \inf_{||\mathbf{t}||=C} \{\mathbf{t}'\mathbf{S}_n(\mathbf{t})\} + \inf_{||\mathbf{t}||=C} \left\{ - \mathbf{t}'n^{-\frac{1}{2}} \sum_{i=1}^{n} \boldsymbol{\psi}(X_i, \boldsymbol{\theta}_0) + \mathbf{t}'\boldsymbol{\Gamma}(\boldsymbol{\theta}_0)\mathbf{t}\right\} \\
&\geq -C \inf_{||\mathbf{t}||=C} ||\mathbf{S}_n(\mathbf{t})|| + C^2\lambda_1(\boldsymbol{\theta}_0) - C||n^{-\frac{1}{2}} \sum_{i=1}^{n} \boldsymbol{\psi}(X_i, \boldsymbol{\theta}_0)||.
\end{aligned}$$

Hence

$$\begin{aligned}
&P\Big(\inf_{||\mathbf{t}||=C} \left\{\mathbf{t}'n^{-\frac{1}{2}} \sum_{i=1}^{n} \boldsymbol{\psi}(X_i, \boldsymbol{\theta}_0 + n^{-\frac{1}{2}}\mathbf{t})\right\} > 0\Big) \\
&\geq P\Big\{||n^{-\frac{1}{2}} \sum_{i=1}^{n} \boldsymbol{\psi}(X_i, \boldsymbol{\theta}_0)|| + \inf_{||\mathbf{t}||=C} ||\mathbf{S}_n(\mathbf{t})|| < C\lambda_1(\boldsymbol{\theta}_0)\Big\} \\
&\geq 1 - \epsilon \quad \text{for } n \geq n_0.
\end{aligned} \tag{5.2.11}$$

By Theorem 6.4.3 of Ortega and Rheinboldt (1970), we conclude from (5.2.11) that for $n \geq n_0$ the system of equations

$$\sum_{i=1}^{n} \psi_j(X_i, \boldsymbol{\theta}_0 + n^{-\frac{1}{2}}\mathbf{t}) = 0, \quad j = 1, \ldots, p,$$

has a root $\mathbf{T}_n$ that lies in the sphere $||t|| \leq C$ with probability exceeding $1 - \epsilon$ for $n \geq n_0$. Then $\mathbf{M}_n = \boldsymbol{\theta}_0 + n^{-\frac{1}{2}}\mathbf{T}_n$ is a solution of (5.2.5) satisfying $P(||n^{\frac{1}{2}}(\mathbf{M}_n - \boldsymbol{\theta}_0)|| \leq C) \geq 1 - \epsilon$ for $n \geq n_0$. Inserting $\mathbf{t} \to n^{\frac{1}{2}}(\mathbf{M}_n - \boldsymbol{\theta}_0)$ in (5.2.8), we obtain the representation (5.2.7).

To prove the uniform asymptotic linearity (5.2.8), we refer to the Appendix (Section A.2) and note that by **A1** and **A2**,

$$\sup_{||\mathbf{t}||\leq C} |n^{-\frac{1}{2}} \sum_{i=1}^{n} \left[\psi_j(X_i, \boldsymbol{\theta}_0 + n^{-\frac{1}{2}}\mathbf{t}) - \psi_j(X_i, \boldsymbol{\theta}_0) - n^{-\frac{1}{2}} \sum_{k=1}^{p} t_k\dot{\psi}_{jk}(X_i, \boldsymbol{\theta}_0)\right]|$$

$$\leq C^2 n^{-\frac{3}{2}} \sum_{i=1}^{n} \sum_{k,l=1}^{p} M_{jkl}(X_i, \boldsymbol{\theta}_0) = O_p(n^{-\frac{1}{2}}), \quad j = 1, \ldots, p.$$

Moreover, uniformly for $||\mathbf{t}|| \leq C$,

$$\mathbb{E}\Big[n^{-1} \sum_{i=1}^{n} \sum_{k=1}^{p} t_k(\dot{\psi}_{jk}(X_i, \boldsymbol{\theta}_0) - \gamma_{jk}(\boldsymbol{\theta}_o))\Big]^2$$

$$= n^{-1}\mathrm{Var}\Big\{\sum_{k=1}^{p} t_k \dot{\psi}_{jk}(X_1, \boldsymbol{\theta}_0)\Big\} \leq K/n$$

for some $K > 0$, $j = 1, \ldots, p$. Combining the last two statements, we arrive at (5.2.8), and this completes the proof of the theorem. □

If $\rho(x, \mathbf{t})$ is convex in $\mathbf{t}$, then the solution of (5.2.2) is uniquely determined. However, since we did not assume ρ to be convex, there may exist more roots of the system (5.2.5) satisfying (5.2.6). Let $\mathbf{M}_n$, $\mathbf{M}_n^\star$ be two such sequences of roots. The following corollary of Theorem 5.2.1 shows that then $\mathbf{M}_n$ and $\mathbf{M}_n^\star$ are asymptotically equivalent up to $O_p(n^{-1})$:

COROLLARY 5.2.2. Let $\mathbf{M}_n$ and $\mathbf{M}_n^\star$ be two sequences of roots of (5.2.5) satisfying (5.2.6). Then, under the conditions of Theorem 5.2.1,

$$||\mathbf{M}_n - \mathbf{M}_n^\star|| = O_p(n^{-1}) \text{ as } n \to \infty. \tag{5.2.12}$$

In the rest of this section we will deal with the case $p = 1$, the case of a single one-dimensional parameter. Theorem 5.2.1 and its corollaries apply naturally to this special case. Here we are able not only to prove the representation (5.2.7) but also to supplement it by the asymptotic (nonnormal) distribution of the remainder term (SOADR).

In particular, assume that $\theta_0 \in \Theta$, Θ being an open interval in $\mathbb{R}_1$. Let $\rho(x, t) : \mathbb{R}_1 \times \Theta \mapsto \mathbb{R}_1$ be a function such that the function

$$h(t) = \mathbb{E}_{\theta_0}\rho(X_1, t) \tag{5.2.13}$$

exists for $t \in \Theta$ and has a unique minimum at $t = \theta_0$. Assume that $\rho(x, t)$ is absolutely continuous in t with the derivatives $\psi(x, t)$, $\dot{\psi}(x, t)$, and $\ddot{\psi}(x, t)$ with respect to t. As in the general case, define the M-estimator M_n of θ_0 as a solution of the minimization (5.2.2) in $t \in \Theta \subset \mathbb{R}_1$; then M_n is a root of the equation

$$\sum_{i=1}^{n} \psi(X_i, t) = 0, \quad t \in \Theta. \tag{5.2.14}$$

We will impose the following conditions on ρ and on F:

B1. *Moments of derivatives.* There exists $K > 0$ and $\delta > 0$ such that

$$\mathbb{E}[\psi(X_1, \theta_0 + t)]^2 \leq K, \quad \mathbb{E}[\dot{\psi}(X_1, \theta_0 + t)]^2 \leq K,$$

and

$$\mathbb{E}[\ddot{\psi}(X_1, \theta_0 + t)]^2 \leq K, \text{ for } |t| \leq \delta.$$

B2. *Fisher's consistency.* $0 < \gamma_1(\theta_0) < \infty$, where $\gamma_1(\theta) = \mathbb{E}_\theta \dot{\psi}(X_1, \theta)$.

B3. *Uniform continuity in the mean.* There exist $\alpha > 0$, $\delta > 0$, and a function $M(x, \theta_0)$ such that $m = \mathbb{E}M(X_1, \theta_0) < \infty$ and

$$|\ddot{\psi}(x, \theta_0 + t) - \ddot{\psi}(x, \theta_0)| \leq |t|^\alpha M(x, \theta_0) \text{ for } |t| \leq \delta, \text{a.s. } [F(x, \theta_0)].$$

Theorem 5.2.2 *(SOADR) Let $\{X_i, i \geq 1\}$ be a sequence of i.i.d.r.v.'s with the d.f. $F(x, \theta_0), \theta_0 \in \Theta$, Θ being an open interval in $\mathbb{R}_1$. Let $\rho(x, t)$: $\mathbb{R}_1 \times \Theta \mapsto \mathbb{R}_1$ be a function such that $h(t)$ of (5.2.13) has a unique minimum at $t = \theta_0$. Then, under the conditions* **B1-B3**, *there exists a sequence $\{M_n\}$ of roots of the equation (5.2.14) such that, as $n \to \infty$,*

$$n^{\frac{1}{2}}(M_n - \theta_0) = O_p(1), \tag{5.2.15}$$

$$M_n = \theta_0 - (n\gamma_1(\theta_0))^{-1} \sum_{i=1}^{n} \psi(X_i, \theta_0) + R_n, \tag{5.2.16}$$

and

$$nR_n \xrightarrow{\mathcal{D}} (\xi_1 - \xi_2\{\gamma_2(\theta_0)/2\gamma_1(\theta_0)\} \cdot \xi_2, \tag{5.2.17}$$

where

$$\gamma_2(\theta) = \mathbb{E}_\theta \ddot{\psi}(X_1, \theta), \tag{5.2.18}$$

and $(\xi_1, \xi_2)'$ is a random vector with normal $\mathcal{N}(\mathbf{0}, \mathbf{S})$ distribution, with $\mathbf{S} = [s_{ij}]_{i,j=1,2}$ and

$$s_{11} = (\gamma_1(\theta_0))^{-2}\text{Var}_{\theta_0}\dot{\psi}(X_1, \theta_0), \quad s_{22} = (\gamma_1(\theta_0))^{-2}\mathbb{E}_{\theta_0}\psi^2(X_1, \theta_0),$$
$$s_{12} = s_{21} = (\gamma_1(\theta_0))^{-2}\text{Cov}_{\theta_0}(\dot{\psi}(X_1, \theta_0), \psi(X_1, \theta_0). \tag{5.2.19}$$

Proof. For notational simplicity, denote $\gamma_j(\theta_0)$ by γ_j, $j = 1, 2$. Consider the random process $Y_n = \{Y_n(t),\ t \in [-B, B]\}$ defined by

$$Y_n(t) = \gamma_1^{-1} \sum_{i=1}^{n} [\psi(X_i, \theta_0 + n^{-\frac{1}{2}}t) - \psi(X_i, \theta_0)] - n^{\frac{1}{2}}t, \quad |t| \leq B,\ 0 < B < \infty. \tag{5.2.20}$$

The realizations of Y_n belong to the space $D[-B, B]$. We will first show that, as $n \to \infty$, Y_n converges to some Gaussian process in the Skorokhod J_1-topology on $D[-B, B]$, and then return to the proof of the theorem.

Lemma 5.2.1 *Under the conditions of Theorem 5.2.2, Y_n converges weakly (in the Skorokhod J_1-topology on $D[-B, B]$) to the Gaussian process $Y = \{Y(t),\ t \in [-B, B]\}$ where*

$$Y(t) = t\xi_1 - \{\gamma_2 t^2/(2\gamma_1)\}, \quad t \in [-B, B], \tag{5.2.21}$$

for any fixed B, $0 < B < \infty$, and ξ_1 is a random variable with normal $\mathcal{N}(0, s_{11})$ distribution.

Proof of Lemma 5.2.1. Denote

$$Z_n(t) = \gamma_1^{-1} \sum_{i=1}^{n} A_n(X_i, t), \quad Z_n^0(t) = Z_n(t) - \mathbb{E} Z_n(t), \tag{5.2.22}$$

where

$$A_n(X_i, t) = \psi(X_i, \theta_0 + n^{-\frac{1}{2}} t) - \psi(X_i, \theta_0), \quad t \in \mathbb{R}_1, \ i = 1, \ldots, n. \tag{5.2.23}$$

We will first prove that $Z_n^0 \Longrightarrow Y$ as $n \to \infty$. From the results in Section 2.5.10, it follows that we are to prove the weak convergence of finite dimensional distribution of Z_n^0 to those of Y and the tightness of the sequence of distributions of Z_n^0, $n = 1, 2, \ldots$. To prove the former, we show that, for any $\boldsymbol{\lambda} \in \mathbb{R}_p$, $\mathbf{t} \in \mathbb{R}_p$, $p \geq 1$,

$$\operatorname{Var}\Big\{ \sum_{j=1}^{p} \lambda_j Z_n(t_j) \Big\} \to s_{11} (\boldsymbol{\lambda}' \mathbf{t})^2 \quad \text{as } n \to \infty, \tag{5.2.24}$$

and then the weak convergence follows by the classical central limit theorem. To prove (5.2.24), we write

$$\operatorname{Var}\Big\{ \sum_{j=1}^{p} \lambda_j Z_n(t_j) \Big\} = \gamma_1^{-2} \sum_{i=1}^{n} \sum_{j,k=1}^{m} \lambda_j \lambda_k \operatorname{Cov}(A_n(X_i, t_j), A_n(X_i, t_k))$$

and it suffices to study $\operatorname{Cov}(A_n(X_1, t_1), A_n(X_1, t_2))$, $t_1, t_2 \in [0, B]$ (the considerations are analogous when (t_1, t_2) belong to another quadrant). Then we have, for $n \geq n_0$,

$$\begin{aligned}
& |\mathbb{E}\{A_n(X_1, t_1) A_n(X_1, t_2) - n^{-1} t_1 t_2 (\dot{\psi}(X_1, \theta_0))^2\}| \\
\leq\ & |\mathbb{E}\{[A_n(X_1, t_1) - n^{-\frac{1}{2}} t_1 (\dot{\psi}(X_1, \theta_0)] A_n(X_1, t_2)\}| \\
& + |\mathbb{E} n^{-\frac{1}{2}} t_1 \dot{\psi}(X_1, \theta_0) [A_n(X_1, t_2) - n^{-\frac{1}{2}} t_2 (\dot{\psi}(X_1, \theta_0)]\}| \\
\leq\ & |\mathbb{E}\Big\{ \int_0^{n^{-\frac{1}{2}} t_1} \int_0^u \ddot{\psi}(X_1, \theta_0 + v) dv \int_0^{n^{-\frac{1}{2}} t_2} \dot{\psi}(X_1, \theta_0 + w) dw \Big\}| \\
& + |\mathbb{E}\Big\{ n^{-\frac{1}{2}} t_1 \dot{\psi}(X_1, \theta_0) \int_0^{n^{-\frac{1}{2}} t_2} \int_0^u \ddot{\psi}(X_1, \theta_0 + v) dv du \Big\}| \\
\leq\ & (K/2) n^{-\frac{3}{2}} |t_1 t_2| (|t_1| + |t_2|).
\end{aligned}$$

Similarly

$$\begin{aligned}
& |\mathbb{E} A_n(X_1, t_1) \mathbb{E} A_n(X_1, t_2) - n^{-1} t_1 t_2 (\mathbb{E} \dot{\psi}(X_1, \theta_0))^2| \\
\leq\ & |t_1 t_2| (|t_1| + |t_2|) \mathrm{O}(n^{-\frac{3}{2}})
\end{aligned}$$

and thus we obtain (5.2.24); this further implies

$$\begin{aligned} & \mathbb{E}\Big(|Z_n^0(t) - Z_n^0(t_1)||Z_n^0(t_2) - Z_n^0(t)|\Big) \\ \leq & \ \frac{1}{2}\Big\{\mathbb{E}(Z_n^0(t) - Z_n^0(t_1))^2 + \mathbb{E}(Z_n^0(t_2) - Z_n^0(t))^2\Big\} \\ \to & \ (s_{11}/2)\Big\{(t-t_1)^2 + (t_2-t)^2\Big\} \leq (s_{11}/2)(t_2-t_1)^2, \end{aligned}$$

as $n \to \infty$ for every $t_1, t, t_2, \quad -B \leq t \leq t_2 \leq B$. Consequently, by (2.5.138), we conclude that $\{Z_n^0\}$ is tight; and this completes the proof of the weak convergence of Z_n^0 to Y. To prove the weak convergence of the process Y_n, it remains to show that

$$\mathbb{E}Z_n(t) - n^{-\frac{1}{2}}t - (\gamma_2 t^2/(2\gamma_1)) \to 0 \text{ as } n \to \infty, \tag{5.2.25}$$

uniformly in $t \in [-B, B]$. For that, it suffices to prove the uniform convergence

$$n|\mathbb{E}\Big\{A_n(X_1,t) - n^{-\frac{1}{2}}t\dot{\psi}(X_1,\theta_0) - (t^2/(2n))\ddot{\psi}(X_1,\theta_0)\Big\}| \to 0.$$

But the left-hand side is majorized by $m|n^{-\frac{1}{2}}t|^{\alpha}(t^2/2)$ by $(B.3)$; this tends to 0 uniformly over $[-B, B]$ and thus the lemma is proved. □

The propositions (5.2.15) and (5.2.16) of Theorem 5.2.2 with $R_n = \mathrm{O}_p(n^{-1})$ follow from the general case in Theorem 5.2.1. Hence it remains to prove that nR_n has the asymptotic distribution as described in (5.2.17). The main idea is to make a random change of time $t \to n^{1/2}(M_n - \theta_0)$ in the process Y_n. This will be accomplished in several steps. First, we will show that the two-dimensional process

$$\mathbf{Y}_n^{\star} = \Big\{\mathbf{Y}_n^{\star}(t) = (Y_n(t), n^{\frac{1}{2}}(M_n - \theta_0))', \quad t \in [-B, B]\Big\} \tag{5.2.26}$$

which belongs to $(D[-B, B])^2$, converges weakly to the Gaussian process

$$\mathbf{Y}^{\star} = \Big\{(t\xi_1 + (\gamma_2 t^2/(2\gamma_1)), \xi_2)', \quad t \in [-B, B]\Big\}. \tag{5.2.27}$$

Notice that the second component of $\mathbf{Y}_n^{\star}(t)$ does not depend on t. It follows from (5.2.16) that $\mathbf{Y}_n^{\star}$ is asymptotically equivalent [up to the order of $\mathrm{O}_p(n^{-1/2})$] to the process

$$\mathbf{Y}_n^{0\star} = \Big\{\mathbf{Y}_n^{0\star}(t) = (Y_n(t), -n^{-\frac{1}{2}}\gamma_1^{-1}\sum_{i=1}^{n}\psi(X_i,\theta_0))', \quad |t| \leq B\Big\} \tag{5.2.28}$$

and this process converges to $\mathbf{Y}^{\star}$ by Lemma 5.2.1 and by the classical central limit theorem.

Second, denoting $\big[a\big]_B = a \cdot I[-B \leq a \leq B], \ a \in \mathbb{R}_1, \ B > 0$, we consider the process

$$\Big[\mathbf{Y}_n^{\star}\Big]_B = \Big\{\Big[\mathbf{Y}_n^{\star}(t)\Big]_B = (Y_n(t), \big[n^{\frac{1}{2}}(M_n - \theta_0)\big]_B)', \quad t \in [-B, B]\Big\}. \tag{5.2.29}$$

Then, by Lemma 5.2.1,

$$\left[\mathbf{Y}_n^\star\right]_B \Longrightarrow \mathbf{Y}^\star \tag{5.2.30}$$

for every fixed $B > 0$, where $\mathbf{Y}^\star$ is the Gaussian process given in (5.2.27), which has continuous sample paths. Hence, as in Section 2.5.10, we can make the random change of time $t \to \left[n^{1/2}(M_n - \theta_0)\right]_B$ in the process $\left[\mathbf{Y}_n^\star\right]_B$. By (5.2.30), and referring to Billingsley (1968), Section 17, pp. 144-145, we conclude that for every fixed $B > 0$, as $n \to \infty$,

$$Y_n(\left[n^{1/2}(M_n - \theta_0)\right]_B) \to \xi_1\left[\xi_2\right]_B - \left[\gamma_2(\left[\xi_2\right]_B)^2/(2\gamma_1)\right]. \tag{5.2.31}$$

Now, given $\epsilon > 0$, there exists $B_0 > 0$ such that, for every $B \geq B_0$,

$$\mathbb{P}\left\{\left[\xi_2\right]_B \neq \xi_2\right\} < \epsilon \quad \text{and} \quad \mathbb{P}\left\{\xi_1\xi_2 \neq \xi_1\left[\xi_2\right]_B\right\} < \epsilon. \tag{5.2.32}$$

Moreover, by (5.2.15), there exists B_1 ($\geq B_0$) and an integer n_0 such that

$$\mathbb{P}(n^{\frac{1}{2}}|M_n - \theta_0)| > B) < \epsilon \quad \text{for } B \geq B_1 \text{ and } n \geq n_0. \tag{5.2.33}$$

Combining (5.2.31)-(5.2.33), we obtain

$$\begin{aligned}
&\limsup_{n\to\infty} \mathbb{P}\left\{Y_n(n^{\frac{1}{2}}(M_n - \theta_0)) \leq y\right\} \\
\leq\ &\limsup_{n\to\infty} \mathbb{P}\left\{Y_n(\left[n^{\frac{1}{2}}(M_n - \theta_0))\right]_B \leq y\right\} + \epsilon \\
=\ &\mathbb{P}\left\{\xi_1\left[\xi_2\right]_B - (2\gamma_1)^{-1}\gamma_2(\left[\xi_2\right]_B)^2 \leq y\right\} + \epsilon \\
\leq\ &\mathbb{P}\left\{\xi_1\xi_2 - (2\gamma_1)^{-1}\gamma_2\xi_2^2 \leq y\right\} + 3\epsilon.
\end{aligned}$$

Similarly

$$\begin{aligned}
&\limsup_{n\to\infty} \mathbb{P}\left\{Y_n(n^{\frac{1}{2}}(M_n - \theta_0)) > y\right\} \\
\leq\ &\limsup_{n\to\infty} \mathbb{P}\left\{Y_n(\left[n^{\frac{1}{2}}(M_n - \theta_0))\right]_B > y\right\} + \epsilon \\
\leq\ &\mathbb{P}\left\{\xi_1\xi_2 - (2\gamma_1)^{-1}\gamma_2\xi_2^2 > y\right\} + 3\epsilon,
\end{aligned}$$

and this completes the proof of (5.2.17) and hence of the theorem. □

5.3 M-ESTIMATION OF LOCATION: FIXED SCALE

Although the results of the preceding section apply to the simple location model, it is possible to obtain deeper results in this model under somewhat simpler regularity conditions. Hence we find it convenient to deal with the location model separately for the following reasons:

1. The M-estimators were originally developed for the classical location model.

2. The consistency, asymptotic normality and Bahadur-type representations for M-estimators of location can be proved under weaker regularity conditions on ρ and F than in the general case. In fact Bahadur (1966) developed the asymptotic representation for the sample quantiles which coincide with M-estimators generated by step-functions ψ (see also Section 4.7 concerning regression quantiles).

3. Unlike the L-estimators, the M-estimators of location may not be scale-equivariant, although they are translation-equivariant. To obtain a scale-equivariant M-estimator of location, one may studentize M_n by an appropriate scale-statistic S_n and define M_n as a solution of the minimization

$$\sum_{i=1}^{n} \rho((X_i - t)/S_n) := \min \quad \text{(with respect to } t \in \mathbb{R}_1). \tag{5.3.1}$$

The methodology of the preceding section may not apply to this studentized case. In the present section, we will consider M-estimators with fixed scale not involving the studentization, which will be studied in the next section.

Let $X_1, X_2, \ldots$ be i.i.d. r.v.'s with the d.f. $F(x-\theta)$. Let $\rho : \mathbb{R}_1 \mapsto \mathbb{R}_1$ be an absolutely continuous function with the derivative ψ and such that the function

$$h(t) = \int \rho(x-t)dF(x) \tag{5.3.2}$$

has a unique minimum at $t = 0$. Let us first consider the case when ψ is an absolutely continuous function which could be decomposed as

$$\psi(t) = \psi_1(t) + \psi_2(t), \quad t \in \mathbb{R}_1, \tag{5.3.3}$$

where ψ_1 has an absolutely continuous derivative ψ_1' and ψ_2 is a piecewise linear continuous function, constant outside a bounded interval. More precisely, we shall impose the following conditions on ψ_1, ψ_2 and F:

A1. *Smooth component* ψ_1. ψ_1 is absolutely continuous with an absolutely continuous derivative ψ_1' such that

$$\int (\psi_1'(x+t))^2 dF(x) < K_1 \text{ for } |t| \leq \delta,$$

and ψ_1' is absolutely continuous and $\int |\psi_1''(x+t)|dF(x) < K_2$ for $|t| \leq \delta$, where δ, K_1 and K_2 are positive constants.

A2. *Piecewise linear component* ψ_2. ψ_2 is absolutely continuous with the derivate

$$\psi_2'(x) = \alpha_\nu \quad \text{for } r_\nu < x \leq r_{\nu+1},\ \nu = 1, \ldots, k,$$

where $\alpha_0, \alpha_1, \ldots \alpha_k$ are real numbers, $\alpha_0 = \alpha_k = 0$, and $-\infty = r_0 < r_1 < \ldots < r_k < r_{k+1} = \infty$.

A3. *F smooth around the difficult points.* F has two bounded derivatives f and f' and $f > 0$ in a neighborhood of $r_1, \ldots r_k$.

A4. *Fisher's consistency.* $\gamma = \gamma_1 + \gamma_2 > 0$, where $\gamma_i = \int \psi_i'(x)dF(x)$ for $i = 1, 2$, and $\int \psi^2(x)dF(x) < \infty$.

The M-estimator M_n of θ is then defined as a solution of the minimization

$$\sum_{i=1}^{n} \rho(X_i - t) := \min \text{ with respect to } t \in \mathbb{R}_1. \tag{5.3.4}$$

Under the conditions **A1-A4**, M_n coincides with a root of the equation

$$\sum_{i=1}^{n} \psi(X_i - t) = 0. \tag{5.3.5}$$

if ρ is convex, and hence ψ nondecreasing, M_n can be defined uniquely in the form

$$M_n = \frac{1}{2}(M_n^+ + M_n^-), \tag{5.3.6}$$

where

$$M_n^- = \sup\{t : \sum_{i=1}^{n} \psi(X_i - t) > 0\}, \quad M_n^+ = \inf\{t : \sum_{i=1}^{n} \psi(X_i - t) < 0\}. \tag{5.3.7}$$

If ρ is not convex, then the equation (5.3.5) may have more roots. The conditions **A1-A4** guarantee that there exists at least one root of (5.3.5) that is $\sqrt{n}$-consistent estimator of θ and that admits an asymptotic representation. This is formally expressed in the following theorem:

Theorem 5.3.1 *Let $X_1, X_2, \ldots$ be i.i.d. r.v.'s with d.f. $F(x - \theta)$. Let $\rho : \mathbb{R}_1 \mapsto \mathbb{R}_1$ be an absolutely continuous function whose derivative ψ can be decomposed as (5.3.3) and such that the function $h(t)$ of (5.3.2) has a unique minimum at $t = 0$. Then under the conditions (A.1) - (A.4), there exists a sequence $\{M_n\}$ of roots of the equation (5.3.5) such that*

$$n^{\frac{1}{2}}(M_n - \theta) = O_p(1) \tag{5.3.8}$$

and

$$M_n = \theta + (n\gamma)^{-1} \sum_{i=1}^{n} \psi(X_i - \theta) + R_n, \quad \text{where } R_n = O_p(n^{-1}). \tag{5.3.9}$$

The asymptotic representation (5.3.9) immediately implies that the sequence $\{n^{1/2}(M_n - \theta)\}$ is asymptotically normally distributed as $n \to \infty$; hence we have the following corollary:

COROLLARY 5.3.1 *Under the conditions of Theorem 5.3.1, there exists a sequence $\{M_n\}$ of solutions of the equation (5.3.5) such that $n^{1/2}(M_n - \theta)$ has asymptotically normal distribution $\mathcal{N}(0, \sigma^2(\psi, F))$ with $\sigma^2(\psi, F) = \gamma^{-2} \int \psi^2(x) dF(x)$.*

Proof of Theorem 5.3.1. We will first prove the uniform asymptotic linearity of the form

$$\sup_{|t| \leq C} |\sum_{i=1}^{n} [\psi_j(X_i - \theta - n^{-\frac{1}{2}} t) - \psi_j(X_i - \theta)] + n^{\frac{1}{2}} t \gamma_j| = O_p(1), \tag{5.3.10}$$

as $n \to \infty$, for any fixed $C > 0$ and $j = 1, 2$. If $j = 1$ and hence $\psi = \psi_1$ is the smooth component of ψ, we may refer to the proof of Lemma 5.2.1: Denoting

$$\begin{aligned} Z_{nj}(t) &= \sum_{i=1}^{n} [\psi_j(X_i - n^{-\frac{1}{2}} t) - \psi_j(X_i)] \\ Z_{nj}^0(t) &= Z_{nj}(t) - \mathbb{E}_0 Z_{nj}(t), \quad j = 1, 2; \end{aligned} \tag{5.3.11}$$

(we could put $\theta = 0$ without loss of generality), we get from the mentioned proof that for $t_1 \leq t \leq t_2$,

$$\limsup_{n \to \infty} \mathbb{E}_0(|Z_{n1}^0(t) - Z_{n1}^0(t_1)||Z_{n1}^0(t_2) - Z_{n1}^0(t)|) \leq K(t_2 - t_1)^2, \tag{5.3.12}$$

where $K > 0$ is a constant. Hence by (2.5.138), we have

$$\sup_{|t| \leq C} |Z_{n1}^0| = O_p(1). \tag{5.3.13}$$

Moreover, by **A1**,

$$|n \mathbb{E}_0[\psi_1(X_1 - n^{-\frac{1}{2}} t) - \psi_1(X_1) + n^{-\frac{1}{2}} t \psi_1'(X_1)]| \leq K^\star \tag{5.3.14}$$

for $n \geq n_0$ and uniformly for $|t| \leq C$; $K^\star > 0$ is a constant. Combining (5.3.13) and (5.3.14) gives (5.3.10) for the case $j = 1$.

Let now ψ_2 be the piecewise linear continuous function of **A2**. Without loss of generality, we may consider the particular case when for some $-\infty < r_1 < r_2 < \infty$,

$$\psi_2(x) = r_1, x \text{ or } r_2 \quad \text{according as} x \text{ is} \ < r_1, \ r_1 \leq x \leq r_2 \text{ or } > r_2. \tag{5.3.15}$$

Note that ψ_2 is first-order Lipschitz, so that for any $t_1 \leq t_2$,

$$\begin{aligned} &\mathrm{Var}_0[Z_{n2}(t_2) - Z_{n2}(t_1)] \\ &\leq \sum_{i=1}^{n} \mathbb{E}_0[\psi_2(X_i - n^{-\frac{1}{2}} t_2) - \psi_2(X_i - n^{-\frac{1}{2}} t_1)]^2 \\ &\leq K(t_2 - t_1)^2, \quad 0 < K < \infty. \end{aligned} \tag{5.3.16}$$

Moreover, by **A3**,

$$\begin{aligned}|n\mathbb{E}_0[\psi_2(X_i - n^{-\frac{1}{2}}t_2) - \psi_2(X_i - n^{-\frac{1}{2}}t_1) + n^{-\frac{1}{2}}\psi_2'(X_1)(t_2 - t_1)]| \\ \leq \quad n\mathbb{E}_0 \int_{n^{-\frac{1}{2}}t_1}^{n^{-\frac{1}{2}}t_2} \int_0^u \int_{\boldsymbol{R}_1} |\psi_2'(x) f'(x+v)| dx dv du \\ \leq \quad K(t_2 - t_1) \quad \text{for } n \geq n_0, \ 0 < K < \infty. \qquad (5.3.17)\end{aligned}$$

Combining (5.3.16) and (5.3.17), we arrive at (5.3.10) for the case $j = 2$. Hence we conclude that

$$\sup_{|t| \leq C} |\sum_{i=1}^{n} [\psi(X_i - \theta - n^{-\frac{1}{2}}t) - \psi(X_i - \theta) + n^{\frac{1}{2}}\gamma t| = \mathrm{O}_p(1), \qquad (5.3.18)$$

and, following the arguments in (5.2.9)-(5.2.11), we further conclude that given $\epsilon > 0$, there exist $C > 0$ and an integer n_0 so that the equation $\sum_{i=1}^{n} \psi(X_i - \theta n^{-\frac{1}{2}}t) = 0$ has a root T_n in the interval $[-C, C]$ with probability exceeding $1 - \epsilon$ for $n \geq n_0$. Then $M_n = \theta + n^{-\frac{1}{2}}T_n$ is a root of (5.3.5), and it satisfies (5.3.8). Finally, it is sufficient to insert $t \rightarrow n^{\frac{1}{2}}(M_n - \theta)$ in (5.3.18) in order to obtain the asymptotic representation (5.3.9). This completes the proof of the theorem. □

Theorem 5.3.1 guarantees an existence of a $\sqrt{n}$-consistent M-estimator even in the case where ρ is not convex and hence ψ is not monotone. In such a case there may exist more roots of equation (5.3.5) that are $\sqrt{n}$-consistent estimators of θ. It then follows from (5.3.18) that all of them admit the asymptotic representation (5.3.9), and that they are asymptotically equivalent up to $\mathrm{O}_p(n^{-1})$. Hence we have the following corollary:

COROLLARY 5.3.2. *Let M_n and $M_n^\star$ be two roots of (5.3.5) satisfying (5.3.8). Then, under the conditions of Theorem 5.3.1,*

$$|M_n - M_n^\star| = \mathrm{O}_p(n^{-1}) \ \text{ as } n \rightarrow \infty. \qquad (5.3.19)$$

The above result may not exclude existence of other roots that are not consistent estimators of θ. This begs the question which roots to choose as an estimator. If we know some $\sqrt{n}$-consistent estimator T_n (e.g., the sample median is $\sqrt{n}$-consistent in the symmetric model under mild conditions), then we may look for the root nearest to T_n. Another possibility is to look for a one-step or k-step version of M_n starting again with a $\sqrt{n}$-consistent initial estimator of θ. These estimators deserve a special attention and will be considered in Chapter 7. However, we may remark that all results hold under the crucial assumption that the function $h(t)$ of (5.3.2) has a unique minimum at $t = 0$. This condition deserves a careful examination not only in the location model but also in the general case. In general as in the case with the MLE or some other estimators, the answer to the question of consistency without this condition may not be affirmative. Some problems are set at the end of this chapter, and these show that the M-estimator may be inconsistent when $h(t)$ does not have the global minimum at $t = 0$.

5.3.1 Possibly Discontinuous but Monotone ψ

We now turn to the important case that while ρ is absolutely continuous, its derivative ψ may have jump discontinuities. This case could typically appear in the location and regression models where it covers the estimation based on L_1-norm. More precisely, we will assume that ρ is absolutely continuous with the derivative ψ which can be written as the sum

$$\psi = \psi_c + \psi_s, \tag{5.3.20}$$

where ψ_c is an absolutely continuous function satisfying the conditions **A1-A4** and ψ_s is a step-function

$$\psi_s = \beta_j \quad \text{for } q_j < x < q_{j+1},\ j = 0, 1, \ldots, m, \tag{5.3.21}$$

where $\beta_0, \beta_1, \ldots, \beta_m$ are real numbers (not all equal) and $-\infty = q_0 < q_1 < \ldots < q_m < q_{m+1} = \infty$, m being a positive integer. We assume that the d.f. F is absolutely continuous with the density f which has a bounded derivative f' in a neighborhood of $q_1, \ldots .q_m$. We denote

$$\gamma_s = \sum_{j=1}^{m} (\beta_j - \beta_{j-1}) f(q_j). \tag{5.3.22}$$

The M-estimator M_n of θ is defined as a solution of the minimization (5.3.4). We assume that $h(t)$ of (5.3.2) has a unique minimum at $t = 0$. Notice that the equation (5.3.5) may not have any roots.

Following the proof of Theorem 4.7.1 (for the case $x_{i1} = 1,\ x_{ij} = 0,\ i = 1, \ldots, n;\ j = 2, \ldots, p$), we obtain

$$\sup_{|t| \le C} |n^{-\frac{1}{2}} \sum_{i=1}^{n} [\psi_s(X_i - \theta - n^{-\frac{1}{2}} t) - \psi_s(X_i - \theta)] + \gamma_s t| = O_p(n^{-\frac{1}{4}}), \tag{5.3.23}$$

for any fixed $C > 0$. Combined with (5.3.18), this gives

$$\sup_{|t| \le C} |n^{-\frac{1}{2}} \sum_{i=1}^{n} [\psi(X_i - \theta - n^{-\frac{1}{2}} t) - \psi(X_i - \theta)] + \gamma t| = O_p(n^{-\frac{1}{4}}), \tag{5.3.24}$$

with $\gamma = \gamma_c + \gamma_s$ and

$$\gamma_c = \int_{R_1} \psi_c'(x) dF(x). \tag{5.3.25}$$

If ρ is convex and thus ψ nondecreasing and if $\gamma > 0$, then M_n can be uniquely determined as in (5.3.6) and (5.3.7). In this case, following the proof of

(4.7.17), we could prove that

$$n^{-\frac{1}{2}} \sum_{i=1}^{n} \psi(X_i - M_n) = \mathrm{O}_p(n^{-\frac{1}{2}}); \tag{5.3.26}$$

see Problem 5.3.1. Moreover, following the proof of (4.7.20), we obtain that M_n is $\sqrt{n}$-consistent estimator of θ:

$$n^{\frac{1}{2}}(M_n - \theta) = \mathrm{O}_p(1). \tag{5.3.27}$$

Substituting $n^{1/2}(M_n - \theta)$ for t in (5.3.24) and using (5.3.26), we come to the asymptotic representation

$$M_n = \theta + (n\gamma)^{-1} \sum_{i=1}^{n} \psi(X_i - \theta) + \mathrm{O}_p(n^{-\frac{3}{4}}). \tag{5.3.28}$$

The results are summarized in the following theorem:

Theorem 5.3.2 *Let $X_1, X_2, \ldots$ be i.i.d. r.v.'s with the d.f. $F(x - \theta)$. Let $\rho : \mathbb{R}_1 \mapsto \mathbb{R}_1$ be an absolutely continuous function such that $h(t)$ of (5.3.2) has a unique minimum at $t = 0$. Assume that $\psi = \rho' = \psi_c + \psi_s$ where ψ_c is absolutely continuous, nondecreasing function satisfying* **A1-A4** *and ψ_s is a nondecreasing step function of (5.3.21). Let F have the second derivative f' in a neighborhood of the jump points of ψ_s and let γ of (5.3.25) be positive. Then the M-estimator M_n defined in (5.3.6) and (5.3.7) is a $\sqrt{n}$-consistent estimator of θ that admits the asymptotic representation (5.3.28).*

5.3.2 Possible Discontinuous and Nonmonotone ψ

Let us look at the case where ρ is not convex, and then M_n can be defined as the point of global minimum in (5.3.4). Consider the steps which led to the proof of consistency and of asymptotic representation, and determine which claims can be still proved under weaker conditions. First, notice that the asymptotic linearity in (5.3.24) does not need the monotonicity of ψ, whereas the monotonicity of ψ is crucial for the proof of (5.3.26) [by the method used for (4.7.17)]. Hence we cannot claim the $\sqrt{n}$-consistency unless the step component ψ_s vanishes. However, if for any reason we assume an existence of a $\sqrt{n}$-consistent solution of (5.3.4), we could immediately claim the same representation (5.3.28).

The estimators generated by non-monotone and possibly discontinuous ψ-functions appear in the literature (e.g., Andrews et al. 1972) and some even have a special name, like the *skipped mean* generated by

$$\rho(x) = \begin{cases} x^2/2, & \text{if } |x| \leq k, \\ k^2/2, & \text{if } |x| > k, \end{cases} \qquad \psi(x) = \begin{cases} x, & \text{if } |x| < k, \\ 0, & \text{if } |x| > k, \end{cases}$$

or the *skipped median* corresponding to

$$\rho(x) = \begin{cases} |x|, & \text{if } |x| \leq k, \\ k, & \text{if } |x| > k, \end{cases} \qquad \psi(x) = \begin{cases} \text{sign}x, & \text{if } |x| < k, \\ 0, & \text{if } |x| > k. \end{cases}$$

Hence such estimators deserve a special study, and their existence should be justified at least by their consistency (a simulation study of their behavior can be found in the above monograph). We will provide some regularity conditions, under which are such estimators consistent; although not exhausting, they cover also other redescending ψ-functions (a collection of such functions could be found, e.g., in Hampel et al. 1986).

B1. *Shape of* ρ. $\rho : \mathbb{R}_1 \mapsto \mathbb{R}_1$ is nonnegative, absolutely continuous, symmetric about zero, and nondecreasing on $[0, \infty)$.

B2. *Shape of derivative.* The derivative ψ of ρ could be decomposed in $\psi = \psi_c + \psi_s$, where ψ_c is a bounded absolutely continuous function of (5.3.4) satisfying conditions **A1-A3** behind (5.3.3), and ψ_s is the step function of (5.3.21).

B3. *Tails and smoothness of* F. F is an absolutely continuous d.f., $F(x) + F(-x) = 1$, $x \in \mathbb{R}_1$ satisfying $1 - F(x) = O(x^{-\nu})$ as $x \to \infty$, $\nu > 0$. The density f has a bounded derivative in a neighborhood of $q_1, \ldots, q_m$.

B4. *Fisher's consistency.* γ defined by (5.3.25) is positive.

Theorem 5.3.3 *Let $X_1, X_2, \ldots$ be i.i.d. r.v.'s with the d.f. $F(x - \theta)$, F satisfying* **B3**. *Let ρ be a function satisfying* **B1**, **B2**, *and* **B4**, *such that $h(t)$ of (5.3.2) has a unique minimum at $t = 0$. Then the point M_n of global minimum of (5.3.4) is a $\sqrt{n}$-consistent estimator of θ which admits the asymptotic representation (5.3.28).*

Proof. We will first show that for any $\tau \leq 1/2$ and $C > 0$,

$$\sup_{|t| \leq C} |n^{-1+\tau} \sum_{i=1}^{n} [\rho(X_i - \theta - n^{-\tau}t) - \rho(X_i - \theta) - h(n^{-\tau}t) + h(0)]| = O_p(n^{-\frac{1}{2}}), \tag{5.3.29}$$

and that for $0 < \tau \leq 1/2$,

$$\sup_{|t| \leq C} |n^{-1+\tau} \sum_{i=1}^{n} [\rho(X_i - \theta - n^{-\tau}t) - \rho(X_i - \theta)] - n^{-\tau}\gamma(\frac{1}{2}t^2)|$$

$$= O_p(n^{-\frac{1}{2}}) + O_p(n^{-2\tau}).$$

Actually for $0 \leq t \leq C$ (the case $-C \leq t \leq 0$ is analogous),

$$\text{Var}_\theta \Big\{ n^{-1+\tau} \sum_{i=1}^{n} [\rho(X_i - \theta - n^{-\tau}t) - \rho(X_i - \theta)] \Big\}$$

$$\leq \; n^{-1+2\tau} \Big\{ \int_0^{n^{-\tau}t} (\mathbb{E}_0 \psi^2(X_1 - v))^{\frac{1}{2}} dv \Big\}^2 \leq Kn^{-1}t^2, \;\; K > 0.$$

Analogously we get for $-C \leq t \leq u \leq C$,

$$\mathrm{Var}_\theta\Big\{n^{-1+\tau}\sum_{i=1}^{n}[\rho(X_i-\theta-n^{-\tau}u)-\rho(X_i-\theta-n^{-\tau}t)]\Big\} \leq Kn^{-1}(u-t)^2,$$

and this implies (5.3.29) with the aid of (2.5.138).

Let M_n be the point of global minimum of $\sum_{i=1}^n \rho(X_i-t)$. Assume that $M_n-\theta=\mathrm{O}_p(n^{-\tau})$, $M_n-\theta$ is not of the order $\mathrm{o}_p(n^{-\tau})$, for some $\tau \leq 1/2$. Substituting $n^\tau(M_n-\theta)$ for t in (5.3.29), we get from the assumptions

$$\begin{aligned} 0 &\geq n^{-1+\tau}\sum_{i=1}^{n}[\rho(X_i-M_n)-\rho(X_i-\theta)] \\ &= n^\tau[h(M_n-\theta)-h(0)]+\mathrm{O}_p(n^{-\frac{1}{2}}). \end{aligned}$$

Hence

$$0 \leq h(M_n-\theta)-h(0)=\mathrm{O}_p(n^{-\frac{1}{2}-\tau}). \tag{5.3.30}$$

On the other hand, we obtain for $M_n-\theta \geq 0$ (the case $M_n-\theta<0$ is analogous),

$$\begin{aligned} 0 &\leq h(M_n-\theta)-h(0)=\int_0^{M_n-\theta}h'(u)du=\int_0^{M_n-\theta}\int_0^u h''(v)dvdu \\ &= \int_{\boldsymbol{R}_1}\int_0^{M_n-\theta}\int_0^u f(x+v)dvdud\psi(x) \\ &= \int_0^{M_n-\theta}\int_0^u\int_{\boldsymbol{R}_1} f(x)d\psi(x)dvdu \\ &\quad +\int_{\boldsymbol{R}_1}\int_0^{M_n-\theta}\int_0^u (f(x+v)-f(x))dvdud\psi(x) \\ &= (M_n-\theta)^2+\mathrm{O}(|M_n-\theta)|^3) \end{aligned}$$

which is $\mathrm{O}_p(n^{-\tau-1/2})$ by (5.3.30). This is a a contradiction unless $M_n-\theta=\mathrm{O}_p(n^{-1/2})$. Hence there is no $\tau<1/2$ such that $M_n-\theta=\mathrm{O}_p(n^{-\tau})$.

Finally, assume that $M_n-\theta \neq \mathrm{O}_p(n^{-1/2})$; then by the above, $M_n-\theta$ cannot be $\mathrm{O}_p(n^{-\tau})$ for any $\tau<1/2$ either. Hence there exists an $\epsilon>0$ such that given any $K>0$, there exists a sequence $\{n_k\}_{k=1}^\infty$ for which

$$\mathbb{P}(|M_{n_k}-\theta|>3Kn_k^{1/\nu})>3\epsilon \quad \text{for} \quad k=1,2,\ldots. \tag{5.3.31}$$

with ν given in **B3**. On the other hand, by assumption **B3**,

$$\max_{1\leq i\leq n}|X_i-\theta|=\mathrm{O}_p(n^{1/\nu}). \tag{5.3.32}$$

More precisely, there exists $C > 0$ and n_0 to a given $\epsilon > 0$ such that, for $n \geq n_0$,

$$\begin{aligned} \mathbb{P}(\max_{1 \leq i \leq n} |X_i - \theta| > Cn^{1/\nu}) &= 1 - [2F(Cn^{1/\nu}) - 1]^n \leq 2n[1 - F(Cn^{1/\nu})] \\ &\leq K_1 C^{-\nu} \leq \epsilon, \end{aligned} \tag{5.3.33}$$

with some $K_1 > 0$. Substituting C for K in (5.3.31), we obtain

$$\begin{aligned} &\mathbb{P}\Big\{\frac{1}{n_k}\sum_{i=1}^{n_k} \rho(X_i - M_{n_k}) \geq \rho(2C(n_k)^{1/\nu})\Big\} \\ &\geq \mathbb{P}\Big\{\min_{1 \leq i \leq n_k} |X_i - M_{n_k}| \geq 2C(n_k)^{1/\nu}\Big\} \\ &\geq \mathbb{P}\Big\{|M_{n_k} - \theta| > 3C(n_k)^{1/\nu}, \ \max_{1 \leq i \leq n} |X_i - \theta| < C(n_k)^{1/\nu}\Big\} \\ &\geq 2\epsilon. \end{aligned} \tag{5.3.34}$$

Finally, $\rho(C(n_k)^{1/\nu}) > 0$ for sufficiently large n, and

$$\begin{aligned} &\mathbb{P}\Big\{\frac{1}{n}\sum_{i=1}^{n} \rho(X_i - X_{n:1}) \geq \rho(Cn^{1/\nu})\Big\} \\ &\leq \mathbb{P}\Big\{\rho(X_{n:n} - X_{n:1}) \geq \rho(Cn^{1/\nu})\Big\} \\ &= \mathbb{P}\Big\{\rho(X_{n:n} - X_{n:1}) \geq \rho(Cn^{1/\nu}), \ X_{n:n} - X_{n:1} \geq Cn^{1/\nu}\Big\} \\ &\leq \epsilon. \end{aligned} \tag{5.3.35}$$

Combining (5.3.34) and (5.3.35), we obtain

$$\begin{aligned} &\mathbb{P}\Big\{\frac{1}{n_k}\sum_{i=1}^{n_k} \rho(X_i - M_{n_k}) > \frac{1}{n_k}\sum_{i=1}^{n_k} \rho(X_i - X_{n_k:1})\Big\} \\ &\geq \mathbb{P}\Big\{\frac{1}{n_k}\sum_{i=1}^{n_k} \rho(X_i - M_{n_k}) \geq \rho(2Cn_k^{1/\nu}), \\ &\qquad \frac{1}{n_k}\sum_{i=1}^{n_k} \rho(X_i - X_{n_k:1}) < \rho(Cn_k^{1/\nu})\Big\} \geq \epsilon \end{aligned}$$

for sufficiently large k, and this is a contradiction to the fact that M_n minimizes $\sum_{i=1}^{n} \rho(X_i - t)$.

We conclude that $M_n - \theta = \mathrm{O}_p(n^{-1/2})$ as $n \to \infty$. Substituting $n^{1/2}(M_n - \theta)$ for t in the asymptotic linearity (5.3.24) (which does not need the monotonicity of ψ), we get the asymptotic representation (5.3.28). □

5.3.3 Second-Order Distributional Representations

The representations in (5.3.9) and (5.3.28) convey a qualitative difference: Jump discontinuities in the score function ψ lead to a slower rate of convergence for the remainder term R_n. The qualitative difference is even more apparant when the respective asymptotic distributions of the normalized R_n are compared. In addition to the difference in these normalizing factors, the distributions also differ in their functional forms. This reveals that M-estimators based on ψ-functions having jump discontinuities may entail more elaborate iterative solutions (compared to smooth ψ-functions).

The SOADR in the location model, for smooth ψ, follows from Theorem 5.2.2. In the symmetric case, where $F(x)+F(-x)=1$, and $\rho(-x)=\rho(x)$, $x \in \mathbb{R}_1$, this result simplifies so much that it deserves a separate and special treatment.

Theorem 5.3.4 *Let $\{X_i, i \geq 1\}$ be a sequence of i.i.d. r.v.'s with the d.f. $F(x-\theta)$, F symmetric about origin. Let ρ : $\mathbb{R}_1 \mapsto \mathbb{R}_1$ be an absolutely continuous symmetric function. Assume the conditions* **A1-A4** *of Section 5.3. Then there exists a sequence $\{M_n\}$ of roots of the equation (5.3.5) that admits representation (5.3.9) wherein*

$$nR_n \xrightarrow{\mathcal{D}} \xi_1^\star \xi_2^\star \text{ as } n \to \infty; \tag{5.3.36}$$

the random vector $(\xi_1^\star, \xi_2^\star)$ has bivariate normal distribution $\mathcal{N}_2(\mathbf{0}, \mathbf{S}^\star)$ with $\mathbf{S}^\star = [s_{i,j}]_{i,j=1,2}$ and

$$s_{11}^\star = \gamma^{-2} \int (\psi'(x))^2 dF(x) - 1, \quad s_{12}^\star = s_{21}^\star = 0$$

$$s_{22}^\star = \gamma^{-2} \int \psi^2(x) dF(x). \tag{5.3.37}$$

Let us now consider the situation described in Theorem 5.3.2, where $\psi = \psi_c + \psi_s$, a sum of absolutely continuous and step function components. If ψ_s does not vanish (this practically means that γ_s in (5.3.22) is positive) then the asymptotic distribution of $n^{3/4}R_n$ is considerably different from that of nR_n in the last theorem.

Theorem 5.3.5 *Assume the conditions of Theorem 5.3.2. Moreover, assume that F satisfies the following conditions:*

F has an absolutely continuous density f and finite Fisher's information

$$I(f) = \int_{\mathbf{R}_1} \left(\frac{f'(x)}{f(x)}\right)^2 dF(x) < \infty. \tag{5.3.38}$$

f' is bounded and continuous in a neighborhood of jump points $q_1, \ldots, q_s$ of ψ_s, (ψ_s nondecreasing) and

$$\gamma_s = \sum_{j=1}^{m} (\beta_j - \beta_{j-1}) f(q_j) > 0. \tag{5.3.39}$$

Denote by

$$\nu_s^2 = \sum_{j=1}^{m} (\beta_j - \beta_{j-1})^2 f(q_j) \ (> 0). \tag{5.3.40}$$

Then M_n defined in (5.3.6) and (5.3.7) admits the asymptotic representation (5.3.28) and

$$n^{\frac{3}{4}} R_n \xrightarrow{\mathcal{D}} \xi \ \textit{as}\ n \to \infty, \tag{5.3.41}$$

where

$$\mathbb{P}(\xi \le x) = 2\int_0^\infty \Phi\big(x/(w\sqrt{t})\big)\, d\Phi(t), \quad x \in \mathbb{R}_1, \tag{5.3.42}$$

Φ being the standard normal d.f. and

$$w = \gamma^{-\frac{3}{2}} \Big(\int_{\mathbf{R}_1} \psi^2(x) dF(x)\Big)^{\frac{1}{2}} \nu_s. \tag{5.3.43}$$

Remark. If we put $\psi_c \equiv 0$ and $\psi_x(x) = p - I[x \le 0]$ for some $0 < p < 1$, then $\int \psi(x-t)dF(x) = 0$ for $t = F^{-1}(p)$ (assume that $f(F^{-1}(p)) \neq 0$). Replacing $\psi(x)$ by $\psi^\star(x) = \psi(x - F^{-1}(p))$ we have $\gamma = \gamma_s = \nu_s^2 = f(F^{-1}(p))$. The M-estimator generated by $\psi^\star$ is then $X_{n:[np]} - F^{-1}(p)$ and we have the following corollary of Theorem 5.3.5, which coincides with the result of Kiefer (1967):

COROLLARY 5.3.5.1. *Assume that F is twice differentiable at $F^{-1}(p)$ and $f(F^{-1}(p)) > 0$. Then*

$$n^{\frac{1}{2}}(X_{n:[np]} - F^{-1}(p)) = n^{-\frac{1}{2}} (f(F^{-1}(p)))^{-1} \sum_{i=1}^{n} (p - I[X_i \le F^{-1}(p)]) + R_n(p) \tag{5.3.44}$$

and

$$\lim_{n\to\infty} \mathbb{P}(n^{\frac{1}{4}} f(F^{-1}(p)) R_n(p) \le x) = 2\int_0^p \Phi\Big(\frac{x}{(p(1-p))^{\frac{1}{4}}\sqrt{t}]}\Big]) d\Phi(t). \tag{5.3.45}$$

Proof of Theorem 5.3.5. Denote by

$$Z_n(t) = n^{-\frac{1}{4}} \sum_{i=1}^{n} A_n(X_i, t), \tag{5.3.46}$$

and

$$Z_n^0(t) = Z_n(t) + n^{\frac{1}{4}} \gamma_s t, \quad t \in \mathbb{R}_1 \tag{5.3.47}$$

where

$$A_n(X_i, t) = \psi_s(X_i - \theta - n^{-\frac{1}{2}} t) - \psi_s(X_i - \theta), \quad i = 1, \ldots, n.$$

We will prove that

$$\nu_s^{-1} Z_n^0 \xrightarrow{\mathcal{D}} W \tag{5.3.48}$$

in the Skorokhod J_1-topology on $D[-B, B]$ for any fixed $B > 0$, where $W = \{W(t) : t \in [-B, B]\}$ is a centered Gaussian process with

$$\mathbb{E}W(s)W(t) = \begin{cases} |s| \wedge |t|, & \text{if } s \cdot t > 0, \\ 0, & \text{otherwise.} \end{cases} \tag{5.3.49}$$

Considering the identity

$$\psi = \beta_0 + \sum_{j=1}^{m} (\beta_j - \beta_{j-1}) \psi_j \tag{5.3.50}$$

where for each $j (= 1, \ldots, m)$

$$\psi_j(x) = \begin{cases} 0, & \text{if } x \leq q_j, \\ 1, & \text{if } x > q_j, \end{cases} \tag{5.3.51}$$

we can prove (5.3.48) only for a single function ψ_i, say, for the function $\psi_s = \psi$ with $q_j = q$. Moreover we can put $\theta = 0$. Then $\nu_s^2 = \gamma_s = f(q) > 0$, and for $-B \leq t_1 < \ldots t_a < 0 \leq t_{a+1} < \ldots < t_b \leq B$ (a, b integers, $0 \leq a \leq b$) and for $\boldsymbol{\lambda} \in \mathbb{R}_b$, we can easily show that as $n \to \infty$,

$$\text{Var} \sum_{j=1}^{b} \lambda_j Z_n(t_j) \to \nu_s^2 \boldsymbol{\lambda}' \mathbf{C}_b \boldsymbol{\lambda},$$

where $\mathbf{C}_b = (c_{jk})_{j,k=1}^{b}$ and for each $j, k (= 1, \ldots, b)$,

$$c_{jk} = \begin{cases} |t_j| \wedge |t_k|, & \text{if } t_j t_k > 0, \\ 0, & \text{otherwise.} \end{cases}$$

By the central limit theorem we conclude that the finite-dimensional distribution of $Z_n - \mathbb{E}Z_n$ converge to those of W.

To prove the tightness of $Z_n - \mathbb{E}Z_n$, let us write the inequalities for $t \le u$

$$\begin{aligned}
&\limsup_{n\to\infty} \mathbb{E}[Z_n(u) - Z_n(t) - \mathbb{E}(Z_n(u) - Z_n(t))]^4 \\
&\le \limsup_{n\to\infty} \Big\{ 11\mathbb{E}[\psi_s(X_1 - n^{-\frac{1}{2}}u) - \psi_s(X_1 - n^{-\frac{1}{2}}t)]^4 \\
&\quad +(n-1)(\mathbb{E}[\psi_s(X_1 - n^{-\frac{1}{2}}u) - \psi_s(X_1 - n^{-\frac{1}{2}}t)]^2)^2 \Big\} \\
&= \limsup_{n\to\infty} \Big\{ 11|F(q + n^{-\frac{1}{2}}u) - F(q + n^{-\frac{1}{2}}t)| \\
&\quad +(n-1)|F(q + n^{-\frac{1}{2}}t) - F(q + n^{-\frac{1}{2}}u)|^2 \Big\} \\
&= \nu_s(u-t)^2.
\end{aligned}$$

This implies tightness by (2.5.138) and altogether gives the weak convergence of $\nu_s^{-1}(Z_n - \mathbb{E}Z_n)$ to W. To prove (5.3.48), we could approximate $\mathbb{E}Z_n(t)$ by $-n^{-\frac{1}{2}}\gamma_s t$ uniformly in $[-B, B]$. But

$$\begin{aligned}
&\sup_{0\le t\le B} |\mathbb{E}Z_n(t) + n^{\frac{1}{4}}\gamma_s t| \\
&\le n^{\frac{3}{4}} \sup_{0\le t\le B} |\mathbb{E}A_n(X_1, t) + n^{-\frac{1}{2}}tf(q)| \\
&= n^{\frac{3}{4}} \sup_{0\le t\le B} |F(q + n^{-\frac{1}{2}}t) - F(q) - n^{-\frac{1}{2}}tf(q)| \\
&\le n^{\frac{3}{4}} \sup_{0\le t\le B} \int_0^{n^{-\frac{1}{2}}t} \int_0^u |f'(q+v)|dvdu \\
&\le KB^2 n^{-\frac{1}{4}},
\end{aligned}$$

and we get analogous inequalities for $\sup_{-B\le t\le 0}$. This proves the weak convergence (5.3.48).

Now we consider the process $Y_n = \{Y_n(t) : |t| \le B\}$ with

$$Y_n(t) = n^{-\frac{1}{4}} \sum_{i=1}^n [\psi_c(X_i - \theta - n^{-\frac{1}{2}}t) - \psi_c(X_i - \theta)] + n^{\frac{1}{4}}t\gamma_c. \tag{5.3.52}$$

By Lemma 5.2.1,

$$\sup\{|Y_n(t)| : |t| \le B\} = o_p(n^{-\frac{1}{4}}). \tag{5.3.53}$$

Combining (5.3.48) and (5.3.53), we obtain

$$W_n \xrightarrow{\mathcal{D}} W \text{ as } n \to \infty \tag{5.3.54}$$

in Skorokhod J_1-topology on $D[-B, B]$ for any fixed $B > 0$, where

$$W_n(t) = \nu_s^{-1}\{n^{-\frac{1}{4}} \sum_{i=1}^n [\psi(X_i - \theta - n^{-\frac{1}{2}}t) - \psi(X_i - \theta)] + n^{\frac{1}{4}}\gamma t\}, \quad t \in \mathbb{R}_1. \tag{5.3.55}$$

As in the proof of Lemma 5.2.1, we will make a random change of time $t \to \left[n^{\frac{1}{2}}(M_n - \theta))\right]_B$ in the process (5.3.47). By (5.3.28),

$$n^{\frac{1}{2}}(M_n - \theta) = \sum_{i=1}^{n} U_{ni} + o_p(1)$$

where $U_{ni} = n^{-\frac{1}{2}}\gamma^{-1}\sum_{i=1}^{n}\psi(X_i - \theta)$, $\mathbb{E}U_{ni} = 0$, $\text{Var}U_{ni} = n^{-1}\sigma^2\gamma^{-2}$, $\sigma^2 = \int \psi^2(x)dF(x)$, $i = 1, \ldots, n$. By (5.3.46) and (5.3.48),

$$\text{Cov}(Z_n(t), \sum_{i=1}^{n} U_{ni}) = O(n^{-\frac{1}{4}}|t|), \quad \text{Cov}(Y_n(t), \sum_{i=1}^{n} U_{ni}) = O(n^{-\frac{1}{4}}|t|), \tag{5.3.56}$$

Hence the finite dimensional distributions of the two-dimensional process

$$\left\{(W_n(t), n^{\frac{1}{2}}(M_n - \theta))) : |t| \le B\right\} \tag{5.3.57}$$

converge to those of the process

$$\left\{(W(t), \xi^0) : |t| \le B\right\} \tag{5.3.58}$$

where ξ^0 is a random variable with normal $\mathcal{N}(0, \sigma^2/\gamma^2)$ distribution, independent of W. Also, $n^{1/2}(M_n - \theta)$ is relatively compact, while the weak convergence (5.3.48) ensures the relative compactness of (5.3.52). This implies the weak convergence of (5.3.57) to (5.3.58). Applying then the random change of time $t \to \left[n^{\frac{1}{2}}(M_n - \theta)\right]_B$ and then letting $B \to \infty$, we obtain

$$\begin{aligned} W_n(n^{\frac{1}{2}}(M_n - \theta)) &= \nu_s^{-1}\left\{n^{-\frac{1}{4}}\sum_{i=1}^{n}[\psi(X_i - M_n) - \psi(X_i - \theta)] + n^{\frac{3}{4}}\gamma(M_n - \theta)\right\} \\ \to \quad & W(\xi^0) = I[\xi^0 > 0]W_1(\xi^0) + I[\xi^0 < 0]W_2(|\xi^0|), \end{aligned} \tag{5.3.59}$$

where W_1 and W_2 are independent copies of the standard Wiener process. By (5.3.28) and (5.3.59),

$$W_n(n^{\frac{1}{2}}(M_n - \theta)) = \gamma\nu_s^{-1}n^{\frac{3}{4}}R_n + \nu_s^{-1}n^{-\frac{1}{4}}\sum_{i=1}^{n}\psi(X_i - M_n), \tag{5.3.60}$$

and the last term is negligible by (5.3.26). Combining (5.3.59) and (5.3.60), we come to the proposition of the theorem. □

5.4 STUDENTIZED M-ESTIMATORS OF LOCATION

Let $X_1, \ldots, X_n$ be independent observations with a joint d.f. $F(x-\theta)$. Consider some scale statistic $S_n = S_n(X_1, \ldots, X_n)$ such that

$$S_n(\mathbf{x}) > 0 \ \text{ a.e. } \mathbf{x} \in \mathbb{R}_n$$

$$S_n(\mathbf{x}+c) = S_n(\mathbf{x}), \quad c \in \mathbb{R}_1, \ \mathbf{x} \in \mathbb{R}_n \ \text{(translation invariance)},$$

$$S_n(c\mathbf{x}) = cS_n(\mathbf{x}), \quad c > 0, \ \mathbf{x} \in \mathbb{R}_n, \ \text{(scale equivariance)}. \tag{5.4.1}$$

Assume that there exists a functional $S = S(F) > 0$ such that

$$n^{\frac{1}{2}}(S_n - S) = O_p(1) \quad \text{as } n \to \infty. \tag{5.4.2}$$

The M-estimator of θ studentized by S_n is then defined as a solution of the minimization

$$\sum_{i=1}^{n} \rho\Big(\frac{X_i - t}{S_n}\Big) := \min \tag{5.4.3}$$

with respect to $t \in \mathbb{R}_1$, where ρ is an appropriate function. We will assume that ρ is absolutely continuous with the derivative ψ. Moreover, we will assume throughout that the function

$$h(t) = \int_{-\infty}^{\infty} \rho\Big(\frac{x-t}{S(F)}\Big) dF(x) \tag{5.4.4}$$

has the only minimum at $t = 0$.

Notice that the studentized estimator is equivariant both in location and scale. We will study the asymptotic distribution and the asymptotic representation of M_n. Typically the asymptotic behavior of M_n depends on the studentizing statistic S_n, unless we assume the symmetry of both F and ρ. In other respects, the situation is similar as in the fixed-scale case: The remainder term in the asymptotic representation is typically of order $O_p(n^{-1})$ or $O_p(n^{-3/4})$, depending on whether $\psi = \rho'$ is smooth or whether it has jump discontinuities. We should require F to be smooth in a neighborhood of S-multiples of "difficult points" of ψ. The continuous component of ψ does not need to be monotone; however, for the sake of simplicity, we will consider only nondecreasing step functions ψ.

Summarizing, we can distinguish three sets of conditions on ψ and on F:

A1. $\psi : \mathbb{R}_1 \mapsto \mathbb{R}_1$ is a step function,

$$\psi(x) = \beta_j \quad \text{for } x \in (q_j, q_{j+1}], \ j = 0, 1, \ldots, m,$$

where

$$-\infty < \beta_0 \leq \beta_1 \leq \ldots \leq \beta_m < \infty$$

and

$$-\infty = q_0 < \ldots < q_m < q_{m+1} = \infty.$$

A2. F has two bounded derivatives f and f' in a neighborhood of $Sq_1, \ldots, Sq_m$ and

$$f(Sq_j) > 0, \quad j = 1, \ldots, m.$$

B1. ψ is absolutely continuous with derivative ψ', which is a step function,

$$\psi'(x) = \alpha_\nu \quad \text{for } \mu_\nu < x < \mu_{\nu+1}, \quad \nu = 0, 1, \ldots, k,$$

where

$$\alpha_0, \alpha_1, \ldots, \alpha_k \in \mathbb{R}_1, \quad \alpha_0 = \alpha_k = 0,$$

and

$$-\infty = \mu_0 < \mu_1 < \ldots < \mu_k < \mu_{k+1} = \infty.$$

B2. F has a bounded derivative f in a neighborhood of $S\mu_1, \ldots, S\mu_k$.

C1. ψ is an absolutely continuous function with an absolutely continuous derivative ψ', and there exists a $\delta > 0$ such that

$$\mathbb{E}_0 \sup\left\{|\psi''(\frac{X_1 - t}{Se^u})|(X_1 - t)^2 : |t|, |u| \leq \delta\right\} < \infty,$$

$$\mathbb{E}_0 \sup\left\{|\psi'(\frac{X_1 - t}{Se^u})(X_1 - t)| : |t|, |u| \leq \delta\right\} < \infty,$$

and

$$\mathbb{E}_0(X_1\psi'(\frac{X_1}{S}))^2 < \infty.$$

The following main theorem of the section gives the asymptotic representation of M_n.

Theorem 5.4.1 *Let M_n be a solution of the minimization (5.4.3) such that*

$$\sqrt{n}(M_n - \theta) = O_p(1) \quad \text{as } n \to \infty. \tag{5.4.5}$$

Let S_n satisfy (5.4.1) and (5.4.2), and let $h(t)$ in (5.4.4) have a unique minimum at $t = 0$. Then

1. Under conditions **A1** *and* **A2**, *$M_n = (M_n^+ + M_n^-)/2$, where*

$$M_n^- = \sup\left\{t : \sum_{i=1}^n \psi(\frac{X_i - t}{S_n}) > 0\right\}, \; M_n^+ = \inf\left\{t : \sum_{i=1}^n \psi(\frac{X_i - t}{S_n}) < 0\right\}, \tag{5.4.6}$$

and it admits the representation

$$M_n = \theta + (n\gamma_1)^{-1}\sum_{i=1}^n \psi(\frac{X_i - \theta}{S}) - \frac{\gamma_2}{\gamma_1}(\frac{S_n}{S} - 1) + O_p(n^{-\frac{3}{4}}), \tag{5.4.7}$$

where

$$\gamma_1 = \sum_{\nu=1}^{m} (\beta_\nu - \beta_{\nu-1} f(Sq_\nu), \quad \gamma_2 = \sum_{\nu=1}^{m} Sq_\nu(\beta_\nu - \beta_{\nu-1} f(Sq_\nu), \tag{5.4.8}$$

2. Under the condition **B1** *and* **B2** *or* **C1**,

$$M_n = \theta + (n\gamma_1)^{-1} \sum_{i=1}^{n} \psi(\frac{X_i - \theta}{S}) - \frac{\gamma_2}{\gamma_1}(\frac{S_n}{S} - 1) + O_p(n^{-1}), \tag{5.4.9}$$

where

$$\gamma_1 = S^{-1} \int_{-\infty}^{\infty} \psi'(\frac{x}{S}) dF(x), \quad \gamma_2 = S^{-1} \int_{-\infty}^{\infty} x\psi'(\frac{x}{S}) dF(x). \tag{5.4.10}$$

Theorem 5.4.1 has several interesting corollaries.

First, whereas the minimization (5.4.3) has a unique solution when ρ is convex (and ψ monotone), it may have more solutions under **B1***and***B2** or under **C1**, where such conditions are not imposed. Any pair $M_n^{(1)}$ and $M_n^{(2)}$ of solutions are close to each other in the following sense (see Problem 5.4.3):

COROLLARY 5.4.1. *Let $M_n^{(1)}$ and $M_n^{(2)}$ be any pair of solutions of (5.4.3). If both $M_n^{(1)}$ and $M_n^{(2)}$ satisfy (5.4.5) then, under* **B1** *and* **B2** *or* **C1**, *respectively,*

$$M_n^{(1)} - M_n^{(2)} = O_p(n^{-1}). \tag{5.4.11}$$

Second, while the representations (5.4.7) and (5.4.9) do not directly imply the asymptotic normality of M_n, they do imply the asymptotic normality of a linear combination of M_n and S_n (see Problem 5.4.3):

COROLLARY 5.4.2. *Assume that*

$$\sigma^2 = \int_{-\infty}^{\infty} \psi^2(\frac{x}{S}) dF(x) < \infty. \tag{5.4.12}$$

Then, under the conditions of Theorem 5.4.1,

$$n^{\frac{1}{2}} \Big\{ \gamma_1(M_n - \theta) + \gamma_2(\frac{S_n}{S}) \Big\} \tag{5.4.13}$$

has asymptotically normal distribution $\mathcal{N}(0, \sigma^2)$.

However, γ_2 vanishes under the symmetry of F and ρ; this considerably simplifies the asymptotic representation and distribution of M_n: (see Problem 5.4.4):

COROLLARY 5.4.3. *If* $F(x) + F(-x) = 1$ and $\rho(-x) = \rho(x)$, $x \in \mathbb{R}_1$, *then*

$$M_n = \theta + (n\gamma_1)^{-1} \sum_{i=1}^{n} \psi(\frac{X_i - \theta}{S}) + R_n, \tag{5.4.14}$$

where $R_n = O_p(n^{-3/4})$ *under* **A1** *and* **A2**, *and* $R_n = O_p(n^{-1})$ *under* **B1** *and* **B2** *or* **C1**, *respectively. Moreover,* $\sqrt{n}(M_n - \theta)$ *is then asymptotically normally distributed* $\mathcal{N}(0, \sigma^2/\gamma_1^2)$.

Without the symmetry conditions, but assuming that S_n itself admits an asymptotic representation, we could still get a representation and the asymptotic normality for M_n itself:

COROLLARY 5.4.4. *Assume that* S_n *admits the asymptotic representation* (see Problem 5.4.5):

$$\frac{S_n}{S} - 1 = n^{-1} \sum_{i=1}^{n} \phi(X_i - \theta) + o_p(n^{-\frac{1}{2}}), \tag{5.4.15}$$

where $\int_{-\infty}^{\infty} \phi(x)dF(x) = 0$, $\int_{-\infty}^{\infty} \phi^2(x)dF(x) < \infty$. *Then*

$$M_n = \theta + (n\gamma_1)^{-1} \sum_{i=1}^{n} \Big\{\psi(\frac{X_i - \theta}{S}) - \gamma_2 \phi(X_i - \theta)\Big\} + o_p(n^{-\frac{1}{2}}), \tag{5.4.16}$$

and $n^{1/2}(M_n - \theta)$ *is asymptotically normally distributed with expectation 0 and with the variance*

$$\gamma_1^{-2} \int_{-\infty}^{\infty} (\psi(\frac{x}{S}) - \gamma_2 \phi(x))^2 dF(x). \tag{5.4.17}$$

The proof of Theorem 5.4.1 is based on the following asymptotic linearity lemma:

Lemma 5.4.1 *For every* $t, u : \ |t|, |u| < C$, *let*

$$Z_n(t, u) = n^{-\frac{1}{2}} \sum_{i=1}^{n} \Big\{\psi(\frac{X_i - \theta - tn^{-\frac{1}{2}}}{Se^{un^{-\frac{1}{2}}}}) - \psi(\frac{X_i - \theta}{S})\Big\}. \tag{5.4.18}$$

1. Under **A1** *and* **A2**, *for any fixed* $C > 0$,

$$\sup\Big\{|Z_n(t, u) + t\gamma_1 + u\gamma_2| : \ |t|, |u| \le C\Big\} = O_p(n^{-\frac{1}{4}}). \tag{5.4.19}$$

2. Under **B1** *and* **B2** *or* **C1** $\ \forall C > 0$,

$$\sup\Big\{|Z_n(t, u) + t\gamma_1 + u\gamma_2| : \ |t|, |u| \le C\Big\} = O_p(n^{-\frac{1}{2}}). \tag{5.4.20}$$

Proof of Lemma 5.4.1 Without loss of generality, we assume $\theta = 0$.

1.We assume, without loss of generality, that $\psi(x) = 0$ or 1 according to $x \le q$ or $x > q$, respectively. Then

$$Z_n(t, u) - Z_n(0, u) = -n^{\frac{1}{2}}[F_n(Sqe^{un^{-\frac{1}{2}}} + n^{-\frac{1}{2}}t) - F_n(Sq)], \tag{5.4.21}$$

where $F_n(x) = n^{-1}\sum_{i=1}^{n} I[X_i \leq x]$ is the empirical d.f. corresponding to $X_1, \ldots, X_n$. Then, by Komlós, Májor and Tusnády (1975),

$$\begin{aligned} Z_n(t,u) &= n^{-\frac{1}{2}}[F(Sqe^{n^{-\frac{1}{2}}u} + n^{-\frac{1}{2}}t) - F(Sq)] + \mathrm{O}_p(n^{-\frac{1}{4}}) \\ &= -t\gamma_1 - u\gamma_2 + \mathrm{O}_p(n^{-\frac{1}{4}}) \end{aligned} \tag{5.4.22}$$

uniformly in $|t|, |u| \leq C$.

2. Assume the conditions **B1** and **B2**. If we denote

$$\psi_\nu(x) = \begin{cases} \mu_\nu, & \text{for } x < \mu_\nu, \\ x, & \text{for } \mu_\nu \leq x \leq \mu_{\nu+1}, \\ \mu_{\nu+1}, & \text{for } x > \mu_{\nu+1}, \quad \text{for } \nu = 1, \ldots, k, \end{cases} \tag{5.4.23}$$

then $\psi \equiv \sum_{\nu=1}^{k} \alpha_\nu \psi_\nu$. Hence we can consider only $\psi \equiv \psi_1$. Then

$$\gamma_1 = S^{-1}(F(\mu_2 S) - F(\mu_1 S)), \quad \gamma_2 = S^{-1}\int_{\mu_1 S}^{\mu_2 S} x dF(x), \tag{5.4.24}$$

and, if $t > 0$ (the case $t < 0$ is treated analogously),

$$\begin{aligned} &|Z_n(t,u) - Z_n(0,u) + \frac{t}{n}\sum_{i=1}^{n}\psi'(\frac{X_i}{Se^{un^{-\frac{1}{2}}}})(Se^{un^{-\frac{1}{2}}})^{-1}| \\ &= |(Se^{un^{-\frac{1}{2}}})^{-1} n^{-\frac{1}{2}} \sum_{i=1}^{n}\int_0^{tn^{-\frac{1}{2}}}[-\psi'(\frac{X_i - T}{Se^{un^{-\frac{1}{2}}}}) + \psi'(\frac{X_i}{Se^{un^{-\frac{1}{2}}}})]dT| \\ &= |(Se^{un^{-\frac{1}{2}}})^{-1} n^{\frac{1}{2}} \int_0^{tn^{-\frac{1}{2}}} \Big\{ - F_n(S\mu_2 e^{un^{-\frac{1}{2}}} + T) + F_n(S\mu_2 e^{un^{-\frac{1}{2}}}) \\ &\quad + F_n(S\mu_1 e^{un^{-\frac{1}{2}}} + T) - F_n(S\mu_1 e^{un^{-\frac{1}{2}}})\Big\} dT| \\ &\leq |(Se^{un^{-\frac{1}{2}}})^{-1} n^{\frac{1}{2}} \int_0^{tn^{-\frac{1}{2}}}\int_0^{T}\Big\{ - f(S\mu_2 e^{un^{-\frac{1}{2}}} + V) \\ &\quad + f(S\mu_1 e^{un^{-\frac{1}{2}}} + V)\Big\} dV dT| \\ &\quad + (Se^{un^{-\frac{1}{2}}})^{-1}\int_0^{tn^{-\frac{1}{2}}}\Big\{ - q_n(S\mu_2 e^{un^{-\frac{1}{2}}} + T) + q_n(S\mu_2 e^{un^{-\frac{1}{2}}}) \\ &\quad + q_n(S\mu_1 e^{un^{-1/2}} + T) - q_n(S\mu_1 e^{un^{-1/2}})\Big\} dT| \end{aligned} \tag{5.4.25}$$

where

$$q_n(x) = n^{1/2}(F_n(x) - F(x)), \; x \in \mathbb{R}_1.$$

Again by Komlós, Májor and Tusnády (1975),

$$q_n(x) = B_n(F(x)) + \mathrm{O}_p(n^{-1/2}\log n) \;\; \text{uniformly in } x \in \mathbb{R}_1, \tag{5.4.26}$$

where the Brownian bridge $B_n(.)$ depends on $X_1, \ldots, X_n$. Hence, by (5.4.25) and (5.4.26),

$$\begin{aligned} &|Z_n(t,u) - Z_n(0,u) + \frac{t}{n}\sum_{i=1}^{n}\psi'(\frac{X_i}{Se^{un^{-\frac{1}{2}}}})(Se^{un^{-\frac{1}{2}}})^{-1}| \\ = \quad & C^2 O_p(n^{-\frac{1}{2}}) + O_p(n^{-\frac{3}{4}}) \end{aligned} \tag{5.4.27}$$

uniformly in $|t|, |u| \le C$. On the other hand,

$$(Se^{un^{-\frac{1}{2}}})^{-1}\frac{t}{n}\sum_{i=1}^{n}\psi'(\frac{X_i}{Se^{un^{-\frac{1}{2}}}}) - t\gamma_1 = O_p(n^{-\frac{1}{2}}) \tag{5.4.28}$$

uniformly in $|t|, |u| \le C$.

Similarly, for $u > 0$ (the case $u < 0$ is analogous),

$$\begin{aligned} &|Z_n(0,u) + \frac{u}{n}\sum_{i=1}^{n}\frac{X_i}{S}\psi'(\frac{X_i}{S})| \\ = \quad & |n^{-\frac{1}{2}}\sum_{i=1}^{n}\int_0^{un^{-\frac{1}{2}}}\frac{X_i}{S}\Big\{ - e^{-U} I[S\mu_1 e^U < X_i < S\mu_2 e^U] \\ & + I[S\mu_1 < X_i < S\mu_2]\Big\} dU| \\ = \quad & S^{-1}n^{\frac{1}{2}}\int_0^{un^{-\frac{1}{2}}} x\Big\{I[S\mu_1 < X_i < S\mu_2] - e^{-U}I[S\mu_1 e^U < X_i < S\mu_2 e^U]\Big\} \\ & d(F_n - F)| + S^{-1}O_p(n^{-\frac{1}{2}}) \\ = \quad & O_p(n^{-\frac{1}{2}}) \text{ uniformly in } |u| \le C. \end{aligned} \tag{5.4.29}$$

Moreover

$$|\frac{u}{n}\sum_{i=1}^{n}\frac{X_i}{S}\psi'(\frac{X_i}{S}) - u\gamma_2| = |u\int_{\mu_1 S}^{\mu_2 S} x d(F_n - F)| = O_p(n^{-\frac{1}{2}}) \tag{5.4.30}$$

uniformly in $|u| \le C$. Combining (5.4.25)-(5.4.30), we arrive at (5.4.20) under conditions **B1** and **B2**.

Assume now the condition **C1** and denote

$$Z_n^{\star}(t,u) = \sum_{\epsilon_1=0,1}\sum_{\epsilon_2=0,1}(-1)^{\epsilon_1+\epsilon_2}Z_n(t - \epsilon_1 t, u - \epsilon_2 u). \tag{5.4.31}$$

Then, obviously,

$$Z_n(t,u) = Z_n^{\star}(t,u) + Z_n^{\star}(t,0) + Z_n^{\star}(0,u), \tag{5.4.32}$$

and for $t > 0$ and $u > 0$ (other quadrants are treated analogously),

$$\begin{aligned} Z_n^\star(t,u) &= n^{-\frac{1}{2}}\sum_{i=1}^n \int_0^{tn^{-\frac{1}{2}}}\int_0^{un^{-\frac{1}{2}}} \Big\{\psi''(\frac{X_i - T}{Se^U})\frac{X_i - T}{(Se^U)^2} \\ &\quad +\psi'(\frac{X_i - T}{Se^U})\frac{1}{Se^U}\Big\}dU\,dT. \end{aligned}$$

Thus under **C1**,

$$\mathbb{E}_0 \sup\Big\{|Z_n^\star(t,u)| : \ |t|,|u| \le C\Big\} \le C^2Kn^{-\frac{1}{2}} \tag{5.4.33}$$

for some K, $0 < K < \infty$ and $n \ge n_0$.

Similarly, for $t > 0$ (the case $t < 0$ is analogous),

$$\begin{aligned} Z_n^\star(t,u) + t\gamma_1 &= n^{-\frac{1}{2}}S^{-1}\sum_{i=1}^n \int_0^{tn^{-\frac{1}{2}}} \Big\{-\psi'(\frac{X_i - T}{S}) + \mathbb{E}_0\psi'(\frac{X_i}{S})\Big\}dT \\ &= n^{-\frac{1}{2}}S^{-1}\sum_{i=1}^n \int_0^{tn^{-\frac{1}{2}}}\int_0^T \psi''(\frac{X_i - V}{S})dV\,dT \\ &\quad -\frac{t}{nS}\sum_{i=1}^n[\psi'(\frac{X_i}{S}) - \mathbb{E}_0\psi'(\frac{X_i}{S})]. \end{aligned} \tag{5.4.34}$$

By **C1**,

$$\begin{aligned} &\mathbb{E}_0\Big\{\sup|\frac{1}{S\sqrt{n}}\sum_{i=1}^n \int_0^{tn^{-\frac{1}{2}}}\int_0^T \psi''(\frac{X_i - V}{S})dV\,dT| : \ |t| \le C\Big\} \\ &\qquad \le S^{-2}C^2Kn^{-\frac{1}{2}} \end{aligned} \tag{5.4.35}$$

for some K, $0 < K < \infty$ and $n \ge n_0$. On the other hand,

$$\begin{aligned} &\mathbb{E}_0 \sup\Big\{|\frac{t}{nS}\sum_{i=1}^n \Big[\psi'(\frac{X_i - V}{S}) - \mathbb{E}_0\psi'(\frac{X_i - V}{S})\Big]| : \ |t| \le C\Big\}^2 \\ &\qquad \le S^{-2}C^2n^{-1}\mathbb{E}_0(\phi'(\frac{X_1}{S}))^2. \end{aligned} \tag{5.4.36}$$

Finally, for $u > 0$ (the case $u < 0$ is analogous),

$$\begin{aligned} &Z_n^\star(0,u) + u\gamma_2 \\ &= n^{-\frac{1}{2}}\sum_{i=1}^n \int_0^{un^{-\frac{1}{2}}} \Big\{(-\frac{X_1}{Se^U})\psi'(\frac{X_1}{Se^U}) + \mathbb{E}_0[(\frac{X_1}{Se^U})\psi'(\frac{X_1}{Se^U})]\Big\}dU \\ &= n^{-\frac{1}{2}}S^{-1}\sum_{i=1}^n \int_0^{un^{-\frac{1}{2}}}\int_0^U \Big\{(\frac{X_i}{Se^V})^2\psi''(\frac{X_i}{Se^V}) + (\frac{X_i}{Se^V})\psi'(\frac{X_i}{Se^V})\Big\}dV\,dU \\ &\quad -\frac{u}{n}\sum_{i=1}^n \Big\{\frac{X_i}{S}\psi'(\frac{X_i}{S}) - \mathbb{E}_0(\frac{X_i}{S}\psi'(\frac{X_i}{S}))\Big\}. \end{aligned} \tag{5.4.37}$$

By **C1**,

$$\mathbb{E}_0 \sup \Big\{ |n^{-\frac{1}{2}} S^{-1} \sum_{i=1}^{n} \int_0^{un^{-\frac{1}{2}}} \int_0^{U} \{ (\frac{X_i}{Se^V}) \psi''(\frac{X_i}{Se^V}) + (\frac{X_i}{Se^V}) \psi'(\frac{X_i}{Se^V}) \} dV\, dU | \, : \, |u| \le C \Big\} \le C^2 K n^{-\frac{1}{2}} \tag{5.4.38}$$

for some K, $0 < K < \infty$ and $n \ge n_0$, and

$$\mathbb{E}_0 \sup \Big\{ \frac{u}{n} \sum_{i=1}^{n} \Big[\frac{X_i}{S} \psi'(\frac{X_i}{S}) - \mathbb{E}_0 (\frac{X_i}{S} \psi'(\frac{X_i}{S})) \Big] \, : \, |u| \le C \Big\}^2$$

$$\le S^{-2} C^2 n^{-1} \mathbb{E}_0 [X_1 \psi'(\frac{X_1}{S})]^2. \tag{5.4.39}$$

Combining (5.4.32)-(5.4.39), we arrive at (5.4.20) under the condition **C1** with the aid of the Markov inequality. The lemma is proved. □

Proof of Theorem 5.4.1. Inserting $u \to n^{1/2} \ln(S_n/S) = O_p(1))$ in (5.4.19) and (5.4.20) and recalling (5.4.2) we obtain

$$\sup_{|t| \le C} \Big\{ |n^{-\frac{1}{2}} \sum_{i=1}^{n} [\psi(\frac{X_i - \theta - n^{-\frac{1}{2}}}{S_n}) - \psi(\frac{X_i - \theta}{S})] + t\gamma_1 + n^{\frac{1}{2}}(\frac{S_n}{S} - 1)\gamma_2 | \Big\}$$

$$= \begin{cases} O_p(n^{-\frac{1}{4}}), & \text{under} \mathbf{A1} - \mathbf{A2}, \\ O_p(n^{-\frac{1}{2}}), & \text{under} \mathbf{B1} - \mathbf{B2} \text{ or } \mathbf{C1}. \end{cases} \tag{5.4.40}$$

1. Under **A1** and **A2**, where ψ is a nondecreasing step function, we could follow the proof of (4.7.20) and obtain the $\sqrt{n}$-consistency of M_n. Moreover, analogously, as in the proof of (4.7.17), we can conclude that

$$n^{-\frac{1}{2}} \sum_{i=1}^{n} \psi(\frac{X_i - M_n}{S_n}) = O_p(n^{-\frac{1}{2}}). \tag{5.4.41}$$

Inserting $u \to n^{1/2}(M_n - \theta)$ in (5.4.40), we obtain the first proposition of the theorem.

2. Under conditions **B1** and **B2** and under **C1**, we assume $n^{1/2}(M_n - \theta) = O_p(1)$. Inserting $u \to n^{1/2}(M_n - \theta)$ in (5.4.40), we obtain the the second proposition. □

The corollaries of Theorem 5.4.1 follow immediately, and we leave their proofs as an exercise.

One open question was tacitly avoided in Theorem 5.4.1: If ρ is not convex then the existence of a solution of (5.4.3) satisfying (5.4.5) may be doubtful. The following theorem deals with this problem.

Theorem 5.4.2 *Let S_n satisfy (5.4.1) and (5.4.2), and let $h(t)$ in (5.4.4) have a unique minimum at $t = 0$ and $\gamma_1 > 0$.*

1. under **B1** *and* **B2** *or under* **C1**, *there exists a solution M_n of the equation*

$$\sum_{i=1}^{n} \psi(\frac{X_i - t}{S_n}) = 0 \tag{5.4.42}$$

such that $n^{1/2}(M_n - \theta) = O_p(1)$. M_n then admits the representation (5.4.9) and Corollaries 5.4.1 - 5.4.4 apply to M_n.

2. Moreover, if **B1** *and* **B2** *or* **C1**, *respectively, apply also to ψ' in the role of ψ, then there exists a local minimum $M_n^\star$ of $\sum_{i=1}^{n} \rho((X_i - t)/S_n)$ satisfying $n^{1/2}(M_n - \theta) = O_p(1)$.*

Proof. Under the conditions of the theorem, $\gamma_1 > 0$ and

$$n^{-\frac{1}{2}} \sum_{i=1}^{n} \psi(\frac{X_i - \theta}{S_n}) = O_p(1), \quad n^{\frac{1}{2}}(\frac{S_n}{S} - 1) = O_p(1),$$

Hence, given $\epsilon > 0$, there exist $C > 0$ and n_0 so that for $n \geq n_0$,

$$I\!P\Big\{|n^{-\frac{1}{2}} \sum_{i=1}^{n} \psi(\frac{X_i - \theta}{S_n})| > C\gamma_1/3\Big\} < \epsilon/3,$$

$$I\!P\Big\{|n^{\frac{1}{2}}(\frac{S_n}{S} - 1)\gamma_2| > C\gamma_1/3\Big\} < \epsilon/3. \tag{5.4.43}$$

By (5.4.40),

$$\begin{aligned} & n^{-\frac{1}{2}} \sum_{i=1}^{n} \psi\Big(\frac{X_i - \theta - n^{-\frac{1}{2}}C}{S_n}\Big) \\ = \; & n^{-\frac{1}{2}} \sum_{i=1}^{n} \psi(\frac{X_i - \theta}{S}) - C\gamma_1 - n^{\frac{1}{2}}(\frac{S_n}{S} - 1)\gamma_2 + O_p(n^{-\frac{1}{2}}) \\ < \; & 0 \text{ for } n \geq n_0 \text{ with probability } \geq 1 - \epsilon. \end{aligned} \tag{5.4.44}$$

Analogously,

$$n^{-\frac{1}{2}} \sum_{i=1}^{n} \psi\Big(\frac{X_i - \theta + n^{-\frac{1}{2}}C}{S_n}\Big) > 0 \text{ for } n \geq n_0 \text{ with probability } \geq 1 - \epsilon. \tag{5.4.45}$$

Due to the continuity of ψ, (5.4.44) and (5.4.45) imply that there exists T_n such that

$$\sum_{i=1}^{n} \psi\Big(\frac{X_i - \theta - n^{-\frac{1}{2}}T_n}{S_n}\Big) = 0$$

and $I\!P(|T_n| < C) \geq 1 - 2\epsilon$ for $n \geq n_0$. Put $M_n = n^{-\frac{1}{2}}T_n + \theta$. Then $\sum_{i=1}^{n} \psi((X_i - M_n)/S_n) = 0$ and $n^{1/2}(M_n - \theta) = O_p(1)$.

2. Let M_n be the solution of (5.4.42) from part 1. We may virtually repeat the proof of Lemma 5.4.1 for

$$n^{-\frac{1}{2}}\sum_{i=1}^{n}\psi'\Big(\frac{X_i - tn^{-\frac{1}{2}}}{Se^{un^{-\frac{1}{2}}}}\Big)$$

and obtain for some γ_1', γ_2' ($|\gamma_1'|, |\gamma_2'| < \infty$)

$$\begin{aligned} & n^{-1}\sum_{i=1}^{n}\psi'(\frac{X_i - M_n}{S_n}) \\ = \; & n^{-1}\sum_{i=1}^{n}\psi'(\frac{X_i-\theta}{S_n}) - \gamma_1'(M_n-\theta) - \gamma_2'(\frac{S_n}{S}-1) + O_p(n^{-1}) \\ = \; & \gamma_1 S + o_p(1). \end{aligned}$$

Hence, given $\epsilon > 0$, there exists n_0, and for $n \geq n_0$,

$$\mathbb{P}(A_n) \geq 1-\epsilon$$

where $A_n = \Big\{w \in \Omega : \; n^{-1}\sum_{i=1}^{n}\psi'(\frac{X_i - M_n}{S_n}) > 0\Big\}$. Put $M_n^\star = M_n$ for $w \in A_n$; let $M_n^\star$ be any point of local minimum of $\sum_{i=1}^{n}\rho(X_i - t)/S_n)$ at the other points. Then $M_n^\star$ is a local minimum of the above function which is a $\sqrt{n}$-consistent estimator of θ. □

Remark. Under additional conditions on ρ, S_n and on the tails of F, we could apparently prove that also the point of global minimum of $\sum_{i=1}^{n}\rho((X_i - t)/S_n)$ is a $\sqrt{n}$-consistent estimator of θ analogously as in the case of nonstudentized M-estimator.

5.5 M-ESTIMATION IN LINEAR REGRESSION MODEL

Consider the linear model

$$\mathbf{Y} = \mathbf{X}\boldsymbol{\beta} + \mathbf{E}, \tag{5.5.1}$$

where $\mathbf{Y} = (Y_1, \ldots, Y_n)'$ is the vector of observations, $\mathbf{X} = \mathbf{X}_n$ is a (known or observable) design matrix of order $(n \times p)$, $\boldsymbol{\beta} = (\beta_1, \ldots, \beta_p)'$ is an unknown parameter, and $\mathbf{E} = (E_1, \ldots, E_n)'$ is a vector of i.i.d. errors with a distribution function F.

The M-estimator of location parameter extends to the model (5.5.1) in a straightforward way: Given an absolutely continuous $\rho : \mathbb{R}_1 \mapsto \mathbb{R}_1$ with derivative ψ, we define an M-estimator of $\boldsymbol{\beta}$ as a solution of the minimization

$$\sum_{i=1}^{n} \rho(Y_i - \mathbf{x}_i'\mathbf{t}) := \min \tag{5.5.2}$$

with respect to $\mathbf{t} \in \mathbb{R}_p$, where $\mathbf{x}_i'$ is the ith row of $\mathbf{X}_n$, $i = 1, \ldots, n$. Such M-estimator $\mathbf{M}_n$ is regression equivariant:

$$\mathbf{M}_n(\mathbf{Y} + \mathbf{X}\mathbf{b}) = \mathbf{M}_n(\mathbf{Y}) + \mathbf{b} \quad \text{for} \quad \mathbf{b} \in \mathbb{R}_p, \tag{5.5.3}$$

but $\mathbf{M}_n$ is generally not scale equivariant: It does not satisfy

$$\mathbf{M}_n(c\mathbf{Y}) = c\mathbf{M}_n(\mathbf{Y}) \quad \text{for} \quad c > 0. \tag{5.5.4}$$

On the other hand, the studentization leads to estimators that are scale as well as regression equivariant. The studentized M-estimator is defined as a solution of the minimization

$$\sum_{i=1}^{n} \rho((Y_i - \mathbf{x}_i'\mathbf{t})/S_n) := \min, \tag{5.5.5}$$

where $S_n = S_n(\mathbf{Y}) \geq 0$ is an appropriate scale statistic. For the best results S_n should be regression invariant and scale equivariant:

$$S_n(c(\mathbf{Y} + \mathbf{X}\mathbf{b})) = cS_n(\mathbf{Y}) \quad \text{for} \quad \mathbf{b} \in \mathbb{R}_p \quad \text{and} \quad c > 0. \tag{5.5.6}$$

The minimization (5.5.5) should be supplemented by a rule how to define $\mathbf{M}_n$ if $S_n(\mathbf{Y}) = 0$. However, in typical cases it appears with probability zero, and the specific rule does not effect the asymptotic properties of $\mathbf{M}_n$.

For example, defining $\mathbf{H} = \mathbf{X}(\mathbf{X}'\mathbf{X})^{-1}\mathbf{X}'$ as the projection matrix and $\mathbf{I}_n$ as the identity matrix, the square root of the residual sum of squares

$$S_n(\mathbf{Y}) = (\mathbf{Y}'(\mathbf{I}_n - \mathbf{H})\mathbf{Y})^{\frac{1}{2}} \tag{5.5.7}$$

satisfies (5.5.6). However, it is not used in robust procedures. We will rather use the scale statistics based on regression quantiles. Another possibility are statistics of the type $||\mathbf{Y} - \mathbf{X}\hat{\boldsymbol{\beta}}(1/2)||$ where $\hat{\boldsymbol{\beta}}(\alpha)$ is α-regression quantile (L_1-estimator of $\boldsymbol{\beta}$ (and $||\cdot||$ is an appropriate norm; (see Problem 5.5.1).

We will deal only with the studentized M-estimators. The nonstudentized M-estimators are covered as a special case (though of course they typically need weaker regularity conditions). Similarly, as in the location model, we will assume that $\psi = \rho'$ can be decomposed into the sum

$$\psi = \psi_a + \psi_c + \psi_s, \tag{5.5.8}$$

where ψ_a is absolutely continuous function with absolutely continuous derivative, ψ_c is a continuous, piecewise linear function that is constant in a neighborhood of $\pm\infty$, and ψ_s is a nondecreasing step function.

More precisely, we will impose the following conditions on (5.5.5):

M1. $S_n(\mathbf{Y})$ is regression invariant and scale equivariant, $S_n > 0$ a.s. and

$$n^{\frac{1}{2}}(S_n - S) = O_p(1)$$

for some functional $S = S(F) > 0$.

M2. The function $h(t) = \int \rho((z-t)/S)dF(z)$ has the unique minimum at $t = 0$.

M3. For some $\delta > 0$ and $\eta > 1$,

$$\int_{-\infty}^{\infty} \left\{|z| \sup_{|u|\leq\delta} \sup_{|v|\leq\delta} |\psi_a''(e^{-v}(z+u)/S)|\right\}^{\eta} dF(z) < \infty$$

and

$$\int_{-\infty}^{\infty} \left\{|z|^2 \sup_{|u|\leq\delta} |\psi_a''(z+u)/S)|\right\}^{\eta} dF(z) < \infty,$$

where $\psi_a'(z) = (d/dz)\psi_a(z)$, and $\psi_a''(z) = (d^2/dz^2)\psi_a(z)$.

M4. ψ_c is a continuous, piecewise linear function with knots at $\mu_1, \ldots, \mu_k$, which is constant in a neighborhood of $\pm\infty$. Hence the derivative ψ_c' of ψ_c is a step function

$$\psi_c'(z) = \alpha_\nu \text{ for } \mu_\nu < z < \mu_{\nu+1},\ \nu = 0, 1, \ldots, k,$$

where $\alpha_0, \alpha_1, \ldots, \alpha_k \in \mathbb{R}_1$, $\alpha_0 = \alpha_k = 0$ and $-\infty = \mu_0 < \mu_1 < \ldots < \mu_k < \mu_{k+1} = \infty$. We assume that $f(z) = \frac{dF(z)}{dz}$ is bounded in neighborhoods of $S\mu_1, \ldots S\mu_k$.

M5. $\psi_s(z) = \lambda_\nu$ for $q_\nu < z \leq q_{\nu+1}$, $\nu = 1, \ldots, m$ where $-\infty = q_0 < q_1 < \ldots q_m < q_{m+1} = \infty$, $-\infty < \lambda_0 < \lambda_1 < \ldots < \lambda_m < \infty$. We assume that $0 < f(z) = (d/dz)F(z)$ and $f'(z) = (d^2/dz^2)F(z)$ are bounded in neighborhoods of $Sq_1, \ldots, Sq_m$.

The asymptotic representation for $\mathbf{M}_n$ will involve the functionals

$$\gamma_1 = S^{-1} \int_{-\infty}^{\infty} (\psi_a'(z/S) + \psi_c'(z/S))dF(z), \tag{5.5.9}$$

$$\gamma_2 = S^{-1} \int_{-\infty}^{\infty} z(\psi_a'(z/S) + \psi_c'(z/S))dF(z), \tag{5.5.10}$$

$$\gamma_1^\star = \sum_{\nu=1}^{m} (\lambda_\nu - \lambda_{\nu-1}) f(Sq_\nu), \tag{5.5.11}$$

and

$$\gamma_2^\star = S \sum_{\nu=1}^{m} (\lambda_\nu - \lambda_{\nu-1}) q_\nu f(Sq_\nu). \tag{5.5.12}$$

Condition **M3** is essentially a moment condition that holds, for example, if ψ_a'' is bounded and either (1) $\psi_a''(z) = 0$ for $z < a$ or $z > b$, $-\infty <$

$a < b < \infty$, or (2) $\int_{-\infty}^{\infty} |z|^{2+\epsilon} dF(z) < \infty$ for some $\epsilon > 0$. If S is known or if we consider the nonstudentized M-estimator, we can omit **M1** and replace **M3** by

M3'. For some $\delta > 0$ and $\eta > 1$,

$$\int_{-\infty}^{\infty} \Big\{ \sup_{|u| \le \delta} |\psi_a''((z+a)/S)| \Big\}^{\eta} dF(z) < \infty.$$

Conditions **M4** and **M5** depict explicitly the trade-off between the smoothness of ψ and smoothness of F. The class of functions ψ_c covers the usual Huber's and Hampel's proposals.

Moreover we will impose the following conditions on the matrix $\mathbf{X}_n$:

X1. $x_{i1} = 1,\ i = 1, \ldots, n.$

X2. $n^{-1} \sum_{i=1}^{n} ||\mathbf{x}_i||^4 = O_p(1).$

X3. $\lim_{n \to \infty} \mathbf{Q}_n = \mathbf{Q}$,

where $\mathbf{Q}_n = n^{-1} \mathbf{X}_n' \mathbf{X}_n$ and $\mathbf{Q}$ is a positive definite $p \times p$ matrix.

Let $\mathbf{M}_n$ be a solution of the minimization (5.5.5). If $\psi = \rho'$ is continuous (i.e., $\psi_s \equiv 0$), then $\mathbf{M}_n$ is a solution of the system of equations

$$\sum_{i=1}^{n} \mathbf{x}_i \psi\Big(\frac{Y_i - \mathbf{x}_i' \mathbf{t}}{S_n}\Big) = \mathbf{0}. \tag{5.5.13}$$

However, this system may have more roots, while only one of them leads to a global minimum of (5.5.5). We will prove that there exists at least one root of (5.5.13) which is a $\sqrt{n}$-consistent estimator of $\boldsymbol{\beta}$.

If ψ is a nondecreasing step function, $\psi_a = \psi_c \equiv 0$, then $\mathbf{M}_n$ is a point of minimum of the convex function $\sum_{i=1}^{n} \rho((Y_i - \mathbf{x}_i' t)/S_n)$ of $\mathbf{t} \in \mathbb{R}_p$, and its consistency and asymptotic representation may be proved using a different argument. The basic results on studentized M-estimators of regression are summarized in the following three theorems.

Theorem 5.5.1 *Consider the model (5.5.1) and assume the conditions* **M1-M4, X1-X3**, *and that γ_1 defined in (5.5.9) is different from zero. Then, provided $\psi_s \equiv 0$, there exists a root $\mathbf{M}_n$ of the system (5.5.13) such that*

$$n^{\frac{1}{2}} ||\mathbf{M}_n - \boldsymbol{\beta}|| = O_p(1) \quad \text{as } n \to \infty. \tag{5.5.14}$$

Moreover any root $\mathbf{M}_n$ of (5.5.13) satisfying (5.5.14) admits the representation

$$\mathbf{M}_n - \boldsymbol{\beta} = (n\gamma_1)^{-1} \mathbf{Q}_n^{-1} \sum_{i=1}^{n} \mathbf{x}_i \psi(E_i/S) - \frac{\gamma_2}{\gamma_1}\Big(\frac{S_n}{S} - 1\Big)\mathbf{e}_1 + \mathbf{R}_n, \tag{5.5.15}$$

where $||\mathbf{R}_n|| = O_p(n^{-1})$ and $\mathbf{e}_1 = (1, 0, \ldots, 0)' \in \mathbb{R}_p$.

Theorem 5.5.2 *Consider the linear model (5.5.1) and assume the conditions* **M1**, **M2**, **M5**, *and* **X1-X3**. *Let* $\mathbf{M}_n$ *be the point of global minimum of (5.5.5). Then, provided that* $\psi_a = \psi_s = 0$,

$$n^{\frac{1}{2}}||\mathbf{M}_n - \boldsymbol{\beta}|| = O_p(1) \ \textit{as}\ n \to \infty, \tag{5.5.16}$$

and $\mathbf{M}_n$ *admits the representation*

$$\mathbf{M}_n - \boldsymbol{\beta} = (n\gamma_1^\star)^{-1}\mathbf{Q}_n^{-1}\sum_{i=1}^{n}\mathbf{x}_i\psi(E_i/S) - \frac{\gamma_2^\star}{\gamma_1^\star}(\frac{S_n}{S} - 1)\mathbf{e}_1 + \mathbf{R}_n, \tag{5.5.17}$$

where $||\mathbf{R}_n|| = O_p(n^{-3/4})$ *and* $\mathbf{e}_1 = (1, 0, \ldots, 0)' \in \mathbb{R}_p$.

Remark. Notice that only the first (intercept) components of the second terms in representations (5.5.15) and (5.5.17) are different from zero; the slope components of $\mathbf{M}_n$ are not affected by S_n.

Combining the above results, we immediately obtain the following theorem for the general class of M-estimator:

Theorem 5.5.3 *Consider the model (5.5.1) and assume the conditions* **M1-M4**, *and* **X1-X3**. *Let* ψ *be either continuous or monotone, and let* $\gamma_1 + \gamma_1^\star \neq 0$. *Then, for any M-estimator* $\mathbf{M}_n$ *satisfying* $n^{1/2}(\mathbf{M}_n - \boldsymbol{\beta}) = O_p(1)$,

$$\begin{aligned}\mathbf{M}_n - \boldsymbol{\beta} &= (n(\gamma_1 + \gamma_1^\star))^{-1}\mathbf{Q}_n^{-1}\sum_{i=1}^{n}\mathbf{x}_i\psi(E_i/S) \\ &- (\gamma_1 + \gamma_1^\star)^{-1}(\gamma_2 + \gamma_2^\star)(\frac{S_n}{S} - 1)\mathbf{e}_1 + \mathbf{R}_n\end{aligned} \tag{5.5.18}$$

where

$$||\mathbf{R}_n|| = \begin{cases} O_p(n^{-1}), & \textit{if } \psi_s \equiv 0, \\ O_p(n^{-3/4}), & \textit{otherwise.}\end{cases} \tag{5.5.19}$$

Theorems 5.5.1-5.5.3 have several interesting corollaries, parallel to those in the location model; we leave the proofs as Problems 5.5.1-5.5.4.

COROLLARY 5.5.1 Under the conditions of Theorem 5.5.1, let $\mathbf{M}_n^{(1)}$ and $\mathbf{M}_n^{(2)}$ be any pair of roots of the system of equations (5.5.13), both satisfying (5.5.14). Then

$$||\mathbf{M}_n^{(1)} - \mathbf{M}_n^{(2)}|| = O_p(n^{-1}). \tag{5.5.20}$$

COROLLARY 5.5.2 Assume that

$$\sigma^2 = \int_{-\infty}^{\infty}\psi^2(z/S)dF(z) < \infty. \tag{5.5.21}$$

Then, under the conditions of Theorems 5.5.1-5.5.3, respectively, the sequence

$$n^{\frac{1}{2}}\Big\{\hat{\gamma}_1(\mathbf{M}_n - \boldsymbol{\beta}) + \hat{\gamma}_2(\frac{S_n}{S} - 1)\mathbf{e}_1\Big\} \tag{5.5.22}$$

has the asymptotic p-dimensional normal distribution $\mathcal{N}_p(\mathbf{0}, \sigma^2\mathbf{Q}^{-1})$; here $\hat{\gamma}_i$ stands for $\gamma_i, \gamma_i^{\star}$ or $\gamma_i + \gamma_i^{\star}$, respectively, $i = 1, 2$.

COROLLARY 5.5.3 Let $F(z) + F(-z) = 1$, $\rho(-z) = \rho(z), z \in \mathbb{R}_1$. Then, under the conditions of either of Theorems 5.3.1 – 5.3.3, respectively,

$$\mathbf{M}_n - \boldsymbol{\beta} = (n\hat{\gamma}_1)^{-1}\mathbf{Q}_n^{-1}\sum_{i=1}^{n}\mathbf{x}_i\psi(E_i/S) + \mathbf{R}_n \tag{5.5.23}$$

where $||\mathbf{R}_n|| = O_p(n^{-1})$ provided that $\psi_s \equiv 0$ and $||\mathbf{R}_n|| = O_p(n^{-3/4})$ otherwise. Moreover, if $\sigma^2 < \infty$ for σ^2 in (5.5.21), $n^{1/2}(\mathbf{M}_n - \boldsymbol{\beta})$ is asymptotically normally distributed

$$\mathcal{N}_p(\mathbf{0}, (\sigma^2/\hat{\gamma}_1^2)\mathbf{Q}^{-1}). \tag{5.5.24}$$

In some contexts we may consider the *restricted M-estimator* $\mathbf{M}_n$ of $\boldsymbol{\beta}$ defined as a solution of the minimization (5.5.5) under the linear constraint

$$\mathbf{A}\beta = \mathbf{c}, \tag{5.5.25}$$

where $\mathbf{A}$ is a $q \times p$ matrix of the full rank and $c \in \mathbf{R}$. The relation (5.5.25) may be an hypothesis H_0 of interest. Similarly as in the unrestricted case, we get the following representation for restricted M-estimator:

COROLLARY 5.5.4 *Let* $\mathbf{M}_n$ *be the restricted M-estimator of* $\boldsymbol{\beta}$ *under the constraint* (5.5.25). *Then, under the conditions of Theorem 5.5.3,* $\mathbf{M}_n$ *admits the asymptotic representation*

$$\mathbf{M}_n - \beta = \hat{\gamma}_1^{-1}\left[\mathbf{I}_p - \mathbf{Q}^{-1}\mathbf{A}'\left(\mathbf{A}\mathbf{Q}^{-1}\mathbf{A}'\right)^{-1}\mathbf{A}\right].\left(\mathbf{Q}^{-1}\zeta_n - \hat{\gamma}_2\mathbf{e}_1\left(\frac{S_n}{S} - 1\right)\right), \tag{5.5.26}$$

where $\zeta_n = n^{-1}\sum_{i=1}^{n}\mathbf{x}_i\psi(E_i/S)$, *and* $\hat{\gamma}_\nu = \gamma_\nu + \gamma_\nu^{*}$, $\nu = 1, 2$.

Theorems 5.5.1-5.5.3 will be proved with the aid of the following asymptotic linearity lemma:

Lemma 5.5.1 *Consider the model (5.5.1) with* $\mathbf{X}_n$ *satisfying* **X1-X3**. *Let* $\psi : \mathbb{R}_1 \mapsto \mathbb{R}_1$ *be a function of the form* $\psi = \psi_a + \psi_c + \psi_s$. *(1). If* $\psi_s \equiv 0$, *then, under the conditions* **M3-M4**,

$$\sup_{||\mathbf{t}||\leq C, |u|\leq C} ||n^{-\frac{1}{2}}\sum_{i=1}^{n}\mathbf{x}_i\{\psi(e^{-n^{-\frac{1}{2}}u}(E_i - n^{-\frac{1}{2}}\mathbf{x}_i'\mathbf{t})/S)$$

$$- \psi(E_i/S) + n^{-\frac{1}{2}}(\gamma_1 \mathbf{x}_i'\mathbf{t} + \gamma_2 u)\}|| = O_p(n^{-\frac{1}{2}}), \qquad (5.5.27)$$

for any fixed $C > 0$ as $n \to \infty$. (2) If $\psi_1 - \psi = 0$ then under the conditions **M5**,

$$\sup_{||\mathbf{t}|| \leq C, |u| \leq C} ||n^{-\frac{1}{2}} \sum_{i=1}^{n} \mathbf{x}_i \{\psi(e^{-n^{-\frac{1}{2}}u}(E_i - n^{-\frac{1}{2}}\mathbf{x}_i'\mathbf{t})/S)$$

$$- \psi(E_i/S) + n^{-\frac{1}{2}}(\gamma_1^{\star} \mathbf{x}_i'\mathbf{t} + \gamma_2^{\star} u)\}|| = O_p(n^{-\frac{1}{4}}) \qquad (5.5.28)$$

for any fixed $C > 0$ as $n \to \infty$.

The proof of Lemma 5.5.1 is relegated to the Appendix (see Section A.2).

Proof of Theorem 5.5.1. Notice that $\gamma_1 \geq 0$ by condition **M2**; hence $\gamma_1 > 0$ under the conditions of the theorem. Inserting $u \to n^{1/2} \ln(S_n/S)$ $(= O_p(1))$ in (5.5.27), we obtain (in view of **M1**)

$$\sup_{||\mathbf{t}|| \leq C} ||\mathbf{E}_n(\mathbf{t}, S_n) - \mathbf{E}_n(\mathbf{0}, S) + \gamma_1 \mathbf{Q}\mathbf{t} + \gamma_2 \mathbf{q}_1 n^{\frac{1}{2}}(\frac{S_n}{S} - 1)|| = O_p(n^{-\frac{1}{2}}), \quad (5.5.29)$$

where

$$\mathbf{E}_n(\mathbf{t}, s) = n^{-\frac{1}{2}} \sum_{i=1}^{n} \mathbf{x}_i \psi((E_i - n^{-\frac{1}{2}}\mathbf{x}_i'\mathbf{t})/s), \quad \mathbf{t} \in \mathbb{R}_p, \ s > 0, \qquad (5.5.30)$$

and $\mathbf{q}_1$ is the first column of the matrix $\mathbf{Q}$ from the condition **X3**. Notice that

$$\mathbf{E}_n(n^{\frac{1}{2}}(\mathbf{M}_n - \boldsymbol{\beta}), S_n) = n^{-\frac{1}{2}} \sum_{i=1}^{n} \mathbf{x}_i \psi((Y_i - \mathbf{x}_i'\mathbf{M}_n)/S_n)$$

and thus $n^{1/2}(\mathbf{M}_n - \boldsymbol{\beta})$ is a root of $\mathbf{E}_n(\mathbf{t}, S_n) = \mathbf{0}$ if and only if $\mathbf{M}_n$ is a root of (5.5.13). The $\sqrt{n}$-consistency of $\mathbf{M}_n$ will follow from proposition 6.3.4 in Ortega and Rheinboldt (1973, p. 163) if we can show that

$$\mathbf{t}'\mathbf{E}_n(\mathbf{t}, S_n) < 0 \qquad (5.5.31)$$

for all $\mathbf{t}$, $||\mathbf{t}|| = C$ in probability as $n \to \infty$ for sufficiently large $C > 0$. To prove this, notice that we may write that, given $\eta > 0$ and $\epsilon > 0$,

$$\mathbb{P}\{\sup_{||\mathbf{t}||=C} \mathbf{t}'\mathbf{E}_n(\mathbf{t}, S_n) \geq 0\}$$

$$\leq \mathbb{P}\{\sup_{||\mathbf{t}||=C} \mathbf{t}'\mathbf{E}_n(\mathbf{t}, S_n) \geq 0, \ \sup_{||\mathbf{t}|| \leq C} ||\mathbf{E}_n(\mathbf{t}, S_n) - \mathbf{E}_n(\mathbf{0}, S)$$

$$+ \gamma_1 \mathbf{Q}\mathbf{t} + \gamma_2 \mathbf{q}_1 n^{\frac{1}{2}}(\frac{S_n}{S} - 1)|| < \eta\} + \frac{\epsilon}{2}$$

for $n \geq n_0$ and for any fixed $C > 0$, as it follows from (5.5.29). Hence

$$\begin{aligned}
&\mathbb{P}\{\sup_{||\mathbf{t}||=C} \mathbf{t}'\mathbf{E}_n(\mathbf{t}, S_n) \geq 0\} \\
&\leq \mathbb{P}\{\sup_{||\mathbf{t}||=C} [\mathbf{t}'\mathbf{E}_n(\mathbf{0}, S) - \gamma_1 \mathbf{t}'\mathbf{Q}\mathbf{t} - \gamma_2 \mathbf{t}'\mathbf{q}_1 n^{\frac{1}{2}}(\frac{S_n}{S} - 1)] \geq -C\eta\} + \frac{\epsilon}{2} \\
&\leq \mathbb{P}\Big\{C||\mathbf{E}_n(\mathbf{0}, S_n)|| + |\gamma_2|C||\mathbf{q}_1||n^{\frac{1}{2}}|\frac{S_n}{S} - 1| \geq \gamma_1 C^2 \lambda_{\min} - C\eta\Big\} + \frac{\epsilon}{2} \leq \epsilon
\end{aligned}$$

for $n \geq n_1$, and for sufficiently large $C > 0$ because both $||\mathbf{E}_n(\mathbf{0}, S)||$ and $n^{1/2}|S_n/S - 1|$ are $O_p(1)$ as $n \to \infty$; $\lambda_{\min}$ is the minimal eigenvalue of $\mathbf{Q}$. Hence, given $\epsilon > 0$, there exist $C > 0$ and n_1 such that

$$\mathbb{P}\Big\{\sup_{||\mathbf{t}||=C} \mathbf{t}'\mathbf{E}_n(\mathbf{t}, S_n) < 0\Big\} > 1 - \epsilon \quad \text{for } n \geq n_1,$$

and this in turn implies that, with probability $> 1 - \epsilon$ and for $n \geq n_1$, there exists a root of the system $\mathbf{E}_n(\mathbf{t}, S_n) = \mathbf{0}$ inside the ball $||\mathbf{t}|| \leq C$. This completes the proof of (5.5.14).

Inserting $n^{1/2}(\mathbf{M}_n - \boldsymbol{\beta})$ for $\mathbf{t}$ in (5.5.29), we obtain

$$||\gamma_1 \mathbf{Q}_n n^{\frac{1}{2}}(\mathbf{M}_n - \boldsymbol{\beta}) + \gamma_2 \mathbf{q}_1^{(n)} n^{\frac{1}{2}}(\frac{S_n}{S} - 1) - n^{-\frac{1}{2}} \sum_{i=1}^{n} \mathbf{x}_i \psi(E_i/S)|| = O_p(n^{-\frac{1}{2}}),$$

with $\mathbf{q}_1^{(n)}$ being the first column of $\mathbf{Q}_n$. This gives the representation (5.5.15). □

Proof of Theorem 5.5.2. Inserting $n^{1/2}\ln(S^{-1}S_n - 1)$ for u in (5.5.28), we obtain

$$\begin{aligned}
\sup_{||\mathbf{t}||\leq C} ||n^{-\frac{1}{2}} \sum_{i=1}^{n} \mathbf{x}_i \{\psi((E_i - n^{-\frac{1}{2}}\mathbf{x}_i'\mathbf{t})/S_n) - \psi(E_i/S)\} \\
+\gamma_1^{\star}\mathbf{Q}\mathbf{t} + \gamma_2^{\star}\mathbf{q}_1 n^{\frac{1}{2}}(\frac{S_n}{S} - 1)|| = O_p(n^{-\frac{1}{4}}).
\end{aligned}$$

This in turn implies, through the integration over $\mathbf{t}$ and using **M1**, that

$$\begin{aligned}
\sup_{||\mathbf{t}||\leq C} |\sum_{i=1}^{n} \big[\rho((E_i - n^{-\frac{1}{2}}\mathbf{x}_i'\mathbf{t})/S_n) - \rho(E_i/S)\big] \\
+ \mathbf{t}'\mathbf{Z}_n - a\mathbf{t}'\mathbf{Q}\mathbf{t}| = o_p(1) \quad \text{as } n \to \infty,
\end{aligned} \tag{5.5.32}$$

where $a = \frac{\gamma_1^{\star}}{2S} > 0$ by (5.5.11) and $(M.5)$ and where

$$\mathbf{Z}_n = S^{-1}\Big\{n^{-\frac{1}{2}} \sum_{i=1}^{n} \mathbf{x}_i \psi(E_i/S) - \gamma_2^{\star}\mathbf{q}_1 n^{\frac{1}{2}}(\frac{S_n}{S} - 1)\Big\}; \tag{5.5.33}$$

obviously $||\mathbf{Z}_n|| = O_p(1)$. If $\mathbf{M}_n$ minimizes (5.5.5), then $n^{1/2}(\mathbf{M}_n - \boldsymbol{\beta})$ minimizes the convex function

$$G_n(\mathbf{t}) = \sup_{||\mathbf{t}|| \le C} |\sum_{i=1}^{n} [\rho((E_i - n^{-\frac{1}{2}}\mathbf{x}_i'\mathbf{t})/S_n) - \rho(E_i/S)] \tag{5.5.34}$$

with respect to $\mathbf{t} \in \mathbb{R}_p$. By (5.5.32),

$$\min_{||\mathbf{t}|| \le C} G_n(\mathbf{t}) = \min_{||\mathbf{t}|| \le C} \{ -\mathbf{t}'\mathbf{Z}_n + a\mathbf{t}'\mathbf{Q}\mathbf{t} \} + o_p(1) \tag{5.5.35}$$

for any fixed $C > 0$. Denoting

$$\mathbf{U}_n = \arg\min_{\mathbf{t} \in \mathbb{R}_p} \{ -\mathbf{t}'\mathbf{Z}_n + a\mathbf{t}'\mathbf{Q}\mathbf{t} \}$$

we get $\mathbf{U}_n = \frac{a}{2}\mathbf{Q}^{-1}\mathbf{Z}_n$ $(= O_p(1))$ and

$$\min_{\mathbf{t} \in \mathbb{R}_p} \{ -\mathbf{t}'\mathbf{Z}_n + a\mathbf{t}'\mathbf{Q}\mathbf{t} \} = \frac{1}{4a}\mathbf{Z}_n'\mathbf{Q}\mathbf{Z}_n.$$

Hence we can write

$$-\mathbf{t}'\mathbf{Z}_n + a\mathbf{t}'\mathbf{Q}\mathbf{t} = a\{(\mathbf{t} - \mathbf{U}_n)'\mathbf{Q}(\mathbf{t} - \mathbf{U}_n) - \mathbf{U}_n'\mathbf{Q}\mathbf{U}_n)\}$$

and then rewrite (5.5.32) in the form

$$\sup_{||\mathbf{t}|| \le C} |G_n(\mathbf{t}) - a[(\mathbf{t} - \mathbf{U}_n)'\mathbf{Q}(\mathbf{t} - \mathbf{U}_n) - \mathbf{U}_n\mathbf{Q}\mathbf{U}_n)]| \xrightarrow{P} \text{ as } n \to \infty. \tag{5.5.36}$$

Substituting $\mathbf{U}_n$ for $\mathbf{t}$, we further obtain

$$G_n(\mathbf{U}_n) + a\mathbf{U}_n'\mathbf{Q}\mathbf{U}_n = o_p(1). \tag{5.5.37}$$

We want to show that

$$||n^{\frac{1}{2}}(\mathbf{M}_n - \boldsymbol{\beta}) - \mathbf{U}_n|| = o_p(1). \tag{5.5.38}$$

Consider the ball $\mathcal{B}_n$ with center $\mathbf{U}_n$ and radius $\delta > 0$. This ball lies in a compact set with probability exceeding $(1 - \epsilon)$ for $n \ge n_0$ because $||\mathbf{t}|| \le ||\mathbf{t} - \mathbf{U}_n|| + ||\mathbf{U}_n|| \le \delta + C$ for $\mathbf{t} \in \mathcal{B}_n$ and for some $C > 0$ (with probability exceeding $1 - \epsilon$ for $n \ge n_0$). Hence by (5.5.32),

$$\Delta_n = \sup_{\mathbf{t} \in \mathcal{B}_n} |\sum_{i=1}^{n} [\rho((E_i - n^{-\frac{1}{2}}\mathbf{x}_i'\mathbf{t})/S_n) - \rho(E_i/S)]$$

$$+ \mathbf{t}'\mathbf{Z}_n - a\mathbf{t}'\mathbf{Q}\mathbf{t}| = o_p(1). \tag{5.5.39}$$

Let $\mathbf{t} \notin \mathcal{B}_n$; then $\mathbf{t} = \mathbf{U}_n + k\mathbf{v}$ where $k > \delta$ and $||\mathbf{v}|| = 1$. Let $\mathbf{t}^\star$ be the boundary point of $\mathcal{B}_n$ that lies on the line from $\mathbf{U}_n$ to $\mathbf{t}$ (i.e., $\mathbf{t}^\star = \mathbf{U}_n + \delta\mathbf{v}$). Then $\mathbf{t}^\star = (1 - (\delta/k))\mathbf{U}_n + (\delta/k)\mathbf{t}$, and, by (5.5.36)-(5.5.37),

$$a\delta^2\lambda_{\min} + G_n(\mathbf{U}_n) - 2\Delta_n \leq G_n(\mathbf{t}^\star) \leq \frac{\delta}{k}G_n(\mathbf{t}) + (1 - \frac{\delta}{k})G_n(\mathbf{U}_n),$$

where $\lambda_{\min}$ is the minimal eigenvalue of $\mathbf{Q}$. Hence

$$\inf_{||\mathbf{t}-\mathbf{U}_n|| \geq \delta} G_n(\mathbf{t}) \geq G_n(\mathbf{U}_n) + \frac{k}{\delta}(a\delta^2\lambda_{\min} - 2\Delta_n)$$

and this implies that, given $\delta > 0$ and $\epsilon > 0$, there exist n_0 and $\eta > 0$ such that for $n \geq n_0$

$$I\!P\Big\{\inf_{||\mathbf{t}-\mathbf{U}_n|| \geq \delta} G_n(\mathbf{t}) - G_n(\mathbf{U}_n) > \eta\Big\} > 1 - \epsilon.$$

Thus

$$I\!P(||n^{\frac{1}{2}}(\mathbf{M}_n - \boldsymbol{\beta}) - \mathbf{U}_n)|| \leq \delta) \to 1$$

for any fixed $\delta > 0$, as $n \to \infty$, and this is (5.5.38). We obtain the asymptotic representation (5.5.17) after inserting $t \to n^{1/2}(\mathbf{M}_n - \boldsymbol{\beta})$ and $u \to n^{1/2}\ln(S_n/S)$ into (5.5.28). □

Proof of Theorem 5.5.3. If ψ is continuous, then $\psi_s \equiv 0$, and the proposition follows from Theorem 5.5.1.

Hence, let $\psi_s \not\equiv 0$ but let ψ be nondecreasing. First, combining (1) and (2) of Lemma 5.2.1, we obtain

$$\sup_{||\mathbf{t}|| \leq C, |u| \leq C} ||n^{-\frac{1}{2}}\sum_{i=1}^{n}\mathbf{x}_i\{\psi(e^{-n^{-\frac{1}{2}}u}(E_i - n^{-\frac{1}{2}}\mathbf{x}_i'\mathbf{t})/S)$$

$$- \psi(E_i/S) + n^{-\frac{1}{2}}(\tilde{\gamma}_1\mathbf{x}_i'\mathbf{t} + \tilde{\gamma}_2 u)\}|| = O_p(n^{-\frac{1}{4}}) \qquad (5.5.40)$$

for any fixed $C > 0$ where $\tilde{\gamma}_i = \gamma_i + \gamma_i^\star$, $i = 1, 2$. Let $\mathbf{M}_n$ be the global solution of the minimization $\sum_{i=1}^n \rho((Y_i - \mathbf{x}_i'\mathbf{t})/S_n) = \min$ of a function that is convex in $\mathbf{t}$ (because ψ is nondecreasing). Then, following step by step the proof of Theorem 5.3.2, we arrive at $n^{1/2}||\mathbf{M}_n - \boldsymbol{\beta}|| = O_p(1)$. Inserting $\mathbf{t} \to n^{1/2}(\mathbf{M}_n - \boldsymbol{\beta})$ and $u \to n^{1/2}\ln(S_n/S)$ in (5.3.40), we receive the representation of $n^{1/2}(\mathbf{M}_n - \boldsymbol{\beta})$. □

5.6 SOME STUDENTIZING SCALE STATISTICS FOR LINEAR MODELS

In Section 5.4 we considered, for the simple location model, some studentizing scale statistics that are translation invariant and scale equivariant [see (5.4.2)].

For linear models the concept of translation invariance extends to that of the affine invariance. This can be stated as follows: Consider the linear model $\mathbf{Y} = \mathbf{X}\boldsymbol{\beta} + \mathbf{E}$, as introduced in (5.5.1). Consider the transformation

$$\mathbf{Y}^\star(\mathbf{b}) = \mathbf{Y} + \mathbf{X}\mathbf{b}, \ \mathbf{b} \in \mathbb{R}_p. \tag{5.6.1}$$

Let then $S_n = S_n(\mathbf{Y})$ be a scale statistic such that

$$S_n(\mathbf{Y} + \mathbf{X}\mathbf{b}) = S_n(\mathbf{Y}) \ \ \forall \mathbf{b} \in \mathbb{R}_p, \mathbf{Y} \in \mathbb{R}_n. \tag{5.6.2}$$

Then S_n is said to be affine invariant. On the other hand, as in (5.4.2) the scale equivariance means that $S_n(c\mathbf{Y}) = cS_n(\mathbf{Y}) \ \forall c \in \mathbb{R}^+, \ \mathbf{Y} \in \mathbb{R}_n$. Combining this with (5.6.2), we say that S_n is affine invariant and scale equivariant if

$$S_n(c(\mathbf{Y} + \mathbf{X}\mathbf{b})) = cS_n(\mathbf{Y}) \ \ \forall \mathbf{b} \in \mathbb{R}_p, \ c > 0, \ \mathbf{Y} \in \mathbb{R}_n. \tag{5.6.3}$$

If we consider the usual LSE $\hat{\boldsymbol{\beta}}$ of $\boldsymbol{\beta}$ [ref: (5.5.1)], then the square root of the residual sum of squares defined in (5.5.7):

$$||\mathbf{Y} - \mathbf{X}\hat{\boldsymbol{\beta}}|| = [(\mathbf{Y} - \mathbf{X}\hat{\boldsymbol{\beta}})'(\mathbf{Y} - \mathbf{X}\hat{\boldsymbol{\beta}})]^{\frac{1}{2}} \tag{5.6.4}$$

possesses the invariance equivariance property in (5.6.3). But this is a highly nonrobust statistic, and hence, in robust inference, there is a genuine need to explore alternative scale statistics that are basically robust and satisfy (5.6.3). In this section we present a broad review of such scale statistics in linear models.

1. *Median absolute deviation from the median*(MAD). An extension of MAD to model (5.5.1) was proposed by Welsh (1986): Let $\hat{\boldsymbol{\beta}}^0$ be an initial $\sqrt{n}$-consistent and affine and scale equivariant estimator of $\boldsymbol{\beta}$, and denote by

$$Y_i(\hat{\boldsymbol{\beta}}^0) = Y_i - \mathbf{x}_i'\hat{\boldsymbol{\beta}}^0, \ \ i = 1, \ldots, n, \tag{5.6.5}$$

$$\xi_{\frac{1}{2}}(\hat{\boldsymbol{\beta}}^0) = \text{med}_{1 \leq i \leq n} Y_i(\hat{\boldsymbol{\beta}}^0), \tag{5.6.6}$$

and

$$S_n = \text{med}_{1 \leq i \leq n} |Y_i(\hat{\boldsymbol{\beta}}^0) - \xi_{\frac{1}{2}}(\hat{\boldsymbol{\beta}}^0)|. \tag{5.6.7}$$

Then S_n satisfies (5.6.2) and (5.6.3) provided that $\hat{\boldsymbol{\beta}}^0$ is affine and scale equivariant. Moreover, under some regularity conditions on F, S_n is $\sqrt{n}$-consistent estimator of the population median deviation. Welsh (1986) also derived its asymptotic representation of the type (5.4.12), hence Theorem 5.5.2 applies.

2. *L-statistics based on regression quantiles.* The α-regression quantile $\hat{\boldsymbol{\beta}}(\alpha)$ for model (5.5.1) was introduced by Koenker and Bassett (1978) as a solution of the minimization problem

$$\sum_{i=1}^{n} \rho_\alpha(Y_i - \mathbf{x}_i\mathbf{t}) := \min \quad \text{with respect to } \mathbf{t} \in \mathbb{R}_p, \tag{5.6.8}$$

where

$$\rho_\alpha(z) = |z|\{\alpha I(z > 0] + (1-\alpha)I[z < 0]\}, \quad z \in \mathbb{R}_1. \tag{5.6.9}$$

The Euclidean distance of two regression quantiles

$$S_n = ||\hat{\boldsymbol{\beta}}_n(\alpha_2) - \hat{\boldsymbol{\beta}}_n(\alpha_1)||, \quad 0 < \alpha_1 < \alpha_2 < 1,$$

satisfies (5.6.2) and (5.6.3) and $S_n \xrightarrow{P} S(F) = F^{-1}(\alpha_2) - F^{-1}(\alpha_1)$. Its asymptotic representation follows, for example, from that for $\hat{\boldsymbol{\beta}}_n(\alpha)$ derived by Ruppert and Carroll (1980). The Euclidean norm may be replaced by L_p-norm or by another appropriate norm. An alternative statistic is the deviation of the first components of regression quantiles, $S_n = \hat{\beta}_n(\alpha_2) - \hat{\beta}_n(\alpha_1)$, $0 < \alpha_1 < \alpha_2 < 1$, with the same population counterpart.

More generally, Bickel and Lehmann (1979) proposed various measures of spread of the distribution F, that could also serve as the scale functional $S(F)$. The corresponding scale statistic is then an estimator of $S(F)$ based on regression quantiles. As an example, we could take

$$S(F) = \Big\{ \int_{\frac{1}{2}}^{1} [F^{-1}(u) - F^{-1}(1-u)]^2 d\Lambda(u) \Big\}^{\frac{1}{2}} \tag{5.6.10}$$

where Λ is the uniform distribution on $(1/2, 1-\delta)$, $0 < \delta < 1/2$; then

$$S(F) = \Big\{ \int_{\frac{1}{2}}^{1} ||\hat{\boldsymbol{\beta}}(u) - \hat{\boldsymbol{\beta}}(1-u)||^2 \, d\Lambda(u) \Big\}^{\frac{1}{2}}. \tag{5.6.11}$$

3. Falk (1986) proposed a histogram and kernel type estimators of the value $(1/f(F^{-1}(\alpha)))$, $0 < \alpha < 1$, in the location model, $f(x) = \frac{dF(x)}{dx}$. Dodge and Jurečková (1993) extended Falk's estimators to the linear model in the following way: First let

$$H_n^{(\alpha)} = \{\hat{\boldsymbol{\beta}}_{n1}(\alpha + \nu_n) - \hat{\boldsymbol{\beta}}_{n1}(\alpha - \nu_n)\}/(2\nu_n) \tag{5.6.12}$$

where

$$\nu_n = o(n^{-\frac{1}{3}}) \text{ and } n\nu_n \to \infty \text{ as } n \to \infty, \tag{5.6.13}$$

be a histogram type $(n\nu_n)^{1/2}$-consistent estimator of $1/f(F^{-1}(\alpha))$ satisfying (5.6.2) and (5.6.3). Second, considering the kernel function $k: \mathbb{R}_1 \mapsto \mathbb{R}_1$ with a compact support, which is continuous on its support and satisfies

$$\int k(x)dx = 0 \text{ and } \int xk(x)dx = -1, \tag{5.6.14}$$

we construct the following kernel type estimator of $1/f(F^{-1}(\alpha))$:

$$\chi_n^{(\alpha)} = \nu_n^{-2} \int_0^1 \hat{\beta}_{n1}(u) k\Big(\frac{\alpha - u}{\nu_n}\Big) du, \tag{5.6.15}$$

where

$$\nu_n \to 0, \quad n\nu_n^2 \to \infty \quad \text{and} \quad n\nu_n^3 \to 0 \quad \text{as } n \to \infty. \tag{5.6.16}$$

Again, $\chi_n^{(u)}$ is a $(n\nu_n)^{1/2}$-consistent estimator of $1/f(F^{-1}(\alpha))$, whose asymptotic variance may be less than that of (5.6.12) for some kernels. Due to their lower rates of consistency, we will not use the above estimators for a simple studentization but rather in an inference on the population quantiles based on the regression data.

4. Jurečková and Sen (1994) constructed a scale statistics based on *regression rank scores*, which are dual to the regression quantiles in the linear programming sense and represent an extension of the rank scores to the linear model (we recommend Gutenbrunner and Jurečková 1992 for a detailed account of this concept). More precisely, regression rank scores $\hat{a}_n(\alpha) = (\hat{a}_{n1}(\alpha), \ldots, \hat{a}_{nn}(\alpha))'$, $0 < \alpha < 1$, for the model (5.5.1) are defined as a solution of the maximization

$$\mathbf{Y}'\hat{a}_n(\alpha) := \max \tag{5.6.17}$$

under the restriction

$$\mathbf{X}'(\hat{a}_n(\alpha) - \mathbf{1}_n'(1-\alpha)) = \mathbf{0}, \quad \hat{a}_n(\alpha) \in [0,1]^n, \quad 0 < \alpha < 1. \tag{5.6.18}$$

In the location model, they reduce to the rank scores considered in Hájek and Šidák (1967). The proposed scale statistic is of the form

$$S_n = n^{-1} \sum_{i=1}^{n} Y_i \hat{b}_{ni}, \tag{5.6.19}$$

where

$$\hat{b}_{ni} = -\int_{-\alpha_0}^{1-\alpha_0} \phi(\alpha) d\hat{a}_{ni}(\alpha), \quad 0 < \alpha_0 < 1/2, \tag{5.6.20}$$

and the score-generating function $\phi : [0,1] \mapsto \mathbb{R}_1$ is nondecreasing, square integrable and skew-symmetric, standardized so that $\int_{\alpha_0}^{1-\alpha_0} \phi^2(\alpha) d\alpha = 1$. Then S_n satisfies (5.6.3) and is a $\sqrt{n}$-consistent estimator of $S(F) = \int_{\alpha_0}^{1-\alpha_0} \phi(\alpha) F^{-1}(\alpha) d\alpha$. In the location model, S_n reduces to the Jaeckel (1972) measure of dispersion of ranks of residuals. Some of these details are provided in the last section of Chapter 6, which deals with regression rank scores in depth.

5.7 HADAMARD DIFFERENTIABILITY APPROACHES FOR M-ESTIMATORS IN LINEAR MODELS

The general asymptotics developed in this chapter mostly rest on the premise that M-estimators in regression parameters have uniform asymptotic linearity, for which different sets of (sufficient) regularity conditions have been considered in the earlier sections. An alternative derivation of this basic result based on Hadamard (or compact) differentiability approaches is considered in this section. These results are mostly the generalization of those in Sections 4.5 (dealing with i.i.d. observations), and they have been studied at length by Ren and Sen (1991, 1994, 1995a,b), among others. We provide here only a general motivation and a broad outline, for more technical treatments of this subject matter, we refer the reader to the papers cited above.

For the linear model (5.5.1), the M-estimation procedure based on the norm $\rho(.) : \mathbb{R} \mapsto \mathbb{R}$, given in (5.5.2), may as well be formulated in terms of the estimating equation

$$\sum_{i=1}^{n} \mathbf{x}_i \psi(\mathbf{Y}_i - \mathbf{x}_i' \boldsymbol{\beta}) = \mathbf{0}, \tag{5.7.1}$$

whenever $\rho(.)$ has a derivative ψ (a.e.); we have designated ψ as the score function. In the preceding section, under diverse regularity conditions on ψ, $\mathbf{x}_i$ and the d.f. F (of the error component), the "uniform asymptotic linearity" result was established. In this context the basic statistic is

$$\mathbf{M}_n(\mathbf{u}) = \sum_{i=1}^{n} \mathbf{x}_i \psi(e_i - \mathbf{x}_i' \mathbf{u}), \quad \mathbf{u} \in \mathbb{R}_p. \tag{5.7.2}$$

For the location model, $\mathbf{x}_i = 1 \ \forall i$, $u \in \mathbb{R}$, and the approach of Section 4.5, clearly extends to the M-estimators (which are, however, defined implicitly and hence require extra care in the treatment of their asymptotics). For linear models, although the errors e_i are i.i.d. r.v.'s, the Y_i are not i.d., and hence a somewhat different approach is needed. For this study, we define

$$\mathbf{C}_n = \mathbf{X}'\mathbf{X} = \sum_{i=1}^{n} \mathbf{x}_i \mathbf{x}_i' = ((r_{nij}))_{i,j=1,\ldots,p} \tag{5.7.3}$$

$$\mathbf{C}_n^0 = \text{Diag}(r_{n11}^{\frac{1}{2}}, \ldots, r_{npp}^{\frac{1}{2}})', \quad 1 \leq i \leq n, \tag{5.7.4}$$

$$\mathbf{c}_{ni} = \left(\mathbf{C}_n^0\right)^{-1} \mathbf{x}_i = (c_{ni1}, \ldots, c_{nip})', \quad 1 \leq i \leq n, \tag{5.7.5}$$

and

$$\mathbf{Q}_n = \sum_{i=1}^{n} \mathbf{c}_{ni} \mathbf{c}_{ni}' = (\mathbf{C}_n^0)^{-1} \mathbf{C}_n (\mathbf{C}_n^0)^{-1}. \tag{5.7.6}$$

With these notations introduced, we take a varied form of $\mathbf{M}_n(\mathbf{u})$ in (5.7.2); namely we let

$$\mathbf{M}_n(\mathbf{u}) = \sum_{i=1}^{n} \mathbf{c}_{ni} \psi(e_i - \mathbf{c}_{ni}' \mathbf{u}), \quad \mathbf{u} \in \mathbb{R}_p. \tag{5.7.7}$$

Our goal is to study the following result: As $n \to \infty$,

$$\sup_{||u|| \leq K} ||\mathbf{M}_n(\mathbf{u}) - \mathbf{M}_n(\mathbf{0}) + \mathbf{Q}_n \mathbf{u}\gamma|| \xrightarrow{P} 0, \tag{5.7.8}$$

for every $K : 0 < K < \infty$, where

$$\gamma = \int_{\boldsymbol{R}} f(x) d\psi(x). \tag{5.7.9}$$

Whenever $I(f) < \infty$, we write

$$\gamma = \int_{\boldsymbol{R}} \psi(x)\{-f'(x)/f(x)\} dF(x). \tag{5.7.10}$$

Let us introduce the empirical function (vector)

$$\mathbf{S}_n^\star(t, \mathbf{u}) = \sum_{i=1}^{n} \mathbf{c}_{ni} I(e_i \leq F^{-1}(t) + \mathbf{c}'_{ni}\mathbf{u}), \quad \text{for } t \in [0,1], \ \mathbf{u} \in \mathbb{R}_p. \tag{5.7.11}$$

Then, we have from (5.7.7) and (5.7.11),

$$\mathbf{M}_n(\mathbf{u}) = \int_{\boldsymbol{R}} \psi(F^{-1}(t)) d\mathbf{S}_n^\star(t, \mathbf{u}), \ \ \mathbf{u} \in \mathbb{R}_p, \tag{5.7.12}$$

which is a linear function of $\mathbf{S}_n^\star(.)$.

Referring back to Section 4.5, we may note that for $\tau(F)$ a functional of F, whenever $\tau(.)$ is Hadamard (or compact) differentiable, its Hadamard derivative is also a linear functional [see (4.5.1)-(4.5.2)]. This provides an intuitive justification for adopting a plausible Hadamard differentiability approach for the study of (5.7.8), it and is explored here.

We keep in mind the same notations and definitions of first- (and second-)order Hadamard derivatives, as introduced in Section 4.5 [see (4.5.1)-(4.5.8)]. For the current investigation, we consider as in Ren and Sen (1991) some extended statistical functionals. For simplicity of presentation take first the case of $p = 1$, when the c_{ni} and u are both real numbers. We write

$$c_{ni} = c_{ni}^+ - c_{ni}^-; \ \ c_{ni}^+ = \max(0, c_{ni}), \ \ c_{ni}^- = -\min(0, c_{ni}). \tag{5.7.13}$$

We assume that

$$\sum_{i=1}^{n} c_{ni}^2 = 1 \quad \text{and} \quad \max_{1 \leq i \leq n} c_{ni}^2 \to 0 \quad \text{as } n \to \infty. \tag{5.7.14}$$

Also we assume that the F is absolutely continuous and has a positive and continuous derivative $F' = f$ with limits $\pm\infty$. We define $\mathbf{S}_n^\star(t, u)$ as in before (for $p = 1$), and replacing c_{ni} by c_{ni}^+ and c_{ni}^-, we obtain parallel expressions for $S_n^{+\star}(t, u)$ and $S_n^{-\star}(t, u)$.

Let U_n be the empirical d.f. of the $F(e_i)$, $1 \le i \le n$, and let U be the classical uniform (0,1) d.f. Let $\tau(.)$ be a right continuous functions hav-ing left-hand limits). Then the following result is due to Ren and Sen (1991; Theorem 3.2):
Suppose that $\tau: D[0,1] \mapsto \mathbb{R}$ is a functional, and it is Hadamard differentiable at U. Also assume that the c_{ni} and F satisfy the conditions mentioned before. Then, for any $K > 0$, as $n \to \infty$,

$$\sup_{|u|<K} \Big|\sum_{i=1}^{n} c_{ni}^{+}\tau\Big(\frac{S_n^{+\star}(.,u)}{\sum_{i=1}^{n} c_{ni}^{+}}\Big) - \sum_{i=1}^{n} c_{ni}^{-}\tau\Big(\frac{S_n^{+\star}(.u)}{\sum_{i=1}^{n} c_{ni}^{-}}\Big) - \tau(U)\sum_{i=1}^{n} c_{ni}$$

$$- \tau_U'\Big(S_n^{\star}(.,u) - U(.)\sum_{i=1}^{n} c_{ni}\Big)\Big| \to 0, \text{ in probability.} \tag{5.7.15}$$

Now identifying a potential $\tau(.)$ for which $M_n(u)$ may be Hadamard derivative, it is possible to show that for any $0 < K < \infty$, as $n \to \infty$,

$$\sup_{|u|\le K} \Big|\sum_{i=1}^{n} c_{ni}\Big\{\tau\big(\frac{S_n^{\star}(.,u)}{\sum_{i=1}^{n} c_{ni}}\big) - \tau\big(\frac{S_n^{\star}(.,0)}{\sum_{i=1}^{n} c_{ni}}\big)\Big\}$$

$$- [M_n(u) - M_n(0)]\Big| \xrightarrow{P} 0, \tag{5.7.16}$$

and

$$\sup_{|u|\le K} \Big|\sum_{i=1}^{n} c_{ni}\Big\{\tau\big(\frac{S_n^{\star}(.,u)}{\sum_{i=1}^{n} c_{ni}}\big) - \tau\big(\frac{S_n^{\star}(.,0)}{\sum_{i=1}^{n} c_{ni}}\big)\Big\} + u\gamma\Big| \xrightarrow{P} 0. \tag{5.7.17}$$

This implies the asymptotic linearity.

Further we define

$$\mathbf{D}_p = \text{Diag}\,\Big(\sum_{i=1}^{n} c_{ni1}, \dots, \sum_{i=1}^{n} c_{nip}\Big) \tag{5.7.18}$$

$$\boldsymbol{\tau}(\mathbf{S}_n^{\star}(.\mathbf{u})) = \Big(\tau\big(\frac{S_{n1}^{\star}(.,\mathbf{u})}{\sum_{i=1}^{n} c_{ni1}}\big), \dots, \tau\big(\frac{S_{np}^{\star}(.,\mathbf{u})}{\sum_{i=1}^{n} c_{nip}}\big)\Big)'. \tag{5.7.19}$$

We desire to show that under appropriate regularity conditions, as $n \to \infty$,

$$\sup_{\|\mathbf{u}\|\le K} \|\mathbf{D}_p\Big\{\tau(S_n^{\star}(.,\mathbf{u}) - \tau(S_n^{\star}(.,\mathbf{0}))\Big\} - [M_n(u) - M_n(0)]\|\| \xrightarrow{P} 0, \tag{5.7.20}$$

and further

$$\sup_{\|\mathbf{u}\|<K} \|\mathbf{D}_p\Big\{\tau(S_n^{\star}(.,\mathbf{u}) - \tau(S_n^{\star}(.,\mathbf{0}))\Big\} + \mathbf{Q}_n\mathbf{u}\gamma\| \xrightarrow{P} 0, \tag{5.7.21}$$

as $n \to \infty$. These two in turn imply that

$$\sup_{\|\mathbf{u}\|<K} \|\mathbf{M}_n(\mathbf{u}) - \mathbf{M}_n(\mathbf{0}) + \mathbf{Q}_n\mathbf{u}\gamma\| \xrightarrow{P} 0, \tag{5.7.22}$$

so the uniform asymptotic linearity holds.

This Hadamard differentiability approach provides a good method for the study of the asymptotic properties of M-estimators of regression through asymptotic linear approximations for the estimators. We may remark, in passing, that Pollard (1991) used the convexity lemma to derive the uniform asymptotic linearity through convexity property of the criterion function and pointwise linearity results, which are comparatively easier to verify. The Hadamard differentiability approach does not depend on any such convexity property. Rather, it allows one to derive the asymptotic approximation of the estimating equations [i.e., $\mathbf{M}_n(\mathbf{u})$] uniformly over a compact set. The regularity conditions appear to be less stringent. We pose some of these as problems at the end of the chapter.

As in the proof of Lemma 5.5.1, we relegate the derivations of (5.7.15)-(5.7.16) and (5.7.19)-(5.7.21) to the Appendix [see Section A.3]. Earlier, in Section 4.5, for i.i.d. r.v.'s, we considered a general SOADR result for a second-order Hadamard differentiable functional. Such a result also holds for M-estimators of regression parameters. The relevant methodology has been developed by Ren and Sen (1995b) wherein the second-order Hadamard differentiability of M-functionals plays a basic role. For a complete discussion of the development, the reader is invited to look into this and other appropriate references.

5.8 PROBLEMS

5.2.1. Consider a generalized Laplace density

$$f(x;\theta) = \text{constant} \exp\{-|x-\theta|^\alpha\}, \quad x \in I\!R,$$

where $\alpha \in (0,1)$ and $\theta \in \Theta \subset I\!R$. Show that

$$\frac{\partial}{\partial\theta} \log f(x;\theta) = -\alpha|x-\theta|^{\alpha-1}\text{sign}(x-\theta)$$

so that

$$|\frac{\partial}{\partial\theta} \log f(x;\theta)|dF(x;\theta) < \infty, \quad \forall \alpha \in (0,1),$$

but that for $\alpha \le 1/2$,

$$\{\frac{\partial}{\partial\theta} \log f(x;\theta)\}^2 dF(x;\theta) \not< \infty.$$

Hence, or otherwise, comment on the asymptotic properties of the MLE of θ (when α is known and $0 < \alpha \le 1/2$).

5.2.2. (Continuation). Consider the same model when $\alpha \in [1,2]$. Comment on the asymptotic properties of the MLE of θ. What are the main differences in the two cases: (a) $0 < \alpha \le 1/2$ and (b) $\alpha \ge 1$?

5.2.3. (Continuation). For the same model, let $\psi(x;\theta) = -f'_x(x;\theta)/f(x;\theta)$, $x \in \mathbb{R}$. Verify whether the regularity conditions, **A1-A2** in Section 5.2 hold or not? Consider the three cases separately: (a) $0 < \alpha < 1$, (b) $\alpha = 1$, (c) $1 < \alpha < 2$? What happens when $\alpha = 2$ or $\alpha > 2$ (but, not an integer)?

5.2.4. Let $g(x)$ be an absolutely continuous, bounded, and skew-symmetric function, and let $\psi(x,t) = g(x-t)$, $x \in \mathbb{R}, t \in \mathbb{R}$. Show that (5.2.5) can then be incorporated for the estimation of θ in Problem 5.2.1. What happens when $g(-x) = -g(x)$ for all x, where $g(0+) \neq 0$? Can **A1-A2** be modified to include such a jump discontinuity of $\psi(.)$?

5.2.5. Consider the two-parameter Laplace density

$$f(x,\boldsymbol{\theta}) = \frac{1}{2}\theta_2^{-1}\exp\{-|x-\theta_1|/\theta_2\}, \quad x \in \mathbb{R}, \theta_1 \in \mathbb{R}, \theta_2 \in \mathbb{R}^+.$$

Let $\boldsymbol{\psi}(x,\mathbf{t}) = (\psi_1(x;\mathbf{t}), \psi_2(x;\mathbf{t}))' = (\text{sign}(x-t_1), |x-t_1|/t_2 - 1)'$. Show that **A1-A2** may not hold here, although if one uses the median-estimator of θ_1, for the estimation of θ_2, $\psi_2(.)$ can be tuned to satisfy both **A1** and **A2**.

5.3.1. Verify (5.3.26).

5.3.2. Consider the function $\rho(y) = \log(1+y^2)$, $y \in \mathbb{R}$, and define $h(t)$ as in (5.3.2). Assume further that the d.f. F has a pdf f, symmetric about 0 such that f has heavy tails [i.e., $f(x)$ converges to 0 as $x \to \infty$ at a rate slower than the normal or exponential pdf.] Show then that $h(t) = 0$ may not have a unique minimum at $t = 0$.

5.3.3. (Continuation). Verify the last problem when $f(x)$ is a Cauchy density. Is $\rho(.)$ in Problem 5.3.2 convex?

5.3.4. Consider the following redescending score function:
(a) $\psi_1(x) = \sin(x/a)$ for $-a\pi \le x \le a\pi$, and 0 otherwise.
(b) $\psi_2(x) = x(k^2 - x^2)^p$, $-k \le x \le k$, and 0 otherwise, where k and p are positive numbers.
(c) $\psi_3(x) = x$, for $-k \le x \le k$; $k\text{sign}x$, for $(0<)k \le |x| \le k+a$, $a > 0$, and $(ke^{k+a-|x|})\text{sign}x$, for $x > k+a$.
In each case, find $h(t)$, defined by (5.3.2), and verify whether a unique root for $h(t) = 0$ exists or not?

5.3.5. For (5.3.60) show that the last term is $o_p(1)$ as $n \to \infty$.

5.4.1. For $h(t)$ defined by (5.4.4), show that when ψ is nonmonotone (as in Problem 5.3.4), there may not be a unique solution of $h(t) = 0$.

5.4.2. Provide a formal proof of Corollary 5.4.1.

5.4.3. Provide a formal proof of Corollary 5.4.2.

5.4.4. Provide a formal proof of Corollary 5.4.3.

5.4.5. Provide a formal proof of Corollary 5.4.4.

5.5.1. For $0 < \alpha < 1$, let $\hat{\boldsymbol{\beta}}_\alpha$ be the α-regression quantile (estimator) of $\boldsymbol{\beta}$. Use the results in Section 5.6, and study its properties.

5.5.2. Provide a formal proof of Corollary 5.5.1.

5.5.3. Provide a formal proof of Corollary 5.5.2.

5.5.4. Provide a formal proof of Corollary 5.5.3.

Chapter 6

Asymptotic Representations for R-Estimators

6.1 INTRODUCTION

Like the M-estimators in Chapter 5, the R-estimators are implicitly defined statistical functionals. Although for R-estimators a differentiable statistical functional approach can be formulated, it may call for bounded score functions that do not include the important cases of normal scores, log-rank scores, and other unbounded ones. Moreover the ranks are integer valued random variables, so that even for smooth score functions, the rank scores may not possess smoothness to a very refined extent. For this reason the treatment of Chapter 5 may not go through entirely for R-estimators of location or regression parameters. Hence, to encompass a larger class of rank-based estimators, we will follow the tracks based on the asymptotic linearity of rank statistics in shift or regression parameters: This approach has been popularized in the past two decades, and some accounts of the related developments are also available in some other contemporary text books on nonparametrics. However, our aim is to go beyond these reported developments onto the second-order representations and to relax some of the currently used regularity conditions. In this way we aim to provide an up-to-date and unifying treatment of the asymptotic representations for R-estimators of location and regression parameters. In Section 6.2 we start with the usual signed rank statistics and present the first-order asymptotic representations for allied R-estimators of location of a symmetric distribution. The next section deals with second-order representations for such R-estimators of location (under additional regularity conditions). Asymptotic representations for R-estimators of the slope in a simple linear regression model are studied in Section 6.4, and the second-order results are then presented in Section 6.5. Section 6.6 deals with the general multiple regression models and contains the first-order asymptotic representation results for the usual R-estimators of the regression and intercept parameters. The last two sections are devoted to regression rank scores estimators which have been developed only in the recent past.

6.2 FIRST ORDER REPRESENTATIONS FOR R-ESTIMATORS OF LOCATION

In Section 3.4 [i.e., (3.4.9)] we have introduced R-estimators of location of a symmetric distribution based on appropriate signed rank statistics. Hence we will first study the asymptotic representation for signed-rank statistics. As in (3.4.7), we define a signed rank statistic $S_n^0(b)$ as

$$n^{-1/2}\sum_{i=1}^{n}\text{sign}(X_i-b)a_n^0(R_{ni}^+(b)),\quad b\in\mathbb{R}_1,$$

where $R_{ni}^+(b)$ = rank of $|X_i-b|$ among $|X_1-b|,\ldots,|X_n-b|$, for $i=1,\ldots,n$. Here we take the *scores* as

$$a_n^0(i)=\mathbb{E}[\phi(U_{n:i})],\quad 1\le i\le n;\ n\ge 1,\tag{6.2.1}$$

where $U_{n:1}<\ldots<U_{n:n}$ stand for the ordered r.v.'s of a sample of size n from the uniform (0,1) distribution, and the score function $\phi=\{\phi(u),\ u\in(0,1)\}$ is taken to be nondecreasing. We let

$$\phi(u)=\phi^\star\Big(\frac{1+u}{2}\Big),\ u\in(0,1),\ \phi^\star\nearrow\text{ in }u,\tag{6.2.2}$$

$$\phi^\star(u)+\phi^\star(1-u)=0\ \ \forall u\in[0,1]\tag{6.2.3}$$

so that $\phi^\star$ is skew-symmetric about 1/2. We assume that $X_1,\ldots,X_n$ have the symmetric d.f. $F(x-\theta)$ where $F(x)+F(-x)=1\ \forall x\in\mathbb{R}_1$. We may assume that $\theta=0$ without loss of generality, and for notational simplicity, we let $R_{ni}^+(0)=R_{ni}^+$, $\forall i,n$, and $F^+(x)=F(x)-F(-x)=2F(x)-1,\ x\ge 0$. Finally, we let

$$\begin{aligned}Q_n^0 &= S_n^0(0)-n^{-\frac{1}{2}}\sum_{i=1}^{n}\phi^\star(F(X_i))\\ &= n^{-\frac{1}{2}}\sum_{i=1}^{n}\text{sign}X_i\{a_n^0(R_{ni}^+)-\phi(F^+(|X_i|))\}.\end{aligned}\tag{6.2.4}$$

We now have the following:

Theorem 6.2.1 *Let X_i, $i > 1$ be i.i.d. r.v.'s with a continuous and symmetric d.f. F, defined on $\mathbb{R}_1$. Let $\phi : (0,1) \mapsto \mathbb{R}_1^+$ be $\nearrow$ such that $\phi(0) = 0$ and $\int_0^1 \phi^2(u)du = A_\phi^2 < \infty$. Then*

$$Q_n^0 \to 0, \quad \textit{in probability, as } n \to \infty. \tag{6.2.5}$$

Proof. Let $\mathcal{B}_n$ be the sigma-field generated by the vector of signs and the vector of the ranks [i.e., $\mathcal{B}_n = \mathcal{B}(\text{sign}X_1, \dots, \text{sign}X_n; R_{n1}^+, \dots, R_{nn}^+)$] for $n \geq 1$. Note that by (6.2.2) and (6.2.3), $\phi^\star(F(X_i)) = (\text{sign}X_i)\phi(F^+(|X_i|))$, for every $i = 1, \dots, n$. Hence, invoking the stochastic independence of the two vectors $(\text{sign}X_1, \dots, \text{sign}X_n)$ and $(R_{n1}^+, \dots, R_{nn}^+)$ along with the independence of the R_{ni}^+ and the ordered r.v.'s corresponding to the $\phi(F^+(|X_i|))$, it follows that

$$\begin{aligned}
\mathbb{E}\Big[n^{-\frac{1}{2}}\sum_{i=1}^n \phi^\star(F(X_i))|\mathcal{B}_n\Big] &= \mathbb{E}\Big[n^{-\frac{1}{2}}\sum_{i=1}^n (\text{sign}X_i)\phi(F(X_i))|\mathcal{B}_n\Big] \\
&= n^{-\frac{1}{2}}\sum_{i=1}^n (\text{sign}X_i)\mathbb{E}\Big[\phi(F(X_i))|R_{ni}^+\Big] \\
&= n^{-\frac{1}{2}}\sum_{i=1}^n (\text{sign}X_i)a_n^0(R_{ni}^+) \\
&= S_n^0(0), \quad n \geq 1.
\end{aligned} \tag{6.2.6}$$

Therefore, by (6.2.4) and (6.2.6), we obtain

$$\begin{aligned}
\mathbb{E}[Q_n^0]^2 &= \mathbb{E}\Big[\phi^2(F^+(|X_i|)\Big] - \mathbb{E}\Big[(S_n^0(0))^2\Big] \\
&= \int_0^1 \phi^2(u)du - n^{-1}\sum_{i=1}^n \{a_n^0(i)\}^2,
\end{aligned} \tag{6.2.7}$$

where, by (6.2.1),

$$\begin{aligned}
n^{-1}\sum_{i=1}^n \{a_n^0(i)\}^2 &= n^{-1}\sum_{i=1}^n \Big\{\mathbb{E}[\phi(U_{n:i})]\Big\}^2 \\
&\leq n^{-1}\sum_{i=1}^n \mathbb{E}[\phi^2(U_{n:i})] = n^{-1}\sum_{i=1}^n \mathbb{E}[\phi^2(U_i)] \\
&= \int_0^1 \phi^2(u)du = A_\phi^2, \quad n \geq 1.
\end{aligned} \tag{6.2.8}$$

Defining

$$\phi_n^0(u) = a_n^0(i) \quad \text{for } (i-1)/n < u \leq i/n, \ i = 1, \dots, n, \tag{6.2.9}$$

we obtain by using (6.2.2) and (6.2.3) that

$$\lim_{n\to\infty} \phi_n^0(u) = \phi(u), \quad \text{for every (fixed) } u \in [0,1); \tag{6.2.10}$$

the proof is left as Problem 6.2.1. Hence, noting that

$$n^{-1}\sum_{i=1}^{n}\{a_n^0(i)\}^2 = \int_0^1 \{\phi_n^0(u)\}^2 du, \tag{6.2.11}$$

we obtain by using (6.2.8), (6.2.9) and the Fatou lemma (Section 2.5) that (6.2.7) converges to 0 as $n \to \infty$, and hence, by the Chebyshev inequality, $Q_n \to 0$, in probability, as $n \to \infty$. □

It is not necessary to define the scores by (6.2.1). For (6.2.5) to hold for suitably defined scores $a_n(i)$, $i = 1, \ldots, n$, all we need is to assume that as $n \to \infty$,

$$n^{-1}\sum_{i=1}^{n}[a_n(i) - a_n^0(i)]^2 \to 0. \tag{6.2.12}$$

Toward this end, note that

$$\begin{aligned}
\mathbb{E}\Big[(S_n(0) - S_n^0(0))^2\Big] &= n^{-1}\mathbb{E}\Big[\sum_{i=1}^{n}\text{sign}X_i\{a_n(R_{ni}^+) - a_n^0(R_{ni}^+)\}\Big]^2 \\
&= n^{-1}\sum_{i=1}^{n}\mathbb{E}\Big[a_n(R_{ni}^+) - a_n^0(R_{ni}^+)\Big]^2 \\
&= n^{-1}\sum_{i=1}^{n}\Big[a_n(i) - a_n^0(i)\Big]^2
\end{aligned} \tag{6.2.13}$$

(where we have invoked the independence of the signs and the ranks), so (6.2.12) ensures the stochastic convergence of $S_n(0) - S_n^0(0)$ to 0, and this ensures that (6.2.5) also holds for $S_n^0(0)$ being replaced by $S_n(0)$. In this context, note that

$$\begin{aligned}
n^{-1}\sum_{i=1}^{n}a_n^2(i) &= n^{-1}\sum_{i=1}^{n}\{a_n^0(i)\}^2 + n^{-1}\sum_{i=1}^{n}\Big[a_n(i) - a_n^0(i)\Big]^2 \\
&\quad -2n^{-1}\sum_{i=1}^{n}a_n^0(i)\Big[a_n^0(i) - a_n(i)\Big].
\end{aligned} \tag{6.2.14}$$

Using the Cauchy-Schwarz inequality on the last term and (6.2.14), we conclude that (6.2.12) ensures

$$n^{-1}\sum_{i=1}^{n}a_n^2(i) \to A_\phi^2 = \int_0^1 \phi^2(u)du, \quad \text{as } n \to \infty. \tag{6.2.15}$$

Alternatively, if we write

$$\phi_n(u) = a_n(i), \quad \text{for } (i-1)/n < u \le i/n,\ i = 1, \ldots, n;\ n \ge 1, \tag{6.2.16}$$

then using (6.2.9), (6.2.16), and the simple triangular inequality, we may put (6.2.12) in the form:

$$\int_0^1 \left\{\phi_n(u) - \phi(u)\right\}^2 du \to 0 \quad \text{as } n \to \infty; \tag{6.2.17}$$

we leave the proof as Problem 6.2.2. This form is often more convenient to verify in practice. For example, we often take the scores as

$$a_n(i) = \phi(i/(n+1)), \;\; i = 1, \ldots, n; \; n \geq 1, \tag{6.2.18}$$

where the score-generating function $\phi(.)$ satisfies (6.2.2) and (6.2.3). Then, by the fundamental theorem of integral calculus, we obtain

$$\lim_{n\to\infty} \left\{ \int_0^1 \phi_n^2(u)du \right\} = \lim_{n\to\infty} \left\{ n^{-1} \sum_{i=1}^{n} \phi^2\Big(\frac{i}{n+1}\Big) \right\} = A_\phi^2 = \int_0^1 \phi^2(u)du. \tag{6.2.19}$$

To verify (6.2.17), all we need to show that

$$\lim_{n\to\infty} \left\{ \int_0^1 \phi_n(u)\phi(u)du \right\} = A_\phi^2, \tag{6.2.20}$$

and this can be easily done by writing

$$\int_0^1 \phi_n(u)\phi(u)du = \sum_{i=1}^{n} \phi\Big(\frac{i}{n+1}\Big) \int_{(i-1)/n}^{i/n} \phi(u)du$$

and invoking the monotonicity of $\phi(.)$ along with its square integrability; we leave the proof of (6.2.20) as Problem 6.2.3.

We will now incorporate Theorem 6.2.1 in our basic study of the asymptotic representations for R-estimators of location. As we mentioned before, we use an asymptotic linearity approach that enables us to approximate $S_n(n^{-1/2}t) - S_n(0)$ by $(-\gamma t)$ uniformly in $|t| \leq C$, with a suitable γ depending on ϕ and F. We define $\{Z_n = Z_n(t), \;\; t \in [-C, C]\}$, $C > 0$ an arbitrary fixed constant, by letting

$$Z_n(t) = S_n(n^{-\frac{1}{2}}t) - S_n(0), \;\; t \in [-C, C]. \tag{6.2.21}$$

We will consider the scores of the form

$$a_n(i) = \phi\Big(\frac{i}{n+1}\Big), \;\; i = 1, \ldots, n.$$

Theorem 6.2.2 *Let $X_1, \ldots, X_n, \ldots$ be a sequence of i.i.d. r.v.'s with a symmetric d.f. F. Assume that F possesses an absolutely continuous density f and finite Fisher's information*

$$I(f) = \int_{\mathbf{R}_1} \Big(f'(x)/f(x)\Big)^2 dF(x) \ (< \infty). \tag{6.2.22}$$

Let $\phi : [0,1] \mapsto \mathbb{R}^+$ be a nondecreasing and square integrable function such that $\phi(0) = 0$. Then, for any fixed $C > 0$ as $n \to \infty$,

$$\sup\Big\{|Z_n(t) + \gamma t| : |t| \leq C\Big\} \xrightarrow{P} 0, \tag{6.2.23}$$

where

$$\gamma = -\int_{\mathbf{R}_1} \phi^\star(F(x))f'(x)dx = \int_{\mathbf{R}_1} \phi^\star(u)\psi_f(u)du = \langle\phi^\star, \psi_f\rangle, \tag{6.2.24}$$

where

$$\psi_f(u) = -f'(F^{-1}(u))/f(F^{-1}(u)), \ \ 0 < u < 1, \tag{6.2.25}$$

is the Fisher score function and $\phi^\star$ is related to ϕ by (6.2.2). [Note that $\langle\psi_f, \psi_f\rangle = \int_0^1 \psi_f^2(u)du = \int_{\mathbf{R}_1}(f'(x)/f(x))^2dF(x) = I(f)$].

We carry out the proof in several steps. First, we prove that for every $n(\geq 1)$ and every nondecreasing $\phi(.)$, $S_n(b)$ is nonincreasing in $b \in \mathbb{R}_1$; this result has independent interest of its own.

Lemma 6.2.1 *Under the hypothesis of Theorem 6.2.2, the statistic $S_n(b)$ is nonincreasing in b ($\in \mathbb{R}_1$) with probability one.*

Proof. Note that the sign$(X_i - b)$ are nonincreasing in b ($\in \mathbb{R}_1$), and assume the values $+1, 0$, or -1. On the other hand, for $X_i > b$, $|X_i - b|$ is nonincreasing in b, while for $X_i < b$, $|X_i - b|$ is nondecreasing. Hence, for the $X_i > b$, $R^+_{ni}(b)$ are nonincreasing in b, and for the $X_i < b$, $R^+_{ni}(b)$ are nondecreasing in b. Since sign$(X_i - b)$ is +1 or -1 according as X_i is $>$ or $<$ b, from the above monotonicity results we immediately conclude that $S_n(b)$ is nonincreasing in $b \in \mathbb{R}_1$. □

Returning to the proof of Theorem 6.2.2, in the second step, we prove a pointwise convergence result.

Lemma 6.2.2 *Let $t \in \mathbb{R}_1$ be an arbitrary fixed number. Then, under the hypothesis of Theorem 6.2.2, $Z_n(t) + \gamma t \to 0$, in probability, as $n \to \infty$.*

Proof. Let us denote by

$$V_n(t) = n^{-\frac{1}{2}} \sum_{i=1}^{n} \text{sign}(X_i - t)\phi(F^+(|X_i - t|)), \ \ t \in \mathbb{R}_1. \tag{6.2.26}$$

Then, by Theorem 6.2.1, (6.2.12) and (6.2.17),

$$S_n(0) - V_n(0) \to 0, \quad \text{in probability, as } n \to \infty. \tag{6.2.27}$$

Let $\{P_n\}$ and $\{P_n^\star\}$ denote the probability distributions with the densities $P_n = \prod_{i=1}^n f(X_i)$ and $P_n^\star = \prod_{i=1}^n f(X_i - n^{-1/2}t)$, respectively, for $n \geq 1$. Then, by (2.5.119), the sequence $\{P_n^\star\}$ is *contiguous* to $\{P_n\}$ (see Problem 6.2.4) so that (6.2.27) implies that

$$S_n(0) - V_n(0) \to 0, \quad \text{under } \{P_n^\star\} \text{ as well, as } n \to \infty. \tag{6.2.28}$$

We rewrite (6.2.28) as

$$S_n(n^{-\frac{1}{2}}t) - V_n(n^{-\frac{1}{2}}t) \to 0, \quad \text{under } \{P_n\} \text{ measure, as } n \to \infty. \tag{6.2.29}$$

Hence it suffices to show that for any (fixed) $t \in \mathbb{R}_1$

$$V_n(n^{-\frac{1}{2}}t) - V_n(0) + \gamma t \to 0, \quad \text{in probability, as } n \to \infty. \tag{6.2.30}$$

Note that apart from the monotonicity and square integrability of $\phi^\star$, we are not assuming any other regularity conditions (and in general $\phi^\star$ is unbounded at the two extremities). Hence, for our subsequent manipulations, we consider the following approximation to $\phi^\star$. For each k (≥ 1), let

$$\begin{aligned}\phi_k^\star(u) &= \phi^\star\Big(\frac{1}{k+1}\Big)I\Big(u < \frac{1}{k}\Big) + \phi^\star(u)I\Big(\frac{1}{k} \leq u \leq \frac{k-1}{k}\Big) \\ &\quad + \phi^\star\Big(\frac{k}{k+1}\Big)I\Big(u > \frac{k-1}{k}\Big).\end{aligned} \tag{6.2.31}$$

Then we have

$$\lim_{k\to\infty}\Big\{\int_0^1 \{\phi_k^\star(u) - \phi^\star(u)\}^2 du\Big\} = 0. \tag{6.2.32}$$

Noting that $(\text{sign}y)\phi(F^+(|y|) = \phi^\star(F(y))$, $\forall y \in \mathbb{R}_1$, we obtain from (6.2.26) and (6.2.31) that the parallel version for $V_n(t)$ corresponding to $\phi_k^\star$ is

$$V_{nk}(t) = n^{-\frac{1}{2}}\sum_{i=1}^n \phi_k^\star(F(X_i - t)),\ t \in \mathbb{R}_1. \tag{6.2.33}$$

As in (6.2.24), we let $\gamma_k = \langle\phi_k^\star, \psi_f\rangle$, $k \geq 1$, so that by (6.2.32),

$$\begin{aligned}(\gamma_k - \gamma)^2 &= \langle(\phi_k^\star - \phi^\star), \psi_f\rangle^2 \\ &\leq ||\phi_k^\star - \phi^\star||^2||\psi_f||^2 \\ &= I(f)||\phi_k^\star - \phi^\star||^2 \to 0, \quad \text{as } k \to \infty.\end{aligned} \tag{6.2.34}$$

Further, note that by (6.2.32),

$$\begin{aligned}&Var[V_{nk}(0) - V_n(0)] \\ =\ & \mathbb{E}[V_{nk}(0) - V_n(0)]^2 \\ =\ & \mathbb{E}\Big[\phi^\star(F(X_1)) - \phi_k^\star(F(X_1))\Big]^2 \\ =\ & \int_0^1 \Big[\phi^\star(u) - \phi_k^\star(u)\Big]^2 du \to 0, \text{ as } k \to \infty,\end{aligned} \tag{6.2.35}$$

so that $V_{nk}(0) - V_n(0)$ converges to 0 in probability, as $k \to \infty$ uniformly in $n = 1, 2, \ldots$. Using the contiguity of $\{P_n^\star\}$ to $\{P_n\}$, it follows as in (6.2.28) and (6.2.29) that for any fixed t,

$$V_{nk}(n^{-\frac{1}{2}}t) - V_n(n^{-\frac{1}{2}}t) \xrightarrow{P} 0, \quad \text{as } n \to \infty,\ k \to \infty. \tag{6.2.36}$$

Thus from (6.2.30), (6.2.34), (6.2.35), and (6.2.36), it follows that we have to show that for every (fixed) k and any (fixed) $t \in [-C, C]$, as $n \to \infty$,

$$V_{nk}(n^{-\frac{1}{2}}t) - V_{nk}(0) + t\gamma_k \xrightarrow{P} 0. \tag{6.2.37}$$

About this proof we may note that

$$\begin{aligned}
&\mathbb{E}\Big[V_{nk}(n^{-\frac{1}{2}}t) - V_{nk}(0) + t\gamma_k\Big] \\
= \ & n^{\frac{1}{2}}\Big[\int_{\mathbf{R}_1} \phi_k^\star(F(x))d\Big[F(x + n^{-\frac{1}{2}}t) - F(x)\Big] \\
& -n^{-\frac{1}{2}}t\Big\{\int_{\mathbf{R}_1} \phi_k^\star(F(x))f'(x)dx\Big\}\Big] \\
= \ & n^{\frac{1}{2}}\Big[\int_{\mathbf{R}_1} \phi_k^\star(F(x))d\Big\{F(x + n^{-\frac{1}{2}}t) - F(x) - n^{-\frac{1}{2}}tf(x)\Big\}\Big]
\end{aligned} \tag{6.2.38}$$

Since, for every (fixed) k, $\phi_k^\star$ is monotone (skew-symmetric) and is bounded from below and above by $\phi^\star(1/(k+1))$ and $\phi^\star(k/(k+1))$, respectively, while $f(x)$ is absolutely continuous, integrating the right hand side of (6.2.38) by parts and noting that

$$F(x + n^{-\frac{1}{2}}t) - F(x) - n^{-\frac{1}{2}}tf(x) = \mathrm{O}(n^{-1})$$

a.e. (for all x : $|\phi_k^\star(F(x))| \le |\phi^\star(k/(k+1))|$), we obtain that (6.2.38) converges to 0 as $n \to \infty$. Finally, we note that

$$\mathrm{Var}\Big[V_{nk}(n^{-\frac{1}{2}}t) - V_{nk}(0)\Big] = \mathrm{Var}\Big[\phi_k^\star(F(X_1 - n^{-\frac{1}{2}}t)) - \phi_k^\star(F(X_1))\Big], \tag{6.2.39}$$

It follows by similar arguments, that (6.2.39) converges to 0 as $n \to \infty$. Again, using the Chebyshev inequality and (6.2.38), we conclude that (6.2.37) holds. □

Let us now return to the third phase of the proof of Theorem 6.2.2. We want to prove that the stochastic convergence in Lemma 6.2.2 is uniform in $t \in [-C, C]$ for every fixed C ($0 < C < \infty$). Let C, ϵ, and η be arbitrary positive numbers. Consider a partition of the interval $[-C, C]$: $-C = t_0 < t_1 < \ldots < t_p = C$ such that

$$t_i - t_{i-1} \le \frac{\eta}{2|\gamma|}, \quad \text{for every } i = 1, \ldots, p. \tag{6.2.40}$$

Then, by Lemma 6.2.2., for every positive ϵ and η, there exists a positive integer n_0, such that for every i $(= 0, 1, \ldots, p)$,

$$\mathbb{P}\Big\{|Z_n(t_i) + \gamma t_i| \geq \eta/4\Big\} \leq \frac{\epsilon}{p+1}, \quad \forall n \geq n_0. \tag{6.2.41}$$

On the other hand, by Lemma 6.2.1, for every $t \in [t_{i-1}, t_i]$,

$$|Z_n(t) + \gamma t| \leq |Z_n(t_{i-1}) + \gamma t_{i-1}| + |Z_n(t_i) + \gamma t_i| + |\gamma(t_i - t_{i-1})|$$

for every $i = 1, 2, \ldots, p$. Hence

$$\begin{aligned} &\mathbb{P}\Big\{\sup\{|Z_n(t) + t\gamma| : t \in [-C, C]\} \geq \eta\Big\} \\ &\leq \sum_{i=0}^{p} \mathbb{P}\Big\{|Z_n(t_i) + \gamma t_i| \geq \eta/4\Big\} < \epsilon \quad \text{for every } n \geq n_0. \end{aligned} \tag{6.2.42}$$

This completes the proof of Theorem 6.2.2. □

We will now employ Theorem 6.2.2 in the derivation of the first order asymptotic representation for R-estimators of location. We first want to prove that a general R-estimator is $\sqrt{n}$-consistent.

Lemma 6.2.3 *Let $X_1, X_2, \ldots$ be a sequence of i.i.d. r.v.'s with a d.f. $F(x-\theta)$ with unknown θ and such that F has an absolutely continuous symmetric density f having a finite Fisher information $I(f)$. Let $\phi : [0,1) \mapsto \mathbb{R}^+$ be a nondecreasing square integrable score function such that $\phi(0) = 0$. Let γ be defined as (6.2.24), and assume that $\gamma \neq 0$.*

Then for the R-estimator T_n of θ based on $S_n(.)$ through the relation (3.4.9),

$$n^{\frac{1}{2}}|T_n - \theta| = O_p(1), \quad \textit{as } n \to \infty. \tag{6.2.43}$$

Proof. Note that by definition of T_n, for any (fixed) $C > 0$,

$$\begin{aligned} \mathbb{P}_\theta\Big\{n^{\frac{1}{2}}|T_n - \theta| > C\Big\} &= \mathbb{P}_0\Big\{T_n > n^{-\frac{1}{2}}C\Big\} + \mathbb{P}_0\Big\{T_n < -n^{-\frac{1}{2}}C\Big\} \\ &\leq \mathbb{P}_0\Big\{S_n(n^{-\frac{1}{2}}C) \geq 0\Big\} + \mathbb{P}_0\Big\{S_n(-n^{-\frac{1}{2}}C) \leq 0\Big\}. \end{aligned} \tag{6.2.44}$$

Further, as ϕ is assumed to be nondecreasing, $\gamma = \int_0^1 f(F^{-1}(u))d\phi^\star(u)$ is > 0, while by Theorem 6.2.1, $S_n(0) \overset{P}{\sim} n^{-\frac{1}{2}} \sum_{i=1}^n \phi^\star(F(X_i)) \overset{\mathcal{D}}{\sim} \mathcal{N}(0, \langle\phi^\star, \phi^\star\rangle^2)$. Thus there exists a positive $C^\star$ and an integer n_0 such that for $n \geq n_0$,

$$\mathbb{P}_0\Big\{|S_n(0)| > C^\star\Big\} < \epsilon/2 \tag{6.2.45}$$

For the first term on the right-hand side of (6.2.44), we have for any $\eta > 0$,

$$\mathbb{P}_0\Big\{S_n(n^{-\frac{1}{2}}C) \geq 0\Big\} = \mathbb{P}_0\Big\{S_n(n^{-\frac{1}{2}}C) - S_n(0) + \gamma C \geq -S_n(0) + \gamma C\Big\}$$
$$\leq \mathbb{P}_0\Big\{Z_n(C) + \gamma C \geq \eta\Big\} + \mathbb{P}_0\Big\{S_n(0) > \gamma C - \eta\Big\}. \tag{6.2.46}$$

Thus, choosing C so large that $\gamma C > C^{\star} + \eta$, we obtain by using Lemma 6.2.2 and (6.2.45) that (6.2.46) can be made smaller than ϵ for every $n \geq n_0$. A similar case holds for the second term on the right-hand side of (6.2.44). □

Now we are in a position to formulate the main theorem of this section.

Theorem 6.2.3 *Let $\{X_i;\ i \geq 1\}$ be a sequence of i.i.d. r.v.'s with a d.f. $F(x-\theta)$ where F has an absolutely continuous, symmetric density function f having a finite Fisher information $I(f)$. Let $\phi : [0,1) \mapsto \mathbb{R}^+$ be a nondecreasing, square integrable score function such that $\phi(0) = 0$, and assume that γ, defined by (6.2.24), is different from 0. Then, for the R-estimator T_n, defined by (3.4.7)-(3.4.9), as $n \to \infty$,*

$$T_n - \theta = (n\gamma)^{-1} \sum_{i=1}^{n} \phi^{\star}(F(X_i - \theta)) \ + \ o_p(n^{-\frac{1}{2}}). \tag{6.2.47}$$

Proof. By Theorem 6.2.2 and Lemma 6.2.3, we conclude that when θ holds,

$$n^{\frac{1}{2}}(T_n - \theta) - \gamma^{-1} S_n(\theta) \xrightarrow{P} 0, \quad \text{as } n \to \infty. \tag{6.2.48}$$

On the other hand, by Theorem 6.2.1, when θ holds,

$$S_n(\theta) - n^{-\frac{1}{2}} \sum_{i=1}^{n} \phi^{\star}(F(X_i - \theta)) \xrightarrow{P} 0, \quad \text{as } n \to \infty, \tag{6.2.49}$$

so (6.2.47) is a direct consequence of (6.2.48) and (6.2.49). □

We conclude this section with a remark that the representation in (6.2.47) is very weak in the sense that the remainder term is $o_p(n^{-1/2})$ (although in terms of the score function ϕ it encompasses a large class). In the next section we will study this remainder term in a more precise manner, albeit bringing in additional regularity conditions on the score function ϕ and the d.f. F.

6.3 SOADR FOR R-ESTIMATORS OF LOCATION

In this section we would like to refine Theorems 6.2.1, 6.2.2 and 6.2.3 and supplement them with the exact rates of convergence whenever possible. This will naturally be possible only under some additional regularity conditions on ϕ and on F. We will need some of the following conditions:

$\mathbf{A_q}$. *Chernoff – Savage* Condition. $\phi(u)$ has a second derivative $\phi''(u)$ a.e. on $(0,1)$, where

$$|\phi''(u)| \leq M(1-u)^{-q} \quad \text{for some } q \geq 0,\ M > 0,\ \forall u \in (0,1). \tag{6.3.1}$$

$\mathbf{A}^{\star}$. *Decomposition* of ϕ. The score function $\phi(u)$ can be expressed as a sum

$$\phi_1(u) + \phi_2(u),\ u \in (0,1), \tag{6.3.2}$$

where

$$\phi_1(u) = \alpha_j \text{ for } s_j < u \leq s_{j+1},\ \ j = 0,1,\ldots,p, \tag{6.3.3}$$

$$0 = s_0 < s_1 < \ldots < s_p < s_{p+1},\ 0 = \alpha_0 \leq \alpha_1 \leq \cdots \leq \alpha_p, \tag{6.3.4}$$

with at least two α being different, and ϕ_2 satisfies $\mathbf{A_q}$ for some $0 < q < 2$. Parallel to (6.2.2), we denote by

$$\phi_i^{\star}(u) = \begin{cases} \phi_i(2u-1), & \frac{1}{2} \leq u < 1, \\ -\phi_i(1-u), & 0 < u < \frac{1}{2},\ i = 1,2. \end{cases} \tag{6.3.5}$$

We will impose either of the following regularity conditions on the d.f. F:

B1. *Conitinuity* Condition. F is continuous everywhere.

B2. *Differentiability* Condition. F has two bounded derivatives f and f' in a neighborhood of each of the points $F^{-1}(s_j')$, $s_j' = (1+s_j)/2$, for $j = 0,1,\ldots,p$, and

$$\int_0^1 f(F^{-1}(u))d\phi_1^{\star}(u) = \sum_{j=1}^{p}(\alpha_j - \alpha_{j-1})f(F^{-1}(s_j')) > 0. \tag{6.3.6}$$

First, we consider the case of a smooth score function $\phi = \phi_2$.

Theorem 6.3.1 *If ϕ $(= \phi_2)$ satisfies $\mathbf{A_q}$ for some $0 < q < 2$ and F satisfies* **B1**, *then, defining Q_n as in (6.2.4) [with the scores $a_n(i) = \phi(i/(n+1))$, $i = 1,\ldots,n$],*

$$Q_n = O_p(n^{-\frac{1}{2}}) \quad \text{as } n \to \infty. \tag{6.3.7}$$

In the same setup, for $q = 2$, we have

$$Q_n = O_p(n^{-\frac{1}{2}}\log n) \quad \text{as } n \to \infty. \tag{6.3.8}$$

Proof We let $U_i = F^+(|X_i|)$, $i \geq 1$, and

$$u_n(t) = (n+1)^{-1} \sum_{i=1}^{n} I(U_i \leq t), \quad t \in [0,1]; \tag{6.3.9}$$

$$D_n^{(\eta)} = \sup\left\{ n^{\frac{1}{2}} |u_n(t) - t| / \{t(1-t)\}^{\frac{1}{2}-\eta} : 0 < t < 1 \right\} \ (\eta > 0). \tag{6.3.10}$$

Note that by (2.5.179), for every η $(0 < \eta \leq 1/2)$ and $\epsilon > 0$, there exist a positive C $(< \infty)$ and an integer n_0, such that

$$\mathbb{P}\left\{ D_n^{(\eta)} > C \right\} \leq \epsilon/2, \quad \text{for every } n \geq n_0. \tag{6.3.11}$$

Let us also denote by $H_i = H_i^+ \cup H_i^-$, where

$$H_i^+ = \left[0 \leq U_{n:i} - \frac{i}{n+1} \leq n^{-\frac{1}{2}} C (U_{n:i}(1 - U_{n:i}))^{\frac{1}{2}-\eta} \right],$$

and

$$H_i^- = \left[0 \geq U_{n:i} - \frac{i}{n+1} \geq -n^{-\frac{1}{2}} C (U_{n:i}(1 - U_{n:i}))^{\frac{1}{2}-\eta} \right],$$

for $i = 1, \ldots, n$. Then, for the first case (of $q < 2$), we have

$$\begin{aligned}
\mathbb{P}\left\{ n^{\frac{1}{2}} |Q_n| > K \right\} &= \mathbb{P}\left\{ n^{\frac{1}{2}} |Q_n| > K,\ D_n^{(\eta)} \leq C \right\} + \epsilon/2 \\
&\leq \mathbb{P}\left\{ |\sum_{i=1}^{n} \text{sign} X_i \left[\phi(u_n(U_i)) - \phi(U_i) \right] I(H_i)| > K \right\} + \epsilon/2 \\
&\leq K^{-2} \sum_{i=1}^{n} \mathbb{E}\left\{ \left[\phi((U_{n:i}) - \phi(i/(n+1)) \right]^2 I(H_i) \right\} + \epsilon/2 \\
&\leq K^{-2} \sum_{i=1}^{n} \mathbb{E}\left\{ \left[\int_{i/(n+1)}^{U_{n:i}} M(1-t)^{1-q} dt \right]^2 I(H_i^+) \right\} \\
&+ K^{-2} \sum_{i=1}^{n} \mathbb{E}\left\{ \left[\int_{U_{n:i}}^{i/(n+1)} M(1-t)^{1-q} dt \right]^2 I(H_i^-) \right\} + \epsilon/2.
\end{aligned} \tag{6.3.12}$$

Since $q < 2$ (i.e., $1 - q > -1$), we use the definition of H_i^+ and H_i^-, and obtain, on choosing $\eta < 2 - q$, that the right-hand side of (6.3.12) is bounded from above by

$$n^{-1} C^2 M^2 K^{-2} \sum_{i=1}^{n} \mathbb{E}\left\{ U_{n:i}^{1-2\eta} (1 - U_{n:i})^{3-2q-2\eta} \right\} +$$

$$(n+1)^{-1} M^2 K^{-2} \sum_{i=1}^{n} \left[\left(\frac{i}{n+1} \right)^{1-2\eta} \left(1 - \frac{i}{n+1} \right)^{3-2q} \right] + \epsilon/2$$

$$= K^{-2}[\mathrm{O}(1)] + \epsilon/2 < \epsilon, \tag{6.3.13}$$

(by letting K adequately large). For $q = 2$ we rewrite

$$\begin{aligned} n^{\frac{1}{2}} Q_n &= \sum_{\{i: R_{ni}^+ \le n-k\}} \mathrm{sign} X_i \left[\phi(R_{ni}^+/(n+1)) - \phi(U_{n:R_{ni}^+}) \right] \\ &+ \sum_{\{i: R_{ni}^+ > n-k\}} \mathrm{sign} X_i \left[\phi(R_{ni}^+/(n+1)) - \phi(U_{n:R_{ni}^+}) \right] \\ &= Q_{n1}^\star + Q_{n2}^\star, \quad \text{say}, \end{aligned} \tag{6.3.14}$$

where k (≥ 2) is a positive integer. Note that under (6.3.1) for $q = 2$, $|\phi(u)|$ is bounded from above by $C + (-\log(1-u))$ for all $u \in [0,1]$, where C is a nonnegative constant, so $|\phi(i/(n+1))|$ is bounded from above by $C + \log(n/(n-i+1))$, for every $i = 1, \ldots, n$. It can be shown similarly [see Problem 6.3.1] that $\mathbb{E}\phi(U_{n:i})$ is bounded from above by $C + c\log n$, for all i ($= 1, \ldots, n$), where c is a positive constant. Hence, we obtain that

$$\mathbb{E}|Q_{n2}^\star| \le k[\mathrm{O}(\log n)] = \mathrm{O}(\log n), \quad \text{for every fixed } k. \tag{6.3.15}$$

By (6.3.15) and the Chebyshev inequality, we have $Q_{n2}^\star = \mathrm{O}_p(\log n)$. Further, using the stochastic independence of the $\mathrm{sign}X_i$ and R_{ni}^+, we obtain

$$\begin{aligned} \mathbb{E}(Q_{n1}^{\star 2}) &= \sum_{i: i \le n-k} \mathbb{E}\left[\phi(U_{n:i}) - \phi(i/(n+1))\right]^2 \\ &= \sum_{i: i \le n-k} M^2 \mathbb{E}\left\{ \frac{\left[(U_{n:i}) - (i/(n+1))\right]^2}{\left\{1 - (U_{n:i} \vee i/(n+1))\right\}^2} \right\}, \end{aligned} \tag{6.3.16}$$

where the last step follows by using (6.3.1) for $q = 2$. Note that for $U_{n:i} \le i/(n+1)$, $1 - U_{n:i} \vee i/(n+1)$ is $\ge 1 - (i/(n+1))$, while $\mathbb{E}[(U_{n:i} - (i/(n+1)))^2 I(U_{n:i} \le i/(n+1))] \le i(n+1-i)/[n(n+1)^2] \le 1/(4n)$, for every i ($\le n$). Thus, for each $i \le n-k$; $k \ge 2$, we have

$$\begin{aligned} &\mathbb{E}\left[(U_{n:i} - (i/(n+1)))^2 / \left\{1 - (U_{n:i} \vee (i/(n+1)))\right\}^2\right] \\ \le\ & 2\mathbb{E}\left[(U_{n:i} - (i/(n+1)))^2 / \{1 - U_{n:i}\}^2\right] \\ =\ & \frac{n(n-1)}{(n-i)(n-i-1)} 2\mathbb{E}\left[(U_{n-2:i} - i/(n+1))^2\right] \\ =\ & \frac{2n(n-1)}{(n-i)(n-i-1)} \left[\mathbb{E}\left\{(U_{n-2:i} - i/(n-1))^2\right\} + 4[i/((n+1)(n-1))]^2\right] \\ =\ & \frac{2n(n-1)}{(n-i)(n-i-1)} \times \frac{i(n-i-1)}{n(n-1)^2} + 4[i/((n+1)(n-1))]^2 \\ =\ & \frac{2i}{(n-i)(n-1)}[1 + \frac{4ni}{(n+1)^2(n-i-1)}]. \end{aligned} \tag{6.3.17}$$

Note that for $i \leq n-k$, $k \geq 2$, $i/(n-2) \leq 1$, and $n-i \geq k \geq 2$. Thus, if we sum over i from 1 to $n-k$, the right-hand side of (6.3.16) is bounded by a term of the order $\log(n-2)$. Hence $\mathbb{E}(Q_{n1}^{\star 2} = \mathrm{O}(\log n)$, as $n \to \infty$, so $Q_{n1}^{\star} = \mathrm{O}_p((\log n)^{1/2})$ as $n \to \infty$. Hence (6.3.8) follows from (6.3.14) and the stochastic orders of $Q_{n1}^{\star}$ and $Q_{n2}^{\star}$. □.

Next we consider the case where the score-generating function ϕ is a step function [i.e., $\phi = \phi_1$, in (6.3.2)]. In this case we have a slower rate.

Theorem 6.3.2 *If the score function $\phi = \phi_1$ satisfies* $\mathbf{A}^{\star}$ *and the d.f. F satisfies* **B2**, *then*

$$Q_n = \mathrm{O}(n^{-\frac{1}{4}}), \quad \textit{in probability, as } n \to \infty. \tag{6.3.18}$$

Proof. Without any loss of generality, we consider ϕ of the form

$$\phi(u) = \begin{cases} 0, & \text{if } u < s, \\ \frac{1}{2}, & \text{if } u = s, \\ 1, & \text{if } u > s, \end{cases} \tag{6.3.19}$$

for some $0 < s < 1$. From (6.2.4) and (6.3.19), we obtain that $\mathbb{E}[Q_n] = 0$ and

$$\begin{aligned} \mathrm{Var}(Q_n) &= n^{-1}\sum_{i=1}^{n}\Big\{\mathbb{P}\Big(U_i \leq s,\ R_{ni}^{+}/(n+1) > s\Big) \\ &\quad + \mathbb{P}\Big(U_i > s,\ R_{ni}^{+}/(n+1) \leq s\Big)\Big\}. \end{aligned} \tag{6.3.20}$$

Note that

$$\begin{aligned} &\mathbb{P}\Big\{R_{ni}^{+}/(n+1) > s,\ U_i \leq s\Big\} \\ &= \int_0^s \mathbb{P}\Big\{R_{ni}^{+}/(n+1) > s | U_i = u\Big\} du \\ &= \int_0^s \mathbb{P}\Big\{\sum_{\substack{j=1 \\ \neq i}}^{n} I(U_j \leq u) - (n-1)u > (n+1)s - 1 - (n-1)u | U_i = u\Big\} du \\ &\leq \int_0^s \exp\Big\{-2(n-1)\Big[\frac{n+1}{n-1}(s - \frac{1}{n+1}) - u\Big]^2\Big\} du = \mathrm{O}(n^{-\frac{1}{2}}), \end{aligned} \tag{6.3.21}$$

where we have used the Hoeffding inequality (2.5.25) for the bounded r.v.'s $I(U_j \leq u)$. In a similar manner it follows that

$$\mathbb{P}\Big\{R_{ni}^{+}/(n+1) < s,\ U_i \geq s\Big\} = \mathrm{O}(n^{-\frac{1}{2}}), \quad \text{as } n \to \infty. \tag{6.3.22}$$

Hence, (6.3.20) is $\mathrm{O}(n^{-\frac{1}{2}})$, so (6.3.18) follows by the the Chebyshev inequality. Under (6.3.3) and (6.3.4), Q_n can be expressed as a sum of p components for each of which (6.3.18) holds, and hence the proof of the theorem is complete. □

Looking at (6.3.7), (6.3.8), (6.3.18), we may observe that if both the components ϕ_1 and ϕ_2 are present in (6.3.2), then (6.3.18) dominates the situation, and hence we arrive at the following.

Theorem 6.3.3 *If the score function ϕ satisfies $\mathbf{A}^\star$ and the d.f. F satisfies both* **B1** *and* **B2**, *then defining Q_n as in (6.2.4), we have*

$$Q_n = \mathrm{O}_p(n^{-\frac{1}{4}}) \quad \text{as } n \to \infty.$$

The above stochastic order clearly reflects the influence of discontinuities of the score generating function with respect to the second order approximation of $S_n(0)$ by $V_n(0)$. Next we consider the impact of such discontinuities on the second-order properties of $Z_n(t)$, $t \in [-C, C]$, defined by (6.2.21).

Theorem 6.3.4 *Let ϕ $(= \phi_1)$ satisfy $\mathbf{A}^\star$ [namely, (6.3.3) and (6.3.4)] and the d.f. F satisfy* **B1** *and* **B2**. *Then, for any $C > 0$,*

$$\sup\Big\{|Z_n(t) + \gamma_1 t| : -C \leq t \leq C\Big\} = \mathrm{O}_p(n^{-\frac{1}{4}}) \quad \text{as } n \to \infty, \tag{6.3.23}$$

where

$$\gamma_1 = \int_0^1 f(F^{-1}(u))d\phi_1^\star(u) \;\; (> 0), \tag{6.3.24}$$

and $\phi_1^\star$ is defined by (6.3.5).

Proof. Without loss of generality, we define ϕ as in (6.3.19), and let $s' = (1+s)/2$. Note that $\gamma_1 = 2f(F^{-1}(s')) > 0$ (by assumption) in this special case. Let then $H_n(y)$, $y \geq 0$, be the empirical d.f. of the $|X_i|$, $i = 1, \dots, n$. For every $t \in \mathbb{R}_1$, we may write

$$S_n(n^{-\frac{1}{2}}t) = n^{\frac{1}{2}} \int_0^\infty \text{sign}(x - n^{-\frac{1}{2}}t) I\big[\frac{n}{n+1} H_n(|x - n^{-\frac{1}{2}}t|) > s\big] dF_n(x). \tag{6.3.25}$$

Thus

$$\begin{aligned}
& S_n(n^{-\frac{1}{2}}t) - S_n(0) \\
= \;& n^{\frac{1}{2}} \int_{-\infty}^\infty [\text{sign}(x - n^{-\frac{1}{2}}t) - \text{sign}x] I\big[H_n(|x - n^{-\frac{1}{2}}t|) > \frac{n+1}{n}s\big] dF_n(x) \\
+ \;& n^{\frac{1}{2}} \int_{-\infty}^\infty \text{sign}x \big\{ I\big[H_n(|x - n^{-\frac{1}{2}}t|) > \frac{n+1}{n}s\big] \\
& \quad - I\big[H_n(|x|) > \frac{n+1}{n}s\big]\big\} dF_n(x)
\end{aligned} \tag{6.3.26}$$

Note that for $t > 0$, $\text{sign}(x - n^{-1/2}t) - \text{sign}x$ is -1 if $0 < x \le n^{-1/2}t$, and it is equal to 0, otherwise; a similar case holds for $t < 0$. Also, by the Hoeffding inequality,

$$\begin{aligned} & \mathbb{P}\Big\{H_n(2n^{-\frac{1}{2}}C) > \frac{n+1}{n}s\Big\} \\ & \le \Big\{H_n(2n^{-\frac{1}{2}}C) - F^+(2n^{-\frac{1}{2}}C) > s - F^+(2n^{-\frac{1}{2}}C)\Big\} \\ & \le \exp\Big\{-2n[s - F^+(2n^{-\frac{1}{2}}C)]^2\Big\} \quad \text{for every } s > 0. \end{aligned} \tag{6.3.27}$$

[We choose n so large that $F^+(2n^{-\frac{1}{2}}C) < s$.] Thus we have

$$\begin{aligned} & \sup\Big\{\Big|n^{\frac{1}{2}}\int_{-\infty}^{\infty}[\text{sign}(x - n^{-\frac{1}{2}}t) - \text{sign}x]I[H_n(|x - n^{-\frac{1}{2}}t|) > \\ & \qquad \frac{n+1}{n}s]dF_n(x)\Big| : \ |t| \le C\Big\} \\ & \le n^{\frac{1}{2}}\big[F_n(n^{-\frac{1}{2}}C) - F_n(-n^{-\frac{1}{2}}C)\big]I\big[H_n(2n^{-\frac{1}{2}}C) > s\big] \\ & = n^{\frac{1}{2}}H_n(n^{-\frac{1}{2}}C)I\big[H_n(2n^{-\frac{1}{2}}C) > s\big]. \end{aligned} \tag{6.3.28}$$

Hence, using (6.3.27) along with the fact that $F^+(n^{-\frac{1}{2}}C) = O(n^{-\frac{1}{2}})$, we obtain that (6.3.28) can be made $O_p(n^{-r})$, for any $r > 0$ (and we choose $r > 1/4$). On the other hand, for the second term on the right-hand side of (6.3.26), we virtually follow the proof of Theorem 4.2.3 (with x localized to $F^{-1}(s')$ and $F^{-1}(1 - s')$, and conclude that

$$\begin{aligned} & \sup_{|t| \le C}\Big\{\Big|n^{\frac{1}{2}}\int_{-\infty}^{\infty}\text{sign}x\big[I(H_n(|x - n^{-\frac{1}{2}}t|) > \frac{n+1}{n}s) \\ & \qquad - I(H_n(|x|) > \frac{n+1}{n}s)\big]dF_n(x) + \gamma_1 t\Big|\Big\} \\ & = O_p(n^{-\frac{1}{4}}) \quad \text{as } n \to \infty; \end{aligned} \tag{6.3.29}$$

we leave the details as Problem 6.3.2. Thus, as $n \to \infty$,

$$\sup\Big\{|S_n(n^{-\frac{1}{2}}t) - S_n(0) + \gamma_1 t| : \ |t| \le C\Big\} = O_p(n^{-\frac{1}{4}}), \tag{6.3.30}$$

and the proof of the theorem is complete. □.

Next we consider the behavior of $Z_n(t)$, $t \in [-C, C]$ when $\phi = \phi_2$. In this case, we can obtain a better rate under an additional regularity condition:

B3. (*Chernoff-Savage condition on the Fisher score function*). The density f of F is absolutely continuous, and for some $w > 0$, $K < \infty$,

$$|f'(F^{-1}(1+u)/2)f(F^{-1}((1+u)/2)| \le K(1-u)^{-\frac{1}{2}+w}, \quad u \in (0,1), \tag{6.3.31}$$

and for every $\eta > 0$, there exists a positive δ such that

$$\int_0^{\infty}\{1 - F^+(x)\}^{-1+\eta}dF^+(x^+ - \delta) < \infty. \tag{6.3.32}$$

Note that (6.3.31) entails more than the square integrability of the Fisher score function.

Theorem 6.3.5 *Let ϕ satisfy $\mathbf{A_q}$ for some $q \in [0, 2)$, and F satisfy* **B3**. *Then, for any finite C (> 0), as $n \to \infty$,*

$$\sup\left\{|Z_n(t) + t\gamma_2| : |t| \leq C\right\} = O_p(n^{-\frac{1}{2}}), \tag{6.3.33}$$

where

$$\gamma_2 = \int_{-\infty}^{\infty} \phi^\star(F(x))f'(x)dx, \tag{6.3.34}$$

and $\phi^\star$ is defined by (6.3.5).

The proof is quite technical in nature; the general outline may be attempted by the more adventurous reader as an exercise (see Problems 6.3.3-6.3.4).

Theorems 6.3.1 through 6.3.5 provide us with the necessary tools for the study of the asymptotic representations of R-estimators involving suitable orders for the remainder terms.

Theorem 6.3.6 *Let $\{X_i;\ i \geq 1\}$ be i.i.d. r.v.'s with a d.f. $F(x - \theta)$, $x \in \mathbb{R}_1$ and θ real (unknown), such that $F(x) + F(-x) = 1$, $x \in \mathbb{R}_1$. Let T_n be the R-estimator of θ, defined as in (3.7) and (3.8), and assume that the score generating function ϕ satisfies (6.2.2) and (6.2.3). Further, assume that*

$$\gamma = \int_0^1 f(F^{-1}(u))d\phi^\star(u) \quad \textit{is positive.} \tag{6.3.35}$$

Then, the following representations hold:

1. *Under* $\mathbf{A}^\star$, **B1**, *and* **B2** **as n → ∞**,

$$T_n - \theta = (n\gamma)^{-1}\sum_{i=1}^{n}\phi^\star(F(X_i - \theta)) + O_p(n^{-\frac{3}{4}}). \tag{6.3.36}$$

2. *Under* $\mathbf{A_1}$ *and* **B3** *as $n \to \infty$,*

$$T_n - \theta = (n\gamma)^{-1}\sum_{i=1}^{n}\phi^\star(F(X_i - \theta)) + O_p(n^{-1}\log n). \tag{6.3.37}$$

3. *Under* $\mathbf{A_q}$ *with $0 \leq q < 2$ and* **B3**, *as* $n \to \infty$,

$$T_n - \theta = (n\gamma)^{-1}\sum_{i=1}^{n}\phi^\star(F(X_i - \theta)) + O_p(n^{-1}). \tag{6.3.38}$$

Proof. By Theorem 6.2.3, $n^{-1/2}(T_n - \theta)$ is asymptotically normal with 0 mean and variance $\gamma^{-2}||\phi^\star||^2$ whenever $||\phi^\star||^2$ is positive and finite, and hence $n^{1/2}|T_n - \theta| = \mathrm{O}_p(1)$, as $n \to \infty$. Therefore, inserting $t = n^{1/2}(T_n - \theta) + \epsilon$ or $t = n^{1/2}(T_n - \theta) - \epsilon$ $(\epsilon > 0)$ in (6.3.23), we obtain that under $\mathbf{A}^\star$, **B1**, and **B2**,

$$S_n(T_n^+ - n^{-\frac{1}{2}}\epsilon) - S_n(T_n^+ + n^{-\frac{1}{2}}\epsilon) = 2\gamma\epsilon + \mathrm{O}_p(n^{-\frac{1}{4}}), \qquad (6.3.39)$$

uniformly in $0 \leq \epsilon \leq C$. On the other hand, by definition of T_n,

$$S_n(T_n + n^{-\frac{1}{2}}\epsilon) \leq 0 \leq S_n(T_n - n^{-\frac{1}{2}}\epsilon), \qquad (6.3.40)$$

and by the monotonicity of S_n stated in Lemma 6.2.1,

$$S_n(T_n) \leq S_n(T_n - n^{-\frac{1}{2}}\epsilon) - S_n(T_n + n^{-\frac{1}{2}}\epsilon). \qquad (6.3.41)$$

Hence, letting $\epsilon \downarrow 0$, we obtain from (6.3.39) and (6.3.41),

$$S_n(T_n) = \mathrm{O}_p(n^{-\frac{1}{4}}) \quad \text{as } n \to \infty. \qquad (6.3.42)$$

Now, inserting $t = n^{1/2}(T_n - \theta)$ in (6.3.23), we get

$$n^{\frac{1}{2}}(T_n - \theta) = \gamma^{-1}S_n(\theta) + \mathrm{O}_p(n^{-\frac{1}{4}}) \quad \text{as } n \to \infty, \qquad (6.3.43)$$

and this together with (6.3.18) ensures (6.3.36), and thus the proposition 1.

Proceeding analogously and using Theorem 6.3.5, we obtain that under $\mathbf{A_q}$, with $0 \leq q < 2$ and under $[B3]$,

$$n^{\frac{1}{2}}(T_n - \theta) = \gamma^{-1}S_n(\theta) + \mathrm{O}_p(n^{-\frac{1}{2}}) \quad \text{as } n \to \infty, \qquad (6.3.44)$$

and then we arrive at the propositions 2 and 3 using (6.3.8) and (6.3.7), respectively.

The remainder term $\mathrm{O}_p(n^{-\frac{1}{2}}\log n)$ in (6.3.37) is not surprising. A similar order appears in the approximation of the empirical process $\{n^{\frac{1}{2}}(F_n - F)\}$ by a Brownian bridge, and rank statistics are functionals related to such empirical processes. Notice that the case $q = 2$ corresponds to the normal scores statistic, and thus (6.3.37) corresponds to the R-estimator with the normal scores.

6.4 REPRESENTATIONS FOR R-ESTIMATORS OF REGRESSION

Consider the simple regression model

$$X_i = \theta + \beta t_i + e_i, \quad i = 1, \ldots, n, \qquad (6.4.1)$$

where $\mathbf{X}_n = (X_1, \ldots, X_n)'$ is an n-vector of observable r.v.'s, $\mathbf{t}_n = (t_1, \ldots, t_n)'$ is the vector of known (given) regression constants (not all equal), θ and β are the unknown *intercept* and *slope* parameters, respectively, and the e_i are assumed to be i.i.d. random errors with a continuous d.f. F, defined on $\mathbb{R}_1$ (it may not be necessary to assume that F is symmetric about 0). As in Section 3.4 we consider estimators of θ and β based on suitable rank statistics, and these are termed *R-estimators* of *regression*. Toward this end, as in (3.4.20), we let

$$L_n(b) = \sum_{i=1}^{n} (t_i - \bar{t}) a_n(R_{ni}(b)), \quad b \in \mathbb{R}_1, \ n \geq 2, \tag{6.4.2}$$

where $a_n(1) \leq \ldots \leq a_n(n)$ are given scores (to be defined more precisely later on), $\bar{t}_n = n^{-1}\sum_{i=1}^{n} t_i$, and $R_{ni}(b)$ is the rank of $X_i - bt_i$ among the $X_j - bt_j$, $j = 1, \ldots, n$; for $i = 1, \ldots, n$. Recall that the ranks are invariant under shift, so subtraction of a common a from the $X_i - bt_i$ would not affect their relative ranks. Further note that $X_i - \beta t_i$, $i = 1, \ldots, n$ are i.i.d. r.v.'s with the d.f. $F(e - \theta)$, $e \in \mathbb{R}_1$, so the $R_{ni}(\beta)$ are all exchangeable (for each $n \geq 2$). This leads to

$$\mathbb{E}[L_n(\beta)|\beta] = \bar{a}_n \sum_{i=1}^{n} (t_i - \bar{t}_n) = 0, \tag{6.4.3}$$

where $\bar{a}_n = n^{-1}\sum_{i=1}^{n} a_n(i)$. We now have a motivation for obtaining an estimator of β by equating $L_n(b)$ to 0 (in a well-defined manner). Toward this end, first, we consider the following:

Lemma 6.4.1 *If $a_n(1) \leq \ldots \leq a_n(n)$, and the t_i are not all equal, then $L_n(b)$ is nonincreasing in $b \in \mathbb{R}_1$ for every $n \geq 2$.*

Proof. Consider two lines $X_i - bt_i$ and $X_j - bt_j$ $(i \neq j)$, $b \in \mathbb{R}_1$; if $t_i = t_j$, the lines are parallel, while for $t_i \neq t_j$ they intersect at point, say, b_{ij}. Thus, considering the set of all pairs $\{(t_i, t_j) : \ t_i \neq t_j, \ 1 \leq i < j \leq n\}$, we arrive at a set of $N^\star$ points where these n lines intersect. We denote these points in ascending order by $b_1^\star < \ldots < b_{N^\star}^\star$. By assumption, $N^\star$ is bounded from below and above by $(n-1)$ and $\binom{n}{2}$, respectively. Also, without any loss of generality, we take $t_1 \leq \ldots \leq t_n$, with at least one strict inequality. Then for every $b \in \mathbb{R}_1$ we may rewrite $L_n(b)$ as

$$L_n(b) = n^{-1} \sum_{\{1 \leq i < j \leq n\}} (t_j - t_i)\Big[a_n(R_{nj}(b)) - a_n(R_{ni}(b))\Big]. \tag{6.4.4}$$

Note that between two successive $b^\star$, say, $b_m^\star$ and $b_{m+1}^\star$, the ranks $R_{ni}(b)$ do not change. At $b_{m+1}^\star = b_{ij}$ only the pair (i, j) undergoes a change of ranks, while the remainings $(n-2)$ remain the same. At this point, as t_j is

$\geq t_i$, $R_{nj}(b_{ij}+) = R_{nj}(b_{ij}-) - 1$ and $R_{ni}(b_{ij}+) = R_{ni}(b_{ij}-) + 1 = R_{nj}(b_{ij}-)$. Thus we have

$$\begin{aligned} L_n(b_{ij}+) &\\ = \ & L_n(b_{ij}-) - (t_j - t_i)\Big[a_n(R_{nj}(b_{ij}-)) - a_n(R_{ni}(b_{ij}-) - 1)\Big] \\ \leq \ & L_n(b_{ij}-), \end{aligned} \tag{6.4.5}$$

since the $a_n(r)$ are nondecreasing in r $(1 \leq r \leq n)$. Letting $b_0^\star = -\infty$ and $b_{N^\star+1}^\star = +\infty$, we readily obtain from (6.4.5) that $L_n(b)$ is nonincreasing in $b \in \mathbb{R}_1$. □

Observe that the scores $a_n(k)$, $k = 1, \ldots, n$ are all discrete, so $L_n(b)$ is a nonincreasing step function of $b \in \mathbb{R}_1$. Thus there may not be a unique point at which $L_n(b)$ is identically equal to 0. We eliminate this difficulty in the same way as in the case of the R-estimators of location. We let

$$\hat{\beta}_n^- = \sup\{b : L_n(b) > 0\} \quad \text{and} \quad \hat{\beta}_n^+ = \sup\{b : L_n(b) < 0\} \tag{6.4.6}$$

and define R-estimator $\hat{\beta}_n$ of β based on the linear rank statistic $L_n(.)$ as

$$\hat{\beta}_n = \frac{1}{2}(\hat{\beta}_n^+ + \hat{\beta}_n^-). \tag{6.4.7}$$

Having obtained $\hat{\beta}_n$, we may consider the residuals

$$\hat{X}_i = X_i - \hat{\beta}_n t_i, \quad i = 1, \ldots, n. \tag{6.4.8}$$

Based on these residuals, we may define a signed-rank statistic as in Section 6.2, and consider an R-estimator $\hat{\theta}_n$ of θ as in (3.4.7) and (3.4.8).

In this section we are primarily interested in the asymptotic representations for the estimators $\hat{\beta}_n$ and $\hat{\theta}_n$, when suitable regularity conditions are imposed on the scores $a_n(k)$ as well as on the underlying d.f. F. Observe that if the t_i in (6.4.1) can only assume the values 0 and 1, the model reduces to the two-sample location model, and β is the difference of the location parameters. Thus the findings of this section would remain pertinent to the two-sample location model as well.

Let us introduce some of the notations we need in the sequel. Let $T_n^2 = \sum_{i=1}^n (t_i - \bar{t}_n)^2$, $n \geq 2$, and define the normalized constants

$$c_{ni} = (t_i - \bar{t}_n)/T_n, \ i = 1, \ldots, n. \tag{6.4.9}$$

(so that $\sum_{i=1}^n c_{ni} = 0$ and $\sum_{i=1}^n c_{ni}^2 = 1$). Also, denote the score-generating function $\phi = \{\phi(u), \ u \in (0,1)\}$, and assume that $\phi(u)$ is nondecreasing in $u \in (0,1)$. Because of the translation-invariance of $L_n(.)$, we may assume without any loss of generality that $\bar{\phi} = \int_0^1 \phi(u)du = 0$. Parallel to (6.2.1), we define here

$$a_n^0(i) = \mathbb{E}[\phi(U_{n:i})], \ i = 1, \ldots, n, \tag{6.4.10}$$

although we may as well take

$$a_n(i) = \phi(\mathbb{E}U_{n:i}) = \phi(i/(n+1)), \quad i = 1, \ldots, n.$$

Let then

$$L_n^{0\star}(0) = T_n^{-1} L_n^0(0) = \sum_{i=1}^{n} c_{ni} a_n^0(R_{ni}(0)), \tag{6.4.11}$$

$$V_n(0) = \sum_{i=1}^{n} c_{ni} \phi(F(X_i)), \quad Q_n^0 = L_n^{0\star}(0) - V_n(0). \tag{6.4.12}$$

Theorem 6.4.1 *Let $\{X_i;\ i \geq 1\}$ be i.i.d. r.v's with a continuous d.f. F, and let ϕ be a nondecreasing function such that $\int_0^1 \phi^2(u)du < \infty$. Then*

$$Q_n^0 \to 0, \quad \textit{in probability, as } n \to \infty. \tag{6.4.13}$$

Proof. Note that by (6.4.10),

$$\begin{aligned} \bar{a}_n &= n^{-1} \sum_{i=1}^{n} \mathbb{E}\phi(U_{n:i}) \\ &= \int_0^1 \phi(u) \Big\{ \sum_{i=1}^{n} \binom{n-1}{i-1} u^{i-1}(1-u)^{n-i} \Big\} du \\ &= \int_0^1 \phi(u)du = \bar{\phi} = 0, \end{aligned} \tag{6.4.14}$$

for every $n \geq 1$. Let $\mathcal{B}_n = \mathcal{B}(R_{n1}(0), \ldots, R_{nn}(0))$ be the sigma-field generated by the vector of ranks at the nth stage, $n \geq 1$. Note that when $\beta = 0$, $F(X_i)$, $i = 1, \ldots, n$ are i.i.d. r.v.'s having the uniform (0, 1) d.f., and moreover the vector of ranks and order statistics are stochastically independent. Thus $\mathbb{E}[\phi(F(X_i))|\mathcal{B}_n] = \mathbb{E}[\phi(U_{n:R_{ni}})|R_{ni} \text{ given}] = a_n^0(R_{ni})$, for every $i = 1, \ldots, n$ [here R_{ni} stands for $R_{ni}(0)$]. Hence we obtain

$$\begin{aligned} \mathbb{E}[V_n(0)|\mathcal{B}_n] &= \sum_{i=1}^{n} c_{ni} \mathbb{E}[\phi(F(X_i))|\mathcal{B}_n] \\ &= \sum_{i=1}^{n} c_{ni} a_n^0(R_{ni}(0)) = L_n^{0\star}(0), \quad \text{for every } n \geq 1. \end{aligned} \tag{6.4.15}$$

Now we have

$$\begin{aligned} \mathbb{E}[(Q_n^0)^2] &= \mathbb{E}[V_n^2(0)] - \mathbb{E}[(L_n^{0\star}(0))^2] \\ &= \int_0^1 \phi^2(u)du - \frac{n}{n-1}\Big[n^{-1}\sum_{i=1}^{n}\{a_n^0(i)\}^2\Big]. \end{aligned} \tag{6.4.16}$$

Proceeding as in the proof of Theorem 6.2.1, we find that the right-hand side of (6.4.16) converges to 0 as $n \to \infty$, and hence (6.4.13) holds. □

Note that under (6.2.12) or (6.2.17), (6.4.13) also applies to other forms of score functions, and we omit these duplications. As in Section 6.2, we will incorporate the *asymptotic linearity of linear rank statistics in regression parameter* in the derivation of our main results. Toward this end, define $L_n^\star(b) = T_n^{-1}L_n(b)$, $b \in \mathbb{R}_1$, and define $Z_n = \{Z_n(t),\ t \in [-K, K]\}$ for an arbitrary positive number K by letting

$$Z_n(t) = L_n^\star(T_n^{-1}t) - L_n^\star(0), \quad -K \leq t \leq K, \tag{6.4.17}$$

where, for simplicity of presentation, we may take $a_n(k) = \phi(k/(n+1))$, $k = 1, \ldots, n$.

Theorem 6.4.2 *Let $\{X_i;\ i \geq 1\}$ be a sequence of i.i.d. r.v.'s with a continuous d.f. F (defined on $\mathbb{R}_1$) and assume that F possesses an absolutely continuous probability density function f with a finite Fisher information $I(f)$ defined by (6.2.22). Let ϕ : $(0,1) \mapsto \mathbb{R}_1$ be a nondecreasing and square integrable function (and we let $\bar{\phi} = 0$). Then for every (fixed) $K > 0$ as $n \to \infty$,*

$$\sup\Big\{|Z_n(t) + \gamma t : |t| \leq K\Big\} \xrightarrow{P} 0, \tag{6.4.18}$$

where

$$\gamma = \int_{-\infty}^{\infty} \phi(F(x))\{-f'(x)\}dx = \int_0^1 \phi(u)\psi_f(u)du = \langle \phi, \psi_f \rangle, \tag{6.4.19}$$

and $\psi_f(.)$ is defined by (6.2.25). We assume that

$$\max\Big\{|c_{ni}| : 1 \leq i \leq n\Big\} \to 0, \quad \textit{as } n \to \infty \tag{6.4.20}$$

Proof. We closely follow the proof of Theorem 6.2.2. First, Lemma 6.4.1 is the counterpart of Lemma 6.2.1 for the linear rank statistics. Second, in (6.4.12), we replace the X_i by $X_i - bt_i$, and denote the statistic by $V_n(b)$, $b \in \mathbb{R}_1$. We let $\bar{t}_n = 0$ at this stage for an unessential simplification. Then, since $X_i - bt_i = X_i - bT_n c_{ni}$, $i \geq 1$, we know that $V_n(t/T_n)$ corresponds to the statistic in (6.4.12) where X_i are replaced by $X_i - tc_{ni}$, $i = 1, \ldots, n$. Let $\{P_n\}$ and $\{P_n^\star\}$ denote the probability distributions with the densities $p_n = \prod_{i=1}^n f(X_i)$ and $p_n^\star = \prod_{i=1}^n f(X_i - tc_{ni})$, respectively, $n \geq 1$. Then, by (2.5.179), the sequence $\{P_n^\star\}$ is contiguous to $\{P_n\}$ (for any fixed t), and (6.4.13) implies that

$$L_n^\star(0) - V_n(0) \xrightarrow{P} 0 \quad \text{under } \{P_n^\star\}, \quad \text{as } n \to \infty. \tag{6.4.21}$$

Therefore, proceeding as in (6.2.28)-(6.2.29), we conclude that

$$L_n^\star(T_n^{-1}t) - V_n(T_n^{-1}t) \xrightarrow{P} 0 \quad \text{under } \{P_n\} \text{ measure as } n \to \infty. \tag{6.4.22}$$

To show that $Z_n(t) + \gamma t \to 0$ in probability for any fixed $t: |t| \leq K$ as $n \to \infty$, it suffices to show that for any fixed t,

$$V_n(T_n^{-1}t) - V_n(0) + \gamma t \xrightarrow{P} 0 \quad \text{uniformly in } t \in [-K, K]. \tag{6.4.23}$$

Toward this end, we may repeat the proof of Lemma 6.2.2 [particularly (6.2.30) through (6.2.39)] and obtain (6.4.23); we therefore leave the details as Problem 6.4.1. Observe that (6.4.20) is needed in establishing the contiguity of $\{P_n^\star\}$ to $\{P_n\}$ and hence in the proof of (6.4.23) as well. Problem 6.4.2 is therefore posed to bring it out clearly. Finally, because of the monotonicity property established in Lemma 6.4.1, we may proceed precisely along the same line as in (6.2.40)-(6.2.42) and claim that the uniform pointwise stochastic convergence $Z_n(t) + \gamma t$ (to 0) implies (6.4.18). The details are left as Problem 6.4.3. This completes the proof of Theorem 6.4.2. □

Theorem 6.4.3 *For the simple regression model in (6.4.1), assume that the error distribution F has an absolutely continuous probability density function f having a finite Fisher information $I(F)$. Let $\phi : (0,1) \mapsto \mathbb{R}_1$ be a nondecreasing, square integrable score function, and assume that (1) γ, defined by (6.4.19), is different from 0, and (2) the regression constants t_i satisfy (6.4.20). Then, for the R-estimator $\hat{\beta}_n$, defined by (6.4.7), we have for $n \to \infty$,*

$$T_n(\hat{\beta}_n - \beta) = \gamma^{-1} \sum_{i=1}^{n} c_{ni}\phi(F(X_i - \theta - \beta t_i)) + \mathrm{O}_p(1). \tag{6.4.24}$$

[Note that the $X_i - \theta - \beta t_i = e_i$ are i.i.d. r.v.'s with the d.f. F, so $F(X_i - \theta - \beta t_i)$ are i.i.d. r.v.'s with the uniform distribution on $(0,1)$.]

Proof. The proof parallels that of Theorem 6.2.3. First, we show that

$$T_n|\hat{\beta}_n - \beta| = \mathrm{O}_p(1), \quad \text{as } n \to \infty. \tag{6.4.25}$$

For this note that by definition in (6.4.7) and the monotonicity property in Lemma 6.4.1, for any fixed C (> 0),

$$\begin{aligned} & \mathbb{P}_\beta\{T_n|\hat{\beta}_n - \beta| > C\} \\ = \; & \mathbb{P}_0\{T_n|\hat{\beta}_n| > C\} \\ \leq \; & \mathbb{P}_0\{L_n(C/T_n) \geq 0\} + \mathbb{P}_0\{L_n(-C/T_n) \leq 0\} \\ = \; & \mathbb{P}_0\{L_n^\star(C/T_n) \geq 0\} + \mathbb{P}_0\{L_n^\star(-C/T_n) \leq 0\}, \end{aligned} \tag{6.4.26}$$

where $\mathbb{P}_0$ refers to the model when $\beta = 0$. At this stage we make use of (6.4.21) along with the classical central limit theorem for $V_n(0)$ [involving independent summands for which the Lindeberg condition is ensured by (6.4.20) and the square integrability of the score function; see Problem 6.4.4). Thus $L_n^\star(0)$ is asymptotically normal with 0 mean and variance $||\phi||^2$. The rest of the proof of (6.4.25) follows precisely as in (6.2.45) and (6.2.46). Having obtained (6.4.25), we directly obtain (6.4.24) by an appeal to (6.4.18) [which yields that $T_n(\hat{\beta}_n - \beta) - L_n^\star(\beta) \xrightarrow{P} 0$, as $n \to \infty$, when β holds] and the stochastic equivalence of $\mathbf{L}_n^\star(\beta)$ and $V_n(\beta)$. □

We may remark that (6.4.24), known as the *first-order asymptotic representation of R-estimator of regression*, demands minimal conditions on the score function and the underlying d.f. A natural question arising in this context is whether the remainder term $[\mathrm{O}_p(1)]$ in (6.4.24) can be made more precise as a function of the sample size n or T_n? As in the location model in Section 6.3, an answer to this query would involve more precise regularity conditions on ϕ as well as F. For example, jump discontinuities in ϕ may entail a slower order than in the case where ϕ is sufficiently smooth. We will therefore try to probe into this problem in a manner similar to that in Section 6.3. Before that, we consider the first order representation results for the estimator of θ. For this we rewrite (6.4.1) as

$$X_i = \theta + \beta\bar{t}_n + \beta(t_i - \bar{t}_n) + e_i = \theta_n^0 + \beta T_n c_{ni} + e_i, \quad i = 1, \dots, n, \quad (6.4.27)$$

where $\theta_n^0 = \theta + \beta\bar{t}_n$. Then the residuals in (6.4.8) may be written as

$$\begin{aligned} \hat{X}_i &= \theta_n^0 + \beta T_n c_{ni} + e_i - \hat{\beta}_n\bar{t}_n - \hat{\beta}_n T_n c_{ni} \\ &= \theta_n^0 - \hat{\beta}_n\bar{t}_n + T_n(\beta - \hat{\beta}_n)c_{ni} + e_i \\ &= \theta + e_i + T_n(\beta - \hat{\beta}_n)\{(\bar{t}_n/T_n) + c_{ni}\}, \quad i = 1, \dots, n. \quad (6.4.28) \end{aligned}$$

We make additional assumption that

$$\limsup_{n\to\infty}\{|\bar{t}_n|/T_n\} = 0. \quad (6.4.29)$$

Then, by (6.4.20) and (6.4.25), the last term on the right-hand side of (6.4.28) will be $\mathrm{O}_p(1)$, uniformly in i $(\leq n)$, as $n \to \infty$. To study the asymptotic properties of the R-estimator of θ based on the residuals in (6.4.28), we need to consider first some extensions of the results in Section 6.2 when the observations are subject to "small' perturbations. Specifically, we consider the following extension of Theorem 6.2.2: We start with a two-dimensional time parameter stochastic process $\{Z_n(s,t),\ (s,t) \in [-C,C]^2\}$, where C is an arbitrary positive number, and

$$Z_n(s,t) = S_n(n^{-\frac{1}{2}}s, t) - S_n(0,0), \ (s,t) \in [-C,C]^2, \quad (6.4.30)$$

where $S_n(a,b)$ is the usual signed-rank statistic considered in Section 6.2 but based on the pseudo-observations $X_i - a - bc_{ni}$, $i = 1, \dots, n$, and the c_{ni} are defined as in (6.4.9). Suppose further that γ is defined as in (6.2.24) and that

$$\max\{|c_{ni}| : 1 \leq i \leq n\} = \mathrm{O}(n^{-\frac{1}{2}}) \quad \text{as } n \to \infty. \quad (6.4.31)$$

Theorem 6.4.4 *Let $\{X_i;\ i \geq 1\}$ be a sequence of i.i.d. r.v.'s with a symmetric d.f. F and assume that F possesses an absolutely continuous density function f with a finite Fisher information $I(f)$. Further, assume that the score function ϕ satisfies the conditions in Theorem 6.2.2, and the constants c_{ni} satisfy (6.4.31). Then for every (fixed) $C > 0$ as $n \to \infty$,*

$$\sup\left\{|Z_n(s,t) + s\gamma| : -C \leq s \leq C,\ -C \leq t \leq C\right\} \xrightarrow{P} 0, \tag{6.4.32}$$

Note that ϕ refers to the score-generating function for the signed- rank statistic $S_n(.)$, F is assumed to be symmetric (not needed for the earlier theorems in this section), and by definition $\sum_{i=1}^{n} c_{ni} = 0$, so its contribution does not arise in the drift function in (6.4.32).

Proof. We may assume, without any loss of generality, that the c_{ni} are nondecreasing in i $(1 \leq i \leq n)$. First using arguments similar to Lemma 6.4.1 we verify that Lemma 6.2.1 directly extends to the following (Problem 6.4.5):

$$S_n(a,b) \text{ is nonincreasing in } a \text{ and is a difference of two nonincreasing functions of } b,\ (a,b) \in \mathbb{R}_2. \tag{6.4.33}$$

Second, we let $\{P_n\}$ and $\{P_n^\star\}$ denote the probability distributions with the densities $\prod_{i=1}^{n} f(X_i)$ and $\prod_{i=1}^{n} f(X_i - n^{-1/2}s - tc_{ni})$, respectively, for $n \geq 1$. Then the sequence $\{P_n^\star\}$ is contiguous to $\{P_n\}$ (Problem 6.4.6). As such, Lemma 6.2.2 directly extends to the following: For any (fixed) $C > 0$ as $n \to \infty$,

$$Z_n(s,t) + \gamma s \xrightarrow{P} 0 \quad \text{for } (s,t) \in [-C, C]^2. \tag{6.4.34}$$

The rest of the proof follows the same track as in (6.2.40) through (6.2.42) with a two-dimensional gridding, and hence the details are omitted. □

Theorem 6.4.4 provides us with the necessary tools for studying the first order asymptotic representation for $\hat{\theta}_n$, the R-estimator of θ, based on the residuals in (6.4.28). In this context we find it convenient to assume that

$$\lim_{n\to\infty} \{n^{-1}T_n^2\} = T^2 : 0 < T < \infty \text{ exists}. \tag{6.4.35}$$

While it may suffice to assume that $\liminf_{n\to\infty}\{n^{-1}T_n^2\} > 0$, (6.4.35) seems to be much more handy. It does not impose any serious restriction on the model.

By virtue of (6.4.25), (6.4.28), and Theorem 6.4.4, it follows that whenever $\limsup_{n\to\infty}\{n^{\frac{1}{2}}|\bar{t}_n|/T_n)$ is bounded, then

$$n^{\frac{1}{2}}(\hat{\theta}_n - \theta) = \gamma^{-1}S_n(\theta,\beta) - n^{\frac{1}{2}}\bar{t}_n(\hat{\beta}_n - \beta) + o_p(1) \quad \text{as } n \to \infty. \tag{6.4.36}$$

At this stage we assume that the score function ϕ for the linear rank statistics $L_n(b)$ is the same as the score function $\phi^\star$ appearing in (6.2.2) for the signed-rank statistics. We need to replace ϕ by $\phi^\star$ in (6.4.24), although the two γ

in (6.2.24) and (6.4.24) are the same. With this adjustment we find from (6.4.36) and (6.4.24) that

$$\begin{aligned}
n^{\frac{1}{2}}(\hat{\theta}_n - \theta) &= \gamma^{-1} S_n(\theta, \beta) - \gamma^{-1}\{n^{\frac{1}{2}}\bar{t}_n / T_n\} \\
&\quad \times \Big\{ \sum_{i=1}^{n} c_{ni} \phi^\star(F(X_i - \theta - \beta t_i)) \Big\} + o_p(1) \\
&= \gamma^{-1} \Big\{ n^{-\frac{1}{2}} \sum_{i=1}^{n} \phi^\star(F(X_i - \theta - \beta t_i)) \\
&\quad - \Big[n^{\frac{1}{2}} \bar{t}_n / T_n \Big] \sum_{i=1}^{n} c_{ni} \phi^\star(F(X_i - \theta - \beta t_i)) \Big\} + o_p(1) \\
&= \gamma^{-1}\{ n^{-\frac{1}{2}} \sum_{i=1}^{n} \phi^\star(F(e_i)) - [n^{\frac{1}{2}}\bar{t}_n / T_n] \sum_{i=1}^{n} c_{ni}\phi^\star(F(e_i))\} \\
&\quad + o_p(1) \qquad (6.4.37)
\end{aligned}$$

where the e_i are i.i.d. r.v's with the d.f. F. If we assume further that $\lim_{n\to\infty} \bar{t}_n = \bar{t}$, then $n^{\frac{1}{2}}\bar{t}_n / T_n$ converges to $\bar{t}/T$ under (6.4.35). For large n, the right-hand side of (6.4.37) may also be written

$$n^{\frac{1}{2}}(\hat{\theta}_n - \theta) = \gamma^{-1} n^{-\frac{1}{2}} \sum_{i=1}^{n} \Big\{ 1 - T^{-1} \bar{t} c_{ni} \Big\} \phi^\star\Big(F(e_i)\Big) + o_p(1). \qquad (6.4.38)$$

Further, if $n^{-1}T_n^2$ goes to $+\infty$ but $\bar{t}_n$ still goes to $\bar{t}$ (finite), then (6.4.38) still holds, with a further simplification that $T^{-1}\bar{t}c_{ni} = 0$ for all i, which is the same representation as in (6.2.47). Finally, if the score function in $L_n(b)$ is different from $\phi^\star$, we may still use (6.2.47), (6.4.24), and (6.4.36) to obtain a first-order representation involving two γ coefficients and two different score functions. Hence we arrive at the following:

Theorem 6.4.5 *Assume that the errors e_i in the model (6.4.1) are i.i.d. r.v.'s with a d.f. F, symmetric about 0 and such that $f = F'$ is absolutely continuous and has a finite Fisher information $I(f)$. Let the scores and the signed scores be generated by a skew-symmetric and square integrable function $\phi^\star : (0,1) \mapsto \mathbb{R}_1$ [the signed scores by (6.2.2)]. Let $\gamma \neq 0$ with γ defined in (6.2.24), let c_{ni} satisfy (6.4.31) and let $\lim_{n\to\infty}(n^{\frac{1}{2}}\bar{t}_n/T_n) = \bar{t}/T$. Then the R-estimator $\hat{\theta}_n$ of θ admits the first order asymptotic representation (6.4.38) and the simultaneous R-estimator $(\hat{\theta}_n, \hat{\beta}_n)$ of (θ, β) admits the representation*

$$\gamma n^{\frac{1}{2}} \begin{pmatrix} \hat{\theta}_n - \theta \\ \hat{\beta}_n - \beta \end{pmatrix} = n^{-\frac{1}{2}} \sum_{i=1}^{n} \begin{pmatrix} 1 - n^{\frac{1}{2}} c_{ni}\bar{t}/T \\ n^{\frac{1}{2}} c_{ni}/T \end{pmatrix} \phi^\star(F(e_i)) + o_p(1). \qquad (6.4.39)$$

Both results extend to the case of nonisomorphic score-generating functions via (6.4.24) and (6.4.36).

We have already noticed that for the estimation of the regression coefficient β, it is not necessary to assume that the underlying error distribution F is symmetric about 0, while this assumption is a part of the location parameter estimation. Since the usual ranks $R_{ni}(a,b)$ [of the $X_i - a - bt_i$, $i = 1, \ldots, n$] do not depend on the shift a ($\in \mathbb{R}_1$), the $L_n(a,b)$ may not be used to provide an estimator of θ, and the estimator $\hat{\beta}_n$ is translation-invariant. On the other hand, being translation-invariant, linear rank statisticsl do not provide an estimator of θ. We therefore consider an alternative criterion whose gradient yields a vector of (weighted) signed-rank statistics that are capable of providing simultaneous R-estimators of (θ, β). We adopt the same notation as in around (6.4.9), and for simplicity of presentation we consider a reparametrization: $(\theta,\beta) \mapsto (\theta^\star, \beta^\star)$, where $\theta^\star = \theta + \beta\bar{t}_n$ and $\beta^\star = T_n\beta$. Then we have $\theta + \beta t_i = \theta^\star + \beta^\star c_{ni}$ for $i = 1, \ldots, n$, where $\sum_{i=1}^n c_{ni} = 0$ and $\sum_{i=1}^n c_{ni}^2 = 1$. Also we denote by $R_{ni}^+(a,b)$ the rank of $|X_i - a - bc_{ni}|$ among the $|X_r - a - bc_{nr}|$, $1 \le r \le n$, for $i = 1, \ldots, n$. Let

$$D_n(a,b) = \sum_{i=1}^n a_n(R_{ni}^+(a,b))|X_i - a - bc_{ni}|, \quad (a,b) \in \mathbb{R}_2. \tag{6.4.40}$$

Note that the scores $a_n(k)$ are nonnegative and monotone (in $k : 1 \le k \le n$) so that for any $\lambda : 0 \le \lambda \le 1$ and $(a_j, b_j) \in \mathbb{R}_2$, $j = 1, 2$,

$$\begin{aligned}
&D_n(\lambda(a_1,b_1) + (1-\lambda)(a_2,b_2)) \\
&\quad \le \sum_{i=1}^n a_n(R_{ni}^+\{\lambda(a_1,b_1) + (1-\lambda)(a_2,b_2)\}) \\
&\quad\quad \times\{\lambda|X_i - a_1 - b_1c_{ni}| + (1-\lambda)|X_i - a_2 - b_2c_{ni}|\} \\
&\quad \le \lambda\sum_{i=1}^n a_n(R_{ni}^+(a_1,b_1))|X_i - a_1 - b_1c_{ni}| \\
&\quad\quad +(1-\lambda)\sum_{i=1}^n a_n(R_{ni}^+(a_2,b_2))|X_i - a_2 - b_2c_{ni}|\} \\
&\quad = \lambda D_n(a_1,b_1) + (1-\lambda)D_n(a_2,b_2).
\end{aligned} \tag{6.4.41}$$

Therefore $D_n(a,b)$ is a convex, continuous, and piecewise linear function of (a,b), and we define an estimator $(\hat{\theta}^\star, \hat{\beta}^\star)$ of (θ,β) as

$$(\hat{\theta}^\star, \hat{\beta}^\star) = \text{Arg min}\{D_n(a,b) : (a,b) \in \mathbb{R}_2\}. \tag{6.4.42}$$

It is easy to verify that at the points of continuity of $D_n(a,b)$, we have

$$\begin{aligned}
(\partial/\partial a)D_n(a,b) &= -\sum_{i=1}^n \text{sign}(X_i - a - bc_{ni})a_n(R_{ni}^+(a,b)) \\
&= -n^{\frac{1}{2}}S_n(a,b),
\end{aligned} \tag{6.4.43}$$

$$\begin{aligned}(\partial/\partial b)D_n(a,b) &= -\sum_{i=1}^{n} c_{ni}\text{sign}(X_i - a - bc_{ni})a_n(R_{ni}^+(a,b)) \\ &= -S_n^\star(a,b), \quad \text{say.}\end{aligned} \tag{6.4.44}$$

This result provides the motivation and justification for incorporation of (weighted) signed-rank statistics in deriving simultaneous (R-) estimators of $(\theta^\star, \beta^\star)$ (or (θ, β), based on (6.4.42). We may proceed as in the proof of Theorem 6.2.2 (or 6.4.2), and obtain that for any (fixed) $(a,b) \in \mathbb{R}_2$, as $n \to \infty$,

$$|S_n(n^{-\frac{1}{2}}a, b) - S_n(0,0) + a\gamma| \xrightarrow{P} 0, \tag{6.4.45}$$

$$|S_n^\star(n^{-\frac{1}{2}}a, b) - S_n^\star(0,0) + b\gamma| \xrightarrow{P} 0, \tag{6.4.46}$$

where γ has been introduced in (6.2.24). Hence

$$\begin{aligned}&D_n(n^{-\frac{1}{2}}a, b) - D_n(0,0) + aS_n(0,0) + bS_n^\star(0,0) \\ &\quad + (\gamma/2)(a^2 + b^2) \xrightarrow{P} 0, \quad \text{as } n \to \infty.\end{aligned} \tag{6.4.47}$$

The afore mentioned convexity of $D_n(a,b)$ and the pointwise stochastic convergence in (6.4.47) imply that for every (fixed) $C:\ 0 < C < \infty$, the convergence in (6.4.47) is uniform in $(a,b): a^2 + b^2 \leq C^2$, and the same uniform stochastic convergence holds for (6.4.45) and (6.4.46). Proceeding then as in the proof of Theorem 5.5.2, we conclude that as $n \to \infty$,

$$\left\| \begin{pmatrix} n^{\frac{1}{2}}(\hat{\theta}_n^\star - \theta^\star) \\ \hat{\beta}_n^\star - \beta^\star \end{pmatrix} - \text{Arg min}_{(a,b)'} \left\{ \begin{array}{c} (\gamma/2)(a^2+b^2) - \\ aS_n(0,0) - bS_n^\star(0,0) \end{array} \right\} \right\| \xrightarrow{P} 0, \tag{6.4.48}$$

so

$$\begin{pmatrix} n^{\frac{1}{2}}(\hat{\theta}_n^\star - \theta^\star) \\ \hat{\beta}_n^\star - \beta^\star \end{pmatrix} - \gamma^{-1} \begin{pmatrix} S_n(0,0) \\ S_n^\star(0,0) \end{pmatrix} \xrightarrow{P} \mathbf{0}. \tag{6.4.49}$$

Finally, the first-order representation [as in (6.4.39)] follows from (6.4.49) by incorporating the same for the vector of weighted signed-rank statistics. Because of the translation- and scale- equivariance properties of R-estimators, we have from (6.4.49) and some standard steps that as $n \to \infty$,

$$\begin{pmatrix} n^{\frac{1}{2}}(\hat{\theta}_n - \theta) \\ T_n(\hat{\beta}_n - \beta) \end{pmatrix} - \gamma^{-1} \begin{pmatrix} 1 & -\sqrt{n}\bar{t}_n/T_n \\ 0 & 1 \end{pmatrix} \begin{pmatrix} S_n(0,0) \\ S_n^\star(0,0) \end{pmatrix} \xrightarrow{P} \mathbf{0}, \tag{6.4.50}$$

where $(\hat{\theta}_n, \hat{\beta}_n)$ refer to the R-estimator of (θ, β) based on $D_n(a,b)$. Problem 6.4.7 is setup to verify (6.4.50) from (6.4.49). As in the case of R-estimators of β based on $L_n(b)$ and θ based on the residual signed-rank statistics, in general an iterative procedure may be needed here to solve for $(\hat{\theta}_n, \hat{\beta}_n)$. A $\sqrt{n}$-consistent initial estimator of (θ, β) may be used in a Newton-Raphson method for this purpose. Kraft and van Eeden (1972) proposed some related estimators (under additional regularity conditions), which Mckean and Schrader (1980) established as a "norm" characterization of $D_n(a, b)$ in proposing some related estimators. The convexity approach sketched here (along the lines of Hössjer 1994) not only provides a simpler derivation of the first-order representation in (6.4.50) but also demonstrates the existence of a global minimum for (6.4.42).

6.5 SOADR FOR REGRESSION R-ESTIMATORS

This section runs parallel to Section 6.4 and deals with the second-order representations for R-estimator of regression and intercept parameters considered in the previous section. In this connection we may define Q_n (or Q_n^0) as in (6.4.12), and our first task is to improve the order of approximation in Theorem 6.4.1. This will be done under additional regularity conditions on the score function ϕ as well as the regression constants t_i. In this setup we define the scores $a_n^0(i)$ as in (6.4.10), and note that

$$n^{-1}\sum_{i=1}^{n}\{a_n^0(i)\}^2 = n^{-1}\sum_{i=1}^{n}\mathbb{E}[\phi^2(U_{n:i})] - n^{-1}\sum_{i=1}^{n}\{\mathbb{E}[\phi^2(U_{n:i}) - [\mathbb{E}\phi(U_{n:i})]^2\}$$

$$= \int_0^1 \phi^2(u)du - n^{-1}\sum_{i=1}^{n}\mathrm{Var}\{\phi(U_{n:i})\}.$$

Thus, for every $n \geq 2$, the right hand side of (6.4.16) reduces to

$$\mathbb{E}[(Q_n^0)^2] = (n-1)^{-1}\Big(\sum_{i=1}^{n}\mathrm{Var}\{\phi(U_{n:i})\} - \int_0^1 \phi^2(u)du\Big). \tag{6.5.1}$$

We may note further that for each i $(= 1, \ldots, n)$,

$$\mathrm{Var}\{\phi(U_{n:i})\} \leq \mathbb{E}\Big\{[\phi(U_{n:i}) - \phi(i/(n+1))]^2\Big\}, \quad \forall n \geq 1, \tag{6.5.2}$$

so that

$$\mathbb{E}[(Q_n^0)^2] \leq (n-1)^{-1}\sum_{i=1}^{n}\mathbb{E}\Big\{[\phi(U_{n:i}) - \phi(i/(n+1))]^2\Big\}, \quad \forall n \geq 2. \tag{6.5.3}$$

Now, as in Section 6.3, we introduce the following regularity conditions:

$\mathbf{A_q}$. *Chernoff-Savage-type condition on* ϕ. $\phi(u)$ has a second derivative $\phi''(u)$ a.e. on $u \in (0,1)$, where

$$|\phi''(u)| \leq M\{u(1-u)\}^{-q} \ \forall u \in (0,1), \text{ for some } q \geq 0, M > 0. \quad (6.5.4)$$

$\mathbf{A}^\star$. *Decomposition of* ϕ. The same as in (6.3.2) through (6.3.4).

B1. *Continuity of* F. The same as in after (6.3.5).

B2. *Differentiability of* F. F has two bounded derivatives f and f' in a neighborhood of each of the points $F^{-1}(s_j)$, $j = 0, \ldots, p$, and (6.3.6) holds (with the s'_j replaced by the s_j).

Theorem 6.5.1 *Suppose that the score function* ϕ *satisfies* $\mathbf{A_q}$ *and the d.f.* F *satisfies* **B1**. *Then for every* $q: \ 0 \leq q < 2$,

$$Q_n^0 = \mathrm{O}_p(n^{-\frac{1}{2}}) \quad \text{as } n \to \infty, \quad (6.5.5)$$

while for $q = 2$,

$$Q_n^0 = \mathrm{O}_p(n^{-\frac{1}{2}} \log n) \quad \text{as } n \to \infty, \quad (6.5.6)$$

Finally, if the score function ϕ *satisfies* $\mathbf{A}^\star$ *and* F *satisfies* **B2**, *then*

$$Q_n^0 = \mathrm{O}_p(n^{-\frac{1}{4}}) \quad \text{as } n \to \infty. \quad (6.5.7)$$

Outline of the proof. Note that

$$\mathbb{E}[\Big(U_{n:i} - (i/(n+1))\Big)^2] = i(n+1-i)/\{n(n+1)^2\}, \ \forall i = 1, \ldots, n, n \geq 1.$$

Thus, whenever ϕ has a bounded first derivative ϕ', it follows from (6.5.3) that $\mathbb{E}[(Q_n^0)^2] = \mathrm{O}(n^{-1})$, so $Q_n^0 = \mathrm{O}_p(n^{-\frac{1}{2}})$ as $n \to \infty$. This eliminates the need to prove (6.5.5) for $q = 0$, and hence we consider only the case $q > 0$. For $q < 0$, we have by (6.5.4) and double integration that

$$|\phi(u)| \leq M^\star\{u(1-u)\}^{2-q}, \quad \text{for every } u \in (0,1),$$

where $2 - q > 0$ and $M^\star < \infty$. Thus $\phi(u)$ is a bounded and differentiable function of $u \in (0,1)$. As such, we set problem 6.5.1 to verify that

$$\mathbb{E}[\phi(U_{n:i}) - \phi(i/(n+1))]^2 \leq M^0 n^{-1}, \quad \text{uniformly in } i \ (= 1, \ldots, n), \quad (6.5.8)$$

where M^0 is a positive number depending on $M^\star$. [The case of $q = 1$ needs a little more delicate attention: $|\phi'(u)|$ is bounded by $M'\{-\log[u(1-u)]\}$, for all $u \in (0,1)$, where M' is a finite positive number, so that $|\phi(u)|$ is bounded by

$$M'' + M^\star\{-u(1-u)\log[u(1-u)]\},$$

which is also a bounded and differentiable function of $u \in (0,1)$.] Thus, (6.5.3), (6.5.8) and the Chebyshev inequality imply (6.5.5) for every $q < 2$. For $q = 2$, consider the partition of $\{1,\ldots,n\}$ in two sets, namely $\{k \leq i \leq n-k\}$ and $\{[i < k] \cup [i > n-k]\}$, analogously as in (6.3.14); for the first set, we repeat the proofs of (6.3.16) -(6.3.17) while for the second set, we observe that

$$\mathbb{E}[\phi(U_{n:i}) - \phi(i/(n+1))]^2 \leq 2[\mathbb{E}\phi^2(U_{n:i}) + \phi^2(i/(n+1))] = \mathrm{O}((\log n)^2),$$

where the last step follows by using Problem 6.3.1. This completes the proof of (6.5.6). Finally, the proof of (6.5.7) follows precisely along the same linen of the proof of Theorem 6.3.2 [see (6.3.19)-(6.3.21)], and hence the details are omitted. □

Observe that if we use the scores $a_n(i) = \phi(i/(n+1))$ instead of the scores $a_n^0(i)$, $i = 1,\ldots,n$, and define Q_n as in (6.4.12), we have

$$\mathbb{E}[Q_n^2] \leq 2\mathbb{E}[(Q_n^0)^2] + 2\mathbb{E}[L_n^\star(0) - L_n^{0\star}(0)^2]$$

$$= 2\mathbb{E}[(Q_n^0)^2] + 2(n-1)^{-1}\sum_{i=1}^{n}[a_n^0(i) - \phi(i/(n+1))]^2, \qquad (6.5.9)$$

so Theorem 6.5.1 will again lead to the same stochastic order for Q_n. Thus, (6.5.5) through (6.5.7) also pertain to this alternative definition of the scores. Let us now define the $Z_n(t)$ and $V_n(t)$ as in Section 6.4 [see (6.4.17) and (6.4.22)]. Then we have the following:

Theorem 6.5.2 *If the score function ϕ $(\equiv \phi_1)$ satisfies* $\mathbf{A}^\star$ *and the d.f. F satisfies* **B1** *and* **B2**, *then for any $C > 0$,*

$$\sup\Big\{|Z_n(t) + t\gamma| : -C \leq t \leq C\Big\} = \mathrm{O}_p(n^{-\frac{1}{4}}), \quad \textit{as } n \to \infty, \qquad (6.5.10)$$

where $\gamma = \int_0^1 f(F^{-1}(u))d\phi(u) > 0$.

Proof. As in the proof of Theorem 6.3.4, we may take without any loss of generality that $\phi(u)$ is equal to 0 or 1 according as u is $\leq$ or $>$ s for some $s \in (0,1)$. Then, note that by definition

$$L_n^\star(T_n^{-1}t) = \sum_{i=1}^{n} c_{ni}\big[I\big(\sum_{j=1}^{n} I(X_i - X_j \geq t(c_{ni} - c_{nj})) > ns\big)\big]$$

$$= \sum_{i=1}^{n} c_{ni}\big[I\big(\sum_{\substack{j=1\\ j\neq i}}^{n} I(X_i - X_j \geq t(c_{ni} - c_{nj})) > ns - 1\big)\big]. \qquad (6.5.11)$$

Now for each $i =$ $(1,\ldots,n)$ the sum $\sum_{j=1(j\neq i)}^{n} I(X_i - X_j \geq t(c_{ni} - c_{nj})$ (given X_i) involves $(n-1)$ independent zero-one-valued r.v.'s on which the Hoeffding inequality can again be applied analogously as in (6.3.27). Consequently, noting that $\sum_{j=1}^{n} c_{nj} = 0$ and that the d.f. F has an absolutely continuous density f, we obtain that for every positive ϵ_n,

$$\mathbb{P}\Big\{|\sum_{\substack{j=1\\ \neq i}}^{n} I(X_i - X_j \geq t(c_{ni} - c_{nj})) - (n-1)F(X_i - tc_{ni})| \geq (n-1)\epsilon_n\Big\}$$
$$\leq 2\exp\{-2n\epsilon_n^2\}, \quad \text{for every } n \geq n_0. \tag{6.5.12}$$

Choose $\epsilon_n = n^{-\frac{1}{2}}(\log n)^{\frac{1}{2}}$ so that $\exp\{-2n\epsilon_n^2\} = n^{-2}$. Hence, using (6.5.12), for each i $(= 1, \ldots, n)$, we obtain that we may replace the right-hand side of (6.5.11) in probability by

$$\sum_{i=1}^{n} c_{ni}\Big\{I[F(X_i - tc_{ni})) \geq s + \mathrm{O}(\epsilon_n))]\Big\}. \tag{6.5.13}$$

A similar case holds for $L_n^\star(0)$. For $Z_n(t)$ we may likewise take

$$\sum_{i=1}^{n} c_{ni}\Big\{I[F(X_i - tc_{ni})) \geq s + \mathrm{O}(\epsilon_n))] - I[F(X_i) \geq s + \mathrm{O}(\epsilon_n)]\Big\}. \tag{6.5.14}$$

Note that $\sum_{i=1}^{n} c_{ni}^2 = 1$ and we have assumed that $\max\{|c_{ni}| : 1 \leq i \leq n\} = \mathrm{O}(n^{-\frac{1}{2}})$. If we compute the mean of (6.5.14), we will have it equal to $-f(F^{-1}(s)t + \mathrm{O}(n^{-\frac{1}{2}})$ and variance equal to $\sum_{i=1}^{n} c_{ni}^2[\mathrm{O}(tc_{ni}|)] = \mathrm{O}(n^{-\frac{1}{2}})$, which would then imply that for every (fixed) $t \in [-C, C]$,

$$Z_n(t) + tf(F^{-1}(s)) = \mathrm{O}_p(n^{-\frac{1}{4}}) \quad \text{as } n \to \infty. \tag{6.5.15}$$

For the particular form of ϕ chosen, we also note that $\gamma = f(F^{-1}(s))$. Hence, the rest of the proof follows by (6.5.15) and the monotonicity of $Z_n(t)$ in t. In this connection recall that by (6.5.12), we can replace $\mathrm{O}_p(n^{-\frac{1}{4}})$ in (6.5.15) by the statement that the left-hand side is bounded in absolute value by $c_1 n^{-\frac{1}{4}}$, with a probability greater than $1-n^{-1}$ for every $n \geq n_0$. Choosing a set of $n^{\frac{1}{4}}$ grid points for t on $[-C, C]$ and noting that within each of these subintervals (of width $2Cn^{-\frac{1}{4}}$), we finf that the increment of $t\gamma$ is $\mathrm{O}(n^{-\frac{1}{4}})$, while by virtue of Lemma 6.4.1, $Z_n(t)$ is monotone nonincreasing in t. We conclude that (6.5.15) holds simultaneously for all $t \in [-C, C]$, and this completes the proof of the theorem. □

Next we consider the case of a smooth and differentiable score function $\phi = \phi_2$, where we may have better rates under some additional regularity conditions:

B3. *Chernoff-Savage condition on the Fisher score function.* The density f of the d.f. F is absolutely continuous, and for some $w > 0$ and $K < \infty$,

$$\Big|\frac{f'(F^{-1}(u))}{f(F^{-1}(u))}\Big| \leq K[u(1-u)]^{-\frac{1}{2}+w} \quad \text{for every } u \in (0,1), \tag{6.5.16}$$

and for every positive $\eta > 0$ there exists a positive δ such that

$$\int_{\mathbb{R}_1} \{[1 - F(x)]F(x)\}^{-1+\eta} dF(x \pm \delta) < \infty. \tag{6.5.17}$$

Then we have the following:

Theorem 6.5.3 *Let ϕ satisfy $\mathbf{A_q}$ for some $q \in [0, 2]$ and let F satisfy* **B3**. *Then for any (fixed) $C > 0$,*

$$\sup\{|Z_n(t) + t\gamma| : |t| \le C\} = O_p(n^{-\frac{1}{2}}), \quad \text{as } n \to \infty, \tag{6.5.18}$$

where

$$\gamma = -\int_{\mathbb{R}_1} \phi(F(x))f'(x)dx. \tag{6.5.19}$$

The theorem is analogous to Theorem 6.3.5, and its proof is therefore relegated to the exercises (see problems 6.5.2-6.5.3).

Theorems 6.5.1, 6.5.2, and 6.5.3 lead us to the following representations for R-estimators of the regression coefficient β.

Theorem 6.5.4 *Consider the simple regression model in (6.4.1) with i.i.d. errors e_i having the d.f. F, defined on $\mathbb{R}_1$. Let $\hat{\beta}_n$ be R-estimator of β, defined by (6.4.7) and (6.4.8). Assume that the score generating function ϕ is nondecreasing and square integrable, and further that*

$$\gamma = \int_0^1 f(F^{-1}(t))d\phi(t) \text{ is positive and finite.} \tag{6.5.20}$$

Finally, assume that the regression constants t_i satisfy (6.4.31) and (6.4.35). Then, the following representations hold:

1. Under $\mathbf{A}^\star$ *(for ϕ) and* **B1** *and* **B2** *(for the d.f.) as $n \to \infty$,*

$$T_n(\hat{\beta}_n - \beta) = \gamma^{-1} \sum_{i=1}^{n} c_{ni}\phi(F(X_i - \theta - \beta t_i)) + O_p(n^{-\frac{1}{4}}). \tag{6.5.21}$$

2. Under $\mathbf{A_2}$ *and* **B3**, *as $n \to \infty$,*

$$T_n(\hat{\beta}_n - \beta) = \gamma^{-1} \sum_{i=1}^{n} c_{ni}\phi(F(X_i - \theta - \beta t_i)) + O_p(n^{-\frac{1}{2}} \log n). \tag{6.5.22}$$

3. Under $\mathbf{A_q}$ *with $q \in [0, 2)$ and* **B3**, *as $n \to \infty$,*

$$T_n(\hat{\beta}_n - \beta) = \gamma^{-1} \sum_{i=1}^{n} c_{ni}\phi(F(X_i - \theta - \beta t_i)) + O_p(n^{-\frac{1}{2}}). \tag{6.5.23}$$

The proof is parallel to that of Theorem 6.3.6, and is left for the reader to attempt Problem 6.5.4.

6.6 REPRESENTATIONS FOR R-ESTIMATORS IN LINEAR MODEL

Let us consider the univariate linear model

$$X_i = \theta + \boldsymbol{\beta}'\mathbf{c}_i + e_i, \quad i = 1, \ldots, n. \tag{6.6.1}$$

Here $\mathbf{X} = (X_1, \ldots, X_n)'$ is an n-vector of observable r.v.'s, $\mathbf{c}_i = (c_{i1}, \ldots, c_{ip})'$, $i \geq 1$ are vectors of given regression constants (not all equal), θ is an intercept parameter, $\boldsymbol{\beta} = (\beta_1, \ldots, \beta_p)'$ is the (unknown) vector of regression parameters, $p \geq 1$, and the e_i are i.i.d. errors having a continuous (but unknown) d.f. F defined on $\mathbb{R}_1$. Note that for $p = 1$, (6.6.1) reduces to the simple regression model (6.4.1). As before for the estimation of $\boldsymbol{\beta}$,we do not need to assume that F is symmetric about 0, although we may find it necessary to do so for the estimation of θ. We intend to employ suitable linear rank statistics for the estimation of $\boldsymbol{\beta}$ and aligned signed-rank statistic for θ.

First, consider the estimation of $\boldsymbol{\beta}$. For every $\mathbf{b} \in \mathbb{R}_p$ let

$$R_{ni}(\mathbf{b}) = \text{ Rank of } (X_i - \mathbf{b}'\mathbf{c}_i) \tag{6.6.2}$$

among the $X_j - \mathbf{b}'\mathbf{c}_j$ $(1 \leq j \leq n)$, for $i = 1, \ldots, n$. For each n (≥ 1) we consider a set of scores $a_n(1) \leq \ldots \leq a_n(n)$ (not all equal) as in Section 6.4. Define a vector of (aligned) linear rank statistics

$$\begin{aligned} \mathbf{L}_n(\mathbf{b}) &= (L_{n1}(\mathbf{b}), \ldots, L_{np}(\mathbf{b}))' \\ &= \sum_{i=1}^{n} (\mathbf{c}_i - \bar{\mathbf{c}}_n) a_n(R_{ni}(\mathbf{b})) \ \ \mathbf{b} \in \mathbb{R}_p, \end{aligned} \tag{6.6.3}$$

where $\bar{\mathbf{c}}_n = n^{-1} \sum_{i=1}^{n} \mathbf{c}_i$. Note that the $R_{ni}(\mathbf{b})$ in (6.6.2) are translation invariant, and hence there is no need for adjustment for θ in (6.6.3). Further, under $\boldsymbol{\beta} = \mathbf{0}$, $R_{n1}(\mathbf{0}), \ldots R_{nn}(\mathbf{0})$ are interchangeable r.v.'s [assuming each permutation of $1, \ldots, n$ with the common probability $(n!)^{-1}$], so we have

$$\mathbb{E}_{\boldsymbol{\beta}=0} \mathbf{L}_n(\mathbf{0}) = \mathbf{0}, \quad \forall n \geq 1. \tag{6.6.4}$$

From this perspective the situation is quite comparable to the simple regression case in Section 6.4, and we may be tempted to "equate" $\mathbf{L}_n(\mathbf{b})$ in order to $\mathbf{0}$ to obtain suitable estimators of $\boldsymbol{\beta}$. However, the simple monotonicity argument of Lemma 6.4.1 does not generally hold for $p \geq 2$. Jurečková (1971) laid down the foundation of R-estimation of $\boldsymbol{\beta}$ based on the L_1-norm

$$||\mathbf{L}_n(\mathbf{b})|| = ||\mathbf{L}_n(\mathbf{b}_1)||_1 = \sum_{j=1}^{p} |L_{nj}(\mathbf{b})|, \ \mathbf{b} \in \mathbb{R}_p \tag{6.6.5}$$

and defining

$$\mathbf{b}^0 : || \ \mathbf{L}_n(\mathbf{b}^0)|| = \inf_{\mathbf{b} \in \mathbb{R}_p} ||\mathbf{L}_n(\mathbf{b})||. \tag{6.6.6}$$

Such $\mathbf{b}^0$ may not be uniquely determined. Let

$$\mathcal{D}_n = \text{ set of all } \mathbf{b}^0 \text{ satisfying (6.6.6).} \tag{6.6.7}$$

The justification of $\mathcal{D}_n$ in estimation of $\boldsymbol{\beta}$ rests in establishing that $\mathcal{D}_n$ represents a sufficiently shrinking neighborhood of $\boldsymbol{\beta}$, in probability as $n \to \infty$. We then can eliminate the arbitrariness in (6.6.6) by letting

$$\hat{\boldsymbol{\beta}}_n = \text{ center of gravity of } \mathcal{D}_n. \tag{6.6.8}$$

Of course, instead of the L_1-norm in (6.6.6) we could have chosen an L_2 (or, in general, an L_r) norm (for an arbitrary $r > 0$) and arrived in a similar formulation. We will see later on that the choice of r in such an L_r-norm is not that crucial. Jaeckel (1972) eliminated this arbitrariness by using a slightly different formulation. He introduced the measure of *rank dispersion*

$$D_n(\mathbf{b}) = \sum_{i=1}^{n}(X_i - \mathbf{b}'\mathbf{c}_i)a_n(R_{ni}(\mathbf{b})), \quad \mathbf{b} \in \mathbb{R}_p, \tag{6.6.9}$$

[where $a_n(1), \ldots, a_n(n)$ are nondecreasing scores] and proposed to estimate $\boldsymbol{\beta}$ by minimizing $D_n(\mathbf{b})$ with respect to $\mathbf{b} \in \mathbb{R}_p$. If we set $\bar{a}_n = n^{-1}\sum_{i=1}^{n} a_n(i) = 0$ (what we could without loss of generality), we could show that $D_n(\mathbf{b})$ is translation-invariant. Actually, noting that $a_n(R_{ni}(\mathbf{b}))$ are translation-invariant and denoting $D_n(\mathbf{b}, k)$ the measure (6.6.9) calculated for the pseudo-observations $X_i + k - \mathbf{b}'\mathbf{c}_i$, $i = 1, \ldots, n$, we could write

$$\begin{aligned} D_n(\mathbf{b}, k) &= \sum_{i=1}^{n}(X_i + k - \mathbf{b}'\mathbf{c}_i)(a_n(R_{ni}(\mathbf{b})) - \bar{a}_n) \\ &= D_n(\mathbf{b}) + k\sum_{i=1}^{n}(a_n(R_{ni}(\mathbf{b})) - \bar{a}_n) = D_n(\mathbf{b}). \end{aligned} \tag{6.6.10}$$

Moreover it can be shown (see Problem 6.6.2) that

$$D_n(\mathbf{b}) \text{ is nonnegative, continuous, piecewise linear,}$$
$$\text{and convex function of } \mathbf{b} \in \mathbb{R}_p. \tag{6.6.11}$$

Note that (6.6.11) ensures that $D_n(\mathbf{b})$ is differentiable in $\mathbf{b}$ almost everywhere and

$$(\partial/\partial\mathbf{b})D_n(\mathbf{b})|_{\mathbf{b}^0} = -L_n(\mathbf{b}^0) \tag{6.6.12}$$

at any point $\mathbf{b}^0$ of differentiability of D_n. If D_n is not differentiable in $\mathbf{b}^0$, we can work with the subgradient $\nabla D_n(\mathbf{b}_0)$ of D_n at $\mathbf{b}_0$ defined as the operation satisfying

$$D_n(\mathbf{b}) - D_n(\mathbf{b}^0) \geq (\mathbf{b} - \mathbf{b}^0)\nabla D_n(\mathbf{b}_0) \tag{6.6.13}$$

for all $\mathbf{b} \in \mathbb{R}_p$. The following formulation incorporates an extension of the original Jurečková (1971a) construction, due to Heiler and Willers (1988), where we make the following assumptions:

1. *Smoothness of* F. The errors e_i in (6.6.1) are i.i.d. r.v.'s with a d.f. F (not depending on n) such that F possesses an absolutely continuous density f with a finite Fisher information $I(f)$ [defined by (6.2.22)].

2. *Score-generating function.* This function $\phi : (0,1) \mapsto \mathbb{R}_1$, is assumed to be nonconstant, nondecreasing and square integrable on $(0,1)$, so

$$0 < A_\phi^2 = \int_0^1 \phi^2(u)du < \infty. \tag{6.6.14}$$

The scores $a_n(i)$ are defined either by (6.4.10) or as $\phi(i/(n+1))$, $1 \le i \le n$, for $n \ge 1$.

3. *Generalized Noether condition.* Assume that

$$\lim_{n\to\infty} \max_{1\le i\le n} (\mathbf{c}_i - \bar{\mathbf{c}}_n)'\mathbf{V}_n^{-1}(\mathbf{c}_i - \bar{\mathbf{c}}_n) = 0, \tag{6.6.15}$$

where

$$\mathbf{V}_n = \sum_{i=1}^{n} (\mathbf{c}_i - \bar{\mathbf{c}}_n)(\mathbf{c}_i - \bar{\mathbf{c}}_n)', \quad n > p. \tag{6.6.16}$$

Note that for any vector $\mathbf{a}$ and a positively definite matrix $\mathbf{B}$,

$$\mathbf{a}'\mathbf{B}^{-1}\mathbf{a} = \lambda_{\max}(\mathbf{a}\mathbf{a}'\mathbf{B}^{-1}) = \text{trace}(\mathbf{a}\mathbf{a}'\mathbf{B}^{-1}). \tag{6.6.17}$$

Hence we could verify (see Problem 6.6.3) that the generalized Noether condition in (6.6.15) implies that the Noether condition in (6.4.20) holds for each of the p-coordinates.

It will be convenient to make an orthonormal transformation on the model in (6.6.1): Put

$$\mathbf{c}_i^\star = \mathbf{V}_n^{-\frac{1}{2}} \mathbf{c}_i, \tag{6.6.18}$$

and introduce the reparametrization

$$\boldsymbol{\beta}_n^\star = \boldsymbol{\beta}^\star = \mathbf{V}_n^{\frac{1}{2}} \boldsymbol{\beta}, \quad n = 1, 2, \ldots \tag{6.6.19}$$

Then

$$\mathbf{V}_n^\star = \sum_{i=1}^{n} (\mathbf{c}_i^\star - \bar{\mathbf{c}}_n^\star)(\mathbf{c}_i^\star - \bar{\mathbf{c}}_n^\star)' = \mathbf{I}_p \tag{6.6.20}$$

and

$$\boldsymbol{\beta}'\mathbf{c}_i = \boldsymbol{\beta}^{\star\prime}\mathbf{c}_i^\star, \quad 1 \le i \le n. \tag{6.6.21}$$

The reparametrization (6.6.19) accompanied by a conjugate transformation (6.6.18) leave the estimation problem invariant and this simplification does not reduce the generality of the model.

We denote the reduced linear rank statistics by

$$\mathbf{L}_n^\star(\mathbf{b}) = \sum_{i=1}^{n}(\mathbf{c}_i^\star - \bar{\mathbf{c}}_n^\star)a_n(R_{ni}^\star(\mathbf{b})), \quad \mathbf{b} \in \mathbb{R}_p, \tag{6.6.22}$$

where $R_{ni}^\star(\mathbf{b})$ are defined as in (6.6.2) with the $\mathbf{c}_i$ replaced by $\mathbf{c}_i^\star$. For $p = 1$, (6.6.21) reduces to the parallel case of $L_n^\star(\mathbf{b})$ in (6.4.16) and it follows from Theorem 6.4.2 that, under $\boldsymbol{\beta}^\star = \mathbf{0}$ and under assumption 2,

$$||\mathbf{L}_n^\star(\mathbf{0}) - \mathbf{S}_n^\star(\mathbf{0})|| \xrightarrow{P} 0 \quad \text{as } n \to \infty, \tag{6.6.23}$$

where

$$\mathbf{S}_n^\star(\mathbf{b}) = \sum_{i=1}^{n}(\mathbf{c}_i^\star - \bar{\mathbf{c}}_n^\star)\phi(F(X_i - \theta - \mathbf{b}'\mathbf{c}_n^\star)), \quad \mathbf{b} \in \mathbb{R}_p. \tag{6.6.24}$$

Similarly let P_n and $P_n^\star$ be the probability distributions with the respective densities $p_n = \prod_{i=1}^n f(X_i - \theta)$ and $p_n^\star = \prod_{i=1}^n f(X_i - \theta - \boldsymbol{\lambda}'\mathbf{c}_i^\star)$, $\boldsymbol{\lambda} \in \mathbb{R}_p$ fixed. Then $\{P_n^\star\}$ are contiguous to $\{P_n\}$, and proceeding as in (6.2.28)-(6.2.29), we conclude that under $\boldsymbol{\beta}^\star = \mathbf{0}$ and for any (fixed) $\boldsymbol{\lambda} \in \mathbb{R}_p$,

$$||\mathbf{L}_n^\star(\boldsymbol{\lambda}) - \mathbf{S}_n^\star(\boldsymbol{\lambda})|| \xrightarrow{P} 0 \quad \text{as } n \to \infty. \tag{6.6.25}$$

On the other hand, by (6.2.24),

$$\begin{aligned} & \mathbf{S}_n^\star(\boldsymbol{\lambda}) - \mathbf{S}_n^\star(\mathbf{0}) + \gamma\boldsymbol{\lambda} \\ = & \sum_{i=1}^{n}(\mathbf{c}_i^\star - \bar{\mathbf{c}}_n^\star)\big[\phi(F(X_i - \theta - \boldsymbol{\lambda}'\mathbf{c}_i^\star)) - \phi(F(X_i - \theta)) \\ & +\gamma(\mathbf{c}_i^\star - \bar{\mathbf{c}}_n^\star)'\boldsymbol{\lambda}\big], \end{aligned} \tag{6.6.26}$$

where γ is defined by (6.4.19) [with $\psi_f(.)$ as in (6.2.25)]. Notice that the right-hand side of (6.6.26) is a sum of independent summands and the multivariate central limit theorem, with its degeneracy, implies its stochastic convergence to $\mathbf{0}$. Thus, parallel to (6.4.23), we have under $\boldsymbol{\beta}^\star = \mathbf{0}$ and under all three assumptions, that for any (fixed) $\boldsymbol{\lambda} \in \mathbb{R}_p$,

$$||\mathbf{S}_n^\star(\boldsymbol{\lambda}) - \mathbf{S}_n^\star(\mathbf{0}) + \gamma\boldsymbol{\lambda}|| \xrightarrow{P} 0 \quad \text{as } n \to \infty, \tag{6.6.27}$$

see Problem 6.6.4. Hence

$$\begin{aligned} & ||\mathbf{L}_n^\star(\boldsymbol{\lambda}) - \mathbf{L}_n^\star(\mathbf{0}) + \gamma\boldsymbol{\lambda}|| \\ \le & \ ||\mathbf{L}_n^\star(\boldsymbol{\lambda}) - \mathbf{S}_n^\star(\boldsymbol{\lambda})|| + ||\mathbf{S}_n^\star(\boldsymbol{\lambda}) - \mathbf{S}_n^\star(\mathbf{0}) + \gamma\boldsymbol{\lambda}|| \\ & +||\mathbf{L}_n^\star(\mathbf{0}) - \mathbf{S}_n^\star(\mathbf{0})|| \xrightarrow{P} 0 \quad \text{as } n \to \infty. \end{aligned} \tag{6.6.28}$$

under all three assumptions and for fixed $\boldsymbol{\lambda} \in \mathbb{R}_p$.

It remains to prove that the convergence in (6.2.28) is uniform over any compact set $\mathbb{K} = \{\boldsymbol{\lambda} : \ ||\boldsymbol{\lambda}|| \leq C\}$, $0 < C < \infty$. In the case $p = 1$ the uniformity result follows from the monotonicity of $L_n^\star(b)$ in b (see Theorem 6.4.2). For $p \geq 2$ Jurečková (1971) proved the uniformity under a rather restrictive concordance-discordance condition on $\mathbf{c}_{ni}^\star$, $i = 1, \ldots, n$. As pointed out by Heiler and Willers (1988), this extra condition is not really necessary. Toward this end, we consider the following lemma:

Lemma 6.6.1 *(Heiler and Willers 1988). For monotone scores (see assumption 2), the pointwise stochastic convergence in (6.6.28) is equivalent to the uniform stochastic convergence on a compact subset $\mathbb{K}$. That is, under the same regularity conditions as demanded for (6.6.28),*

$$\sup\left\{||\mathbf{L}_n^\star(\boldsymbol{\lambda}) - \mathbf{L}_n^\star(\mathbf{0}) + \gamma\boldsymbol{\lambda}|| : \ ||\boldsymbol{\lambda}|| \leq C\right\} \xrightarrow{P} 0 \tag{6.6.29}$$

as $n \to \infty$ for any fixed C, $0 < C < \infty$.

The proof of (6.6.29) rests on (6.6.12), (6.6.13), the convexity of $D_n(\mathbf{b})$ in (6.6.11) and a convexity argument, included in (2.5.142)-(2.5.143), we leave the details as Problem 6.6.5.

Recall that, by virtue of (6.6.18)-(6.6.21),

$$R_{ni}^\star(\boldsymbol{\lambda}) = R_{ni}(\mathbf{V}_n^{-\frac{1}{2}}\boldsymbol{\lambda}), \ \mathbf{c}_i^\star = \mathbf{V}_n^{-\frac{1}{2}}\mathbf{c}_i, \quad i = 1, \ldots, n,$$

so that

$$\mathbf{L}_n^\star(\boldsymbol{\lambda}) = \mathbf{V}_n^{-\frac{1}{2}}\mathbf{L}_n(\mathbf{V}_n^{-\frac{1}{2}}\boldsymbol{\lambda}), \quad \forall\boldsymbol{\lambda} \in \mathbb{R}_p. \tag{6.6.30}$$

We may rewrite (6.6.29) in the form that for $n \to \infty$,

$$\sup\left(||\mathbf{V}_n^{-\frac{1}{2}}\left\{\mathbf{L}_n(\mathbf{V}_n^{-\frac{1}{2}}\boldsymbol{\lambda}) - \mathbf{L}_n(\mathbf{0}) + \mathbf{V}_n^{\frac{1}{2}}\boldsymbol{\lambda}\gamma\right\}|| : \ ||\boldsymbol{\lambda}|| \leq C\right) \xrightarrow{P} 0 \tag{6.6.31}$$

under assumptions 1 to 3 and under $\boldsymbol{\beta} = \mathbf{0}$. This can be further rewritten for $n \to \infty$ as

$$\sup\left(||\mathbf{V}_n^{-\frac{1}{2}}\left\{\mathbf{L}_n(\mathbf{b}) - \mathbf{L}_n(\mathbf{0}) + \gamma\mathbf{V}_n\mathbf{b}\right\}|| : \ ||\mathbf{V}_n^{\frac{1}{2}}\mathbf{b}|| \leq C\right) \xrightarrow{P} 0. \tag{6.6.32}$$

Finally, translating (6.6.32) to the case of true $\boldsymbol{\beta}$, we have under the three assumptions, under $\boldsymbol{\beta}$, and for any fixed C, $0 < C < \infty$,

$$\sup\left(||\mathbf{V}_n^{-\frac{1}{2}}\left\{\mathbf{L}_n(\mathbf{b}) - \mathbf{L}_n(\boldsymbol{\beta}) + \gamma\mathbf{V}_n(\mathbf{b} - \boldsymbol{\beta})\right\}|| : \ ||\mathbf{V}_n^{\frac{1}{2}}(\mathbf{b} - \boldsymbol{\beta})|| \leq C\right) \xrightarrow{P} 0. \tag{6.6.33}$$

The uniform asymptotic linearity in (6.6.33), along with the monotonicity of ϕ, enables us to conclude that the diameter of $\mathcal{D}_n$, defined in (6.6.6)-(6.6.8), is $O_p(||\mathbf{V}_n^{-\frac{1}{2}}||)$, and that under the three assumptions,

$$\sup\left(||\mathbf{V}_n^{-\frac{1}{2}}\left\{\mathbf{L}_n(\boldsymbol{\beta}) - \gamma\mathbf{V}_n(\mathbf{b} - \boldsymbol{\beta})\right\}|| : \ \mathbf{b} \in \mathcal{D}_n\right) \xrightarrow{P} 0, \quad \text{as } n \to \infty. \tag{6.6.34}$$

For every $\hat{\boldsymbol{\beta}}_n$ belonging to $\mathcal{D}_n$, we have

$$\begin{aligned}\mathbf{V}_n^{\frac{1}{2}}(\hat{\boldsymbol{\beta}}_n - \boldsymbol{\beta}) &= \mathbf{V}_n^{-\frac{1}{2}}\mathbf{L}_n(\boldsymbol{\beta}) + \mathbf{o}_p(1)\\ &= \mathbf{L}_n^{\star}(\boldsymbol{\beta}^{\star}) + \mathbf{o}_p(1)\\ &= \mathbf{S}_n^{\star}(\boldsymbol{\beta}^{\star}) + \mathbf{o}_p(1), \end{aligned} \tag{6.6.35}$$

where, by (6.6.24), $\mathbf{S}_n^{\star}(\boldsymbol{\beta}^{\star})$ under $\boldsymbol{\beta}^{\star}$ has the same distribution as $\mathbf{S}_n^{\star}(\mathbf{0})$ under $\boldsymbol{\beta}^{\star} = \mathbf{0}$. For the latter case, we can readily use the multivariate central limit theorem and claim that under $\boldsymbol{\beta}^{\star} = \mathbf{0}$, and assumptions 2 and 3,

$$\mathbf{S}_n^{\star}(\mathbf{0}) \sim \mathcal{N}_p(\mathbf{0}, A_{\phi}^2\mathbf{I}_p), \tag{6.6.36}$$

where A_{ϕ}^2 is defined by (6.6.15). By (6.6.35) and (6.6.36), we obtain that when $\boldsymbol{\beta}$ holds, under all three assumptions,

$$\mathbf{V}_n^{\frac{1}{2}}(\hat{\boldsymbol{\beta}}_n - \boldsymbol{\beta}) \sim \mathcal{N}_p(\mathbf{0}, \gamma^{-2}A_{\phi}^2\mathbf{I}_p). \tag{6.6.37}$$

An analogous argument may be based on the convexity of the Jaeckel measure of dispersion (6.6.9), which by the asymptotic linearity, (6.6.29) and (6.6.12) can be approximated by a quadratic function of $\mathbf{b}$ uniformly over $||\mathbf{b}|| \leq C$. We have similar arguments used already in in connection with regression quantiles (Chapter 4) and M-estimators generated by a step function (Chapter 5).

Let us now write,

$$\mathbf{d}_{ni} = \mathbf{V}_n^{-\frac{1}{2}}(\mathbf{c}_i^{\star} - \bar{\mathbf{c}}_n^{\star}) = \mathbf{V}_n^{-1}(\mathbf{c}_i - \bar{\mathbf{c}}_n), \ \ 1 \leq i \leq n. \tag{6.6.38}$$

Then, by virtue of (6.6.38), we have under all three assumptions,

$$(\hat{\boldsymbol{\beta}}_n - \boldsymbol{\beta}) = \gamma^{-1}\sum_{i=1}^{n}\mathbf{d}_{ni}\phi(F(e_i)) + \mathbf{o}_p(||\mathbf{V}_n^{-\frac{1}{2}}||), \tag{6.6.39}$$

and this leads to the main theorem of this section.

Theorem 6.6.1 *Let $\hat{\boldsymbol{\beta}}_n$ be the R-estimator of $\boldsymbol{\beta}$ in the linear model (6.6.1), defined by (6.6.8). Then, under assumptions 1, 2 and 3, $\hat{\boldsymbol{\beta}}_n$ admits the (first-order) asymptotic representation (6.6.39).*

Let us consider a similar representation for an R-estimator of the intercept parameter θ. In this context, we rewrite (6.6.1) as

$$\begin{aligned} X_i &= \theta + \boldsymbol{\beta}'\mathbf{c}_i + e_i\\ &= (\theta - \boldsymbol{\beta}'\bar{\mathbf{c}}_n) + \boldsymbol{\beta}'(\mathbf{c}_i - \bar{\mathbf{c}}_n) + e_i\\ &= \theta_n^0 + \boldsymbol{\beta}'(\mathbf{c}_i - \bar{\mathbf{c}}_n) + e_i, \ \ i = 1, \ldots, n, \end{aligned} \tag{6.6.40}$$

where $\theta_n^0 = \theta + \boldsymbol{\beta}'\bar{\mathbf{c}}_n$. Replace $\boldsymbol{\beta}$ by its R-estimator $\hat{\boldsymbol{\beta}}_n$, and consider the residuals

$$\begin{aligned}\hat{X}_i &= X_i - \hat{\boldsymbol{\beta}}_n'(\mathbf{c}_i - \bar{\mathbf{c}}_n) - \hat{\boldsymbol{\beta}}_n'\bar{\mathbf{c}}_n \\ &= \theta + e_i + (\boldsymbol{\beta} - \hat{\boldsymbol{\beta}}_n)'(\bar{\mathbf{c}}_n + (\mathbf{c}_i - \bar{\mathbf{c}}_n)) \\ &= \theta + e_i + (\boldsymbol{\beta} - \hat{\boldsymbol{\beta}}_n)'\mathbf{V}_n^{\frac{1}{2}}\left\{\mathbf{V}_n^{-\frac{1}{2}}(\bar{\mathbf{c}}_n + (\mathbf{c}_i - \bar{\mathbf{c}}_n))\right\}, \qquad (6.6.41)\end{aligned}$$

$i = 1, \ldots, n$. Note that by (6.6.37),

$$\|(\boldsymbol{\beta} - \hat{\boldsymbol{\beta}}_n)'\mathbf{V}_n^{\frac{1}{2}}\| = \mathrm{O}_p(1), \qquad (6.6.42)$$

while by (6.6.15),

$$\max_{1 \le i \le n} \|\mathbf{V}_n^{-\frac{1}{2}}(\mathbf{c}_i - \bar{\mathbf{c}}_n)\| = \mathrm{o}_p(1). \qquad (6.6.43)$$

Hence the last term on the right-hand side of (6.6.41) is negligible in probability uniformly in $\{i :\ 1 \le i \le n\}$, while the result of Section 6.2 apply for $\theta + e_i$, $i = 1, \ldots, n$. As such, an R-estimator $\hat{\theta}_n$ of θ, based on the residuals $\hat{X}_i$, $1 \le i \le n$, may be defined as in (3.4.9). Following Section 6.2, we should assume the symmetry of the d.f. F about 0 (otherwise $\hat{\theta}_n$ is generally biased). Moreover, following the condition (6.4.34), we assume that

$$\lim_{n \to \infty} n^{\frac{1}{2}}\bar{\mathbf{c}}_n'\mathbf{V}_n^{-\frac{1}{2}} = \boldsymbol{\xi}' \text{ exists.} \qquad (6.6.44)$$

The estimator will be based on the signed-rank statistic with the score function $\phi(u) = \phi^\star((u+1)/2)$, $0 \le u \le 1$ where $\phi^\star$ is skew-symmetric about 1/2 [i.e., $\phi^\star(u) + \phi^\star(1-u) = 0,\ 0 < u < 1$]. The unknown $\boldsymbol{\beta}$ in this signed-rank statistic is replaced by $\hat{\boldsymbol{\beta}}$ (aligned signed-rank statistic based on the $\hat{X}_i$), and we can prove the asymptotic linearity in α, $\boldsymbol{\lambda}$, parallel to (6.6.28). Again, Lemma 6.6.1 enables to show that this asymptotic linearity holds uniformly over an arbitrary compact set. As such, parallel to (6.4.35), we have

$$\begin{aligned}& n^{\frac{1}{2}}(\hat{\theta}_n - \theta) \\ &= \gamma^{-1}n^{-\frac{1}{2}}\Big\{\sum_{i=1}^n \phi^\star(F(e_i)) - \boldsymbol{\xi}'\sum_{i=1}^n \mathbf{V}_n^{-\frac{1}{2}}(\mathbf{c}_i - \bar{\mathbf{c}}_n)\phi^\star(F(e_i))\Big\} + \mathrm{o}_p(1) \\ &= n^{-\frac{1}{2}}\gamma^{-1}\sum_{i=1}^n \Big[1 - \boldsymbol{\xi}'\mathbf{V}_n^{-\frac{1}{2}}(\mathbf{c}_i - \bar{\mathbf{c}}_n)\Big]\phi^\star(F(e_i)) + \mathrm{o}_p(1). \qquad (6.6.45)\end{aligned}$$

This (first-order) asymptotic representation implies the asymptotic normality of $n^{\frac{1}{2}}(\hat{\theta}_n - \theta)$. Combining (6.6.39) and (6.6.45), we conclude that, under the pertaining regularity conditions as $n \to \infty$,

$$n^{\frac{1}{2}}\begin{bmatrix}\hat{\theta}_n - \theta \\ \hat{\boldsymbol{\beta}}_n - \boldsymbol{\beta}\end{bmatrix} = \gamma^{-1}n^{-\frac{1}{2}}\sum_{i=1}^n \begin{bmatrix}1 - \boldsymbol{\xi}'\mathbf{d}_{ni} \\ n^{\frac{1}{2}}\mathbf{d}_{ni}\end{bmatrix}\phi^\star(F(e_i)) + \mathrm{o}_p(1) \qquad (6.6.46)$$

with the $\mathbf{d}_{ni}$ given by (6.6.38).

The exact order of the remainder terms in the asymptotic representations (6.6.39), (6.6.45), and (6.6.46) would be of interest. If we want to follow the methods of Section 6.5, we should consider two steps:

1. Derive the pointwise confidence rate for $\mathbf{L}_n^\star(\mathbf{b}) - \mathbf{L}_n^\star(\mathbf{0}) - \gamma\mathbf{b}$ for fixed $\mathbf{b} \in I\!\!K$.

2. Prove that this convergence rate is uniform for $\mathbf{b} \in I\!\!K$.

In the general linear model treated in this section, the pointwise convergence rates hold under the same regularity conditions as in Section 6.5. Thus we have a pointwise rate $O_p(n^{-\frac{1}{4}})$, $O_p(n^{-\frac{1}{2}})$, or $O_p(n^{-\frac{1}{2}} \log n)$ [e.g., $= O_p(a_n)$] depending on the nature of the score function ϕ. Then, we would have for every fixed $\mathbf{b} \in I\!\!K$,

$$(a_n \log n)^{-1} \| \mathbf{L}_n^\star(\mathbf{b}) - \mathbf{L}_n^\star(\mathbf{0}) + \gamma\mathbf{b} \| \xrightarrow{P} 0, \quad \text{as } n \to \infty. \tag{6.6.47}$$

Hence, using the same convexity arguments as in Lemma 6.6.1, we can show that (6.6.47) holds uniformly in $\mathbf{b} \in I\!\!K$. In this way we obtain that under the regularity conditions of Section 6.5, (6.6.39) can be refined as

$$\hat{\boldsymbol{\beta}}_n - \boldsymbol{\beta} = \gamma^{-1} \sum_{i=1}^{n} \mathbf{d}_{ni} \phi(F(e_i)) + R_n \tag{6.6.48}$$

where $R_n = o_p(n^{-\frac{1}{2}} \log n)$, $R_n = o_p(n^{-\frac{1}{4}} \log n)$, and $R_n = o_p(n^{-\frac{1}{2}} (\log n)^2)$, respectively, depending on the score function ϕ.

The exact rates of R_n would apparently coincide with those in the simple regression model of Section 6.5. The difficulties with the proof of this fact are primarily due to the lack of ordering of the $\mathbf{c}_i$. Also (6.4.50) extends directly to the linear model case (see Problem 6.6.6).

6.7 REGRESSION RANK SCORES

In the context of L-estimators in linear (regression) models, in Section 4.7, regression quantiles (RQ) have been introduced and studied. The concept of dual regression quantiles extends, rather surprisingly, the concept of rank scores from the location model to the linear regression model. Hence, instead of dual regression quantiles, we will use the name regression rank (RR) scores. The duality of RQ's to RR's is not only in the linear programming sense, but it also extends the duality of order statistics and ranks from the location to the linear regression model.

As in previous section, we consider the linear model:

$$X_i = \boldsymbol{\beta}' \mathbf{c}_i + e_i, \quad i = 1, \ldots, n, \tag{6.7.1}$$

where the X_i are the observable r.v.'s, $\mathbf{c}_i = (c_{i1}, \ldots, c_{ip})'$, $i \geq 1$ are given p-vectors of (known) regression constants (with $c_{i1} = 1 \; \forall i \geq 1$), $\boldsymbol{\beta} = (\beta_1, \ldots, \beta_p)'$ is a p-vector of unknown parameters, and the e_i are i.i.d. r.v.'s having a continuous d.f. F defined on $\mathbb{R}_1$. We take $p \geq 2$, and note that β_1 is an intercept, while $\beta_2, \ldots, \beta_p$ are slopes. The case of $p = 1$ relates to the location model treated in Sections 6.2 and 6.3.

The vector of RR is denoted by $\hat{\mathbf{a}}_n(\alpha) = (\hat{a}_{n1}(\alpha), \ldots, \hat{a}_{nn}(\alpha))'$ where $\alpha \in (0,1)$. This is defined as the optimal solution of the linear programming problem:

$$\begin{aligned} &\sum_{i=1}^{n} X_i \hat{a}_{ni}(\alpha) = \max, \\ &\sum_{i=1}^{n} c_{ij} \hat{a}_{ni}(\alpha) = (1-\alpha) \sum_{i=1}^{n} c_{ij}, \quad j = 1, \ldots, p; \\ &\quad \hat{a}_{ni}(\alpha) \in [0,1], \; \forall 1 \leq i \leq n, \; 0 \leq \alpha \leq 1. \end{aligned} \tag{6.7.2}$$

This concept is better understood in the simple case of $p = 1$ (i.e., location model) where all c_i are scalars equal to 1. Then we have $\hat{a}_{ni}(\alpha) = a_n^\star(R_i, \alpha)$, where R_i is the rank of X_i among $X_1, \ldots, X_n$ and

$$a_n^\star(R_i, \alpha) = \begin{cases} 0, & \text{if } R_i/n < \alpha, \\ R_i - n\alpha, & \text{if } (R_i - 1)/n \leq \alpha \leq R_i/n, \\ 1, & \text{if } \alpha < (R_i - 1)/n, \end{cases} \tag{6.7.3}$$

$i = 1, \ldots, n$, are scores considered in Hájek (1965), as a starting point for an extension of the Kolmogorov-Smirnov test to regression alternatives.

As in the location model the regression quantiles are suitable mainly for estimation, while the regression rank scores are suitable also for testing hypothesis in the linear model, specially when the hypothesis concerns only some component of $\boldsymbol{\beta}$ while the other components are considered as nuisance. Linear RR-tests are constructed by Gutenbrunner et al. (1993), and the R-tests of Kolmogorov-Smirnov type by Jurečková (1992a). One may also construct estimators (RR-estimators) of subvectors of $\boldsymbol{\beta}$, based on RR, analogous to R-estimators (Jurečková (1992c)). Let us illustrate some finite-sample properties of regression rank scores in order to see how far the analogy with the ordinary ranks goes and where the deviations start. In the model (6.7.1), denote $\mathbf{C} = \mathbf{C}_n$ the $n \times p$ matrix with components c_{ij}, $1 \leq i \leq n$ and $1 \leq j \leq p$, and assume that $\mathbf{C}$ is of rank p and satisfies the standardization

$$c_{i1} = 1, \quad i = 1, \ldots, n \tag{6.7.4}$$

and

$$\sum_{i=1}^{n} c_{ij} = 0, \quad j = 2, \ldots, p. \tag{6.7.5}$$

While (6.7.4) is crucial for the existence of RR's (otherwise, the RQ's are not distinguishable for different α's and thus RR's are not well defined), the assumption (6.7.5) is rather for mathematical convenience.

Let $\beta(\alpha)$ be the α-regression quantile and let $\hat{\mathbf{a}}_n(\alpha)$ be the vector of regression rank scores, $0 < \alpha < 1$. Some finite-sample properties of these statistics are summarized in the following lemma:

Lemma 6.7.1 *(1)* $\hat{\mathbf{a}}_n(\alpha, \mathbf{X} + \mathbf{Cb}) = \hat{\mathbf{a}}_n(\alpha, \mathbf{X})\ \forall \mathbf{b} \in \mathbb{R}_p$, *(i.e., the regression rank scores are invariant to the regression with matrix* $\mathbf{C}$ *).*

(2) If $\mathbf{C}$ *satisfies (6.7.4) and (6.7.5), then*

$$\hat{\beta}_1(\alpha_1) \leq \hat{\beta}_2(\alpha_2) \quad \text{for any } \alpha_1 \leq \alpha_2$$

(3) If, on the contrary, $\sum_{i=1}^n c_{ij} = 0$ *for* $j = 1, \ldots, p$, *then*

$$\hat{\beta}_1(\alpha_1) = \hat{\beta}_2(\alpha_2) \quad \text{for any } 0 < \alpha_1, \alpha_2 < 1$$

(i.e., the regression quantiles corresponding to different α*'s are not distinguishable).*

Remark. By proposition 1, every statistical procedure based on RR's is invariant with respect to the $\mathbf{C}$-regression. This is important mainly in the models where we have a nuisance $\mathbf{C}$-regression. Moreover the monotonicity of $\hat{\beta}_1(\alpha)$ claimed in prop[osition 2 is in correspondence with the fact that the first component of the regression quantile represents the location quantile, while the other components estimate the slope parameters.

Proof. Proposition 1 follows directly from (6.7.2). Moreover by (6.7.1), for every $\mathbf{t} \in \mathbb{R}_p$,

$$\sum_{i=1}^n [\rho_{\alpha_2}(X_i - \mathbf{c}_i'\mathbf{t}) - \rho_{\alpha_1}(X_i - \mathbf{c}_i'\mathbf{t})] = (\alpha_2 - \alpha_1)\sum_{i=1}^n (X_i - \mathbf{c}_i'\mathbf{t}),$$

where $\rho_\alpha(x) = |x|\{\alpha I(x > 0) + (1-\alpha)I[x < 0]\}$. This immediately implies 3. This further implies under (6.7.4) and (6.7.5) that

$$\begin{aligned}\sum_{i=1}^n \rho_{\alpha_2}(X_i - \mathbf{c}_i'\hat{\boldsymbol{\beta}}(\alpha_2)) \quad &\geq \quad \sum_{i=1}^n \rho_{\alpha_1}(X_i - \mathbf{c}_i'\hat{\boldsymbol{\beta}}(\alpha_1)) \\ &\quad +n(\alpha_2 - \alpha_1)(\bar{X} - \hat{\boldsymbol{\beta}}(\alpha_2)), \qquad (6.7.6)\end{aligned}$$

where $\bar{X} = n^{-1}\sum_{i=1}^n X_i$. Similarly

$$\begin{aligned}\sum_{i=1}^n \rho_{\alpha_1}(X_i - \mathbf{c}_i'\hat{\boldsymbol{\beta}}(\alpha_1)) \quad &\geq \quad \sum_{i=1}^n \rho_{\alpha_2}(X_i - \mathbf{c}_i'\hat{\boldsymbol{\beta}}(\alpha_2)) \\ &\quad -n(\alpha_2 - \alpha_1)(\bar{X} - \hat{\boldsymbol{\beta}}(\alpha_1)), \qquad (6.7.7)\end{aligned}$$

Combining (6.7.6) and (6.7.7), we arrive at proposition 2. □

Lemma 6.7.2 *Under (6.7.4) and (6.7.5) and for* $0 < \alpha_1 \leq \alpha_2 < 1$,

$$
\begin{aligned}
(\alpha_2 - \alpha_1)\hat{\boldsymbol{\beta}}_1(\alpha_1) &\leq n^{-1}\sum_{i=1}^{n} X_i[\hat{a}_i(\alpha_1) - \hat{a}_i(\alpha_2)] \\
&\leq (\alpha_2 - \alpha_1)\hat{\boldsymbol{\beta}}_1(\alpha_2)
\end{aligned} \tag{6.7.8}
$$

Moreover, if α *is a point of continuity of* $\hat{\beta}_1(.)$, *then*

$$
\hat{\boldsymbol{\beta}}_1(\alpha_1) = -n^{-1}\sum_{i=1}^{n} X_i \hat{a}_i'(\alpha) \tag{6.7.9}
$$

where $\hat{a}_i'(\alpha) = (\partial)/(\partial\alpha)a_i(\alpha)$.

Remark 1. For a fixed $\alpha \in (0,1)$, let $I(\alpha) = \{i : 0 < \hat{a}_i(\alpha) < 1\}$; then $X_i = \mathbf{c}_i'\hat{\boldsymbol{\beta}}(\alpha)$ for $i \in I(\alpha)$ and $\#I(\alpha) = p$. The identity (6.7.9) means that while, in the location model, $\hat{\beta}_1(\alpha)$ coincides just with one order statistic (i.e., with one observation), in the linear regression model $\hat{\beta}_1(\alpha)$ coincides with a linear combination of those X_i's which have an exact fit for α. The weight of X_i is positive or negative according as whether $\hat{a}_i(\alpha)$ is decreasing or increasing in α for all $i \in I(\alpha)$. The weights $\hat{a}_i(\alpha)$, $i \in I(\alpha)$, satisfy

$$
-n^{-1}\sum_{i=1}^{n} \hat{a}_i'(\alpha) = 1, \sum_{i \in I(\alpha)} c_{ij}\hat{a}_i'(\alpha) = 0, \ j = 2, \ldots, p. \tag{6.7.10}
$$

Remark 2. The sets $\{\hat{\boldsymbol{\beta}}(\alpha) : 0 < \alpha < 1\}$ and $\{I(\alpha) : 0 < \alpha < 1\}$ form a sufficient statistic for the model (6.7.1). In fact, with probability 1, every i_0 belongs to some $I(\alpha_{i_0})$, $i_0 = 1, \ldots, n$ by the continuity of $\hat{a}_i(\alpha)$ and due to the fact that $\hat{a}_i(0) = 1$, $\hat{a}_i(1) = 0$, $i = 1, \ldots, n$. Fixing the corresponding α_0, we have $X_{i_0} = \mathbf{c}_{i0}'\hat{\boldsymbol{\beta}}(\alpha_{i_0})$ with probability 1, $i_0 = 1, \ldots, n$.

Proof. The duality of $\hat{\boldsymbol{\beta}}(\alpha)$ and $\hat{\mathbf{a}}(\alpha)$ implies that

$$
\sum_{i=1}^{n} \rho_\alpha(X_i - \mathbf{c}_i'\hat{\boldsymbol{\beta}}(\alpha)) = \sum_{i=1}^{n} X_i(\hat{a}_i(\alpha) - (1-\alpha)). \tag{6.7.11}
$$

Hence (6.7.8) follows from (6.7.6) and (6.7.7). Moreover $\hat{\beta}_1(\alpha)$ is a step function of α, and $\hat{a}_i(\alpha)$ is a continuous, piecewise linear function of α, $0 < \alpha < 1$. The breakpoints of two functions coincide (e.g., see Gutenbrunner and Jurečková 1992). Thus (6.7.9) follows from (6.7.8). □.

6.7.1 Asymptotic Representation of RR-Process

The statistical inference based on RR's is typically based on the functionals of the *regression rank scores process*

$$
\{n^{\frac{1}{2}}\sum_{i=1}^{n} d_{ni}\hat{a}_{ni}(\alpha) : 0 \leq \alpha \leq 1\} \tag{6.7.12}
$$

with appropriate coefficients $(d_{n1}, \ldots, d_{nn})$. In the location model such process was studied in of Hájek (1965) who among others proved its weak convergence to Brownian Bridge in the uniform topology on [0, 1]. We are able to approximate the process (6.7.12) by an empirical process as well as by the Hájek process, uniformly in the segment $[\alpha_n^\star, 1-\alpha_n^\star]$, with $\alpha_n^\star$ defined in (6.7.19) below.

As in the location model most of the statistical inference will be based on *linear regression rank score statistics* (linear RR statistics) which are constructed in the following way: Let us choose a *score-generating function* $\phi: (0,1) \mapsto \mathbb{R}_1$, nondecreasing and square integrable on $(0,1)$ and put

$$\phi_n(\alpha) = \begin{cases} \phi(\alpha_n^\star), & \text{if } 0 \leq \alpha < \alpha_n^\star, \\ \phi(\alpha), & \text{if } \alpha_n^\star \leq \alpha \leq 1-\alpha_n^\star, \\ \phi(1-\alpha_n^\star), & \text{if } 1-\alpha_n^\star < \alpha \leq 1. \end{cases} \tag{6.7.13}$$

Let $\hat{\mathbf{a}}_n(\alpha) = (\hat{a}_{n1}(\alpha), \ldots, \hat{a}_{nn}(\alpha))'$ be the RR's corresponding to the model (6.7.1); calculate the *scores* $\hat{\mathbf{b}} = (\hat{b}_{n1}, \ldots, \hat{b}_{nn})'$ generated by ϕ, in the following way

$$\hat{b}_{ni} = -\int_0^1 \phi_n(\alpha) d\hat{a}_{ni}(\alpha), \ i = 1, \ldots, n. \tag{6.7.14}$$

The *linear RR-statistic* is then

$$S_{nn} = n^{\frac{1}{2}} \sum_{i=1}^{n} d_{ni} \hat{b}_{ni}. \tag{6.7.15}$$

Note that (6.7.14) can be rewritten as

$$\hat{b}_{ni} = -\int_0^1 \phi_n(\alpha) \hat{a}'_{ni}(\alpha) d\alpha, \ \ i = 1, \ldots, n. \tag{6.7.16}$$

and the derivatives $\hat{a}'_{ni}(\alpha)$ are step-functions on $[0,1]$. In the location model this reduces to

$$\hat{b}_{ni} = n \int_{n^{-1}(R_i-1)}^{n^{-1}R_i} \phi_n(\alpha), \ \ i = 1, \ldots, n. \tag{6.7.17}$$

Typical choices of ϕ are in correspondence with the classical rank-tests theory (Wilcoxon, normal scores, median scores).

We will formulate (without proofs) asymptotic representations of RR-process and of linear RR-statistics. The proofs, which go beyond the scope of this book, can be found in Jurečková (1992d); under more restrictive conditions, some of the results are proved in Gutenbrunner and Jurečková (1992) and in Gutenbrunner et al. (1993).

The propositions stated below hold under the following regularity conditions on the distribution of the errors, on $\mathbf{C}_n$ and on $\mathbf{d}_n$: We will consider the linear model (6.7.1) with i.i.d. errors $e_1, \ldots, e_n$, distributed according to a d.f. F satisfying the conditions:

F1. F is absolutely continuous with absolutely continuous, positive, and bounded density $f(x)$, $A < x < B$. The derivative f' of f is bounded a.e. in (A, B), where

$$A = \sup\{x : F(x) = 0\} \quad \text{and} \quad B = \inf\{x : F(x) = 1\}.$$

F2. The density f is monotonically decreasing to 0 as $x \to A+$ and $x \to B-$, and, for some $a > 0$,

$$\sup_{0<\alpha<1} \left\{\alpha(1-\alpha)\frac{|f'(F^{-1}(\alpha))|}{f^2(F^{-1}(\alpha))}\right\} \leq 1 + a.$$

Comments on the conditions F1 and F2 . Parzen (1979) calls the density f satisfying **F2** *tail monotone with exponent* $1 + a$. This conditions was first used by Csörgö and Révész (1978) who showed that such distribution admits a strong approximation of the quantile process. Parzen (1980) also shows that if there exist limits at $\alpha \to 0$ and $\alpha \to 1$ equal to $1 + a_0$ and $1 + a_1$, respectively, of the expression

$$\alpha(1-\alpha)\frac{|f'(F^{-1}(\alpha))|}{f^2(F^{-1}(\alpha))},$$

then $f(F^{-1}(\alpha)) = \alpha^{1+a_0} L_0(\alpha)$ as $\alpha \to 0$ and $= (1-\alpha)^{1+a_1} L_1(\alpha)$ as $\alpha \to 1$, respectively, where L_0 and L_1 are slowly varying functions at $\alpha = 0$ and $\alpha = 1$, respectively (Parzen 1970, Seneta 1976, Csörgö et al. 1982). Then for any $\epsilon > 0$ there exist K_1, $K_2 > 0$ and $\alpha_0 \in (0, 1)$ such that (e.g., Feller 1966)

$$K_1 \alpha^{1+a_0+\epsilon} < f(F^{-1}(\alpha)) < K_2 \alpha^{1+a_0-\epsilon} \tag{6.7.18}$$

for $0 < \alpha < \alpha_0$, and similarly in a neighborhood of 1. We refer to Parzen (1979) for a detailed exposition of the tail-behavior of distributions.

Denote

$$\alpha_n^\star = n^{-1/(1+2b)} \tag{6.7.19}$$

for some fixed b satisfying $0 < \delta \leq b - a \leq a + \delta$ for some $\delta > 0$, and

$$\sigma_\alpha = \frac{(\alpha(1-\alpha))^{\frac{1}{2}}}{f(F^{-1}(\alpha))}, \quad 0 < \alpha < 1. \tag{6.7.20}$$

The following regularity conditions will be imposed on the design matrix $\mathbf{C}_n$:

C1. $c_{i1} = 1, \quad i = 1, \ldots, n.$

C2. $\lim_{n\to\infty} \mathbf{D}_n = \mathbf{D}$, where $\mathbf{D}_n = n^{-1}\mathbf{C}_n'\mathbf{C}_n$, and $\mathbf{D}$ is a positive definite $p \times p$ matrix.

C3. $\max_{1\leq i\leq n}\|c_{ni}\| = O(n^{\triangle})$ as $n\to\infty$, where

$$\wedge = \frac{b-a-\delta}{1+2b} \wedge \frac{1}{4}.$$

C4. $n^{-1}\sum_{i=1}^{n}\|c_{ni}\|^4 = O(1)$ as $n\to\infty$.

Finally, we assume that $\mathbf{d}_n = (d_{n1},\ldots,d_{nn})'$ satisfy two conditions:

D1. $\mathbf{C}_n'\mathbf{d}_n = \mathbf{0}$.

D2. $n^{-1}\sum_{i=1}^{n} d_{ni}^2 \to \Gamma^2,\ 0<\gamma^2<\infty; n^{-1}\sum_{i=1}^{n} d_{ni}^4 = O(1)$ and

$$\max_{1\leq i\leq n}|d_{ni}| = O(n^{\triangle}) \text{ as } n\to\infty.$$

Theorem 6.7.1 *Under* $\mathbf{F1}-\mathbf{F2}, \mathbf{C1}-\mathbf{C4}$, *and* $\mathbf{D1}-\mathbf{D3}$,
(1) $\sup\left\{n^{-\frac{1}{2}}|\sum_{j=1}^{n} d_{ni}[\hat{a}_{ni}(\alpha)-\bar{a}_i(\alpha)]|:\ \alpha_n^\star\leq\alpha\leq 1-\alpha_n^\star\right\}$

$$= o_p(n^{-\frac{1}{2}+((b-\delta)/(1+2b))}\log n) \quad as\ n\to\infty, \tag{6.7.21}$$

where

$$\bar{a}_i(\alpha) = I[e_i > F^{-1}(\alpha)],\ \ 0<\alpha<1,\ i=1,\ldots,n. \tag{6.7.22}$$

(2) Let $R_{n1},\ldots,R_{nn}$ *denote the ranks of errors* $e_1,\ldots,e_n$, *and let* $a_n^\star(R_i,\alpha)$, $i=1,\ldots,n$, *be the Hájek scores defined in (6.7.3). Then*

$$\sup\left\{|n^{-\frac{1}{2}}\sum_{j=1}^{n} d_{ni}[\hat{a}_{ni}(\alpha)-a_n^\star(R_i,\alpha)]|:\ \alpha_n^\star\leq\alpha\leq 1-\alpha_n^\star\right\}\xrightarrow{P} 0. \tag{6.7.23}$$

The following theorem extends the uniform approximation of RR process to the whole segment [0,1] under a more stringent condition on the distribution tails.

Theorem 6.7.2 *Assume the conditions* $\mathbf{F1}-\mathbf{F2}, \mathbf{C1}-\mathbf{C4}$ *and* $\mathbf{D1}-\mathbf{D3}$ *with the constants* a, b *and* δ *satisfying the additional restrictions*

$$0<a<b<\frac{1}{2} \quad and \quad 0\leq b-a-\delta<\frac{1}{2}-b. \tag{6.7.24}$$

(1) Then as $n\to\infty$,

$$\sup_{0\leq\alpha\leq 1}|n^{-\frac{1}{2}}\sum_{i=1}^{n} d_{ni}(\hat{a}_{ni}(\alpha)-\bar{a}_i(\alpha)|\xrightarrow{P} 0, \tag{6.7.25}$$

and the process

$$\{\Gamma^{-1}n^{-\frac{1}{2}}\sum_{i=1}^{n} d_{ni}\hat{a}_{ni}(\alpha) : 0 \leq \alpha \leq 1\} \tag{6.7.26}$$

converges in law to the Brownian bridge in the Prokhorov topology on $[0, 1]$.

(2) Moreover as $n \to \infty$,

$$\sup_{0\leq\alpha\leq1} |n^{-\frac{1}{2}}\sum_{i=1}^{n} d_{ni}[\hat{a}_{ni}(\alpha) - a_n^{\star}(R_i, \alpha]| \xrightarrow{P} 0. \tag{6.7.27}$$

The following theorem gives an asymptotic representation of simple linear rank score statistic (6.7.15) for the class of score functions satisfying some Chernoff-Savage-type condition.

Theorem 6.7.3 *Assume that the score-generating function* $\phi : (0, 1) \to \mathbb{R}_1$ *is nondecreasing, square integrable, and such that the derivative* $\phi'(t)$ *exists for* $0 < t < \alpha_0$ *and* $1 - \alpha_0 < t < 1$, $0 < \alpha < \frac{1}{2}$, *In this domain it satisfies*

$$|\phi'(t)| \leq c[t(1-t)]^{-1-\delta^{\star}} \tag{6.7.28}$$

with $c > 0$ *and* $0 < \delta < \delta^{\star}$. *Let* S_{nn} *be the linear RR-statistic corresponding to* ϕ. *Then, under* $\mathbf{F1} - \mathbf{F2}, \mathbf{C1} - \mathbf{C4}$, *and* $\mathbf{D1} - \mathbf{D3}$,

$$S_{nn} = T_n + o_p(1) \quad \textit{as } n \to \infty, \tag{6.7.29}$$

where

$$T_n = n^{-\frac{1}{2}}\sum_{i=1}^{n} d_{ni}\phi(F(e_i)). \tag{6.7.30}$$

Note that Theorem 6.7.3 provides the asymptotic (normal) distribution of S_{nn}. This distribution coincides with that of simple linear rank statistics studied in Section 6.2. We proceed now to present the RR-estimators of $\boldsymbol{\beta}$ whose properties can be studied with the aid of the above theorems.

6.7.2 Asymptotics for RR-Estimators

Recall that for the classical R-estimators of $\boldsymbol{\beta}$, even if we are to estimate a component (e.g., $\boldsymbol{\beta}^{(2)}$ of $\boldsymbol{\beta}$), we need to solve for the entire vector [see (6.6.6)-(6.6.10)]. However, the situation is more flexible with RR-estimators. To illustrate this point, we proceed as in Jurečková (1991) and write

$$\boldsymbol{\beta}' = (\beta_0, \boldsymbol{\beta}^{(1)}, \boldsymbol{\beta}^{(2)\prime}), \tag{6.7.31}$$

where $\boldsymbol{\beta}^{(1)}$ is $(p-q-1)$ -vector, and $\boldsymbol{\beta}^{(2)}$ is q -vector for some $q:\ 1 \leq q \leq p-1$; if we let $q = p-1$, we have $\boldsymbol{\beta}^{(1)} = 0$. For $\mathbf{t} \in \mathbb{R}_q$ we consider the pseudo-observations

$$X_i - \mathbf{t}'\mathbf{c}^{(2)}, \quad i = 1, \ldots, n, \tag{6.7.32}$$

where we let $\mathbf{c}_i = (c_{i1}, \mathbf{c}^{(1)'}, \mathbf{c}^{(2)'})'$, $i = 1, \ldots, n$, partitioned in concordance with (6.7.31). We also partition the design matrix $\mathbf{C}_n$ as $(\mathbf{1}_n, \mathbf{C}_{n1}, \mathbf{C}_{n2})$ where $\mathbf{C}_{nj}$ is a $n \times p_j$ matrix, $p_1 = p - q - 1$, $p_2 = q$, $j = 1, 2$. Thus $\mathbf{c}_i^{(2)'}$ is the ith row of $\mathbf{C}_{n2}$. Let $\hat{a}_{ni}(\alpha, \mathbf{X}_n - \mathbf{C}_{n2}\mathbf{t})$, $i = 1, \ldots, n$, be the regression rank scores derived from (6.7.2) when $\mathbf{X}_n$ and $\mathbf{C}_n$ are replaced by $(\mathbf{X}_n - \mathbf{C}_{n2}\mathbf{t}$ and $(\mathbf{1}_n, \mathbf{C}_{n1})$ respectively. Note that this relates to the model with the parameter $(\beta_0, \boldsymbol{\beta}^{(1)'})'$. Define the $\hat{b}_{ni}$ as in (6.7.16), and denote them as

$$\hat{b}_{ni}(\mathbf{X}_n - \mathbf{C}_{n2}\mathbf{t}), \quad i = 1, \ldots, n. \tag{6.7.33}$$

Note that by virtue of (6.7.2) for every $n \geq 1$,

$$\frac{1}{n}\sum_{i=1}^{n} \hat{b}_{ni}(\mathbf{X}_n - \mathbf{C}_{n2}\mathbf{t}) = \bar{\phi} = \int_0^1 \phi(u)du. \tag{6.7.34}$$

Let then

$$D_n(\mathbf{t}) = \sum_{i=1}^{n}(X_i - \mathbf{t}'\mathbf{c}_i^{(2)})[\hat{b}_{ni}(\mathbf{X}_n - \mathbf{C}_{n2}\mathbf{t}) - \bar{\phi}]. \tag{6.7.35}$$

Then the RR-estimator $\bar{\boldsymbol{\beta}}_n^{(2)}$ of $\boldsymbol{\beta}^{(2)}$, proposed by Jurečková (1991), is defined by

$$\bar{\boldsymbol{\beta}}_n^{(2)} = \text{Arg min}\{D_n(\mathbf{t}) :\ \mathbf{t} \in \mathbb{R}_q\}. \tag{6.7.36}$$

The function $D_n(\mathbf{t})$, $\mathbf{t} \in \mathbb{R}_q$, corresponds to Jaeckel's (1972) measure of dispersion (see Section 6.6) but based on regression rank scores. It is continuous, piecewise linear, and convex in $\mathbf{t} \in \mathbb{R}_q$. It is also differentiable with respect to $\mathbf{t}$ a.e., and the vector of derivatives, whenever it exists, coincides with a particular vector of linear rank scores statistics, as studied in Section 6.7.1. This enables us to use the uniform asymptotic linearity (in $\mathbf{t}$) of such RR statistics (see Jurečková 1992) to derive the asymptotic representation for RR-estimators in the same fashion as in Section 6.6 (dealing with R-estimators of $\boldsymbol{\beta}$). Therefore, omitting such derivations, we arrive at the following theorem.

Theorem 6.7.4 *Under the regularity conditions in Section 6.7.1, whenever*

$$0 < \gamma = -\int_0^1 \phi(\alpha)df(F^{-1}(\alpha)) < \infty,$$

we have for $n \to \infty$,

$$n^{\frac{1}{2}}(\bar{\boldsymbol{\beta}}_n^{(2)} - \boldsymbol{\beta}^{(2)}) = (\gamma\sqrt{n})^{-1}(\mathbf{D}^{(2)})^{-1}\sum_{i=1}^{n}(\mathbf{c}_i^{(2)} - \hat{\mathbf{c}}_i^{(2)})\phi(F(e_i)) + o_p(1), \quad (6.7.37)$$

where

$$\hat{\mathbf{C}}_{n2} = \mathbf{C}_{n1}(\mathbf{C}'_{n1}\mathbf{C}_{n1})^{-1}\mathbf{C}'_{n1}\mathbf{C}_{n2} \quad (6.7.38)$$

is the projection of $\mathbf{C}_{n2}$ on the space spanned over the columns of $\mathbf{C}_{n1}$, $(\hat{\mathbf{c}}_i^{(2)})'$ is the ith row of $\hat{\mathbf{C}}_{n2}$ and $\mathbf{D}^{(2)} = \lim_{n\to\infty}\mathbf{D}_n^{(2)}$, where

$$\mathbf{D}_n^{(2)} = n^{-1}(\mathbf{C}_{n2} - \hat{\mathbf{C}}_{n2})'(\mathbf{C}_{n2} - \hat{\mathbf{C}}_{n2}). \quad (6.7.39)$$

This theorem provides a basis for the study of asymptotic properties of RR-estimators in linear models. In passing, we may observe that for $q = p-1$ (i.e., $\boldsymbol{\beta}^{(1)} = \mathbf{0}$), we have $\mathbf{C}_{n1} = \mathbf{0}$, so $\hat{\mathbf{C}}_{n2} = \mathbf{0}$. Hence, $\mathbf{D}^{(2)}$ reduces to $\mathbf{D}$, defined by **C2**. Looking back at (6.7.37), we note that the $\phi(F(e_i))$ are i.i.d. r.v.'s with 0 mean and finite variance A_ϕ^2, while the $\mathbf{c}_i^{(2)} - \hat{\mathbf{c}}_i^{(2)}$ satisfy the generalized Noether condition. Therefore by the Hájek-Šidák CLT [see (2.5.75)], extended to the multivariate case, we have for $n \to \infty$,

$$n^{\frac{1}{2}}(\bar{\boldsymbol{\beta}}_n^{(2)} - \boldsymbol{\beta}^{(2)}) \xrightarrow{\mathcal{D}} \mathcal{N}_q(\mathbf{0}, \gamma^{-2}A_\phi^2(\mathbf{D}^{(2)})^{-1}); \quad (6.7.40)$$

we leave the proof as Problem 6.7.4. Moreover, as a special case of $q = p - 1$, we have

$$n^{\frac{1}{2}}(\bar{\boldsymbol{\beta}}_n - \boldsymbol{\beta}) \xrightarrow{\mathcal{D}} \mathcal{N}_{p-1}(\mathbf{0}, \gamma^{-2}A_\phi^2\mathbf{D}^{-1}). \quad (6.7.41)$$

In the above development, β_0 is treated as a nuisance parameter; its estimation can be accomplished by the use of (aligned-)signed-ranks, as was discussed in Section 6.6. In Chapters 8 and 9, we will make use of these results in drawing statistical conclusions on the regression parameter $\boldsymbol{\beta}$.

Jurečková and Sen (1993) have shown that under suitable regularity conditions, the R-estimator and RRS-estimator of $\boldsymbol{\beta}^{(2)}$ (or $\boldsymbol{\beta}$), based on a common score function ϕ, are asymptotically equivalent, and that they share the same asymptotic relative efficiency (A.R.E.) properties. Thus the choice between them should be based on mainly on computational and diagnostic aspects.

Let us refer to Theorem 6.6.1, where a first-order asymptotic distributional representation for the R-estimator is given. Under the regularity conditions of this section, incorporating (6.6.39) and considering only q components (for $\boldsymbol{\beta}^{(2)}$), it follows that for the R-estimator $\hat{\boldsymbol{\beta}}_n^{(2)}$ of $\boldsymbol{\beta}^{(2)}$, based on the score function ϕ, we have the same representation as in (6.7.37) (for the RR-estimator). Therefore we obtain as $n \to \infty$,

$$n^{\frac{1}{2}}\|\hat{\boldsymbol{\beta}}_n^{(2)} - \bar{\boldsymbol{\beta}}_n^{(2)}\| \xrightarrow{\boldsymbol{P}} 0, \quad (6.7.42)$$

and this asymptotic equivalence holds for all q ($\leq p-1$) and all subsets of $\boldsymbol{\beta}$.

We may take advantage of computational convenience of R-estimator in the case of low dimension of $\boldsymbol{\beta}^{(2)}$ (e.g., if $q = 1$). While the RR are calculated with the aid of linear programming techniques, the minimization in (6.7.36) is only with respect to a q-dimensional vector.

6.8 STUDENTIZING SCALE STATISTIC AND REGRESSION RANK SCORES

In dealing with studentized M-estimators in linear models in Section 5.5, we were looking for appropriate studentizing scale statistics. Recall that in order to obtain a regression- and scale-equivariant M-estimator, we should studentize it by a regression-invariant and scale-equivariant scale statistic S_n, satisfying

$$S_n(\mathbf{X} + \mathbf{C}\mathbf{b}) = S_n(\mathbf{X}), \quad \mathbf{b} \in I\!R_p$$
$$S_n(a\mathbf{X}) = aS_n(\mathbf{X}), \quad c > 0. \tag{6.8.1}$$

Surprisingly, not many scale-statistics satisfying (6.8.1) are studied in the literature. The robust scale statistics proposed in Huber (1981), Hampel et al. (1986), Croux (1992), among others, are only translation and not regression invariant. The root of the *residual sum of squares*,

$$S_n = \{(\mathbf{X} - \hat{\mathbf{X}})'(\mathbf{X} - \hat{\mathbf{X}})\}^{\frac{1}{2}}, \tag{6.8.2}$$

where $\hat{\mathbf{X}} = \mathbf{C}\hat{\boldsymbol{\beta}}$ with $\hat{\boldsymbol{\beta}}$ being the LSE, satisfies (6.8.1); however, such S_n is closely connected with the normal distribution and is quite nonrobust.

In Section 5.5 we mentioned a class of L-statistics based on regression quantiles. As an alternative we will propose another class of scale statistics, based on *regression rank scores*. Jurečková (1992a) considered such statistics as criterion for estimators based on regression rank scores. These statistics represent an extension of Jaeckel's (1972) rank dispersion of the residuals. However, Jaeckel's dispersion is only translation- invariant and not regression-invariant. Hence it does not satisfy (6.7.31). The proposed statistics are applicable in the studentization and in various other contexts.

We will define the scale statistic in the following way: Select a score function $\phi:\ (0,1) \mapsto I\!R_1$, nondecreasing, nonconstant, square integrable on (0,1) and such that $\phi(\alpha) = -\phi(1-\alpha)$, $0 < \alpha < 1$. Fix a number α_0, $0 < \alpha_0 < 1/2$, and assume, without loss of generality, that ϕ is standardized so that

$$\int_{\alpha_0}^{1-\alpha_0} \phi^2(\alpha)d\alpha = 1. \tag{6.8.3}$$

Calculate the *regression scores* $\hat{\mathbf{b}} = (\hat{b}_{n1}, \ldots, \hat{b}_{nn})'$ generated by ϕ, for the same linear model as in Section 6.7. Then

$$\hat{b}_{ni} = -\int_{\alpha_0}^{1-\alpha_0} \phi(\alpha) d\hat{a}_{ni}(\alpha), \quad i = 1, \ldots, n, \tag{6.8.4}$$

where the $\hat{a}_{ni}(\alpha)$ are obtained as in (6.7.2). The proposed scale statistic S_n has the form

$$S_n = S_n(\mathbf{X}) = n^{-1} \sum_{i=1}^{n} X_i \hat{b}_{ni} = n^{-1} \mathbf{X}' \hat{\mathbf{b}}_n. \tag{6.8.5}$$

S_n is regression-invariant because

$$S_n(\mathbf{X} + \mathbf{Cd}) = n^{-1}(\mathbf{X} + \mathbf{Cd})' \hat{\mathbf{b}}_n(\mathbf{X} + \mathbf{Cd}) = n^{-1} \mathbf{X}' \hat{\mathbf{b}}_n = S_n(\mathbf{X}). \tag{6.8.6}$$

Since $\hat{\mathbf{a}}_n(\alpha)$ and $\hat{\mathbf{b}}_n$ are scale-invariant, S_n is also scale-equivariant. Furthermore, because $n^{-1} \sum_{i=1}^{n} \hat{b}_{ni} = 0$ and S_n is regression-invariant and hence also translation-invariant, (6.8.5) implies that $S_n \geq 0$ and $S_n > 0$ with probability 1.

Under some regularity conditions on F, ϕ, and $\mathbf{C}$, the statistic S_n converges in probability to the functional $S(F) = \int_{\alpha_0}^{1-\alpha_0} \phi(\alpha) F^{-1}(\alpha) d\alpha$, and $n^{1/2}(S_n - S(F)$ is asymptotically normal. The asymptotic behavior is summarized in the following theorem (due to Jurečková and Sen 1994):

Theorem 6.8.1 *Let S_n be the scale statistic defined in (6.7.34) and (6.7.35), generated by a nondecreasing score function ϕ, $\phi(1-\alpha) = -\phi(\alpha)$, $0 < \alpha < 1$, satisfying (6.7.33). Then, under the conditions* **F1**, **C1**, **C2**, *and* **C4**,

$$S_n \xrightarrow{P} S(F) \quad \text{as } n \to \infty, \tag{6.8.7}$$

where

$$S(F) = \int_{\alpha_0}^{1-\alpha_0} \phi(\alpha) F^{-1}(\alpha) d\alpha. \tag{6.8.8}$$

Second, $n^{\frac{1}{2}}(S_n - S(F)) \xrightarrow{\mathcal{D}} \mathcal{N}(0, \sigma^2)$, *where*

$$\begin{aligned} \sigma^2 &= \sigma^2(\phi, F, \alpha_0) \\ &= \int_{F^{-1}(\alpha_0)}^{F^{-1}(1-\alpha_0)} \{F(x \wedge y) - F(x)F(y)\} \phi(F(x)) \phi(F(y)) dx dy \end{aligned} \tag{6.8.9}$$

Finally, S_n admits the following asymptotic representation

$$\begin{aligned} S_n - S(F) &= n^{-1} \sum_{i=1}^{n} \psi(e_i) + o_p(n^{-\frac{1}{2}}) \\ &= \int_{F^{-1}(\alpha_0)}^{F^{-1}(1-\alpha_0)} \phi(F(z)) \{F(z) - F_n(z)\} dz + o_p(n^{-\frac{1}{2}}) \end{aligned} \tag{6.8.10}$$

where

$$\psi(z) = \int_{F^{-1}(\alpha_0)}^{F^{-1}(1-\alpha_0)} \left\{\frac{\alpha - I[F(z) \leq \alpha]}{f(F^{-1}(\alpha))}\right\}\phi(\alpha)d\alpha, \quad z \in \mathbb{R}_1 \tag{6.8.11}$$

and $F_n(z)$ *is the empirical distribution function of* $e_1, \ldots, e_n$.

For the proof, we refer to Theorem 3.1 in Gutenbrunner and Jurečková (1992), who deal with a more general setup.

Remark. The scale statistic S_n is robust in the sense that its influence function, and hence its global sensitivity is bounded, as can be deduced from (6.8.10). We could also admit $\alpha_0 = 0$ in the definition of S_n; then the asymptotic properties would be retained but S_n would not be robust.

6.9 PROBLEMS

6.2.1. Show that for every $i: \ 1 \leq i \leq n$ and $n \geq 1$,

$$a_n^{(0)}(i) = \mathbb{E}\phi(U_{n:i}) = i\binom{n}{i}\int_0^1 \phi(u)u^{i-1}(1-u)^{n-i}du.$$

Hence, or otherwise, show that for every (fixed) $u \in [0,1)$, $a_n^0([nu]) \to \phi(u)$ as $n \to \infty$. Further show that for this pointwise convergence to hold, $\int_0^1 |\phi(u)|du < \infty$ is sufficient but not necessary.

6.2.2. Show that for every $n \geq 1$,

$$n^{-1}\sum_{i=1}^{n}\{a_n(i) - a_n^0(i)\}^2 \leq \int_0^1 \{\phi_n(u) - \phi_n^0(u)\}^2 du$$

$$\leq 2\{\int_0^1 \{\phi_n(u) - \phi(u)\}^2 du + \int_0^1 \{\phi_n^0(u) - \phi(u)\}^2 du\},$$

where by (6.2.7) and (6.2.11), the second term on the right hand side converges to 0 as $n \to \infty$. Hence show that (6.2.17) entails (6.2.12).

6.2.3. Show that for a nondecreasing ϕ,

$$\phi_n(u)\phi(u)du = n^{-1}\sum_{i=1}^{n}\phi(\frac{1}{n+1})\phi(\xi_{ni}),$$

where $\xi_{ni} \in ((i-1)/n, i/n)$, $1 \leq i \leq n$. Show further that for ϕ square integrable, $\phi^2(\frac{n}{n+1})\frac{1}{n} \to 0$ and $n^{-1}\phi(\xi_{nn}) = o(\frac{1}{\sqrt{n}}$, as $n \to \infty$. Hence, or otherwise, verify (6.2.20).

6.2.4. Define $\{P_n\}$ and $\{P_n^\star\}$ as in after (6.2.27), and show that under $\{P_n\}$ as $n \to \infty$,

$$\log(P_n^\star/P_n) \xrightarrow{\mathcal{D}} \mathcal{N}(-\frac{1}{2}t^2 I(f), I(f)).$$

Hence, invoking (2.5.119), show that $\{P_n^\star\}$ is contiguous to $\{P_n\}$.

6.3.1. Show [by repeated integration over $(0,u)$, $u \in [0,1)$] that under $\mathbf{A_2}$, $\phi(\mathbf{u}) \leq \mathbf{c_1} - \mathbf{c_2}\log(\mathbf{1}-\mathbf{u}), \forall \mathbf{u} \in [\mathbf{0},\mathbf{1}]$, where c_1, c_2 are nonnegative constants. Hence use Problem 6.2.1 to verify that

$$\max\{a_n^0(i) : 1 \leq i \leq n\} = a_n^0(n) \leq C + c\log n.$$

6.3.2. Use Theorem 4.2.3 to verify (6.3.29).

6.3.3. Define $Z_n(t)$ and γ_2 as in Theorem 6.3.5, and let

$$W_n(t) = n^{\frac{1}{2}}\{Z_n(t) + t\gamma_2\},\ t \in [-C, C].$$

Show that for (6.3.33) to hold, it suffices that $W_n = \{W_n(t),\ -C \leq t \leq C\}$ converges in law, in the Skorokhod J_1-topology on $[-C, C]$ to a Gaussian function W. Note that by the results in Section 2.5.10, for the above weak convergence result to hold, it suffices to show that: (a) The finite-dimensional distributions of W_n converge to the corresponding ones of W, and (b) $\{W_n\}$ is *tight* or relatively compact.

6.3.4. (Continuation.) Show that for the Problem 6.3.3, W can be expressed as

$$W(t) = tZ^\star,\ \ -C \leq t \leq C,$$

where $Z^\star$ has the normal distribution with 0 mean and variance δ^2 with

$$\delta^2 = \mathrm{Var}\{\phi'(F(|X_1|))f(|X_1|) - \int_{|X_1|}^{\infty} \phi'(F(x))f'(x)dx\}.$$

[Hint. (Hušková and Jurečková, 1983): Use a decomposition of $Z_n(t)$ into 6 terms, each involving an empirical process for which fluctuation bounds were developed earlier.]

6.4.1. Verify (6.4.23) [along the lines of Lemma 6.2.2].

6.4.2. Show that under (6.4.20) the contiguity of $\{P_n^\star\}$ to $\{P_n\}$, as established in the location model in Problem 6.2.4, holds for the simple regression model too.

6.4.3. Show that the inequality in (6.2.42) also holds for the regression model, and hence uniform pointwise convergence implies (6.4.23).

6.4.4. Define $V_n(0)$ as in (6.4.12). Verify the Hájek-Šidák CLT in (2.5.75) for $V_n(0)$.

6.4.5. Extend the result in Lemma 6.2.1 to the simple regression model involving two parameters (s, t).

6.4.6. With the setup following (6.4.33), verify that $\{P_n^\star\}$ is contiguous to $\{P_n\}$.

6.4.7. Provide a proof of Lemma 6.4.2.

6.4.8. Verify the contiguity of $\{P_n^\star\}$ to $\{P_n\}$ in the context of a two-parameter model in simple regression.

6.4.9. For the two-parameter simple regression model, verify that the likelihood function is locally quadratic in the MLE, and hence, maximization of the likelihood leads to the MLE as a linear combination of the two score statistics plus a remainder term that converge to $\mathbf{0}$ at faster rate.

6.4.10. Use Theorem 6.4.6 to verify (6.4.46).

6.5.1. Show that for a bounded and differentiable $\phi(.)$, (6.5.8) holds, uniformly in i, $1 \leq i \leq n$. [Hint: Use the bound for uniform order statistics.]

6.5.2. Show that Problem 6.3.3 also applies to the regression model.

6.5.3. Consider the analogue of Problem 6.3.4 for the regression model.

6.5.4. Provide a formal proof of Theorem 6.5.4.

6.6.1. Verify (6.6.4).

6.6.2. Verify (6.6.11).

6.6.3. Verify (6.6.17). Hence, or otherwise, show that the generalized Noether condition in (6.6.15) ensures that for each coordinate the usual Noether condition holds.

6.6.4. Verify (6.6.27).

6.6.5. Verify (6.6.29).

6.6.5. Verify that (6.4.50) holds for linear model as well.

6.7.1. Provide a formal proof of Theorem 6.7.1.

6.7.2. Provide a formal proof of Theorem 6.7.2.

6.7.3. Verify the representation result in (6.7.29).

6.7.4. Verify (6.7.40).

6.8.1. Provide a proof of Theorem 6.7.3 using Theorem 3.1 of Gutenbrunner and Jurečková (1992).

Chapter 7

Asymptotic Interrelations of Estimators

7.1 INTRODUCTION

We have seen in the preceding chapters that a parameter can be estimated in various ways. This begs a natural question, how to compare the competing estimators. In the finite sample case, such a comparison of estimators is based on their *risks*, which are computed by reference to a suitable *loss function*. The choice of loss function is a basic concept in this respect, but there may not be a unique choice which can be judged the most appropriate one. Usually there are some technical difficulties in an exact computation of such a risk, especially when the sample size is not so small. Fortunately, in an asymptotic setup, by allowing the sample size to increase indefinitely, it is often possible to induce various approximations and/or simplifications by which either the limit of a risk or the value of risk from the asymptotic distribution can be obtained in a closed form; this renders a convenient way of (asymptotic) comparisons of competing estimators. While we could consider various loss criteria and their risk counterparts, asymptotic normality considerations usually lead to the choice of squared error loss, and thus to the *mean square error* (MSE) for a real-valued estimator and to *quadratic risks* in the multiparameter case.

Let $X_1, X_2, \ldots$ be a sequence of i.i.d. observations with a d.f. $F(x-\theta)$, $\theta \in \Theta$, and let there be two sequences $\{T_{1n}\}$ and $\{T_{2n}\}$ of estimators of the function $g(\theta)$. If $n^{\frac{1}{2}}(T_{in} - g(\theta)) \xrightarrow{\mathcal{D}} \mathcal{N}(0, \sigma_i^2)$ as $n \to \infty$, $i = 1, 2$, then the asymptotic relative efficiency of $\{T_{2n}\}$ with respect to $\{T_{1n}\}$ is formulated as

$$e_{2,1} = \sigma_1^2/\sigma_2^2. \tag{7.1.1}$$

If, alternatively, $\{T_{2n'}\}$ is based on n' observations, then $n^{\frac{1}{2}}(T_{2n'} - g(\theta))$ has the same asymptotic distribution $\mathcal{N}(0, \sigma_1^2)$ as $n^{\frac{1}{2}}(T_{1n} - g(\theta))$ if and only if $n' = n'(n)$ is so chosen that $\lim_{n\to\infty}[n/n'(n)]$ exists and

$$\lim_{n\to\infty}[n/n'(n)] = e^{\star}_{2,1} = \sigma_1^2/\sigma_2^2 = e_{2,1}. \tag{7.1.2}$$

The equality $e_{2,1} = 1$ means that $\{T_{1n}\}$ and $\{T_{2n}\}$ are equally asymptotically efficient. We could continue with a more refined comparison based on the *deficiency* of $\{T_{2n}\}$ with respective to $\{T_{1n}\}$: Assume that

$$\mathbb{E}[n(T_{in} - g(\theta)^2] = \tau^2 + \frac{a_i}{n} + \mathrm{o}(n^{-1}), \quad i = 1, 2; \tag{7.1.3}$$

then the deficiency of $\{T_{2n}\}$ with respect to $\{T_{1n}\}$ is defined as

$$d = \frac{a_2 - a_1}{\tau^2}. \tag{7.1.4}$$

On the other hand, d is also equal to $\lim_{n\to\infty} d(n) = \lim_{n\to\infty}[n'(n) - n]$, where $n'(n)$ is chosen in such a way that $\mathbb{E}[n(T_{2n'} - g(\theta))^2]$ coincides with $\mathbb{E}[n(T_{1n} - g(\theta))^2]$ up to terms of order n^{-1}. Similar considerations apply also to a general (asymptotic) risk replacing the (asymptotic) mean square error.

We often want not only to compare two or more estimators, but also to study whether there may exist closer relations of their functional and/or other characteristics. This may be possible with the aid of asymptotic representations derived in the preceding chapters, using the asymptotic representations of the difference $n^{\frac{1}{2}}(T_{2n} - T_{1n})$. If the leading terms in the respective representations of $n^{\frac{1}{2}}(T_{1n} - g(\theta))$ and $n^{\frac{1}{2}}(T_{2n} - g(\theta))$ differ from each other, then $n^{\frac{1}{2}}(T_{2n} - T_{1n})$ has a nondegenerate asymptotic normal distribution. However, it may be more interesting to enquire that whether this asymptotic distribution is degenerate and under what conditions? This would mean that $\{T_{1n}\}$ and $\{T_{2n}\}$ are asymptotically equivalent; the order of this equivalence coincides with the order of the remainder terms of the respective representations. We can speak about the *equivalence of* $\{T_{1n}\}$ *and* $\{T_{2n}\}$ *of the first order* if we are able to prove

$$n^{\frac{1}{2}}(T_{2n} - T_{1n}) = \mathrm{o}_p(1) \text{ as } n \to \infty, \tag{7.1.5}$$

while we speak about the asymptotic equivalence of the second order if we know the exact order of the right-hand side of (7.1.5). Typically, if $\|n^{\frac{1}{2}}(T_{2n} - T_{1n})\| \xrightarrow{P} 0$ as $n \to \infty$, then the order of this distance is either $\mathrm{O}_p(n^{-\frac{1}{2}})$ or $\mathrm{O}_p(n^{-\frac{1}{4}})$, respectively, depending on the smoothness of the score functions generating the estimators. Moreover, after an appropriate standardization (i.e., being multiplied by a positive power of n), $n^{\frac{1}{2}}(T_{2n} - T_{1n})$ may again have a nondegenerate asymptotic (nonnormal) distribution; this provides an additional information on the interrelation of the sequences of estimators.

The asymptotic interrelations of estimators will be studied in the present chapter; although we will concentrate on robust estimators; analogous results hold in the parametric setup. Considering, for instance, the classes of

M-, R-, and L-estimators (of location or regression parameters), we know that either ofthese classes contains an asymptotically efficient element for a given distribution. One naturally expects that the optimal elements are asymptotically equivalent; if it is the case, then what is the order of the equivalence? If we have, for instance, an M-estimator of θ, what is its asymptotically equivalent R-estimation counterpart, if any? We will consider such and similar questions.

The approach based on asymptotic representations was used by Ibragimov and Hasminskii (1970, 1971, 1972, 1979, sec. 3.1 and 3.2), who proved an asymptotic equivalence of Bayesian and maximum likelihood estimators in a locally asymptotic normal family of distributions. Jaeckel (1971) established a close relation of M-, L-, and R-estimators of location. The asymptotic relations of M-, L-, and R-estimators in the location and in the linear regression models were studied in Jurečková (1977, 1983a, b, 1985, 1986), Hušková and Jurečková (1981, 1985), Rivest (1982) and by van Eeden (1983), among others. Hanousek (1988, 1990) established the asymptotic equivalence of M-estimators to P-estimators (Pitman-type estimators) and to B-estimator (Bayes-type estimators) in the location model.

For many estimators, defined implicitly, we may construct one-step or k- step versions starting with an initial estimator. If the initial estimator is $\sqrt{n}$-consistent then already the one-step version is asymptotically equivalent to the noniterative estimator, and the order of this equivalence is increasing with k. Approximations of estimators by their one- or k-step versions were studied by Bickel (1973), Janssen, Jurečková, and Veraverbeke (1987), Jurečková (1983, 1986), Jurečková and Portnoy (1987), Jurečková and Welsh (1990), Malý (1991), among others; the effect of the initial estimator on the quality of the one-step version was studied by Jurečková and Sen (1990).

7.2 ASYMPTOTIC INTERRELATIONS OF LOCATION ESTIMATORS

Let $X_1, X_2, \ldots, X_n$ be independent observations with a d.f. $F(x - \theta)$, where F is an unknown member of a family $\mathcal{F}$ of symmetric distribution functions and θ is the parameter to be estimated. Among various estimators of θ, the M-, L-, and R-estimators play the most important role. Either of these three types is defined with the aid of some score (or weight) function that determines its robustness and efficiency properties. Selecting these score functions properly, we may obtain asymptotically equivalent estimators. However, we should emphasize that the relation between the score functions, leading to asymptotically equivalent estimators, depend on the underlying distribution function F which is considered as unknown. Hence we cannot calculate one estimator numerically once we have a value of another estimator. However, with the knowledge of such interrelations, we may select the estimator that best fits the specific situation. The asymptotic equivalence may also carry some specific features from one type of estimator to another. Each of these classes has some advantages as well as disadvantages: M-estimators have attractive minimax properties, but they are generally not scale-equivariant and have to be supplemented by an additional estimator of scale, or they have to be studentized, which may violate their minimax properties. L-estimators are computationally appealing, mainly in the location model. R-estimators retain the advantages and disadvantages of the rank tests on which they are based.

The close relation of M-, L-, and R-estimators of location was first noticed by Jaeckel (1971) who found sufficient conditions for their asymptotic equivalence. These results were then extended to the linear model and to the second-order asymptotic equivalence by the other authors referred to in Section 7.1. Let us present some of the most interesting results of this kind.

7.2.1 Asymptotic Relations of Location M- and L-Estimators

Let $X_1, X_2, \ldots$ be independent random variables, identically distributed according to d.f. $F(x-\theta)$ such that $F(x)+F(-x)=1,\ x \in I\!R_1$; let $X_{n:1} \leq \ldots X_{n:n}$ be the order statistics corresponding to $X_1, \ldots, X_n$. Let M_n be the M-estimator of θ generated by a nondecreasing step-function ψ of the form

$$\psi(x) = \alpha_j, \quad \text{for} \quad s_j < x < s_{j+1}, \quad j = 1, \ldots, k, \tag{7.2.1}$$

where $-\infty = s_0 < s_1 < \ldots < s_k < s_{k+1} = \infty$, $-\infty < \alpha_0 \leq \alpha_1 \leq \ldots \leq \alpha_k < \infty$, $\alpha_j = -\alpha_{k-j+1}$, $s_j = -s_{k-j+1}$, $j = 1, \ldots, k$ (at least two α's different). It means that M_n is a solution of the minimization

$$\sum_{i=1}^{n} \rho(X_i - t) := \min \text{ with respect to } t \in I\!R_1, \tag{7.2.2}$$

where ρ is a continuous convex symmetric piecewise linear function with the derivative $\rho' = \psi$. The first theorem gives an L-estimator counterpart of M_n, which will be a linear combination of several quantiles.

Theorem 7.2.1 *Let M_n be an M-estimator of θ defined in (7.2.1)-(7.2.2). Assume that d.f. F has two bounded derivatives f and f', and that f is positive in neighborhoods of $s_1, \ldots, s_k$. Then*

$$M_n - L_n = O_p(n^{-\frac{3}{4}}) \tag{7.2.3}$$

where L_n is the L-estimator, $L_n = \sum_{i=1}^{n} a_j X_{n:[np_j]}$ with

$$p_j = F(s_j),\ a_j = \gamma^{-1}(\alpha_j - \alpha_{j-1}) f(s_j),\ j = 1, \ldots, k \tag{7.2.4}$$

and

$$\gamma = \sum_{j=1}^{k} (\alpha_j - \alpha_{j-1}) f(s_j)\ (> 0). \tag{7.2.5}$$

Proof. By Theorem 5.3.2, M_n admits the asymptotic representation (5.3.28). On the other hand, let F_n be the empirical d.f. corresponding to $X_1, \ldots, X_n$:

$$F_n(x) = n^{-1} \sum_{i=1}^{n} I[X_i \leq x], \quad x \in \mathbb{R}_1, \tag{7.2.6}$$

and let $Q_n(t)$ be the corresponding empirical quantile function, $0 < t < 1$:

$$Q_n(t) = \begin{cases} X_{n:i} & \text{if } (i-1)/n < t < i/n, \\ 0 & \text{if } t = 0; \quad i = 1, \ldots, n. \end{cases} \tag{7.2.7}$$

We may set $\theta = 0$ without loss of generality. Using the representation (5.3.27) and (7.2.3)-(7.2.7), we obtain that

$$\begin{aligned} n^{\frac{1}{2}}(M_n - L_n) &= n^{\frac{1}{2}}\gamma^{-1} \sum_{j=1}^{k} (\alpha_j - \alpha_{j-1})[F_n(s_j) - F(s_j)] \\ &\quad + f(s_j)(Q_n(F(s_j)) - s_j) + \mathrm{O}_p(n^{-\frac{1}{4}}) \end{aligned} \tag{7.2.8}$$

and this is of order $\mathrm{O}_p(n^{-\frac{1}{4}})$ by Kiefer (1967). □

Remark. The condition that the first derivative of F is positive and finite in a neighborhood of each of $s_1, \ldots, s_k$ is crucial for (7.2.3). In a nonregular case where there are singular points at which the above assumption does not hold, M- and $L-$ estimators are not asymptotically equivalent any more, and they may have diffeent rates of consistency. The behavior of estimators in such nonregular cases was studied by Akahira (1975a,b) Akahira and Takeuchi (1981), Chanda (1975), de Haan and Taconis-Haantjes (1979), Ghosh and Sukhatme (1981), Ibragimov and Hasminskii (1979) and Jurečková (1983c) among others.

We will now consider the asymptotic interrelation of two important estimators: Huber's M-estimator and the trimmed mean. There has been some confusion in the history of this problem: One might intuitively expect that the Winsorized rather than the trimmed mean resembles Huber's estimator (see Huber 1964). Bickel (1965) was apparently the first who recognized the close connection between Huber's estimator and the trimmed mean. Jaeckel (1971) found the L-estimator counterpart to the given M-estimator such that the difference between the two is $\mathrm{O}_p(n^{-1})$, as $n \to \infty$, provided that the function ψ generating the M-estimator (and hence the function J generating the L-estimator) is sufficiently smooth. However, Huber's ψ-function is not smooth enough to be covered by Jaeckel's proof. This important special case, situated on the border of smooth and nonsmooth cases and yet attaining the order $\mathrm{O}_p(n^{-1})$, needs to be proved in a more delicate way.

Theorem 7.2.2 *Let M_n be the Huber M-estimator of θ generated by the ψ-function*

$$\psi(x) = \begin{cases} x & \text{if } |x| \le c \\ c\,\mathrm{sign}x & \text{if } |x| > c, \end{cases} \tag{7.2.9}$$

for some $c > 0$, and let L_n be the α-trimmed mean,

$$L_n = \frac{1}{n - 2[n\alpha]} \sum_{i=[n\alpha]+1}^{n-[n\alpha]} X_{n:i}. \tag{7.2.10}$$

Assume the following:
1. $c = F^{-1}(1-\alpha)$.
2. F has absolutely continuous symmetric density f and positive and finite Fisher information,

$$0 < I(f) = \int \left(\frac{f'(x)}{f(x)}\right)^2 dF(x) < \infty. \tag{7.2.11}$$

3. $f(x) > a > 0$ for all x satisfying

$$\alpha - \epsilon \le F(x) \le 1 - \alpha + \epsilon, \;\; 0 < \alpha < \frac{1}{2}, \;\; \epsilon > 0. \tag{7.2.12}$$

4. $f'(x)$ exists in interval $(F^{-1}(\alpha - \epsilon),\; F^{-1}(1 - \alpha + \epsilon))$.
Then

$$L_n - M_n = O_p(n^{-1}) \quad \text{as } n \to \infty. \tag{7.2.13}$$

Proof. We may put $\theta = 0$ without loss of generality. It follows from Theorem 5.3.1 that

$$\begin{aligned} M_n &= (1-2\alpha)^{-1}\Big\{n^{-1}\sum_{i=1}^{n} X_{n:i} I[F^{-1}(\alpha) \le X_{n:i} \le F^{-1}(1-\alpha)] \\ &+ n^{-1}F^{-1}(\alpha)\sum_{i=1}^{n} I[X_{n:i} \le F^{-1}(\alpha)] \\ &+ n^{-1}F^{-1}(1-\alpha)\sum_{i=1}^{n} I[X_{n:i} > F^{-1}(1-\alpha)]\Big\} + O_p(n^{-1}) \\ &= (1-2\alpha)^{-1}\Big\{n^{-1}\sum_{i=i_n}^{j_n} X_{n:i} + F^{-1}(1-\alpha)\Big[1 - F_n(F^{-1}(1-\alpha)) \\ &\quad -F_n(F^{-1}(\alpha)-)\Big]\Big\} + O_p(n^{-1}), \end{aligned} \tag{7.2.14}$$

where

$$i_n = nF_n(F^{-1}(\alpha)-) + 1, \;\; j_n = nF_n(F^{-1}(1-\alpha)). \tag{7.2.15}$$

By (7.2.15),

$$X_{n:i} = X_{n:n-[n\alpha]} + O_p(n^{-\frac{1}{2}}) \tag{7.2.16}$$

for every integer i between $n - [n\alpha]$ and j_n; analogously

$$X_{n:i} = X_{n:[n\alpha]+1} + O_p(n^{-\frac{1}{2}}) \tag{7.2.17}$$

for every integer i between $[n\alpha]+1$ and i_n. Hence, combining (7.2.14)-(7.2.17), we obtain

$$\begin{aligned}
&(1-2\alpha)(M_n - L_n)\\
&= F^{-1}(1-\alpha)\Big[1 - F_n(F^{-1}(1-\alpha)) - F_n(F^{-1}(\alpha)-)\Big]\\
&\quad -\Big[X_{n:n-[n\alpha]+1} + O_p(n^{-\frac{1}{2}})\Big]\Big[F_n(F^{-1}(\alpha)-) - \alpha + O_p(n^{-1})\Big]\\
&\quad +\Big[X_{n:n-[n\alpha]} + O_p(n^{-\frac{1}{2}})\Big]\Big[F_n(F^{-1}(1-\alpha)) - (1-\alpha) + O_p(n^{-1})\Big]\\
&\quad + O_p(n^{-1}),
\end{aligned} \tag{7.2.18}$$

and this, by Bahadur's representation of sample quantiles, studied in Chapter 4, is equal to

$$\begin{aligned}
& F^{-1}(1-\alpha)[1 - F_n(F^{-1}(1-\alpha)) - F_n(F^{-1}(\alpha)-)]\\
- & \{F^{-1}(\alpha) + [f(F^{-1}(\alpha))]^{-1}[\alpha - F_n(F^{-1}(\alpha)-] + O_p(n^{-\frac{1}{2}})\}\\
& \quad \cdot\{F_n(F^{-1}(\alpha)-) - \alpha + O_p(n^{-\frac{1}{2}})\}\\
+ & \{F^{-1}(1-\alpha) + [f(F^{-1}(\alpha))]^{-1}[1 - \alpha - F_n(F^{-1}(1-\alpha)]\\
+ & O_p(n^{-\frac{1}{2}})\}\{F_n(F^{-1}(1-\alpha)) - (1-\alpha) + O_p(n^{-1})\} + O_p(n^{-1})\\
= & O_p(n^{-1}).
\end{aligned} \tag{7.2.19}$$

This completes the proof. □

Combining Theorems 4.2.1 and 7.2.2, we immediately obtain the M-estimator counterpart of the α-Winsorized mean

$$L_n = n^{-1}\Big\{[n\alpha]X_{n:[n\alpha]} + \sum_{i=[n\alpha]+1}^{n-[n\alpha]} X_{n:ii} + [n\alpha]X_{n:n-[n\alpha]+1}\Big\}. \tag{7.2.20}$$

COROLLARY. *Under the conditions of Theorem 7.2.2, defining L_n, the α-Winsorized mean as in (7.2.20), we have*

$$L_n - M_n = O_p(n^{-\frac{3}{4}}) \tag{7.2.21}$$

where M_n is the M-estimator generated by the function

$$\psi(x) = \begin{cases} F^{-1}(\alpha) - [\alpha/f(F^{-1}(\alpha))] & \text{if } x < F^{-1}(\alpha),\\ x & \text{if } F^{-1}(\alpha) \le x \le F^{-1}(1-\alpha),\\ F^{-1}(1-\alpha) + [\alpha/f(F^{-1}(\alpha))] & \text{if } x > F^{-1}(1-\alpha). \end{cases} \tag{7.2.22}$$

Let us now consider general L-estimator of the form

$$L_n = \sum_{i=1}^{n} c_{ni} X_{n:i}, \tag{7.2.23}$$

where the coefficients c_{ni}, $i = 1, \ldots, n$, are generated by a function $J : (0,1) \mapsto \mathbb{R}_1$, $J(u) = J(1-u)$, $0 < u < 1$, according to (4.3.2) or (4.3.3), respectively. We will find an M-estimator counterpart M_n to L_n in two situations: (1) J trimmed, that is, $J(u) = 0$ for $0 \leq u < \alpha$ and $1 - \alpha < u \leq 1$ and (2) J possibly untrimmed but under more restrictive conditions on F. In either case the corresponding M-estimator is generated by the following continuous and skew-symmetric ψ-function:

$$\psi(x) = -\int_{\mathbb{R}_1} (I[y \geq x] - F(y)) J(F(y)) dy, \quad x \in \mathbb{R}_1. \tag{7.2.24}$$

Theorem 7.2.3 *For an L-estimator L_n in (7.2.23) suppose that the c_{ni}, generated by a function $J : (0,1) \mapsto \mathbb{R}_1$, satisfy*
1.

$$J(u) = J(1-u),\ 0 < u < 1, \quad \int_0^1 J(u) du = 1. \tag{7.2.25}$$

2. $J(u) = 0$ for $u \leq \alpha$ and $u \geq 1 - \alpha$ for some $\alpha \in (0, 1/2)$, J is continuous on (0,1) up to a finite number of points $s_1, \ldots, s_m$, $\alpha < s_1 < \ldots < s_m < 1-\alpha$ and J is Lipschitz of order $\nu \leq 1$ in intervals (α, s_1), $(s_1, s_2), \ldots (s_m, 1-\alpha)$.
3. F is absolutely continuous with symmetric density f and $F^{-1}(u) = \inf\{x : F(x) \geq u\}$ satisfies the Lipschitz condition of order 1 in neighborhoods of $s_1, \ldots, s_m$.
4.

$$\int_{-A}^{A} f^2(x) dx < \infty \text{ where } A = F^{-1}(1 - \alpha + \epsilon),\ \epsilon > 0. \tag{7.2.26}$$

Then

$$L_n - M_n = O_p(n^{-\nu}) \tag{7.2.27}$$

where M_n is the M-estimator generated by the function (7.2.24).

Proof. Assume that $\theta = 0$ without loss of generality. ψ-function (7.2.24) satisfies the condition **A1-A4** of Section 5.3 for trimmed J satisfying 1-2 and for F satisfying 3-4. Moreover the derivative $h'(t)$ of the function h in (5.3.3) is decreasing at $t = 0$. Hence (7.2.27) follows by combining the asymptotic representation (4.3.4) of L_n (see Theorem 4.3.1) with the asymptotic representation (5.3.9) of M_n (Theorem 5.3.1). □

For an untrimmed J, more restrictive conditions on J and F are needed; better results will be obtained for the c_{ni} in (4.3.3):

$$c_{ni} = \int_{(i-1)/n}^{i/n} J(u) du, \quad i = 1, \ldots, n. \tag{7.2.28}$$

Theorem 7.2.4 *Let L_n be an L-estimator (7.2.23) with the coefficients c_{ni} defined in (7.2.28) and the following conditions:*
1. $J(u)$ is continuous up to a finite number of points $s_1, \ldots, s_m$, $0 < s_1 < \ldots < s_m < 1$ and satisfies the Lipschitz condition of order $\nu > 0$ in each of the intervals $(0, s_1)$, $(s_1, s_2), \ldots (s_m, 1)$;
2. $F^{-1}(s) = inf\{x : F(x) \geq s\}$ satisfies the Lipschitz condition of order 1 in a neighborhood of $s_1, \ldots, s_m$.
3. F is absolutely continuous with symmetric density f, and

$$\sup\{|x|^{\beta} F(x)(1 - F(x)) : x \in \mathbb{R}_1\} < \infty \tag{7.2.29}$$

for some $\beta > 2/(\nu + 1 - \triangle)$, $0 < \triangle < 1$.
4. J and F satisfy

$$\int_{\mathbb{R}_1} J^2(F(x)) f(x + t) dx \leq K_1, \tag{7.2.30}$$

and

$$\int_{\mathbb{R}_1} |J'(F(x))| f(x) f(x + t) dx \leq K_2, \tag{7.2.31}$$

for $|t| \leq \delta$, $K_1, K_2, \delta > 0$. then

$$L_n - M_n = \mathrm{O}_p(n^{-r}) \quad as \quad n \to \infty, \tag{7.2.32}$$

where M_n is the M-estimator generated by the ψ-function of (7.2.24) and

$$r = \min\{(\nu + 1)/2, 1\}. \tag{7.2.33}$$

Proof. The proposition follows from the representation of L_n and M_n derived in Theorems 4.3.1 and 5.3.1; the assumptions 1 - 4 of Theorem 7.2.4 imply the conditions of the respective theorems (see Problem 7.2.1). □

L-estimators in the location model are both location and scale equivariant, while M-estimators are generally only location equivariant. This shortcoming of M-estimators can be removed by a studentization using an appropriate scale statistic. Studentized M-estimators are studied in detail in Section 5.4, where their asymptotic representations are also derived. Hence starting with an L-estimator L_n with the weight function J, we are naturally interested in its M-estimator counterpart. Recall that the studentized M-estimator of θ is defined as a solution of the minimization (5.4.1) with the scale statistics $S_n = S_n(X_1, \ldots, X_n)$ being translation invariant and scale equivariant in the sense of (5.4.2). If $\psi = \rho'$ is continuous, then M_n is also a root of the equation

$$\sum_{i=1}^{n} \psi((X_i - t)/S_n) = 0,\ t \in \mathbb{R}_1. \tag{7.2.34}$$

It is assumed [see (5.4.3)] that S_n is $\sqrt{n}$-consistent estimator of some positive functional $S(F)$. Then typically the studentized M-estimator M_n, asymptotically equivalent to L_n, will be a root of (7.2.34) with

$$\psi(x) = -\int_{\mathbb{R}_1} (I[y \geq xS(F)] - F(y))J(F(y))dy, \; x \in \mathbb{R}_1. \tag{7.2.35}$$

The question is which conditions are needed for this asymptotic equivalence? The following theorem gives sufficient conditions for the asymptotic equivalence of a trimmed L-estimator and its studentized M-estimation counterpart.

Theorem 7.2.5 *Let L_n be the L-estimator generated by the function J; assume conditions 1 - 3 of Theorem 7.2.3 and condition 4 with A replaced by $A = a + bF^{-1}(1-\alpha)$ and with $a \in \mathbb{R}_1$, $b > 0$, hold. Assume further that there exists a scale statistic S_n satisfying (5.4.2) and (5.4.3). Then*

$$L_n - M_n = O_p(n^{-\nu}) \tag{7.2.36}$$

where M_n is the M-estimator generated by ψ of (7.2.35) and studentized by S_n.

Proof. The result follows from the asymptotic representations of L_n and M_n derived in Theorems 4.3.1 and 5.4.1, respectively, whose conditions are easily verified in the trimmed case. □

Finally, consider possibly untrimmed L-estimator L_n and a scale statistic S_n satisfying (5.4.2) and (5.4.3). We expect that L_n is asymptotically equivalent to an M-estimator M_n generated by ψ of (7.2.35). The following theorem gives one set of sufficient conditions for this asymptotic equivalence.

Theorem 7.2.6 *Let L_n be an L-estimator of (7.2.23) with the coefficients c_{ni} defined in (7.2.29), generated by J satisfying (7.2.25) and conditions 1 and 2 of Theorem 7.2.4; further assume that F satisfies condition 3 of Theorem 7.2.4 and that together J and F satisfy the following conditions:*

$$\int_{\mathbb{R}_1} x^2 J^2(F(x)) f(xe^u + t)dx \leq K_1, \tag{7.2.37}$$

$$\int_{\mathbb{R}_1} |J'(F(x))| f(x) f(xe^u + t)dx \leq K_2, \tag{7.2.38}$$

$$\int_{\mathbb{R}_1} x^2 |J'(F(x))| f(x) f(xe^u + t)dx \leq K_3, \tag{7.2.39}$$

for $|t|$, $|u| \leq \delta$, $K_1, K_2, K_3, \delta > 0$. Assume that there exists a scale statistic S_n satisfying (5.4.2) and (5.4.3). Then

$$L_n - M_n = O_p(n^{-r}), \tag{7.2.40}$$

with r given in (7.2.33) where M_n is the estimator generated by ψ of (7.2.35) and studentized by S_n.

Proof. The proposition follows from the asymptotic representation of L_n and M_n given in Theorems 4.3.1 and 5.4.1. □

7.2.2 Asymptotic Relations of Location M- and R-Estimators

Let $\{X_i;\ i \geq 1\}$ be a sequence of independent observations from a population with a d.f. $F(x-\theta)$, where $F(x)+F(-x) = 1$, $x \in I\!R_1$, and F has an absolutely continuous density f and finite and positive Fisher's information $I(f)$. Let $\phi:\ [0,1) \mapsto I\!R^+$ be a nondecreasing score function such that

$$\phi(0) = 0 \text{ and } \int_0^1 \phi^2(u)du < \infty, \tag{7.2.41}$$

and assume that γ, defined in (6.2.4), is nonzero. Let R_{ni}^+ denote the rank of $|X_i - t|$ among $|X_1 - t|, \ldots, |X_n - t|$, $i = 1, \ldots, n$, $t \in I\!R_1$, and consider the linear signed-rank statistic

$$S_n(\mathbf{X}_n - t\mathbf{1}_n) = n^{-1/2}\sum_{i=1}^{n} \operatorname{sign}(X_i - t)a_n(R_{ni}^+(t)) \tag{7.2.42}$$

with the scores $a_n(1) \leq \ldots a_n(n)$ satisfying (6.2.16) and (6.2.17). The R-estimator T_n of θ was defined in (3.4.8)-(3.4.9) as

$$T_n = (T_n^- + T_n^+)/2 \tag{7.2.43}$$

where

$$T_n^- = \sup\{t:\ S_n(t) > 0\},\ T_n^+ = \inf\{t:\ S_n(t) < 0\}. \tag{7.2.44}$$

The following theorem gives an M-estimator counterpart that is first order asymptotically equivalent to T_n.

Theorem 7.2.7 *Let $\{X_i:\ i \geq 1\}$ be a sequence of i.i.d. r.v.'s with a d.f. $F(x-\theta)$ where F has an absolutely continuous, symmetric density f and positive and finite Fisher's information $I(f)$. Let T_n be an R-estimator of θ defined in (7.2.43) and (7.2.44), generated by a nondecreasing function ϕ : $[0,1) \mapsto I\!R^+$ satisfying (7.2.41). Assume that*

$$0 < \gamma = -\int_{I\!R_1} \phi^\star(F(x))f'(x)dx < \infty, \tag{7.2.45}$$

where

$$\phi^\star(u) = \begin{cases} \phi(2u-1), & 1/2 \le u < 1, \\ -\phi^\star(1-u), & 0 < u \le 1/2. \end{cases} \tag{7.2.46}$$

Then

$$n^{\frac{1}{2}}(T_n - M_n) = o_p(1) \quad \text{as } n \to \infty, \tag{7.2.47}$$

where M_n is the M-estimator of θ generated by the ψ-function

$$\psi(x) = c\phi^\star(F(x)), \quad \forall x \in \mathbb{R}_1, \text{ for some } c > 0. \tag{7.2.48}$$

Proof. By Theorem 6.2.6,

$$\begin{aligned} n^{\frac{1}{2}}(T_n - \theta) &= (n\gamma)^{-1} \sum_{i=1}^{n} \phi^\star(F(X_i - \theta)) + o_p(1) \\ &= (n\gamma)^{-1} \sum_{i=1}^{n} \psi(X_i - \theta) + o_p(1). \end{aligned} \tag{7.2.49}$$

Hence the theorem follows from Theorems 6.2.6 and 5.3.2. □

Let us keep the notation of the last theorem. Taking Theorems 6.3.6 and 5.3.2, respectively, into account, we obtain the second-order asymptotic relations of R-estimator and M-estimator under more refined conditions on F and on the score-generating functions. Namely we could formulate the results in the following theorem.

Theorem 7.2.8 *Let $\{X_i;\ i \ge 1\}$ be i.i.d. r.v.'s with a d.f. $F(x-\theta)$, $x \in \mathbb{R}_1$, $\theta \in \mathbb{R}_1$, such that $F(x) + F(-x) = 1$, $x \in \mathbb{R}_1$. Let T_n be an R-estimator of θ. Assume that the score-generating function ϕ is nondecreasing, square integrable on (0,1) and satisfies (7.2.41), and that both ϕ and F satisfy (7.2.45). Let M_n be an estimator of θ generated by the ψ-function (7.2.48).*
1. Under the conditions $\mathbf{A}^\star, \mathbf{B1}$, *and* $\mathbf{B2}$ *of Theorem 6.3.6,*

$$T_n - M_n = O_p(n^{-\frac{3}{4}}) \quad \text{as } n \to \infty; \tag{7.2.50}$$

2. Under $\mathbf{A_1}$ *and* $\mathbf{B3}$ *of Theorem 6.3.6,*

$$T_n - M_n = O_p(n^{-1} \log n). \tag{7.2.51}$$

3. Under $\mathbf{A_q}$ *with $0 < q < 2$ and* $\mathbf{B3}$,

$$T_n - M_n = O_p(n^{-1}). \tag{7.2.52}$$

We relegate the proof of this theorem to Problem 7.2.5.

7.2.3 Asymptotic Relations of Location L- and R-Estimators

The asymptotic interrelations of L- and R-estimators of location follow from the results in Sections 7.2.1 and 7.2.2. Indeed, given an L-estimator generated by a symmetric function J, the M-estimation counterpart is generated by the function ψ in (7.2.24). On the other hand, the R-estimation counterpart is generated by the score-generating function $\phi(u) = \phi^\star((u+1)/2)$, $0 < u < 1$, where $\phi^\star$ is determined by the relation $\phi^\star(u) = \psi(F^{-1}(u))$, $0 < u < 1$. Hence we get the following relation between J and $\phi^\star$:

$$\phi^\star(u) = -\int_0^1 (I[v \geq u] - v) J(v) dv, \quad 0 < u < 1. \tag{7.2.53}$$

The respective L-, M-, and R-estimators, generated by this triple of functions, are then asymptotically equivalent to each other, but the conditions under which this asymptotic equivalence holds need to be formulated, combining the above results. We leave these conditions as Problem 7.2.6 and rather make several observations concerning the asymptotic interrelations.

As mentioned earlier, such asymptotic relations do not enable us to calculate the estimators numerically from each other because the relations between the pertaining weight functions involve the unknown F. The relations are rather between the classes of estimators than between the individual members. On the other hand, the knowledge of these relations gives us a better picture of the related classes of estimators. For instance, the Wilcoxon score R-estimator has the score function $\phi^\star(u) = u - 1/2$, $0 < u < 1$. Hence the related class of M-estimators is generated by the functions $\psi(x) = F(x) - 1/2$, $x \in \mathbb{R}_1, F \in \mathcal{F}$. Further the corresponding class of L-estimators is generated by functions $f(F^{-1}(u)/\int_{\mathbb{R}} f^2(x)dx)$, $0 < u < 1$, $F \in \mathcal{F}$. These estimators have bounded influence functions, and it makes no sense to winsorize $\phi^\star$ and in this way obtain the trimmed Hodges-Lehmann estimator, as some authors recommend.

Another example is the normal scores R-estimator generated by $\phi^\star(u) = \Phi^{-1}(u)$ [Φ being the N(0,1) d.f.], $0 < u < 1$. Then

$$\psi(x) = \Phi^{-1}(F(x)), \; x \in \mathbb{R}_1 \tag{7.2.54}$$

and

$$J(u) = \left\{\frac{f(F^{-1}(u))}{\Phi'(\Phi^{-1}(u))}\right\}\left\{\int_{\mathbb{R}_1} f(F^{-1}(\Phi(x)))dx\right\}^{-1}, \; 0 < u < 1. \tag{7.2.55}$$

Such estimators are considered as nonrobust because they have unbounded influence functions. However, looking at (7.2.54) and (7.2.55), we see that the normal scores estimator is sensitive to the tails of the underlying (unknown) distribution. Indeed, considering the Parzen (1979) classification of tails, we see that $J(0) = J(1) = 0$ for a heavy-tailed F, $J(0) = J(1) = 1$ for an exponentially tailed F and $\lim_{u \to 0,1} J(u) = \infty$ for a short tailed F.

R-estimation counterpart of the α-trimmed mean corresponds to the score function

$$\phi^{\star}(u) = \begin{cases} F^{-1}(\alpha), & 0 < u < F^{-1}(\alpha), \\ F^{-1}(u), & F^{-1}(\alpha) \leq u \leq F^{-1}(1-\alpha), \\ F^{-1}(1-\alpha), & F^{-1}(1-\alpha) < u < 1. \end{cases} \tag{7.2.56}$$

Taking into account the Huber M-estimator and its L- and R-estimator counterparts, we come to a conclusion that not only the class of M-estimators but also L-estimators and R-estimators contain an element, asymptotically minimax over a family of contaminated normal distributions (see Problems 7.2.7 - 7.2.8). More about the asymptotic interrelations and their consequences may be found in Jurečková (1989).

7.3 ASYMPTOTIC RELATIONS IN THE LINEAR MODEL

7.3.1 M- and R-Estimators

Using the asymptotic representations derived in Chapters 5 and 6, we may derive the asymptotic interrelations of M- and R-estimators in the linear regression model. Consider the linear model

$$\mathbf{Y} = \mathbf{X}\boldsymbol{\beta} + \mathbf{E}, \tag{7.3.1}$$

where $\mathbf{Y} = (Y_1, \ldots, Y_n)'$ is the vector of observations, $\mathbf{X} = \mathbf{X}_n$ is a known design matrix of order $n \times p$, $\boldsymbol{\beta} = (\beta_1, \ldots, \beta_p)'$ is a parameter and $\mathbf{E} = (E_1, \ldots, E_n)'$ is vector of i.i.d. errors with a d.f. F. M-estimator of $\boldsymbol{\beta}$ is defined as a solution of the minimization (5.5.2) with the function ρ. The asymptotic representation of $\mathbf{M}_n$ was derived in Section 5.5 in the form

$$\mathbf{M}_n - \boldsymbol{\beta} = \gamma_M^{-1} \mathbf{Q}_n^{-1} \sum_{i=1}^{n} \mathbf{x}_i \psi(E_i) + o_p(||\mathbf{Q}_n^{-1/2}||), \tag{7.3.2}$$

with

$$\gamma_M = \int_{\boldsymbol{R}} f(x) d\psi(x) \tag{7.3.3}$$

and

$$\mathbf{Q}_n = \sum_{i=1}^{n} \mathbf{x}_i \mathbf{x}_i'. \tag{7.3.4}$$

The R-estimator $\mathbf{T}_n$ of $\boldsymbol{\beta}$ is defined as a solution of the minimization of the measure of *rank dispersion* [see (6.6.9)]

$$D_n(\mathbf{b}) = \sum_{i=1}^{n}(Y_i - \mathbf{x}_i'\mathbf{b})[a_n(R_{ni}(\mathbf{b})) - \bar{a}_n], \tag{7.3.5}$$

where $R_{ni}(\mathbf{b})$ is the rank of the residual $Y_i - \mathbf{x}_i\mathbf{b}$ among $Y_1 - \mathbf{x}_1'\mathbf{b}, \ldots, Y_n - \mathbf{x}_n'\mathbf{b}$, $\mathbf{b} \in \mathbb{R}_p$, $i = 1, \ldots, n$; $a_n(1), \ldots, a_n(n)$ are scores generated by a nondecreasing, square-integrable score function $\phi : (0,1) \mapsto \mathbb{R}_1$ and $\bar{a}_n = n^{-1}\sum_{i=1}^{n} a_n(i)$. Notice that $D_n(\mathbf{b})$ is invariant to the translation; hence the above procedure is not able to estimate the intercept, which should be estimated by signed-rank statistics. Thus, for simplicity of comparison, we will consider the model without an intercept:

$$\sum_{i=1}^{n} x_{ij} = 0, \; j = 1, \ldots, p. \tag{7.3.6}$$

Under the regularity conditions of Section 6.6, $\mathbf{T}_n$ admits the asymptotic representation

$$\mathbf{T}_n - \boldsymbol{\beta} = \gamma_R^{-1}\sum_{i=1}^{n}\mathbf{Q}_n^{-1}\mathbf{x}_i\phi(F(E_i)) + o_p(||\mathbf{Q}_n^{-1/2}||) \tag{7.3.7}$$

with γ_R defined in (7.2.45).

Combining the representations (7.3.2) and (7.3.7), and regarding their regularity conditions, we obtain the following (first order) asymptotic relations of $\mathbf{M}_n$ and $\mathbf{T}_n$.

Theorem 7.3.1 *1. Assume that the errors E_i in the model (7.3.1) are i.i.d. r.v.'s with a d.f. F which has an absolutely continuous density f with a positive and finite Fisher's information $I(f)$.*
2. Let the design matrix $\mathbf{X}_n$ satisfy (7.3.6) and the generalized Noether condition

$$\lim_{n\to\infty}\max_{1\le i\le n}\mathbf{x}_i'\mathbf{Q}_n^{-1}\mathbf{x}_i = 0. \tag{7.3.8}$$

3. Let $\mathbf{T}_n$ be the R-estimator of $\boldsymbol{\beta}$ defined through the minimization (7.3.5), generated by the score function $\phi : (0,1) \mapsto \mathbb{R}_1$, nonconstant, nondecreasing and square integrable and such that

$$\phi(1-u) = -\phi(u), \;\; 0 < u < 1. \tag{7.3.9}$$

Then

$$\mathbf{M}_n - \mathbf{T}_n = o_p(||\mathbf{Q}_n^{-1/2}||), \tag{7.3.10}$$

where $\mathbf{M}_n$ is the M-estimator of $\boldsymbol{\beta}$ generated by the function

$$\psi(x) = a + b\phi(F(x)), \;\; \forall x \in \mathbb{R}_1, \tag{7.3.11}$$

with $a \in \mathbb{R}_1$, $b > 0$ being arbitrary numbers.

Remark. In Section 6.5, on the simple linear regression model (i.e., the model of regression line), we derived the second-order asymptotic representations of R-estimator of slope parameter. These more accurate asymptotic representations, when combined with those for M-estimators, should lead to the second-order asymptotic relations of M- and R-estimators in the linear model. We leave their derivation as Problem 7.3.2.

7.3.2 Asymptotic Relations of Huber's M-Estimator and Trimmed LSE in the Linear Model

The trimmed least squares estimator (TLSE), introduced in (4.7.12), is a straightforward extension of the trimmed mean to the linear regression model. Noting the close relation of the class of trimmed means and that of Huber's M-estimators of location, we expect an analogous phenomenon in the linear regression model. Actually the relation of the trimmed LSE and of Huber's estimator is quite analogous; the only difference may appear in the first components when F is asymmetric because the population functionals then do not coincide, and both estimators may be biased in the first components. This bias appears in the asymmetric location model too.

Consider the model (7.3.1), but unlike in the last section, assume that the model contains an intercept:

$$x_{i1} = 1, \ i = 1, \ldots, n. \tag{7.3.12}$$

The (α_1, α_2)-trimmed LSE $\mathbf{T}_n(\alpha_1, \alpha_2)$ of $\boldsymbol{\beta}$ is defined in (4.7.12); it can also be described as a solution of the minimization

$$\sum_{i=1}^{n} a_i (Y_i - \mathbf{x}_i \mathbf{t})^2 := \min \quad \text{with respect to } \mathbf{t} \in I\!R_p, \tag{7.3.13}$$

and the a_i defined as in (4.7.11).

The following theorem provides one set of regularity conditions leading to the asymptotic equivalence of trimmed LSE and of the Huber estimator of order $O_p(n^{-1})$, up to a bias that may appear in the first components in case of asymmetric F.

Theorem 7.3.2 *In the linear model (7.3.1) assume the following*
1. $\mathbf{X}_n$ *satisfies (7.3.12) and the following two*

$$(a) \qquad \lim_{n\to\infty} n^{-1}\mathbf{X}_n'\mathbf{X}_n = \mathbf{Q} \quad \textit{is p.d.}, \tag{7.3.14}$$

$$(b) \qquad \max\{|x_{ij}| : 1 \le i \le n; \ 1 \le j \le p\} = O(1) \ \textit{as} \ n \to \infty. \tag{7.3.15}$$

2. F *is absolutely continuous with density* f *such that* $0 < f(x) < \infty$ *for* $F^{-1}(\alpha_1) - \epsilon \le x \le F^{-1}(\alpha_2) + \epsilon$ *with some* $\epsilon > 0$ *and that* f *has a bounded*

derivative f' in a neighborhood of $F^{-1}(\alpha_1)$ and $F^{-1}(\alpha_2)$.
Then

$$\|\mathbf{T}_n(\alpha_1, \alpha_2) - \mathbf{M}_n - \mathbf{e}_1\eta(\alpha_2 - \alpha_1)^{-1}\| = O_p(n^{-1}), \tag{7.3.16}$$

where

$$\eta = (1 - \alpha_2)F^{-1}(\alpha_2) + \alpha_1 F^{-1}(\alpha_1), \;\; \mathbf{e}_1 = (1, 0, \ldots, 0)' \in \mathbb{R}_p, \tag{7.3.17}$$

and $\mathbf{M}_n$ is the M-estimator generated by the ψ-function

$$\psi(x) = \begin{cases} F^{-1}(\alpha_1), & x < F^{-1}(\alpha_1), \\ x, & F^{-1}(\alpha_1) \le x \le F^{-1}(\alpha_2), \\ F^{-1}(\alpha_2), & x > F^{-1}(\alpha_2). \end{cases} \tag{7.3.18}$$

Proof. Theorem follows from the asymptotic representations of $T_n(\alpha_1, \alpha_2)$ (Theorem 4.7.2) and of M_n (Theorem 5.5.1). □

The following corollary concerns the symmetric case.

COROLLARY. *If F is symmetric around 0 and $\alpha_1 = \alpha$, $\alpha_2 = 1 - \alpha$, $0 < \alpha < 1/2$, then under the conditions of Theorem 7.3.2,*

$$\|\mathbf{T}_n(\alpha) - \mathbf{M}_n\| = O_p(n^{-1}). \tag{7.3.19}$$

7.4 APPROXIMATION BY ONE-STEP VERSIONS

Many estimators, like maximum likelihood, M- and R-estimators, are defined implicitly as a solution of a minimization problem or of a system of equations. Sometimes it may be difficult to obtain this solution algebraically, and there may exist more solutions among which one is consistent and/or efficient, and so on. We have already touched on this problem in the context of M-estimators generated by a nonconvex ρ-function. In the case of continuously differentiable ρ, there exists a root of the pertaining system of equations that is a $\sqrt{n}$-consistent estimator of the parameter; however, there is no criterion distinguishing the right one among the multiple roots.

The asymptotic equivalence of an M-estimator M_n with a one-step Newton-Raphson adjustment starting with an appropriate consistent initial estimate is well known in the context of likelihood estimation with a sufficiently smooth kernel (e.g., Lehmann 1983, sec.6.3). This smoothness precludes the standard Huber and Hampel kernels. Moreover, unlike in the MLE case, we can still estimate some functional of the unknown distribution function. Bickel (1975) introduced the one-step version of the M-estimator M_n in the linear model, which is asymptotically equivalent to M_n in probability as $n \to \infty$. Jurečková (1983) and Jurečková and Portnoy (1987) studied the order of approximation of M_n by its one-step version in the linear model; the latter paper considered also the effect of the initial estimator on the resulting breakdown point. Welsh (1987) constructed a one-step version of L-estimator in the linear model; its relation to the one-step M-estimator was studied by Jurečková and Welsh (1990). Janssen, Jurečková, and Veraverbeke (1985) studied the order of an approximation of M-estimator of a general parameter by its one- and two-step versions. Jurečková and Sen (1990) studied the effect of the initial estimator on the second order asymptotics of the one-step version of M-estimator of general parameter. Malý (1991) studied the orders of approximations of studentized M-estimators of location by their one-step versions.

In the context of M-estimators, we generally observe an interesting phenomenon relating to multiple (k-step) iterations: If the ρ-function is sufficiently smooth then the order of approximation of M_n by its k-step version increases rapidly with k [more precisely, it is $O_p(n^{-(k-1)/2})$]. However, if $\psi = \rho'$ has jump discontinuities, this order is only $O_p(n^{-1-2^{-k-1}})$ (see Malý 1991).

7.4.1 One-step Version of General M-Estimator

Let $X_1, X_2, \ldots$ be a sequence of i.i.d. r.v.'s with a d.f. $F(x, \theta_0)$, where $\theta_0 \in \Theta$, an open interval in $\mathbb{R}_1$. Let $\rho: \mathbb{R}_1 \times \Theta \mapsto \mathbb{R}_1$ be a function, absolutely continuous in θ with the derivative $\psi(x, \theta) = \partial\rho(x, \theta)/\partial\theta$. We assume that $\mathbb{E}_{\theta_0}\rho(X_1, \theta)$ exists for all $\theta \in \Theta$ and has a minimum at $\theta = \theta_0$.

The M-estimator M_n of θ_0 is defined as a solution of the minimization (5.2.2). Let us first consider the case that the function $\psi(x, \theta)$ is also absolutely continuous in θ. Then M_n is a solution of the equation

$$\sum_{i=1}^{n} \psi(X_i, t) = 0. \tag{7.4.1}$$

By Theorem 5.2.2, there exists a sequence $\{M_n\}$ of roots of (7.4.1) satisfying $n^{\frac{1}{2}}(M_n - \theta_0) = O_p(1)$ as $n \to \infty$, which admits the asymptotic representation

$$M_n = \theta_0 - (n\gamma_1(\theta_0))^{-1} \sum_{i=1}^{n} \psi(X_i, \theta_0) + R_n, \tag{7.4.2}$$

where $R_n = O_p(n^{-1})$ as $n \to \infty$,

$$\gamma_1(\theta) = \mathbb{E}_\theta \dot{\psi}(X_1, \theta), \quad \text{and } \dot{\psi}(x, \theta) = (\partial/\partial\theta)\psi(x, \theta) \tag{7.4.3}$$

Often, it is difficult to find an explicit and consistent solution of (7.4.1). Therefore a standard technique is to look at an iterative solution of this equation. Starting with an initial consistent estimator $M_n^{(0)}$, we may consider the successive estimators $\{M_n^{(k)}; k \geq 1\}$ defined recursively by $M_n^{(k)} = M_n^{(k-1)}$ if $\gamma_n^{(k-1)} = 0$, and if $\gamma_n^{(k-1)} \neq 0$, as

$$M_n^{(k)} = M_n^{(k-1)} - (n\hat{\gamma}_n^{(k-1)})^{-1} \sum_{i=1}^{n} \psi(X_i, M_n^{(k-1)}), \qquad (7.4.4)$$

for $k = 1, 2, \ldots$, where we take

$$\hat{\gamma}_n^{(k)} = n^{-1} \sum_{i=1}^{n} \dot{\psi}(X_i, M_n^{(k)}), \quad k = 0, 1, 2, \ldots. \qquad (7.4.5)$$

We will show that, under some regularity conditions,

$$n(M_n^{(1)} - M_n) = \mathrm{O}_p(1), \quad n(M_n^{(k)} - M_n) = \mathrm{o}_p(1), \ k \geq 2 \qquad (7.4.6)$$

provided that

$$\sqrt{n}(M_n^{(0)} - \theta_0) = \mathrm{O}_p(1), \quad \text{as } n \to \infty. \qquad (7.4.7)$$

Together with (7.4.2) this implies that

$$M_n^{(1)} = \theta_0 - (n\gamma_1(\theta_0))^{-1} \sum_{i=1}^{n} \psi(X_i, \theta_0) + R_n^{(1)}, \qquad (7.4.8)$$

where

$$R_n^{(1)} = \mathrm{O}_p(n^{-1}) \quad \text{as } n \to \infty. \qquad (7.4.9)$$

A closer look at the behavior of the remainder $R_n^{(1)}$ reveals the role of the initial estimator $M_n^{(0)}$.

The *second order distributional representation* (SOADR) of M_n, which is the asymptotic representation (7.4.2) supplemented with the asymptotic (nonnormal) distribution of nR_n is given by Theorem 5.2.2. We will now derive the asymptotic distribution of $nR_n^{(1)}$, the SOADR for the one-step version. Its most important corollary will be the asymptotic distribution of $n(M_n^{(1)} - M_n)$, a properly standardized difference of M_n and its one-step version. Whenever this asymptotic distribution is degenerate for some initial $M_n^{(0)}$, we can obtain an approximation of M_n of order $\mathrm{o}_p(n^{-1})$.

We will confine ourselves to the situations where the initial estimator $M_n^{(0)}$ itself is asymptotically linear in the sense that it admits the asymptotic representation

$$M_n^{(0)} = \theta_0 + n^{-1} \sum_{i=1}^{n} \phi(X_i, \theta_0) + \mathrm{o}_p(n^{-1/2}) \qquad (7.4.10)$$

with a suitable function $\phi(x, \theta)$ on $\mathbb{R}_1 \times \Theta$ such that

$$0 < \mathbb{E}_\theta \phi^2(X_1, \theta_0) < \infty \quad \text{in neighborhood of } \theta_0. \tag{7.4.11}$$

We will impose the following conditions on F and ϕ:

1.

$$h(\theta) = \mathbb{E}_{\theta_0} \psi(X_1, \theta) \text{ exists } \forall \theta \in \Theta \tag{7.4.12}$$

and has a unique root at $\theta = \theta_0$.

2. $\dot{\psi}(x, \theta)$ is absolutely continuous in θ, and there exist $\delta > 0$, $K_1, K_2 > 0$, such that

$$\mathbb{E}_{\theta_0} |\dot{\psi}(X_1, \theta_0 + t)|^2 \leq K_1, \quad \mathbb{E}_{\theta_0} |\ddot{\psi}(X_1, \theta_0 + t)|^2 \leq K_2 \ \forall |t| \leq \delta, \tag{7.4.13}$$

where

$$\dot{\psi}(x, \theta) = \frac{\partial}{\partial \theta} \psi(x, \theta), \quad \ddot{\psi}(x, \theta) = \frac{\partial}{\partial \theta} \dot{\psi}(x, \theta).$$

3. $\gamma_1(\theta_0)$ as defined in (7.4.3) is nonzero and finite.

4.

$$0 < \mathbb{E}_{\theta_0} \psi^2(X_1, \theta) < \infty \quad \text{in a neighborhood of } \theta_0. \tag{7.4.14}$$

5. There exist $\alpha > 0$, $\delta > 0$, and a function $H(x, \theta_0)$ such that $\mathbb{E}_{\theta_0} H(X_1, \theta_0) < \infty$ and

$$|\ddot{\psi}(x, \theta_0 + t) - \ddot{\psi}(x, \theta_0)| \leq |t|^\alpha H(x, \theta_0) \quad \text{a.e. } [F(x, \theta_0)] \tag{7.4.15}$$

for $|t| \leq \delta$. Let

$$U_{n1} = n^{1/2}\Big(n^{-1} \sum_{i=1}^{n} \dot{\psi}(X_1, \theta_0) - \gamma_1(\theta_0)\Big), \tag{7.4.16}$$

$$U_{n2} = n^{-1/2} \sum_{i=1}^{n} \psi(X_i, \theta_0), \tag{7.4.17}$$

$$U_{n3} = n^{-1/2} \sum_{i=1}^{n} \phi(X_i, \theta_0), \tag{7.4.18}$$

and let $\mathbf{U}_n = (U_{n1}, U_{n2}, U_{n3})'$. Under the above conditions $\mathbf{U}_n$ is asymptotically normally distributed,

$$\mathbf{U}_n \xrightarrow{\mathcal{D}} \mathcal{N}_3(\mathbf{0}, \mathbf{S}), \tag{7.4.19}$$

where $\mathbf{S}$ is a (3×3) matrix with the elements

$$\begin{aligned} s_{11} &= \mathrm{Var}_{\theta_0} \dot{\psi}(X_1, \theta_0), \quad s_{22} = \mathbb{E}_{\theta_0} \psi^2(X_1, \theta_0), \\ s_{33} &= \mathbb{E}_{\theta_0} \phi^2(X_1, \theta_0), \quad s_{12} = \mathrm{Cov}_{\theta_0}(\dot{\psi}(X_1, \theta_0), \psi(X_1, \theta_0)), \\ s_{13} &= \mathrm{Cov}_{\theta_0}(\dot{\psi}(X_1, \theta_0), \phi(X_1, \theta_0)) \quad s_{23} = \mathrm{Cov}_{\theta_0}(\psi(X_1, \theta_0), \phi(X_1, \theta_0)) \\ s_{21} &= s_{12}, \quad s_{31} = s_{13} \quad s_{32} = s_{23}. \end{aligned} \tag{7.4.20}$$

Theorem 7.4.1 *Assume that $\rho(x,\theta)$ and $\psi(x,\theta) = \partial\rho(x,\theta)/\partial\theta$ are absolutely continuous in θ and the above conditions 1 - 4 are satisfied. Let $M_n^{(1)}$ be the one-step estimator defined in (7.4.4) and (7.4.5) with $M_n^{(0)}$ satisfying (7.4.10). Then $M_n^{(1)}$ admits the asymptotic representation (7.4.8) and*

$$nR_n^{(1)} \xrightarrow{\mathcal{D}} U^\star \ \text{as } n \to \infty, \tag{7.4.21}$$

where

$$U^\star = \gamma_1^{-2} U_2 \Big(U_1 - \frac{U_2\gamma_2}{2\gamma_1} \Big) + \gamma_2 (2\gamma_1^3)^{-1} (U_2 + \gamma_1 U_3)^2 \tag{7.4.22}$$

with

$$\gamma_1 = \gamma_1(\theta_0) \ \text{and} \ \gamma_2 = \gamma_2(\theta_0) = \mathbb{E}_{\theta_0} \ddot{\psi}(X_1, \theta_0). \tag{7.4.23}$$

Notice that, by Theorem 5.2.2, the distribution of the first term on the right-hand side of (7.4.22) coincides with the asymptotic distribution of nR_n, where R_n is the remainder term in (7.4.2). Hence the effect of the initial estimate $M_n^{(0)}$ appears only in the second term of the right-hand side of (7.4.22). In this way we get the following corollary:

COROLLARY. *Under the conditions of Theorem 7.4.1,*

$$n(M_n^{(1)} - M_n) \xrightarrow{\mathcal{D}} \frac{\gamma_2}{2\gamma_1^3} (U_2 + \gamma_1 U_3)^2 \ \text{as } n \to \infty. \tag{7.4.24}$$

Consequently

$$M_n^{(1)} - M_n = o_p(n^{-1}) \tag{7.4.25}$$

if and only if either

$$\phi(x,\theta) = -\gamma_1^{-1} \psi(x,\theta) \ \forall x \in \mathbb{R}_1, \ \theta \in \Theta, \tag{7.4.26}$$

for the function ϕ in the representation (7.4.10) or

$$\gamma_2(\theta_0) = \mathbb{E}_{\theta_0} \ddot{\psi}(X_1, \theta_0) = 0. \tag{7.4.27}$$

Remarks. 1. The formula (7.4.25) tells us that M_n and $M_n^{(1)}$ are asymptotically equivalent up to the order n^{-1} if the initial estimator $M_n^{(0)}$ has the same influence function as M_n.
2. The asymptotic distribution of $n(M_n^{(1)} - M_n)$ is the central χ^2 with 1 d.f., up to the multiplicative factor $\sigma^2\gamma_2/(2\gamma_1^3)$, where

$$\begin{aligned} \sigma^2 &= \mathbb{E}(U_2 + \gamma_1 U_3)^2 \\ &= \mathbb{E}\psi^2(X_1,\theta_0) + \gamma_1^2 \mathbb{E}\phi^2(X_1,\theta_0) \\ &+ 2\gamma_1 \mathbb{E}(\psi(X_1,\theta_0)\phi(X_1,\theta_0)). \end{aligned} \tag{7.4.28}$$

The asymptotic distribution is confined to the positive or negative part of $\mathbb{R}_1$ depending on whether or not γ_2/γ_1 is positive.
3. The asymptotic relative efficiency of $M_n^{(1)}$ to M_n is equal to 1. On the other hand, the second moment of the asymptotic distribution of $n(M_n^{(1)} - M_n)$ in (7.2.24) may be considered as a measure of deficiency of $M_n^{(1)}$ with respect to M_n; hence

$$d(M_n^{(1)}, M_n) = \frac{3}{16}\sigma^4(\gamma_2\gamma_1^{-3})^2 \tag{7.4.29}$$

with σ defined by (7.4.28).
4.If $k \geq 2$, then

$$M_n^{(k)} - M_n = o_p(n^{-1}). \tag{7.4.30}$$

5. In the location model $\psi(x,t) = \psi(x-t)$ and $F(x,\theta) = F(x-\theta)$. In the symmetric case when $F(x) + F(-x) = 1$, $\psi(-x) = -\psi(x)$, $x \in \mathbb{R}_1$ is $\gamma_2 = 0$, and hence $M_n^{(1)} - M_n = o_p(n^{-1})$.
6. Unless $M_n^{(0)}$ and M_n have the same influence functions, the ratio of the second moment of the limiting distribution of $n^{1/2}(M_n^{(0)} - M_n)$ and of the first absolute moment of that of $n(M_n^{(1)} - M_n)$ is equal to $\gamma_2/(2\gamma_1)$ and hence is independent of the choice of $M_n^{(0)}$.

Proof of Theorem 7.4.1. By Theorem 5.2.2 and by virtue of the conditions, we have

$$\begin{aligned} n^{-1}\sum_{i=1}^{n}\ddot{\psi}(X_i,\theta_0) &= \gamma_2 + o_p(1), && (7.4.31)\\ n^{1/2}(\hat{\gamma}_n - \gamma_1) &= U_{n1} + \gamma_2 U_{n3} + o_p(1), && (7.4.32)\\ n^{1/2}(\hat{\gamma}_n^{-1} - \gamma_1^{-1}) &= -\gamma_1^{-2}U_{n1} + \gamma_2\gamma_1^{-2}U_{n3} + o_p(1), && (7.4.33)\\ n^{-1/2}\sum_{i=1}^{n}\psi(X_i, M_n^{(0)}) &= U_{n2} + \gamma_1 U_{n3} + O_p(n^{-1/2}). && (7.4.34) \end{aligned}$$

Hence,

$$nR_n^{(1)} = \gamma_1^{-2}\Big\{\frac{1}{2}\gamma_1\gamma_2 U_{n3}^2 + \gamma_2 U_{n2}U_{n3} + U_{n1}U_{n2}\Big\} + o_p(1), \tag{7.4.35}$$

and this gives the desired result. □

7.4.2 One-Step Version of M-Estimator of Location

The location model can be considered as a special case of the general linear model. Theorem 7.4.1 provides the orders of approximations of estimators by their one-step versions also in the location model, at least under some conditions. We will leave the precise formulation of these results to the reader as an exercise and rather turn our attention to the cases not covered by the general model, namely to the studentized estimators and to less smooth ρ-functions.

Let $X_1, \ldots, X_n$ be independent observations with a common d.f. $F(x-\theta)$. The studentized M-estimator M_n of θ was defined in Section 5.4 as a solution of the minimization (5.4.1) with the studentizing scale statistic S_n satisfying the conditions (5.4.2) and (5.4.3). Theorem 5.4.1 provided the asymptotic representation of M_n which can be written in a unified form as

$$M_n = \theta + (n\gamma_1)^{-1} \sum_{i=1}^{n} \psi\Big(\frac{X_i-\theta}{S}\Big) - \frac{\gamma_2}{\gamma_1}\Big(\frac{S_n}{S}-1\Big) + \mathrm{O}_p(n^{-1/2-x}), \quad (7.4.36)$$

where γ_1 and γ_2 are suitable functionals of F, dependent on $S = S(F)$ but not on θ and $0 < x \leq 1/2$; specific values of x in Theorem 5.4.1 were $x = 1/4$ or $x = 1/2$ according to the smoothness of $\psi = \rho'$.

On the other hand, in Chapter 5 we proved the asymptotic linearity of the empirical process (Lemma 5.4.1)

$$Z_n(t,u) = n^{-1/2} \sum_{i=1}^{n} \Big\{\psi\Big(\frac{X_i - \theta - n^{-\frac{1}{2}}t}{Se^{n^{-\frac{1}{2}}u}}\Big) - \psi\Big(\frac{X_i-\theta}{S}\Big)\Big\}. \quad (7.4.37)$$

More precisely, for any $C > 0$, as $n \to \infty$,

$$\sup\Big\{|Z_n(t,u) + t\gamma_1 + u\gamma_2| : \ |t|, |u| \leq C\Big\} = \mathrm{O}_p(n^{-x}). \quad (7.4.38)$$

Inserting $t \to n^{1/2}(M_n^{(0)} - \theta)$ $[= \mathrm{O}_p(1)]$ and $u \to n^{1/2}(Z_n - Z)$, $Z_n = \ln S_n, Z = \ln S$ in (7.4.38), where $M_n^{(0)}$ is an initial estimator of θ, we express M_n as

$$M_n = M_n^{(0)} + (n\gamma_1)^{-1} \sum_{i=1}^{n} \psi\Big(\frac{X_i - M_n^{(0)}}{S_n}\Big) + \mathrm{O}_p(n^{-1/2-x}). \quad (7.4.39)$$

Hence we can define the one-step version of the studentized M-estimator M_n as

$$M_n^{(1)} = \begin{cases} M_n^{(0)} + (n\hat{\gamma}_1^{(1)})^{-1} \sum_{i=1}^{n} \psi\big(\frac{X_i - M_n^{(0)}}{S_n}\big), & \hat{\gamma}_1^{(1)} \neq 0, \\ M_n^{(0)}, & \hat{\gamma}_1^{(1)} = 0, \end{cases} \quad (7.4.40)$$

where $\hat{\gamma}_1^{(1)}$ is a consistent estimator of γ_1. The order of approximation of M_n by its one-step version (7.4.40) is given in the following theorem:

Theorem 7.4.2 *Assume the conditions of Theorem 5.4.1, and let $M_n^{(1)}$ be the one-step version of M_n with an estimator $\hat{\gamma}_1^{(1)}$ of $\gamma_1 > 0$ such that*

$$\hat{\gamma}_1^{(1)} - \gamma_1 = \mathrm{O}_p(n^{-x}), \quad (7.4.41)$$

where $x = 1/4$ or $1/2$ under the conditions of parts 1 or 2) of Theorem 5.4.1, respectively. Then

$$M_n - M_n^{(1)} = \mathrm{O}_p(n^{-x-1/2}) \quad as\ n \to \infty. \tag{7.4.42}$$

Remark. If ψ' exists and is continuous, we may take

$$\hat{\gamma}_1^{(1)} = (nS_n)^{-1} \sum_{i=1}^{n} \psi'((X_i - M_n^{(0)})/S_n). \tag{7.4.43}$$

Another possible choice is (for some $-\infty < t_1 < t_2 < \infty$)

$$\begin{aligned} \hat{\gamma}_1^{(1)} &= n^{-1/2}(t_2 - t_1)^{-1} \sum_{i=1}^{n} \Big\{ \psi\Big(\frac{X_i - M_n^{(0)} + n^{-1/2}t_2}{S_n}\Big) \\ &\quad - \psi\Big(\frac{X_i - M_n^{(0)} + n^{-1/2}t_1}{S_n}\Big) \Big\}. \end{aligned} \tag{7.4.44}$$

Actually (7.4.41) applies to (7.4.44) as well (Problem 7.4.3).

Proof of Theorem 7.4.2. Since $\gamma_1 > 0$, we have

$$(\gamma_1/\hat{\gamma}_1)^{-1} = \mathrm{O}_p(n^{-x}) \text{ as } n \to \infty. \tag{7.4.45}$$

Combining (7.4.36) and (7.4.39), we obtain

$$\begin{aligned} n^{1/2}(M_n - M_n^{(1)}) &= -(\gamma_1/\hat{\gamma}_1^{(1)})(n^{1/2}\gamma_1)^{-1} \sum_{i=1}^{n} [\psi((X_i - M_n^{(0)})/S_n) \\ &- \psi((X_i - \theta)/S)] + n^{1/2}(M_n^{(0)} - \theta) + (\gamma_2/\gamma_1)n^{1/2}((S_n/S) - 1)) \\ &- ((\gamma_1/\hat{\gamma}_1^{(1)}) - 1)\Big\{(n^{1/2}\gamma_1)^{-1} \sum_{i=1}^{n} \psi((X_i - \theta)/S) - n^{1/2}(M_n^{(0)} - \theta) \\ &- (\gamma_2/\gamma_1)n^{1/2}((S_n/S) - 1)\Big\} + \mathrm{O}_p(n^{-x}) \\ &= \mathrm{O}_p(n^{-x}). \end{aligned} \tag{7.4.46}$$

□

Similarly, we could define the k-step version of M_n ($k \geq 1$),

$$M_n^{(k)} = \begin{cases} M_n^{(k-1)} + (n\hat{\gamma}_1^{(k)})^{-1} \sum_{i=1}^{n} \psi\big(\frac{X_i - M_n^{(k-1)}}{S_n}\big), & \text{if } \hat{\gamma}_1^{(k)} \neq 0, \\ M_n^{(k-1)} & \text{if } \hat{\gamma}_1^{(k)} = 0, \end{cases} \tag{7.4.47}$$

where $\hat{\gamma}_1^{(k)}$ is a consistent estimator of γ_1. For instance, we may take the estimators (7.4.43) and (7.3.44) with $M_n^{(0)}$ being replaced $M_n^{(k-1)}$, $k = 1, 2, \ldots$.

The asymptotic properties of $M_n^{(k)}$ were studied by Malý (1991); here, for the sake of simplicity, we will restrict ourselves to the special case $S_n = 1$, namely to the non-studentized M-estimators.

The following asymptotic linearity lemma provides the key to our asymptotic considerations:

Lemma 7.4.1 *Let $X_1, \ldots, X_n$ be i.i.d. r.v.'s with the d.f. F. Consider the following empirical process with time parameter $t, u \in \mathbb{R}_1$, and an index $\tau \geq 1/2$:*

$$Q_n = Q_n(t,u) = \sum_{i=1}^{n} \{\psi(X_i - n^{-1/2}t - n^{-\tau}u) - \psi(X_i - n^{-1/2}t)\}. \quad (7.4.48)$$

1. If either ψ and ψ' are absolutely continuous,

$$\mathbb{E}|\psi_\delta''(X_1)| < \infty \quad \text{for } 0 < \delta \leq \delta_0, \quad (7.4.49)$$

$$\psi_\delta''(x) = \sup_{|y| \leq \delta} |\psi''(x+y)|, \; x \in \mathbb{R}_1, \quad (7.4.50)$$

$$\mathbb{E}(\psi'(X_1))^2 < \infty, \quad (7.4.51)$$

or ψ is absolutely continuous with the step-function derivative

$$\psi'(x) = \beta_j, \;\; r_j < x < r_{j+1}, \;\; j = 1, \ldots, l, \quad (7.4.52)$$

$\beta_0, \ldots, \beta_l \in \mathbb{R}_1$, $\beta_0 = \beta_l = 0$, $-\infty = r_0 < r_1 < \ldots < r_l < r_{l+1} = \infty$, and F has a bounded derivative $f(x)$ in a neighborhood of $r_1, \ldots, r_l$, then for any $\tau \geq 1/2$ and any $C > 0$ whenever

$$\gamma_1 = \mathbb{E}\psi'(X_1) > 0, \quad (7.4.53)$$

we have

$$\sup_{|t|,|u| \leq C} \left\{ |n^{-1/2}Q_n(t,u) + n^{1/2-\tau}\gamma_1 u| \right\} = O_p(n^{-\tau}). \quad (7.4.54)$$

2. If ψ is a step-function,

$$\psi(x) = \alpha_j \text{ if } q_j < x < q_{j+1}, \;\; j = 0, 1, \ldots, m, \quad (7.4.55)$$

and F is continuous with two bounded derivatives f and f', $f > 0$ in a neighborhood of $q_1, \ldots q_m$, then, for any $\tau \geq 1/2$,

$$\sup_{|t|,|u| \leq C} \left\{ |n^{-1/2}Q_n(t,u) + n^{1/2-\tau}\gamma_1 u| \right\} = O_p(n^{-\tau/2}). \quad (7.4.56)$$

with

$$\gamma_1 = \sum_{j=1}^{m} (\alpha_j - \alpha_{j-1}) f(q_j) > 0. \quad (7.4.57)$$

Proof. Let first ψ and ψ' be absolutely continuous. Using the Taylor expansion, we get with some $0 < h_1, \; h_2 < 1$ and $K > 0$,

$$
\begin{aligned}
&\sup_{|t|,|u|\le C} |n^{-1/2}Q_n(t,u) + n^{1/2-\tau}\gamma_1 u| \\
= & \sup_{|t|,|u|<C} \Big\{ | - n^{-1/2-\tau} u \sum_{i=1}^{n} [\psi'(X_i) - \mathbb{E}\psi'(X_i)] \\
& + n^{-1-\tau} ut \sum_{i=1}^{n} \psi''(X_i - h_2 n^{-1/2} t) \\
& + \frac{1}{2} n^{-1/2-2\tau} u^2 \sum_{i=1}^{n} \psi''(X_i - n^{-1/2}t - h_1 n^{-\tau} u)| \Big\} \\
& \le n^{-\tau} C |n^{-1/2} \sum_{i=1}^{n} [\psi'(X_i) - \mathbb{E}\psi'(X_i)]| \\
& + n^{-\tau} C^2 K \sup_{|t|\le C} \Big\{ |n^{-1} \sum_{i=1}^{n} \psi''(X_i - h_2 n^{-1/2} t)| \Big\} \\
& + \frac{1}{2} n^{1/2-2\tau} C^2 K \sup_{|t|\le C} \Big\{ |n^{-1} \sum_{i=1}^{n} \psi''(X_i - h_1 n^{-1/2} t)| \Big\} \\
= & \ \mathrm{O}_p(n^{-\tau}),
\end{aligned}
\tag{7.4.58}
$$

due to (7.4.49) – (7.4.51). Under the second condition of proposition 1, it suffices to consider the function

$$
\psi(x) = \begin{cases} r_1 & \text{for } x < r_1, \\ x & \text{for } r_1 \le x \le r_2, \\ r_2 & \text{for } x > r_2, \end{cases} \tag{7.4.59}
$$

with $-\infty < r_1 < r_2 < \infty$, without loss of generality. Assume first that $u > 0$ (the case $u < 0$ is analogous), and denote by

$$
\begin{aligned}
a_n &= n^{-\tau} u, \\
\mathcal{I}_i &= I[X_i - n^{-1/2}t \in (r_1 - a_n, r_2 + a_n) \cup (r_2 - a_n, r_2 + a_n)], \\
\mathcal{J}_i &= I[X_i - n^{-1/2}t \in (r_1 + a_n, r_2 - a_n)]; & (7.4.60) \\
Q_n^i &= \psi(X_i - n^{-1/2}t - a_n) - \psi(X_i - n^{-1/2}t), \ i = 1, \ldots, n. & (7.4.61)
\end{aligned}
$$

Then

$$
Q_n = \sum_{i=1}^{n} Q_n^i \mathcal{I}_i - \sum_{i=1}^{n} a_n \mathcal{J}_i, \tag{7.4.62}
$$

and we have

$$\begin{aligned}
&\sup\Big\{|n^{-1/2}\sum_{i=1}^{n} Q_n^i \mathcal{I}_i| : |t| < C, |u| < C\Big\} \\
&\leq \; n^{1/2-\tau} K \sup\Big\{|\hat{F}_n(R_1^+) - \hat{F}_n(R_1^-) + \hat{F}_n(R_2^+) \\
&\qquad -\hat{F}_n(R_2^-)| : |t| < C, |u| < C\Big\}, \qquad (7.4.63)
\end{aligned}$$

where $\hat{F}_n$ stands for the empirical d.f. based on $X_1, \ldots, X_n$ and

$$R_j^- = r_j + n^{-1/2}t - a_n, \;\; R_j^+ = r_j + n^{-1/2}t + a_n, \; j = 1, 2. \qquad (7.4.64)$$

Using the approximation of the empirical d.f. process by Brownian bridge [see (2.5.171)-(2.5.174)], we come to the conclusion that the right-hand side of (7.4.63) is $\mathrm{O}_p(n^{-\tau})$.

Similarly it can be shown that

$$\sup_{|t|,|u|\leq C} \Big\{n^{-1/2}\sum_{i=1}^{n} |a_n\mathcal{I}_i + n^{1/2-\tau}u\gamma_1|\Big\} = \mathrm{O}_p(n^{-\tau}), \qquad (7.4.65)$$

and this completes the proof of Part 1.

For Part 2, without loss of generality, we can consider the function

$$\psi(x) = \begin{cases} 0 & \text{if } x \leq q, \\ 1 & \text{if } x > q. \end{cases} \qquad (7.4.66)$$

Then

$$n^{-1/2}Q_n(t,u) = -n^{1/2}[\hat{F}_n(q + n^{-1/2}t + n^{-\tau}u) - \hat{F}_n(q + n^{-1/2}t)], \quad (7.4.67)$$

and the proposition follows analogously using the approximation of the empirical d.f. process. □

We are now able to derive the order of the approximation of M_n by its k-step version.

Theorem 7.4.3 *Let $X_1, X_2, \ldots$ be i.i.d. observations with the d.f. $F(x-\theta)$. Let M_n be the M-estimator of θ generated by an absolutely continuous function ρ with the derivative ψ. Let $M_n^{(k)}$ be the k-step version of M_n defined in (7.4.47), $k = 1, 2, \ldots$ with a $\sqrt{n}$-consistent initial estimator $M_n^{(0)}$.*
1. Under the assumptions of part 1 of Lemma 7.4.1, whenever

$$\hat{\gamma}_1^{(k)} - \gamma_1 = \mathrm{O}_p(n^{-1/2}), \; k = 1, 2, \ldots, \qquad (7.4.68)$$

we have

$$n^{1/2}(M_n^{(k)} - M_n) = \mathrm{O}_p(n^{-k/2}), \; k = 1, 2, \ldots. \qquad (7.4.69)$$

2. Under the assumptions of part 2 of Lemma 7.4.1, whenever

$$\hat{\gamma}_1^{(k)} - \gamma_1 = \mathrm{O}_p(n^{-1/4}), \; k = 1, 2, \ldots, \qquad (7.4.70)$$

we have

$$n^{1/2}(M_n^{(k)} - M_n) = O_p(n^{2^{-k-1}-(1/2)}),\ k = 1, 2, \ldots. \tag{7.4.71}$$

Proof. 1. (7.4.69) holds for $k = 1$ by Theorem 7.4.2. Assume that (7.4.69) holds for $k-1$. Inserting $t \to n^{1/2}(M_n - \theta)$ $[= O_p(1)]$ and $u \to n^{\tau}(M_n^{(k-1)} - M_n)$ $[O_p(1)$ for $\tau = k/2]$ in $Q_n(t, u)$ we obtain

$$n^{-1/2}\sum_{i=1}^{n}\psi(X_i - M_n^{(k-1)}) + n^{1/2}(M_n^{(k-1)} - M_n)\gamma_1 = O_p(n^{-k/2}). \tag{7.4.72}$$

Moreover, by definition of $M_n^{(k)}$, we have for $\hat{\gamma}_1^{(k)} \neq 0$,

$$\begin{aligned} n^{1/2}(M_n^{(k)} - M_n) &= (n^{1/2}\hat{\gamma}_1^{(k)})^{-1}\sum_{i=1}^{n}\psi(X_i - M_n^{(k-1)}) \\ &+ n^{1/2}(M_n^{(k-1)} - M_n) \end{aligned} \tag{7.4.73}$$

and, by the induction hypothesis, this leads to the desired result.

The proof of part 2 is quite analogous. □

Remark. Notice that while the approximation of M_n by $M_n^{(k)}$ really improves with increasing k in the case of continuous ψ, the order of approximation of M_n by its k-step version increases very slowly with k if ψ has jump discontinuities. In that case, it is never better than $O_p(n^{-1})$. This surprising conclusion also concerns the estimators based on L_1-norm.

7.4.3 One-Step Versions in Linear Models

For obvious reasons the one- and k-step iterations of estimators have even more applicability in the linear models than in the one-parameter ones.The ideas and properties of one- and k-step versions in the linear regression model are analogous to those in the location model. However, to prove them, we need more technicalities. As a compromise, in order to bring the forms and properties of these estimators as well as the basic references and yet not to consume much space on the technicalities, we will give a brief description of one and k-step versions in the linear regression model along with some basic properties, with only either hints or references to proofs.

Consider the linear regression model (7.3.1) with the design matrix $\mathbf{X}$ of order $n \times p$. For simplicity, we assume that $x_{i1} = 1$, $i = 1, \ldots, n$. The *one-step version of M-estimator* of $\boldsymbol{\beta}$ was first proposed by Bickel (1975) and later investigated by Jurečková (1983), Jurečková and Portnoy (1987), and Jurečková and Welsh (1990). Keeping the notation of Section 7.3, we may define the one-step version $\mathbf{M}_n^{(1)}$ of the M-estimator $\mathbf{M}_n$, generated by an absolutely continuous function ρ with $\psi = \rho'$ and studentized by a scale statistics $S_n = S_n(\mathbf{Y})$, as

$$\mathbf{M}_n^{(1)} = \begin{cases} \mathbf{M}_n^{(0)} + \hat{\gamma}_n^{-1}\mathbf{W}_n & \text{if } \hat{\gamma}_n \neq 0, \\ \mathbf{M}_n^{(0)} & \text{if } \hat{\gamma}_n = 0, \end{cases} \tag{7.4.74}$$

where

$$\mathbf{W}_n = \mathbf{Q}_n^{-1} \sum_{i=1}^{n} \mathbf{x}_i \psi((Y_i - \mathbf{x}_i'\mathbf{M}_n^{(0)})/S_n) \tag{7.4.75}$$

and $\hat{\gamma}_n$ is the estimator of γ_1 based on $Y_1, \ldots, Y_n$. For instance, we may consider

$$\hat{\gamma}_n = \frac{1}{2t} n^{-1/2} \sum_{i=1}^{n} \{\psi(\frac{Y_i - \mathbf{x}_i'\mathbf{M}_n^{(0)} + n^{-1/2}t}{S_n}) - \psi(\frac{Y_i - \mathbf{x}_i'\mathbf{M}_n^{(0)} - n^{-1/2}t}{S_n})\}.$$

By the asymptotic linearity results of Chapter 5, $\hat{\gamma}_n$ is a $n^{1/2}$- or $n^{1/4}$-consistent estimator of γ_1, depending on whether ψ is smooth or it has jump discontinuities. If ψ' exists, we may still consider

$$\hat{\gamma}_n = (nS_n)^{-1} \sum_{i=1}^{n} \psi'((Y_i - \mathbf{x}_i'\mathbf{M}_n^{(0)})/S_n). \tag{7.4.76}$$

Then, under the conditions of Theorem 5.5.1 (for smooth ψ) provided that $n^{\frac{1}{2}}||\mathbf{M}_n^{(0)} - \boldsymbol{\beta}|| = O_p(1)$, we have

$$||\mathbf{M}_n - \mathbf{M}_n^{(1)}|| = O_p(n^{-1}), \tag{7.4.77}$$

while under the conditions of Theorem 5.5.2 (step-function ψ),

$$||\mathbf{M}_n - \mathbf{M}_n^{(1)}|| = O_p(n^{-3/4}). \tag{7.4.78}$$

We expect that also the behavior of the k-step version will be analogous as in the location case.

The possible existence of leverage points in the matrix $\mathbf{X}_n$ leads to the fact that the breakdown point of the M-estimator does not exceed $1/(p+1)$. The situation is not better with L- and R-estimators in the linear model (see Donoho and Huber 1983). The following simple modification of the one-step estimator in (7.4.74)- (7.4.75) yields an estimator $\mathbf{M}_n^{\star}$, satisfying (7.4.75), which has the same breakdown point as the initial estimator $\mathbf{M}_n^{(0)}$:

$$\mathbf{M}_n^{\star} = \begin{cases} \mathbf{M}_n^{(0)} + \hat{\gamma}_n^{-1}\mathbf{W}_n & \text{if } ||\hat{\gamma}_n^{-1}\mathbf{W}_n|| \leq a, \\ \mathbf{M}_n^{(0)} & \text{otherwise}, \end{cases} \tag{7.4.79}$$

for some fixed $a > 0$. We refer to Jurečková and Portnoy (1987) for the proof in case of a smooth ψ. Moreover, if the initial estimator has a high breakdown point but a lower rate of consistency, for some $\tau \in (1/4, 1/2]$,

$$n^{\tau}||\mathbf{M}_n^{(0)} - \boldsymbol{\beta}|| = O_p(1) \text{ as } n \to \infty, \tag{7.4.80}$$

then $\mathbf{M}_n^\star$ inherits the breakdown point from $\mathbf{M}_n^{(0)}$ and

$$\|\mathbf{M}_n^\star - \mathbf{M}_n\| = \mathrm{O}_p(n^{-2\tau}). \tag{7.4.81}$$

Kraft and van Eeden (1972) proposed the *linearized rank estimator* in the linear regression model (7.3.1): Let $\phi: (0,1) \mapsto \mathbb{R}_1$ be a nondecreasing, square-integrable score function, and let $R_{ni}(\mathbf{t})$ denote the rank of the residual $Y_i - \mathbf{x}_i'\mathbf{t},\ \mathbf{t} \in \mathbb{R}_p,\ i = 1, \ldots, n$. Consider the vector of *linear rank statistics* $\mathbf{S}_n(\mathbf{t}) = (S_{n1}(\mathbf{t}), \ldots, S_{np}(\mathbf{t}))'$ where

$$S_{nj}(\mathbf{t}) = \sum_{i=1}^{n} (x_{ij} - \bar{x}_j) a_n(R_{ni}(\mathbf{t})), \quad j = 1, \ldots, p, \tag{7.4.82}$$

and $a_n(i),\ 1 \leq i \leq n$, are the *scores* generated by ϕ, for example,

$$a_n(i) = \phi\Big(\frac{i}{n+1}\Big), \quad i = 1, \ldots, n. \tag{7.4.83}$$

The linearized rank estimator $\mathbf{T}_n^{(l)}$ is then defined with the aid of an initial estimator $\mathbf{T}_n^{(0)}$ as

$$\mathbf{T}_n^{(l)} = \mathbf{T}_n^{(0)} + \alpha^{-2} n^{-1} \mathbf{Q}_n^{-1} \mathbf{S}_n(\mathbf{T}_n^{(0)}), \tag{7.4.84}$$

where

$$\mathbf{Q}_n = \Big(\Big(q_{jk}^{(n)}\Big)\Big)_{j,k=1,\ldots,p},$$

$$q_{jk}^{(n)} = n^{-1} \sum_{i=1}^{n} (x_{ij} - \bar{x}_j)(x_{ik} - \bar{x}_k), \quad |mbox for j, k = 1, \ldots, p, \tag{7.4.85}$$

$\bar{x}_j = n^{-1} \sum_{i=1}^{n} x_{ij},\ j = 1, \ldots, p$, and

$$\alpha^2 = \int_0^1 (\phi(u) - \bar{\phi})^2 du, \quad \bar{\phi} = \int_0^1 \phi(u) du. \tag{7.4.86}$$

If $\mathbf{T}_n^{(0)}$ is $\sqrt{n}$-consistent estimator of $\boldsymbol{\beta}$ and $\phi(u) - \bar{\phi} \equiv \phi(u, f)$, where

$$\phi(u, f) = -\{f'(F^{-1}(u))/f(F^{-1}(u))\}, \ 0 < u < 1, \tag{7.4.87}$$

then $T_n^{(l)}$ is an asymptotically efficient estimator of $\boldsymbol{\beta}$. Under a general ϕ the asymptotic (normal) distribution of $T_n^{(l)}$ depends on $T_n^{(0)}$ and ϕ. The asymptotic relation of $T_n^{(l)}$ to the ordinary R-estimator $\mathbf{T}_n$ of $\boldsymbol{\beta}$ generated by ϕ is described by the following relation: As $n \to \infty$,

$$n^{1/2}(\mathbf{T}_n - \mathbf{T}_n^{(l)}) = \Big(1 - \frac{\gamma}{\alpha^2}\Big) \mathbf{S}_n(\boldsymbol{\beta}) + \mathrm{o}_p(1), \tag{7.4.88}$$

where

$$\gamma = \int_0^1 \phi(u)\phi(u, f) du, \tag{7.4.89}$$

see Kraft and van Eeden (1972) and Jurečková (1989) for detailed proofs.

The *one-step version* $\mathbf{T}_n^{(1)}$ of R-estimator $\mathbf{T}_n$ generated by ϕ may be defined as follows:

$$\mathbf{T}_n^{(1)} = \begin{cases} \mathbf{T}_n^{(0)} + \hat{\gamma}_n^{-1} n^{-1} \mathbf{Q}_n^{-1} \mathbf{S}_n(\mathbf{T}_n^{(0)}) & \text{if } \hat{\gamma}_n \neq 0, \\ \mathbf{T}_n^{(0)} & \text{if } \hat{\gamma}_n = 0, \end{cases} \tag{7.4.90}$$

where $\hat{\gamma}_n$ is an estimator of γ in (7.4.88). The asymptotic linearity of the linear rank statistics offers the following possibility of estimating γ :

$$\hat{\gamma}_n = ||\frac{1}{2} n^{-1/2} \mathbf{Q}_n^{-1} [\mathbf{S}_n(\mathbf{T}_n^{(0)} - n^{-1/2}\mathbf{e}_j) - \mathbf{S}_n(\mathbf{T}_n^{(0)} + n^{-1/2}\mathbf{e}_j)]|| \tag{7.4.91}$$

with $\mathbf{e}_j$ standing for the jth unit vector, $1 \leq j \leq p$. If $\mathbf{T}_n^{(0)}$ is a $\sqrt{n}$-consistent estimator of $\boldsymbol{\beta}$ and the conditions of Chapter 6 leading to the asymptotic linearity of $\mathbf{S}_n$ are satisfied, then

$$n^{1/2} ||\mathbf{T}_n^{(1)} - \mathbf{T}_n|| = o_p(1). \tag{7.4.92}$$

7.5 PROBLEMS

7.2.1. Show that the assumption 1- 4 in Theorem 7.2.4 imply that the regularity assumptions of both Theorems 4.3.1 and 5.3.1 hold.

7.2.2. Provide an outline of the proof of Theorem 7.2.5.

7.2.3. Provide a formal proof of Theorem 7.2.6.

7.2.4. Provide an outline of the proof of Theorem 7.2.7.

7.2.5. Provide a formal proof of Theorem 7.2.8.

7.2.6. For a location parameter of a symmetric d.f., consider the following
a. An L-estimator based on a score function $J(.)$.
b. An M-estimator based on a score function $\psi(.)$.
c. A R-estimator based on a score function $\phi(.)$.
In each case use the appropriate theorems from earlier chapters, and specify the regularity conditions needed on the d.f. F and the score function so that a first-order asymptotic distributional representation holds. Hence, or otherwise, specify the regularity conditions under which they are (pairwise) asymptotically first-order equivalent.

7.2.7. Let Φ be the standard normal d.f., and let P be the set of all symmetric probability measures arising from Φ through ϵ-contamination: $\mathcal{P} = \{F : F =$

$(1-\epsilon)\Phi + \epsilon H,\ H \in \mathcal{M}\}$, where $\mathcal{M}$ is the set of all probability measures on $\mathbb{R}_1$. Show that here the Fisher information $I(f)$ is minimized by the pdf

$$f_0(x) = ((1-\epsilon)/\sqrt{2\pi})\{e^{-\frac{1}{2}x^2} I(|x| \leq k) + e^{\frac{1}{2}k^2 - k|x|} I(|x| > k)\},$$

where

$$2\phi(k)/k - 2\Phi(-k) = \epsilon/(1-\epsilon). \tag{7.5.1}$$

Hence, or otherwise, show that the minimax M-estimator of θ corresponds to the score function $\psi_0 :\ \psi_0(x) = -(\partial/\partial x)\log f_0(x)$. [Huber 1981]

7.2.8. Using the results in Problem 7.2.6, show that for the ϵ-error contaminated model in Problem 7.2.7, the following two estimates are asymptotically minimax (along with the M-estimator considered in Problem 7.2.7):
(a) The α-trimmed mean with

$$\alpha = F_0(-k) = (1-\epsilon)\Phi(-k) + \epsilon/2,$$

and (b) R-estimator of θ based on the score function

$$\phi_0(u) = \begin{cases} -k, & u \leq \alpha, \\ \Phi^{-1}((u - \epsilon/20/(1-\epsilon)), & \alpha \leq t \leq 1-\alpha, \\ k, & u \geq 1-\alpha, \end{cases}$$

with α defined by (a). [Huber 1981]

7.3.1. Use Theorems 4.7.2 and 5.5.1 and outline a proof of Theorem 7.3.1.

7.3.2. Characterize the second order asymptotic relations of M- and R- estimators of the slope of the regression line.

7.4.1. Verify (7.4.19).

7.4.2. Use Theorem 7.4.1 to provide a SOADR equivalence for the simple location model when $\psi = \rho'$ is smooth. State the needed regularity conditions.

7.4.3. Show that (7.4.41) holds for $\hat{\gamma}_1^{(1)}$ defined by (7.4.44).

PART II

Robust Statistical Inference

Ask a mathematician: What is your bread and butter ?
You w'd hear a mutter: Not on a statistics platter!
On the same question to a mathematical statistician,
you w'd hear: showing that everything is Gaussian!
The poor statistician entrapped in the middle-line,
can only think of statistical inference as affine!
The robust monster, in applications, roars in anger:
without bootstrapping, how could I bite my finger?
And, the know-infinite real experts in data-analysis
are the untouchables in the pursuit of neometa analysis!

Chapter 8

Robust Sequential and Recursive Point Estimation

8.1 INTRODUCTION

In a simple estimation problem, given n independent observations $X_1, \ldots, X_n$ from a distribution F $(\in \mathcal{F})$ defining a parameter θ as $\theta(F)$, $F \in \mathcal{F}$, one would estimate θ by a sample statistic (or estimator) $T_n = T(X_1, \ldots, X_n)$. In this conventional setup the sample size is prefixed, although it may be large. In a large sample setup, it is not only customary to check that T_n is consistent (a minimal requirement) but also that it is asymptotically efficient in some well-defined manner. On the top of this, in line with the main theme of this book, we would naturally emphasize robustness aspects of such an estimator. The intricate relationship of robustness and asymptotic efficiency has been explored in earlier chapters. Going further along this avenue, we will consider suitable sequential and recursive versions of robust estimators, both in location and regression models.

Let us look at a well-defined loss function $L(T_n, \theta) : \mathbb{R} \times \mathbb{R} \mapsto \mathbb{R}^+$. In a conventional setup, we would consider an estimator $T_n^\star$ of θ such that

$$\mathbb{E}_F L(T_n^\star, \theta) = \inf\{\mathbb{E}_F L(T_n, \theta) : T_n \in \mathcal{U}\}, \tag{8.1.1}$$

where $\mathcal{U}$ is a class of estimators of θ. Since F is not known and it belongs to a class $\mathcal{F}$, we may require that (8.1.1) hold for all $F \in \mathcal{F}$, namely that $T_n^\star$ be optimal uniformly in $F \in \mathcal{F}$. Such a strong property may not hold in general. In such a case we would choose an estimator T_n such that it remains robust for entire class $\mathcal{F}$ and at the same time it achieves efficiency (at least asymptotically) for a broad subclass of $\mathcal{F}$. However, it is often the case that the cost of sampling must be included in the loss, whereby the loss function $L(T_n, \theta)$ depends also on the sample size n. For example, if c (> 0) is the cost per unit sample, we could take the loss function $L(T_n, \theta)$ as $a[T_n - \theta]^2 + cn$, where $a > 0$ is a suitable constant. We can see that both a and c are given, while θ is unknown and T_n is stochastic in nature. So $\mathbb{E}_F L(T_n, \theta) = cn + a\mathbb{E}_F(T_n - \theta)^2$ depends on (a, c, n) as well as on F, where the dependence on F arises mainly through the mean square error $\mathbb{E}_F(T_n - \theta)^2$. More generally, when the cost of sampling is incorporated in the loss, we can write

$$\mathbb{E}_F L(T_n, \theta) = c(n) + \nu_n(F), \tag{8.1.2}$$

where $c(n)$ is nonnegative and nondecreasing in n, while $\nu_n(F)$ is nonnegative and nonincreasing in n; moreover $\nu_n(F)$ is generally unknown and depends on F. In a parametric setup, $\nu_n(F)$ may depend explicitly on some parameters of F. On the other hand, allowing F to be a member of a general class $\mathcal{F}$, we may at best express $\nu_n(F)$ as a functional of the d.f. F whose domain is $\mathcal{F}$. In many cases the functional form on $\nu_n(F)$ may be complicated or unknown, and hence an exact solution of (8.1.1) may be difficult to obtain. However, under quite general regularity conditions, pertaining to the nondecreasing nature of $c(n)$ and nonincreasing nature of $\nu_n(F)$, there may exist a sample size $n^\star$ such that

$$\mathbb{E}_F L(T_{n^\star}, \theta) = \inf\{\mathbb{E}_F L(T_n, \theta) : n \geq n_0\}, \tag{8.1.3}$$

where n_0 is the minimal sample size for which the expected loss functions are defined properly. If there are more solutions of (8.1.3), then $n^\star$ would be the minimum among them. The value $n^\star$ is called the *optimal sample size* in the context of minimal risk estimation of θ. It is clear that $n^\star$ depends on both functions $c(.)$ and $\nu_n(F)$; hence it also depends on F. Thus no fixed sample size $n^\star$ can be optimal simultaneously for all F belonging to a class $\mathcal{F}$. This feature of the minimal risk estimation problem therefore calls for suitable sequential rules. We will study such sequential estimators based on robust statistics for both location and regression models.

One of the problems cropping up in sequential estimation theory is the computational aspect of an estimator that is typically nonlinear. To compute the estimator afresh at each stage would be a laborious task. In the case of the sample mean, the estimate at stage $n + 1$ could be computed as a linear combination of that at stage n and of the $(n + 1)$-th observation. Thus the sequential estimator could be obtained recursively. The natural question arises: how far could such recursive estimators be considered in the case of robust M-, L-, and R-estimators? We will probe also into this aspect of robust estimation.

8.2 MINIMUM RISK ESTIMATION

As we mentioned in discussing (8.1.1) and (8.1.3), there are two basic minimum risk estimation problems. In the first case the sample size n is fixed and our objective is to choose an optimal estimator (e.g., $T_n^\star$), such that the risk $\mathbb{E}_F L(T_n^\star, \theta)$ of $T_n^\star$ is minimal in the class of competing estimators of θ. In a parametric framework the d.f. F involves only a finite set of unknown parameters (say ξ), and then we need to verify (8.1.1) for all ξ. This requirement may not be satisfied for a finite sample size and thus often one allows n to increase indefinitely, and tries to verify the minimum asymptotic risk property of some regular estimators (see Hájek 1970).

Robust estimators were developed to behave reasonably well in a neighborhood (usually contaminated) of a fixed distribution shape, and in this neighborhood they have the asymptotic minimax properties described earlier. However, in a nonparametric setup the actual d.f. F belongs to a broad class $\mathcal{F}$; hence we have the requirement that (8.1.1) hold for all $F \in \mathcal{F}$ demands additional regularity conditions, which are very seldom met in practical applications (even asymptotically). One possibility is to use a convenient adaptive estimator, if there is any; we will introduce the adaptive estimators in Section 8.5. However, notice that though we admit an indefinitely large n, the estimation problem is still fixed-sample size (i.e., nonsequential) in nature.

The second estimation problem treated in (8.1.3) is basically sequential in nature. In most simple parametric models, where we estimate parameter θ among other (nuisance) parameters, the risk of any estimator T_n of θ depends on the unknown parameter and on the sample size n in a rather involved manner. Hence the optimal sample size $n^\star$ also depends on the nuisance parameter, and to find it can be a complicated ordeal.

In a nonparametric setting, the risk of T_n of θ depends on a functional $\nu(F)$, $F \in \mathcal{F}$. This also calls for an involved solution, and we are only rarely able to establish its form exactly. Nevertheless, often an asymptotic solution exists for a broad class of estimators.

For the sake of simplicity, we will first consider the case of a real parameter $\theta(F)$ admitting a suitable estimator for every $n \geq n_0$. As in (8.1.2), incorporating the cost of sampling in the loss function, we write the risk function as

$$R(T_n, \theta, F) = \mathbb{E}_F L(T_n, \theta) = c(n) + \nu_n(F), \ n \geq n_0, \tag{8.2.1}$$

where $c(n)$, $n \geq n_0$ is nonnegative, increasing in n and $c(\infty) = \infty$; $\nu_n(F)$ is a nonnegative functional of F ($\in \mathcal{F}$), nonincreasing in n ($\geq n_0$), and such that $\lim_{n\to\infty} \nu_n(F) = 0$ for every $F \in \mathcal{F}$. Since $c(n)$ is $\nearrow$ and $\nu_n(F)$ is $\searrow$ in n, there exists an optimal n, say, $n^\star$ such that

$$c(n^\star) + \nu_{n^\star}(F) = \inf\{c(n) + \nu_n(F) : \ n \geq n_0\}. \tag{8.2.2}$$

We call the estimator $T_{n^\star}$, based on the sample size $n^\star$, the *minimum risk estimator* of θ (within the class $\{T_n : \ n \geq n_0\}$). However, $\nu_n(F)$ depends on the d.f. F, which is either unknown or depends on some unknown parameters. On the other hand, the cost function $c(n)$ is independent of F. It is often taken as a linear function $c(n) = c_0 + c_1 n$, where c_0 (> 0) is the overall cost, and c_1 (> 0) is the cost per unit sample. The optimal sample size $n^\star$ thus depends on the nature of the cost function $c(n)$; it is also depends on the underlying F (or its parameters) through $\nu_n(F)$. As a result no fixed sample size $n^\star$ satisfies the equation (8.2.2) simultaneously for all $F \in \mathcal{F}$, and no fixed sample size minimum risk estimator generally exists within the class $\{T_n : n \geq n_0\}$. However, under appropriate regularity conditions, guaranteeing a good asymptotic behavior of the class $\{T_n\}$, suitable sequential procedures exist in an asymptotic setup, and they are asymptotically minimax in the sense of (8.2.2). This asymptotic setup relates to small values of c, the cost of sampling per unit, so that

$$c(n) = cn, \quad n > 0, \text{ where } c \to 0. \tag{8.2.3}$$

In a majority of cases we consider the quadratic loss function $(T_n - \theta)^2$. For a class of asymptotically normal estimators, we have

$$\psi(n)(T_n - \theta) \quad \to \quad \mathcal{N}(0, \nu(F)) \tag{8.2.4}$$

for a suitable nondecreasing function $\psi(n)$, where $\nu(F)$ is a nonnegative functional of F. Typically $\psi(n) = \sqrt{n}$ in the location model and $\psi^2(n)$ is the sum of squares of the regression constants in the simple linear regression model. In the previous text we have mostly assumed that $\psi^2(n) \sim n$ as $n \to \infty$. On the other hand, if we adopt an absolute error loss $|T_n - \theta|$, then it follows from (8.2.4) that $\nu_n(F)$ behaves as $(2n^{-1}\nu(F)/\pi)^{1/2}$ as $n \to \infty$. We may also use the L_p-norm $|T_n - \theta|^p$, $p > 0$; under (8.2.4), $\nu_n(F) \sim n^{-p^\star}\nu^\star(F)$, where $p^\star = p - 1$ or $p/2$ according as $p \in (1, 2]$ or $p \geq 2$, and $\nu^\star(F)$ depends on $\nu(F)$ through the expression for the p-th absolute moment of the normal law. If we consider a binary-valued loss function (zero for $|T_n - \theta| \leq d$, for some $d > 0$, and 1 for $|T_n - \theta| > d$), then $\nu_n(F) = \mathbb{P}\{|T_n - \theta| > d\}$ may be taken as $[\phi(d)]^n$ where $0 < \phi(d) < 1$. This latter result is adapted from the large deviation result for $\{T_n\}$. For the sake of simplicity, let us consider the most frequent case

$$\nu_n(F) \sim n^{-1}\nu(F) \tag{8.2.5}$$

(which is in correspondence with the mean square error), where $\nu(F)$ is a functional of F. Then, as $c \searrow 0$, the risk of T_n satisfies

$$R(T_n, \theta; F) \sim cn + n^{-1}\nu(F), \tag{8.2.6}$$

and the right-hand side is minimized at

$$n_c^\star = n^\star(c, F) \simeq \{c^{-1}\nu(F)\}^{1/2} \quad \text{as } c \searrow 0. \tag{8.2.7}$$

The minimum risk behaves as

$$\lambda_F(c) = 2\{c\nu(F)\}^{1/2} \quad \text{as } c \searrow 0. \tag{8.2.8}$$

Hence the problem of an asymptotically minimax sequential procedure that meets the requirements in (8.2.2) in a well-defined asymptotic sense will be solved if we estimate the parameter (functional) $\nu(F)$ consistently in a sequential fashion. We intend to present such sequential point estimators in a robust setup.

It follows from (8.2.7) and (8.2.8) that

$$\lim_{c\searrow 0} n^{\star}(c, F) = \infty \quad \text{and} \quad \lim_{c\searrow 0} \lambda_F(c) = 0. \tag{8.2.9}$$

Actually, as $c \searrow 0$, $n^{\star}(c, F) \sim O(c^{-1/2})$ and $\lambda_F(c) \sim O(c^{1/2})$, hence $cn + \nu_n(F)$ is dominated by $\nu_n(F)$ for smaller values of n and may not be small. This fact lies in the formulation of a sequential procedure that replaces $\nu(F)$ by an appropriate sequence $\{v_n\}$ of consistent estimators, and then verifies (8.2.7) in a sequential way. Moreover it is imperative that $\{v_n\}$ converges to $\nu(F)$ in a well defined probabilistic sense.

Motivated by these considerations, we define a *stopping variable* N_c for every $c > 0$ by letting

$$N_c = \inf\{n \geq n_0 : \; n^2 \geq c^{-1}(v_n + n^{-h})\}, \tag{8.2.10}$$

where h (> 0) is a suitably chosen constant, mainly adapted to limit N_c from being unduly small. Actually, by the above definition, we have for every $c > 0$,

$$N_c \geq c^{-1/(2+h)}, \quad \text{with probability 1}, \tag{8.2.11}$$

so as $c \searrow 0$, we may as well replace the initial sample size n_0 by $n_{0c} = n_0 c^{-1/(2+h)}$. Given a class of estimators $\{T_n : \; n \geq n_0\}$, we consider the *sequential point estimator* T_{N_c} with $n = N_c$; it has the risk

$$\lambda_F^{\star}(c) = \mathbb{E}_F(cN_c) + \mathbb{E}_F L(T_n, \theta), \; c > 0. \tag{8.2.12}$$

We will call T_{N_c} an *asymptotic minimum risk estimator* (AMRE) provided that

$$\lim_{c\searrow 0}\{\lambda_F^{\star}(c)/\lambda_F(c)\} = 1 \; \forall F \in \mathcal{F}. \tag{8.2.13}$$

This AMRE property is relative to the optimal (nonsequential) estimator $T_{n^{\star}}$ where $n_c^{\star}$ is given by (8.2.7). In the literature the property (8.2.13) is usually referred to as the *first-order asymptotic minimum risk efficiency* (FOAMRE) property (see, Woodroofe 1982), while one speaks on the second-order AMRE (SOAMRE) property when $\lambda_F^{\star}(c)/\lambda_F(c) = 1 + l(c) + o(l(c))$ and $l(c)$ [$= O(c^{1/2})$ as $c \searrow 0$] is minimum with respect to some criterion.

Remarks: As we explained earlier, $L(T_n, \theta)$ based on L_1-norm typically relates to $\nu_n(F) \sim n^{-1/2}\nu^{\star}(F)$, while $L(T_n, \theta)$ based on L_p-norm modifies (8.2.5) to

$$\nu_n(F) \sim n^{-q} v^{\star}(F) \;\; \text{for some } q > 0. \tag{8.2.14}$$

In that case (8.2.7) has to be modified to

$$n_c^\star \sim \{c^{-1}q\nu^\star(F)\}^{1/(q+1)} \quad \text{as } c \searrow 0, \tag{8.2.15}$$

and the corresponding change in (8.2.10) has the form

$$N_c = \inf\{n \geq n_0 : n^{q+1} \geq c^{-1}q(v_n + n^{-h})\},\ c > 0. \tag{8.2.16}$$

The rest of the discussion remains pertinent. This leads us to formulate a general stopping variable N_c, $c > 0$, as

$$N_c = \inf\{n \geq n_0 :\ \psi(n) \geq c^{-1}(v_n + n^{-h})\},\ c > 0, \tag{8.2.17}$$

for an appropriate nonnegative and nondecreasing function $\psi(n)$. A general review of AMRE estimators in a nonparametric setup could be found in Sen (1981a, ch. 10). In the present chapter, we will put the main emphasis on applications of these techniques to the robust estimators.

For simplicity, let us first consider the stopping rule N_c in (8.2.10). The calculation of $\mathbb{E}_F N_c$ and $\mathbb{E}_F L(T_{N_c}, \theta)$ simplifies if the T_n have some reversed martingale properties (e.g., the U-statistics; see Hoeffding 1948). The robust estimators like the L-, M-, and R-estimators typically admit the representations (of the first or second order) with leading terms possessing the reversed martingale structure and the other terms being of lower stochastic order. Hence one possibility would be to try to show that not only the stochastic order but also the L_2-norm of the remaining terms is small; however, this we shall be able to show only under some additional conditions on the moments. In the case of M-estimators, the moment convergence would be proved under a bounded score function; similarly we would prove the moment convergence of trimmed L-estimators. Analogously the moment convergence of R-estimators would be proved in the case that the score-generating function ϕ belongs to $L_r(0,1)$, $r > 4$; this would cover the normal scores estimator, log-rank scores estimator, and Wilcoxon scores estimator, which satisfy this condition for every fixed $r > 0$. These results on the moment convergence have an interest of their own.

Motivated by this feature, we assume that

A1. For some $r > 4$ there exist finite K_r and an integer n_0 such that

$$\mathbb{E}_F\{(n^{1/2}|T_n - \theta|)^r\} \leq K_r < \infty\ \forall n \geq n_0, \tag{8.2.18}$$

where K_r may depend on F but remains finite for all $F \in \mathcal{F}$.

A2. For every $\delta > 0$,

$$\lim_{n\to\infty} \mathbb{E}_F\{\max_{m:|m-n|\leq \delta n} n|T_n - T_m|^2\} = \gamma_\delta \tag{8.2.19}$$

exists and

$$\gamma_\delta \searrow 0 \text{ as } \delta \searrow 0. \tag{8.2.20}$$

Because N_c in (8.2.10) depends on the sequence $\{v_n\}$ of the estimators of $\nu(F)$, it will demand the following conditions on these estimators:

B1. There exists a $\tau > 0$ such that for every $\epsilon > 0,\ n \geq n_0$,

$$\mathbb{P}_F\{|v_n - \nu(F)| > \epsilon\} \leq K(\epsilon, \tau)n^{-\tau-1} \tag{8.2.21}$$

where $K(\epsilon, \tau) < \infty\ \forall F \in \mathcal{F}$.

B2. Either (8.2.21) holds for $\tau \geq 1 + h$ where h is defined as in (8.2.10), or for every $n \geq n_0$,

$$\mathbb{P}_F\{\sup_{m \geq n} |v_n - \nu(F)| > \epsilon\} \leq K(\epsilon, \tau)n^{-\tau} \tag{8.2.22}$$

for some $\tau > 1 + h$ and every $F \in \mathcal{F}$.

Note that while **A1** and **A2** are moment conditions, **B1** and **B2** are probability inequality conditions.

Theorem 8.2.1 *Let $\{T_n\} = \{T_n(X_1, \ldots, X_n)\}$ be a sequence of estimators of θ satisfying the conditions* **A1** *and* **A2** *for all $\mathcal{F}$. Let $\{v_n\}$ be a sequence of estimators of $\nu(F)$ satisfying* **B1** *and* **B2**. *Define the stopping rule as in (8.2.10). Then T_{N_c} is a first-order asymptotically risk efficient estimator of θ as $c \searrow 0$.*

Proof. First, we need to establish some convergence properties of N_c. Define $n_c^\star$ as in (8.2.7), and let

$$n_{1c} = \left[c^{-1/(2+h)}\right] \quad \text{and} \quad n_{2c} = [n_c^\star(1-\epsilon)],\ c > 0, \tag{8.2.23}$$

where $\epsilon > 0$ is arbitrary. Then $N_c \geq n_{1c}$ with probability 1 $\forall c > 0$ by (8.2.11), while by definition in (8.2.10),

$$\begin{aligned}
& \mathbb{P}\{N_c \leq n_{2c}\} \\
\leq\ & \mathbb{P}\{v_n < cn^2 \ \text{ for some } n,\ n_{1c} \leq n \leq n_{2c}\} \\
\leq\ & \mathbb{P}\{v_n < cn_{2c}^2 \text{ for some } n,\ n_{1c} \leq n \leq n_{2c}\} \\
=\ & \mathbb{P}\{v_n - \nu(F) < cn_{2c}^2 - \nu(F) \text{ for some } n,\ n_{1c} \leq n \leq n_{2c}\} \\
\leq\ & \mathbb{P}\{v_n - \nu(F) < \nu(F)[(1-\epsilon)^2 - 1] \text{ for some } n,\ n_{1c} \leq n \leq n_{2c}\} \\
\leq\ & \mathbb{P}\{|v_n - \nu(F)| > \epsilon\nu(F) \text{ for some } n,\ n_{1c} \leq n \leq n_{2c}\} \\
\leq\ & \mathbb{P}\{\sup_{n \geq n_{1c}} |v_n - \nu(F)| > \epsilon\nu(F)\} \\
\leq\ & K(\epsilon, \tau)n_{1c}^{-\tau} = \mathrm{o}\big(c^{(1+h)/(2+h)}\big) = \mathrm{o}\big(c^{(1/2)+\eta}\big),
\end{aligned} \tag{8.2.24}$$

where $\eta = h/(4 + 2h) > 0$. We might also have used **B1** instead of **B2**. Next, note that by (8.2.10),

$$c^{-1/2}\big(v_{N_c}\big)^{1/2} < N_c < n_0 + c^{-1/2}\big(v_{N_c-1} + (N_c - 1)^{-h}\big)^{1/2} \tag{8.2.25}$$

with probability 1. Since $n_{1c} \to 0$ as $c \searrow 0$, (8.2.11) implies that $N_c \to \infty$a.s., while **B1** ensures that $v_n \to \nu(F)$ a.s. as $n \to \infty$. Hence $v_{N_c} \to \nu(F)$ a.s. as $c \searrow 0$, and therefore we obtain by (8.2.7) and (8.2.25) that

$$N_c/n_c^\star \to 1 \text{ a.s. as} c \searrow 0. \tag{8.2.26}$$

Let us further denote $n_{3c} = n_c^\star(1+\epsilon)$ with arbitrary $\epsilon > 0$. Then, by (8.2.10), we have for every $n \geq n_{3c}$,

$$\begin{aligned} \mathbb{P}\{N_c > n\} &= \mathbb{P}\{m^2 \leq c^{-1}(v_m + m^{-h}) \text{ for } n_{1c} \leq m \leq n\} \\ &\leq \mathbb{P}\{cn^2 \leq v_n\} = \mathbb{P}\{cn^2 - \nu(F) \leq v_n - \nu(F)\} \\ &\leq \mathbb{P}\{|v_n - \nu(F)| > \epsilon\nu(F)\} = K(\epsilon, \tau)n^{-1-\tau}, \end{aligned} \tag{8.2.27}$$

so that

$$\mathbb{E}\{(N_c/n_c^\star)I[N_c \geq n_{3c}]\} = (n_c^\star)^{-1} \sum_{n \geq n_{3c}} n\mathbb{P}\{N - c = n\} \to 0 \text{ as } c \searrow 0. \tag{8.2.28}$$

We obtain from (8.2.24), (8.2.26), and (8.2.28) that

$$\lim_{c \searrow 0} \mathbb{E}_F(N_c/n_c^\star) = 1. \tag{8.2.29}$$

Thus, to establish (8.2.13), it suffices to show that

$$\lim_{c \searrow 0} \{c^{-1}(n_c^\star)^{-1}\mathbb{E}_F(T_{N_c} - \theta)^2\} = 1 \; \forall F \in \mathcal{F}. \tag{8.2.30}$$

We write

$$\begin{aligned} \mathbb{E}_F(T_{N_c} - \theta)^2 &= \mathbb{E}_F\{(T_{N_c} - \theta)^2 I[N_c \leq n_{2c}]\} \\ &+ \mathbb{E}_F\{(T_{N_c} - \theta)^2 I[N_c \geq n_{3c}]\} \\ &+ \mathbb{E}_F\{(T_{N_c} - \theta)^2 I[n_{2c} < N_c < n_{3c}]\}. \end{aligned} \tag{8.2.31}$$

Recall that for $r > 2$,

$$\begin{aligned} &\mathbb{E}_F\{(T_{N_c} - \theta)^2 I[N_c \leq n_{2c}]\} \\ &= \sum_{n=n_{1c}}^{n_{2c}} \mathbb{E}_F\{(T_n - \theta)^2 I[N_c = n]\} \\ &\leq \sum_{n=n_{1c}}^{n_{2c}} (\mathbb{E}_F|T_n - \theta|^r)^{2/r}(\mathbb{P}\{N_c = n\})^{1-(2/r)}. \end{aligned} \tag{8.2.32}$$

By **A1** and (8.2.24) the right-hand side of (8.2.32) is $O(c^{(1/2)+\eta'})$ for some η', while $cn_c^\star = O(c^{1/2})$ as $c \searrow 0$. Hence, (8.2.32) is $o(cn_c^\star)$ as $c \searrow 0$. A similar treatment applies to the second term on the right-hand side of (8.2.31). Finally, by **A2** and (8.2.26) we may conclude that

$$\frac{\mathbb{E}_F\{(T_{N_c} - \theta)^2 I[n_{2c} < N_c < n_{3c}]\}}{\mathbb{E}_F(T_{n_c^\star} - \theta)^2} \to 1 \text{ as } c \searrow 0, \tag{8.2.33}$$

and this completes the proof of the theorem. □

Theorem 8.2.2 *Assume the conditions of Theorem 8.2.1 and that*

$$n^{1/2}(v_n - \nu(F)) \xrightarrow{\mathcal{D}} \mathcal{N}(0, \beta^2(F)) \tag{8.2.34}$$

for some $\beta(F)$, $0 < \beta(F) < \infty$, $F \in \mathcal{F}$, and assume that the sequence $\{v_n\}$ satisfies the Anscombe (1952) condition, that is, for every $\epsilon > 0$,

$$\lim_{n\to\infty} \mathbb{P}\{\max_{m:|m-n|<\delta n} |v_m - v_n| > n^{-1/2}\epsilon\} \searrow 0 \quad as\ \delta \searrow 0. \tag{8.2.35}$$

Let N_c be the stopping rule defined in (8.2.10) with $h > 1/2$. Then, as $c \searrow 0$,

$$(n_c^\star)^{-1/2}(N_c - n_c^\star) \xrightarrow{\mathcal{D}} \mathcal{N}(0, \frac{\beta^2(F)}{4\nu^2(F)}). \tag{8.2.36}$$

Proof. By the fact that $n_c^\star \to \infty$ a.s. as $c \searrow 0$, by (8.2.26), and using (8.2.34) and (8.2.35), we obtain that

$$(N_c)^{1/2}(v_{N_c} - \nu(F)) \xrightarrow{\mathcal{D}} \mathcal{N}(0, \beta^2(F)). \tag{8.2.37}$$

as $c \searrow 0$. Using the transformation $v_n \to v_n^{1/2} I[v_n > 0]$, we further obtain

$$(N_c)^{1/2}((v_{N_c})^{1/2} - \nu^{1/2}(F)) \xrightarrow{\mathcal{D}} \mathcal{N}(0, \frac{\beta^2(F)}{4\nu(F)}), \quad \text{as } c \searrow 0. \tag{8.2.38}$$

Both (8.2.37) and (8.2.38) hold similarly for $(N_c - 1)$. Using (8.2.35) along with the fact that $n_c^\star \sim c^{-1/2}\nu^{1/2}(F)$ as $c \searrow 0$, we have $(n_c^\star)^{-1/2}(N_c - n_c^\star) \sim c^{-1/2}(n_c^\star)^{-1/2}(v_{N_c})^{1/2} - \nu^{1/2}(F)) \sim \sqrt{n_c^\star}\nu^{-1/2}(F)(v_{N_c}^{1/2} - \nu^{1/2}(F))$ as $c \searrow 0$, and the rest of the proof follows from the Anscombe condition (8.2.34). □

Let us examine how the regularity conditions **A1**, **A2**, **B1**, and **B2** apply to the estimators considered in Chapters 3-6. We will restrict our consideration to estimators T_n of $\theta \in \mathbb{R}_1$ satisfying

$$T_n - \theta = \sum_{i=1}^{n} Z_{ni} + R_n, \tag{8.2.39}$$

where the Z_{ni} are independent (centered) random variables and R_n is the residual term. In the special case of location model, we further have $Z_{ni} = n^{-1}Z_i$ where the Z_i are i.i.d. r.v.'s with 0 mean and a finite variance, while in the regression model the Z_{ni} depend on the regression constants and are independent but not necessarily identically distributed. If in the location model the Z_i have a finite rth absolute moment for some $r > 4$, one immediately obtains that

$$\mathbb{E}_F\{|n^{\frac{1}{2}} \sum_{i=1}^{n} Z_{ni}|^r\} \leq K_r < \infty \; \forall n \geq n_0. \tag{8.2.40}$$

Therefore, to verify **A1**, it suffices to show that

$$n^{1/2}|R_n| \to 0 \quad \text{in the } r\text{th mean} \quad \text{as } n \to \infty, \tag{8.2.41}$$

for some $r > 4$. This may generally require more stringent regularity conditions. In the case of Hoeffding's (1948) U-statistics, it suffices to assume that the kernel itself has a finite rth-order absolute moment by virtue of the inherent reversed martingale property. If $T_n = T(F_n)$ is a Hadamard-differentiable statistical functional, it suffices to show that the Hadamard derivative of $T(.)$ at F has a finite rth-order absolute moment for some $r > 4$. Since a large class of L- and M-estimators of location satisfies the Hadamard differentiability condition, where the Hadamard derivative is also bounded with probability 1, verification of (8.2.18) poses no serious problem. For R-estimators, this Hadamard differentiability does not hold for unbounded score-generating function ϕ; however, if $|\phi|^r$ is integrable on (0,1), for some $r > 4$, then again (8.2.18) can be verified by some standard techniques [e.g.,see Sen 1981a, ch.10). The manipulations are more tedious for the regression model, and they may demand more stringent regularity conditions on the regression constants.

Concerning condition **A2**, this could be verified by standard manipulations provided that $\{(T_n - \theta) : \ n \geq n_0\}$ is a reversed martingale and $n^{1/2}(T_n - \theta)$ is uniformly square integrable. Without this characteristic, we should impose some extra (uniform integrability) conditions on $\{(T_n - \theta)\}$. However, (8.2.18) generally ensures (8.2.19) [Problem 8.2.4]. Moreover, notice that (8.2.22) is implied by (8.2.21), so that the problem remains to verify (8.2.21) for some $\tau > 1 + h$.

An alternative measure of performance of the sequential estimator is the *asymptotic distributional risk* (ADR). Assume that $\{T_n\}$ is a consistent asymptotically normal sequence of estimators of θ:

$$n^{1/2}(T_n - \theta) \xrightarrow{\mathcal{D}} \mathcal{N}(0, \sigma_T^2) \tag{8.2.42}$$

as $n \to \infty$ for some $\sigma_T > 0$. Then the asymptotic distributional risk is defined as

$$\rho_D(T_n, \theta) = cn + n^{-1}\sigma_T^2. \tag{8.2.43}$$

Let N_c be defined as in (8.2.10), and assume that also

$$N_c^{1/2}(T_{N_c} - \theta) \xrightarrow{\mathcal{D}} \mathcal{N}(0, \sigma_T^2) \quad \text{as } c \searrow 0. \tag{8.2.44}$$

Then

$$N_c/n_c^0 \xrightarrow{P} 1 \quad \text{as } c \searrow 0, \tag{8.2.45}$$

where n_c^0 is defined by

$$\rho_D(T_{n_c^0}, \theta) = \inf\{\rho_D(T_n, \theta) : \ n \geq n_0\}. \tag{8.2.46}$$

Since $n_c^0 \sim c^{-1/2}\sigma_T$ $(\to \infty)$ as $c \searrow 0$, we may argue that the ADR-efficiency of the sequential estimator $\{T_N\}$ with respect to $\{T_{n_c^0}\}$ (which is ADR optimal) is equal to 1.

The main advantage of this ADR efficiency criterion is that it does not require the computation of $\mathbb{E}_F L(T_N, \theta)$, but adopts its natural counterpart from the variance of the asymptotic distribution, which is possible under weaker regularity conditions. In fact, under (8.2.45), the asymptotic normality of T_{N_c} follows from the asymptotic normality of T_n and from the Anscombe condition; and the convergence in probability in (8.2.45) may be verified through the first order asymptotic distributional representation (FOADR) of v_n [see Problem 8.2.5 and Problem 8.2.6].

Theorem 8.2.3 *Suppose that (1) $\{T_n\}$ admits the asymptotic representation (8.2.39) in which the leading term is asymptotically normal and the remainder term R_n satisfies the Anscombe conditions (8.2.35); (2) $\{v_n\}$ is strongly consistent estimator of $\nu(F)$. Then, as $c \searrow 0$,*

$$(a)\ N_c/n_c^0 \xrightarrow{\mathbb{P}} 1, \tag{8.2.47}$$

$$(b)\ \text{the ADR efficiency of } \{T_{N_c}\} \text{ w.r.t. } \{T_{n_c^0}\} = 1; \tag{8.2.48}$$

$$(c)\ T_{N_c} \text{ is asymptotically normally distributed as } c \searrow 0.$$

Proof. Let us write the representation (8.2.39) as

$$T_n - \theta = n^{-1}\sum_{i=1}^{n} \phi(X_i, \theta) + R_n, \tag{8.2.49}$$

where $\phi(X_i, \theta)$ are centered with variance $\sigma_T^2 = \mathbb{E}_F \phi^2(X, \theta)$. Then

$$N_c^{1/2}(T_{n_c^0} - \theta) \xrightarrow{\mathcal{D}} \mathcal{N}(0, \sigma_T^2), \tag{8.2.50}$$

while the Anscombe condition on $\{R_n\}$ yields that for every $\epsilon > 0$

$$\lim_{n\to\infty} \mathbb{P}\{\max_{m:|m-n|<\delta n} n^{1/2}|R_n - R_m| > \epsilon\} \searrow 0 \text{ as } \delta \searrow 0. \tag{8.2.51}$$

Next $N_c \to \infty$ a.s. as $c \searrow 0$ by (8.2.11), so the a.s. convergence of v_n to σ_T^2 implies that both the left- and right-hand sides of (8.2.25), divided by n_c^0, converge a.s. to 1. Thus N_c/n_c^0 satisfies (8.2.47). Applying the central limit theorem to the $\phi(X_i, \theta)$, we conclude from (8.2.51) that also $\{T_n\}$ satisfies the Anscombe conditions. This, along with (8.2.47), imply that T_{N_c} is asymptotically normal. Parts (a) and (c) then imply (8.2.48). □

Theorem 8.2.4 *If v_n admits a first-order asymptotic representation, it is asymptotically normal, and the remainder term satisfies the Anscombe condition and if h in (8.2.10) is chosen greater than 1/2, then*

$$(n_c^0)^{-1/2}(N_c - n_c^0) \xrightarrow{\mathcal{D}} \mathbb{N}(0, \gamma^2), \tag{8.2.52}$$

where $\gamma^2 = \beta^2(F)/(4\sigma_T^2)$ and $\beta^2(F) =$ asymptotic variance of $n^{1/2}(v_n - \sigma_T^2)$.

The proof is analogous to that of Theorem 8.2.2 and is omitted.

Remark. The asymptotic representation (8.2.49) relates specifically to the location model and to some other general parameter model with i.i.d. r.v.'s. In a simple regression model, the first term on the right-hand side of (8.2.49) is typically of the form $\sum_{i=1}^{n} c_{ni}\phi(e_i, \theta)$, where the e_i are the i.i.d. random errors and the c_{ni} are suitable regression constants. The central limit theorem remaining applicable to to this model, the conclusion remains valid. Since a similar modification holds for the v_n, Theorem 8.2.4 remains valid too.

Let us now consider the more general case of a vector-valued parameter $\boldsymbol{\theta} = (\theta_1, \ldots, \theta_p)'$, $p \geq 1$. This situation is pertinent to the simple model of regression line ($p = 2$), the general linear model and the multivariate location model. We assume that $\boldsymbol{\theta}$ is estimable by a suitable estimator $\mathbf{T}_n = (T_{n1}, \ldots, T_{np})'$ for $n \geq n_0$, and that when estimating $\boldsymbol{\theta}$ by $\mathbf{T}_n$, the loss incurred consists of the cost of the sampling and of a suitable distance of $\mathbf{T}_n$ from $\boldsymbol{\theta}$. The typical choice of this distance is based on the quadratic norm:

$$||\mathbf{T}_n - \boldsymbol{\theta}||_Q = \{(\mathbf{T}_n - \boldsymbol{\theta})'\mathbf{Q}(\mathbf{T}_n - \boldsymbol{\theta})\}^{1/2}, \tag{8.2.53}$$

with some nonnegative definite matrix $\mathbf{Q}$. In particular, if the dispersion matrix of $\mathbf{T}_n$ has the form $a(n) \cdot \mathbf{V}$, where $\mathbf{V}$ is a known nonnegative definite matrix and $a(n) > 0$ is a scalar factor, then the natural choice of $\mathbf{Q}$ is $\mathbf{V}^{-1}$. This leads to *Mahalanobis distance* of $\mathbf{T}_n$ and the case $\mathbf{Q} = \mathbf{I}$ leads to the Euclidean distance. There are still other distances, but for the sake of simplicity of presentation, we will stick to a quadratic form with a given $\mathbf{Q}$. Then, similarly as in (8.2.1), we formulate the risk function as

$$R(\mathbf{T}_n, \boldsymbol{\theta}; F) = cn + \mathbb{E}_F||\mathbf{T}_n - \boldsymbol{\theta}||_Q^2 = cn + \mathrm{Trace}(\mathbf{Q}\boldsymbol{\nu}_n), \tag{8.2.54}$$

where $c > 0$ is the cost per unit sample, and

$$\boldsymbol{\nu}_n = \mathbb{E}(\mathbf{T}_n - \boldsymbol{\theta})(\mathbf{T}_n - \boldsymbol{\theta})' = \boldsymbol{\nu}_n(F) \tag{8.2.55}$$

generally depends on n and the form of the d.f. F is unknown. Based on our previous results, we may assume that

$$\boldsymbol{\nu}_n(F) = n^{-1}\boldsymbol{\nu}(F) + \mathrm{o}(n^{-1}) \quad \text{as } n \to \infty \tag{8.2.56}$$

under general regularity conditions, where $\boldsymbol{\nu}(F)$ is a nonnegative definite matrix functional of F. Based on (8.2.54) and (8.2.56), we may argue that as $c \searrow 0$, the right-hand side of (8.2.54) is minimized at

$$n_c^\star \ (\equiv n_c^\star(F)) \cong c^{-1/2}[\text{Trace}(\mathbf{Q}\boldsymbol{\nu}(F))]^{1/2}. \tag{8.2.57}$$

The minimum risk is therefore

$$\lambda_F(c) \simeq 2\{c\text{Trace}(\mathbf{Q}\boldsymbol{\nu}(F)))\}^{1/2} \quad \text{as} \quad c \searrow 0. \tag{8.2.58}$$

Since both $n_c^\star$ and $\lambda_F(c)$ depend on F, an (asymptotically) optimal solution based on a nonrandom sample size may not exist. Hence we take recourse to suitable sequential procedures that yield the asymptotically minimum risk efficiencies (AMRE). A natural extension of (8.2.10) would be a stopping number N_c involving an appropriate estimator v_n of Trace$(\mathbf{Q}\boldsymbol{\nu}(F))$; the result would be the sequential (point) estimator $\mathbf{T}_{N_c}$. A more general loss function leads to the risk

$$R(\mathbf{T}_n, \boldsymbol{\theta}; F) = cn + I\!E_F \rho(\mathbf{T}_n, \boldsymbol{\theta}), \tag{8.2.59}$$

where the metric $\rho(\mathbf{a}, \mathbf{b}) : I\!R_p \times I\!R_p \mapsto I\!R^+$ may depend on n as well as on F through $\boldsymbol{\theta}$ and/or other parameters. As in the first part of this section, we assume that

$$I\!E_F \rho(\mathbf{T}_n, \boldsymbol{\theta}) \ \text{ exists for all } n \geq n_0, \ F \in \mathcal{F},$$

$$I\!E_F \rho(\mathbf{T}_n, \boldsymbol{\theta}) \ \text{ is nonincreasing in } n \ (\geq n_0), \tag{8.2.60}$$

$$n^q I\!E_F \rho(\mathbf{T}_n, \boldsymbol{\theta}) \to \nu(F) \ \ \forall F \in \mathcal{F} \tag{8.2.61}$$

as $n \to \infty$, for some $q \in (0, 1]$ and some $\nu(F)$.

Moreover we assume that there exists a sequence $\{v_n\}$ of consistent estimators of $\nu(F)$. Then we consider a stopping variable

$$N_c = \min\{n \geq n_0 : n^{q+1} \geq c^{-1} q(v_n + n^{-h})\}, \ c > 0, \tag{8.2.62}$$

where h is a suitable positive constant. Hence, for every $c > 0$, T_{N_c} is the resulting sequential point estimator of $\boldsymbol{\theta}$. We want to show that it is AMRE as $c \searrow 0$ [relative to the optimal nonrandom sample size estimator under known $\nu(F)$]. The result will be proved under the conditions analogous to **A1**, **A2**, **B1**, and **B2** in (8.2.18)-(8.2.22) with the only difference that $|T_n - \theta|$ is replaced by $||\mathbf{T}_n - \boldsymbol{\theta}||_Q$. Theorems 8.2.1 and 8.2.2 then extend to the vector parameter case, and similarly for Theorems 8.2.3 and 8.2.4. Therefore we omit those details.

However, the multiparameter estimation problem merits some detailed discussion due to the so-called Stein effect [after Stein (1956)]. When estimating the mean vector $\boldsymbol{\mu}$ of a p-variate normal distribution with $p > 2$, the MLE $\bar{\mathbf{X}}_n$ is not admissible and may be dominated by some other estimators under the quadratic loss. Led by Stein (1956), there has been a steady growth of research in this area, though the developments are confined either to normal F or to an exponential family of F. On the other hand, the estimation of location, regression, or general parameters, which are asymptotically multinormal, are all subject to the Stein-effect. The AMRE problem for the multinormal mean vector (with unknown dispersion matrix) has been treated in a general sequential setup by Ghosh, Nickerson, and Sen (1987), and since then the corresponding asymptotic theory has been extended to encompass a broad class of parametric as well as nonparametric (and robust) estimators.

In the next section, we will give an account of such shrinkage estimators from the point of view of the asymptotic risk efficiency. We will mainly concentrate to the classes of estimators considered in earlier chapters.

8.3 AMRE OF STEIN-RULE ESTIMATORS

Consider a vector valued parameter $\boldsymbol{\theta} = (\theta_1, \ldots, \theta_n)'$ where the θ_j are considered as functionals of a d.f. F. Suppose that there exists a robust estimator $\mathbf{T}_n = (T_{ni}, \ldots, T_{np})'$ that is asymptotically multinormal so that

$$n^{1/2}(\mathbf{T}_n - \boldsymbol{\theta}) \overset{\mathcal{D}}{\to} \mathcal{N}_p(\mathbf{0}, \mathbf{V}) \tag{8.3.1}$$

for some nonnegative definite matrix $\mathbf{V}$ of order $(p \times p)$. Assume that $\mathbf{V}$ is estimable,that there exists a sequence $\{\mathbf{V}_n\}$ of nonnegative definite matrices such that

$$\mathbf{V}_n \to \mathbf{V} \text{ a.s.} \quad \text{as } \to \infty. \tag{8.3.2}$$

We will deal with both *Stein rule estimation* and *preliminary test estimation* methodologies. In both the cases the construction of an estimator rests on the choice of a special value $\boldsymbol{\theta}_0$ of $\boldsymbol{\theta}$, called *pivot*, whose plausibility plays a basic role in the motivation. Then, regarding (8.3.1) and (8.3.2), we will construct an asymptotic test statistic for testing the hypothesis $H_0 : \boldsymbol{\theta} = \boldsymbol{\theta}_0$ vs. $H_1 : \boldsymbol{\theta} \neq \boldsymbol{\theta}_0$ in the form

$$\mathcal{L}_n = n(\mathbf{T}_n - \boldsymbol{\theta}_0)'\mathbf{V}_n^{-1}(\mathbf{T}_n - \boldsymbol{\theta}_0) = n||\mathbf{T}_n - \boldsymbol{\theta}_0||^2_{\mathbf{V}_n^{-1}} \tag{8.3.3}$$

If $\mathbf{V}$ is positive definite, then the asymptotic distribution of $\mathcal{L}_n$ under H_0 is the central chi square with p degrees of freedom, while under H_1 is $\mathcal{L}_n$ stochastically larger; under local (Pitman-type) alternatives, it has asymptotically noncentral chi square distribution with p d.f. and with an appropriate noncentrality parameter. The preliminary test parameter takes on the form

$$\mathbf{T}_n^{PT} = \begin{cases} \boldsymbol{\theta}_0 & \text{if } \mathcal{L}_n \leq l_{n\alpha}, \\ \mathbf{T}_n & \text{if } \mathcal{L}_n > l_{n\alpha}, \end{cases} \tag{8.3.4}$$

where α is the significance level and the critical value $l_{n\alpha}$ is asymptotically equal to $\chi^2_{p\alpha}$, the $(1-\alpha)$-quantile of the chi square distribution with p d.f.

Thus we may write

$$\mathbf{T}_n^{PT} = \boldsymbol{\theta}_0 I[\mathcal{L}_n \le l_{n\alpha}] + \mathbf{T}_n I[\mathcal{L}_n > l_{n\alpha}], \tag{8.3.5}$$

as a convex combination of the pivot $\boldsymbol{\theta}_0$ and the unrestricted estimator $\mathbf{T}_n$, where the zero-one coefficients depend on the result of the preliminary test. $\mathbf{T}_n^{PT}$ performs much better than $\mathbf{T}_n$ when $\boldsymbol{\theta}$ is actually close to the assumed pivot; for $\boldsymbol{\theta} \neq \boldsymbol{\theta}_0$, $\mathbf{T}_n^{PT}$ becomes equivalent to $\mathbf{T}_n$ in probability as $n \to \infty$. However, $\mathbf{T}_n^{PT}$ may not dominate $\mathbf{T}_N$ for all $\boldsymbol{\theta}$ with respect to the asymptotic risk. In fact the asymptotic distributional risk of $\mathbf{T}_n^{PT}$ may be generally strictly greater than that $\mathbf{T}_n$ outside a small neighborhood of the pivot, although this excess is usually small and tends to zero as $\|\boldsymbol{\theta} - \boldsymbol{\theta}_0\|$ increases. In other words, a preliminary test estimator may not be admissible, even asymptotically.

In contrast, the Stein-rule estimator $\mathbf{T}_n^S$ generally has the desired dominance property, exact for normal or some exponential models and asymptotic for a larger class of models. Let us consider a quadratic loss in (8.2.53) and put

$$d_n = \text{ smallest characteristic root of } \mathbf{QV}_n \tag{8.3.6}$$

and select a shrinkage factor $k : 0 < k \le 2(p-2)$, $p \ge 3$. The typical shrinkage version of $\mathbf{T}_n$ is then

$$\mathbf{T}_n^S = (\mathbf{I} - k d_n \mathcal{L}_n^{-1} \mathbf{Q}^{-1} \mathbf{V}_n^{-1}) \mathbf{T}_n. \tag{8.3.7}$$

The role of the pivot $\boldsymbol{\theta}_0$ is hidden in the test statistic $\mathcal{L}_n$, but the dichotomy of the test in (8.3.4) is here replaced by a smoother mixture of $\boldsymbol{\theta}_0$ and $\mathbf{T}_n$. Also the matrix $\mathbf{I}_n - k d_n \mathcal{L}_n^{-1} \mathbf{Q}^{-1} \mathbf{V}_n^{-1}$ may not always be nonnegative definite, and there exist alternative versions of $\mathbf{T}_n^S$ such as the *positive-rule part*. For our purpose the form (8.3.7) will suffice.

The asymptotic distributional risk of $\mathbf{T}_n$ with respect to the quadratic loss $\|\cdot\|_{\mathbf{Q}}^2$ will be, due to (8.3.2),

$$ADR(\mathbf{T}_n, \boldsymbol{\theta}) = \text{Trace}(\mathbf{QV}), \tag{8.3.8}$$

and it remains stationary for all $\boldsymbol{\theta} \in \boldsymbol{\Theta}$ provided that $\mathbf{V}$ does not depend on $\boldsymbol{\theta}$. However, the behavior of the risk $\mathbf{T}_n^S$ is quite different, as we will illustrate in the asymptotic behavior of $\mathcal{L}_n$ for $\boldsymbol{\theta}$ far away from the pivot $\boldsymbol{\theta}_0$. Under $\boldsymbol{\theta}$ being the right parameter value, as $n \to \infty$,

$$n^{-1}\mathcal{L}_n \xrightarrow{P} \|\boldsymbol{\theta} - \boldsymbol{\theta}_0\|_{\mathbf{V}^{-1}}^2 > 0 \ \{\Rightarrow \mathcal{L}_n^{-1} = \mathrm{O}_p(n^{-1})\}. \tag{8.3.9}$$

Also by (8.3.2), (8.3.6), and (8.3.9) as $n \to \infty$,

$$d_n \to \delta \text{ a.s.}, \quad \mathbf{V}_n^{-1} \to \mathbf{V}^{-1} \text{ a.s.}, \tag{8.3.10}$$

where

$$\delta \text{ is the smallest characteristic root of } \mathbf{QV}. \tag{8.3.11}$$

So

$$kd_n l_n^{-1}\mathbf{Q}^{-1}\mathbf{V}_n^{-1} = O_p(n^{-1}), \tag{8.3.12}$$

and moreover

$$\sqrt{n}||\mathbf{T}_n - \mathbf{T}_n^S||_Q \xrightarrow{P} 0 \text{ as } n \to \infty. \tag{8.3.13}$$

Therefore the Stein-rule estimator $\mathbf{T}_n^S$ is asymptotically risk-equivalent to $\mathbf{T}_n$ whenever $\boldsymbol{\theta} \neq \boldsymbol{\theta}_0$. However, when we consider the local alternatives of the Pitman type:

$$H_n : \boldsymbol{\theta} = \boldsymbol{\theta}_{(n)} = n^{-1/2}\boldsymbol{\xi},\ \boldsymbol{\xi} \in \mathbb{R}_p, \tag{8.3.14}$$

then under H_n, $\mathcal{L}_n$ asymptotically has a noncentral chi square distribution with p d.f., and with the noncentrality parameter

$$\triangle = \boldsymbol{\xi}'\mathbf{V}^{-1}\boldsymbol{\xi} \tag{8.3.15}$$

and

$$n^{1/2}\mathbf{T}_n = \mathbf{Z}_n \xrightarrow{\mathcal{D}} \mathbf{Z} \simeq \mathcal{N}_p(\boldsymbol{\xi}, \mathbf{V}). \tag{8.3.16}$$

By (8.3.10),

$$\begin{aligned} \mathcal{L}_n &= \mathbf{Z}_n\mathbf{V}_n^{-1}\mathbf{Z}_n = (\mathbf{Z}_n\mathbf{V}^{-1}\mathbf{Z}_n)(\mathbf{Z}_n\mathbf{V}_n^{-1}\mathbf{Z}_n)/(\mathbf{Z}_n\mathbf{V}^{-1}\mathbf{Z}_n)\} \\ &= (\mathbf{Z}_n\mathbf{V}^{-1}\mathbf{Z}_n)\{1 + o_p(1)\} = (\mathbf{Z}_n\mathbf{V}^{-1}\mathbf{Z}_n) + o_p(1). \end{aligned} \tag{8.3.17}$$

Hence by (8.3.7), (8.3.9), (8.3.10), (8.3.16), and (8.3.17), under $\{H_n\}$,

$$\begin{aligned} n^{1/2}(\mathbf{T}_n^S - \boldsymbol{\theta}_{(n)}) &= n^{1/2}(\mathbf{T}_n - \boldsymbol{\theta}_{(n)}) - \frac{k\delta\mathbf{Q}^{-1}\mathbf{V}^{-1}(n^{1/2}\mathbf{T}_n)}{(\mathbf{Z}_n\mathbf{V}^{-1}\mathbf{Z}_n)} + o_p(1) \\ &= (\mathbf{Z}_n - \boldsymbol{\xi}) - \frac{k\delta\mathbf{Q}^{-1}\mathbf{V}^{-1}\mathbf{Z}_n}{(\mathbf{Z}_n\mathbf{V}^{-1}\mathbf{Z}_n)} + o_p(1) \\ &\xrightarrow{\mathcal{D}} \mathbf{Z} - \boldsymbol{\xi} - (\mathbf{Z}'\mathbf{V}^{-1}\mathbf{Z})^{-1}\mathbf{Q}^{-1}\mathbf{V}^{-1}\mathbf{Z}. \end{aligned} \tag{8.3.18}$$

Therefore the ADR of $\mathbf{T}_n^S$ under $\{H_n\}$, is given by

$$\begin{aligned} ADR(\mathbf{T}_n^S, \boldsymbol{\theta}_{(n)}) &= \mathbb{E}\{(\mathbf{Z} - \boldsymbol{\xi})'\mathbf{Q}(\mathbf{Z} - \boldsymbol{\xi})\} \\ &- 2k\delta\mathbb{E}\{(\mathbf{Z} - \boldsymbol{\xi})'\mathbf{Q}\mathbf{Q}^{-1}\mathbf{V}^{-1}\mathbf{Z}(\mathbf{Z}\mathbf{V}^{-1}\mathbf{Z})^{-1}\} \\ &+ k^2\delta^2\mathbb{E}\{(\mathbf{Z}'\mathbf{V}^{-1}\mathbf{Q}^{-1}\mathbf{Q}\mathbf{Q}^{-1}\mathbf{V}^{-1}\mathbf{Z}(\mathbf{Z}\mathbf{V}^{-1}\mathbf{Z})^{-2}\}. \end{aligned} \tag{8.3.19}$$

Note that $\mathbb{E}\{(\mathbf{Z}-\boldsymbol{\xi})'\mathbf{Q}(\mathbf{Z}-\boldsymbol{\xi})\} = \text{Trace}(\mathbf{QV})$; the other terms may be treated with the aid of the following lemma:

Lemma 8.3.1 *(Stein identity). Let* $\mathbf{W} \sim \mathcal{N}_p(\boldsymbol{\theta}, \mathbf{I})$ *and* $\phi : \mathbb{R}^+ \mapsto \mathbb{R}^+$ *be a real function. Then*

$$\mathbb{E}\{\phi(||\mathbf{W}||^2)\mathbf{W}\} = \boldsymbol{\theta}\mathbb{E}\{\phi(\chi^2_{p+2}(\triangle))\}, \tag{8.3.20}$$

and for any positive definite symmetric matrix $\mathbf{A}$,

$$\mathbb{E}\{\phi(||\mathbf{W}||^2)\mathbf{W}'\mathbf{A}\mathbf{W}\} = \text{Trace}(\mathbf{A})\mathbb{E}\{\phi(\chi^2_{p+2}(\triangle))\} + \boldsymbol{\theta}'\mathbf{A}\boldsymbol{\theta}\mathbb{E}\{\phi(\chi^2_{p+4}(\triangle))\}, \tag{8.3.21}$$

where $\chi^2_q(\delta)$ *stands for a random variable having the noncentral chi-square distribution with* q *d.f. and noncentrality parameter* δ, *and* $\triangle = \boldsymbol{\theta}'\boldsymbol{\theta} = ||\boldsymbol{\theta}||^2$.

Proof of Lemma 8.3.1. Note that, by definition,

$$\int_{\boldsymbol{R}_p}\int \phi(||\mathbf{w}||^2)(2\pi)^{-p/2}\exp\{-\frac{1}{2}||\mathbf{w}-\boldsymbol{\theta}||^2\}d\mathbf{w} = \mathbb{E}\{\phi(\mathbf{W}'\mathbf{W})\}$$

$$= e^{-\triangle/2}\sum_{r\geq 0}\frac{(\triangle/2)^r}{r!}2^{-(p/2+r)}\big(\Gamma(p/2+r)\big)^{-1}\int_0^\infty \phi(y)e^{-y/2}y^{(p/2+r-1)}dy. \tag{8.3.22}$$

Differentiating with respect to $\boldsymbol{\theta}$, we have

$$\int_{\boldsymbol{R}_p}\int \phi(||\mathbf{w}||^2)(\mathbf{w}-\boldsymbol{\theta})(2\pi)^{-p/2}\exp\{-\frac{1}{2}||\mathbf{w}-\boldsymbol{\theta}||^2\}d\mathbf{w}$$

$$= -\boldsymbol{\theta}\mathbb{E}\{\phi(\mathbf{W}'\mathbf{W})\} + \boldsymbol{\theta}e^{-\triangle/2}\sum_{r\geq 0}\frac{(\triangle/2)^r}{r!}2^{-(\frac{p}{2}+r+1)}$$

$$\cdot\big(\Gamma(\frac{p}{2}+r+1)\big)^{-1}\int_0^\infty \phi(y)e^{-y/2}y^{(\frac{p}{2}+r)}dy, \tag{8.3.23}$$

which yields

$$\begin{aligned}\mathbb{E}\{\phi(||\mathbf{W}||^2)\mathbf{W}\} - \boldsymbol{\theta}\mathbb{E}\{\phi(||\mathbf{W}||^2)\}\\ = \quad -\boldsymbol{\theta}\mathbb{E}\{\phi(||\mathbf{W}||^2)\} + \boldsymbol{\theta}\mathbb{E}\{\phi(\chi^2_{p+2}(\triangle))\}\end{aligned} \tag{8.3.24}$$

and this directly leads to (8.3.20). Further note that

$$\mathbb{E}\{\phi(||\mathbf{W}||^2)\mathbf{W}'\mathbf{A}\mathbf{W}\} = \text{Trace}(\mathbf{A}\mathbb{E}\{\phi(||\mathbf{W}||^2)\mathbf{W}\mathbf{W}'\}). \tag{8.3.25}$$

Differentiating (8.3.23) with respect to $\boldsymbol{\theta}$ and using (8.3.25), we obtain (8.3.21) along parallel lines. □

Recall that for every $k > 0$, $q \geq 2k+1$, $\triangle = \boldsymbol{\xi}'\mathbf{V}^{-1}\boldsymbol{\xi}$,

$$\begin{aligned}\mathbb{E}\{\chi_q^{-2k}(\triangle)\} &= e^{-\triangle/2}\sum_{r\geq 0}\frac{(\delta/2)^r}{r!}\mathbb{E}\{\chi^{-2k}_{q+2r}(0)\}\\ &= \mathbb{E}_1\{\mathbb{E}_2[\chi^{-2k}_{q+2\mathcal{K}}(0))|\mathcal{K}\},\end{aligned} \tag{8.3.26}$$

where E_1 is the expectation with respect to $\mathcal{K}$, and

$$\mathcal{K} \text{ is a Poisson r.v. with mean } = \triangle/2. \tag{8.3.27}$$

Further note that for every $r \geq 1$ and $q > 2r$,

$$\mathbb{E}\{\chi_q^{-2r}(0)\} = \{(q-2)\dots(q-2r)\}^{-1}. \tag{8.3.28}$$

Note that from (8.3.27) and (8.3.28), we have for $\mathcal{K} \sim \text{Poisson}(\triangle/2)$,

$$\mathbb{E}\{\chi_q^{-2r}(\triangle) = \mathbb{E}\{[(q-2+2\mathcal{K})\dots(q-2r+2\mathcal{K})]^{-1}\}, \tag{8.3.29}$$

for every $q > 2r$ and $r \geq 1$. Also

$$\begin{aligned} &\mathbb{E}\{\chi_q^{-2r}(\triangle) - \chi_{q+2}^{-2r}(\triangle)\} \\ &= \mathbb{E}\Big\{\frac{1}{(q-2+2\mathcal{K})\dots(q-2r+2\mathcal{K})} - \frac{1}{(q+2\mathcal{K})\dots(q-2r+2+2\mathcal{K})}\Big\} \\ &= 2r\mathbb{E}\{[(q+2\mathcal{K})\dots(q-2r+2\mathcal{K})]^{-1}\} = 2r\mathbb{E}(\chi_{q+2}^{-2r-2}(\triangle)), \end{aligned} \tag{8.3.30}$$

$r \geq 1$. Similarly for $p > 2$,

$$\begin{aligned} \triangle\mathbb{E}(\chi_{p+2}^{-2}(\triangle)) &= e^{-\triangle/2}\sum_{r\geq 0}\frac{(\triangle/2)^r}{r!}\frac{\triangle}{p+2r} \\ &= 2e^{-\triangle/2}\sum_{r\geq 0}\frac{(\triangle/2)^{r+1}}{(r+1)!}\frac{\triangle}{p+2r} \\ &= (p-2)\Big\{e^{-\triangle/2}\sum_{r\geq 0}\frac{(\triangle/2)^r}{r!}\Big[\frac{1}{p-2} - \frac{1}{p+2r-2}\Big]\Big\} \\ &= 1-(p-2)\mathbb{E}(\chi_p^{-2}(\delta)). \end{aligned} \tag{8.3.31}$$

Moreover, letting $\triangle^\star = \boldsymbol{\xi}'\mathbf{V}^{-1}\mathbf{Q}^{-1}\mathbf{V}^{-1}\boldsymbol{\xi}$, we have

$$\begin{aligned} \frac{\triangle^\star}{\triangle} &= \frac{(\boldsymbol{\xi}'\mathbf{V}^{-1}\mathbf{Q}^{-1}\mathbf{V}^{-1}\boldsymbol{\xi})}{(\boldsymbol{\xi}'\mathbf{V}^{-1}\boldsymbol{\xi})} \leq \text{Ch}_1(\mathbf{Q}^{-1}\mathbf{V}^{-1}) \\ &= \{\text{Ch}_p(\mathbf{Q},\mathbf{V})\}^{-1} = 1/\delta,\ \boldsymbol{\xi} \in \mathbb{R}_p. \end{aligned} \tag{8.3.32}$$

Thus, by virtue of Lemma 8.3.1, we can simplify (8.3.19) to

$$\begin{aligned} ADR(\mathbf{T}_n^S, \mathbf{Q}_n) &= \text{Trace}(\mathbf{QV}) - 2k\delta + 2k\delta(\boldsymbol{\xi}'\mathbf{V}^{-1}\boldsymbol{\xi})\mathbb{E}(\chi_{p+2}^{-2}(\triangle)) \\ &+ k^2\delta^2\triangle(\mathbf{Q}^{-1}\mathbf{V}^{-1})\mathbb{E}(\chi_{p+2}^{-4}(\triangle)) \\ &+ (\boldsymbol{\xi}'\mathbf{V}^{-1}\mathbf{Q}^{-1}\mathbf{V}^{-1}\boldsymbol{\xi})\mathbb{E}(\chi_{p+2}^{-4}(\triangle)) \\ &= \text{Trace}(\mathbf{QV}) - 2k\delta(p-2)\mathbb{E}(\chi_p^{-2}(\triangle)) \\ &+ k^2\delta^2\triangle(\mathbf{Q}^{-1}\mathbf{V}^{-1})\mathbb{E}(\chi_{p+2}^{-4}(\triangle)) \\ &+ k^2\delta^2\triangle^\star\mathbb{E}(\chi_{p+4}^{-4}(\triangle)), \end{aligned} \tag{8.3.33}$$

where the last term on the right-hand side of (8.3.33) is bounded from above by

$$k^2\delta\triangle\mathbb{E}(\chi_{p+4}^{-4}(\triangle))$$

$$
\begin{aligned}
&= \frac{1}{2}k^2\delta\Delta[\mathbb{E}(\chi_{p+2}^{-2}(\Delta)) - \mathbb{E}(\chi_{p+4}^{-2}(\Delta))] \\
&= \frac{1}{2}k^2\delta\{[1-(p-2)\mathbb{E}(\chi_p^{-2}(\Delta))] - [1-p\mathbb{E}(\chi_{p+2}^{-2}(\Delta))]\} \\
&= \frac{1}{2}k^2\delta\{2\mathbb{E}(\chi_{p+2}^{-2}(\Delta)) - (p-2)\mathbb{E}(\chi_p^{-2}(\Delta) - \chi_{p+2}^{-2}(\Delta))\} \\
&= k^2\delta\{\mathbb{E}(\chi_{p+2}^{-2}(\Delta)) - (p-2)\mathbb{E}(\chi_{p+2}^{-4}(\Delta))\}. \qquad (8.3.34)
\end{aligned}
$$

Therefore $ADR(\mathbf{T}_n^S, \boldsymbol{\theta}_{(n)})$ is bounded from above by

$$
\begin{aligned}
&\text{Trace}(\mathbf{QV}) - k\delta[2(p-2)-k]\mathbb{E}(\chi_p^{-2}(\Delta))] \\
&-\mathbb{E}(\chi_{p+2}^{-4}(\Delta))\{k^2\delta[p - \delta\text{Trace}(\mathbf{Q}^{-1}\mathbf{V}^{-1})]\}, \qquad (8.3.35)
\end{aligned}
$$

where

$$
\begin{aligned}
\delta\text{Trace}(\mathbf{Q}^{-1}\mathbf{V}^{-1}) &= \delta(\sum_{j=1}^{p}\text{Ch}_j(\mathbf{Q}^{-1}\mathbf{V}^{-1})) \\
&= \left\{\frac{\sum_{j=1}^{p}\text{Ch}_j(\mathbf{Q}^{-1}\mathbf{V}^{-1})}{\text{Ch}_{\max}(\mathbf{Q}^{-1}\mathbf{V}^{-1})}\right\} \\
&\leq p. \qquad (8.3.36)
\end{aligned}
$$

Consequently $ADR(\mathbf{T}_n^S, \boldsymbol{\theta}_{(n)})$ is dominated by Trace$(\mathbf{QV})$ whenever $0 \leq k \leq 2(p-2)$; this leads to the following theorem:

Theorem 8.3.1 *Under the hypothesis $\{H_n\}$ defined in (8.3.14), and for any value of the shrinkage factor k satisfying $0 \leq k \leq 2(p-2)$,*

$$
ADR(\mathbf{T}_n^S, \boldsymbol{\theta}_{(n)}) \leq ADR(\mathbf{T}_n, \boldsymbol{\theta}_{(n)}), \qquad (8.3.37)
$$

with the equality holding for the nonlocal alternatives (when $n||\boldsymbol{\theta}_{(n)}||^2 \to \infty$ as $n \to \infty$ or $||\boldsymbol{\xi}|| \to \infty$). Hence $\mathbf{T}_n$ is asymptotically dominated by the Stein-rule estimator $\mathbf{T}_n^S$.

In the case of $\mathbf{V} = \mathbf{Q}^{-1}$ [i.e., corresponding to the Mahalanobis norm when Trace$(\mathbf{Q}^{-1}\mathbf{V}^{-1}) = p$ and $\delta = 1$], the reduction in (8.3.35) is maximal for the choice of $k = p-2$. If, however, generally $\mathbf{V} \neq \mathbf{Q}^{-1}$, the reduction depends on δ, on Trace$(\mathbf{Q}^{-1}\mathbf{V}^{-1})$ and on Δ. Theorem 8.3.1 relates to the simple case of a multivariate location model or of a general parameter model based on i.i.d. r.v.'s. Let us now turn to a more complex situation of a linear regression model, which has been studied by Sen and Saleh (1987).

Consider the usual linear model

$$
\mathbf{Y}_n = (Y_1, \ldots, Y_n)' = \mathbf{X}_n\boldsymbol{\beta} + \mathbf{e}_n, \qquad (8.3.38)
$$

$$
\mathbf{e}_n = (e_1, \ldots, e_n)', \quad \boldsymbol{\beta} = (\beta_1, \ldots, \beta_p)'. \qquad (8.3.39)
$$

$\mathbf{X}_n$ is an $n \times p$ design matrix of known regression constants, $p \geq 1$, $n > p$, and $\boldsymbol{\beta}$ is a vector of unknown regression parameters. The errors e_i are i.i.d. r.v.'s with a continuous d.f. F, defined on $\mathbb{R}_1$. Without loss of generality assume that that Rank$(\mathbf{X}_n) = p$. Writing $p = p_1 + p_2$, $p_1 \geq 0$, $p_2 \geq 0$, consider a partition

$$\boldsymbol{\beta}' = (\boldsymbol{\beta}_1' \, , \, \boldsymbol{\beta}_2') \text{ and } \mathbf{X}_n = (\mathbf{X}_{n1} \, , \, \mathbf{X}_{n2}), \tag{8.3.40}$$

with $\boldsymbol{\beta}_1, \boldsymbol{\beta}_2, \mathbf{X}_{n1}$ and $\mathbf{X}_{n2}$ of order $p_1 \times 1$, $p_2 \times 1$, $n \times p_1$, and $n \times p_2$, respectively. Then rewrite (8.3.38) as

$$\mathbf{Y}_n = \mathbf{X}_{n1}\boldsymbol{\beta}_1 + \mathbf{X}_{n2}\boldsymbol{\beta}_2 + \mathbf{e}_n. \tag{8.3.41}$$

We are primarily interested in the estimation of $\boldsymbol{\beta}_1$ when it is plausible that $\boldsymbol{\beta}_2$ is "close" to some specified $\boldsymbol{\beta}_2^0$ (which, without loss of generality, we may set to be $\mathbf{0}$). For example, in the case of a multifactor design, we may be interested in estimating the vector of maineffects $\boldsymbol{\beta}_1$, while there is a question whether the vector of interaction effects $\boldsymbol{\beta}_2$ may be ignored. We may treat this problem analogously as in (8.3.1) through (8.3.7), but we should be aware of more complexities due to the following two reasons: (1) The pivot $\boldsymbol{\beta}_2 = \mathbf{0}$ relates to the nuisance parameter $\boldsymbol{\beta}_2$, while $\boldsymbol{\beta}_1$ remains unknown, and (2) the scale factor $n^{1/2}$ in (8.3.1) has to be replaced by a matrix $\mathbf{D}_n$ defined by

$$\mathbf{D}_n(\mathbf{X}_n'\mathbf{X}_n)^{-1}\mathbf{D}_n' = \mathbf{I}_p. \tag{8.3.42}$$

We will assume that as $n \to \infty$,

$$n^{-1}\mathbf{C}_n \to \mathbf{Q} \text{ (p.d.)}; \tag{8.3.43}$$

$$\mathbf{C}_n = \mathbf{X}_n'\mathbf{X}_n, \text{ and } n^{-1/2}\mathbf{D}_n \to \mathbf{D} \text{ (p.d.)}. \tag{8.3.44}$$

Then, parallel to (8.3.1), we will assume that we have an estimator $\mathbf{T}_n$ of $\boldsymbol{\beta}$ satisfying

$$\mathbf{D}_n(\mathbf{T}_n - \boldsymbol{\beta}) \xrightarrow{\mathcal{D}} \mathcal{N}_p(\mathbf{0}, \mathbf{V}) \text{ as } n \to \infty, \tag{8.3.45}$$

and that there exists a sequence $\{\mathbf{V}_n\}$ of n.n.d. matrices satisfying (8.3.2). The partition (8.3.41) induces the following decomposition of the matrix $\mathbf{C}_n$:

$$\mathbf{C}_n = \begin{pmatrix} \mathbf{C}_{n11} & \mathbf{C}_{n12} \\ \mathbf{C}_{n21} & \mathbf{C}_{n22} \end{pmatrix} = \begin{pmatrix} \mathbf{X}_{n1}'\mathbf{X}_{n1} & \mathbf{X}_{n1}'\mathbf{X}_{n2} \\ \mathbf{X}_{n2}'\mathbf{X}_{n1} & \mathbf{X}_{n2}'\mathbf{X}_{n2} \end{pmatrix}. \tag{8.3.46}$$

Regarding (8.3.43), we have

$$\mathbf{Q} = \begin{pmatrix} \mathbf{Q}_{11} & \mathbf{Q}_{12} \\ \mathbf{Q}_{21} & \mathbf{Q}_{22} \end{pmatrix}, \quad \mathbf{Q}_{ij} = \lim_{n \to \infty} n^{-1}\mathbf{C}_{nij}, \; i, j = 1, 2. \tag{8.3.47}$$

Finally, we will assume that

$$n^{-1}\sum_{i=1}^{n} \mathbf{x}_i'\mathbf{x} = n^{-1}\sum_{i=1}^{n} \|\mathbf{x}_i\|^2 = O(1) \text{ as } n \to \infty, \tag{8.3.48}$$

where $\mathbf{x}_i$ is the ith row of $\mathbf{X}_n$, $i = 1, \ldots n$. Note that by (8.3.43) and (8.3.48),

$$\max_{1 \le i \le n} \{\mathbf{x}_i' \mathbf{C}_n^{-1} \mathbf{x}_i)^{1/2} = O(n^{-1/2}) \text{ as } n \to \infty. \tag{8.3.49}$$

As in (8.3.3), we will consider the class of test statistics for testing the null hypothesis

$$H_0 : \boldsymbol{\beta}_2 = \mathbf{0} \text{ vs. } H_1 : \boldsymbol{\beta}_2 \neq \mathbf{0}, \ (\boldsymbol{\beta}_1 \text{ nuisance}). \tag{8.3.50}$$

Such a test statistic $\mathcal{L}_n$ can be constructed from aligned test statistics or derived estimators. For example, we may partition $\mathbf{T}_n$ satisfying (8.3.45) as

$$\mathbf{T}_n' = (\mathbf{T}_{n1}' \ , \ \mathbf{T}_{n2}'). \tag{8.3.51}$$

Then, if we put

$$\mathbf{D}_n^{-1} \mathbf{V}_n \mathbf{D}_n'^{-1} = \mathbf{J}_n = \begin{pmatrix} \mathbf{J}_{n11} & \mathbf{J}_{n12} \\ \mathbf{J}_{n21} & \mathbf{J}_{n22} \end{pmatrix}, \tag{8.3.52}$$

we may consider the test statistic

$$\mathcal{L}_n = \mathbf{T}_{n2}' \mathbf{J}_{n22}^{-1} \mathbf{T}_{n2}. \tag{8.3.53}$$

Alternatively, we may use aligned test statistics based on M- or R-estimators of $\boldsymbol{\beta}_1$. Let us illustrate this approach on the M-tests; the case of R-tests is quite analogous. For every $\mathbf{b} \in \mathbb{R}_p$, define

$$\mathbf{M}_n(\mathbf{b}) = \sum_{i=1}^{n} \mathbf{x}_i \psi(Y_i - \mathbf{x}_i' \mathbf{b}). \tag{8.3.54}$$

Then the unrestricted M-estimator $\tilde{\boldsymbol{\beta}}_n = (\tilde{\boldsymbol{\beta}}_{n1}', \tilde{\boldsymbol{\beta}}_{n2}')'$ of $\boldsymbol{\beta}$ is a solution of the system of p equations

$$\mathbf{M}_n(\mathbf{b}) := \mathbf{0} \quad \text{with respect to } \mathbf{b} \in \mathbb{R}_p. \tag{8.3.55}$$

Partitioning further $\mathbf{b}'$ as $(\mathbf{b}_1' \ , \ \mathbf{b}_2')$ and $\mathbf{M}_n'(.)$ as $(\mathbf{M}_{n1}'(.) \ \ \mathbf{M}_{n2}'(.))$, we get the restricted M-estimator $\hat{\boldsymbol{\beta}}_{n1}$ of $\boldsymbol{\beta}_1$ as a solution of the system of p_1 equations

$$\mathbf{M}_{n1}((\mathbf{b}_1', \mathbf{0}')') = \mathbf{0} \text{ with respect to } \mathbf{b} \in \mathbb{R}_{p_1}. \tag{8.3.56}$$

Let then

$$S_n^2 = (n - p_1)^{-1} \sum_{i=1}^{n} \psi^2(T_i - \mathbf{x}_{i1}' \hat{\boldsymbol{\beta}}_{n1}), \tag{8.3.57}$$

where $\mathbf{x}_i' = (\mathbf{x}_{i1}' \ \ \mathbf{x}_{i2}')$, $i = 1, \ldots, n$, and

$$\mathbf{C}_{nrr:s} = \mathbf{C}_{nrr} - \mathbf{C}_{nrs} \mathbf{C}_{nss}^{-1} \mathbf{C}_{nsr}, \ r \neq s = 1, 2. \tag{8.3.58}$$

Now an appropriate aligned M-test statistic is

$$\mathcal{L}_n = S_n^{-2}\{(\hat{\mathbf{M}}_{n2})'\mathbf{C}_{n22.1}(\hat{\mathbf{M}}_{n2})\}, \tag{8.3.59}$$

where

$$(\hat{\mathbf{M}}_{n2}) = \mathbf{M}_{n2}(\hat{\boldsymbol{\beta}}_{n1}, \mathbf{0}). \tag{8.3.60}$$

Based on the uniform asymptotic linearity results in Chapters 4, 5, and 6, we may conclude that under $H_0 : \boldsymbol{\beta}_2 = \mathbf{0}$, $\mathcal{L}_n$ has asymptotically the central chi square distribution with p_2 d.f.. Hence, as in (8.3.4), we may formulate this preliminary test estimator of $\boldsymbol{\beta}_1$ as follows:

$$\begin{aligned} \hat{\boldsymbol{\beta}}_{n1}^{PT} &= \begin{cases} \hat{\boldsymbol{\beta}}_{n1} & \text{if } \mathcal{L}_n < \chi^2_{p_2,\alpha}, \\ \tilde{\boldsymbol{\beta}}_{n1} & \text{if } \mathcal{L}_n \geq \chi^2_{p_2,\alpha}, \end{cases} \\ &= \hat{\boldsymbol{\beta}}_{n1} I[\mathcal{L}_n < \chi^2_{p_2,\alpha}] + \tilde{\boldsymbol{\beta}}_{n1} I[\mathcal{L}_n \geq \chi^2_{p_2,\alpha}]. \end{aligned} \tag{8.3.61}$$

On the other hand, the Stein-rule estimator $\hat{\boldsymbol{\beta}}_{n1}^{S}$ of $\boldsymbol{\beta}_1$ is defined as

$$\hat{\boldsymbol{\beta}}_{n1}^{S} = \hat{\boldsymbol{\beta}}_{n1} + (\mathbf{I}_{p_1} - kd_n n^{-1}\mathcal{L}_n^{-1}\mathbf{W}^{-1}\mathbf{C}_{n11.2})(\tilde{\boldsymbol{\beta}}_{n1} - \hat{\boldsymbol{\beta}}_{n1}), \tag{8.3.62}$$

where

$$d_n = \text{ smallest characteristic root of } \; n\mathbf{W}\mathbf{C}_{n11.2}^{-1}, \tag{8.3.63}$$

k (> 0) is a shrinkage factor, and the other notations have been introduced before.

The natural choice of $\mathbf{W}$ based on (8.3.46) and the classical Mahalanobis distance is given by

$$\mathbf{W} = n^{-1}\mathbf{C}_{n11.2} \text{ (or } \mathbf{C}_{11.2}). \tag{8.3.64}$$

In this case the Stein-rule estimator (8.3.61) reduces to

$$\hat{\boldsymbol{\beta}}_{n1}^{S} = \hat{\boldsymbol{\beta}} n1 + (1 - k\mathcal{L}_n^{-1})(\tilde{\boldsymbol{\beta}}_{n1} - \hat{\boldsymbol{\beta}}_{n1}), \tag{8.3.65}$$

and this in turn suggests that the *positive-rule version*:

$$\hat{\boldsymbol{\beta}}_{n1}^{S+} = \hat{\boldsymbol{\beta}}_{n1} + (1 - k\mathcal{L}_n^{-1})^+(\tilde{\boldsymbol{\beta}}_{n1} - \hat{\boldsymbol{\beta}}_{n1}), \tag{8.3.66}$$

where $a^+ = \max\{0, a\}$, $a \in \mathbb{R}$.

We have seen before [see (8.3.7)-(8.3.13)] that under a fixed alternative to H_0, the preliminary test estimator and the Stein-rule estimator become asymptotically isomorphic to the classical version $\mathbf{T}_n$ as $n \to \infty$. This is also the case in the linear regression model: If $\boldsymbol{\beta}_2 \neq \mathbf{0}$, then $\hat{\boldsymbol{\beta}}_{n1}^{PT}$ and $\hat{\boldsymbol{\beta}}_{n1}^{S}$ (or $\hat{\boldsymbol{\beta}}_{n1}^{S+}$) become asymptotically equivalent to the unrestricted estimator $\hat{\boldsymbol{\beta}}_{n1}$.

The difference appear in a "shrinking neighborhood" of the pivot $\boldsymbol{\beta}_2 = \mathbf{0}$. Hence, we will consider a sequence $\{H_n^\star\}$ of alternatives with

$$H_n^\star : \boldsymbol{\beta}_2 = \boldsymbol{\beta}_{2(n)} = n^{-1/2}\boldsymbol{\xi},\ \boldsymbol{\xi} \in \mathbb{R}_{p_2}. \tag{8.3.67}$$

and the null hypothesis H_0 may be then expressed as $H_0 : \ \boldsymbol{\xi} = \mathbf{0}$. For a suitable estimator $\boldsymbol{\beta}_{n1}^\star$ of $\boldsymbol{\beta}$, denote

$$G^\star(x) = \lim_{n\to\infty} \mathbb{P}\{n^{1/2}(\hat{\boldsymbol{\beta}}_{n1}^\star - \boldsymbol{\beta}_1) \leq \mathbf{x}|H_n^\star\},\ \mathbf{x} \in \mathbb{R}_{p_1}, \tag{8.3.68}$$

and assume that $G^\star(.)$ is nondegenerate. Then the asymptotic distributional risk of $\hat{\boldsymbol{\beta}}_{n1}^\star$ corresponding to the quadratic loss function $n\|\hat{\boldsymbol{\beta}}_{n1}^\star - \boldsymbol{\beta}_1\|_{\mathbf{W}}^2$ (with a given p.d. matrix $\mathbf{W}$) is defined as

$$R(\boldsymbol{\beta}_1^\star; \mathbf{W}) = \text{Trace}(\mathbf{W})\int\cdots\int_{\mathbb{R}_{p_1}} \mathbf{x}\mathbf{x}' dG^\star(\mathbf{x}) = \text{Trace}(\mathbf{W}\mathbf{V}^\star), \tag{8.3.69}$$

where $\mathbf{V}^\star$ is the dispersion matrix of the d,f, $G^\star$. Denote

$$\sigma_\psi^2 = \int_{\mathbb{R}} \psi^2(y)dF(y) = \mathbb{E}_F(\psi^2(e_i)), \tag{8.3.70}$$

$$\gamma = \gamma(\psi, F) = \int_{\mathbb{R}} \psi(x)\{-f'(x)/f(x)\}dF(x), \tag{8.3.71}$$

$$\nu^2 = \sigma_\psi^2/\gamma^2. \tag{8.3.72}$$

As in Chapter 5, we assume either that the density f has a finite Fisher information $I(f) = \int_{\mathbb{R}}\{-f'(x)/f(x)\}^2 dF(x)$ and $\psi(.)$ is square integrable with respect to F or that $\psi(x)$ has a derivative $\psi'(x)$ a.e. So $I(f) < \infty$ may be relaxed. However, we assume that $\gamma(\psi, F) > 0$ in either case. This is satisfied for example if ψ is nondecreasing and nonconstant and if $f(.)$ is log-convex (see Chapters 4, 5, and 6, for details).

Let $\Phi(\mathbf{x}, \boldsymbol{\mu}, \boldsymbol{\Sigma})$ denote the d.f. of the multivariate normal distribution with the mean vector $\boldsymbol{\mu}$ and with the dispersion matrix $\boldsymbol{\Sigma}$, and let $H_p(x, \triangle)$, $x \in \mathbb{R}^+$, denote the d.f. of the noncentral chi-square distribution with p d.f. and with the noncentrality parameter $\triangle$. The asymptotic behavior of the estimators of $\boldsymbol{\beta}_1$ is described in the following theorem (we relegate the proof as an exercise):

Theorem 8.3.2 *1. Let $\tilde{\boldsymbol{\beta}}_{n1}$, $\hat{\boldsymbol{\beta}}_{n1}$ and $\hat{\boldsymbol{\beta}}_{n1}^{PT}$ be the versions of the M-estimators of $\boldsymbol{\beta}_1$ as described above, and let $\mathcal{L}_n$ be the test statistic defined in (8.3.59). Then, under $H_n^\star$,*

$$\lim_{n\to\infty} \mathbb{P}\{\mathcal{L}_n \leq x|H_n^\star\} = H_{p_2}(x, \triangle),\ \ x \geq 0,\ \ \triangle = \nu^{-2}(\boldsymbol{\xi}'\mathbf{Q}_{22.1}\boldsymbol{\xi}), \tag{8.3.73}$$

$$\lim_{n\to\infty} \mathbb{P}\{n^{1/2}(\hat{\boldsymbol{\beta}}_{n1} - \boldsymbol{\beta}_1) \leq x|H_n^\star\} = \Phi_{p_1}(\mathbf{x} + \mathbf{Q}_{11}^{-1}\mathbf{Q}_{12}\boldsymbol{\xi}; \mathbf{0}, \nu^2\mathbf{C}_{11}^{-1}), \tag{8.3.74}$$

$$\lim_{n\to\infty} \mathbb{P}\{n^{\frac{1}{2}}(\hat{\boldsymbol{\beta}}_{n1}^{PT} - \boldsymbol{\beta}_1) \le x | H_n^\star\}$$
$$= H_{p_2}(\chi^2_{p_2}, \alpha; \triangle)\Phi_{p_2}(\mathbf{x} + \mathbf{Q}_{11}^{-1}\mathbf{Q}_{12}\boldsymbol{\xi}; \mathbf{0}, \nu^2\mathbf{Q}_{11}^{-1})$$
$$+ \int_{\boldsymbol{E}(\boldsymbol{\xi})} \Phi_{p_1}(\mathbf{x} - \mathbf{B}_{12}\mathbf{B}_{22}^{-2}\mathbf{Z}; \mathbf{0}, \nu^2\mathbf{B}_{11.2}) d\Phi_{p_2}(\mathbf{z}; \mathbf{0}, \nu^2\mathbf{B}_{22}), \quad (8.3.75)$$

where $\mathbf{B} = \mathbf{Q}^{-1}$*(so that* $\mathbf{DBD}' = \mathbf{I}$*), the* $\mathbf{B}_{ij}$ *and* $\mathbf{B}_{ii.j}$ *are defined as in (8.3.46) and (8.3.58), and*

$$\mathbb{E}_{\boldsymbol{\xi}} = \{\mathbf{z} \in \mathbb{R}_{p_2} : (\mathbf{z} + \boldsymbol{\xi})'\mathbf{Q}_{22.1}(\mathbf{z} + \boldsymbol{\xi}) \ge \nu^2\chi^2_{p,\alpha}\}. \quad (8.3.76)$$

(ii) Let $\hat{\boldsymbol{\beta}}_{n1}^{S}$ *be the Stein-rule M-estimator of* $\boldsymbol{\beta}_1$*. Then, under* $H_n^\star$*,*

$$n^{1/2}(\hat{\boldsymbol{\beta}}_{n1}^{S} - \boldsymbol{\beta}_1) \xrightarrow{\mathcal{D}} \mathbf{B}_1\mathbf{U} + \frac{\nu^2 k \mathbf{Q}_{11}^{-1}\mathbf{Q}_{12}(\mathbf{B}_2\mathbf{U} + \boldsymbol{\xi})}{(\mathbf{B}_2\mathbf{U} + \boldsymbol{\xi})'\mathbf{Q}_{22.1}(\mathbf{B}_2\mathbf{U} + \boldsymbol{\xi})}, \quad (8.3.77)$$

where k *is the shrinkage factor,* $\mathbf{U}$ *is a random vector with the* $\mathcal{N}_p(\mathbf{0}, \nu^2\mathbf{Q})$ *distribution, and* $\mathbf{B} = (\mathbf{B}_1', \mathbf{B}_2')'$ *with* $\mathbf{B}_1$ *and* $\mathbf{B}_2$ *of order* $p_1 \times p$ *and* $p_2 \times p$*, respectively.*

We now appeal to the Stein identity (Lemma 8.3.1), to (8.3.69) and to Theorem 8.3.2. We arrive at the following theorem (whose proof is again left as an exercise):

Theorem 8.3.3 *Under* $\{H_n^\star\}$ *and under the assumed regularity conditions,*

$$R(\tilde{\boldsymbol{\beta}}_1; \mathbf{W}) = \nu^2 \text{Trace}(\mathbf{W}\mathbf{Q}_{11.2}^{-1}), \quad (8.3.78)$$
$$R(\hat{\boldsymbol{\beta}}_1; \mathbf{W}) = \nu^2 \text{Trace}(\mathbf{W}\mathbf{Q}_{11}^{-1}) + \boldsymbol{\xi}'\mathbf{M}\boldsymbol{\xi}, \quad (8.3.79)$$

where $\mathbf{M} = \mathbf{Q}_{21}\mathbf{Q}_{11}^{-1}\mathbf{W}\mathbf{Q}_{11}^{-1}\mathbf{Q}_{12}$*;*

$$R(\hat{\boldsymbol{\beta}}_1^{PT}; \mathbf{W}) = \nu^2\{\text{Trace}(\mathbf{W}\mathbf{Q}_{11.2}^{-1})[1 - H_{p_2+2}(\chi^2_{p_2,\alpha}; \triangle)]$$
$$+ \boldsymbol{\xi}'\mathbf{M}\boldsymbol{\xi}[2H_{p_2+2}(\chi^2_{p_2,\alpha}; \triangle) - H_{p_2+4}(\chi^2_{p_2,\alpha}; \triangle)]\} \quad (8.3.80)$$

and

$$R(\hat{\boldsymbol{\beta}}_1^{S}; \mathbf{W}) = \nu^2\{\text{Trace}(\mathbf{W}\mathbf{Q}_{11.2}^{-1} - k\text{Trace}(\mathbf{M}\mathbf{Q}_{22.1}^{-1})[2\mathbb{E}(\chi^{-2}_{p_2+2}(\triangle))$$
$$-k\mathbb{E}(\chi^{-4}_{p_2+2}(\triangle))] + k(k+4)(\boldsymbol{\xi}'\mathbf{M}\boldsymbol{\xi})\mathbb{E}(\chi^{-4}_{p_2+4}(\triangle))\}, \quad (8.3.81)$$

where $\triangle$*,* $H_q(.)$ *and the other notations were introduced in Theorem 8.3.2.*

In the special case $\mathbf{Q}_{12} = \mathbf{0}$ we have $\mathbf{Q}_{11.2} = \mathbf{Q}_{11}$ and $\mathbf{M} = \mathbf{0}$. Then (8.3.78)-(8.3.81) are all reduced to $\nu^2\text{Trace}(\mathbf{W}\mathbf{Q}_{11}^{-1})$ and hence all these versions are equivalent to each other with respect to the asymptotic distributional risk. Because of this fact, we will concentrate on the case $\mathbf{Q}_{12} \ne \mathbf{0}$. Further the special choice $\mathbf{W} = \mathbf{Q}_{11.2}\nu^{-2}$, corresponding to Mahalanobis distance, deserves special attention. The risks in Theorem 8.3.3 take on considerably simpler forms that enables us to study the dominance of the estimators and to compare their efficiencies. First, (8.3.78) reduces to p_1, which is a constant independent of $\boldsymbol{\xi} \in \mathbb{R}_{p_2}$. Further, denoting

$$\mathbf{M}^0 = \mathbf{Q}_{12}\mathbf{Q}_{22}^{-1}\mathbf{Q}_{21}\mathbf{Q}_{11}^{-1}, \tag{8.3.82}$$

we obtain by using (8.3.77)-(8.3.79)

$$R(\tilde{\boldsymbol{\beta}}_1, |\mathbf{w}) \underset{<}{>} R(\hat{\boldsymbol{\beta}}_1, \mathbf{W}) \text{ according as } \boldsymbol{\xi}'\mathbf{M}\boldsymbol{\xi} \underset{>}{<} \text{Trace}(\mathbf{M}^0), \tag{8.3.83}$$

and

$$\begin{aligned} R(\tilde{\boldsymbol{\beta}}_1, |\mathbf{w}) \quad & \underset{<}{\geq} R(\hat{\boldsymbol{\beta}}_1^{PT}, \mathbf{W}) \quad \text{according as} \\ \boldsymbol{\xi}'\mathbf{M}\boldsymbol{\xi} \quad & \underset{>}{\leq} \frac{\text{Trace}(\mathbf{M})H^0_{p_2+2}(\chi^2_{p_2,\alpha};\triangle)}{2H_{p_2+2}(\chi^2_{p_2,\alpha};\triangle) - H_{p_2+4}(\chi^2_{p_2,\alpha};\triangle)}. \end{aligned} \tag{8.3.84}$$

In the special case of $\boldsymbol{\xi} = \mathbf{0}$ both $\hat{\boldsymbol{\beta}}_1$ and $\hat{\boldsymbol{\beta}}_{PT}$ obviously dominate $\tilde{\boldsymbol{\beta}}_1$ in their ADR. In that case we also have

$$\mathbf{M} = \nu^2\mathbf{Q}_{21}(\mathbf{I} - \mathbf{M}^\star)\mathbf{Q}_{11}^{-1}\mathbf{Q}_{12} \text{ with } \mathbf{M}^\star = \mathbf{Q}_{11}^{-1}\mathbf{Q}_{12}\mathbf{Q}_{22}^{-1}\mathbf{Q}_{21}. \tag{8.3.85}$$

So $\text{Trace}(\mathbf{M}^\star) = \text{Trace}(\mathbf{M}^0)$. In the general case,

$$\begin{aligned} &\text{Trace}(\mathbf{W}\mathbf{Q}_{11}^{-1}) = \nu^{-2}\text{Trace}(\mathbf{Q}_{11.2}\mathbf{Q}_{11}^{-1}) \\ &= \nu^{-2}\text{Trace}(\mathbf{I}_{p_2} - (\mathbf{Q}_{12}\mathbf{Q}_{22}^{-1}\mathbf{Q}_{21}\mathbf{Q}_{11}^{-1}) \\ &= \nu^{-2}[p_1 - \text{Trace}(\mathbf{M}^\star)] = \nu^2[p_1 - \text{Trace}(\mathbf{M}^0)], \end{aligned} \tag{8.3.86}$$

which is a positive number independent of $\boldsymbol{\xi}$. Hence under the nonlocal alternative, when $\boldsymbol{\xi}$ eventually moves away from $\mathbf{0}$, the risk of $\hat{\boldsymbol{\beta}}_1$ in (8.3.79) monotonically increases and goes to ∞ as $\boldsymbol{\xi}'\mathbf{M}\boldsymbol{\xi} \to \infty$. Thus the restricted estimator $\hat{\boldsymbol{\beta}}_1$ may not perform well when the assumed pivot is different form the tried value of $\boldsymbol{\beta}_2$. Concerning the preliminary test estimator $\hat{\boldsymbol{\beta}}_1^{PT}$, the last term in its risk in (8.3.80) is bounded away from infinity even when $\triangle \to \infty$. Actually $H_{p_2}(x, \triangle) > H_{p_2+4}(x, \triangle)$, $\forall x > 0$ and $\triangle \geq 0$ and $H_{p_2}(x, \triangle) \to 0$ as $\triangle \to \infty$. Hence, even if $R(\hat{\boldsymbol{\beta}}_1^{PT}, \mathbf{W})$ exceeds $R(\tilde{\boldsymbol{\beta}}_1, \mathbf{W})$ for $\boldsymbol{\xi}$ lying outside a closed neighborhood of the pivot, the excess is bounded, and it converges to 0 as $\triangle \to \infty$. From this point of view, though inadmissible, $\hat{\boldsymbol{\beta}}_1^{PT}$ is quite robust with respect to ADR and does not demand any extra condition on $\mathbf{M}^0$ besides the basic one that $\mathbf{M}^0 \neq \mathbf{0}$.

Let us consider the Stein-rule estimator in case of the Mahalanobis distance. It follows from (8.3.79) and (8.3.81) that for $\hat{\boldsymbol{\beta}}_1^S$ to dominate $\tilde{\boldsymbol{\beta}}_1$, that is for $R(\tilde{\boldsymbol{\beta}}_1; \mathbf{W}) - R(\hat{\boldsymbol{\beta}}_1^S; \mathbf{W})$ to be nonnegative, we need that

$$\begin{aligned} &k\text{Trace}(\mathbf{M}^0)[2\mathbb{E}(\chi_{p_2+2}^{-2}(\Delta)) - k\mathbb{E}(\chi_{p_2+2}^{-4}(\Delta))] \\ &\geq k(k+4)(\boldsymbol{\xi}'\mathbf{M}\boldsymbol{\xi})\mathbb{E}(\chi_{p_2+2}^{-4}(\Delta)), \ \forall \boldsymbol{\xi}. \end{aligned} \tag{8.3.87}$$

This sufficient conditions can be rewritten in the following way:

$$2\mathbb{E}(\chi_{p_2+2}^{-2}(\Delta)) - k\mathbb{E}(\chi_{p_2+2}^{-4}(\Delta)) \geq h(k+4)\Delta\mathbb{E}(\chi_{p_2+2}^{-4}(\Delta)) \tag{8.3.88}$$

for all $\Delta \geq 0$, where

$$h = \frac{\text{Ch}_{\max}(\mathbf{M}^0)}{\text{Trace}(\mathbf{M}^0)} \in (0, 1]. \tag{8.3.89}$$

It follows from (8.3.85) that

$$(\boldsymbol{\xi}'\mathbf{M}\boldsymbol{\xi})/\Delta \leq \nu^{-2}\text{Ch}_{\max}(\mathbf{M}\mathbf{Q}_{22.1}^{-1}) = \text{Ch}_{\max}(\mathbf{M}^0) = h\text{Trace}(\mathbf{M}^0). \tag{8.3.90}$$

The condition (8.3.88) in turn requires that

$$p_2 \geq 3, \ \ 0 < k < 2(p_2 - 2), \ \text{and} \ h(k+4) \leq 2 \tag{8.3.91}$$

or, equivalently, $\hat{\boldsymbol{\beta}}_1^S$ dominates $\tilde{\boldsymbol{\beta}}_1$ in ADR whenever

$$p^\star = \min(p_1, p_2) \geq 3, \ \ h < \frac{1}{2}, \ \ 0 \leq k \leq \min\{\frac{2}{h} - 4, 2(p_2 - 2)\}. \tag{8.3.92}$$

Actually by (8.3.22),

$$\begin{aligned} \text{Trace}(\mathbf{M}^0) &= \text{Trace}([\mathbf{Q}_{11} - \mathbf{Q}_{11.2}]\mathbf{Q}_{11}^{-1}) \\ &= p_1 \text{Trace}(\mathbf{Q}_{11.2}\mathbf{Q}_{11}^{-1}) \\ (&= p_2 \text{Trace}(\mathbf{Q}_{22.1}\mathbf{Q}_{22}^{-1}) \ ) \\ &\leq p^\star. \end{aligned} \tag{8.3.93}$$

Since (8.3.90) implies that $h < 1/2$, hence $\text{Rank}(\mathbf{M}^0) \leq 2$.

Finally, (8.3.80) and (8.3.92) lead us to the following conclusion on the relation of $\hat{\boldsymbol{\beta}}_1^{PT}$ and $\hat{\boldsymbol{\beta}}_S$:

Under the regularity conditions pertaining to (8.3.80) and (8.3.81), the preliminary test estimator $\hat{\boldsymbol{\beta}}_1^{PT}$ fails to dominate the Stein-rule estimator $\hat{\boldsymbol{\beta}}_1^S$ (in the light of their ADR). Further, if the significance level α of the preliminary test satisfies the condition

$$H_{p_2}(\chi_{p_2+2}^2; 0) \geq \frac{q(2(p_2 - 2) - q)}{p_2(p_2 - 2)}, \tag{8.3.94}$$

then the Stein-rule estimator fails to dominate the preliminary test estimator.

We leave the proof as an exercise (Problem 8.3.3).

Notice that (8.3.92) holds for a wide range of values p_2 and q because α is usually small and thus $\chi^2_{p_2,\alpha}$ large. For example, in case $\alpha = 0.05$, it holds for all $p_2 \in [3, 11]$ (and even for higher p_2 when q is small), while for $\alpha = 0.05$, it holds for all $p_2 \leq 21$. However, the maximum excess of the risk of the PTE over that of the SRE is generally very small, and hence it is acceptable in practical applications.

A parallel study can be made for the positive-rule version $\hat{\boldsymbol{\beta}}_1^{S+}$ of the Stein-rule estimate in (8.3.66); namely (8.3.62) can be modified as follows: Denoting

$$\mathbf{U}_n^\star = \begin{cases} \mathbf{0} & \text{if } \mathcal{L}_n \leq k, \\ \mathbf{I} - kd_n n^{-1}\mathcal{L}_n^{-1}\mathbf{W}^{-1}\mathbf{c}_{n11.2} & \text{if } \mathcal{L}_n \leq k, \end{cases} \tag{8.3.95}$$

we define

$$\hat{\boldsymbol{\beta}}_{n1}^{S\star} = \hat{\boldsymbol{\beta}}_{n1} + \mathbf{U}_n^\star(\tilde{\boldsymbol{\beta}}_{n1} - \hat{\boldsymbol{\beta}}_{n1}). \tag{8.3.96}$$

The risk of $\hat{\boldsymbol{\beta}}_{n1}^{S\star}$ could be studied on parallel lines.

We will now use Theorem 8.3.3 in the construction of the asymptotic minimum risk efficiency (AMRE) estimation in a sequential setup. As in (8.2.59), the optimal sample size $n_c^0 = \mathrm{O}(c^{-1/2})$ as $c \searrow 0$. However, there is no pivot in Section 8.2. Hence n_c^0 is independent of $\boldsymbol{\xi}$ appearing in the local alternative $H_n^\star$. On the other hand, the ADR of shrunken estimators depend on $\boldsymbol{\xi}$. Hence some version of the local alternative should appear also in the sequential setup. For this purpose, define a sequence $\{A(c) : c > 0\}$ of nested neighborhoods

$$A(c) = \{\boldsymbol{\theta} : ||\boldsymbol{\theta} - \boldsymbol{\theta}_0|| \leq Ac^{1/4}\},\ c > 0, \tag{8.3.97}$$

with the pivot $\boldsymbol{\theta}_0$. Then, defining the stopping number N_c as in (8.2.63), we can show that the random vector $c^{-1/4}(\mathbf{T}_{N_c}^S - \mathbf{T}_{N_c})$ has a nondegenerate asymptotic distribution for $\boldsymbol{\theta} \in A(c)$, while it can be degenerate for $\boldsymbol{\theta} \notin A(c)$. Hence $\mathbf{T}_{N_c}^S$ and $\mathbf{T}_{N_c}$ are not asymptotically (distributional risk) equivalent when $\theta \in A(c)$, while they may be so for $\boldsymbol{\theta} \notin A(c)$. We have an analogous conclusion for $\mathbf{T}_{N_c}^{S\star}$. Since $N_c/n_c^0 \xrightarrow{P} 1$ as $c \searrow 0$, we may conclude from Theorem 8.3.3 that $\mathbf{T}_n^S$ dominates $\mathbf{T}_N$ when $\boldsymbol{\theta} \in A(c)$. Hence the reduction of the ADR is possible by "shrinking" the estimators in the case that the assumed "pivot" is "close to " the true situation. This is in agreement with the general results on sequential shrinkage estimation for the normal model (studied in detail by Ghosh, Nickerson, and Sen 1987, and others).

There is still a question that applies to both normal and nonnormal models: v_n in (8.2.62) estimates $\nu(F)$. However, in replacing $\mathbf{T}_n$ by a shrunken estimator $\mathbf{T}_n^S$, we must also replace $\nu(F)$ by $\nu^S(F)$ which depends on $\boldsymbol{\theta}$ as well as on $\nu(F)$ and is $\leq \nu(F)\ \forall\boldsymbol{\theta}$. Thus, if it is possible to estimate $\nu^S(F)$ consistently by, say, v_n^S which is then used in (8.2.62) to define the stopping number N_c^S, we may readily conclude that $N_c^S \leq N_c$ whenever

$v_n^S \le v_n \; \forall n$. This question may be formulated as follows: Does $\mathbf{T}^S_{N^S_c}$ have a smaller risk than $\mathbf{T}_{N_c}$ for all $\boldsymbol{\theta}$? A more critical question is: Does there exist a stopping rule $\{N_c^\star;\; c > 0\}$, such that (1) $N_c^\star$ is stochastically smaller than N_c while (2) the ADR of $\mathbf{T}^S_{N_c^\star}$ is not grater than the ADR of $\mathbf{T}^S_{N_c}$? The answer is affirmative, and the problem is solved by an iterative process that can be quite cumbersome (see Sen 1990). We omit the details. The reader might well consider N_c, $c \ge 0$, in (8.2.62) as a good working rule, since the extra gain is generally very small.

Remark. A challenging alternative to the procedures developed in this section is the class of procedures based on regression rank scores (RR) defined in Section 6.7. An estimator of $\boldsymbol{\beta}_1$ based on RR, proposed by Jurečková (1992b), is invariant to $\boldsymbol{\beta}_2$, and hence we may not need to test the hypothesis that $\boldsymbol{\beta}_2 = \mathbf{0}$. If, however, we wish to make a preliminary test, we could again use a test based on linear regression rank scores statistics (6.7.15) which is invariant with respect to $\boldsymbol{\beta}_1$ and hence avoids a necessity to estimate it.

The procedures based on the RR go beyond the scope of this book, and they are still undergoing intensive development. We refer to Gutenbrunner and Jurečková (1992), Jurečková (1991, 1992 a,b, 1993), Gutenbrunner, Jurečková, Koenker and Portnoy (1993), Jurečková and Sen (1994), and Sen (1995b).

8.4 RECURSIVE AND K-STEP ESTIMATORS

We have observed that, typically, robust M- and R-estimators are nonlinear, although they have FOADR under fairly general regularity conditions. In regard to the fixed-sample size estimation procedure, we suggested in earlier chapters that a one-step estimation procedure (based on $\sqrt{n}$-consistent estimator at the initial step) greatly simplifies the computational scheme and simultaneously retains the (first order) efficiency properties in an asymptotic sense. The situation is admittedly more complex for a sequential estimation rule. If for every sample size (n) an estimator has to be computed afresh, and if an iteration procedure is needed to do so, it generally becomes very laborious. Therefore we present here some recursive and k-step estimators that are appealing from the computational point of view. Note that for a linear estimator this problem is not acute. In fact, for the LSE of location, $\bar{X}_n$, we have $\bar{X}_{n+1} = n\bar{X}_n/(n+1) + X_{n+1}/(n+1)$, $\forall n \ge 1$, indicating that estimators can be obtained recursively at successive stages. Such an exact recursive relation may not generally hold for nonlinear estimators; see, for example, the case of the median (posed in Problem 8.4.1). Nevertheless, motivated by our asymptotic representations, considered in Chapters 4-6, we are able to unify some recursive estimators in a common vein.

We represent this methodology first for an M-estimator of location, and use the same for motivating the case of R-estimators of location. Then we consider a broad outline of parallel results for the linear model. Let $\{X_i;\ i \geq 1\}$ be a sequence of i.i.d. r.v.'s with a continuous d.f. $F(x, \theta) = F(x - \theta)$, where θ ($\in \mathbb{R}$) is the location parameter and F is assumed to be symmetric about 0. Based on a sample of size n and a score function $\psi : \mathbb{R} \mapsto \mathbb{R}$, $\hat{\theta}_n$, an M-estimator of θ, has then been defined as a solution of the estimating equation (w.r.t. $t \in \mathbb{R}$):

$$M_n(t) = \int \psi(x - t)dF_n(x) = 0. \tag{8.4.1}$$

In this context the regularity conditions assumed earlier ensures that

$$\int \psi dF = 0, \quad \sigma_\psi^2 = \int \psi^2 dF < \infty, \quad \gamma = \int f d\psi \neq 0. \tag{8.4.2}$$

From (8.4.1) we have

$$\begin{aligned} 0 &= M_n(\hat{\theta}_n) = M_n(\hat{\theta}_{n-1}) + \{M_n(\hat{\theta}_n) - M_n(\hat{\theta}_{n-1})\} \\ &= n^{-1}\psi(X_n - \hat{\theta}_{n-1}) + \int [\psi(x - \hat{\theta}_n) - \psi(x - \hat{\theta}_{n-1})]dF_n(x) \\ &= n^{-1}\psi(X_n - \hat{\theta}_{n-1}) - (\hat{\theta}_n - \hat{\theta}_{n-1})V_n, \end{aligned} \tag{8.4.3}$$

where

$$V_n = \begin{cases} 0, & \text{if } \hat{\theta}_n = \hat{\theta}_{n-1}, \\ \frac{[M_n(\hat{\theta}_n) - M_n(\hat{\theta}_{n-1})]}{\hat{\theta}_{n-1} - \hat{\theta}_n} & \text{if } \hat{\theta}_n \neq \hat{\theta}_{n-1}. \end{cases} \tag{8.4.4}$$

Then we write

$$\hat{\theta}_n = \begin{cases} \hat{\theta}_{n-1}, & \text{if } M_n(\hat{\theta}_{n-1}) = 0, \\ \hat{\theta}_{n-1} + V_n^{-1} n^{-1}\psi(X_n - \hat{\theta}_{n-1}), & \text{otherwise}. \end{cases} \tag{8.4.5}$$

But V_n itself depend on $\hat{\theta}_n$. So we need to replace V_n by a recursive estimator. For this purpose we refer back to the FOADR result in Chapter 5:

$$\hat{\theta}_n - \theta = \gamma^{-1} M_n(\theta) + R_n, \tag{8.4.6}$$

where $M_n(\theta) = \mathrm{O}_p(n^{-1/2})$ and $R_n = \mathrm{O}_p(n^{-\nu})$, for some $\nu > 1/2$, depending on the nature of ψ and F. In fact, under appropriate regularity assumptions leading to a SOADR result, $R_n = \mathrm{O}_p(n^{-3/4})$ or $\mathrm{O}_p(n^{-1})$, depending on whether ψ admits jump discontinuities or it is absolutely continuous. Moreover, $M_n(\theta)$ is a linear statistic, and hence (Problem 8.4.2),

$$n(n-1)\mathbb{E}\{[M_{n-1}(\theta) - M_n(\theta)]^2\} = \sigma_\psi^2; \tag{8.4.7}$$

then $M_{n-1}(\theta) - M_n(\theta) = \mathrm{O}_p(n^{-1})$. By (8.4.6) and (8.4.7) we have

$$\hat{\theta}_n - \hat{\theta}_{n-1} = \mathrm{o}_p(n^{-1/2}), \quad \text{as } \to \infty. \tag{8.4.8}$$

Further, by (8.4.1), we have

$$M_n(\hat{\theta}_n) - M_n(\hat{\theta}_{n-1}) = \int_{\boldsymbol{R}} [\psi(x - \hat{\theta}_n) - \psi(x - \hat{\theta}_{n-1})] dF_n(x). \tag{8.4.9}$$

If $\psi(.)$ admits jump-discontinuities, writing (8.4.9) as

$$\int_{\boldsymbol{R}} \Big[\int_{\hat{\theta}_{n-1}}^{\hat{\theta}_n} d\psi(x - t) \Big] dF_n(x), \tag{8.4.10}$$

we observe that V_n may not behave well (we refer back to Problem 8.4.1). On the other hand, if ψ is absolutely continuous, we gather from (8.4.4) and (8.4.10) that

$$V_n \xrightarrow{\boldsymbol{P}} \gamma, \quad \text{as } n \to \infty. \tag{8.4.11}$$

In such a case we may use a convenient estimator of γ and provide a good approximation of $\hat{\theta}_n$ by a recursive one. We may consider a sequence $\{n^{-1/2}t = t_n,\ n \geq 1\}$, where t ($\in \mathbb{R}$) is fixed and let

$$\hat{V}_n = -t_n^{-1}\{M_n(\hat{\theta}_{n-1} + t_n) - M_n(\hat{\theta}_{n-1})\}. \tag{8.4.12}$$

Then we consider the following recursive M-estimator of θ:

$$\hat{\theta}_n^\star = \begin{cases} \hat{\theta}_{n-1}, & \text{if } M_n(\hat{\theta}_{n-1}) = 0, \\ \hat{\theta}_{n-1} + \hat{V}_n^{-1} n^{-1} \psi(X_n - \hat{\theta}_{n-1}) & \text{if } M_n(\hat{\theta}_{n-1}) \neq 0. \end{cases} \tag{8.4.13}$$

In a sequential scheme, in (8.4.12) and (8.4.13), we replace $\hat{\theta}_{n-1}$ by $\hat{\theta}_{n-1}^\star$, for every $n \geq n_0$.

Basically, motivated by (8.4.5), (8.4.11), and (8.4.13), we may consider a general class of recursive M-estimators of location based on the score function ψ. Let

$$\tilde{\theta}_n^\star = \begin{cases} \tilde{\theta}_{n-1}^\star, & \text{if } M_n(\tilde{\theta}_{n-1}^\star) = 0, \\ \tilde{\theta}_{n-1}^\star + \hat{\gamma}_{n-1}^{-1} n^{-1} \psi(X_n - \tilde{\theta}_{n-1}^\star), & \text{otherwise}, \end{cases} \tag{8.4.14}$$

where $\hat{\gamma}_{n-1}$ is a robust estimator of γ based on the sample of size $n-1$. A very convenient form of $\hat{\gamma}_{n-1}$ (based on the asymptotic uniform linearity results in Chapter 5) is

$$\hat{\gamma}_{n-1} = (2t_n)^{-1}\{M_{n-1}(\tilde{\theta}_{n-1}^\star - t_n) - M_{n-1}(\tilde{\theta}_{n-1}^\star + t_n)\}, \tag{8.4.15}$$

where $t_n = n^{-1/2}t$, $t \in \mathbb{R}$ (fixed). Stochastic (or a.s.) convergence $\hat{\gamma}_{n-1}$ to γ (along with suitable rates of convergence) can be obtained by incorporating the basic results in Section 5.3. Hence (see Problem 8.4.3), it can be verified that under suitable regularity condition,

$$\tilde{\theta}_n^\star = \theta + \gamma^{-1} M_n(\theta) + o_p(n^{-1/2}), \quad \text{as } n \to \infty, \tag{8.4.16}$$

which is the same FOADR result applicable to $\hat{\theta}_n$.

This recursive estimation scheme eliminates to a greater extent the computational complexities of $\hat{\theta}_n$, which is usually based on an iterative procedure. In Section 7.4 we considered a one-step estimator of θ, based on $\sqrt{n}$-consistent estimator $\tilde{\theta}_n$ of θ. Note that

$$\begin{aligned} 0 &= M_n(\hat{\theta}_n) = M_n(\tilde{\theta}_n) + \{M_n(\hat{\theta}_n) - M_n(\tilde{\theta}_n)\} \\ &= M_n(\tilde{\theta}_n) + (\hat{\theta}_n - \tilde{\theta}_n)\{[M_n(\hat{\theta}_n) - M_n(\tilde{\theta}_n)]/(\hat{\theta}_n - \tilde{\theta}_n)\}. \end{aligned}$$

As in (8.4.5) we have for $M_n(\tilde{\theta}_n) \neq 0$,

$$\hat{\theta}_n = \tilde{\theta}_n - M_n(\tilde{\theta}_n)/\hat{V}_n, \tag{8.4.17}$$

where

$$\hat{V}_n = \{[M_n(\hat{\theta}_n) - M_n(\tilde{\theta}_n)]/(\hat{\theta}_n - \tilde{\theta}_n)\}. \tag{8.4.18}$$

It has been shown that $\hat{V}_n + \gamma \xrightarrow{P} 0$ as $n \to \infty$. Thus, parallel to (8.4.15), we define

$$\hat{\gamma}_n^{(0)} = (2t_n)^{-1}\{M_n(\tilde{\theta}_n - t_n) - M_n(\tilde{\theta}_n + t_n)\}, \tag{8.4.19}$$

and define a one-step M-estimator

$$\hat{\theta}_n^{(1)} = \begin{cases} \tilde{\theta}_n, & \text{if } M_n(\tilde{\theta}_n) = 0, \\ \tilde{\theta}_n + (\hat{\gamma}_n^{(0)})^{-1} M_n(\tilde{\theta}_n), & \text{otherwise.} \end{cases} \tag{8.4.20}$$

In the second step, we replace $\tilde{\theta}_n$ and $\hat{\gamma}_n^{(0)}$ by $\tilde{\theta}_n^{(1)}$ and $\hat{\gamma}_n^{(1)}$ respectively, where $\hat{\gamma}_n^{(1)}$ is defined as in (8.4.19) with $\tilde{\theta}_n$ replaced by $\tilde{\theta}_n^{(1)}$. This leads to the two-step estimator $\tilde{\theta}_n^{(2)}$. This process may be continued if desired, and we have the k-step M-estimator of θ:

$$\hat{\theta}_n^{(k)} = \begin{cases} \hat{\theta}_n^{(k-1)}, & \text{if } M_n(\hat{\theta}_n^{(k-1)}) = 0, \\ \hat{\theta}_n^{(k-1)} + (\hat{\gamma}_n^{(k-1)})^{-1} M_n(\hat{\theta}_n^{(k-1)}), & \text{otherwise.} \end{cases} \tag{8.4.21}$$

for $k = 1, 2, \ldots$. Detailed properties of such estimators have been studied by Jurečková and Sen (1991), and we relegate a depiction of this to Problem 8.4.4.

We now incorporate this idea for modifying the recursive estimator in (8.4.14). Our objective to eliminate the small bias that crops up due to recursions done sequentially (over n) for a long range. We denote $\tilde{\theta}_n^\star$ in (8.4.14) as $\tilde{\theta}_n^{\star(0)}$ to indicate that it is an initial recursive estimator, and (8.4.16) establishes its $\sqrt{n}$-consistency. Then, as in (8.4.20), we consider the one-step recursive estimator $\tilde{\theta}_n^{\star(1)}$ where for $\tilde{\theta}_n$ and $\hat{\gamma}_n^{(0)}$, we take $\tilde{\theta}_n^{\star(0)}$ and $\hat{\gamma}_{n-1}$, respectively. The rest of the procedure is the same as in (8.4.21). Thus we have a k-step recursive M-estimator of θ:

$$\tilde{\theta}_n^{\star(k)} = \begin{cases} \tilde{\theta}_n^{\star(k-1)}, & \text{if } M_n(\tilde{\theta}_n^{\star(k-1)}) = 0, \\ \tilde{\theta}_n^{\star(k-1)} + (\hat{\gamma}_n^{(k-1)})^{-1} M_n(\tilde{\theta}_n^{\star(k-1)}), & \text{otherwise.} \end{cases} \tag{8.4.22}$$

for $k = 1, 2, \ldots$; here also for all practical purposes, $k = 2$ may prove to be adequate.

Let us now turn to recursive R-estimators of location. Excepting the case of Wilcoxon signed rank statistic on the sign statistic, the derived estimator $\hat{\theta}_n$ of θ may not have a closed form; it has to be obtained by iteration. This makes k-step estimation rules very appealing. Moreover, in a sequential setup, recursive estimation remains as appealing as in the case of M-estimators. The theory of recursive R-estimation of location runs almost parallel to that of M-estimators treated earlier. Instead of $M_n(t)$ in (8.4.1), we have to incorporate an aligned signed rank statistic $S_n(t)$, $t \in \mathbb{R}$, as introduced in (3.4.9). We may then write for $\hat{\theta}_n \neq \hat{\theta}_{n-1}$,

$$\begin{aligned} S_n(\hat{\theta}_n) &= 0 = S_n(\hat{\theta}_{n-1}) + \{S_n(\hat{\theta}_n) - S_n(\hat{\theta}_{n-1})\} \\ &= S_n(\hat{\theta}_{n-1}) + (\hat{\theta}_n - \hat{\theta}_{n-1}) V_n, \end{aligned} \tag{8.4.23}$$

where

$$V_n = \{S_n(\hat{\theta}_n) - S_n(\hat{\theta}_{n-1})\}/(\hat{\theta}_n - \hat{\theta}_{n-1}). \tag{8.4.24}$$

Then the uniform asymptotic linearity of signed rank statistic (in shift) and FOADR results for $\hat{\theta}_n$ imply that $V_n \xrightarrow{P} -\gamma$ as $n \to \infty$, where an estimator $\hat{\gamma}_n$ of γ can again be constructed as in Chapter 6. Thus we have for $S_n(\hat{\theta}_{n-1}) \neq 0$,

$$\hat{\theta}_n^{\star} = \hat{\theta}_{n-1} + \hat{\gamma}_n^{-1} S_n(\hat{\theta}_{n-1})$$

[where $S_n(t)$ is taken as an average rather than a sum statistic]; for $S_n(\hat{\theta}_{n-1}) = 0$, we have $\hat{\theta}_n^{\star} = \hat{\theta}_{n-1}$. The rest of the discussion including the k-step recursive versions follows the same lines as M-estimators; Problems 8.4.5-8.4.6 are set up to verify some of these details.

From the above discussion pertaining to recursive M- and R-estimators, it is quite clear that if $\{T_n\}$ is a sequence of (possibly) nonlinear estimators, such that T_n admits a FOADR and T_n is derived through an estimating equation, then whenever a consistent estimator of the scale factor appearing in the asymptotic distribution of $\sqrt{n}(T_n - \theta)$ is available, recursive k-step versions of T_n can be obtained, and computationally these will be simpler. Asymptotically they share the properties with T_n. The case of MLE's in a parametric setup is included in this formulation (see Problem 8.4.7). In passing, we may remark that M-estimators of location are not generally scale-equivariant, and hence in Section 5.4 we considered some studentized M-estimators of location. Theorem 5.4.1 pertains to a FOADR result for such studentized M-estimators of location. It follows from the above discussion that under the regularity conditions of Theorem 5.4.1, recursive k-step versions of studentized M-estimators of location can be computed (involving estimators of both θ and the scale factor s), and they share the FOADR result with the estimator considered in Section 5.4. We pose Problem 8.4.8 to illustrate this point. This modification does not arise with R-estimators of location which are scale-equivariant by definition.

Let us note that for the multivariate location model [with d.f. $F(\mathbf{x}, \boldsymbol{\theta})$ = $F(\mathbf{x}-\boldsymbol{\theta})$, $\boldsymbol{\theta} \in \mathbb{R}_p, \mathbf{x} \in R_p$], we may consider M- or R-estimators of $\boldsymbol{\theta}$ on a coordinatewise basis so that their recursive versions can also be obtained along the same line as in before.

We proceed to consider the linear model (5.5.1). We formulate suitable recursive versions of the usual M- and R-estimators of $\boldsymbol{\beta}$, the regression parameter vector. For simplicity of presentation, we consider the fixed-scale case first. With the notations borrowed from Chapter 5, we have the estimating equations yielding an M-estimator of $\hat{\boldsymbol{\beta}}_n$ of $\boldsymbol{\beta}$, based on the score function $\psi : \mathbb{R} \mapsto \mathbb{R}$, given by

$$\mathbf{M}_n(\hat{\boldsymbol{\beta}}_n) = \sum_{i=1}^{n} \mathbf{x}_i \psi(Y_i - \hat{\boldsymbol{\beta}}_n' \mathbf{x}_i) = \mathbf{0}. \tag{8.4.25}$$

Thus, as in the location model, we write

$$\begin{aligned} \mathbf{0} &= \mathbf{M}_n(\hat{\boldsymbol{\beta}}_n) = \mathbf{M}_n(\hat{\boldsymbol{\beta}}_{n-1}) + [\mathbf{M}_n(\hat{\boldsymbol{\beta}}_n) - \mathbf{M}_n(\hat{\boldsymbol{\beta}}_{n-1})] \\ &= \mathbf{x}_n \psi(Y_n - \hat{\boldsymbol{\beta}}_{n-1}' \mathbf{x}_n) + \mathbf{H}_n(\hat{\boldsymbol{\beta}}_n - \hat{\boldsymbol{\beta}}_{n-1}), \end{aligned} \tag{8.4.26}$$

where

$$\mathbf{H}_n = \frac{\partial}{\partial \mathbf{t}'} \mathbf{M}_n(\mathbf{t})|_{\mathbf{t}=\alpha\hat{\boldsymbol{\beta}}_n + (1-\alpha)\hat{\boldsymbol{\beta}}_{n-1}}, \quad 0 < \alpha < 1. \tag{8.4.27}$$

Then we have

$$\hat{\boldsymbol{\beta}}_n - \hat{\boldsymbol{\beta}}_{n-1} = \mathbf{H}_n^{-1} \mathbf{x}_n \psi'(Y_n - \hat{\boldsymbol{\beta}}_{n-1}' \mathbf{x}_n). \tag{8.4.28}$$

Again, $\mathbf{H}_n$ depends both on $\hat{\boldsymbol{\beta}}_n$, and $\hat{\boldsymbol{\beta}}_{n-1}$, and hence the crux of the problem is to replace $\mathbf{H}_n$ by a version depending on $\hat{\boldsymbol{\beta}}_{n-1}$ and $Y_1, \ldots, Y_{n-1}$, $\mathbf{x}_1, \ldots, \mathbf{x}_{n-1}$. An intuitive version is

$$\tilde{\mathbf{H}}_n = \sum_{i=1}^{n} \mathbf{x}_i \mathbf{x}_i' \psi'(Y_i - \hat{\boldsymbol{\beta}}_{n-1}' \mathbf{x}_i). \tag{8.4.29}$$

A recursive version of $\hat{\boldsymbol{\beta}}_n$, denoted by $\hat{\boldsymbol{\beta}}_n^\star$, is given by

$$\hat{\boldsymbol{\beta}}_n^\star = \begin{cases} \hat{\boldsymbol{\beta}}_{n-1}, & \text{if } \mathbf{M}_n(\hat{\boldsymbol{\beta}}_{n-1}) = \mathbf{0}, \\ \hat{\boldsymbol{\beta}}_{n-1} + \tilde{\mathbf{H}}_n^{-1} \mathbf{x}_n \psi(Y_n - \hat{\boldsymbol{\beta}}_{n-1}' \mathbf{x}_n), & \text{otherwise.} \end{cases} \tag{8.4.30}$$

To cut down the influence of $\mathbf{H}_n$, we may also consider a k-step version of (8.4.30) that is analogous to the location model. We denote by $\tilde{\mathbf{H}}_n^\star$ the solution of

$$\mathbf{M}_n(\hat{\boldsymbol{\beta}}_n^\star) - \mathbf{M}_n(\hat{\boldsymbol{\beta}}_{n-1}) = \tilde{\mathbf{H}}_n^\star(\hat{\boldsymbol{\beta}}_n^\star - \hat{\boldsymbol{\beta}}_{n-1}). \tag{8.4.31}$$

In (8.4.30) we replace $\mathbf{H}_n^{-1}$ by $(\tilde{\mathbf{H}}_n^\star)^{-1}$ and denote the resulting estimator by $\hat{\boldsymbol{\beta}}_n^{\star(1)}$. We can repeat this process, k times ($k \geq 1$) and denote the resulting estimator $\hat{\boldsymbol{\beta}}_n^{\star(k)}$ as the k-step recursive estimator of $\boldsymbol{\beta}$. Let us denote

$$\mathbf{H}_n^0 = \sum_{i=1}^n \mathbf{x}_i\mathbf{x}_i'\psi(Y_i - \boldsymbol{\beta}'\mathbf{x}_i). \tag{8.4.32}$$

Since the $Y_i - \boldsymbol{\beta}'\mathbf{x}_i$ are i.i.d. r.v.'s, it follows (see Problem 8.4.9) that

$$\mathbf{C}_n^{-1}\mathbf{H}_n^0 \xrightarrow{P} \mathbf{I}_p \cdot \gamma, \tag{8.4.33}$$

where $\mathbf{C}_n = \sum_{i=1}^n \mathbf{x}_i\mathbf{x}_i'$, γ is defined by (8.4.2) and $\mathbf{I}_p$ is the $p \times p$ identity matrix. On the other hand, looking at (8.4.29) and (8.4.32), we may gather that under appropriate regularity conditions

$$\mathbf{C}_n^{-1}[\tilde{\mathbf{H}}_n - \mathbf{H}_n^0] \xrightarrow{P} \mathbf{0}, \quad \text{as } n \to \infty; \tag{8.4.34}$$

Problem 8.4.10 is set up to verify this. In the preceding development, it is tacitly assumed that $\psi'(.)$ is continuous almost everywhere (i.e., the points of discontinuities of ψ' have null measure). It is possible to rewrite γ in an alternative form (requiring finite Fisher information) where this condition on ψ' may be eliminated. In this case the recursive M-estimation theory works out well. In Chapter 5 we have assumed that $\sum_{i=1}^n x_{ji} = 0$, for $j = 2, \ldots, p$ and $x_{1i} = 1$, $i = 1, \ldots, n$. In a sequential setup we may not be able to impose this condition, since $\sum_{i=1}^n x_{ji} = 0$, $\forall n \Rightarrow x_{ji} = 0\ \forall n$. However, the modifications needed in Theorems 5.5.1-5.5.3 to eliminate this restriction are quite straightforward, and hence we will not cover them.

The situation is much more complicated with studentized M-estimators treated in Section 5.5. In this case we have

$$\mathbf{M}_n(\hat{\boldsymbol{\beta}}_n) = \sum_{i=1}^n \mathbf{x}_i\psi((Y_i - \hat{\boldsymbol{\beta}}_n'\mathbf{x}_i)/S_n) = \mathbf{0}, \tag{8.4.35}$$

where S_n is an appropriate scale statistic. Thus, if we write as in (8.4.26),

$$\begin{aligned} \mathbf{0} &= \mathbf{M}_n(\hat{\boldsymbol{\beta}}_n) = \mathbf{M}_n(\hat{\boldsymbol{\beta}}_{n-1}) + [\mathbf{M}_n(\hat{\boldsymbol{\beta}}_n) - \mathbf{M}_n(\hat{\boldsymbol{\beta}}_{n-1})] \\ &= \sum_{i=1}^n \mathbf{x}_i\psi(\frac{Y_i - \hat{\boldsymbol{\beta}}_{n-1}'\mathbf{x}_i}{S_n}) \\ &\quad + \sum_{i=1}^n \mathbf{x}_i\Big[\psi(\frac{Y_i - \hat{\boldsymbol{\beta}}_n'\mathbf{x}_i}{S_n}) - \psi(\frac{Y_i - \hat{\boldsymbol{\beta}}_{n-1}'\mathbf{x}_i}{S_n})\Big], \end{aligned} \tag{8.4.36}$$

then the first term on the right-hand side of (8.4.36) may not reduce to $\mathbf{x}_n\psi((Y_i - \hat{\boldsymbol{\beta}}'_{n-1}\mathbf{x}_n)/S_n)$, since S_{n-1} and S_n may not be identical or $\psi(.)$ is not linear everywhere. Nevertheless, this term is a linear statistic. On the other hand, for the second term we may proceed as in (8.4.27)-(8.4.28), and as in (8.4.29) take

$$\tilde{\mathbf{H}}_n = S_n^{-1}\sum_{i=1}^{n}\mathbf{x}_i\mathbf{x}_i'\psi'\Big(\frac{Y_i - \hat{\boldsymbol{\beta}}'_{n-1}\mathbf{x}_i}{S_n}\Big). \tag{8.4.37}$$

Thus it may be more convenient to estimate S_n recursively first, and then to use the same in (8.4.36)-(8.4.37) to obtain the recursive version of $\hat{\boldsymbol{\beta}}_n$. It is desirable to use a robust S_n, although the (strong) consistency of S_n is sufficient in this context.

We conclude this section with some remarks on recursive R-estimation in linear models. Although these estimators are scale equivariant (so that studentization is not necessary), computationally they are generally more involved. Whereas $\mathbf{M}_n(\mathbf{t})$ in (8.4.25) is a (vector of) linear statistic, $\mathbf{L}_n(\mathbf{t})$, as defined in Chapter 6, yielding the usual R-estimates of $\hat{\boldsymbol{\beta}}_n$ of $\boldsymbol{\beta}$, is not a linear statistic (but a linear rank statistic) vector. Unfortunately, such (aligned) rank scores often do not behave that linearly, and hence an iterative procedure is needed to solve for $\hat{\boldsymbol{\beta}}_n$. This makes the k-step versions more appealing. The basic uniform asymptotic linearity of rank statistics for the regression parameter, studied in Section 6.6, provides the necessary links, and the methodology is quite similar to the one for recursive M-estimators; Problems 8.4.10-8.4.12 are set up to verify the details. In the concluding sections of Chapter 6, we introduced regression rank scores estimators of $\boldsymbol{\beta}$ and discussed their relative merits and demerits. Note that some linear programming algorithm is involved in their solution, albeit we may not have a closed expression (in general). However, based on the asymptotic equivalence of classical R-estimators and regression rank score estimators, the methodology of recursive estimation also extends to regression rank scores estimators; Problem 8.4.13 is set up to figure out the details.

8.5 ROBUST ADAPTIVE ESTIMATION

The basic results in Chapter 7 convey a general theme on the asymptotic equivalence of L-, M-, and R-estimators of location or regression parameters when they are based on conjugate score functions. The results in the earlier sections of this chapter extend this general asymptotic equivalence result to sequential and recursive estimation problems. This picture is more closely studied when for each of these estimators, the corresponding influence functions are taken to be isomorphic. To illustrate this point, consider the location model. For an R-estimator based on a score function $\phi^+(t)$ [or equivalently $\phi(t)$], $t \in (0, 1)$, the influence function is given by

$$IC(x; F)_R \propto \phi(F(x - \theta)), \quad x \in \mathbb{R}. \tag{8.5.1}$$

For an M-estimator based on a score function $\psi : \mathbb{R} \mapsto \mathbb{R}$, the corresponding influence function is

$$IC(x; F)_M \propto \psi(x - \theta), \; x \in \mathbb{R}, \tag{8.5.2}$$

while for an L-estimator, expressible as a differentiable statistical functional, we have

$$IC(x; F)_L = T_1(x; F), \quad x \in \mathbb{R}, \tag{8.5.3}$$

where $T_1(x; F)$ is the Hadamard (or compact) derivative of the functional at F. Thus, whenever

$$\phi(F(x - \theta)) \equiv \psi(x - \theta) \equiv T_1(x; F), \quad x \in \mathbb{R}, \tag{8.5.4}$$

the three estimators are asymptotic $\sqrt{n}$-equivalent, in probability, and this asymptotic equivalence holds even for the sequential setup where n is replaced by a stopping time N, satisfying some mild restrictions. In a parametric setup, for F belonging to a parametric family, it is possible to choose ϕ in such a way that

$$\mathbb{E}_F \phi^2(F(x - \theta)) = I(f), \tag{8.5.5}$$

where $I(f)$ is the Fisher information (on θ). Then, by the Cramér-Rao information inequality, the corresponding R-estimator is asymptotically optimal (or efficient). [Actually, in (8.5.1), we have $(I(f))^{-1}\phi(F(x - \theta))$.] A similar result holds for the other estimators as well. Thus there arises a pertinent question:

If the d.f. F is not known but is assumed to be a member of a class $\mathcal{F}$, is it possible to choose a score-generating function (probably adaptively) such that (8.5.5) holds for all $F \in \mathcal{F}$?

The answer is in the affirmative. Note that for

$$\psi(F(x - \theta)) \equiv \{-f'(x - \theta)/f(x - \theta)\}, \; x \in \mathbb{R}, \tag{8.5.6}$$

the Cramér-Rao bound is attained. Intuitively one needs to estimate the density and its first derivative in an adaptive manner. This method is, however, not very practical, since it generally requires an enormously large sample size to have the full impact of this adaptive solution (see Hájek and Šidák 1967). A similar situation arises with L- and M-estimators for being asymptotically optimal (in terms of their asymptotic mean square errors). In the sequential estimation problem relating to the AMRE property (see Section 8.2), one has an additional complication: not only to choose such an efficient score function adaptively but also the sample size such that the AMRE property holds with respect to a larger class of (sequential) point estimators location. Specifically one needs to choose adaptively a score function as well as a sample size, such that as $c \downarrow 0$, the (asymptotic) risk of the estimator is equal to

$$2c^{\frac{1}{2}}(I(f))^{-1/2}. \tag{8.5.7}$$

Adaptive (i.e., data-dependent) procedures based on a Fourier series representation of a score function involving an orthonormal system have been developed over the past 15 years, and they are applicable for comparatively moderate sample sizes. For simplicity of presentation, we consider adaptive R-estimators of location/regression parameters, and briefly append the case of other estimators. Most of these developments are adopted from Hušková and Sen (1985, 1986).

Note that by (8.5.6), an optimal score function $\phi = \{\phi(u), u \in (0,1)\}$ is given by

$$\phi_F(u) = -\{f'(F^{-1}(u))/f(F^{-1}(u))\},\ 0 < u < 1. \tag{8.5.8}$$

Then, whenever $I(f) < \infty$, we have by (8.5.8)

$$\int_0^1 \phi_F(u)du = 0 \text{ and } 0 < \int_0^1 \phi_F^2(u)du = I(f) < \infty. \tag{8.5.9}$$

Because $\phi_F \in L^2(0,1)$, we are tempted to use an orthonormal system $\{[P_k(u), u \in (0,1)],\ k \geq 0\}$ in a Fourier series representation

$$\phi_f(u) \sim \sum_{k \geq 0} \gamma_k P_k(u),\ u \in (0,1), \tag{8.5.10}$$

where

$$\int_0^1 P_k(u)du = 0,\ \forall k \geq 0, \tag{8.5.11}$$

$$\int_0^1 P_k(u)P_q(u)du = \delta_{kq} = \begin{cases} 1, & k = q, \\ 0, & k \neq q, \end{cases} \tag{8.5.12}$$

for $k, q \geq 0$, and the γ_k are the Fourier coefficients. In this context, the Legendre polynomial system is particularly useful. For this system we have

$$P_k(u) = (2k+1)^{1/2}(-1)^k(k!)^{-1}[\frac{d^k}{du^k}\{u(1-u)\}^k] \tag{8.5.13}$$

for $u \in (0,1)$ and $k = 0, 1, \ldots, \infty$. Note that for $k = 0, 1$, $P_0(u) \equiv 1$ and $P_1(u) = \sqrt{12}(u - 1/2)$ correspond to the sign statistic and Wilcoxon signed rank statistic score functions, and in general, $P_k(u)$ is a polynomial in u of degree k (≥ 0). Further for every $k \geq 0$,

$$P_{2k}(u) \equiv P_{2k}(1-u) \text{ and } P_{2k+1}(u) = -P_{2k+1}(1-u),\ \ u \in (0,1). \tag{8.5.14}$$

By virtue of (8.5.10)-(8.5.12), we have

$$\langle \phi_F, \phi_F \rangle = \int_0^1 \phi_F^2(t)dt = \sum_{k\geq 0} \gamma_k^2 = I(f), \tag{8.5.15}$$

$$\langle \phi_F, P_k \rangle = \int_0^1 \phi_F(t) P_k(t)dt = \sum_{q\geq 0} \gamma_q \langle P_k, P_q \rangle = \gamma_k, \tag{8.5.16}$$

for evey $k \geq 0$. For the location model, by virtue of the assumed symmetry of F (around 0), $\phi_F(u)$ is skew-symmetric, so by (8.5.14) we have

$$\langle \phi_F, P_{2k} \rangle = 0 \ \forall k \geq 0. \tag{8.5.17}$$

For the location model we take

$$\phi_F(u) \sim \sum_{k\geq 0} \gamma_{2k+1} P_{2k+1}(u), \ u \in (0,1). \tag{8.5.18}$$

The first term on the right-hand side of (8.5.18) is the classical Wilcoxon score function $P_1(.)$, so γ_1 represent the contribution of the same in an L_2-representation for ψ_F, while other terms represent the nonlinearity of $\phi_F(.)$.

Motivated by the above representation, our goal is to choose an adaptive score function

$$\hat{\phi}_{F,M}(u) = \sum_{k\leq M} \hat{\gamma}_{kn} P_k(u), \ u \in (0,1), \tag{8.5.19}$$

where M is a positive integer (r.v.) to be chosen from the dataset, and the $\hat{\gamma}_{k,n}$ are the estimates of the Fourier coefficients γ_k. The results of Chapter 6 enable us to consider suitable nonparametric estimates of the γ_k (based on the asymptotic uniform linearity of the derived rank statistics). We illustrate this procedure for the location model, although a very similar case holds for the linear model.

Consider a R-estimator of θ based on the Wilcoxon signed-rank statistic (i.e., $\phi(u) \equiv P_1(u), \ u \in (0,1)$). We denote this estimator by

$$\tilde{\theta}_{n1} = \tilde{\theta}(X_1, \ldots, X_n; P_1(.)), \ n \geq 1. \tag{8.5.20}$$

$\tilde{\theta}_{n1}$ is the Wilcoxon-score estimator of θ. Hence

$$\tilde{\theta}_{n1} = \text{median}\{\frac{1}{2}(X_i + X_j) : \ 1 \leq i \leq j \leq n\}, \tag{8.5.21}$$

and this closed form easily leads to a recursive version. Next we denote by

$$S_{nk}(t) = \sum_{i=1}^{n} \text{sign}(X_i - t) a_{nk}(R_{ni}^+(t)), \ t \in \mathbb{R}, \tag{8.5.22}$$

where $a_{nk}(i)$ is the score based on the score-function $\phi \equiv P_k$ for $i = 1, \ldots, n, \ k \geq 0$, and $R_{ni}^+(t)$ = rank of $|X_i - t|$ among the $|X_\alpha - t|, \ 1 \leq \alpha \leq n$, for

$i = 1, \ldots, n$. Note that the $P_k(u)$ are bounded and continuously differentiable functions on (0,1). We are in a position to make use of the results in Chapter 6 corresponding to the so-called smooth case [where $\phi''(.)$ is bounded]. As such, we propose the following estimator of γ_k: Let a (> 0) be a fixed positive member, and let

$$\hat{\gamma}_{k,n} = (2a\sqrt{n})^{-1}\{S_{nk}(\hat{\theta}_{n1} - a/\sqrt{n}) - S_{nk}(\tilde{\theta}_{n1} + a/\sqrt{n})\} \tag{8.5.23}$$

for $k \geq 1$. Note that $\tilde{\theta}_{n1}$ is a $\sqrt{n}$-consistent, median-unbiased robust estimator of θ, and hence by the uniform asymptotic linearity result in Chapter 6,

$$\hat{\gamma}_{k,n} \stackrel{P}{\to} \gamma_k, \quad \text{as } n \to \infty. \tag{8.5.24}$$

This implies that for finitely many k (≥ 1), we can make use of (8.5.24) to show that for a fixed M ($< \infty$),

$$\begin{aligned} \hat{\phi}_{F,M}(u) &= \sum_{k \leq M} \hat{\gamma}_{k,n} P_k(u) \\ &\stackrel{P}{\to} \sum_{k \leq M} \gamma_k P_k(u) = \phi_{F,M}(u), \quad \text{say.} \end{aligned} \tag{8.5.25}$$

On the other hand, by (8.5.15) and the finiteness of $I(f)$, we claim that for every $\epsilon > 0$, there exists an M ($= M(\epsilon, F)$) such that

$$\sum_{k > M} \gamma_k^2 < \epsilon. \tag{8.5.26}$$

We iterate that this M may depend on F. For example, if F is logistic, $M = 1$ for every $\epsilon > 0$, but for a normal F, M depends on ϵ and the d.f. intricately. Note that (8.5.26) implies that

$$\begin{aligned} ||\phi_F - \phi_{F,M}||^2 &= \langle \phi_F - \phi_{F,M}, \phi_F - \phi_{F,M} \rangle \\ &= \sum_{k > M} \gamma_k^2 < \epsilon. \end{aligned} \tag{8.5.27}$$

Moreover for every r (≥ 1) and $K > M$,

$$\sum_{k=K+1}^{K+r} \gamma_k^2 < \epsilon. \tag{8.5.28}$$

Actually, in view of the boundedness of the $P_k(.)$, $k \geq 0$, some precise rates of convergence of the $\hat{\gamma}_{k,n}$ (to the γ_k) can be obtained; we refer to Problems 8.5.1-8.5.2 for verification of such results. This enables us to replace a fixed ϵ by $\{\epsilon_n\}$, $\epsilon_n \to 0$ as $n \to \infty$, and M by M_n : $M_n \to \infty$ as $n \to \infty$. Motivated by this, Hušková and Sen (1986) suggested a sequence $\{\epsilon_n\}$ of positive numbers and another (slowly increasing) sequence $\{r_n\}$ of positive integers such that $r_n \to \infty$, but $n^{-1}r_n \to 0$ as $n \to \infty$. They also proposed the following stopping rule:

For every n ($\geq n_0$, an initial sample size), let K_n be a stopping number defined by

$$K_n = \min\{k \geq k_0 : \sum_{j=k+1}^{k+r_n} \hat{\gamma}_{j,n}^2 \leq \epsilon_n\}, \tag{8.5.29}$$

where k_0 (≥ 1) is a predetermined positive integer. Then let

$$\hat{\phi}_n(u) = \sum_{k \leq K_n + r_n} \hat{\gamma}_{k,n} P_k(u), \; u \in (0,1), \tag{8.5.30}$$

$$\hat{I}_n = \sum_{k \leq K_n + r_n} \hat{\gamma}_{k,n}^2, \tag{8.5.31}$$

$$\hat{S}_n(t) = S_n(t) \text{ for } \phi \equiv \hat{\phi}_n. \tag{8.5.32}$$

The adaptive estimator $\hat{\theta}_{nA}$ of θ is then based on the adaptive signed rank statistic $\hat{S}_n(t)$, $t \in \mathbb{R}$. We pose Problems 8.5.3-8.5.5 to verify that

$$||\hat{\phi}_n - \phi_F|| \xrightarrow{P} 0, \quad \text{as } n \to \infty, \tag{8.5.33}$$

$$\hat{I}_n \xrightarrow{P} I(f), \quad \text{as } n \to \infty, \tag{8.5.34}$$

and

$$n^{1/2\}}\hat{I}_n^{1/2}(\hat{\theta}_{nA} - \theta) \xrightarrow{\mathcal{D}} \mathcal{N}(0,1). \tag{8.5.35}$$

These results in turn imply that the adaptive estimator $\hat{\theta}_{nA}$ is asymptotically optimal.

Let us now look into recursive and sequential versions of $\hat{\theta}_{nA}$. For the sequential AMRE problem, it follows that we may consider a stopping number N_c defined by

$$N_c = \min\{n \geq n_0 : \; n \geq (ac^{-1})^{1/2}(\hat{I}_n^{1/2} + n^{-d})\}, \; c > 0,$$

where a, c, d, and n_o are all defined as in Section 8.2, and $\hat{I}_n$ by (8.5.31). In this setup the sequential adaptive version of the classical R-estimator of location is

$$\hat{\theta}_{N_c A}, \text{ defined for every } \; c > 0. \tag{8.5.36}$$

The asymptotic risk of $\hat{\theta}_{N_c A}$ is given by

$$a\mathbb{E}_F\{(\hat{\theta}_{N_c A} - \theta)^2\} + c\mathbb{E}_F N_c, \; c \downarrow 0. \tag{8.5.37}$$

First, we may strengthen (8.5.33) to convergence in the second mean and work with the true score function ϕ_F. It is then possible to show that as $c \downarrow 0$, (8.5.37) behaves as

$$2(ac)^{1/2}\{I(f)\}^{1/2} + \mathrm{o}(\sqrt{c}), \tag{8.5.38}$$

and we relegate the proof as Problem 8.5.6.

Recursive versions can be formulated as in Section 8.4 [by considering the sequence $\hat{S}_n(t),\ n \geq n_0$].

8.6 PROBLEMS

8.2.1 Verify (8.2.41) for U-statistics when the kernel has a finite rth moment for some $r > 4$.

8.2.2. If $T(F_n)$ is first order Hadamard differentiable, show that (8.2.41) hold for every $r > 0$, while (8.2.40) holds whenever the compact derivative of $T(F)$ has a finite rth moment, $r > 4$.

8.2.3. For R-estimators of location/regression, show that (8.2.18) holds whenever the score function $\in L_r(0,1)$, for some $r > 4$.

8.2.4. Show that for a square integrable reversed martingale sequence $\{T_n\}$, (8.2.19) holds. Moreover, if $T_n = T_n^0 + R_n$, where $\{T_n^0\}$ is a reversed martingale, and R_n may be expressed as a linear combination of reversed martingales with the coefficients all converging to 0 (as $n \to \infty$) at suitable rates, then (8.2.19) holds. Verify this with the von Mises functionals as well as for Hadamard differentiable functionals.

8.2.5. Assume that the v_n admit a FOADR, where the principal term is an average over i.i.d. r.v's with finite expectation, while the remainder term is either $O_p(n^{-1})$ or converges to 0 in probability in the mode of the Anscombe condition. Then (8.2.45) holds.

8.2.6. Compare (8.2.10) with (8.2.43) and (8.2.46), and verify that (8.2.45) holds whenever $\max\{|v_m - v_n| : |m-n| \leq \delta n\} \xrightarrow{P} 0$ as $n \to \infty$. Show that the a.s. convergence of v_n (to σ_T^2) ensures the above.

8.3.1. Provide a proof of Theorem 8.3.2.

8.3.2. Provide a proof of Theorem 8.3.3.

8.3.3. Show that under (8.3.94), the Stein-rule estimator fails to dominate the PTE.

8.4.1. For the sample median $\tilde{X}_n = \text{med}(X_1, \ldots, X_n)$, show that $(n+1)\tilde{X}_{n+1}$ may not be equal to $n\tilde{X}_n + X_{n+1}$.

8.4.2. Use the reverse martingale property of $M_n(t)$ in (8.4.1), and verify (8.4.7).

8.4.3. State appropriate regularity conditions under which (8.4.16) holds.

8.4.4. Obtain a stochastic order of convergence of $\hat{\theta}_n^{(k)} - \hat{\theta}_n$.

8.4.5. For R-estimators of location/regression, obtain (8.4.16) under appropriate regularity assumptions.

8.4.6. Extend the results in Problem 8.4.4 to R-estimators of location and regression.

8.4.7. Show that k-step MLE's in a parametric model share the same asymptotic properties as the classical MLE.

8.4.8. For studentized M-estimators of location, obtain the stochastic order of convergence of $\hat{\theta}_n^{(k)} - \hat{\theta}_n$.

8.4.9. Verify (8.4.33).

8.4.10. For a rank estimator $\hat{\boldsymbol{\beta}}_n$ of the regression parameter $\boldsymbol{\beta}$, show that the FOADR results in Chapter 6 enables one to obtain (8.4.16).

8.4.11. Extend this result to the estimator of θ (in a general linear model) based on residual signed rank statistics.

8.4.12. Study the of convergence of $\hat{\boldsymbol{\beta}}_n^{(k)} - \hat{\boldsymbol{\beta}}_n$ for rank estimators.

8.4.13. Use the general equivalence results in the last section of Chapter 6, and obtain results parallel to that in the preceding three problems for regression rank scores estimators.

8.5.1. Define $v_{n,k} = \max\{(2k+1)^{5/2} n^{\delta-1/2},\ (2k+1)^{13/2} n^{-1} (\log n)^{3/2}\}$, for $k \geq 0$, $n \gg k$, where $\delta \in (0, 1/4)$. Then show that based on $P_{2k+1}(.)$, for every $s > 0$, there exists an n_s ($< \infty$) such that

$$\mathbb{P}\{|\hat{\gamma}_{k,n} - \gamma_k| > c v_{n,k}\} \leq d n^{-s},\ c > 0, d > 0,$$

for every $k \leq k_n = \mathrm{o}(n^{2/13} (\log n)^{-2})$.

8.5.2. Define $\phi_{F,M}(u)$ and $\hat{\phi}_{F,M}(u)$ as in (8.5.25), and use the preceding problem to obtain a similar probability inequality for $||\hat{\phi}_{F,M} - \phi_{F,M}||$, for a fixed M.

8.5.3. Verify (8.5.33).

8.5.4. Verify (8.5.34).

8.5.5. Verify (8.5.35).

Chapter 9

Robust Confidence Sets and Intervals

9.1 INTRODUCTION

We have so far laid down the main emphasis on the point estimation problem and its robustness aspects. The point estimators of a parameter may not, by themselves, convey full information on their chance fluctuations, although in many cases (especially when the estimators are asymptotically normal) their (estimated) standard errors provide a good deal of information in this respect. For example, consider the problem of estimating the mean (θ) of a population (admitting finite second moment), and let σ^2 be the variance parameter. Based on a sample $X_1, \ldots, X_n$ of size n, natural estimators of θ and σ^2 are $\bar{X}_n = n^{-1}\sum_{i=1}^n X_i$ and $S_n^2 = (n-1)^{-1}\sum_{i=1}^n (X_i - \bar{X}_n)^2$, respectively. Note that $\bar{X}_n$ is an unbiased estimator of θ and

$$\text{standard error of } \bar{X}_n = n^{-1/2}\sigma, \tag{9.1.1}$$

which can be estimated by $n^{-1/2}S_n$ $(= \{\sum_{i=1}^n (X_i - \bar{X}_n)^2/n(n-1)\}^{1/2})$. $\bar{X}_n$ is an optimal estimator of θ when the d.f. of the X_i is normal. We have considered a more general setup wherein θ is taken as location parameter, and presented various other robust competitors of $\bar{X}_n$. In Chapter 3 we have discussed the vulnerability of $\bar{X}_n$ to error contaminations, outliers, and possible departures from distributional assumptions. These drawbacks are shared by S_n to a greater extent. The confidence intervals for θ, based on asymptotic normality (or even Edgeworth expansion) of $n^{1/2}(\bar{X}_n - \theta)/S_n$, are even more vulnerable to such departures. Robustness considerations are even more important for confidence sets and intervals. Hence robust statistics deserve proper attention, and the current chapter is fully devoted to such robust confidence sets.

Typically, based on a sample $X_1, \ldots, X_n$, one wants to formulate two statistics $L_n = L(X_1, \ldots, X_n)$ and $U_n = U(X_1, \ldots, X_n)$, termed the *lower and upper confidence limits*, respectively, such that when θ holds, L_n and U_n cover θ with a probability $\geq 1 - \alpha$ (= *coverage probability* or *confidence coefficient*), where $0 < \alpha < 1$ (and α is generally taken to be small). That is,

$$\mathbb{P}_\theta\{L_n \leq \theta \leq U_n\} \geq 1 - \alpha, \ \forall \theta \in \Theta. \tag{9.1.2}$$

In a nonparametric setup, $\theta = \theta(F)$, $F \in \mathcal{F}$, is a functional of the d.f. F, so we need to rewrite (9.1.2) as

$$\mathbb{P}_F\{L_n \leq \theta(F) \leq U_n\} \geq 1 - \alpha, \ \forall F \in \mathcal{F}. \tag{9.1.3}$$

Then $I_n = [L_n, U_n]$ is termed a *confidence interval.* If $\boldsymbol{\theta}(F)$ is vector valued, then we may conceive of a subspace I_n (of $\mathbb{R}_p$), usually closed such that

$$\mathbb{P}_F\{\boldsymbol{\theta}(F) \in I_n\} \geq 1 - \alpha, \ \forall F \in \mathcal{F}. \tag{9.1.4}$$

In this case I_n is termed a *confidence region.* A further modification is also possible to cover the case where $\theta(F)$ is possibly infinite-dimensional. However, without unnecessarily going into such abstractions, we may introduce the term *confidence set* to include I_n in (9.1.3), (9.1.4), as well as more general cases. Confidence sets are therefore characterized by (random) subsets of the parameter space, for which the coverage probability is preassigned, and some other desirable or optimal properties hold. For the specific one- parameter problem referred to in (9.1.2) or (9.1.3), the crux of the problem is therefore to choose the two confidence limits, L_n and U_n, in such a way that the coverage probability condition holds, and the length of the interval ($l_n = U_n - L_n$) is as small as possible (this concept is essentially related to the idea of *shortest confidence* intervals). The procedure for choosing such a shortest confidence interval [in a fixed (finite) sample setup] is generally highly parametric in nature, although in an asymptotic setup, whenever an estimator $\hat{\theta}_n$ of $\theta(F)$ is asymptotically normal (AN) and its standard error can be estimated, L_n and U_n may be approximated by the usual normal theory formula $\hat{\theta}_n \pm \tau_{\alpha/2}$ (estimated standard error of $\hat{\theta}_n$), where $\Phi(\tau_{\alpha/2}) = 1 - \alpha/2$, and $\Phi(.)$ is the standard normal d.f. In this formulation it is required that both $\hat{\theta}_n$ and its estimated standard error be robust, and this dominates our deliberations in the subsequent sections. R- and M-estimation of location (and regression) may be obtained by setting some *estimating equation* involving suitable *robust statistics.* In principle, this idea extends to confidence sets, in our case to robust ones. Section 9.2 deals such type I and II confidence intervals in the one parameter case. Section 9.3 extends the considerations to multiparameter case. Both these sections are devoted to the usual *fixed-sample size* (i.e., nonsequential) schemes.

Referred to (9.1.2), we may want to set a further restriction that the length (l_n) of the confidence interval I_n has to be bounded from above by a predetermined $\triangle$ (> 0). Similarly as in models of Chapter 8, no fixed sample size procedure may be optimal simultaneously for all $F \in \mathcal{F}$ (or $\theta \in \Theta$), and hence one may need to take recourse to a sequential scheme. Such procedures are studied in Sections 9.4 and 9.5. The last section deals with some general remarks pertaining to confidence sets in a robust mold.

9.2 TYPE I AND II CONFIDENCE INTERVALS

Based on a sample of n i.i.d.r.v.'s $X_1, \ldots, X_n$ from the d.f. F, let $T_n = T(X_1, \ldots, X_n)$ be an estimator of a parameter $\theta = \theta(F)$, and let

$$G_n(t; F) = I\!\!P_F\{T_n - \theta \le t\},\ t \in I\!\!R, \tag{9.2.1}$$

be the d.f. of $T_n - \theta$. Given a coverage probability $1 - \alpha$ $(0 < \alpha < 1)$, consider a partition $\alpha = \alpha_1 + \alpha_2$, $0 \le \alpha_1, \alpha_2 \le \alpha$, and let

$$t_{n\alpha_1}^- = \sup\{t : G_n(t; F) \le \alpha_1\};\ \ t_{n\alpha_2}^+ = \inf\{t : G_n(t; F) \ge 1 - \alpha_2\}. \tag{9.2.2}$$

Then, by definition,

$$I\!\!P_F\{t_{n\alpha_1}^- \le T_n - \theta \le t_{n\alpha_2}^+\} \ge 1 - \alpha. \tag{9.2.3}$$

If the d.f. of $T_n - \theta$ is symmetric about 0, then one may set $\alpha_1 = \alpha_2 = \alpha/2$ and $t_{n\alpha_1}^- = -t_{n\alpha_2}^+ = -t_{n\alpha/2}^+$; otherwise, some other considerations are taken into account in this partitioning of α. In the simplest situation, if $t_{n\alpha_1}^-$ and $t_{n\alpha_2}^+$ do not depend on θ (or F when F is not specified), then, by inversion, we obtain from (9.2.3) that

$$I\!\!P_F\{T_n - t_{n\alpha_2}^+ \le \theta \le T_n + t_{n\alpha_1}^-\} \ge 1 - \alpha\ \forall F \in \mathcal{F}, \tag{9.2.4}$$

and this provides the desired confidence interval of θ. In practice, this simple prescription may not work out even for the simplest parametric models (e.g., θ binomial parameter, Poisson parameter, or the normal mean when the variance σ^2 is not known), although in some cases this drawback may be overcome by using a studentized form [e.g., $(T_n - \theta)/Z_{nT}^{1/2}$, for suitable Z_{nT}; e.g., normal mean problem, σ^2 unknown]. In an asymptotic setup we have under general regularity assumptions

$$n^{1/2}(T_n - \theta) \xrightarrow{\mathcal{D}} \mathcal{N}(0, \sigma_T^2), \tag{9.2.5}$$

$$Z_{nT}/\sigma_T^2 \xrightarrow{P} 1. \tag{9.2.6}$$

By the Slutsky theorem, as n increases,

$$n^{1/2}(T_n - \theta)/Z_{nT}^{1/2} \xrightarrow{\mathcal{D}} \mathcal{N}(0,1). \tag{9.2.7}$$

In such a case we have

$$\lim_{n\to\infty} P_F\{|n^{1/2}(T_n - \theta)/Z_{nT}^{1/2}| \leq \tau_{\alpha/2}\} = 1 - \alpha, \tag{9.2.8}$$

so

$$I_n = \{T_n - n^{-1/2}\tau_{\alpha/2}Z_{nT}^{1/2},\ T_n + n^{-1/2}\tau_{\alpha/2}Z_{nT}^{1/2}\} \tag{9.2.9}$$

provides an asymptotic confidence interval for θ with coverage probability $1 - \alpha$. This we term a type I confidence interval for θ. The length l_{nT} of interval (9.2.9) satisfies

$$n^{1/2}l_{nT} = 2\tau_{\alpha/2}Z_{nT}^{1/2} \xrightarrow{P} 2\tau_{\alpha/2}\sigma_T, \text{ as } n \to \infty.$$

Hence an optimal choice will be a T_n for which σ_T^2 is a minimum, that is, T_n is BAN (best asymptotically normal). On the other hand, one may want to choose some T_n that are globally or locally robust in a certain manner. A desirable confidence interval would be based on robust T_n accompanied by a robust choice of Z_{nT}, leading to a minimum σ_T^2 within a class. Asymptotic optimality of such a confidence interval can be studied through the optimality properties of Z_{nT}. In this context the optimality of T_n (retaining its robustness) and robustness of Z_{nT} therefore are the basic considerations. In a more general framework, where $\theta = \theta(F)$ is treated as a functional of the underlying d.f. F, this prescription works out well for the general class of estimators discussed in earlier chapters. From practical applications point of view, one must examine the adequacy of asymptotic normality (9.2.8) for moderate sample sizes. In the parametric case, we would often derive an exact formulation of the distribution of $n^{1/2}(T_n - \theta)/Z_{nT}^{1/2}$, and for moderate sample sizes, we would find an adequate solution through Edgeworth-type expansions. Such simple second-order approximation may not be readily adaptable in a robustness setup, although some of the SOADR results discussed in earlier chapters may be very pertinent in this context. There remains considerable scope for further work in this area.

If (9.2.8) holds, then the basic problem is to choose an appropriate $\{Z_{nT}\}$ that consistently (and robustly) estimates σ_T^2 and at the same time makes the asymptotic result tenable (to a satisfactory degree) for finite to moderate sample sizes as well. In a parametric setup we often use *transformations of statistics*; the arc sine transformation on the binomial proportion, the log-transformation for the variance parameter, the square-root transformation for Poisson and chi square variates and the tan hyperbolic inverse transformation for the simple correlation coefficient are the classical examples. However, such a *variance stabilizing* transformation depends explicitly on the structure of the variance σ_T^2 (as a function of the unknown parameters), and in a nonparametric (or robustness) setup such an exact formulation may not be possible. Even when it is so, it may not have the simple form for which the Bartlett transformation methodology can be readily adopted. For this reason a robust choice of Z_{nT} or even estimating the d.f. of $n^{1/2}(T_n - \theta)$ is often made with the aid of *resampling plans.* Among these resampling plans *jackknifing* and *bootstrapping* deserve special mention, and we will study their roles in such robust interval estimation of a real parameter.

First, let us consider the jackknife method. Based on the sample $\{X_1, \ldots, X_n\}$ of size n, let T_n be a suitable estimator of θ. We can assume that (9.2.5) holds, although σ_T^2 may be generally unknown and T_n may not be unbiased for θ. To motivate jackknifing, suppose that

$$\mathbb{E}_F(T_n) = \theta + n^{-1}a + n^{-2}b + \ldots, \tag{9.2.10}$$

where $a, b, \ldots$ are unknown quantities and may as well depend on the underlying F. Let T_{n-1} be the same estimator of θ but based on a sample of size $n-1$. Then by (9.2.10),

$$\mathbb{E}_F(T_{n-1}) = \theta + (n-1)^{-1}a + (n-1)^{-2}b + \ldots, \tag{9.2.11}$$

so

$$\mathbb{E}_F\{nT_n - (n-1)T_{n-1}\} = \theta - \frac{1}{n(n-1)}b + \mathrm{O}(n^{-3}). \tag{9.2.12}$$

Hence the effective bias is $\mathrm{O}(n^{-2})$ instead of $\mathrm{O}(n^{-1})$. Motivated by this, we drop the ith observation (X_i) from the base sample $(X_1, \ldots, X_n)$ and denote the estimator based on this subsample of size $n-1$ by $T_{n-1}^{(i)}$, for $i = 1, \ldots, n$. Following (9.2.12), we define the *pseudovariables* as

$$T_{n,i} = nT_n - (n-1)T_{n-1}^{(i)}, \; i = 1, \ldots, n. \tag{9.2.13}$$

Then the jackknifed version T_n^J of T_n is defined by

$$T_n^J = n^{-1}\sum_{i=1}^{n} T_{n,i}. \tag{9.2.14}$$

Note that by (9.2.12), (9.2.13), and (9.2.14) we have

$$\mathbb{E}_F(T_n^J) = \theta - b/(n(n-1)) + \mathrm{O}(n^{-3}). \tag{9.2.15}$$

So the bias of T_n^J is $\mathrm{O}(n^{-2})$ instead of $\mathrm{O}(n^{-1})$ for T_n. This was the primary reason for introducing the jackknifed version, albeit there are several other important features that we want to present. Let us define

$$\begin{aligned} V_n^J &= (n-1)^{-1}\sum_{i=1}^{n}(T_{n,i}-T_n^J)^2 \\ &= (n-1)\sum_{i=1}^{n}(T_{n-1}^{(i)}-T_n^\star)^2, \end{aligned} \tag{9.2.16}$$

where

$$T_n^\star = n^{-1}\sum_{i=1}^{n} T_{n-1}^{(i)}. \tag{9.2.17}$$

V_n^J is termed the jackknifed (or Tukey) variance estimator. If in particular $T_n = \bar{X}_n$, then $T_{n,i} = X_i$, $i = 1,\dots,n$, so $T_n^\star = \bar{X}_n$, and hence $V_n^J = (n-1)^{-1}\sum_{i=1}^n (X_i - \bar{X}_n)^2 = S_n^2$. This led Tukey (1958) to conjecture that V_n^J consistently estimates σ_T^2 for more general statistics as well, and we will see that this is true (under fairly general regularity conditions). Let $\mathcal{C}_n$ be the sigma-field generated by the unordered collection $\{X_1,\dots,X_n\}$ and by X_{n+j}, $j \geq 1$, for $n \geq 1$. If the X_j are real valued, then $\mathcal{C}_n = \mathcal{C}(X_{n:1},\dots X_{n:n}; X_{n+j}, j \geq 1)$, where $X_{n:1} \leq \dots \leq X_{n:n}$ are the order statistics corresponding to $X_1,\dots,X_n$. Then $\mathcal{C}_n$ is nonincreasing in n (≥ 1). Further note that by (9.2.17),

$$T_n^\star = \mathbb{E}\{T_{n-1}|\mathcal{C}_n\},\ n \geq n_0, \tag{9.2.18}$$

where $n_0 = \inf\{m:\ T_{m-1}$ is well defined $\}$. As a result

$$\begin{aligned} T_n^J &= T_n + (n-1)\mathbb{E}\{(T_n - T_{n-1})|\mathcal{C}_n\} \\ &= T_n - (n-1)\mathbb{E}\{(T_{n-1} - T_n)|\mathcal{C}_n\},\ n \geq n_0. \end{aligned} \tag{9.2.19}$$

Whenever $\{T_n, \mathcal{C}_n;\ n \geq n_0\}$ is a reverse martingale, we have $\mathbb{E}(T_{n-1}|\mathcal{C}_n) = T_n$, so $T_n^J = T_n$, $\forall n \geq n_0$. Thus, for the entire class of reversed martingales, T_n^J agree with T_n for every $n \geq n_0$. By (9.2.16) and (9.2.18) we have

$$\begin{aligned} V_n^J &= n(n-1)\mathbb{E}\{(T_{n-1} - T_n^\star)^2|\mathcal{C}_n\} \\ &= n(n-1)\mathbb{E}\{[T_{n-1} - \mathbb{E}(T_{n-1}|\mathcal{C}_n)]^2|\mathcal{C}_n\} \\ &= n(n-1)\mathrm{Var}(T_{n-1}|\mathcal{C}_n),\ n \geq n_0. \end{aligned} \tag{9.2.20}$$

Thus convergence properties of the conditional variance in (9.2.20) imply the convergence of V_n^J to σ_T^2. If T_n is a reversed martingale, then a sufficient condition for the asymptotic normality (9.2.5) of T_n is that

$$n(n-1)\mathrm{Var}((T_{n-1} - T_n)|\mathcal{C}_n) \to \sigma_T^2 \text{ a.s.}, \quad \text{as } n \to \infty \tag{9.2.21}$$

[see (2.5.85)]. For V_n^J to converge a.s. to σ_T^2 (as $n \to \infty$), it suffices to show that (9.2.21) holds and that $\{T_n\}$ is reversed martingale; see Problem

9.2.1. Generally $\{T_n\}$ may not be a reversed martingale, although U-statistics and L-estimators are reversed martingales under general regularity conditions. Nevertheless, the convergence $V_n^J \to \sigma_T^2$ a.s. may be proved with the aid of asymptotic representations of robust estimators, derived in earlier chapters. Assume that T_n admits the representation

$$T_n - \theta = n^{-1} \sum_{i=1}^{n} \phi(X_i; \theta) + R_n, \tag{9.2.22}$$

where $\mathbb{E}_F \phi(X; \theta) = 0$, $\mathbb{E}_F[\phi(X_i, \theta)]^2 = \sigma_T^2 : 0 < \sigma_T^2 < \infty$, and R_n is the remainder term. Define $R_{n,i}$ and $R_n^\star$ as in (9.2.13) and (9.2.17) (with T_n replaced by R_n). Then by (9.2.22) we have

$$T_{n,i} - \theta = \phi(X_i, \theta) + R_{n,i}, \; i = 1, \ldots, n. \tag{9.2.23}$$

Let us denote

$$V_n^\star = (n-1)^{-1} \sum_{i=1}^{n} \{\phi(X_i; \theta) - \frac{1}{n} \sum_{j=1}^{n} \phi(X_j, \theta)\}^2, \tag{9.2.24}$$

$$V_{n,2} = (n-1)^{-1} \sum_{i=1}^{n} (R_{n,i} - R_n^\star)^2, \tag{9.2.25}$$

and

$$V_{n,12} = (n-1)^{-1} \sum_{i=1}^{n} \Big[\phi(X_i; \theta) - \frac{1}{n} \sum_{j=1}^{n} \phi(X_j; \theta)\Big] (R_{n,i} - R_n^\star). \tag{9.2.26}$$

By (9.2.16), (9.2.23), (9.2.24), (9.2.25), and (9.2.26), we have

$$V_n^J = V_n^\star + V_{n,2} + 2V_{n,12}. \tag{9.2.27}$$

Now $V_n^\star$ is a U-statistic corresponding to a kernel of degree 2 (and hence, is a reversed martingale), so

$$V_n^\star \to \mathbb{E} V_n^\star = \sigma_T^2 \text{ a.s., as} n \to \infty. \tag{9.2.28}$$

The convergence $V_{n,2} \to 0$ in probability /a.s. as $n \to \infty$, would further imply that $V_{n,12} \to 0$ in probability /a.s., as $n \to \infty$. Therefore the crux of the problem is to verify that as $n \to \infty$,

$$\begin{aligned} V_{n,2} &= n(n-1)\mathbb{E}\{(R_{n-1} - \mathbb{E}(R_{n-1}|\mathcal{C}_n))^2|\mathcal{C}_n\} \\ &\to 0, \text{ in probability /a.s.} \end{aligned} \tag{9.2.29}$$

Recall that

$$\begin{aligned} &\mathbb{E}\{(R_{n-1} - \mathbb{E}(R_{n-1}|\mathcal{C}_n))^2|\mathcal{C}_n\} \\ &= \text{Var}(R_{n-1} - R_n)|\mathcal{C}_n) \\ &\leq \mathbb{E}\{(R_{n-1} - R_n)^2|\mathcal{C}_n\}, \; \forall n \geq n_0, \end{aligned} \tag{9.2.30}$$

so

$$\begin{aligned} \mathbb{E}V_{n,2} &\leq \mathbb{E}\{\mathbb{E}\{(R_{n-1}-R_n)^2|\mathcal{C}_n\}\} \cdot n(n-1) \\ &= n(n-1)\mathbb{E}(R_{n-1}-R_n)^2. \end{aligned} \tag{9.2.31}$$

Whenever $\mathbb{E}(R_{n-1}-R_n)^2 = o(n^{-2})$ as $n \to \infty$, (9.2.31) converges to 0 as $n \to \infty$, and hence $V_{n,2} \to 0$, in probability, as $n \to \infty$. Consequently for the weak consistency of V_n^J it suffices to assume that (9.2.22) holds for $\{T_n\}$ and $\mathbb{E}(R_{n-1}-R_n) = o(n^{-2})$ as $n \to \infty$. The a.s. convergence of V_n^J may need some additional (although mild) regularity conditions. Rewrite the representation (9.2.22) in the form

$$T_n - \theta = n^{-1/2}Z_n^{(1)} + n^{-1}Z_n^{(2)}, \tag{9.2.32}$$

where

$$Z_n^{(1)} = n^{-1/2}\sum_{i=1}^{n}\phi(X_i;\theta) \sim \mathcal{N}(0,\sigma_T^2); \tag{9.2.33}$$

$$Z_n^{(2)} = nR_n. \tag{9.2.34}$$

Then we have by (9.2.25) and (9.2.32),

$$\begin{aligned} V_{n,2} &= (n-1)^{-1}\sum_{i=1}^{n}(Z_{n-1}^{i(2)} - Z_n^{(2)})^2 \\ &= n(n-1)^{-1}\mathbb{E}\{[Z_{n-1}^{(2)} - \mathbb{E}(Z_{n-1}^{(2)}|\mathcal{C}_n)]^2|\mathcal{C}_n\} \\ &\leq n(n-1)^{-1}\mathbb{E}\{(Z_{n-1}^{(2)} - Z_n^{(2)})^2|\mathcal{C}_n\}. \end{aligned} \tag{9.2.35}$$

Therefore

$$\begin{aligned} \mathbb{E}\{(Z_{n-1}^{(2)} - Z_n^{(2)})^2|\mathcal{C}_n\} &\to 0 \text{ a.s.,} \\ \Rightarrow V_{n,2} &\to 0 \text{ a.s. as } n \to \infty. \end{aligned} \tag{9.2.36}$$

For U-statistics or the von Mises functional, the classical Hoeffding (H-)dec-om-po-sition of T_n (into orthogonal terms) yields (9.2.32) along with (9.2.36). In general (9.2.29) holds (a.s.) under (9.2.30) provided that

$$R_n = R_{n0} + R_{n1}, \ n \geq n_0, \tag{9.2.37}$$

where $\{R_{n0}, \mathcal{C}_n\}$ is a reversed martingale with

$$n^2\mathbb{E}(R_{n0}) \to c \ (0 \leq c < \infty), \text{ as } n \to \infty, \tag{9.2.38}$$

while

$$nR_{n1} \to 0 \text{ a.s., as } n \to \infty. \tag{9.2.39}$$

Note that (9.2.37)-(9.2.39) hold for second-order Hadamard differentiable statistical functionals (see Problem 9.2.2) as well as for L-estimators with smooth weight functions (see Problem 9.2.3). This leads to the following conclusion:

Theorem 9.2.1 *Let $\{T_n\}$ satisfy (9.2.22). Then the following conditions are sufficient for $V_n^J \to \sigma_T^2$ in prob/a.s. as $n \to \infty$:*
1. $n(n-1)(R_n - R_{n-1})^2 \to 0$ in prob/a.s., as $n \to \infty$:
2. $\{T_n\}$ is a reversed martingale and

$$n(n-1)\text{Var}(T_{n-1}|\mathcal{C}_n) \to \sigma_T^2 \text{ a.s. } as\ n \to \infty, \tag{9.2.40}$$

3. $\{T_n\}$ satisfy (9.2.37)-(9.2.39).

Let us now consider the nonregular case where (9.2.37)-(9.2.39) may not hold, and hence (9.2.40) may not hold either. As a very simple (but illustrative) example, we consider the simple location model where θ is the population median, the density $f(x, \theta) = f(x - \theta)$ is symmetric, and $f(0) > 0$. Then, for T_n=median$(X_1, \ldots, X_n)$, the SOADR result, studied in Chapter 4, implies that

$$T_n - \theta = n^{-1}\sum_{i=1}^{n}\{I(X_i > \theta) - \frac{1}{2}\} + R_n, \tag{9.2.41}$$

where $R_n = \text{O}_p(n^{-3/4})$, and this order is exact. Thus

$$n^2 \mathbb{E} R_n^2 = \text{O}(n^{\frac{1}{2}}) \not\to 0 \ \text{ as } n \to \infty, \tag{9.2.42}$$

and (9.2.37)-(9.2.38) or even (9.2.40) do not hold in general. In such a case the simple jackknifing technique will be inadequate. But we might consider a modified procedure (called the *delete-k jackknife method*) which works well, when k is not too small. If we confine ourselves to differentiable statistical functionals (introduced in Section 3.6), then we can establish the a.s. convergence of V_n^J under fairly general regularity conditions (see, Sen 1989a). Defining the empirical d.f. by

$$F_n(x) = n^{-1}\sum_{i=1}^{n} I(X_i \le x), \ x \in \mathbb{R}, \tag{9.2.43}$$

we have

$$T_n = T(F_n) \quad \text{and} \quad \theta = T(F), \tag{9.2.44}$$

where $T(.)$ is a suitable functional. For a given k $(1 \le k < n)$, for every $\mathbf{i} = (i_1, \ldots, i_k) \in I = \{\mathbf{i} : 1 \le i_1 < \ldots i_k \le n\}$, we let

$$\mathbf{n} = \{1, \ldots, n\}, \ s_{n,k}^{(\mathbf{i})} = \mathbf{n} \setminus \mathbf{i}, \tag{9.2.45}$$

$$F_{n-k}^{(\mathbf{i})}(x) = \frac{1}{n-k}\sum_{j \in S_{n,k}^{(\mathbf{i})}} I(X_j \le x), \ x \in \mathbb{R}; \tag{9.2.46}$$

$$T_{n-k}^{(\mathbf{i})} = T(F_{n-k}^{(\mathbf{i})}), \ \ \mathbf{i} \in I. \tag{9.2.47}$$

Note that $\#I = \binom{n}{k}$. The pseudovariables generated by the delete-k jackknifing are then defined by

$$T_{n,\boldsymbol{i}}^{(k)} = k^{-1}\{nT_n - (n-k)T_{n-k}^{(\boldsymbol{i})}\}, \quad \mathbf{i} \in I. \tag{9.2.48}$$

The delete-k jackknifed estimator of θ is defined as

$$T_{n,k}^J = \binom{n}{k}^{-1} \sum_{\boldsymbol{i}\in I} T_{n,\boldsymbol{i}}^{(k)}, \ k \geq 1. \tag{9.2.49}$$

Side by side, we can introduce the variance functions

$$V_{n,k}^{\star} = \binom{n}{k}^{-1} \sum_{\boldsymbol{i}\in I} (T_{n,\boldsymbol{i}}^{(k)} - T_{n,k}^J)^2, \ \ k \geq 1. \tag{9.2.50}$$

At this stage we appeal to the conventional definition of Hadamard or compact differentiability of $T(.)$ at F, as formulated in (3.6.15)- (3.6.18), and denote the Hadamard or compact derivative by $T^{(1)}(F;x)$ as in (3.6.19). Further, as in (3.6.21), we note that $T^{(1)}(F;x) = IC(x;F,T)$ is also the influence function of $T(.)$ at F, and that

$$\begin{aligned} \sigma_T^2 &= \mathbb{E}_F\{[T^{(1)}(F;x)]^2\} \\ &= \int_{\boldsymbol{R}} (T^{(1)}(F;x))^2 dF(x) = T_1^{\star}(F), \ \text{say}. \end{aligned} \tag{9.2.51}$$

For a functional $T(.)$ satisfying the condition [see (3.6.16)] that

$$\lim_{t\to 0} \{T(F+tH) - T(F)\} = 0 \tag{9.2.52}$$

uniformly in H belonging to a class $\mathcal{C}$ of compact subsets, we introduce a comparatively weaker notion of *Hadamard (or compact) continuity* of $T(.)$ at F belonging to $\mathcal{A}$ (where $\mathcal{C} \in \mathcal{A}$). Moreover, if (9.2.52) holds uniformly in F belonging to a set $\mathcal{F}$, we refer to this as *uniform Hadamard continuity.* If $\mathcal{F}$ consists only of an infinitesimal neighborhood of a fixed F_o, then we say that it is *locally uniformly Hadamard continuous* (and similarly, *locally uniformly Hadamard differentiable*).

The delete-k jackknifed variance estimator is defined by

$$V_{n,k}^J = k(n-k)^{-1}(n-1)V_{n,k}^{\star}, \ k \geq 1, \tag{9.2.53}$$

where the $V_{n,k}^{\star}$ are defined by (9.2.50). We have the following:

Theorem 9.2.2 *If $T(.)$ is first order (locally uniformly) Hadamard differentiable at F and $T_1^{\star}(.)$, defined by (9.2.51), is locally uniformly Hadamard continuous at F, then as $n \to \infty$,*

$$V_{n,1}^J \to \sigma_T^2 \text{ a.s.}, \tag{9.2.54}$$

and whenever $k = k_n$ is nondecreasing in n, with

$$k_n = O(n^{-1-\eta}), \text{ for some } \eta > 0, \tag{9.2.55}$$

for every $k:\ 1 \le k \le k_n$,

$$V_{n,k}^J - V_{n,1}^J \to 0 \text{ a.s.}. \tag{9.2.56}$$

[**Remark.** The regularity conditions of Theorem 9.2.1 and 9.2.2, mainly adapted from Sen (1988a,b;1989a), are not generally (partially) ordered. For some statistics it may be easier to verify (9.2.22) and (9.2.37), although the locally uniformity condition in Theorem 9.2.2may usually be verified under less restrictive conditions than in (9.2.37).]

Proof. Instead of (9.2.23) we consider an expansion of $T(F_{n-1}^{(i)})$ around $T(F_n)$ (where $||F_n - F|| \to 0$ a.s., as $n \to \infty$), and by an appeal to the assumed uniform (locally) Hadamard differentiability of $T(.)$ at F, we have

$$\begin{aligned} T_{n-1}^{(i)} &= T(F_{n-1}^{(i)}) = T(F_n + (F_{n-1}^{(i)} - F_n)) \\ &= T(F_n) + \int T_1(F_n; x) dF_{n-1}^{(i)}(x) + o(||F_{n-1}^{(i)} - F_n||), \end{aligned} \tag{9.2.57}$$

for every $i : 1 \le i \le n$, where

$$||F_{n-1}^{(i)} - F_n|| = n^{-1}, \quad \text{for every } i = 1, \ldots, n. \tag{9.2.58}$$

We have by (9.2.13), (9.2.57), and (9.2.58),

$$T_{n,i} = T_n + T_1(F_n; X_i) + o(1) \text{ a.s. } \forall i = 1, \ldots, n, \tag{9.2.59}$$

where $n^{-1}\sum_{i=1}^n T_1(F_n, X_i) = \int T_1(F_n; x) dF_n(x) = 0$ a.e., and hence we have

$$\begin{aligned} n^{-1}(n-1)V_n^J &= V_{n,1}^J = n^{-1}\sum_{i=1}^n (T_{n,i} - T_{n,1}^J)^2 \\ &= n^{-1}\sum_{i=1}^n (T_1(F_n; X_i) + o(1))^2 \\ &= \int T_1^2(F_n; x) dF_n(x)[1 + o(1)] \\ &= T_1^\star(F_n)[1 + o(1)], \text{ a.s. as } n \to \infty. \end{aligned} \tag{9.2.60}$$

Note that by the Glivenko-Cantelli lemma, $||F_n - F|| \to 0$ a.s., as $n \to \infty$, so the (local) uniform Hadamard continuity of $T_1^\star(.)$ and (9.2.51) imply that $T_1^\star(F_n) \to \sigma_T^2$. □

A notable example where the uniform Hadamard continuity (or differentiability) condition does not hold is the sample quantile (or median).

Such a functional belongs to the so-called *nonsmooth* class where the classical jackknifing may not yield a consistent variance estimator. Nevertheless, the delete-k jackknifing with k not too small yields a satisfactory solution. We leave the details for Problem 9.2.4. While in the above two theorems it has been tacitly assumed that (9.2.5) holds, there may be some situations where a limiting distribution exists but is not normal (see Problem 9.2.5). Also in such a case jackknifing may not yield a consistent variance estimator. Moreover, since the variance factor may not directly appear as a scale parameter in the asymptotic distribution, estimating it does not serve any real purpose [see (9.2.8)]. The nonregular cases should be treated individually, and for some of them there really exist alternative methods of solutions. Bootstrapping,described below, is one of such possibilities.

Recall that T_n ($= T(X_1, \ldots, X_n)$ or $T(F_n)$) is based on the base sample $(X_1, \ldots, X_n)$ with the corresponding empirical d.f. F_n. F_n is a natural (and very desirable) estimator of the true d.f. F. Draw a sample of n observations from the sample d.f. F_n with replacement (under a simple random sampling (SRS) scheme), and denote them by $X_1^\star, \ldots, X_n^\star$, respectively. Let $F_n^\star$ be the empirical d.f. of these $X_1^\star$, $i \leq n$. Then let

$$T_n^\star = T(X_1^\star, \ldots, X_n^\star) \text{ or } T(F_n^\star). \tag{9.2.61}$$

Note that, by definition, given F_n, the $X_j^\star$ are conditionally independent, and

$$\mathbb{P}\{X_j^\star = X_i | F_n\} = 1/n \text{ for every } i = 1, \ldots, n;\ j \geq 1. \tag{9.2.62}$$

Thus $(X_1^\star, \ldots, X_n^\star)$ can have n^n possible realizations $\{(X_{i_1}, \ldots, X_{i_n}) : i_j = 1, \ldots, n, \text{ for } j = 1, \ldots, n\}$, and each of them has the common (conditional) probability n^{-n}. Therefore

$$\mathbb{E}\{T_n^\star | F_n\} = n^{-n} \sum_{i_1=1}^{n} \cdots \sum_{i_n=1}^{n} T(X_{i_1}, \ldots, X_{i_n}), \tag{9.2.63}$$

and the conditional mean square error of $T_n^\star$ (given F_n) is

$$\mathbb{E}\{(T_n^\star - T_n)^2 | F_n\} = n^{-n} \sum_{i_1=1}^{n} \cdots \sum_{i_n=1}^{n} [T(X_{i_1}, \ldots, X_{i_n}) - T_n]^2. \tag{9.2.64}$$

Under fairly general regularity conditions,

$$n\mathbb{E}\{(T_n^\star - T_n)^2 | F_n\} \to \sigma_T^2, \text{ in probability,} \tag{9.2.65}$$

as $n \to \infty$. On the other hand, as n increases, the exact computation of the conditional mean square error in (9.2.64) may generally become a prohibitively laborious task [even worse than delete-k jackknifing where the computational task is $O(n^k)$]. To eliminate this basic problem, Efron (1979, 1982) proposed the following resampling scheme: Draw (independently) a large number (say, M) of such bootstrap samples [i.e., $(X_1^\star, \ldots, X_n^\star)$] from F_n. For the kth bootstrap sample, denote the corresponding observations by $(X_{k1}^\star, \ldots, X_{kn}^\star)$, for $k = 1, \ldots, M$. Note that these $X_{ki}^\star$ are conditionally i.i.d. r.v.'s with the d.f. F_n, and (9.2.62) applies to each of them. Let

$$T_{nk}^\star = T(X_{k1}^\star, \ldots, X_{kn}^\star) \text{ or } T(F_{kn}^\star),\ k = 1, \ldots, M, \tag{9.2.66}$$

and define

$$V_{nM}^\star = nM^{-1}\sum_{k=1}^{M}(T_{nk}^\star - T_n)^2. \tag{9.2.67}$$

This statistics $V_{nM}^\star$ is termed the *bootstrap estimator* of σ_T^2. The resampling plan in the bootstrap method is thus the simple random sampling with replacement (SRSWR). These M replications not only allow a consistent estimator of σ_T^2, they also generate the bootstrap empirical d.f.,

$$G_n^\star(y) = M^{-1}\sum_{k=1}^{M} I\{n^{1/2}(T_{kn}^\star - T_n) \leq y\},\ y \in \mathbb{R}. \tag{9.2.68}$$

Let $G(y)$ denote the limit law $\lim_{n\to\infty} \mathbb{P}_F\{n^{1/2}(T_n - \theta) \leq y\}$ whenever it exists. Then under certain regularity conditions G_n (uniformly) consistently estimate G. Therefore the percentile points of $G(.)$ may as well be estimated from the parallel quantiles of $G_n^\star(.)$, and hence, one may derive an asymptotic confidence interval for θ by using $G_n^\star(.)$ [and without necessarily assuming that $G(.)$ is normal]. For moderate to large sample sizes, Edgeworth type expansions may also be considered for $G_n^\star(y)$, often providing a better approximation to the estimated percentile points of $G(.)$ or $G_n(y) = \mathbb{P}_F\{n^{1/2}(T_n - \theta) \leq y\}$. Often for the $T_{kn}^\star$ the sample size n is modified to some n' (not necessarily equal to n or $n-1$) so as to yield a better rate of convergence. Most of these details are given in the recent text of Hall (1992), and we shall not go into their detailed presentation. Taking clues from Sen (1988b) and Hall (1992), we may argue that whenever T_n admits a first-order representation [see (9.2.22)], $G(y)$ is a normal d.f. with null mean and variance σ_T^2, so $G_n^\star(.)$ converges to the same normal d.f.. Hence either $V_{nM}^\star$ or $G_n^\star(.)$ may be used to derive the desired (asymptotic) confidence interval for θ. In this context it is not be necessary to assume that $G(.)$ is normal so that technically bootstrap has some advantage over jackknifing [which rests on (9.2.5)]. This point should not be overemphasized. There are various examples of nonnormal $G(.)$ where bootstrapping does not work out well. We leave some of these as exercises (see Problems 9.2.7 and 9.2.8). Finally, there remains some arbitrariness in the choice of M. An ideal choice of $M = n^n$ is unpracticable for large n even under the presumption that modern computer cost for generating such bootstrap samples is insignificant. A sensible comparison of jackknifing and bootstrapping should take into account the relative resampling cost factors.

In this way jackknifing would come out as the real winner in most regular cases where $G(.)$ is Gaussian.

Note that in the regular case, under the conditional setup (given F_n) involving SRSWR, $n^{1/2}(F_n^\star - F_n)$ weakly converges to a Gaussian function, and as $||F_n - F|| \to 0$ a.s., as $n \to \infty$, and F is assumed to be continuous this Gaussian function is easily shown to be a tied-down Wiener process (on a transformed time parameter, resulting on the unit interval [0,1]). This establishes that given F_n, $n^{1/2}(T_n^\star - T_n)$ converges weakly to the same functional of the tied-down Wiener process to which $n^{1/2}(T_n - \theta)$ converges in law (unconditionally). This equivalence result may not generally hold in nonregular cases (see van Zwet 1992).

Having studied the consistency property of jackknifed and bootstrap estimators of σ_T^2, now we discuss their robustness aspects. If the functional $T(.)$ is sufficiently smooth (at F), it can be shown that (see, Problem 9.2.9)

$$(n-1)[T(F_n) - T_n^J] \to v(F) \text{ a.s. as } n \to \infty, \tag{9.2.69}$$

where $v(F)$ is termed the "asymptotic bias" of T_n $(= T(F_n))$. This result has two important bearings: First by the close relationship between T_n and its jackknifed version T_n^J; it shows how jackknifing leads to the elimination of the first-order bias term without structurally affecting the original estimator T_n. Second, it also raises the alarm that if $T(F_n)$ is itself not a very robust estimator, its jackknifed version T_n^J will not be generally robust. From the robustness perspective the choice of a robust initial estimator is crucial. Fortunately this prescription fits well with the general objectives of this book, and we do not need to put emphasis on this aspect of jackknifing. But, if we look at the pseudovariables $(T_{n,i})$, we may observe that they are based on deeper smoothness regularity conditions, and hence they are more vulnerable to outliers and error contaminations (than the original T_n). Therefore the classical jackknifing may not be the right course to preserve or enhance robustness in resampling schemes. The bootstrap method does not have the property in (9.2.69), and even if $T_n^\star$ does (nearly) unbiasedly estimate T_n (given F_n), it is subject to all the lack of robustness properties shared by jackknifing.

The variance estimators V_n^J and $V_{nM}^\star$ are generally even less robust that T_n^J and $T_n^\star$. This is apparent from the fact that V_n^J is the variance of the pseudovalues $T_{n,i}$ (and $V_{nM}^\star$ is the variance of the bootstrap versions $T_{kn}^\star$, $1 \leq k \leq M$), where the $T_{n,i}$ (or $T_{kn}^\star$) may not be that robust and hence their second-order sample moment may even be less robust. To make full use of jackknifing and/or bootstrapping, one needs to pay close attention to the choice of initial estimators which not only must be robust themselves but also their pseudovalues $T_{n,i}^2$ (or the bootstrap versions $T_{kn}^{\star 2}$) must be adequately robust. an adequate extent. Alone, the "bounded influence function" criterion (for $T(F_n)$) may not be good enough. For general U-statistics and related von Mises' functionals (for which the kernel is of finite degree), the classical Hoeffding (H-) decomposition of T_n eliminates the need for a bounded influence function but does not guarantee robustness to that extent. For M- and L-estimators the robustness properties of the $T_{n,i}^2$ (or $T_{kn}^{\star 2}$) can be studied under appropriate smoothness conditions on the score functions. However, for R-estimators, there is no need to incorporate jackknifing and/or bootstrapping for variance estimation, since the "uniform asymptotic linearity" approach (as presented in Chapter 6) leads to alternative estimators of variance functionals that are robust in a much broader sense. In effect, in a robustness setup, V_n^J and $V_{nM}^\star$ are more pertinent for smooth functionals, while there are some alternative procedures for M- and R-estimators that work out better. With this motivation, we now proceed to consider type II confidence intervals.

Let $\{T_n = T(X_1, \ldots, X_n)\}$ be a sequence of realvalued estimators (statistics) of the parameter θ of interest.For example, M- and R-estimators of location/regression studied in earlier chapters are of this type, where suitable M- and R-statistics are used in the form of estimating equations to yield the estimates. We assume that the following conditions hold:

1. There is a suitable null hypothesis (H_0) relating to θ, such that under H_0, T_n has a distribution independent of θ.

2. For every real t there exists a transformation $(X_1, \ldots, X_n)' \mapsto (X_1^{(t)}, \ldots, X_n^{(t)})'$ (e.g., $X_j^{(t)} = X_j - t, \ j \geq 1$) such that when θ holds, $(X_1^{(\theta)}, \ldots, X_n^{(\theta)})'$ has the same d.f. as of $(X_1, \ldots, X_n)'$ under $H_0 : \ \theta = 0$.

Regarding condition 1, we often assume that T_n is distribution-free under H_0, and this provides a distribution-free confidence interval for θ. However, T_n is often asymptotically distribution-free and asymptotically normal, with an asymptotic variance σ_T^2 under H_0, but σ_T^2 can be consistently and robustly estimated from the sample. In this case we obtain an asymptotic confidence interval for θ.

By virtue of condition 1, for a given confidence coefficient $1 - \alpha$ ($0 < \alpha < 1$), there exist two numbers, say, $T_{n,\alpha}^{(1)}$ and $T_{n,\alpha}^{(2)}$, depending on the sample size n such that

$$\mathbb{P}\{T_{n,\alpha}^{(1)} \leq T_n \leq T_{n,\alpha}^{(2)} | H_0\} \geq 1 - \alpha, \tag{9.2.70}$$

and $T_{n,\alpha}^{(1)}$, $T_{n,\alpha}^{(2)}$ do not depend on θ. If the d.f. of T_n, under H_0, is symmetric about 0 (or any known constant, which we take to be equal to 0, by translation), then we can take $T_{n,\alpha}^{(1)} = -T_{n,\alpha}^{(2)}$, and choose $T_{n,\alpha}^{(2)}$ as the smallest value for which the right hand side of (9.2.70) is $\geq 1 - \alpha$. If, however, this d.f. is not symmetric about 0, some other considerations must be taken into account in the arbitration of $T_{n,\alpha}^{(1)}$ and $T_{n,\alpha}^{(2)}$. If T_n is asymptotically normal under H_0, then we can take $T_{n,\alpha}^{(1)} = -T_{n,\alpha}^{(2)}$ for large n. Let us denote

$$T_n(t) = T(X_1^{(t)}, \ldots, X_n^{(t)}),\ t \in \mathbb{R}_1. \tag{9.2.71}$$

Then we can define

$$\begin{aligned} \hat{\theta}_{L,n} &= \inf\{\theta :\ T_n(\theta) \in [T_{n,\alpha}^{(1)}, T_{n,\alpha}^{(2)}]\}, \\ \hat{\theta}_{U,n} &= \sup\{\theta :\ T_n(\theta) \in [T_{n,\alpha}^{(1)}, T_{n,\alpha}^{(2)}]\}, \end{aligned} \tag{9.2.72}$$

and we let

$$I_n = \{\theta : \hat{\theta}_{L,n} \le \theta \le \hat{\theta}_{U,n}\}. \tag{9.2.73}$$

By virtue of (9.2.70) and conditions 1 and 2, we conclude that

$$\mathbb{P}_\theta\{\theta : \hat{\theta}_{L,n} \le \theta \le \hat{\theta}_{U,n}\} \ge 1 - \alpha, \tag{9.2.74}$$

so I_n is a distribution-free confidence interval for θ. For the single and two-sample location/scale models, such confidence intervals were proposed by Lehmann (1963) and Sen (1963). Sen (1968a), Jurečková (1969,1971), Koul (1969), and a host of other research workers extended the methodology to a much wider class of problems (including the regression model). In the present section we will describe confidence intervals based on ranks and their extensions based on M-statistics. Their sequential (bounded-length) versions will be considered in Section 9.4.

Consider the one-sample location model with observations $X_1, \ldots, X_n$, and order statistics $X_{n:1}, \ldots, X_{n:n}$. Let $T_n = n^{-1}r_n$, where r_n = number of X_i's below the median θ; then $nT_n \sim Bin(n, F(\theta))$. Under $H_0 : \theta = 0$, $nT_n \sim Bin(n, 1/2)$. Hence we can always find a constant $r_n^\star$ $(0 \le r_n^\star \le [n/2])$ such that $\mathbb{P}\{r_n^\star \le nT_n \le n - r_n^\star | H_0\} \ge 1 - \alpha$. Recalling that $X_i - \theta$ under θ has the same d.f. as X_i under H_0, we obtain from (9.2.71)-(9.2.73) that $(X_{n:r_n^\star}, X_{n:n-r_n^\star+1})$ provides the desired confidence interval for θ. It follows from the convergence of the binomial d.f. to the normal one, that $r_n^\star = n/2 - \tau_{\alpha/2}\sqrt{n}/2 + o(\sqrt{n})$ and $n - r_n^\star = n/2 + \tau_{\alpha/2}\sqrt{n}/2 + o(\sqrt{n})$ as $n \to \infty$, where $\Phi(\tau_{\alpha/2}) = 1 - \alpha/2$. Clearly, for large n, the central order statistics $\{X_{n:k} : |k - n/2| \le k_n\}$ (where $k_n \sim O(n^{1/2})$) provide the desired solution. Since this confidence interval for θ is unaffected by the extreme order statistics, it is robust like the point estimator median. However, like the median it is not very efficient for nearly normal F.

Consider a signed-rank statistic

$$S_n = S(X_1, \ldots, X_n) = \sum_{i=1}^n (\text{sign} X_i) a_n(R_{ni}^+),$$

where the scores $a_n(k)$ are defined as in Chapters 3 and 6 and R_{ni}^+ =rank of $|X_i|$ among $|X_1|, \ldots, |X_n|$, for $i = 1, \ldots, n$. Then under $H_0 : \theta = 0$ (and F

symmetric) S_n is distribution-free, and it has a symmetric d.f. with median 0. Moreover $S(X_1-\theta,\ldots,X_n-\theta)$ has the same d.f. under θ as S_n has under H_0, and $S(X_1-t,\ldots,X_n-t)$ is nonincreasing in $t\in \mathbb{R}_1$. Consequently we may use (9.2.70)-(9.2.74). Let $C_{n,\alpha}$ be chosen so that $\mathbb{P}\{|S_n|\le C_{n,\alpha}|H_0\}\ge 1-\alpha$, and $C_{n,\alpha}$ is the smallest value for which this inequality holds. Let then $\hat{\theta}_{L,n}=\inf\{t:S(X_1-t,\ldots,X_n-t)>C_{n,\alpha}\}$ and $\hat{\theta}_{U,n}=\sup\{t:S(X_1-t,\ldots,X_n-t)<-C_{n,\alpha}\}$. The desired confidence interval for θ is $(\hat{\theta}_{U,n},\hat{\theta}_{L,n})$. By Theorem 6.2.2 we claim that as n increases

$$n^{1/2}(\hat{\theta}_{U,n}-\hat{\theta}_{L,n})\xrightarrow{P}(2/\gamma)\tau_{\alpha/2}A_\phi, \tag{9.2.75}$$

where

$$A_\phi^2=\int_0^1\phi^2(u)du,\quad \gamma=-\int_{-\infty}^{\infty}\phi(F(x))f'(x)dx, \tag{9.2.76}$$

provided that the scores $a_n(k)$ relate to the score function $\phi^\star(u)=\phi((1+u)/2)$, $0\le u\le 1$, and $\phi(u)$ is skew-symmetric (about 1/2), square integrable, and monotone. [If we choose $\phi(u)=\text{sign}(u-1/2)$, we have the sign statistics treated earlier.] Problem 9.2.10 is set up to verify (9.2.75). In a similar manner it follows that as $n\to\infty$,

$$n^{1/2}(\hat{\theta}_{U,n}-\hat{\theta}_n)\xrightarrow{P}\gamma^{-1}\tau_{\alpha/2}A_\phi,\quad n^{1/2}(\hat{\theta}_{L,n}-\hat{\theta}_n)\xrightarrow{P}-\gamma^{-1}\tau_{\alpha/2}A_\phi,$$

where $\hat{\theta}_n$ is the point (R-)estimator of θ, obtained by equating $S_n(X_1-t,\ldots,X_n-t)$ to 0. As a result we have for large n,

$$\begin{aligned}(\,\hat{\theta}_{L,n},\,\hat{\theta}_{U,n})\;&=\;(\,\hat{\theta}_n-n^{-1/2}\tau_{\alpha/2}(\gamma^{-1}A_\phi)+\mathrm{o}(\frac{1}{\sqrt{n}}),\\ &\qquad\hat{\theta}_n+n^{-1/2}\tau_{\alpha/2}(\gamma^{-1}A_\phi)+\mathrm{o}(\frac{1}{\sqrt{n}})\,),\end{aligned} \tag{9.2.77}$$

which is comparable to (9.2.9). However, in this case we do not need to estimate the variance function A_ϕ^2/γ^2, nor to use the point estimator $\hat{\theta}_n$. On the other hand, the solutions for $\hat{\theta}_{L,n}$ and $\hat{\theta}_{U,n}$ are generally based on iteration. There is no apparent need to incorporate jackknifing/bootstrapping or other resampling methods for the estimation of A_ϕ^2/γ^2, and as a result the regularity conditions on ϕ, F are least stringent this context.

Consider next the simple regression model: $X_i=\theta+\beta t_i+e_i$, $i=1,\ldots,n$, where the t_i are known regression constants (not all equal), θ,β are unknown parameters, and the e_i are i.i.d. r.v.'s with a continuous d.f. F (not necessarily symmetric). Our problem is to provide a distribution-free confidence interval for the regression parameter β. A very simple situation crops up with $S_n=\sum_{1\le i<j\le n}\text{sign}(t_i-t_j)\text{sign}(X_i-X_j)$, known as *Kendall's tau coefficient*. Under $H_0:\beta=0$, S_n is distribution-free with location 0, and $S(X_1-bt_1,\ldots,X_n-bt_n)$, $b\in\mathbb{R}_1$ is nonincreasing in b. Moreover

$S(X_1 - \beta t_1, \ldots, X_n - \beta t_n)$ under β has the same d.f. as S_n has under H_0. Equations (9.2.72)-(9.2.74) already led us to the desired solution. Problem 9.2.11 is posed to verify the details. The two-sample location model is a particular case of the linear model (when the c_i can only assume the values 0 and 1), and S_n then reduces to the classical two-sample Wilcoxon-Mann-Whitney statistic. Note that S_n is not a linear rank statistic, and hence the results to be presented in the sequel may not be directly applicable for S_n. Detailed asymptotic results may be found in Ghosh and Sen (1971); see also Problems 9.2.12 and 9.2.13.

We consider a linear rank statistics $L_n(b)$, defined as in (6.4.2), and note that under $\beta = 0$, L_n $(= L_n(0))$ is distribution-free and that under β, $L_n(\beta)$ has the same d.f. as of L_n under $H_0 : \beta = 0$. Further $L_n(\beta)$ is nonincreasing in $b \in \mathbb{R}_1$, so we can again proceed as in (9.2.70)-(9.2.74) and define $\hat{\beta}_{L,n}$ and $\hat{\beta}_{U,n}$ corresponding to a confidence coefficient $1 - \alpha$. By the regularity conditions of Theorem 6.4.2, we can conclude that as n increases,

$$T_n(\hat{\beta}_{U,n} - \hat{\beta}_{L,n}) \xrightarrow{P} 2\gamma^{-1}\tau_{\alpha/2}A_\phi, \tag{9.2.78}$$

where

$$A_\phi^2 = \int_0^1 (\phi(u) - \bar{\phi})^2 du, \;\; \bar{\phi} = \int_0^1 \phi(u)du, \tag{9.2.79}$$

$\phi(u)$ is the score function generating the $a_n(k)$ and γ is defined as in (9.2.76). Additionally, we set

$$T_n^2 = \sum_{i=1}^{n}(t_i - \bar{t}_n)^2, \;\; \bar{t}_n = n^{-1}\sum_{i=1}^{n} t_i, \tag{9.2.80}$$

and assume that $T_n \to \infty$ as $n \to \infty$, with

$$\max\{T_n^{-2}(t_i - \bar{t}_n)^2 : 1 \le i \le n\} \to 0. \tag{9.2.81}$$

Here we could have also considered the point estimator $\hat{\beta}_n$ and used Theorem 6.4.3 to conclude that as $n \to \infty$,

$$T_n(\hat{\beta}_{U,n} - \hat{\beta}_n) \xrightarrow{P} \gamma^{-1}\tau_{\alpha/2}A_\phi, \;\; T_n(\hat{\beta}_{L,n} - \hat{\beta}_n) \xrightarrow{P} -\gamma^{-1}\tau_{\alpha/2}A_\phi. \tag{9.2.82}$$

This expression gives us an analogue of (9.2.77) with n replaced by T_n^2. Likewise in this setup there is no need to incorporate jackknifing/bootstrapping or some other resampling method to estimate the variance function A_ϕ^2/γ^2. The asymptotic uniform linearity results developed in Chapter 6 gives us direct access to the desired asymptotics in a direct way. Moreover the rank statistics are globally robust, and hence the derived confidence intervals are robust too.

Let us next consider some quasi-distribution free (QDF) confidence intervals. These intervals are particularly useful for the M-estimation problems treated in Chapters 3 and 5. In such problems the statistic $T_n =$

$T(X_1, \ldots, X_n)$ satisfy both conditions 1 and 2, although the null hypothesis distribution of T_n may not be genuinely distribution-free. For example, if $T_n(t) = \sum_{i=1}^{n} \psi(X_i - t)$, $t \in \mathbb{R}_1$, is a typical M-statistic with score function ψ, then for the location model with $\theta = 0$, $T_n = T_n(0)$ has a d.f. symmetric about 0, although its exact d.f. depends on the underlying F. However, if $n^{-1/2}T_n(0)$ is asymptotically normal with mean 0 and variance σ_T^2 under H_0, and if a consistent estimator (V_n) of σ_T^2 is available, then we may set for large values of n,

$$n^{-1/2}T_{n,\alpha}^{(j)} = (-1)^j \tau_{\alpha/2} V_n^{1/2} + o_p(1), \; j = 1, 2. \tag{9.2.83}$$

With this modification we proceed as in (9.2.72)–(9.2.73) and obtain a confidence interval as in (9.2.74). The main difference between the QDF procedure and the type I procedure, considered earlier in this section, is that if T_n and V_n are robust statistics, the approximation in (9.2.83) retains that robustness to a greater extent, and hence the inversion technique [as in (9.2.72)] filters that robustness to the derived limits as well. To see this more clearly, let us explain it using M-estimators.

For the one-sample location model $F(x, \theta) = F(x - \theta), F$ symmetric about 0, consider the M-statistic

$$M_n(t) = \sum_{i=1}^{n} \psi(X_i - t), \; t \in \mathbb{R}_1, \tag{9.2.84}$$

where for monotone $\psi(.)$, $M_n(t)$ is also monotone in t. Let $\hat{\theta}_n$ be the M-estimator of θ based on (9.2.84) and defined by (5.3.5)-(5.3.7). Under H_0 : $\theta = 0$, $n^{-1/2}M_n(0) \sim \mathcal{N}(0, \sigma_\psi^2)$, where $\sigma_\psi^2 = \int_{-\infty}^{\infty} \psi^2(x)dF(x)$. A natural estimator of σ_ψ^2 is

$$V_n = n^{-1} \sum_{i=1}^{n} \psi^2(X_i - \hat{\theta}_n). \tag{9.2.85}$$

If we define

$$\begin{aligned} \hat{\theta}_{L,n} &= \inf\{t : n^{-1/2}M_n(t) > \tau_{\alpha/2}V_n^{1/2}\}, \\ \hat{\theta}_{U,n} &= \sup\{t : n^{-1/2}M_n(t) < -\tau_{\alpha/2}V_n^{1/2}\}, \end{aligned} \tag{9.2.86}$$

then we obtain a robust (asymptotic) confidence interval for θ, where

$$n^{1/2}(\hat{\theta}_{U,n} - \hat{\theta}_{L,n}) \xrightarrow{P} 2\gamma^{-1}\tau_{\alpha/2}\sigma_\phi, \text{ as } n \to \infty, \tag{9.2.87}$$

and

$$\gamma = \gamma_1 + \gamma_2, \text{ is defined by } \mathbf{A4} \text{ in Section 5.3.} \tag{9.2.88}$$

We could have also employed Theorem 5.3.1 [i.e., the asymptotic linearity result in (5.3.10)] to formulate an estimator of γ, say, $\hat{\gamma}_n$, so that σ_ψ/γ might be consistently estimated by $V_n^{1/2}/\hat{\gamma}_n$. However, on robustness grounds the quasi-type II procedure appears to be more appealing. Instead of the fixed scale M-estimator, we could even consider one-step or studentized M-statistics (as in Chapter 5) and derive some robust confidence intervals for θ. The modifications for the simple regression model(but confined to a single parameter) are straightforward, and hence we pose this as an exercise (Problem 9.2.14). In either case the asymptotic uniform linearity results of Chapter 5 provide the basis for asymptotic results such as (9.2.87).

In Section 5.2 we have formulated the M-estimation theory for general parameters (in the single parameter case). As long as the relevant M-statistics satisfy conditions 1 and 2 (as with regard to general T_n), we could construct suitable confidence intervals in the same way. However, their asymptotic properties (concerning robustness) reveal that we need some deeper results (i.e., Theorems 5.2.1 and 5.2.2) rather than more precise universal ones that hold for a general model. The regularity conditions pertaining to these theorems would then be analyzed for basic robustness so that the confidence intervals could be prescribed without reservation. To illustrate this point, let us consider the MLE based solutions (which led to the M-solutions presented earlier).

For a general parameter θ, the MLE $\hat{\theta}_n$ is obtained as a solution of the likelihood estimating equation

$$\frac{\partial}{\partial\theta}\log L_n(\theta)|_{\hat{\theta}_n} = 0. \tag{9.2.89}$$

The likelihood score statistic $(\partial/\partial\theta)\log L_n(\theta) = U_n(\theta)$, say, has mean 0, and $n^{-1/2}U_n(\theta)$ is asymptotically normal with 0 mean and variance $I(\theta)$, whereas $n^{1/2}(\hat{\theta}_n - \theta)$ is asymptotically normal with 0 mean and variance $1/I(\theta)$. Here $I(\theta)$ is the Fisher information on θ per unit of the sample. We could use the observed curvature of the likelihood function at $\hat{\theta}_n$, or $I(\hat{\theta}_n)$, to estimate $I(\theta)$ and then use a type I confidence interval for θ based on (9.2.5)-(9.2.9). Alternatively, we could treat (9.2.89) as a general M-statistic and formulate a QDF (asymptotic) confidence interval for θ along the lines in (9.2.70)-(9.2.74). If the loglikelihood function is not parabolic (in θ) or, equivalently, if $U_n(\theta)$ is highly nonlinear, the QDF procedure, in general, would be more robust than the usual type I procedure, although both are asymptotically equivalent.

We conclude this section with a note on an efficiency criterion for comparing confidence intervals for a common parameter. This is mainly adapted from Sen (1966). Suppose that we have two sequences $\{I_n\}$ and $\{I_n^\star\}$ of confidence intervals for θ, where

$$I_n = (\theta_{L,n}, \theta_{U,n}),\ I_n^\star = (\hat{\theta}_{L,n}^\star, \hat{\theta}_{U,n}^\star), \tag{9.2.90}$$

and

$$\lim_{n\to\infty} \mathbb{P}_\theta\{\theta \in I_n\} = 1 - \alpha = \lim_{n\to\infty} \mathbb{P}_\theta\{\theta \in I_n^\star\}. \tag{9.2.91}$$

Note that I_n and $I_n^\star$ are competing intervals, and they are generally not stochastically independent. Suppose that there exist constants δ and $\delta^\star$: $0 < \delta, \delta^\star < \infty$ such that

$$n^{1/2}(\hat{\theta}_{U,n} - \hat{\theta}_{L,n}) \xrightarrow{P} \delta \text{ and } n^{1/2}(\hat{\theta}^\star_{U,n} - \hat{\theta}^\star_{L,n}) \xrightarrow{P} \delta^\star. \quad (9.2.92)$$

Consider two sequences of sample sizes $\{N_n\}$ and $\{N_n^\star\}$ such that

$$\lim_{n\to\infty} (N_n^\star / N_n) = (\delta^\star/\delta)^2 = \rho^2, \text{ say.} \quad (9.2.93)$$

Then, based on (9.2.92) and (9.2.93), we can conclude that the asymptotic relative efficiency (ARE) of $\{I_n\}$ with respect to $\{I_n^\star\}$ is given by

$$e(I, I^\star) = \rho^2 = (\delta^\star/\delta)^2. \quad (9.2.94)$$

For type I confidence intervals, (9.2.6) and (9.2.9) imply that the ARE in (9.2.94) agrees with the conventional Pitman ARE of the corresponding point estimators (e.g., T_n and $T_n^\star$). The same applies to confidence intervals of type II in their QDF version, as we can see by (9.2.75), (9.2.78), or (9.2.87), which contain σ_ψ^2/γ^2, the asymptotic variance of the corresponding point estimator. Hence the point and interval estimation problems are isomorphic from the ARE point of view. This fact leads to a motivation and formulation of robust sequential confidence intervals; see Section 9.4.

9.3 MULTIPARAMETER CONFIDENCE SETS

Consider now a vector parameter $\boldsymbol{\theta}(F) = (\theta_1(F), \ldots, \theta_p(F))' \in \boldsymbol{\Theta} \subset I\!\!R_p$ for some $p \geq 1$. The sample observations $X_1, \ldots, X_n$ may be real- or vector-valued r.v.'s. Several important problems are included in this setup:

1. *Multivariate location model.* The $\mathbf{X}_i$ are themselves p-variate r.v.'s, with a d.f. $F(\mathbf{x}, \boldsymbol{\theta}) = F(\mathbf{x} - \boldsymbol{\theta})$, $\mathbf{x} \in I\!\!R_p$, $\boldsymbol{\theta} \in \boldsymbol{\Theta} \subset I\!\!R_p$, and the form of $F(\mathbf{y})$ is assumed to be independent of $\boldsymbol{\theta}$. Our goal is to construct some robust and efficient (simultaneous) confidence regions for $\boldsymbol{\theta}$.

2. *General univariate linear model.* Consider the model

$$X_i = \theta_0 + \boldsymbol{\beta}'\mathbf{t}_i + e_i, \quad i = 1, \ldots, n, \quad (9.3.1)$$

where the $\mathbf{t}_i = (t_{i1}, \ldots, t_{iq})'$ are known but not all equal regression vectors, $\boldsymbol{\beta}' = (\beta_1, \ldots, \beta_q)$ is an unknown parametric vector, considered the regression parameters, θ_0 is an unknown (intercept) parameter, $q \geq 1$, and the e_i are i.i.d. r.v.'s with a continuous d.f. F defined on $I\!\!R_1$. Thus $p = q + 1$, and our goal is to provide a simultaneous confidence region for $(\theta_0, \boldsymbol{\beta})$ (or $\boldsymbol{\beta}$ when $q \geq 2$).

3. *Multivariate linear model.* Let the $\mathbf{X}_i, \boldsymbol{\theta}_0$, and the $\mathbf{e}_i$ in (9.3.1) be all r-vectors for some $r \geq 1$, and let $\boldsymbol{\beta}'$ be a $r \times q$ matrix of unknown (regression) parameters. Here the $\mathbf{e}_i$ are i.i.d. r.v.'s with a d.f. F defined on $\mathbb{R}_r$, and the form of F is assumed to be independent of $(\boldsymbol{\theta}_0, \boldsymbol{\beta}')$. We desire to have some robust simultaneous confidence regions for $(\boldsymbol{\theta}_0, \boldsymbol{\beta}')$ (or $\boldsymbol{\theta}_0$ or $\boldsymbol{\beta}$).

4. *Location and scale parameters.* Consider the model:

$$F(x, \boldsymbol{\theta}) = F\big((x - \theta_1)/\theta_2\big), \; x \in \mathbb{R}_1, \tag{9.3.2}$$

where $\boldsymbol{\theta} = (\theta_1, \theta_2) \in \mathbb{R}_1 \times \mathbb{R}_1^+$. Then θ_1 is termed the location parameter and θ_2 the scale parameter, and F belongs to the class of location-scale family of d.f.'s. It is also possible to conceive multivariate extensions of the model in (9.3.2).

In such a multiparameter setup our goal is to construct some robust simultaneous confidence regions for $\boldsymbol{\theta} = (\theta_1, \ldots, \theta_p)'$. As in the preceding section we have type I and type II confidence regions. In (9.2.1)-(9.2.4) it was shown how the actual d.f. of an estimator can be used to provide a confidence interval. Similarly let $\mathbf{T}_n = (T_{n1}, \ldots, T_{np})'$ be a suitable estimator of $\boldsymbol{\theta}$ and let

$$G_n(\mathbf{y}, \boldsymbol{\theta}) = \mathbb{P}_{\boldsymbol{\theta}}\{\mathbf{T}_n \leq \mathbf{y}\}, \; \mathbf{y} \in \mathbb{R}_p, \tag{9.3.3}$$

be the d.f. of $\mathbf{T}_n$ defined on $\mathbb{R}_p$. Since (9.3.3) is defined on $\mathbb{R}_p$, the first and foremost question is how to choose a region (in $\mathbb{R}_p$) as the desired confidence region? For example, we might want a multidimensional rectangle $\mathbf{T}_{L,n} \leq \boldsymbol{\theta} \leq \mathbf{T}_{U,n}$ (where $\mathbf{a} \leq \mathbf{b}$ means coordinatewise inequalities) chosen in such a way that (9.1.4) holds. Alternatively, we might choose an appropriate ellipsoidal (or spherical) region for the same purpose. If $G_n(.)$ in (9.3.3) is itself a multivariate normal distribution, we might be naturally attracted by the (studentized) Mahalanobis (D^2) distance, while for the matrix-valued normal distribution, there are some alternative criteria too. Roy's (1953) largest root criterion is noteworthy in this respect, and it leads to some optimal properties of the associated confidence sets (e.g., see Wijsman 1981). For multivariate normal distributions, such confidence sets remain equivariant under the group of affine transformations, so units of measurements of the different variates do not play any significant role. On the other hand, such equivariance properties may not generally hold in robust estimation theory. For example, M-estimators of location, even in the univariate case may not be scale-equivariant, while in the multivariate location model, M-, L-, and R-estimators may not be affine-equivariant. For this reason it may be difficult to advocate standard normal theory confidence sets for such robust estimators, even though they are asymptotically normal. The procedure to be formulated here is mainly geared by its robustness aspects and their computational manageability properties; its construction is mostly based on the asymptotic (multi-)normality of the allied statistic. We should further consider that in the multivariate case the d.f. $G_n(.)$ in (9.3.3) may not be distribution-free even under null hypothesis (of invariance) and for a nonparametric statistic $\mathbf{T}_n$.

Suppose now that there exists a sequence $\{\mathbf{T}_n\}$ of estimators of $\boldsymbol{\theta}$ such that as n increases,

$$n^{1/2}(\mathbf{T}_n - \boldsymbol{\theta}) \xrightarrow{\mathcal{D}} \mathcal{N}_p(\mathbf{0}, \boldsymbol{\Sigma}), \tag{9.3.4}$$

and there exists a sequence $\{\mathbf{V}_n\}$ of consistent estimators of $\boldsymbol{\Sigma}$. If $\boldsymbol{\Sigma}$ is p.d., we may use (9.3.4), Slutsky's theorem, along with Cochran's theorem, and conclude that as n increases,

$$n(\mathbf{T}_n - \boldsymbol{\theta})'\mathbf{V}_n^{-1}(\mathbf{T}_n - \boldsymbol{\theta}) \xrightarrow{\mathcal{D}} \chi_p^2, \tag{9.3.5}$$

where χ_p^2 stands for a r.v. whose d.f. is the central chi square d.f. with p degrees of freedom (D.F.). Next we note that by the Courant theorem (on the ratio of two quadratic forms),

$$\begin{aligned}
&\sup_{\boldsymbol{\lambda}\neq 0} \left[\{\boldsymbol{\lambda}'(\mathbf{T}_n - \boldsymbol{\theta})\}^2/(\boldsymbol{\lambda}'\mathbf{V}_n\boldsymbol{\lambda})\right] \\
&= \mathrm{Ch}_{\max}\{\mathbf{V}_n^{-1}(\mathbf{T}_n - \boldsymbol{\theta})(\mathbf{T}_n - \boldsymbol{\theta})'\} \\
&= (\mathbf{T}_n - \boldsymbol{\theta})'\mathbf{V}_n^{-1}(\mathbf{T}_n - \boldsymbol{\theta}),
\end{aligned} \tag{9.3.6}$$

since $(\mathbf{T}_n - \boldsymbol{\theta})(\mathbf{T}_n - \boldsymbol{\theta})'$ has rank one (so the largest characteristic root is equal to the trace). We obtain from (9.3.5) and (9.3.6) that as n increases,

$$\begin{aligned}
\mathbb{P}\{&\boldsymbol{\lambda}\mathbf{T}_n - n^{-1/2}\chi_{p,\alpha}(\boldsymbol{\lambda}'\mathbf{V}_n\boldsymbol{\lambda})^{1/2} \le \boldsymbol{\lambda}'\boldsymbol{\theta} \\
&\le \boldsymbol{\lambda}\mathbf{T}_n + n^{-1/2}\chi_{p,\alpha}(\boldsymbol{\lambda}'\mathbf{V}_n\boldsymbol{\lambda})^{1/2}, \forall \boldsymbol{\lambda} \neq \mathbf{0}\} \to 1 - \alpha
\end{aligned} \tag{9.3.7}$$

where $\chi_{p,\alpha}^2$ is the $(1-\alpha)$ quantile of the chi square d.f. with p D.F.; this provides a simultaneous confidence region for $\boldsymbol{\theta}$, and it is formulated in the setup of the classical S-method (Scheffé, 1951) of multiple comparison for the multinormal location/regression model. Looking at (9.3.7), we may note that the width of the interval for $\boldsymbol{\lambda}'\boldsymbol{\theta}$ is equal to

$$2n^{-1/2}\chi_{p,\alpha}(\boldsymbol{\lambda}'\mathbf{V}_n\boldsymbol{\lambda})^{1/2} \tag{9.3.8}$$

which not only depends on $\mathbf{V}_n$ but on $\boldsymbol{\lambda}$. It is not unnatural to normalize $\boldsymbol{\lambda}$ in a convenient way, and we may set this as $||\boldsymbol{\lambda}|| = (\boldsymbol{\lambda}'\boldsymbol{\lambda})^{1/2} = 1$. Then

$$\sup\{(\boldsymbol{\lambda}'\mathbf{V}_n\boldsymbol{\lambda}/\boldsymbol{\lambda}'\boldsymbol{\lambda}) : ||\boldsymbol{\lambda}|| = 1\} = \mathrm{Ch}_{\max}(\mathbf{V}_n), \tag{9.3.9}$$

so by (9.3.7) and (9.3.9) we have

$$\lim_{n\to\infty} \mathbb{P}\Big\{\boldsymbol{\lambda}'\mathbf{T}_n - n^{-1/2}\chi_{p,\alpha}\mathrm{Ch}_{\max}(\mathbf{V}_n)]^{1/2} \le \boldsymbol{\lambda}'\boldsymbol{\theta} \le \boldsymbol{\lambda}'\mathbf{T}_n$$
$$+n^{-1/2}\chi_{p,\alpha}\mathrm{Ch}_{\max}(\mathbf{V}_n)]^{1/2},\ \forall\boldsymbol{\lambda} : \|\boldsymbol{\lambda}\| = 1\Big\} \ge 1-\alpha. \qquad (9.3.10)$$

The diameter of the confidence set in (9.3.10) is equal to

$$2n^{-1/2}\chi_{p,\alpha}\{\mathrm{Ch}_{\max}(\mathbf{V}_n)\}^{1/2}. \qquad (9.3.11)$$

Equation (9.3.11) also specifies the maximum diameter for the confidence set in (9.3.7) when $\boldsymbol{\lambda}'\boldsymbol{\lambda} = 1$. This result will be of importance in the sequential case, treated in the next section. Since $\mathbf{V}_n$ is a consistent estimator of $\boldsymbol{\Sigma}$ [in (9.3.4)], $n^{1/2}$ times the diameter in (9.3.11) stochastically converges to

$$2\chi_{p,\alpha}\{\mathrm{Ch}_{\max}(\boldsymbol{\Sigma})\}^{1/2}. \qquad (9.3.12)$$

Therefore, from the shortest confidence set point of view, it appears that we should choose a sequence $\{\mathbf{T}_n^\star\}$ of estimators for which the asymptotic covariance matrix ($\boldsymbol{\Sigma}_{\mathbf{T}}$, say) satisfies the minimax condition:

$$\mathrm{Ch}_{\max}(\boldsymbol{\Sigma}_{\mathbf{T}^\star}) = \inf\big\{\mathrm{Ch}_{\max}(\boldsymbol{\Sigma}_{\mathbf{T}} : \mathbf{T} \subset \mathcal{C}\big\}, \qquad (9.3.13)$$

where $\mathcal{C}$ is a nonempty class of competing estimators of $\boldsymbol{\theta}$. This explains the relevance of the Roy (1953) largest-root criterion. Thus, from this perspective, to obtain robust multiparameter confidence sets based on point estimators and their estimated covariance matrix, we have to keep in mind the following:

1. The estimator $\mathbf{T}_n$ should be robust, efficient, and asymptotically multi-normal.
2. The estimator $\mathbf{V}_n$ is consistent and robust.

For the multivariate location model, if $\mathbf{T}_n$ is the vector of sample means, $\mathbf{V}_n$ is the corresponding sample covariance matrix, and both are known to be highly nonrobust ($\mathbf{V}_n$ being comparatively worse). We see that there is an even a greater need for robust statistics to provide robust confidence sets in the multiparameter case. Further (9.3.13) dictates the choice of $(\mathbf{T}_n, \mathbf{V}_n)$, such that the corresponding $\boldsymbol{\Sigma}$ has the minimax property in (9.3.13). If we have a class of estimators $\{\mathbf{T}_{nj} : j \in \mathcal{C}\}$ with the corresponding class of asymptotic covariance matrices $\{\boldsymbol{\Sigma}_j;\ j \in \mathcal{C}\}$, such that there is a partial order of the $\mathbf{T}_{nj}$ in the sense that

$$j < j' \Rightarrow \boldsymbol{\Sigma}_j - \boldsymbol{\Sigma}_{j'} \text{ is positive semidefinite}, \qquad (9.3.14)$$

then $\mathrm{Ch}_{\max}(\boldsymbol{\Sigma}_j) \ge \mathrm{Ch}_{\max}(\boldsymbol{\Sigma}_{j'})$, so the diameters in (9.3.12) are also ordered. Actually, if (9.3.14) holds, then the *trace-efficiency, D-efficiency* (D for determinant), and *E-efficiency* (E for extreme root) of the $\{\mathbf{T}_{nj}\}$ are all concordant.

In this case locating the optimal j is not a problem, and then we could proceed to choose some robust versions of such an optimal pair $(\mathbf{T}_n^0, \mathbf{V}_n^0)$. In a parametric setup the MLE $\hat{\boldsymbol{\theta}}_n$ of $\boldsymbol{\theta}$ has the asymptotic covariance matrix $\mathbf{I}_{\boldsymbol{\theta}}^{-1}$, the inverse of the Fisher information matrix (on $\boldsymbol{\theta}$), and the multiparameter Cramér-Rao inequality asserts that for all (asymptotically unbiased) estimates $\{\mathbf{T}_n\}$,

$$\boldsymbol{\Sigma}_{T} - \mathbf{I}_{\boldsymbol{\theta}}^{-1} = \text{ p.s.d.} \tag{9.3.15}$$

the use of the MLE $\hat{\boldsymbol{\theta}}_n$ and $\hat{\mathbf{I}}_n$ $[= -n^{-1}(\partial^2/\partial\boldsymbol{\theta}\partial\boldsymbol{\theta}') \log L_n(\boldsymbol{\theta})|_{\boldsymbol{\theta}=\hat{\boldsymbol{\theta}}_n}]$, the sample counterpart of $\mathbf{I}_{\boldsymbol{\theta}}$, lead to (9.3.14) for all estimators belonging to that class, and hence the confidence set based on $(\hat{\theta}_n, \hat{\mathbf{I}}_n)$ is asymptotically optimal. However, both $\hat{\boldsymbol{\theta}}_n$ and $\hat{\mathbf{I}}_n$ are generally highly nonrobust, and hence this asymptotic prescription is of no real utility from the robustness point of view. It is therefore more imperative to choose an estimator that is not only robust but for which a robust $\mathbf{V}_n$ exists and its population counterpart has the desirable property in the light of (9.3.13). This picture not only depends on the diagonal elements of $\boldsymbol{\Sigma}$ but also on the off-diagonal elements which mainly reflect the nature of stochastic dependence among the coordinate variables. Since robust estimators (in the multiparameter case) may not generally be affine-equivariant, a canonical reduction of $\boldsymbol{\Sigma}$ to a diagonal matrix (whose elements are the characteristic roots of $\boldsymbol{\Sigma}$) is not admissible. Moreover colinearity or other limiting degeneracy can reduce the rank of $\mathbf{I}_{\boldsymbol{\theta}}^{-1}$, and this may effect the robust estimators that are typically nonlinear. For example, consider a bivariate normal d.f. with correlation coefficient ρ. Consider, in particular, the medians of the individual coordinate variables. In this case, $\boldsymbol{\Sigma}_T$ can be reduced to $(\pi/2)\begin{pmatrix} 1 & \rho^\star \\ \rho^\star & 1 \end{pmatrix}$, where $\rho^\star = (2/\pi)\sin^{-1}\rho$, when we take $\boldsymbol{\Sigma} = \begin{pmatrix} 1 & \rho \\ \rho & 1 \end{pmatrix}$. As $\rho \to 1$ $\boldsymbol{\Sigma}$ approaches a singular form and so does $\boldsymbol{\Sigma}_T$ but comparatively slowly, and this can affect the picture in (9.3.13). For $\rho \geq 0$ the largest root of $\boldsymbol{\Sigma}$ is equal to $1+\rho$, while that of $\boldsymbol{\Sigma}_T$ is $\pi/2 + \sin^{-1}\rho$. Thus

$$\frac{\text{Ch}_{\max}(\boldsymbol{\Sigma}_T)}{\text{Ch}_{\max}(\boldsymbol{\Sigma})} = \frac{\pi/2 + \sin^{-1}\rho}{1+\rho}, \tag{9.3.16}$$

and this can be expressed as $\pi/2 + (\pi/2)[(2/\pi)\sin^{-1}\rho - \rho]/(1+\rho)$, where (see Problem 9.3.3) $(2/\pi)\sin^{-1}\rho - \rho$ is ≥ 0 for every $\rho \geq 0$. Thus, as $\rho \to 0$ or 1, (9.3.16) converges to $\pi/2 = 1.57$, while for intermediate values of ρ, (9.3.16) is greater than $\pi/2$, indicating that there may be additional loss of efficiency of the confidence set based on the median vector when ρ is different from zero or ± 1. This picture depends on ρ. In the same vein, we may consider a k-variate intraclass correlation normal model where $\boldsymbol{\Sigma} = \sigma^2[(1-\rho)\mathbf{I} + \rho\mathbf{1}\mathbf{1}']$ for some $\rho: -(k-1)^{-1} \leq \rho \leq 1$. Now $\text{Ch}_{\max}(\boldsymbol{\Sigma}) = \sigma^2[1 + (k-1)\rho]$ if $\rho \geq 0$, and $\sigma^2(1-\rho)$ if $\rho \leq 0$ (the proof is left as Problem 9.3.4). Therefore (9.3.16) extends to

$$\begin{array}{ll} ((\pi/2) + (k-1)\sin^{-1}\rho)/(1+(k-1)\rho), & \rho \geq 0, \\ ((\pi/2) - \sin^{-1}\rho)/(1-\rho), & \rho \leq 0. \end{array} \quad (9.3.17)$$

For $\rho \leq 0$ the picture is the same as in (9.3.16) (and it does not depend on k), while for $\rho \geq 0$ we rewrite (9.3.17) as

$$\frac{\pi}{2} + \frac{\pi}{2}\{(k-1)[(2/\pi)\sin^{-1}\rho - \rho](1+(k-1)\rho)^{-1}\}. \quad (9.3.18)$$

Now we can claim that as $\rho \to 0$ or 1 the second term goes to 0, but for $0 < \rho < 1$, noting that $(k-1)/(1+(k-1)\rho)$ is $\uparrow$ in k (see Problem 9.3.5), we can conclude that (9.3.18) can be made arbitrarily large by choosing k large. Thus, for a highdimensional normal d.f. with a nonnegative intraclass correlation, the confidence set based on the coordinatewise medians seems to be considerably inefficient (albeit robust), and hence we do not advocate.

The multiparameter confidence set estimation problem can be handled more conveniently for the univariate linear model in (9.3.1). For example, if we desire to have a confidence set for the regression parameter (vector) $\boldsymbol{\beta}$, we have a model for which (9.3.4) holds with a further simplification that

$$\boldsymbol{\Sigma} = \sigma_0^2 \mathbf{V}_0, \ \mathbf{V}_0 \text{known (p.d.)}, \quad (9.3.19)$$

where σ_0^2 is a nonnegative parameter, and a convenient estimator of σ_0^2, say, v_n^2 exists. In this case an affine transformation $[\mathbf{t}_i \mapsto \mathbf{t}_i^\star = \mathbf{B}\mathbf{t}_i + \mathbf{b}$, $\mathbf{B}$ nonsingular and $\mathbf{b} \in \mathbb{R}_p$, $(i = 1, \ldots, n)]$ on the regression vectors leave the model equivariant, and hence some of the basic comments made before (concerning the covariance matrix $\mathbf{V}_n$ or $\boldsymbol{\Sigma}_T$) for the general case may not be that important. Moreover robust estimation of σ_0^2 may cause much less a concern than that of an arbitrary $\boldsymbol{\Sigma}$. In this case we have

$$\text{Ch}_{\max}(\boldsymbol{\Sigma}) = \sigma_0^2 \text{Ch}_{\max}(\mathbf{V}_0). \quad (9.3.20)$$

The competing estimators in (9.3.13) are discriminated by σ_T^2. In this respect the situation is isomorphic to the uniparameter model treated in Section 9.2, in which case we could omit the above discussion [see (9.3.14)-(9.3.18)].

Let us proceed to consider type II confidence sets in the multiparameter case. Let us go back to Section 9.2 [to (9.2.70)] and postulate the same conditions 1 and 2 but both $\mathbf{T}_n$ and $\boldsymbol{\theta}$ being p-vectors, for some $p \geq 1$. Instead of the interval in (9.2.70), we will look for a suitable set K_n such that

$$\mathbb{P}_0\{\mathbf{T}_n \in K_n\} \geq 1 - \alpha, \quad (9.3.21)$$

where K_n contains $\mathbf{0}$ as an inner point and does not depend on $\boldsymbol{\theta}$. Then parallel to (9.2.71)-(9.2.73), we define

$$I_n = \{\boldsymbol{\theta} : \mathbf{T}_n \in K_n\}. \quad (9.3.22)$$

Whenever K_n and I_n are well defined, we have

$$\mathbb{P}_{\boldsymbol{\theta}}\{\boldsymbol{\theta} \in I_n\} \geq 1-\alpha, \tag{9.3.23}$$

which can be taken as the desired solution. There are two basic issues involved in the formulation of the confidence set I_n in (9.3.22).

1. How to choose K_n in an optimal and robust way?

2. Is it always possible to get a closed convex subset of $\boldsymbol{\Theta}$ as I_n in the specific context where a solution is sought?

The answer to the first question is relatively smoother than that to the second question. We may consider $\mathbf{T}_n$ as a vector of rank test statistics for testing a null hypothesis $H_0 : \boldsymbol{\theta} = \boldsymbol{\theta}_0$ $(= \mathbf{0}$ WLOG$)$. The test criterion would then be the quadratic form of $\mathbf{T}_n$ with respect to its dispersion matrix $\mathbf{W}_n$:

$$\mathcal{L}_n = \mathbf{T}_n' \mathbf{W}_n^{-1} \mathbf{T}_n, \tag{9.3.24}$$

and under H_0, $\mathcal{L}_n \xrightarrow{\mathcal{D}} \chi_p^2$. This leads to the following choice of K_n in (9.3.21):

$$K_n = \{\mathbf{T}_n : \ \mathbf{T}_n \mathbf{W}_n^{-1} \mathbf{T}_n \leq \chi_{p,\alpha}^2\}, \tag{9.3.25}$$

and this provides us with an ellipsoidal region with the origin $\mathbf{0}$. Although $\mathcal{L}_n$ is analogous to the Hotelling T^2-statistic in the multinormal mean estimation problem, the ellipsoidal shape of K_n in (9.3.25) may not be precisely justified in the finite sample case. Often a rectangular region is also contemplated. For example, we may set

$$K_n = \{\mathbf{T}_n : a_{nj} \leq T_{nj} \leq b_{nj}, \ 1 \leq j \leq p\}, \tag{9.3.26}$$

where the a_{nj}, b_{nj} are so chosen that (9.3.21) holds. For multinormal distributions such rectangular regions have been considered by various workers, and in some cases there may be an advantage to using (9.3.26) over (9.3.25). We will make comment on this later.

The second question requires a more delicate treatment. It depends a lot on the nature of $\mathbf{T}_n(\boldsymbol{\theta})$. Recall that $\mathbf{T}_n(\boldsymbol{\theta})$ is itself a p-vector and that each coordinate of $\mathbf{T}_n(\boldsymbol{\theta})$ may typically involve the vector argument $\boldsymbol{\theta}$ (or suitable subsets of it). These are the most dominant characteristics of these functions and they will dominate our solution. In the simplest situation (e.g., rank statistics for the multivariate one sample location model, Problem 9.3.6), $T_{nj}(\boldsymbol{\theta})$ may depend only on the j-th coordinate θ_j of $\boldsymbol{\theta}$ for $j = 1, \ldots, p$. In such a case, if $T_{nj}(\boldsymbol{\theta})$ is a monotone function of θ_j for $j = 1, \ldots, p$, then in (9.3.26) one can directly use the same technique as in (9.2.73) and obtain the desired confidence set I_n. On the other hand, (9.3.25) requires that one traces every point on the boundary (surface) of the ellipsoid, and computationally this approach is too laborious. Frequently each $T_{nj}(\boldsymbol{\theta})$ depends on more than one coordinate of $\boldsymbol{\theta}$, and its monotonicity in the elements of $\boldsymbol{\theta}$ cannot be taken for granted, or it can go in opposite directions for different components of $\boldsymbol{\theta}$ (see Problems 9.3.7 and 9.3.8 on R-estimators and M-estimators, respectively). Inversions of the region (9.3.21) with respect to $\boldsymbol{\theta}$ may be laborious, if at all possible. For these reasons, in the multiparameter confidence set estimation problems, type II regions, desspite of having good robustness property, have not gained much popularity in usage.

As an interesting compromise, we could replace solving (9.3.23) by calculating the point estimator $\hat{\boldsymbol{\theta}}_n$ of $\boldsymbol{\theta}$ and appeal to asymptotic linearity, which may hold for $\mathbf{T}_n$ in the sense that for any compact $K \subset \mathbb{R}_p$,

$$\sup \left\{ n^{-1/2} \|\mathbf{T}_n(\boldsymbol{\theta} + n^{-1/2}\mathbf{b}) - \mathbf{T}_n(\boldsymbol{\theta}) + \gamma \mathbf{V}_n \mathbf{b}\| : \ \mathbf{b} \in K \right\}$$

$$\rightarrow 0, \text{ in probability, as } n \rightarrow \infty, \tag{9.3.27}$$

where γ is a suitable nonzero constant (may depend on F) and $\mathbf{V}_n$ is a suitable matrix. (For rank statistics and M-statistics, such results have been treated in Chapters 5 and 6.) By (9.3.22) and (9.3.25) we may conclude that for large n,

$$I_n \sim \{\boldsymbol{\theta} : n\gamma^2(\boldsymbol{\theta} - \hat{\boldsymbol{\theta}}_n)'\mathbf{V}_n'\mathbf{W}_n^{-1}\mathbf{V}_n(\boldsymbol{\theta} - \hat{\boldsymbol{\theta}}_n) \leq \chi^2_{p,\alpha}\}. \tag{9.3.28}$$

Clearly, if $\mathbf{V}_n, \mathbf{W}_n$ are known, it suffices to estimate γ^2. The estimate of γ can also be based on (9.3.27) by choosing appropriate $\mathbf{b}$, and hence, often there may not be a genuine need to incorporate jackknifing or bootstrapping for the estimation of the dispersion matrix of $\mathbf{T}_n(\boldsymbol{\theta})$ (under $\boldsymbol{\theta}$). In dealing with R- and M-estimators of location and regression parameters, we mainly advocate this alternative method for constructing confidence sets for $\boldsymbol{\theta}$, and likewise $\hat{\gamma}_n$, derived from (9.3.27).

We could try to apply (9.3.27) and (9.3.28) to the adaptive point estimators, introduced in Section 8.5 and obtain the confidence set I_n with asymptotically smallest volume in the class. For clarification purposes we pose some of these ideas as exercises; see Problems 9.3.8 and 9.3.9.

9.4 BOUNDED-WIDTH CONFIDENCE INTERVALS

First, let us consider the case of location parameters. Let $\{X_n, \ n \geq 1\}$ be a sequence of i.i.d. r.v.'s with a d.f. F, defined on $\mathbb{R}$, and let $\theta = \mathbb{E}_F X$ be the mean of the d.f. F. Denote by $I_n = I(X_1, \ldots, X_n)$ a confidence interval for θ

[e.g., see (9.2.1)-(9.2.4)], and denote its length by L_n. Our goal is to construct an I_n such that

$$L_n \leq 2d, \; d > 0 \text{ prescribed}, \tag{9.4.1}$$

$$\mathbb{P}_\theta\{\theta \in I_n\} \geq 1 - \alpha, \; 0 < \alpha < 1, \; \alpha \text{ prescribed}. \tag{9.4.2}$$

Alternatively, θ may be the location parameter of $F(x-\theta)$, with F symmetric about 0; then θ is also the median of $F(x-\theta)$.

Typically, for an estimator T_n of θ, the d.f. $G_n(y) = \mathbb{P}_\theta\{T_n \leq y\}$ depends on θ as well as the underlying d.f. F (possibly through some other nuisance parameter(s) and sample size n); we may refer to Section 9.2 for details. Even for the most simple case of a normal F with an unknown variance σ^2, $G_n(y)$ $(\equiv \Phi(\sqrt{n}(y-\theta)/\sigma))$ depends on σ, so any prefixed sample size (e.g., $n^\star$) cannot lead to a solution for all σ. This motivated Stein (1945) to formulate a two-stage procedure that satisfies both (9.4.1) and (9.4.2). He used a sample size that is a (positive integer valued) r.v. N_d. While the Stein procedure being an exact one (in the sense that it holds for all $d > 0$), it has some drawbacks:

1. Its scope is somewhat limited to normal or multinormal family of d.f.'s.

2. Generally N_d is large (compared to the ideal sample size if σ were known), so it is not usually fully efficient.

For nearly 20 years following this remarkable work, researchers tried to extend the Stein (1945) procedure in various directions, and a breakthrough occurred through an asymptotic formulation by Chow and Robbins (1965). For almost three decades the Chow-Robbins formulation has led to various extensions in diverse setups, including robust and nonparametric sequential interval estimation problems. We will present the sequential procedures using the Chow-Robbins setup. Typically, for the location model, we have a translation equivariant, consistent, and asymptotically normal estimator T_n of θ, for which (9.2.5) holds. Then as in (9.2.8) and (9.2.9), we have a confidence interval I_n (of convergence probability $\approx 1-\alpha$). If the length of I_n has to be bounded by $2d$, for a preassigned $d > 0$, we let $I_n = [T_n - d, T_n + d]$ and compare with (9.2.9), to obtain

$$n = n_d \doteq d^{-2}\tau^2_{\alpha/2}\sigma^2_T, \tag{9.4.3}$$

where $\doteq$ indicated that this relation holds well when d is small. However, σ^2_T $(= \sigma^2_T(F))$ depends on the unknown F and, hence, is itself unknown. Hence n_d $(= n_d(F))$ depends on F and is not known. It is quite natural to replace the unknown σ^2_T by an estimator $Z_n = Z(X_1, \ldots, X_n)$, where Z_n may be updated by a sequential scheme. This leads us to formulate a *stopping variable* N_d :

$$N_d = \inf\{n \geq n_0 : n \geq d^{-2}\tau^2_{\alpha/2}Z_n\}, \; d > 0, \tag{9.4.4}$$

where n_0 is an initial sample size. Now let

$$I_{N_d} = \{T_{N_d} - d, T_{N_d} + d\}. \tag{9.4.5}$$

Based on the stopping rule N_d and the sequence $\{T_n, Z_n\}$, the (sequential) confidence interval I_{N_d} has, by construction, width $2d$, while it remains to confirm (9.4.2) for small d. Note that by definition in (9.4.4),

$$N_d \text{ is nonincreasing in } d\ (> 0). \tag{9.4.6}$$

We will show that the stochastic (or a.s.) convergence property of Z_n (to σ_T^2) ensures a similar property for N_d.

Note that by (9.4.4), $N_d \to \infty$ a.s. as $d \downarrow 0$, and hence, by the monotone convergence theorem,

$$\mathbb{E}_F(N_d) \to \infty \quad \text{as } d \downarrow 0. \tag{9.4.7}$$

Moreover (9.4.4) implies that for any (fixed) $d > 0$ and every $n \geq n_0$,

$$\mathbb{P}_F\{N_d > n\} \leq \mathbb{P}_F\{n < d^{-2}\tau_{\alpha/2}^2 Z_n\}. \tag{9.4.8}$$

Invoking the a.s. convergence of Z_n to σ_T^2, we have

$$\begin{aligned} \mathbb{P}\{N_d = \infty\} &= \lim_{n\to\infty} \mathbb{P}_F\{N_d > n\} \\ &\leq \lim_{n\to\infty} \mathbb{P}_F\{n < d^{-2}\tau_{\alpha/2}^2\sigma_T^2\} = 0. \end{aligned} \tag{9.4.9}$$

Therefore $\mathbb{P}_F\{N_d < \infty\} = 1\ \forall d > 0$. Next, note that

$$d^{-2}\tau_{\alpha/2}^2 Z_{N_d} \leq N_d \leq n_0 + d^{-2}\tau_{\alpha/2}^2 Z_{N_d-1},\ \forall d > 0, \tag{9.4.10}$$

with probability 1, where $Z_n \to \sigma_T^2$ a.s. as $n \to \infty$, while $N_d \to \infty$ a.s. as $d \downarrow 0$. Hence, $Z_{N_d}/\sigma_T^2 \to 1$ a.s., and $\sigma_T^{-2} Z_{N_d-1} \to 1$ a.s., as $d \downarrow 0$. Using (9.4.10), we obtain that

$$N_d/n_d \to 1 \text{ a.s. as } d \downarrow 0, \tag{9.4.11}$$

where n_d is defined by (9.4.3) and $n_0/n_d \to 0$ as $d \downarrow 0$. An application of the Fatou's lemma on (9.4.11) yields then

$$\liminf_{d\downarrow 0}(n_d^{-1})\mathbb{E}_F N_d \geq \mathbb{E}_F\{\liminf_{d\downarrow 0}(n_d^{-1}N_d)\} = 1, \tag{9.4.12}$$

while by (9.4.10),

$$n_d^{-1}\mathbb{E}_F N_d \leq (n_0/n_d) + \mathbb{E}_F(\sigma_T^{-2} Z_{N_d-1}), \tag{9.4.13}$$

where $n_0/n_d \to 0$ as $d \downarrow 0$. We have seen earlier that $Z_{N_d-1}/\sigma_T^2 \to 1$ a.s. as $d \downarrow 0$, so a uniform integrability condition on Z_{N_d-1} would yield that

$\mathbb{E}_F(\sigma_T^{-2} Z_{N_d-1}) \to 1$ as $d \downarrow 0$. For this $\mathbb{E}_F\{\sup_{n \geq n_0} Z_n\} < \infty$ suffices, and in practice, when Z_n admits a FOADR, uniform negligibility (in the first mean) of the remainder term suffices; we leave these as Problems 9.4.1 and 9.4.2. From the above discussion, we obtain that

$$\lim_{d \downarrow 0} \mathbb{E}_F(N_d/n_d) = 1, \tag{9.4.14}$$

which is known as the *asymptotic efficiency* property in the literature.

Next, we assume that $\{T_n\}$ satisfies the Anscombe (1952) condition, as laid down in (8.2.35). Then we may write

$$\begin{aligned}
& N_d^{1/2}(T_{N_d} - \theta)/Z_{N_d}^{1/2} \\
&= \Big(\frac{\sigma_T}{Z_{N_d}^{1/2}}\Big)\Big\{\frac{N_d}{n_d^{1/2}}\Big\}\Big\{\frac{n_d^{1/2}(T_{N_d} - \theta)}{\sigma_T}\Big\} \\
&= \Big(\frac{\sigma_T}{Z_{N_d}^{1/2}}\Big)\Big\{\frac{N_d}{n_d}\Big\}^{1/2}\Big\{\frac{n_d^{1/2}(T_{n_d} - \theta)}{\sigma_T}) + \frac{n_d^{1/2}(T_{N_d} - T_{n_d})}{\sigma_T)}\Big\} \\
&\xrightarrow{\mathcal{D}} \mathcal{N}(0,1) \text{ as } d \downarrow 0.
\end{aligned} \tag{9.4.15}$$

Therefore

$$\lim_{d \downarrow 0} \mathbb{P}_F\{|T_{N_d} - \theta| \leq \tau_{\alpha/2} Z_{N_d}^{1/2} N_d^{-1/2}\} = 1 - \alpha. \tag{9.4.16}$$

By (9.4.4) and (9.4.16) we have

$$\lim_{d \downarrow 0} \mathbb{P}_F\{T_{N_d} - d \leq \theta \leq T_{N_d} + d\} \geq 1 - \alpha. \tag{9.4.17}$$

In the literature, (9.4.17) is known as the *asymptotic consistency* property of I_{N_d} in (9.4.5).

To sum up, we may conclude that (9.4.14) and (9.4.17) provide a natural extension of the Chow-Robbins (1965) theory of bounded width sequential confidence intervals to a general class of estimation problems. The three important features are the following:

1. Asymptotic normality of $n^{1/2}(T_n - \theta)$.

2. Anscombe (1952) condition for $\{T_n\}$.

3. Uniform integrability of Z_{N_d}.

These properties are shared by a general class of robust estimators, as studied in Part I, and can be verified using the FOADR results. Note that the asymptotic normality of stopping times (Theorem 8.2.2) applies also for N_d

in (9.4.4) and estimators $\{Z_n\}$ considered in Section 9.2; indeed such Z_n were shown as asymptotically normal and satisfying the Anscombe (1952) condition.

The confidence interval I_{N_d} in (9.4.5), based on the stopping rule N_d in (9.4.4), is a natural sequential counterpart of the conventional type I confidence intervals considered in Section 9.2. Let us next consider such sequential analogues of the type II confidence intervals treated in (9.2.71)-(9.2.74). With the same notations as in Section 9.2, we may note that for I_n, defined by (9.2.73), the width of the confidence interval is equal to $\hat{\theta}_{U,n} - \hat{\theta}_{L,n}$ (≥ 0). Hence, we may define a stopping rule N_d, $d > 0$, by letting

$$N_d = \inf\{n \geq n_0 : \hat{\theta}_{U,n} - \hat{\theta}_{L,n} \leq 2d\},\ d > 0, \tag{9.4.18}$$

where n_0 is an initial sample size (≥ 2). We let

$$I_{N_d} = \{\theta : \hat{\theta}_{L,N_d} \leq \theta \leq \hat{\theta}_{U,N_d}\},\ d > 0, \tag{9.4.19}$$

so that by its very construction [and (9.4.18)], the width of the (sequential) confidence interval in (9.4.19) is $\leq 2d$. We still need to establish the asymptotic consistency and asymptotic efficiency of the procedure.

As in Section 9.2 [see (9.2.70)-(9.2.74)], we assume that $T_n(\theta)$ is asymptotically normal with mean 0 under θ and that there exists a positive constant ν $(= \nu(F))$ such that

$$n^{1/2}\{\hat{\theta}_{U,n} - \theta\} - \nu\tau_{\alpha/2} \xrightarrow{\mathcal{D}} \mathcal{N}(0,1); \tag{9.4.20}$$

$$n^{1/2}\{\hat{\theta}_{U,n} - \hat{\theta}_{L,n}\} - 2\nu\tau_{\alpha/2} \xrightarrow{P} 0 \text{ as } n \to \infty. \tag{9.4.21}$$

We further assume that for every $\epsilon > 0$ and $\eta > 0$, there exist a δ (> 0) and a positive integer $n_0(\epsilon, \eta)$ such that for every $n \geq n_0(\epsilon, \eta)$,

$$\mathbb{P}\{\max_{m:|m-n|\leq\delta n} n^{1/2}|(\hat{\theta}_{U,m} - \hat{\theta}_{L,m}) - (\hat{\theta}_{U,n} - \hat{\theta}_{L,n})| > \epsilon\} < \eta. \tag{9.4.22}$$

Note the analogy of (9.4.22) with the usual Anscombe (1952) type condition. This may also be verified individually for $\hat{\theta}_{L,n}$ and $\hat{\theta}_{U,n}$. Moreover (9.4.20) follows directly from the asymptotic normality of $T_n(\theta + n^{-1/2}a)$ under θ, which can often be proven either via the asymptotic linearity of T_n or the contiguity of probability measures. We leave it to Problem 9.4.3 to illustrate this point. Further (9.4.21) is a direct consequence of the "asymptotic uniform linearity" results of $T_n(.)$, as we established in earlier chapters for various T_n. Verification of (9.4.22) needs some additional manipulations. First, instead of the asymptotic normality result in (9.4.20), one may establish suitable weak invariance principles, and the "compactness" part of such weak convergence results ensures the Anscombe (1952) condition. This approach has been worked out in detail in Sen (1981a, 1985). We pose Problem 9.4.4 to illustrate the practicality of this approach. Second, if we are able to strengthen (9.4.21) to

$$n^{\frac{1}{2}}(\hat{\theta}_{U,n} - \hat{\theta}_{L,n}) - 2\nu\tau_{\alpha/2} \to 0 \text{ a.s. as } n \to \infty, \tag{9.4.23}$$

then (9.4.22) follows directly from (9.4.23); see Problem 9.4.5. Such a.s. convergence results have also been established by various workers, and we pose Problems 9.4.5 and 9.4.6 to verify this for some special cases. Note that if $\hat{\theta}_{U,n}$ (and $\hat{\theta}_{L,n}$) admit a FOADR, then, whenever the remainder term is $O_p(n^{-1})$, (9.4.21) and (9.4.22) can be established under quite general conditions on this remainder term. Problem 9.4.7 is set up to verify this in a special case.

Note that by definition in (9.4.18), for every $d > 0$,

$$\hat{\theta}_{U,N_d-1} - \hat{\theta}_{L,N_d-1} > 2d \geq \hat{\theta}_{U,N_d} - \hat{\theta}_{L,N_d}. \tag{9.4.24}$$

By defining n_d by

$$n_d = \inf\{n \geq n_0 : n \geq d^{-2}\nu^{-2}\tau^2_{\alpha/2}\}, \; d > 0, \tag{9.4.25}$$

we have

$$dn_d^{1/2} \sim \nu\tau_{\alpha/2} \text{ as } d \downarrow 0. \tag{9.4.26}$$

Thus by (9.4.18), (9.4.21), (9.4.22), (9.4.24), and (9.4.26) we have for $d \downarrow 0$,

$$dN_d^{1/2} \overset{P}{\sim} dn_d^{1/2} \Rightarrow N_d/n_d \overset{P}{\to} 1. \tag{9.4.27}$$

In view of (9.4.20), (9.4.21), and (9.4.27) we conclude that

$$\lim_{d\downarrow 0} \mathbb{P}\{\theta \in I_{N_d}\} = 1 - \alpha. \tag{9.4.28}$$

Next we note that for every $d > 0$,

$$\mathbb{E}N_d = 1 + \sum_{n\geq 1} \mathbb{P}\{N_d > n\}, \tag{9.4.29}$$

where by (9.4.18),

$$\begin{aligned} \mathbb{P}\{N_d > n\} &\leq \mathbb{P}\{\hat{\theta}_{U,n} - \hat{\theta}_{L,n} > 2d\} \\ &\leq (2d)^{-r}\mathbb{E}\{|\hat{\theta}_{U,n} - \hat{\theta}_{L,n}|^r\}, \; r > 0. \end{aligned} \tag{9.4.30}$$

Therefore, to show that $\mathbb{E}N_d < \infty \; \forall d > 0$, it suffices to show that for some $r > 0$,

$$\sum_{n\geq 1} \mathbb{E}\{|\hat{\theta}_{U,n} - \hat{\theta}_{L,n}|^r\} < \infty. \tag{9.4.31}$$

(Typically we may need to set $r > 2$.) Alternatively, we may note that

$$\begin{aligned} &\mathbb{P}\{\hat{\theta}_{U,n} - \hat{\theta}_{L,n} > 2d\} \\ &\quad = \mathbb{P}\{n^{1/2}(\hat{\theta}_{U,n} - \hat{\theta}_{L,n}) - 2\nu\tau_{\alpha/2} > 2dn^{1/2} - 2\nu\tau_{\alpha/2}\} \end{aligned} \tag{9.4.32}$$

where

$$dn^{1/2} - \nu\tau_{\alpha/2} > 0 \text{ for every } n > n_d, \tag{9.4.33}$$

and it goes to $+\infty$ as $n \to \infty$. Thus, it suffices to show that for some $s > 1$ and for every $\eta > 0$, there exist a C_η $(0 < C_\eta < \infty)$ and an n_η such that

$$\mathbb{P}\{|n^{1/2}(\hat{\theta}_{U,n} - \hat{\theta}_{L,n}) - 2\nu\tau_{\alpha/2}| > \eta\} \leq C_\eta n^{-s} \tag{9.4.34}$$

for every $n \geq n_\eta$. Usually (9.4.34) is comparatively easier to verify then (9.4.31). The rest of the proof is similar to that in (9.4.14), so the asymptotic efficiency holds.

In the above development we have tacitly assumed that the asymptotic mean square error of $\hat{\theta}_n$ is of the form $n^{-1}\nu^2$. As we have seen on various estimators of the slope β in the linear regression model (in Chapters 4, 5, and 6), their asymptotic mean square error is typically of the form

$$Q_n^{-1}\nu^2, \quad Q_n \geq 0. \tag{9.4.35}$$

This does not create a problem because we usually assume that $n^{-1}Q_n \to Q > 0$ as $n \to \infty$. It does, however, raises an interesting question: Is it possible to minimize the asymptotic mean square error or the $\mathbb{E}N_d$ in the sequential case by a skillful choice of the regressors? In other words, how does one maximize Q_n by a proper choice of regressors? This is the classical sequential design problem, and fortunately the solutions worked out for the parametric model (under normality assumptions on the error component) remain asymptotically valid for the general robust/nonparametric setups.

We conclude this section with a caveat that for the M-procedure, and for the location as well as simple regression model, instead of (9.2.71)-(9.2.74) we may have to use approximate (asymptotic) solutions [see (9.2.86)]. Although that may not create any additional problems for the verification of asymptotic consistency and efficiency properties, the procedure does not have the exact distribution-free property, even in a nonsequential setup. Problem 9.4.8 is set to illustrate this subtle point.

9.5 SEQUENTIAL CONFIDENCE SETS

We will briefly describe some sequential analogues of the multiparameter confident sets of Section 9.3. First, consider the type I simultaneous confidence region developed in (9.3.4)-(9.3.13). The diameter of the confidence set in (9.3.10) is given by $2n^{-1/2}\chi_{p,\alpha}\text{Ch}_{\max}^{1/2}(\mathbf{V}_n)$ [see (9.3.11)]. To limit this diameter to a preassigned number $2d$ (> 0), intuitively we would want to consider a stopping number N_d, $d > 0$, by letting

$$N_d = \inf\{n \geq n_0 : n \geq d^{-2}\chi^2_{p,\alpha}\text{Ch}_{\max}(\mathbf{V}_n)\},\ d > 0. \tag{9.5.1}$$

For a given $d > 0$, having defined N_d by (9.5.1), we can then consider the following (simultaneous) confidence set (in a sequential setup):

$$I_{N_d} = \{\boldsymbol{\theta} : \boldsymbol{\lambda}'\mathbf{T}_{N_d} - d \leq \boldsymbol{\lambda}'\boldsymbol{\theta} \leq \boldsymbol{\lambda}'\mathbf{T}_{N_d} + d,\ \forall\boldsymbol{\lambda} : ||\boldsymbol{\lambda}|| = 1.\} \tag{9.5.2}$$

Again note that by definition, N_d is $\searrow$ in d (> 0) and that whenever $\mathbf{V}_n$ converges (a.s.) to a nonnegative definite matrix $\boldsymbol{\Sigma}$ (of rank ≥ 1),

$$N_d \rightarrow +\infty \text{ a.s. as } d \downarrow 0. \tag{9.5.3}$$

Actually, if we define n_d, $d > 0$, by letting

$$n_d = \inf\{n \geq n_0 : n \geq d^{-2}\chi^2_{p,\alpha}\text{Ch}_{\max}(\boldsymbol{\Sigma})\}. \tag{9.5.4}$$

then (9.5.3) and the a.s. convergence of $\mathbf{V}_n$ to $\boldsymbol{\Sigma}$ (as $n \rightarrow \infty$) imply that

$$N_d/n_d \rightarrow 1 \text{ a.s. as } d \downarrow 0. \tag{9.5.5}$$

We now assume that in addition to the asymptotic (multi)-normality result in (9.3.4), the classical Anscombe (1952) condition holds for the $\{\mathbf{T}_n\}$. By using (9.5.5) along with (9.3.4), we have

$$N_d^{\frac{1}{2}}(\mathbf{T}_{N_d} - \boldsymbol{\theta}) \sim \mathcal{N}_p(\mathbf{0}, \boldsymbol{\Sigma}) \text{ as } d \downarrow 0. \tag{9.5.6}$$

Moreover (9.5.3), along with the a.s. convergence of $\mathbf{V}_n$ to $\boldsymbol{\Sigma}$ (as $n \rightarrow \infty$), imply that

$$\mathbf{V}_{N_d} \rightarrow \boldsymbol{\Sigma} \text{ a.s. as } d \downarrow 0. \tag{9.5.7}$$

As such, from (9.5.6) and (9.5.7) we conclude that as $d \downarrow 0$,

$$N_d^{1/2}\mathbf{V}_{N_d}^{-1/2}(\mathbf{T}_{N_d} - \boldsymbol{\theta}) \xrightarrow{\mathcal{D}} \mathcal{N}_p(\mathbf{0}, \mathbf{I}_p). \tag{9.5.8}$$

Incorporating the classical Scheffé (S-)method of constructing a simultaneous confidence region, we obtain from (9.5.8) and (9.5.1)-(9.5.2) that

$$\lim_{d\downarrow 0} \mathbb{P}\{\theta \in I_{N_d}\} \geq 1 - \alpha, \tag{9.5.9}$$

and hence the consistency. In Section 9.4 we considered some probability inequalities of the type that for sufficiently large n for some $s > 1$,

$$\mathbb{P}\{|\frac{V_n}{\nu^2} - 1| > \eta\} \leq C_\eta n^{-s}\ \forall\eta > 0. \tag{9.5.10}$$

In the current context we need to extend this to the following: For sufficiently large n, $\forall\eta > 0$,

$$\mathbb{P}\{|(\frac{\text{Ch}_{\max}(\mathbf{V}_n)}{\text{Ch}_{\max}(\boldsymbol{\Sigma})}) - 1| > \eta\} \leq C_\eta n^{-s} \tag{9.5.11}$$

for some $s > 1$. The proof of $\lim_{d \downarrow 0} \mathbb{E}(N_d/n_d) = 1$, the asymptotic efficiency, follows as in the case of a single parameter; see Problem 9.5.1. Whenever $\mathbf{V}_n$ admits a FOADR, we have

$$n^{1/2}\left\{\frac{\mathrm{Ch}_{\max}(\mathbf{V}_n)}{\mathrm{Ch}_{\max}(\boldsymbol{\Sigma})} - 1\right\} \xrightarrow{\mathcal{D}} Z, \tag{9.5.12}$$

where Z has a normal distribution; Problem 9.5.2 is set to verify this [starting with the asymptotic multinormality of $n^{1/2}(\mathbf{V}_n - \boldsymbol{\Sigma})$]. As such assuming that the classical Anscombe (1952) condition holds, we have

$$n_d^{1/2}\left\{\frac{\mathrm{Ch}_{\max}(\mathbf{V}_{N_d})}{\mathrm{Ch}_{\max}(\boldsymbol{\Sigma})} - 1\right\} \xrightarrow{\mathcal{D}} Z \text{ as } d \downarrow 0, \tag{9.5.13}$$

so (see Problem 9.5.3) as in Section 9.4, for $d \downarrow 0$,

$$(N_d^{1/2} - n_d^{1/2}) \xrightarrow{\mathcal{D}} Z. \tag{9.5.14}$$

Let us consider next some sequential versions of the type II confidence sets treated in Section 9.3. As in (9.3.21) we define K_n for some $n \geq n_0$, and by the alignment procedure described in (9.3.25) or (9.3.26), we obtain a confidence set $I_n(\boldsymbol{\theta})$. This may require a tedious trial and error method. The situation is more complex because the vector $\mathbf{T}_n(\mathbf{t})$ is a (typically nonlinear) function of a vector-valued $(\mathbf{t})$. Once this is done, we define a maximum diameter (d_n) for $I_n(\boldsymbol{\theta})$, and then a stopping number N_d by

$$N_d = \inf\{n \geq n_0 : d_n \leq 2d\},\ d > 0. \tag{9.5.15}$$

We may, with advantage, use (9.3.27), i.e., the asymptotic uniform linearity result, and that provides an asymptotic confidence set of the type (9.3.28) for which the stopping variable can be defined as in (9.5.1). From this data set the parameter γ in (9.3.27) (or its vector analogues) is estimated by an estimator $\hat{\gamma}_n$ that is strongly consistent (or at least satisfies an Anscombe (1952) type condition). In a sense (9.3.27) provides the justification for treating type I and type II sequential confidence sets similarly (since they share the same asymptotic properties). Nevertheless, from computational perspectives, type II confidence sets are often more cumbersome than type I sets. The simplicity in the uniparameter case may not hold for the multiparameter models.

9.6 SOME GENERAL REMARKS

The recursive estimators developed in Chapter 8 play an important role in the interval estimation problems as well. For the sequential estimation problems treated in the last two sections, we not only desire to incorporate robust statistics in the definition of the stopping variables, but we also need to pay attention to the computational convenience of the statistics that appear in the stopping variable's (N_d) expression. These statistics are to be computed for successive values of n, and hence their sequential updating is essentially required. Since the robust/nonparametric statistics are generally nonlinear, their recursive versions would be more computationally convenient. In fact there is also reason for initiating a multiphase version of such a recursive estimator. Let $\{n_j; j \geq 1\}$ be an increasing sequence of positive integers such that $n_j \to \infty$ as $j \to \infty$, but $n_j - n_{j-1} = o(n_{j-1})$ as j increases. One method is to compute the actual estimator T_n for $n = n_j$, $j \geq 1$, and use its recursive version for $n \in (n_{j-1}, n_j)$, $j \geq 2$. The idea is to arrest the error due to recursive approximations when cumulated over a long range of values of n. This is a suitable technique if one wants to replace a sequential procedure by a multistage (usually a three-stage) procedure that has the (asymptotic) efficiency of a sequential procedure. For details, mostly in a parametric case, we may refer to Ghosh, Mukhopadhyay, and Sen (1996, ch. 6).

The adaptive estimators considered in Chapter 8 have full asymptotic efficiency in a nonparametric setup. Naturally it is tempting to incorporate them in the confidence set estimation problems as well. The results of Chapter 8 apply directly to the nonsequential confidence interval problem. Considering the applications of adaptive estimators in a sequential case, we refer the reader to to Hušková and Sen (1985, 1986).

9.7 PROBLEMS

9.2.1. If $\{T_n\}$ is a reversed martingale and admits a first order representation, then (9.2.21) holds, and hence $V_n^J \to \sigma_T^2$ a.s. as $n \to \infty$.

9.2.2. If $T_n = T(F_n)$ is second order Hadamard differentiable at F, then (9.2.37)-(9.2.39) hold, and hence $V_n^J \to \sigma_T^2$ a.s. as $n \to \infty$.

9.2.3. For an L-estimator with smooth weights, verify that (9.2.37) -(9.2.39) hold under appropriate regularity conditions.

9.2.4. For the sample median $\tilde{X}_n$, consider a delete-k_n jackknifing where $k_n \to \infty$, but $n^{-1/2}k_n \to 0$ as $n \to \infty$. Show that V_{n,k_n}^J then converges (a.s.) to σ_T^2.

9.2.5. For sample mean $\bar{X}_n$ from a Cauchy distribution, show that the limit distribution is also Cauchy and that jackknifing is not viable (why?).

9.2.6. Let T_n be a U-statistic and $\theta(F)$ be stationary of order 1 (at F). Then show that $n(T_n - \theta(F))$ has a limiting distribution, which is typically a linear combination of independent chi square r.v.'s. Hence, or otherwise, comment on (9.2.9) for such a problem.

9.2.7. For Problem 9.2.5 examine the performance characteristics of the bootstrap variance estimator $V_{nM}^{\star}$ as well as the bootstrap d.f. $G_n^{\star}(.)$.

9.2.8. For Problem 9.2.4 suppose that the pdf $f(x;\theta)$ has a jump discontinuity at θ (population median). Show that bootstrapping may not work out in that case, although it can be modified to do so.

9.2.9. If $T(G)$ is second order Hadamard differentiable at F, then show that (9.2.69) holds. Show also that (9.2.69) holds for the von Mises functionals (under additional regularity conditions), although the latter may not be Hadamard differentiable.

9.2.10. Use Theorem 6.2.2 and verify (9.2.75).

9.2.11. Show that under $b = \beta$, $S_n(\beta) \overset{D}{=} S_n$ under $\beta = 0$, where S_n is the Kendall tau statistic. Hence verify (9.2.72)-(9.2.74) for the estimator of β based on Kendall's tau statistic.

9.2.12. Obtain an asymptotic linearity result for $S_n(\beta + t) - S_n(\beta)$, in t, for the preceding problem.

9.2.13. Verify that (9.2.77) extends directly to the limits obtained by using the Kendall tau statistic.

9.2.14. Obtain M-confidence intervals for β, the regression parameter in a simple regression model. Comment on its robustness properties.

9.3.1. Verify (9.3.5) and (9.3.6), and derive (9.3.7).

9.3.2. Verify (9.3.15).

9.3.3. Show that $(2/\pi)\sin^{-1}\rho - \rho$ is ≥ 0 for $0 \leq \rho \leq 1$.

9.3.4. For the intraclass correlation model, show that $\text{Ch}_{\max}(\boldsymbol{\Sigma}) = \sigma^2(1 + (k-1)\rho)$ if $\rho > 0$ and $\sigma^2(1-\rho)$ if $\rho < 0$.

9.3.5. Verify that $(k-1)/(1+(k-1)\rho)$ is $\nearrow$ in k (≥ 1) for $\rho > 0$.

9.3.6. For the multivariate location model, show that the signed rank statistics $T_{nj}(\boldsymbol{\theta})$ depend on θ_j for $j = 1, \ldots, p$.

9.3.7. For the univariate linear model, show that for rank statistics as well as M-statistics, the $T_{nj}(\boldsymbol{\beta})$ depend on all the coordinates of $\boldsymbol{\beta}$.

9.3.8. For the location model, consider the adaptive signed rank statistics introduced in (8.5.32). Use these to derive appropriate confidence intervals for the location parameter θ. Comment on its asymptotic efficiency.

9.3.9. Obtain parallel results on adaptive R-confidence intervals for regression.

9.4.1. Show that (9.4.14) hold whenever $\mathbb{E}_F\{\sup_{n\geq n_0} Z_n\} < \infty$.

9.4.2. If, in particular, $Z_n \leq (n-m)^{-1}\sum_{i\leq n} Z_i^\star \; \forall n \geq n_0$, where the $Z_i^\star$ are nonnegative i.i.d. r.v.'s with $\mathbb{E}Z^\star < \infty$, then show that (9.4.14) holds without any other moment condition.

9.4.3. Invoking contiguity, show that $T_n(\theta + n^{-1/2}a)$ (for R-estimators) is asymptotically normal, and use this to prove (9.4.22).

9.4.4. Show that for Problem 9.4.3 weak invariance principles for the $\hat{\theta}_n$ lead to the solution under no extra regularity conditions. Show that the "tightness" part of the weak invariance principle yields a simpler proof of the Anscombe (1952) condition.

9.4.5. Show that (9.4.23) implies (9.4.22).

9.4.6. For the signed-rank statistics, verify (9.4.23) under appropriate conditions on F and the score function.

9.4.7. Verify (9.4.23) for M-estimators as well.

9.4.8. Show that for $\hat{\theta}_{U,n} - \hat{\theta}_{L,n}$ admitting an FOADR with a remainder term $O_p(n^{-1})$, (9.4.21) and (9.4.22) follow directly, so (9.4.23) is not required.

9.5.1. Use (9.5.11) to show that $\lim_{d\downarrow 0} \mathbb{E}N_d/n_d^0 = 1$.

9.5.2. Verify (9.5.12) when $\mathbf{V}_n$ admits an FOADR.

9.5.3. Verify (9.5.14).

Chapter 10

Robust Statistical Tests

10.1 INTRODUCTION

In order to limit the influence of outliers (or gross errors), Huber (1965) appealed to the concept of maximin power of a test (for a simple null hypothesis vs. a simple alternative) and derived some censored likelihood ratio-type tests. Later using the concept of so-called "Choquet capacities", Huber and Strassen (1973) extended the robust test theory to a far more general class of statistical hypothesis testing problems. Their elegant theory, however, does not work out conveniently for a composite null hypothesis against a composite alternative testing problem. In our discussion, in Part I we mostly presented statistical models involving several parameters for which composite hypotheses are far more natural than certain simple hypotheses. We may also note that there is an intricate connection between (robust) confidence set estimation and hypothesis testing problems. The last chapter dealt with a variety of such confidence intervals and sets based on robust statistics. In this chapter these will be incorporated in the motivation and derivation of some appropriate robust tests for suitable (simple as well as composite) hypotheses. The results are therefore to be regarded as complementary those in Chapters 8 and 9. By no means do these chapters constitute a complete treatise of the robust statistical tests subject. A full coverage of robust tests is indeed beyond the scope of this book. Nevertheless, the principal findings of the present chapter aim to provide a comprehensive picture of robustness of some statistical tests, interpreted, as in the case of estimates, in a local as well as global sense. M-tests, to be presented in Section 10.2, are more appealing in a local robustness setup, while rankbased (R-) tests are globally robust. Since details of such R-tests are already provided in the texts of Sen (1981a), Puri and Sen (1985), and Shorack and Wellner (1986), among others, the emphasis in Section 10.3 is laid down mainly on tests based on regression rank scores statistics (which were introduced in Chapter 6). The concluding section pertains to a brief introduction to some other related tests based on studentized M-statistics. Sequential analogues of the Wald (1947) sequential probability ratio tests are also considered in the same vein.

10.2 M-TESTS

In view of our primary emphasis on the location and regression models, we motivate such M-tests through the location, and then present their extensions to some linear models.

Let $X_1, \ldots, X_n$ be n i.i.d. r.v.'s from a d.f. $F_\theta(x) = F(x - \theta)$, $x \in \mathbb{R}$, $\theta \in \Theta \subset \mathbb{R}$. It is assumed that F is a symmetric d.f. If the functional form of F is specified, one may be tempted to use the classical likelihood ratio test which possesses some optimal properties (when the assumed model holds). However, such a parametric test is generally very nonrobust, even to small departures from assumed F. Such departures from the assumed model were studied in earlier chapters with the aid of Levy/Kolmogorov metrics as well as the gross errors model. If a least favorable distribution in such a neighborhood can be identified with respect to the two hypotheses, then one can construct a likelihood ratio type statistic for this least favorable law and show that such a test has a maximin power property (Huber 1965). Basically, one conceives of a class $\mathcal{F}_0$ of d.f.'s that define an appropriate neighborhood of F, and let F_0 ($\in \mathcal{F}_0$) be such a least favorable d.f. with respect to the pair (H_0, H_1) of parametric hypotheses (on θ). Then Huber's suggestion is to construct the usual likelihood ratio test statistic (for H_0 vs. H_1) corresponding to the d.f. F_0. Suppose now that F ($\in \mathcal{F}_0$) admits a density f with respect to some sigma-finite measure μ. Then instead of the likelihood ratio statistic, one may also consider (Rao's) efficient score statistic

$$L_n^0 = \sum_{i=1}^{n} (\frac{\partial}{\partial \theta}) \log f(X_i, \theta)|_{\theta_0}, \tag{10.2.1}$$

where $H_0 : \theta = \theta_0$ and $H_1 : \theta = \theta_1$, $\theta_1 >$ (or $<$ or $\neq$)θ_0. In the light of Huber modifications, a robust alternative is to use the test statistic

$$L_n^{\star 0} = \sum_{i=1}^{n} (\frac{\partial}{\partial \theta}) \log f_0(X_i, \theta)|_{\theta_0}, \tag{10.2.2}$$

where f_0 is the density corresponding to F_o. One may, in general, conceive of a score function $\psi : \mathbb{R} \mapsto \mathbb{R}$, and replace $L_n^{\star 0}$ by an M-statistic

$$M_n^0 = \sum_{i=1}^{n} \psi(X_i - \theta_0). \tag{10.2.3}$$

If it is assumed that under $\theta_0 = 0$, $\int_{\mathbb{R}} \psi(x)dF(x) = 0$ and $\sigma_\psi^2 = \int_{\mathbb{R}} \psi^2(x)dF(x)$: $0 < \sigma_\psi < \infty$, then M_n^0 has zero mean and variance $n\sigma_\psi^2$ when θ_0 holds. On the other hand, if ψ is $\nearrow$, then $\mathbb{E}_\theta M_n^0$ is $\stackrel{\leq}{\underset{>}{=}} 0$ according as θ is $\stackrel{\geq}{\underset{<}{=}} \theta_0$. Moreover M_n^0 involves independent summands, so under θ_0,

$$n^{-1/2}(M_n^0/\sigma_\psi) \xrightarrow{\mathcal{D}} \mathcal{N}(0,1). \tag{10.2.4}$$

Therefore a (one- or two-sided) test for $H_0 : \theta = \theta_0$ against $H_1 : \theta > \theta_0$ (or $\theta < \theta_0$ or $\theta \neq \theta_0$) can be based on (10.2.4) with a critical (asymptotic) level τ_α (or $-\tau_\alpha$, or $\tau_{\alpha/2}$). If $\hat{\theta}_n$ is the usual M-estimator of θ based on $X_1, \ldots, X_n$ and the score function ψ, and if we let

$$\begin{aligned} \hat{\sigma}_n^2 &= \int_{\mathbb{R}} \psi^2(x - \hat{\theta}_n)dF_n(x) \\ &= n^{-1}\sum_{i=1}^{n} \psi^2(X_i - \hat{\theta}_n), \end{aligned} \tag{10.2.5}$$

then a convenient form of the test statistic is

$$T_n = n^{-1/2}(M_n^0/\hat{\sigma}_n). \tag{10.2.6}$$

Under $H_0 : \theta = \theta_0$, T_n is asymptotically normal, so for $H_1 : \theta > \theta_0$, the (asymptotic) critical level is τ_α $(0 < \alpha < 1)$. The main justification of this test statistic T_n is the choice of a suitable robust ψ. Toward this study, we consider a sequence of local (Pitman-type) alternative $H_{(n)} : \theta = \theta_{(n)} = \theta_0 + n^{-1/2}\lambda$ for some $\lambda \in \mathbb{R}$. We denote by

$$\delta_\lambda = \lim_{n\to\infty} \int_{\mathbb{R}} \sqrt{n}\{\psi(x - \theta_0) - \psi(x - \theta_0 - n^{-1/2}\lambda)\}dF(x - \theta_0), \tag{10.2.7}$$

and we assume that this limit exists and is positive for all $\lambda > 0$. The usual conditions on ψ, F imposed in Chapters 3 and 5 are sufficient for this purpose. Then, assuming further that

$$\lim_{\delta\to 0} \int_{\mathbb{R}} \psi^2(x - \delta)dF(x) = \int_{\mathbb{R}} \psi^2(x)dF(x) = \sigma_\psi^2, \tag{10.2.8}$$

we obtain that the asymptotic power of the test based on T_n (for H_0 vs. $H_{(n)}$) is given by

$$\begin{aligned} \beta(\lambda) &= \lim_{n\to\infty} \mathbb{P}\{T_n > \tau_\alpha | H_{(n)}\} \\ &= 1 - \Phi(\tau_\alpha - \delta_\lambda/\sigma_\psi), \ \lambda \in \mathbb{R}. \end{aligned} \tag{10.2.9}$$

Thus, within the class of such score tests, an optimal ψ relates to the maximization of $\delta_\lambda/\sigma_\psi$ for every fixed $\lambda \in \mathbb{R}$.

At this stage we refer to Theorem 5.3.2 and identify δ_λ as $\lambda\gamma$, where γ (> 0) is defined as $\gamma_c + \gamma_s$ and given by (5.3.22) and (5.3.25). Therefore the problem reduces to minimizing σ_ψ^2/γ^2 = asymptotic variance of $\sqrt{n}(\hat{\theta}_n - \theta)$, where $\sigma_\psi^2 = \sigma_\psi^2(F)$, $\gamma = \gamma(\psi, F)$, and $F \in \mathcal{F}_0$. We may then use the adaptive approach developed in Section 8.5 and look for an adaptive M-test that is asymptotically optimal within the class of all scores M-tests. Alternatively, as in Huber (1965), we may look for a particular score function $\psi_0 : \mathbb{R} \to \mathbb{R}$, such that

$$\sup_{F \in \mathcal{F}_0} \sigma_\psi^2/\gamma^2 \text{ is a minimum at } \psi = \psi_0, \tag{10.2.10}$$

and obtain an asymptotically maximin power M-test of the score type with ψ_0 in $\mathcal{F}_0$ (e.g., ψ_0 is the Huber function provided that $\mathcal{F}_0$ is the family of contaminated normal distributions). The test for H_0 against two-sided alternatives follow on parallel lines (see Problem 10.2.1).

Let us refer back to Section 5.3 and identify a candidate score function $\psi : \mathbb{R} \to \mathbb{R}$ as

$$\psi(x) = \rho'(x) \text{ where } \rho : \mathbb{R} \mapsto \mathbb{R}^+; \tag{10.2.11}$$

$$Q_n(\theta) = \sum_{i=1}^{n} \rho(X_i - \theta), \ \theta \in \Theta; \tag{10.2.12}$$

and

$$Q_n^\star = \inf\{Q_n(\theta) : \theta \in \Theta\}; \tag{10.2.13}$$

$$\hat{\theta}_n^\star = \operatorname{Arg\,min}\{Q_n(\theta) : \theta \in \Theta\}. \tag{10.2.14}$$

Let then

$$\begin{aligned} Z_n &= Q_n(\theta_0) - \hat{Q}_n^\star \\ &= \sum_{i=1}^{n} [\rho(X_i - \theta_0) - \rho(X_i - \hat{\theta}_n^\star)], \end{aligned} \tag{10.2.15}$$

where $\hat{\theta}_n^\star$ is the M-estimator of θ based on the score function ψ. Z_n is analogous to the classical likelihood ratio type statistic where for a normal F, $\rho(y) \equiv y^2$, so Z_n is reducible to the classical student t-statistic. Now by Theorem 5.3.3 (with $\tau = 1/2$) we obtain that for large n,

$$\begin{aligned} Z_n &\overset{P}{\sim} \frac{1}{2} n(\hat{\theta}_n^\star - \theta_0)^2 \gamma \\ &= \frac{1}{2}\left(\frac{\sigma_\psi^2}{\gamma}\right)\left\{\frac{n\gamma^2(\hat{\theta}_n^\star - \theta_0)^2}{\sigma_\psi^2}\right\}. \end{aligned} \tag{10.2.16}$$

Under $H_0 : \theta = \theta_0$,

$$2\gamma(Z_n/\sigma_\psi^2) \xrightarrow{\mathcal{D}} \chi_1^2. \tag{10.2.17}$$

One may then estimate γ and σ_ψ^2 by $\hat{\gamma}_n$ and $\hat{\sigma}_n^2$ as before, and construct a test statistic

$$Z_n^\star = 2\hat{\gamma}_n\{Z_n/\hat{\sigma}_n^2\}. \tag{10.2.18}$$

Under $H_{(n)} : \theta = \theta_{(n)} = \theta_0 + n^{-1/2}\lambda,\ \lambda \in I\!R,\ Z_n^\star$ has asymptotically a noncentral chi squared d.f. with 1 DF and noncentrality parameter

$$\Delta^\star = \gamma^2\lambda^2\sigma_\psi^{-2}. \tag{10.2.19}$$

The choice of an (asymptotically) optimal ψ_0 is isomorphic to the case of score M-statistics, and we can go either for an adaptive ψ_0 (as in section 8.5) or for a miximin power choice as in (10.2.10). In either case the choice is driven primarily by the basic robustness considerations. In passing, we may compare T_n in (10.2.6) and $Z_n^\star$ in (10.2.18). Whereas for the construction of T_n one only needs to estimate σ_ψ^2 (by $\hat{\sigma}_n^2$), for $Z_n^\star$, one needs both $\hat{\gamma}_n$ and $\hat{\sigma}_n^2$. From this perspective T_n is preferable to $Z_n^\star$. On the other hand, for not so local alternatives, likelihood ratio-type tests behave better than score-type test [see Hoeffding (1965a,b)] so $Z_n^\star$ has an edge over T_n.

In the preceding development it has been tacitly assumed that F has a fixed scale (i.e., the setup of Section 5.3 is taken for granted). It is possible to make use of the basic results on studentized M-statistics developed in Section 5.4, and to extend such M-type results to the case where the scale factor is unknown. Since the approaches are similar, we pose these as Problems 10.2.2 and 10.2.3.

Let us next consider the case of linear hypotheses pertaining to the linear model (5.5.1), for which M-estimators were studied in detail in Section 5.5. In this setup the regression parameter $\boldsymbol{\beta} = (\beta_1, \ldots, \beta_p)' \in B \subset I\!R_p$ for some $p \geq 1$. A linear hypothesis may be framed as

$$H_0 : \mathbf{D}\boldsymbol{\beta} = \mathbf{0} \text{ vs. } H_1 : \mathbf{D}\boldsymbol{\beta} \neq \mathbf{0}, \tag{10.2.20}$$

where $\mathbf{D}$ is a specified $q \times p$ matrix for some $q \leq p$. We may, without loss of generality, let

$$\mathbf{D}^0_{p\times p} = \begin{bmatrix} \mathbf{D} & q \times p \\ \mathbf{D}^\star & (p-q)\times p \end{bmatrix}, \quad \boldsymbol{\theta} = \mathbf{D}^0\boldsymbol{\beta} = \begin{pmatrix} \boldsymbol{\theta}_1 \\ \boldsymbol{\theta}_2 \end{pmatrix} \begin{matrix} q \times 1 \\ (p-q)\times 1. \end{matrix} \tag{10.2.21}$$

We assume that $\mathbf{D}^0$ is a nonsingular matrix with the inverse $(\mathbf{D}^0)^{-1}$. Then we write

$$\mathbf{X}\boldsymbol{\beta} = \mathbf{X}(\mathbf{D}^0)^{-1}\mathbf{D}^0\boldsymbol{\beta} = \mathbf{X}^\star\boldsymbol{\theta}. \tag{10.2.22}$$

so that by this reparametrization we take (5.5.10) as

$$\mathbf{Y} = \mathbf{X}^\star\boldsymbol{\theta} + \mathbf{E} \tag{10.2.23}$$

and let

$$H_0 : \boldsymbol{\theta}_1 = \mathbf{0} \text{ vs. } H_1 : \boldsymbol{\theta}_1 \neq \mathbf{0}, \tag{10.2.24}$$

treating $\boldsymbol{\theta}_2$ as a nuisance parameter (vector). Under H_0 we have therefore

$$\mathbf{Y} = \mathbf{X}_2^\star \boldsymbol{\theta}_2 + \mathbf{E}, \quad \text{where } \mathbf{X}^\star = \begin{pmatrix} \mathbf{X}_1^\star, & \mathbf{X}_2^\star \\ n \times q & n \times (p-q) \end{pmatrix}. \tag{10.2.25}$$

Parallel to (10.2.12) we let

$$Q_n(\boldsymbol{\theta}) = \sum_{i=1}^{n} \rho(Y_i - \boldsymbol{\theta}'\mathbf{x}_i^\star), \ \boldsymbol{\theta} \in \boldsymbol{\Theta} \subset \mathbb{R}_p, \tag{10.2.26}$$

where $\mathbf{x}_i^{\star\prime}$ is the ith row of $\mathbf{X}^\star$, for $i = 1, \ldots, n$. Let then

$$\hat{\boldsymbol{\theta}}_n^\star = \operatorname{Arg\,min}\{Q_n(\boldsymbol{\theta}) : \boldsymbol{\theta} \in \boldsymbol{\Theta}\}, \tag{10.2.27}$$

$$\hat{\boldsymbol{\theta}}_n^{0\star} = \operatorname{Arg\,min}\{Q_n(\boldsymbol{\theta}) : \boldsymbol{\theta}_1 = \mathbf{0}, \boldsymbol{\theta}_2 \in \boldsymbol{\Theta}_2\}, \tag{10.2.28}$$

where $\boldsymbol{\Theta}_2 = \{\boldsymbol{\theta}_2 : (\boldsymbol{\theta}_1', \boldsymbol{\theta}_2') \in \boldsymbol{\Theta}, \ \boldsymbol{\theta}_1 = \mathbf{0}\}$. Letting $\psi \equiv \rho'$, we have by (10.2.27) and (10.2.28),

$$\sum_{i=1}^{n} \mathbf{x}_i^\star \psi(Y_i - (\hat{\boldsymbol{\theta}}_n^\star)'\mathbf{x}_i^\star) = \mathbf{0} \tag{10.2.29}$$

and

$$\sum_{i=1}^{n} \mathbf{x}_{i2}^\star \psi(Y_i - (\hat{\boldsymbol{\theta}}_{n2}^{0\star})'\mathbf{x}_{i2}^\star) = \mathbf{0}, \tag{10.2.30}$$

where $\mathbf{x}_i^{\star\prime} = (\mathbf{x}_{i1}^{\star\prime}, \mathbf{x}_{i2}^{\star\prime})$, $i = 1, \ldots, n$ and $(\hat{\boldsymbol{\theta}}_n^{0\star})' = (\mathbf{0}', (\hat{\boldsymbol{\theta}}_{n2}^{0\star})')$. Then, as in (10.2.15), we let

$$Z_n = Q_n(\hat{\boldsymbol{\theta}}_n^{0\star}) - Q_n(\hat{\boldsymbol{\theta}}_n^\star). \tag{10.2.31}$$

As a direct corollary to Theorem 5.5.1 (when confined to the fixed scale case), we have

$$\hat{\boldsymbol{\theta}}_n^\star - \boldsymbol{\theta} = \gamma^{-1}(\mathbf{X}^{\star\prime}\mathbf{X}^\star)^{-1} \sum_{i=1}^{n} \mathbf{x}_i^\star \psi(Y_i - \boldsymbol{\theta}'\mathbf{x}_i^\star) + \mathbf{R}_n^\star, \tag{10.2.32}$$

where

$$||\mathbf{R}_n^\star|| = o_p(n^{-1/2}); \tag{10.2.33}$$

(10.2.33) can be improved to $O_p(n^{-1})$ under additional regularity conditions on ψ. Similarly, under $H_0 : \boldsymbol{\theta}_1 = \mathbf{0}$, we have

$$\hat{\boldsymbol{\theta}}_{n2}^{0\star} - \boldsymbol{\theta}_2 = \gamma^{-1}(\mathbf{X}_2^{\star\prime}\mathbf{X}_2^\star)^{-1} \sum_{i=1}^{n} \mathbf{x}_{i2}^\star \psi(Y_i - \boldsymbol{\theta}_2'\mathbf{x}_{i2}^\star) + \mathbf{R}_n^{0\star}, \tag{10.2.34}$$

where

$$||\mathbf{R}_n^{0\star}|| = o_p(n^{-1/2}). \tag{10.2.35}$$

Recall that we have assumed in Section 5.5 that as $n \to \infty$,

$$n^{-1}\mathbf{X}^{\star\prime}\mathbf{X}^{\star} \to \mathbf{C}, \quad \text{where } \mathbf{C} \text{ is p.d.} \tag{10.2.36}$$

We partition $\mathbf{C}$ into $((\mathbf{C}_{rs}))_{r,s=1,2}$ [where $\mathbf{C}_{11}$ is $q \times q$, $\mathbf{C}_{12} = \mathbf{C}_{21}'$ is $q \times (p-q)$, and $\mathbf{C}_{22}$ is $(p-q) \times (p-q)$] and write

$$\mathbf{C}^{\star} = \mathbf{C}_{11} - \mathbf{C}_{12}\mathbf{C}_{22}^{-1}\mathbf{C}_{21}. \tag{10.2.37}$$

From (10.2.32), we have

$$n\gamma^2\sigma_{\psi}^{-2}(\hat{\boldsymbol{\theta}}_n^{\star} - \boldsymbol{\theta})'\mathbf{C}(\hat{\boldsymbol{\theta}}_n^{\star} - \boldsymbol{\theta}) \xrightarrow{\mathcal{D}} \chi_p^2, \tag{10.2.38}$$

and from (10.2.34), we have similarly,

$$n\gamma^2\sigma_{\psi}^{-2}(\hat{\boldsymbol{\theta}}_{n2}^{0\star} - \boldsymbol{\theta}_2)'\mathbf{C}_{22}(\hat{\boldsymbol{\theta}}_{n2}^{0\star} - \boldsymbol{\theta}_2) \xrightarrow{\mathcal{D}} \chi_{p-q}^2, \tag{10.2.39}$$

Further, if we write

$$\begin{aligned} \mathbf{M}_n(\boldsymbol{\theta}) &= n^{-1/2}\sum_{i=1}^{n} \mathbf{x}_i^{\star}\psi(Y_i - \boldsymbol{\theta}'\mathbf{x}_i^{\star}) \\ &= \left(\mathbf{M}_{n1}'(\boldsymbol{\theta}),\ \mathbf{M}_{n2}'(\boldsymbol{\theta})\right)', \end{aligned} \tag{10.2.40}$$

then by virtue of (10.2.32) and (10.2.34) we have

$$n^{1/2}\gamma(\hat{\boldsymbol{\theta}}_n^{\star} - \boldsymbol{\theta}) = \mathbf{C}^{-1}\mathbf{M}_n(\boldsymbol{\theta}) + \mathrm{o}_p(1), \tag{10.2.41}$$

$$n^{\frac{1}{2}}\gamma(\hat{\boldsymbol{\theta}}_{n2}^{0\star} - \boldsymbol{\theta}_2) = \mathbf{C}_{22}^{-1}\mathbf{M}_{n2}(\boldsymbol{\theta}) + \mathrm{o}_p(1). \tag{10.2.42}$$

Hence by (10.2.38), (10.2.39), (10.2.41), (10.2.42), and the classical Cochran theorem, we have

$$\begin{aligned} n\gamma^2\sigma_{\psi}^{-2}\{(\hat{\boldsymbol{\theta}}_n^{\star} - \boldsymbol{\theta})'\mathbf{C}(\hat{\boldsymbol{\theta}}_n^{\star} - \boldsymbol{\theta}) - (\hat{\boldsymbol{\theta}}_{n2}^{0\star} - \boldsymbol{\theta}_2)'\mathbf{C}_{22}(\hat{\boldsymbol{\theta}}_{n2}^{0\star} - \boldsymbol{\theta}_2)\} \\ \xrightarrow{\mathcal{D}} \chi_q^2, \text{ when } H_0 \text{ holds.} \end{aligned} \tag{10.2.43}$$

Finally, as a direct corollary to (5.5.37), we have for any compact $K \subset \mathbb{R}_p$, when $\boldsymbol{\theta}$ holds,

$$\sup_{\mathbf{t}\in K} |Q_n(\boldsymbol{\theta} + n^{-1/2}\mathbf{t}) - Q_n(\boldsymbol{\theta}) + \mathbf{t}'\mathbf{M}_n(\boldsymbol{\theta}) + \frac{1}{2}\gamma\mathbf{t}'(n^{-1}\mathbf{X}^{\star\prime}\mathbf{X}^{\star})\mathbf{t}| = \mathrm{o}_p(1); \tag{10.2.44}$$

we pose the derivation of (10.2.44) as Problem 10.2.4. Since by (10.2.41) and (10.2.42), $n^{1/2}\|\hat{\boldsymbol{\theta}}_n^{\star} - \boldsymbol{\theta})\|$ and $n^{1/2}\|\hat{\boldsymbol{\theta}}_{n2}^{0\star} - \boldsymbol{\theta}_2)\|$ are both $\mathrm{O}_p(1)$, from (10.2.29), (10.2.30), and (10.2.31), we have

$$Z_n = \frac{1}{2}\gamma\{n(\hat{\boldsymbol{\theta}}_n^{\star} - \boldsymbol{\theta})'\mathbf{C}(\hat{\boldsymbol{\theta}}_n^{\star} - \boldsymbol{\theta}) - n(\hat{\boldsymbol{\theta}}_{n2}^{0\star} - \boldsymbol{\theta}_2)'\mathbf{C}_{22}(\hat{\boldsymbol{\theta}}_{n2}^{0\star} - \boldsymbol{\theta}_2)\} + \mathrm{o}_p(1). \tag{10.2.45}$$

Thus under $H_0 : \boldsymbol{\theta}_1 = \mathbf{0}$ we have

$$2\gamma(Z_n/\sigma_\psi^2) \xrightarrow{\mathcal{D}} \chi_q^2. \tag{10.2.46}$$

As in the location model, we denote by $\hat{\gamma}_n$ and $\hat{\sigma}_n^2$ some consistent estimator of γ and σ_ψ^2, respectively, and consider a likelihood ratio type statistic

$$\begin{aligned} Z_n^\star &= 2\hat{\gamma}_n\hat{\sigma}_n^{-2}Z_n \\ &= 2\hat{\gamma}_n\hat{\sigma}_n^{-2}[Q_n(\hat{\boldsymbol{\theta}}_n^{0\star}) - Q_n(\hat{\boldsymbol{\theta}}_n^\star)]. \end{aligned} \tag{10.2.47}$$

This type of statistics were proposed by Schrader and McKean (1977) and Schrader and Hettmansperger (1980). Note that $Z_n^\star$ involves the computation of (1) $\hat{\boldsymbol{\theta}}_n^\star$ as well as $\hat{\boldsymbol{\theta}}_n^{0\star}$, (2) $\hat{\gamma}_n$ and (3) $\hat{\sigma}_n^2$. It may not be necessary to do so, especially for large n. We will turn to that scheme shortly. But we note that the decomposition in (10.2.45) implies that if we consider a sequence $\{H_{1(n)}\}$ of local (Pitman-type) alternatives

$$H_{1(n)} : \boldsymbol{\theta}_1 = n^{-\frac{1}{2}}\boldsymbol{\lambda},\ \boldsymbol{\lambda}\ (\text{fixed})\ \in \mathbb{R}_q, \tag{10.2.48}$$

then under $\{H_{1(n)}\}$, (10.2.43) directly extends to a noncentral chi square with q degrees of freedom and noncentrality parameter

$$\triangle = \gamma^2\sigma_\psi^{-2}\boldsymbol{\lambda}'(\mathbf{C}^\star)^{-1}\boldsymbol{\lambda}. \tag{10.2.49}$$

Therefore we conclude that under $\{H_{1(n)}\}$,

$$Z_n^\star \xrightarrow{\mathcal{D}} \chi_{q,\triangle}^2, \tag{10.2.50}$$

and we leave the proof of (10.2.50) as Problem 10.2.5. In passing, we remark that (10.2.50) enables us to use the measure of the Pitman asymptotic relative efficiency of tests, and it agrees with the same for the point (or interval) estimation problem treated in Chapters 8 and 9.

Let us now consider the score-type M-tests, proposed by Sen (1982). With the same notations as in (10.2.20) through (10.2.25), we consider the usual M-estimator of $\boldsymbol{\theta}_2$ (under H_0) based on the score function $\psi : \mathbb{R} \mapsto \mathbb{R}$. Thus we have the estimator $\hat{\boldsymbol{\theta}}_{n2}^{0\star}$ defined by (10.2.30). Note that this does not involve the component $\boldsymbol{\theta}_1$ and is therefore computationally less involved. Let us then denote these statistics by

$$\tilde{\mathbf{M}}_{n1} = n^{-1/2}\sum_{i=1}^n \mathbf{x}_{i1}\psi(Y_i - (\hat{\boldsymbol{\theta}}_{n2}^{0\star})'\mathbf{x}_{i2}), \tag{10.2.51}$$

$$\hat{\sigma}_n^2 = n^{-1}\sum_{i=1}^n \psi^2(Y_i - (\hat{\boldsymbol{\theta}}_{n2}^{0\star})'\mathbf{x}_{i2}), \tag{10.2.52}$$

and

$$\mathbf{C}_n^\star = n^{-1}\{(\mathbf{X}_1^{\star\prime}\mathbf{X}_1^\star) - (\mathbf{X}_1^{\star\prime}\mathbf{X}_2^\star)'(\mathbf{X}_2^{\star\prime}\mathbf{X}_2^\star)^{-1}(\mathbf{X}_2^{\star\prime}\mathbf{X}_1^\star)\}. \tag{10.2.53}$$

Note that $\mathbf{C}_n^\star \to \mathbf{C}^\star$, defined by (10.2.37), as $n \to \infty$. We define the score M-test statistic by

$$T = [(\tilde{\mathbf{M}}_{n1})'(\mathbf{C}_n^\star)^{-1}(\tilde{\mathbf{M}}_{n1})]/\hat{\sigma}_n^2. \tag{10.2.54}$$

A direct application of Lemma 5.5.1 (with a further simplification that here $u = 0$) yields $\tilde{\mathbf{M}}_{n1}$ can be expressed as a linear combination of $\mathbf{M}_{n1}(\boldsymbol{\theta})$ and $\mathbf{M}_{n2}(\boldsymbol{\theta})$ plus a remainder term, which is $\mathbf{o}_p(1)$. As such, we are in a position to use the Cochran theorem on quadratic forms in (asymptotically) normally distributed random variables. Using the Slutzky theorem (on $\hat{\sigma}_n^2/\sigma_\psi^2$), we therefore get

$$T_n \xrightarrow{\mathcal{D}} \chi_q^2 \text{ under } H_0, \tag{10.2.55}$$

while under $\{H_{1(n)}\}$ in (10.2.48) we have

$$T_n \xrightarrow{\mathcal{D}} \chi_{q,\triangle}^2, \tag{10.2.56}$$

where $\triangle$ is given by (10.2.49). We pose the derivation of (10.2.55) -(10.2.56) as Problem 10.2.6. Thus, T_n and $Z_n^\star$ are asymptotically equivalent (under H_0 as well as local alternatives), while T_n does not involve the estimation of γ and the computation of $\hat{\theta}_n^\star$. From this perspective T_n seems to have an advantage over $Z_n^\star$.

In the above discussion we have considered the conventional case of M-tests when the scale S $[= S(F)]$ is assumed to be known. On the other hand, in Section 5.5 we presented general results on M-estimators in linear models for the studentized case [where $S(F)$ is not known]. In view of the basic results in Section 5.5, we are in a position to incorporate studentized M-statistics (and M-estimators) in the formulation of similar M-tests for linear hypothesis when the scale factor $S(F)$ is not assumed to be given. Again, in view of the scale statistics S_n [estimating $S(F)$] and possible nonlinearity of $M_n(./S_n)$, robustness considerations should be given due importance.

While $Z_n^\star$ resembles the classical likelihood ratio test and $T_n^\star$ the Rao scores tests, it is also possible to construct some Wald (1943) type tests that are directly based on some (robust) estimators of the parameters and their asymptotic dispersion matrices. In this context it is not necessary to consider Wald-type tests based only on M-statistics (and estimators), for R- and L-estimators are equally applicable. For this reason we will include a brief discussion on the Wald-type tests in the concluding section of this chapter. The interrelationships of various classes of estimators studied in Chapter 7 have natural analogues for such Wald-type tests, and we will look at them more closely in this chapter. Finally, although in this section we have only considered M-tests for location/regression models, by virtue of the general linearity results in Section 5.2 (see Theorem 5.2.1), similar M-test work out well for general parameters.

10.3 R-TESTS

Whenever a null hypothesis (H_0) relates to an invariance (of the joint distribution of the sample observations) under an appropriate group of transformations (which map the sample space onto itself), optimal invariant tests exist under quite general regularity conditions. For a group of sign-invariant transformations and for the group of arbitrary (strictly) monotone transformations, rank-based (R-)tests occupy a focal point in this development. For the one-sample location model, two- or multisample location/dispersion models, and in general, for simple linear models, such hypotheses of invariance hold, and LMPR (locally most powerful rank) tests exist. Moreover, even when a rank test may not be LMPR for an unknown F, it is generally globally robust, and, in addition, exactly distribution-free (EDF) under the hypothesis of invariance. An excellent treatment of this subject matter is available in the classic text of Hájek and Šidak (1967: *Theory of Rank Tests*). Their work deals with with EDF tests for various hypotheses of invariance, but often in a general linear model, a null hypothesis is itself composite (violating the basic invariance structure). Analogous R-tests have been worked out that are asymptotically distribution-free (ADF) and have suitable properties. Puri and Sen (1985, ch. 7) contains a detailed account of some of these developments in an even more general setup of possibly multivariate observations. We intend to present here only a brief overview of some recent developments not reported in these earlier texts.

Consider a linear model reduced by reparametrization as in (10.2.20)-(10.2.25). We want to test for $H_0 : \boldsymbol{\theta}_1 = \mathbf{0}$ vs. $H_1 : \boldsymbol{\theta}_1 \neq \mathbf{0}$, treating $\boldsymbol{\theta}_2$ as a nuisance parameter. As in (6.6.2)-(6.6.3) we define a vector of linear rank statistics as

$$\mathbf{L}_n(\mathbf{b}) = \sum_{i=1}^{n} (\mathbf{x}_i^\star - \bar{\mathbf{x}}_n^\star) a_n(R_{ni}(\mathbf{b})), \ \mathbf{b} \in I\!\!R_p. \tag{10.3.1}$$

[Note that the $\mathbf{c}_i$ in Section 6.6 are replaced here by $\mathbf{x}_i^\star$ in order to retain the conformity of notations with Section 10.2.] We also define $\mathbf{V}_n$ as in (6.6.16). Then, for $\boldsymbol{\theta}_2 = \boldsymbol{\theta}$, we have a simple null hypothesis, and we consider an EDF test based on

$$\mathcal{L}_n = (\mathbf{L}_n' \mathbf{V}_n^{-1} \mathbf{L}_n)/A_n^2, \tag{10.3.2}$$

where $\mathbf{L}_n = \mathbf{L}_n(\mathbf{0})$ and $A_n^2 = (n-1)^{-1}\sum_{i=1}^n \{a_n(i) - \bar{a}_n\}^2$. Under $H_0 : \boldsymbol{\theta} = \mathbf{0}$, $\mathcal{L}_n \xrightarrow{\mathcal{D}} \chi_p^2$ as $n \to \infty$, while under local alternatives we have a noncentral chi square d.f. with p DF an appropriate noncentrality parameter. This simple prescription may not work out well for subhypothesis testing problem in (10.2.24) when $\boldsymbol{\theta}_2$ is unknown. However, an aligned rank test, considered by Sen and Puri (1977) and Adichie (1978) works out quite well. In principle, this is similar to the score M test based on T_n in (10.2.54). For the (10.2.25) model (relating to $H_0 : \boldsymbol{\theta}_1 = \mathbf{0}$), we proceed as in (6.6.5)-(6.6.8) (with $\mathbf{b}_2 \in \mathbb{R}_{p-q}$) and consider an R-estimator $\hat{\boldsymbol{\theta}}_{n2}^{0\star}$ of $\boldsymbol{\theta}_2$. We let $\hat{\boldsymbol{\theta}}_n^{0\star} = (\mathbf{0}', \hat{\boldsymbol{\theta}}_{n2}^{0\star})'$. Then we consider the aligned ranks $R_{ni}(\hat{\boldsymbol{\theta}}_n^{0\star}) = \tilde{R}_{ni}$, say, $1 \le i \le n$, and find the subvector of the aligned rank statistic

$$\tilde{\mathbf{L}}_{n(1)} = \sum_{i=1}^{n} (\mathbf{x}_{i1}^\star - \bar{\mathbf{x}}_{n1}^\star) a_n(\tilde{R}_{ni}). \tag{10.3.3}$$

Now we define $\mathbf{C}_n^\star$ as in (10.2.53). Then an aligned test statistic is given by

$$\mathcal{L}_n^{(1)} = n^{-1} A_n^{-2} \{\tilde{\mathbf{L}}_{n(1)}' (\mathbf{C}_n^\star)^{-1} \tilde{\mathbf{L}}_{n(1)}\}. \tag{10.3.4}$$

By Lemma 6.6.1 and Theorem 6.6.1, it follows that under $H_0 : \boldsymbol{\theta}_1 = \mathbf{0}$, $\mathcal{L}_n^{(1)} \xrightarrow{\mathcal{D}} \chi_q^2$, so an ADF test can be based on $\mathcal{L}_n^{(1)}$. For local alternative $\{H_{1(n)}\}$, defined by (10.2.48), $\mathcal{L}_n^{(1)} \xrightarrow{\mathcal{D}} \chi_{q,\Delta^\star}^2$, where $\Delta^\star = \gamma^2 A_\phi^{-2}\{\boldsymbol{\lambda}'(\mathbf{C}^\star)^{-1}\boldsymbol{\lambda}\}$, and γ and A_ϕ^2 are defined by (6.4.19) and (6.6.14) respectively. An interesting feature of this aligned R-test is that it involves only the (R-)estimation of the nuisance parameter ($\boldsymbol{\theta}_2$), and not the parameter ($\boldsymbol{\theta}_1$) under testing nor the constant γ [$= \gamma(\phi, F)$]. In this respect it is quite comparable to T_n in (10.2.54), where $\hat{\sigma}_n^2$ is a consistent estimator of σ_ψ^2 (which is not needed here). This feature is not shared by the Wald-type R-tests presented in the next section. Incidently, if $\boldsymbol{\theta}_1$ involves the intercept parameter, then we need to use (aligned) signed-rank statistics instead of linear rank statistics (see Chapter 6 for the estimation problem). But the rest remains the same. We pose these details as Problems 10.3.1-10.3.3.

Alternatively, the tests of $H_0 : \boldsymbol{\theta}_1 = \mathbf{0}$ against $H_1 : \boldsymbol{\theta}_1 \neq \mathbf{0}$ may be based on regression rank scores, developed by Gutenbrunner et al. (1993) and introduced in Section 6.7. Replacing the $\mathbf{c}_i$ in (6.7.1) by the $\mathbf{x}_i^\star$, we write

$$\mathbf{H}_2 = \mathbf{X}_2^\star(\mathbf{X}_2^{\star\prime}\mathbf{X}_2^\star)^{-1}\mathbf{X}_2^{\star\prime}, \quad \hat{\mathbf{X}}_1^\star = \mathbf{H}_2\mathbf{X}_1^\star, \tag{10.3.5}$$

$$\mathbf{Q}_n = n^{-1}(\mathbf{X}_1^\star - \hat{\mathbf{X}}_1^\star)'(\mathbf{X}_1^\star - \hat{\mathbf{X}}_1^\star). \tag{10.3.6}$$

As in (6.7.14) we define a vector of scores based on a score-function $\phi : [0,1] \mapsto \mathbb{R}$, by $\hat{\mathbf{b}}_n = (\hat{b}_{n1}, \ldots, \hat{b}_{nn})'$. We let then

$$\begin{aligned} \mathbf{S}_n &= n^{-\frac{1}{2}}(\mathbf{X}_{n1}^\star - \hat{\mathbf{X}}_{n1}^\star)'\hat{\mathbf{b}}_n \\ &= n^{-\frac{1}{2}}\mathbf{X}_{n1}^{\star\prime}(\mathbf{I} - \mathbf{H}_2)'\hat{\mathbf{b}}_n, \end{aligned} \tag{10.3.7}$$

and define A_ϕ^2 as in (6.6.14). Then their proposed RRS test statistic is given by

$$\mathbf{T}_n^\star = (\mathbf{S}_n' \mathbf{C}_n^{-1} \mathbf{S}_n)/A_\phi^2. \qquad (10.3.8)$$

Notice the similarity of $\mathcal{L}_n^{(1)}$ in (10.3.4) and (10.3.8). The only difference lies in the replacement of $\mathcal{L}_{n(1)}$ by $S_{(n)}$, and both the (q-)vectors are based on the common score generating function ϕ. The basic representation in Theorem 6.7.3 and Theorem 6.7.4 implies that under $H_0 : \boldsymbol{\theta}_1 = \mathbf{0}$, $T_n^\star \xrightarrow{\mathcal{D}} \chi_q^2$, so an ADF test can be based on $T_n^\star$ as well. See Problem 10.3.4 in this context. An important feature of $T_n^\star$ is that it eliminates the estimation of the nuisance parameter $\boldsymbol{\theta}_2$ by using the projection in (10.3.5) and the basic property of the RRS that they are regression-invariant (see Section 6.7). Moreover Theorems 6.7.3 and 6.7.4 may be called on to verify that under $\{H_{1(n)}\}$ in (10.2.48), $T_n^\star \xrightarrow{\mathcal{D}} \chi_{q,\triangle}^\star$, where $\triangle^\star = \gamma^2 A_\phi^{-2} \boldsymbol{\lambda}'(\mathbf{C}^\star)^{-1}\boldsymbol{\lambda}$ agrees with that of $\mathcal{L}_n^{(1)}$ (see Problem 10.3.5). In fact, in Section 6.7.3, we studied the asymptotic equivalence of R- and RRS-estimators. This basic result provides the asymptotic equivalence of $\mathcal{L}_n^{(1)}$ in (10.3.4) and $T_n^\star$ in (10.3.8) under H_0 as well as $\{H_{1(n)}\}$. From an asymptotic perspective $\mathcal{L}_n^{(1)}$ in (10.3.4) rests on the R-estimation of $\boldsymbol{\theta}$ as well as the uniform linearity of the aligned statistics. This may seriously affect the second-order properties. RRS test are more robust from this perspective.

10.4 CONCLUDING NOTES ON ROBUST TESTS

With the due emphasis laid down on robust estimation of location and regression parameters, it seems quite natural to incorporate such estimators more directly in robust testing procedures. This is possible with the adaptation of the classical Wald (1943) method of test construction. Whereas Wald advocated the use of maximum likelihood estimators (or BAN estimators), from robustness considerations we are naturally tempted in replacing such BAN estimators by their robust L-, M-, or R-estimator versions. Doing so begs a natural question: If such robust estimators are derived from robust statistics (as in the case of M- and R-estimators), is it really necessary to use the estimators instead of their providers? We will discuss this point using the simple location model.

Considerations of local robustness and minimaxity often advocate M-estimators over R-estimators which being globally robust may not match the local optimality to that extent. In both the cases the estimates are derived from appropriate robust statistics which may be directly employed in the construction of test statistics, and hence there may not be any real necessity for using the estimates directly: One penalty for using such estimators is to incorporate some estimator of its (asymptotic) mean square error which may make it somewhat less robust. In passing, we may remark that by virtue of the general equivalence results in Chapter 7, instead of M- or R-estimators we may consider some general estimators that are robust and share the same asymptotic properties. The utility of such robust estimators may be more when we consider suitable sequential testing procedures, and we will explain later.

The appeal of the Wald (1943) method is far more apparant in the case of composite hypotheses testing. We illustrate this by reference to the general linear model treated in Sections 10.2 and 10.3. We take recourse to the reparameterized linear model in (10.2.23) and frame the null hypothesis $H_0 : \boldsymbol{\theta}_1 = \mathbf{0}$ against $H_1 : \boldsymbol{\theta}_1 \neq \mathbf{0}$ ($\boldsymbol{\theta}_2$ nuisance) as in (10.2.24). We start with the usual M-estimator $\hat{\boldsymbol{\theta}}_{n(M)}$ of $\boldsymbol{\theta}$ studied in Chapter 5. By an appeal to Theorem 5.5.3 (or Corollary 5.5.3), we conclude that under the assumed regularity conditions,

$$n^{1/2}(\hat{\boldsymbol{\theta}}_{n(M)} - \boldsymbol{\theta}) = \gamma^{-1} n^{1/2} (\mathbf{X}^{\star\prime}\mathbf{X}^{\star})^{-1} \sum_{i=1}^{n} \mathbf{x}_i^{\star} \psi(Y_i - \boldsymbol{\theta}'\mathbf{x}_i^{\star}) + \mathbf{o}_p(1). \quad (10.4.1)$$

Consequently, if γ and σ_{ψ}^2 are consistently estimated by $\hat{\gamma}_n$ and $\hat{\sigma}_n^2$, respectively, one may directly use a quadratic form in $\tilde{\boldsymbol{\theta}}_{n(M)1}$ (the q-component subvector of $\hat{\boldsymbol{\theta}}_{n(M)}$) and conclude by using the classical Cochran theorem, that the same is asymptotically χ_q^2 under H_0. The noncentral law follows easily for local alternatives. This prescription works out well for L- and R-estimators as well. In each case the (asymptotic) dispersion matrix is given by a known matrix with a multiplication factor, which depends on the unknown F, and one needs to estimate the same consistently in order that a Wald-type test statistic can be applied. We therefore formulate this testing procedure in the following manner: Suppose that there exist a robust estimator $\hat{\boldsymbol{\theta}}_n$ of θ based on $Y_1, \ldots, Y_n$, such that on partitioning $\boldsymbol{\theta}$ and $\hat{\boldsymbol{\theta}}_n$ as $(\boldsymbol{\theta}_1', \ \boldsymbol{\theta}_2')'$ and $(\hat{\boldsymbol{\theta}}_{n1}', \ \hat{\boldsymbol{\theta}}_{n2}')'$, respectively, we have a FOADR result:

$$\hat{\boldsymbol{\theta}}_{n1} - \boldsymbol{\theta}_1 = \nu \mathbf{Z}_{n1} + \mathbf{o}_p(n^{-1/2}), \quad (10.4.2)$$

where

$$\sqrt{n}\mathbf{Z}_{n1} \sim \mathcal{N}_q(\mathbf{0}, \mathbf{D}_n). \quad (10.4.3)$$

$\mathbf{D}_n$ is a $q \times q$ matrix of rank q and ν is an unknown functional or parameter. Suppose also that there exists a consistent estimator $\hat{\nu}_n$ of ν (which is to be reasonably robust to suit the purpose). Then set

$$W_n = n \hat{\nu}_n^{-2} \hat{\boldsymbol{\theta}}_{n1}' \mathbf{D}_n^{-1} \hat{\boldsymbol{\theta}}_{n1}. \quad (10.4.4)$$

We obtain directly that under $H_0 : \boldsymbol{\theta}_1 = \mathbf{0}$,

$$W_n \xrightarrow{\mathcal{D}} \chi^2_q, \tag{10.4.5}$$

and under $\{H_{1(n)}\}$ in (10.2.48),

$$W_n \xrightarrow{\mathcal{D}} \chi^2_{q,\Delta^0}, \tag{10.4.6}$$

where

$$\Delta^0 = \nu^{-2}\boldsymbol{\lambda}'\mathbf{D}^{-1}\boldsymbol{\lambda}; \quad \mathbf{D} = \lim_{n\to\infty} D_n. \tag{10.4.7}$$

We pose Problem 10.4.1 to verify (10.4.5) and (10.4.6). Then W_n is a typical Wald-type test statistic. It is usable whenever robust estimators of ν are available and does not require any reparametrization. In this respect the Wald-type test statistics are quite general, and their robustness properties depend on $\hat{\boldsymbol{\theta}}_{n1}$ and $\hat{\nu}_n$. Let us compare them with the other procedures: the M-procedures require an estimation of $\boldsymbol{\theta}$, γ, and σ^2_ψ; the likelihood ratio-type tests of Section 10.2 besides that require estimation of $\boldsymbol{\theta}_1$, while T_n in Section 10.2 involves the estimation of σ^2_ψ and $\hat{\boldsymbol{\theta}}_2$. Among the tests based on ranks, aligned rank tests require only an estimation of $\boldsymbol{\theta}_2$, while the Wald-type tests rest on R-estimator of $\boldsymbol{\theta}_1$ and estimator of γ. On the other hand, the tests based on regression rank scores involve the parametric linear programming and a calculation of scores $\hat{\mathbf{b}}_n$. The relevant computational algorithm was elaborated by Koenker and d'Orey (1994).

Hampel et al. (1986, ch. 7) have considered some tests that are analogous to the likelihood ratio-type tests considered in Section 10.2, containing $Z_n^\star$ as a special case. They have mostly justified such robust tests on the ground of "bounded influence functionals." Generally we would rather recommend the procedures that avoid an estimator of the structural function γ, and this implies either the aligned rank tests or tests based on regression rank scores. The Wald-type robust tests can also be constructed with the aid of adaptive estimators of regression parameters, considered in Chapter 8, that are based on an estimated (asymptototically optimal) function $\hat{\psi}$.

Robust testing procedures have found their way to sequential analysis as well. In the context of location/regression models, some of the robust tests are reported in Sen (1981a, ch. 9). We present here only a brief overview of such sequential tests. For testing a null hypothesis $H_0 : \theta = \theta_0$ against $H_1 : \theta = \theta_1 = \theta_0 + \Delta$, where θ_0 and Δ (> 0) are specified, consider a robust estimator $T_n = T(X_1, \ldots, X_n)$ of θ, such that T_n admits of a FOADR result with (asymptotic) mean square error $n^{-1}\sigma^2_\theta$, and let s_n^2 be a (strongly) consistent estimator of σ^2_θ. Define the

$$Z_n = n\Delta\{T_n - \frac{1}{2}(\theta_0 + \theta_1)\}/s_n^2, \ n \geq n_0, \tag{10.4.8}$$

where n_0 is an initial sample. Then corresponding to the desired type I and type II error probability bounds $\alpha : 0 < \alpha < 1$, $\beta : 0 < \beta < 1$, $\alpha + \beta < 1$, we define $a = \log A$, $b = \log B$, where $A \leq (1 - \beta)/\alpha$ and $B \geq \beta(1 - \alpha)$; the equality signs hold satisfactorily when $\triangle$ is small. As in the classical sequential probability ratio test(SPRT); Wald (1947), we continue drawing observations as long as

$$b < Z_n < a, \ n \geq n_0. \tag{10.4.9}$$

If, for the first time, for $n = N$, $Z_n \notin (b, a)$, the experiment is terminated along with the

$$\left.\begin{array}{ll} \text{acceptance of } H_0 & \text{if } Z_N \leq b \\ \text{acceptance of } H_1 & \text{if } Z_N \geq a \end{array}\right\}. \tag{10.4.10}$$

If no such N exists, we let $N = +\infty$. Thus N is a stopping variable. Whenever T_n admits of a FOADR and $s_n^2 \xrightarrow{P} \sigma_\theta^2$, then for every (fixed) θ and $\triangle$,

$$\mathbb{P}\{N > n|\theta, \triangle\} \to 0, \text{ as } n \to \infty, \tag{10.4.11}$$

so the process terminates with probability 1 (Problem 10.4.4). Whenever $\{T_n\}$ admits a weak Wiener process representation and $s_n^2 \to \sigma_\theta^2$ a.s. as $n \to \infty$, the limiting (as $\triangle \to 0$) OC (operating characteristic) function of the test based on (10.4.9)-(10.4.10) agrees with that of the SPRT for the same hypotheses testing problem (Problem 10.4.5). Finally, under additional bounds on the probability for $|s_n^2/\sigma_\theta^2 - 1| > \epsilon$, $\epsilon > 0$, and some uniform integrability conditions, the ASN (average sample number) of the test in (10.4.9)-(10.4.10) has the same limiting form as that of SPRT (see Problem 10.4.6). This suggests that in the sequential testing H_0 vs. H_1, the MLE may be replaced by robust estimators. In some simple cases we could directly consider the robust (aligned) statistics, avoiding more estimation (see Sen 1981a, ch. 9 for details).

10.5 PROBLEMS

10.2.1. For an ϵ-contamination normal model, obtain the least favorable d.f. F_0 and the corresponding ψ_0 in (10.2.10).

10.2.2. In (10.2.3), instead of M_n^0, consider a studentized version (involving a scale statistic S_n), modify (10.2.6) accordingly, and verify (10.2.9). [Use the results of Section 5.4.]

10.2.3. In (10.2.12), use $\rho(y/S_n)$ instead of $\rho(y)$ for a suitable scale statistic S_n, obtain the parallel version for $Z_n^\star$ in (10.2.18), and verify (10.2.19). [Hints. Use the results of Section 5.4.]

10.2.4. Verify the uniform quadratic approximation in (10.2.44).

10.2.5. Use (10.2.44) and FOADR for $\hat{\boldsymbol{\theta}}_n^\star$ and $\hat{\boldsymbol{\theta}}_n^{0\star}$ to show that (10.2.50) holds under $\{H_{1(n)}\}$.

10.2.6. Use the uniform asymptotic linearity results in Section 5.5 and express $\tilde{\mathbf{M}}_{n1}$ in (10.2.51) as a linear combination of $\mathbf{M}_{n1}$ and $\mathbf{M}_{n2}$ (plus a remainder term $\mathbf{o}_p(1)$). Hence, or otherwise, verify (10.2.55) and (10.2.56).

10.2.7. Extend the $Z_n^\star$ and T_n tests for studentized M-statistics. [Use the linearity results of Section 5.5.]

10.2.8. Work out a T_n-test for a general parameter θ under the regularity conditions of Theorem 5.2.1.

10.3.1. For the simple regression model $Y_i = \theta_1 + \theta_2 x_i + e_i$, $i \geq 1$, consider an aligned signed-rank test for $H_0 : \theta_1 = 0$ vs. $H_1 : \theta_1 \neq 0$ when θ_2 is treated as a nuisance parameter.

10.3.2. Consider the linear model $Y_i = \theta_0 + \boldsymbol{\theta}_1'\mathbf{x}_{i1} + \boldsymbol{\theta}_2'\mathbf{x}_{i2} + e_i$, $i \geq 1$. Show that for testing $H_0 : \boldsymbol{\theta}_1 = \mathbf{0}$ vs. $H_1 : \boldsymbol{\theta}_1 \neq \mathbf{0}$, the usual (aligned) rank statistics can be used for T_n.

10.3.3. In the model in Problem 10.3.2, consider the null hypothesis $H_0 : (\theta_0, \boldsymbol{\theta}_1')' = \mathbf{0}$ vs. $H_1 : (\theta_0, \boldsymbol{\theta}_1')' \neq \mathbf{0}$. Show that the linear rank statistics are not usable for this testing from but that the signed-rank statistics are usable.

10.3.4. Use the basic results in Section 6.7 and verify that under $H_0 : \boldsymbol{\theta}_1 = \mathbf{0}$, $T_n^\star$, defined by (10.3.8), is asymptotically χ_q^2.

10.3.5. Extend the result of the previous exercise to that under $\{H_{1(n)}\}$ in (10.2.48).

10.4.1. Use the Cochran theorem on quadratic forms in (asymptotically) normal vectors and verify (10.4.5) and (10.4.6).

10.4.2. For the one-sample location model, consider adaptive R-tests based on the adaptive score functions of Section 8.5.

10.4.3. For the linear model, for testing H_0 vs. H_1 in (10.2.24), extend the aligned test statistic T_n to that based on (aligned) adaptive rank statistics, and study its asymptotic optimality.

10.4.4. Verify (10.4.11).

10.4.5. Show that the limiting OC function of a SPRT is the same as with tests based on Z_n in (10.4.8)-(10.4.10).

10.4.6. Extend the results of the previous problem to that of the ASN. [Use the results in Sen (1981a, ch. 9)].

Appendix A

A.1 INTRODUCTION

In Part I (particularly Chapter 2), lengthy derivations of some standard results in real and stochastic analysis were omitted. Most of these proofs are available in contemporary texts, and for the applications discussed in this text, these derivations are not so crucial either. Nevertheless, some less standard results mentioned in Part I (particularly in Chapter 5) are indispensable, and hence their derivations are outlined here in order to provide more insight on two fundamental topics : Asymptotic linearity and Hadamard differentiability of statistical functions.

A.2 UNIFORM ASYMPTOTIC LINEARITY

Lemma 5.5.1 provides the central concept behind the derivation of the basic uniform asymptotic linearity results in Section 5.5. The proof of this lemma which is outlined here is mainly adapted from Jurečková and Sen (1989), with the notations in Section 5.5 used as far as possible. We let $\psi = \psi_a + \psi_c + \psi_s : \mathbb{R} \mapsto \mathbb{R}$, and denote the absolutely continuous, continuous, and step-function components by ψ_a, ψ_c, and ψ_s respectively.

Case I: $\psi \equiv \psi_s$ (i.e., $\psi_a \equiv \psi_c \equiv 0$)

Without loss of generality, we assume that there is a single jump-point. We set $\psi_s(y)$ as 0 or 1 according as y is $\leq$ or > 0. Further we assume the regularity assumptions in Section 5.5 (a subset of **M1** $-$ **M5** and **X1** $-$ **X3**) to be true.

Using (5.5.25) or (5.5.26), we set the scale factor $S = 1$ and write

$$\mathbf{S}_n(\mathbf{t}, u) = \sum_{i=1}^{n} \mathbf{x}_i \{\psi(e^{-u/\sqrt{n}}(E_i - n^{-1/2}\mathbf{x}_i'\mathbf{t})) - \psi(\mathbf{E}_i)\}, \tag{A.2.1}$$

for $(\mathbf{t}, u) \in C = [-K, K]^{p+1}$ for some $K : 0 < K < \infty$. We let

$$\mathbf{S}_n^0(\mathbf{t}, u) = \mathbf{S}_n(\mathbf{t}, u) - \mathbb{E}\mathbf{S}_n(\mathbf{t}, u), \ (\mathbf{t}, u) \in C. \tag{A.2.2}$$

By the vector structure in (5.5.26) and (A.2.1), it suffices to show that for each coordinate of $\mathbf{S}_n^0(\mathbf{t}, u)$, the uniform asymptotic linearity result holds for $(\mathbf{t}, u) \in C$. To simplify the proof, we consider only the first coordinate and drop the subscript 1 in $S_{n1}(.)$ or $S_{n1}^0(.)$:

$$S_n^0(\mathbf{t}, u) = \sum_{i=1}^{n} x_{i1}\{I(E_i \leq 0) - I(E_i \leq n^{-1/2}\mathbf{x}_i't) - F(0) + F(n^{-1/2}\mathbf{x}_i't)\}. \tag{A.2.3}$$

[Because of the special nature of ψ_s, $S_n^0(\mathbf{t}, u) \equiv S_n^0(\mathbf{t}, 0) \forall u$. For a more general ψ_s, we could use a similar expression involving u , and proceed in a similar manner (see Jurečková and Sen 1989).] Recall that (A.2.3) involves independent summands, so by the central limit theorem,

$$n^{-1/4} S_n^0(\mathbf{t}, u) \xrightarrow{\mathcal{D}} \mathcal{N}(0, \gamma^2(\mathbf{t}, u)), \ (\mathbf{t}, u) \in C, \tag{A.2.4}$$

where for each $(\mathbf{t}, u) \in C$,

$$\gamma^2(\mathbf{t}, u) = \lim_{n \to \infty} \{(n^{-1} \sum_{i=1}^{n} x_{i1}^2 |\mathbf{x}_i'\mathbf{t}|)\} f(0), \tag{A.2.5}$$

and by **X1** $-$ **X3**, $\gamma^2(\mathbf{t}, u) < \infty \ \forall (\mathbf{t}, u) \in C$ (see Problem A.2.1). Therefore, if we choose an arbitrary positive integer M and a set $(\mathbf{t}_i, u_i)$, $i = 1, \ldots, M$ of points in C, then by (A.2.4)-(A.2.5),

$$\max_{1 \leq i \leq M} |n^{-1/4} S_n^0(\mathbf{t}_i, u_i)| = \mathrm{O}_p(1) \text{ as } n \to \infty. \tag{A.2.6}$$

Note that (A.2.6) relates to the finite-dimensional distributions, while to establish the uniform asymptotic linearity, we need to establish the compactness or tightness of $n^{-1/4} S_n^0(\mathbf{t}, u)$, $(\mathbf{t}, u) \in C$. This can be done either by an appeal to weak convergence of $n^{-1/4} S_n^0(\mathbf{t}, u)$ (on C) to a Wiener function (as done by Jurečková and Sen 1989)) or by a simpler decomposition, which we present below. Consider the set of signs: $\text{sign}(x_{i1}x_{ij})$, $j = 0, 2, 3, \ldots, p$, where $x_{i0} = 1$. Then consider a set of 2^p subsets of $\{1, \ldots, n\}$ such that within each subset $x_{i1}x_{ij}$, have the same sign (which may differ from j to j). Thus we write $S_n^0(.) = \sum_{j \leq 2^p} S_{n,j}^0(.)$. For each $j (= 1, \ldots, 2^p)$, $S_{n,j}^0(\mathbf{t}, u)$ can be expressed as a difference of two functions, each of which is monotone in each argument $\mathbf{t}, u$, although these may not be concordant. By this monotonicity the Bahadur (1966) representation (studied in Chapter 4) extends directly to $S^0_{n,j}(.)$, and hence it follows that

$$\sup\{n^{-1/4}|S^0_{n,j}(\mathbf{t}, u)| : (\mathbf{t}, u) \in C\} = O_p(1) \tag{A.2.7}$$

for $j = 1, \ldots, 2^p$. Hence (A.2.7) implies that

$$\sup\{n^{-1/4}|S^0_n(\mathbf{t}, u)| : (\mathbf{t}, u) \in C\} = O_p(1), \tag{A.2.8}$$

and this implies (5.5.26).

Case II: ψ absolutely continuous, but ψ' step function

We know that ψ is continuous, piecewise linear, and that it is a constant for $x \leq r_1$ or $x \geq r_k$, where $-\infty < r_1 < r_k < \infty$, and there are k jump points for ψ' (denoted by $r_1, \ldots, r_k$). Again for simplicity, we take $k = 2$ (which corresponds to the classical Huber score function). As in Case I, we set $S = 1$, and also consider only the first coordinate of $S_n(\mathbf{t}, u)$, and so on. Since ψ is bounded and satisfies a Lipschitz condition of order 1, it can be shown (see Problem A.2.2) that

$$\sum_{i=1}^{n} x_{i1}\{\psi[e^{-u/\sqrt{n}}(E_i - n^{-1/2}\mathbf{x}_i'\mathbf{t})] - \psi(E_i - n^{-1/2}(uE_i + \mathbf{x}_i'\mathbf{t}))\} = O_p(1) \tag{A.2.9}$$

uniformly in $(\mathbf{t}, u) \in C$. Hence it is sufficient to prove the proposition for

$$S_n(\mathbf{t}, u) = \sum_{i=1}^{n} x_{i1}\{\psi[E_i - n^{-1/2}(uE_i + \mathbf{x}_i'\mathbf{t})] - \psi(E_i)\}. \tag{A.2.10}$$

Now it can be shown (Problem A.2.3) that for any pair $(\mathbf{t}_1, u_1)$ and $(\mathbf{t}_2, u_2)$ of distinct points,

$$\begin{aligned}
&\mathrm{Var}\{S_n(\mathbf{t}_1, u_1) - S_n(\mathbf{t}_2, u)\} \\
&\leq \sum_{i=1}^{n} x_{i1}^2 \mathbb{E}\{[\psi(E_i - n^{-1/2}(u_1E_i + \mathbf{x}_i'\mathbf{t}_1)) \\
&\qquad -\psi(E_i - n^{-1/2}(u_2E_i + \mathbf{x}_i'\mathbf{t}_2))]^2\} \\
&\leq K^\star\{(u_1 - u_2)^2 + ||\mathbf{t}_1 - \mathbf{t}_2||^2\},\ K^\star < \infty,
\end{aligned} \tag{A.2.11}$$

uniformly in $(\mathbf{t}_i, u_i) \in C$, $i = 1, 2$. Also the boundedness of ψ' implies that (Problem A.2.4) that

$$\begin{aligned}
&|\mathbb{E}[S_n(\mathbf{t}_1, u_1) - S_n(\mathbf{t}_2, u_2) + n^{-1/2}\sum_{i=1}^{n}\{x_{i1}\mathbf{x}_i'(\mathbf{t}_1 - \mathbf{t}_2)\gamma_1 + (u_1 - u_2)\gamma_2\}]| \\
&\qquad \leq K^{\star\star}\{||\mathbf{t}_1 - \mathbf{t}_2|| + |u_1 - u_2|\},
\end{aligned} \tag{A.2.12}$$

and hence by (A.2.11) and (A.2.12),

$$\mathbb{E}\{[S_n(\mathbf{t}_1, u_1) - S_n(\mathbf{t}_2, u_2) + n^{-1/2}\sum_{i=1}^{n} x_{i1}(\gamma_1(\mathbf{t}_1 - \mathbf{t}_2)'\mathbf{x}_i + \gamma_2(u_1 - u_2))]^2\}$$

$$\leq K_0\{||\mathbf{t}_1 - \mathbf{t}_2||^2 + (u_1 - u_2)^2\},\ K_0 < \infty; \qquad \text{(A.2.13)}$$

uniformly in $(\mathbf{t}_i, u_i) \in C,\ i = 1, 2$. Therefore pointwise $S_n^0(\mathbf{t}, u) = O_p(1)$. To prove the compactness, we shall consider increments of $S_n(\mathbf{t}, u)$ over small blocks. We present only the case $p = 1$ because it imparts the full generality of considerations. For $t_2 > t_1$, and $u_2 > u_1$, the increment of $S_n(.)$ over the block $B = B(t_1, u_1; t_2, u_2)$ is

$$\begin{aligned} S_n(B) &= S_n(t_2, u_2) - S_n(t_1, u_2) - S_n(t_2, u_1) + S_n(t_1, u_1) \\ &= \sum_{i=1}^{n} x_i \psi_i(E_i; B), \qquad \text{(A.2.14)} \end{aligned}$$

where

$$\begin{aligned} \psi_i(E_i; B) &= \psi(E_i - n^{-1/2}(u_2 E_i + x_i t_2)) - \psi(E_i - n^{-1/2}(u_2 E_i + x_i t_1)) \\ &- \psi(E_i - n^{-1/2}(u_1 E_i + x_i t_2)) \\ &+ \psi(E_i - n^{-1/2}(u_1 E_i + x_i t_1)) \qquad \text{(A.2.15)} \end{aligned}$$

for $i = 1, \ldots, n$. Note that by **X2** in Section 5.5, $\max\{|x_i| : 1 \leq i \leq n\} = O(n^{1/4})$, while ψ is piecewise linear and bounded (a.e.), so $\psi_i(E_i, B)$ is $O(n^{-1/4})$ a.e.. On the other hand, in view of (A.2.15) and piecewise linearity of ψ, $\psi_i(E_i, B) \equiv 0$ if all the arguments in (A.2.15) lay in the same interval among $(-\infty, \mu_1)$, (μ_1, μ_2), (μ_2, ∞). Thus, setting $t_1 = -K,\ u_1 = -K$, $t_2 \in [-K, K]$, $u_2 \in [-K, K]$, we obtain from (A.2.14) and (A.2.15) that

$$\sup\{|S_n(B(-K, -K; t_2, u_2))| : -K \leq t_2 \leq K,\ -K \leq u_2 \leq K\}$$

$$\leq n^{-1/2}\sum_{i=1}^{n} |x_{i1}| K^{\star}(|r_1| + |r_2| + ||\mathbf{x}_i||) I_i, \qquad \text{(A.2.16)}$$

where the I_i, $i = 1, \ldots, n$, are independent nonnegative indicator variables with

$$\mathbb{E} I_i \leq K^{\star}(|r_1| + |r_2| + ||\mathbf{x}_i||) n^{-1/2},\ i = 1, \ldots, n, \qquad \text{(A.2.17)}$$

and $0 < K^{\star} < \infty$. Hence

$$\text{Var}\big(\sup\{|S_n(B(-K, -K; t_2, u_2))| : \ -K \leq t_2, u_2 \leq K\}\big) = O(1), \quad \text{(A.2.18)}$$

and this, combined with (A.2.12), gives the desired result.

Case III: ψ and ψ' both absolutely continuous

In this proof ψ' or ψ may not be bounded, and hence **M3** is introduced to control their unboundedness. Note that for every $(\mathbf{t}, u) \in C$, by a second-order Taylor's expansion,

$$\begin{aligned}
&\psi(e^{-u/\sqrt{n}}(E_i - n^{-1/2}\mathbf{t}\mathbf{x}_i)) \\
&= \psi(E_i) - n^{-1/2}(uE_i + \mathbf{t}'\mathbf{x}_i)\psi'(E_i) \\
&+\frac{1}{2n}u^2(E_i\psi'(E_i) + E_i^2\psi''(E_i)) + \frac{1}{2n}E_i\psi''(E_i)u(\mathbf{x}_i'\mathbf{t}) \\
&+\frac{1}{2n}(\mathbf{t}'\mathbf{x}_i\mathbf{x}_i'\mathbf{t})\psi''(E_i) + \text{ a remainder term.} \qquad \text{(A.2.19)}
\end{aligned}$$

The remainder term involves ψ' and ψ'' as well as the E_i, $\mathbf{x}_i$, $\mathbf{t}$ and u. With a block B defined as in Case II and $S_n(B)$ as in (A.2.12)-(A.2.13), we have by (A.2.14) and (A.2.17),

$$\begin{aligned}
\psi_i(E_i, B) &= 0 - 0 + \frac{1}{n}E_i\psi''(E_i)(u_2 - u_1)\mathbf{x}_i'(\mathbf{t}_2 - \mathbf{t}_1) \\
&+ \frac{1}{2n}(\mathbf{t}_2 - \mathbf{t}_1)'\mathbf{x}_i\mathbf{x}_i'(\mathbf{t}_2 - \mathbf{t}_1)\psi''(E_i) \\
&+ \text{ a remainder term.} \qquad \text{(A.2.20)}
\end{aligned}$$

Therefore we have

$$\begin{aligned}
S_n(B) &= \frac{1}{n}\sum_{i=1}^{n} x_{i1}[E_i\psi''(E_i)(u_2 - u_1)\mathbf{x}_i'(\mathbf{t}_2 - \mathbf{t}_1) \\
&+ \frac{1}{2}(\mathbf{t}_2 - \mathbf{t}_1)'\mathbf{x}_i\mathbf{x}_i'(\mathbf{t}_2 - \mathbf{t}_1)\psi''(E_i)] \\
&+ \text{ a remainder term.} \qquad \text{(A.2.21)}
\end{aligned}$$

Letting $S_n^\star = \sup\{|S_n(B) : B \subset C\}$, we have

$$\begin{aligned}
S_n^\star &\leq n^{-1}\sum_{i=1}^{n} |x_{i1}|\{|\psi''(E_i)|\}[||\mathbf{x}_i||(2k)^{p+1}|E_i| + \frac{1}{2}(2k)^p||\mathbf{x}_i||^2] \\
&+ \text{ remainder term.} \qquad \text{(A.2.22)}
\end{aligned}$$

By **X2** and **M3** of Section 5.5, and the Markov law of large numbers, the first term on the right hand side of (A.2.20) converges to a finite (nonnegative) limit. A very similar treatment holds for the remainder term (which $\xrightarrow{P} 0$ as $n \to \infty$); we refer to Jurečková and Sen (1989) for details. Therefore we obtain that

$$S_n^\star = O_p(1), \quad \text{as } n \to \infty. \qquad \text{(A.2.23)}$$

On the other hand, as in Case II, for any fixed $(\mathbf{t}, u) \in C$, using (A.2.17), $S_n^0(\mathbf{t}, u) = O_p(1)$, so the desired result follows. This completes the proof of Lemma 5.5.1. □

A.3 HADAMARD DIFFERENTIABILITY OF REGRESSION FUNCTIONAL

The notion of first- and second-order Hadamard differentiability of statistical functionals was introduced in Section 4.5 and incorporated in Section 5.7 to formulate parallel results for regression functionals; (5.7.15)-(5.7.17) and (5.7.19)-(5.7.21) relate to the basic results in this direction. We intend to provide more background here to support these results, although our treatment does not include a complete derivation of these results (which is beyond the scope of this book).

In (5.7.12), $\mathbf{M}_n(\mathbf{u})$, expressed as a linear functional of $\mathbf{S}_n^\star(.)$, provides the motivation of a plausible statistical functional $\boldsymbol{\tau}(.)$ whose first compact or Hadamard derivative may be identified as $\mathbf{M}_n(\mathbf{u})$. The primary task is to construct such a $\boldsymbol{\tau}(.)$. There are besides some other technical points that merit careful consideration. First, the compact differentiability of a statistical functional, as discussed in Section 4.5 for a simple (i.i.d.) model and Section 5.7 for the general regression model, rests on the general notion of weak convergence of probability measures on function spaces. We are interested in the space $D[0,1]$ and its ramifications, and the compactness or tightness part of such weak invariance principles provides the necessary ingredients. There is indeed a genuine need for studying weak convergence results pertaining to $\mathbf{S}_n^\star(.,\mathbf{u})$, $\mathbf{u} \in C$. Although such results are related to those in Section 6.6, they can be studied more conveniently by an appeal to the classical weak convergence principles presented in Section 2.5.10.

Second, $\mathbf{S}_n^\star(.,\mathbf{u})$ may not generally belong to the $D^p[0,1]^q$ space, for some $p \geq 1$, $q \geq 1$. For this reason, the decomposition of the c_{ni} (into c_{ni}^+ and c_{ni}^-) in (5.7.13)-(5.7.14) has been incorporated to express$\mathbf{S}_n^\star(.,\mathbf{u})$ as $\mathbf{S}_n^{+\star}(.,\mathbf{u}) - \mathbf{S}_n^{-\star}(.,\mathbf{u})$ such that these components belong to such $D^p[0,1]^q$ spaces. This technical problem does not arise in the i.i.d. model where the c_{ni} are all equal to 1, and for every $u \in \mathbb{R}$, $\mathbf{S}_n^\star(.,u) \equiv \mathbf{S}_n(.,u)$ belongs to the $D[0,1]$ space. In general, when the $\mathbf{c}_{ni}$ are p-vectors, for some $p \geq 1$, we have $q = p+1$, and for each coordinate of the $\mathbf{c}_{ni}$ a similar decomposition (into c_{nik}^+ and c_{nik}^-, $1 \leq k \leq p$) may be needed. This leads to a decomposition of $\mathbf{S}_n^\star(t,\mathbf{u})$ into 2^p components [e.g., $\mathbf{S}_{nj}^\star(t,\mathbf{u})$, $j = 1,\ldots,2^p$], such that each $\mathbf{S}_{n,j}^\star(.)$ can be amended to some other component, say, $\bar{\mathbf{S}}_{n,j}^\star(.)$, which belongs to the $D^p[0,1]^{p+1}$ space.

Third, as we noted in Section 2.5.10, the (extended-) Skorokhod J_1-topology characterizes the tightness part of the related weak convergence results for functions belonging to the $D[0,1]$ or more generally, $D^p[0,1]^q$ spaces.

The J_1-topology is weaker than the uniform topology associated with the $C[0,1]$ space (or $C^p[0,1]^q$ spaces). However, by virtue of the inequalities of the type (2.5.137), working with the uniform topology even for the $D^p[0,1]^q$ spaces may often simplify the statistical formulations without any essential loss of generality; this has been the main theme of Ren and Sen (1991, 1994 a, b) for dealing with the $\bar{\mathbf{S}}_n^\star(.\mathbf{u})$ in the context of regression models. We denote the uniform (i.e.,supnorm) metric by $||.||$. Also let Γ be a set in $D[0,1]$ and $H \in D[0,1]$. Then we let

$$\text{dist}(H,\Gamma) = \inf\{||H-G|| : G \in \Gamma\}. \tag{A.3.1}$$

Next we let $Q(H,t)$, $H \in D[0,1]$, $t \in \mathbb{R}$; that is, $Q : D[0,1] \times \mathbb{R} \mapsto \mathbb{R}$. We assume that for any compact set Γ in $D[0,1]$,

$$\lim_{t\to 0} Q(H,t) = 0, \text{ uniformly for } H \in \Gamma. \tag{A.3.2}$$

We let $\{\eta_n\}$ and $\{\delta_n\}$ be two sequences of real numbers such that $\lim_{n\to\infty} \eta_n = \lim_{n\to\infty} \delta_n = 0$. Then, for any compact set Γ in $D[0,1]$ and every $\epsilon > 0$, there exists a positive integer N such that

$$\text{dist}(H,\Gamma) = \eta_n \Rightarrow |Q(H,\delta_n)| < \epsilon, \ \forall n \geq N. \tag{A.3.3}$$

This is Lemma 4.4 of Ren and Sen (1991), and we pose the proof of (A.3.3) as Problem A.3.1. Finally, we refer to Section 4.5 and 5.7 for related notations and definitions.

To provide an outline of the derivation of (5.7.15)-(5.7.16), for simplicity we, first, let $c_{ni}^+ = c_{ni}$ (≥ 0) $\forall i$, so $\sum_{i\leq n} c_{ni} \to \infty$ as $n \to \infty$ [by (5.7.14)]. Further we let $U_n(.)$ be the empirical d.f. of the $F(e_i)$, $i \leq n$, and let $U(.)$ be the classical uniform d.f. on [0,1]. For $S_n^\star(.)$, defined by (5.7.11), for $p = 1$, $u \in C = [-K, K]$, $0 < K < \infty$, we consider the following two-parameter version:

$$S_n^\star(t,(2u-1)K), \ (t,u) \in [0,1]^2. \tag{A.3.4}$$

Unfortunately, $S_n^\star(.)$ in (A.3.4) $\notin D[0,1]^2$; Neuhaus (1971) has cited a very simple proof of this, and we pose the proof as Problem A.3.2. Also, we set Problem A.3.3 (Ren and Sen, 1991) to verify that with probability 1 the largest jump (discontinuity) of $S_n^\star(.,.)$ is $\leq 2\{\max_{1\leq i\leq n} c_{ni}\}$ and by (5.7.14) this goes to 0 as $n \to \infty$. Thus one may easily obtain a smooth (continuous) version $\bar{S}_n^\star(t,u)$, which belongs to the $C[0,1]^2$ space (with probability 1) such that as $n \to \infty$,

$$||S_n^\star - \bar{S}_n^\star|| \leq 2\{\max_{1\leq i\leq n} c_{ni}\} \to 0 \text{ a.s.} \tag{A.3.5}$$

Let then $T_n(t,u) = \bar{S}_n^\star(t,u) - t\sum_{i\leq n} c_{ni}$, $(t,u) \in [0,1]^2$, and let $\{P_n\}$ be the sequence of probability measures corresponding to the $\{T_n\}$. Then (Problem A.3.4), using the results of Section 2.5.10, it follows that

$$\{P_n\} \text{ is relatively compact.} \tag{A.3.6}$$

Thus the Prokhorov theorem (Section 2.5.10) implies that for any $\epsilon > 0$ there exists a compact set Γ in $C[0,1]^2$ such that $\mathbb{P}(T_n \in \Gamma) > 1 - \epsilon \ \forall n$, and hence by (A.3.5),

$$\mathbb{P}\{T_n \in \Gamma,\ \|\bar{S}_n^\star - S_n\| \le 2 \max_{1\le i\le n} c_{ni}\} \ge 1 - \epsilon,\ \forall n. \tag{A.3.7}$$

Let $\Gamma_1 = \{T_n(.;u) :\ T_n \in \Gamma,\ u \in [0,1]\}$. Then Γ_1 is a compact set in $C[0.1]$ (and hence in $D[0,1]$), and $T_n \in \Gamma \Rightarrow T_n(.,u) \in \Gamma_1\ \forall u \in [0,1]$. This in turn implies that

$$\text{dist}([S_n^\star(., K(2u-1)) - U(.)\sum_{i\le n} c_{ni}], \Gamma_1) \le 2(\max_{1\le i\le n} c_{ni}). \tag{A.3.8}$$

Thus for every n,

$$\begin{aligned} \mathbb{P}\Big\{&\text{dist}([S_n^\star(., K(2u-1)) - U(.)\sum_{i\le n} c_{ni}], \Gamma_1) \\ &\le 2(\max_{1\le i\le n} c_{ni}),\ \forall u \in [0,1]\Big\} \ge 1 - \epsilon. \end{aligned} \tag{A.3.9}$$

Next (A.3.3) holds for $Q(H,t) = t^{-1}\text{Rem}(F + tH)$ (at $(U(\cdot),)$ by the definition of Hadamard differentiability of $\tau(.)$ [see (4.5.5)-(4.5.6)]. Therefore, by (A.3.7)-(A.3.9), we obtain that

$$|\sum_{i\le n} c_{ni}\ \text{Rem}(F + (\sum_{i\le n} c_{ni})^{-1}H)| < \epsilon\ \forall n \ge N. \tag{A.3.10}$$

This in turn implies that for every $n \ge N$,

$$|\sum_{i\le n} c_{ni}\ \text{Rem}((\sum_{i\le n} c_{ni})^{-1} S_n^\star(., K(2u-1)U(.))| < \epsilon\ \forall u \in [0,1], \tag{A.3.11}$$

so for every $n \ge N$,

$$\mathbb{P}\{\sup_{|u|\le K} |\sum_{i=1}^n c_{ni}\ \text{Rem}(\frac{S_n^\star(.,u)}{\sum_{i\le n} c_{ni}} - U(.))| \le \epsilon\} \ge 1 - \epsilon. \tag{A.3.12}$$

Note then that by (4.5.5)-(4.5.6), (A.3.12) implies (5.7.15) when $c_{ni}^+ = c_{ni}, \forall i$. The modification for the more general case treated in (5.7.13) follows then by using the decomposition in (5.7.13) and similar arguments for each component (see, Ren and Sen 1991), and we pose the derivation as Problem A.3.5. Finally, (5.7.16) is a direct corollary to (5.7.15) whenever $M_n(u)$ can be identified as a Hadamard derivative of a suitable $\tau(.)$. Toward this we consider the following:

Let $M; [0,1] \mapsto \mathbb{R}$ be a function defined by $M(t) = \psi(F^{-1}(t))$, $0 \le t \le 1$, where ψ is assumed to be nondecreasing, right-continuous, and bounded.

Since there may not be a unique way of defining $\tau(.)$, we consider a functional $\tau : D[0,1] \to I\!R$ by letting

$$\tau(G) = \int_0^1 h(G(t))e^t dM(t), \ G \in D[0,1], \tag{A.3.13}$$

where

$$h(y) = \begin{cases} -e^{-y}, & y \geq 0, \\ y^{-1}, & y < 0. \end{cases} \tag{A.3.14}$$

Then $\tau(.)$ can be expressed as a composition of the following Hadamard differentiable transformations:

$$\tau(G) = \gamma_2(\gamma_1(G)), \ G \in D[0,1], \tag{A.3.15}$$

where $\gamma_1 : D[0,1] \mapsto L^1[0,1]$ and $\gamma_2 : L^1[0,1] \cap D[0,1] \mapsto I\!R$, are defined by

$$\gamma_1(S) = h \circ S \ \text{ and } \ \gamma_2(S) = \int_0^1 S(t)e^t dM(t). \tag{A.3.16}$$

Since $h(.)$ is differentiable everywhere with a bounded and continuous derivative, it follows that (see Fernholz 1983) γ_1 is Hadamard differentiable at U; we have set up Problem A.3.6 to verify this. Moreover, γ_2 is linear and continuous, and therefore it is Fréchet (and hence Hadamard-)differentiable. Therefore $\tau(.)$ is also Hadamard differentiable at U and its Hadamard derivative at $U \in D[0,1]$ is

$$\tau'_U(G) = \int_0^1 h'(t)G(t)e^t dM(t), \ G \in D[0,1]. \tag{A.3.17}$$

Note further that by (A.3.14),

$$h'(t) = e^{-t}, \ \forall 0 \leq t \leq 1. \tag{A.3.18}$$

Therefore for each $u : |u| \leq K$,

$$\begin{aligned} \tau'_U(S_n^\star(.,u) &= \int_0^1 h'(t)S_n^\star(t,u)e^t dM(t) \\ &= \sum_{i\leq n} c_{ni} \int_{F(e_i - c_{ni}u)}^1 dM(t) \\ &= \sum_{i\leq n} c_{ni} \int_{e_i - uc_{ni}}^\infty d\psi(x) \\ &= \psi(\infty)\sum_{i\leq n} c_{ni} - M_n(u), \end{aligned} \tag{A.3.19}$$

where $|\psi(\infty)| < \infty$, by assumption. This completes the proof of (5.7.16). □

For the linear model involving the vector $\mathbf{c}_{ni}$, for each coordinate, we have a similar decomposition (i.e, c_{nik}^{+} and c_{nik}^{-}, $1 \leq k \leq p$, $i \geq 1$), and by reference to the 2^p subsets of the regressors, the same proof can be directly extended. We omit the details (see, Ren and Sen 1995a).

A.4 PROBLEMS

A.2.1. Provide the steps leading to the result in (A.2.4)-(A.2.5).

A.2.2. Verify (A.2.9).

A.2.3. Verify (A.2.10).

A.2.4. Verify (A.2.11).

A.2.5. Obtain an expression for the remainder term in (A.2.17) and also for (A.2.22). Hence, or otherwise, verify that the remainder term in (A.2.22) $\xrightarrow{P} 0$ as $n \to \infty$. [Jurečková and Sen 1989]

A.3.1. Establish the inequality in (A.3.3).

A.3.2. Show that $S_n^{\star}(.,.)$ defined by (A.3.4) $\notin D[0,1]^2$. [Neuhaus, 1971]

A.3.3. Show that the largest jump of $S_n^{\star}(.,.)$ in (A.3.4) is less than or equal to $2\{\max_{1 \leq i \leq n} c_{ni}\}$.

A.3.4. Establish the compactness of $\{P_n\}$ in (A.3.6). [Ren and Sen 1991]

A.3.5. Use the decomposition in (5.7.13) and verify that (A.3.12) holds for that case too. [Ren and Sen 1991]

A.3.6. Verify the Hadamard differentiability of γ_1 defined by (A.3.16). [Use Proposition 6.1.2 of Fernholz (1983).]

A.3.7. Extend the result of the previous exercise to the composite function $\gamma_2(\gamma_1(G))$. [Ren and Sen 1995a]

References

Adichie, J. N. (1967). Estimate of regression parameters based on rank tests. *Ann. Math. Statist. 38*, 894-904.

Adichie, J. N. (1978). Rank tests for subhypotheses in the general linear regression. *Ann. Statist. 6*, 1012-1026.

Adichie, J. N. (1984). Rank tests in linear models. *Handbook of Statististics, Vol. 4 : Nonparametric Methods.* (eds. P. R. Krishnaiah and P. K. Sen), North Holland, Amsterdam, pp. 229-257.

Aerts, M., Janssen, P., and Veraverbeke, N. (1994). Asymptotic theory for regression quantile estimators in the heteroscedastic regression model. *Asymptotic Statistics; Proc. 5th Prague Cofer.* (eds. M. Hušková and P. Mandl), Physica-Verlag, Heidelberg, pp.151-161.

Akahira, M. (1975a). Asymptotic theory for estimation of location in nonregular cases I: Order of convergence of consistent estimators. *Rep. Stat. Appl. Res., JUSE, 22*, 8-26.

Akahira, M. (1975b). Asymptotic theory for estimation of location in nonregular cases II: Bounds of asymptotic distributions of consistent estimators. *Rep. Stat. Appl. Res., JUSE, 22*, 101-117

Akahira, M., and Takeuchi, K. (1981). *Asymptotic Efficiency of Statistical Estimators: Concepts and Higher Order Asymptotic Efficiency. Springer Lecture Notes in Statistics, 7*, Springer-Verlag, New York.

Amari, S. (1985). *Differential Geometric Methods in Statistics.* Springer-Verlag, New York.

Andersen, P. K. and Gill, R. D. (1982). Cox's regression model for counting processes: a sample study. *Ann. Statist. 10*, 1100 -1120.

Andersen, P. K., Borgan, O., Gill, R. D., and Keiding, N. (1993). *Statistical Models Based on Counting Processes.* Springer-Verlag, New York.

Anderson, J. R. (1978). Use of M-estimator theory to derive asymptotic results for rank regression estimators. *Biometrics 34* ,151. (Abstract)

Andrews, D. F., Bickel, P. J., Hampel, F. R., Huber, P. J., Rogers, W. H., and Tukey, J. W. (1972) *Robust Estimates of Location. Survey and Advances.* Princeton Univ. Press, Princeton.

Andrews, D. W. K. (1986). A note on the unbiasedness of feasible GLS, quasimaximum likelihood, robust adaptive, and spectral estimators of the linear model. *Econometrica 54*, 687-698.

Anscombe, F. J. (1952). Large sample theory of sequential estimation. *Proc. Cambridge Phil. Soc. 48*, 607-607.

Antille, A. (1972). Linearité asymptotique d'une statistique de rang. *Zeit. Wahrsch. verw. Geb. 32*, 147-164.

Antille, A. (1976). Asymptotic linearity of Wilcoxon signed-rank statistics. *Ann. Statist. 4*, 175-186.

Antoch, J. (1984). *Collection of Programs for Robust Regression Analysis.* Charles Univ., Prague.

Antoch, J., Collomb, G., and Hassani, S. (1984). Robustness in parametric and nonparametric regression estimation: An investigation by computer simulation. *COMPSTAT 1984.* Physica Verlag, Vienna, pp. 49-54.

Antoch, J. and Jurečková, J. (1985). Trimmed least squares estimator resistant to leverage points. *Comp. Statist. Quarterly 4*, 329-339.

Atkinson, A. C. (1985). *Plots, Transformations, and Regression.* Clarendon Press, Oxford.

Aubuchon, J. C. and Hettmansperger, T. P. (1984). On the use of rank tests and estimates in the linear model. *Handbook of Statistics. Vol. 4 : Nonparametric Methods.* (eds. P. R. Krishnaiah and P. K. Sen), North Holland, Amsterdam, pp. 259-274.

Azencott, F. J., Birgé, L., Costa, V., Dacunha-Castelle, D., Deniau, C., Deshayes, P., Huber-Carol , C., Jolivaldt, P., Oppenheim, G., Picard, D., Trécourt, P., and Viano, C. (1977). Théorie de la robustesse et estimation d'un paramétre. *Astérisque,* 43-44.

Bahadur, R. R. (1960). Stochastic comparison of tests. *Ann. Math. Statist. 31*, 276-295.

Bahadur, R. R. (1966). A note on quantiles in large samples. *Ann. Math. Statist. 37*, 557-580.

Bahadur, R. R. (1971). *Some Limit Theorems in Statistics.* SIAM, Philadelphia.

Bai, Z.D., Rao, C.R., and Wu, Y. (1990). Recent contributions to robust estimation in linear models. *Probability, Statistics and Design of Experiment: R.C. Bose Mem. Confer.*, (ed. R. R. Bahadur). Wiley Eastern, New Delhi, pp. 33-50.

Bai, Z. D., Rao, C. R., and Wu, Y. (1992). M-estimation of multivariate linear regression parameter under a convex discrepancy function. *Statist. Sinica 2*, 237-254.

Barndorff-Nielsen, O. E., and Cox, D. R. (1989). *Asymptotic Techniques for Use in Statistics.* Chapman and Hall, New York.

Barrodale, I. and Roberts, F. D. K. (1975). Algorithm 478: Solution of an overdetermined system of equations in the L_1-norm. *Commun. ACM 17*, 319-320.

Bassett, G. and Koenker, R. (1978). Asymptotic theory of least absolute error regression. *J. Amer. Statist. Assoc. 73*, 618-622.

Bassett, G. and Koenker, R. (1982). An empirical quantile function for linear models with iid errors. *J. Amer. Statist. Assoc. 77*, 407-415.

Bednarski, T. (1994). Fréchet differentiability and robust estimation. *Asymptotic Statistics; Proc. 5th Prague Symp.*(eds., M. Hušková and P. Mandl), Physica-Verlag, Heidelberg, pp. 49-58.

Bednarski, T., Clarke, B. R., and Kolkiewicz, W. (1991). Statistical expansions and locally uniform Fréchet differentiability. *J. Austral. Math. Soc. Ser. A 50*,88-97.

Bennett, C. A. (1952). Asymptotic properties of ideal linear estimators. *Ph. D. Dissertation,* Univ. of Michigan, Ann Arbor.

Beran, R. J. (1977). Minimum Hellinger distance estimators for parametric models. *Ann. Statist. 5*, 445-463.

Beran, R. J. (1978). An efficient and robust adaptive estimator of location. *Ann. Statist. 6*, 292-313.

Beran, R. J. (1982). Robust estimation in models for independent non-identically distributed data. *Ann. Statist. 12*, 415-428.

Beran, R. J. (1984). Minimum distance procedures. *Handbook of Statistics, Vol. 4 : Nonparametric Methods* (eds. P. R. Krishnaiah and P. K. Sen), North-Holland, Amsterdam, pp. 741-754.

Bhandary, M. (1991). Robust M-estimation of Dispersion matrix with a structure. *Ann. Inst. Statist. Math. 43* , 689-705.

Bhattacharya, P. K., and Gangopadhyay, A. K. (1990). Kernel and nearest neighbour estimation of a conditional quantile. *Ann. Statist. 17*, 1400-1415.

Bhattacharya, R. N., and Ghosh, J. K. (1978). On the validity of the formal Edgeworth expansion. *Ann. Statist. 6*, 434-451.

Bhattacharya, R. N., and Ghosh, J. K. (1988). On moment conditionsfor valid formal Edgeworth expansions. *J. Multivar. Anal. 27*, 68-79.

Bickel, P. J. (1965). On some robust estimates of location. *Ann. Math. Statist. 36*, 847-858.

Bickel, P. J. (1973). On some analogues to linear combinations of order statistics in the linear model. *Ann. Statist. 1*, 597-616.

Bickel, P. J. (1975). One-step Huber estimates in the linear model. *J. Amer. Statist. Assoc. 70*, 428-433.

Bickel, P. J. (1981). Quelques aspects de la statistque robuste. *Ecole dété de St. Flour, Lecture Notes in Mathematics 876*, Springer-Verlag, New York, pp. 1-72.

Bickel, P. J. (1982). On adaptive estimation. *Ann. Statist. 10*, 647-671.

Bickel, P. J., Klaassen, C. A., Ritov, Y., and Wellner, J. A. (1993). *Efficient and Adaptive Estimation for Semiparametric Models*, Johns Hopkins Univ. Press, Baltimore, MD.

Bickel, P. J. and Lehmann, E. L. (1975a). Descriptive statistics for nonparametric model. I Introduction. *Ann. Statist. 3*, 1038-1044.

Bickel, P. J. and Lehmann, E. L. (1975b). Descriptive statistics for nonparametric model. II Location. *Ann. Statist. 3*, 1045-1069.

Bickel, P. J. and Lehmann, E. L. (1976). Descriptive statistics for nonparametric model. III Dispersion. *Ann. Statist. 4*, 1139-1158.

Bickel, P. J. and Lehmann, E. L. (1979). Descriptive statistics for nonparametric model. IV. Spread. *Contributions to Statistics: Jaroslav Hájek Memorial Volume* (ed: J. Jurečková). Acadmia, Prague, pp 33-40.

Bickel, P. J. and Wichura, M. J. (1971). Convergence criteria for multiparameter stochastic process and some applications. *Ann. Math. Statist. 42*, 1656-1670.

Billingsley, P. (1968). *Convergence of Probability Measures.* Wiley, New York.

Birkes, D., and Dodge, Y. (1993). *Alternative Methods of Regression.* Wiley, New York.

Birnbaum, A., and Laska, E. (1967). Optimal robustness: A general method with applications to linear estimates of location. *J. Amer. Statist. Assoc. 62*, 1230-1240.

Blackman, J. (1955). On the approximation of a distribution function by an empirical distribution. *Ann. Math. Statist. 26*, 256-267.

Blom, G. (1956). On linear estimates with nearly minimum variance. *Ark. Mat., 3 (31)*, 365-369.

Bloomfield, P., and Stieger, W. L. (1983). *Least Absolute Deviations: Theory, Applications and Algorithms.* Birhäuser, Boston.

Boos, D. (1979). A differential for L-statistics. *Ann. Statist. 7*, 955-959.

Boos, D. (1982). A test of symmetry associated with the Hodges-Lehmann estimator. *J. Amer. Statist. Assoc. 77*, 647-651.

Boos, D., and Serfling, R. J. (1979). On Berry-Esseen rates for statistical functions, with application to L-estimates. *Technical report*, Florida State Univ., Tallahassee.

Box, G.E.P. (1953) Non-normality and tests of variance. *Biometrika 40*, 318-335.

Box, G.E.P., and Anderson, S. L. (1955). Permutation theory in the derivation of robust criteria and the study of departures from assumption. *J. Royal Statist. Soc. Ser. B 17*, 1-34.

Brillinger, D. R. (1969). An asymptotic representation of the sample distribution. *Bull. Amer. Math. Soc. 75*, 545-547.

Brown, B. M. (1971). Martingale central limit theorems. *Ann. Math. Statist. 42*, 59-66.

Bustos, O. H. (1982). General M-estimates for contaminated p-th order autoregression processes: consistency and asymptotic normality. *Zeit. Wahrsch. verw. Geb. 59*, 491-504.

Carroll, R. J. (1977a). On the uniformity of sequential procedures. *Ann. Statist. 5*, 1039-1046.

Carroll, R. J. (1977b). On the asymptotic normality of stopping times based on robust estimators. *Sankhyā, Ser.A 39*, 355-377.

Carroll, R. J. (1979). On estimating variances of robust estimators when the errors are asymmetric. *J. Amer. Statist. Assoc. 74*, 674-679.

Carroll, R. J. (1982). Robust estimation in certain heteroscedastic linear models when there are many parameters. *J. Statist. Plan. Infer. 7*, 1-12.

Carroll, R. J., and Gallo, P. P. (1982). Some aspects of robustness in the functional errors-in-variables regression models. *Com. Statist. Theor. Meth. A 11*, 2573-2585.

Carroll, R. J., and Ruppert, D. (1980). A comparison between maximum likelihood and generalized least squares in a heteroscedastic model. *J. Amer. Statist. Assoc. 77*, 878-882.

Carroll, R. J., and Ruppert, D. (1982). Robust estimation in heteroscadastic linear models. *Ann. Statist. 10*, 429-441.

Carroll, R. J., and Ruppert, D. (1988). *Transformation and Weighting in Regression.* Chapman and Hall, London.

Carroll, R.J., and Welsh, A. H. (1988). A note on asymmetry and robustness in linear regression. *Amer. Statist. 42*, 285-287.

Chanda, K. C. (1975). Some comments on the asymptotic probability laws of sample quantiles. *Calcutta Statist. Assoc. Bull. 24*, 123-126.

Chatterjee, S., and Hadi, A.S. (1986). Influential observations, high leverage points and outliers in linear regression (with discussion). *Statist. Sci. 1*, 379-416.

Chatterjee, S. K., and Sen, P. K. (1973). On Kolmogorov-Smirnov type test for symmetry. *Ann. Inst. Statist. Math. 25*, 288-300.

Chaudhuri, P. (1991). Nonparametric estimators of regression quantiles and their local Bahadur representation. *Ann. Statist. 19*, 760-777.

Chaudhuri, P. (1992). Generalized regression quantiles: Forming a useful toolkit for robust linear regression. *L_1-Statistical Analysis and Related Methods.* (ed. Y. Dodge). North-Holland, Amsterdam, pp.169-185.

Chaudhuri, P. (1992) Multivariate location estimation using extension of R-estimators through U-statistics-type approach. *Ann. Statist. 20*, 897-916.

Chaudhuri, P., and Sengupta, D. (1993). Sign tests in multi-dimension: Inference based on the geometry of the data cloud. *J. Amer. Statist. Assoc. 88*, 1363-1370.

Cheng, K. F. (1984). Nonparametric estimation of regression function using linear combinations of sample quantile regression functions. *Sankhyā, Ser. A 46*, 287-302.

Chernoff, H., Gastwirth, J. L., and Johns, M. V. (1967). Asymptotic distribution of linear combinations of order statistics. *Ann. Math. Statist. 38*, 52-72.

Clarke, B. R. (1983). Uniqueness and Freĉhet differentiability of functional solutions to maximum likelihood type equations. *Ann. Statisl. 11*, 1196-1205.

Clarke, B. R. (1986). Nonsmooth analysis and Fréchet differentiability of M-functionals. *Probab. Th. Rel. Fields 73*, 197-209.

Collins, J. R. (1976). Robust estimation of a location parameter in the presence of asymmetry. *Ann. Statist. 4*, 68-85.

Collins, J. R. (1977). Upper bounds on asymtotic variances of M-estimators of location. *Ann. Statist. 5*, 646-657.

Collins, J. R. (1982). Robust M-estimators of location vectors. *J. Mult. Anal. 12*, 480-492.

Collins, J. R., and Portnoy, S. (1981). Maximizing the variance of M-estimators using the generalized method of moment spaces. *Ann. Statist. 9*, 567-577.

Collins, J. R., Sheahan, J. N., and Zheng, Z. (1985). Robust estimation in the linear model with asymmetric error distribution. *J. Mult. Anal. 20*, 220-243.

Cook, R. D., and Weisberg, S. (1982). *Residuals and Influence in Regression.* Chapman and Hall, New York.

Croux, C., and Rousseeuw, P. J. (1992). A class of high-breakdown scale estimators based on subranges. *Comm. Statist. Theor. Meth. A21*, 1935-1951.

Csörgő, M., and Horváth, L. (1993). *Weighted Approximations in Probability and Statistics.* Wiley, New York.

Csörgő, M., and Révész, P. (1981). *Strong Approximations in Probability and Statistics.* Akadémiai Kiadó, Budapest.

Daniell, P. J. (1920). Observations weighted according to order. *Ann. Journ. Math. 42*, 222-236.

David, H. A. (1981). *Order Statistics*, 2nd ed. Wiley, New York.

Davidian, M., and Carroll, R. J. (1987). Variance function estimation. *J. Amer. Statist. Assoc. 82*, 1079-1091.

Davies, L. (1994). Desirable properties, breakdown and efficiency in the linear regression model. *Statist. Probab. Letters 19*, 361-370.

Davies, P. L. (1994). Aspects of robust linear regression. *Ann. Statist. 21*, 1843-1899.

de Haan, L., and Taconis-Haantjes, E. (1979). On Bahadur's representation of sample quantiles. *Ann. Inst. Statist. Math. A31*, 299-308.

Denby, L., and Martin, D. (1979). Robust estimation of the first order autoregressive parameters. *J. Amer. Statist. Assoc. 74*, 140-146.

Dionne, L. (1981). Efficient nonparametric estimators of parameters in the general linear hypothesis. *Ann. Statist. 9*, 457-460.

Dodge, Y., and Jurečková, J. (1995). Estimation of quantile density function based on regression quantiles. *Statist. Probab. Letters 23*, 73-78.

Donoho, D. L., and Gasko, M. (1992). Breakdown propertties of location estimates based on halfspace depth and projected outlyingness. *Ann. Statist. 20*, 1803-1827.

Donoho, D. L., and Huber, P. J. (1983). The notion of breakdown point. *Festschrift for E. L. Lehmann* (eds. P.J.Bickel et al.). Wadsworth, Belmont, Calif., 157-184.

Donoho, D. L., and Liu, R. C. (1988). The "automatic" robustness of minimum distance functionals. *Ann. Statist. 16*, 552-586.

Doob, J. L. (1949). Heurustic approach to the Kolmogorov-Smirnov theorems. *Ann. Math. Statist. 20*, 393-403.

Doob, J. L. (1967). *Stochastic Processes.* Wiley, New York.

Dudley, R. M. (1984). A course on empirical Processes. *Lecture Notes in Mathematics 1097*, Springer-Verlag, New York, pp. 1-142.

Dudley, R. M. (1985). An extended Wichura theorem, definitions of Donsker classes, and weighted empirical distributions. *Lecture Notes in Mathematics 1153*, Springer-Verlag, New York, pp. 141-148.

Dupač, V., and Hájek, J. (1969). Asymptotic normality of linear rank statistics. *Ann. Math. Statist. 40*, 1992-2017.

Dutter, R. (1977). Numerical solutions of robust regression problems: Computational aspects, a comparison. *J. Statist. Comp. Simul. 5*, 207-238.

Dvoretzky, A. (1972). Central limit theorem for dependent random variables. *Proc. 6th Berkeley Symp. Math. Statist. Probab.*, vol. 2 (eds. L. LeCam et al.). Univ. Calif. Press, Los Angeles, pp. 513-555.

Dvoretzky, A., Kiefer, J., and Wolfowitz, J. (1956). Asymptotic minimax character of the sample distribution function and the classical multinomial estimator. *Ann. Math. Statist. 27*, 642-669.

Eicker, F. (1970). A new proof of the Bahadur-Kiefer representation of sample quantiles. *Nonparametric Techniques in Statistical Inference.* (ed. M. L. Puri), Cambridge Univ. Press, pp. 321-342.

Falk, M. (1986). On the estimation of the quantile density function. *Statist. Probab. Letters 4*, 69-73.

Fernholtz, L.T. (1983). *von Mises' Calculus for Statistical Functionals. Lecture notes in Statistics, 19*, Springer-Verlag, New York.

Field, C. A., and Wiens, D. P. (1994). One-step M-estimators in the linear model, with dependent errors. *Canad. J. Statist. 22*, 219-231.

Fillippova, A. A. (1961). Mises' theorem on the asymptotic behavior of functionals of empirical distribution functions and its applications. *Teor. Veroyat. Primen. 7*, 24-57.

Fisher, R. A. (1925). Theory of statistical estimation. *Proc. Cambridge Phil. Soc. 22*, 700-725.

Fu, J. C. (1975). The rate of convergence of consistent point estimators. *Ann. Statist. 3*, 234-240.

Gangopadhyay, A. K., and Sen, P. K. (1990). Bootstrap confidence intervals for conditional quantile functions. *Sankhyā Ser. A 52*, 346-363.

Gangopadhyay, A. K., and Sen, P. K. (1992). Contiguity in nonparametric estimation of a conditional functional. In *Nonparameteric Statistics and Related Topics* (ed. A.K.M.E. Saleh), North Holand, Amsterdam, pp. 141-162.

Gangopadhyay, A. K., and Sen, P. K. (1993). Contiguity in Bahadur-type representations of a conditional quantile and application in conditional quantile process. In *Statistics and Probability: R. R. Bahadur Festschrift* (eds. J. K. Ghosh et al.), Wiley Eastern, New Delhi, pp. 219-232.

Gardiner, J. C., and Sen, P. K. (1979). Asymptotic normality of a variance estimator of a linear combination of a function of order statistics. *Zeit. Wahrsch. verw. Geb. 50*, 205-221.

Geertsema, J. C. (1970). Sequential confidence intervals based on rank tests. *Ann. Math. Statist. 41*, 1016-1026.

Geertsema, J. C. (1987). The behavior of sequential confidence intervals under contaminations. *Sequent. Anal. 6*, 71-91.

Geertsema, J. C. (1992). A comparison of nonparametric sequential procedures for estimating a quantile. *Order Statistics and Nonparametrics: Theory and Applications* (eds. P. K. Sen and I. A. Salama), North Holland, Amsterdam, pp. 101-113.

Ghosh, J. K. (1971). A new proof of the Bahadur representation of quantiles and an application. *Ann. Math. Statist. 42*, 1957-1961.

Ghosh, J. K. (1994). *Higher Order Asymptotics*, NSF-CBMS Reg. Confer. Ser. Probab. Statist., Vol. 4, Inst. Math. Statist., Hayward, Calif.

Ghosh, J. K., Mukerjee, R., and Sen, P. K. (1995). Second-order Pitman admissibility and Pitman closeness: The multiparameter case and Stein-rule estimators. *J. Multivar. Anal.*, to appear.

Ghosh, J. K., and Sen, P. K. (1985). On the asymptotic performance of the log-likelihood ratio statistic for the mixture model and related results. In *Proceedings of the Berkeley Coference in Honor of J. Neyman and J. Kiefer*, Vol.2 (eds. L. M. LeCam and R. A. Olshen), Wadsworth, Belmont, Calif., pp. 789-806.

Ghosh, J. K., Sen, P. K., and Mukerjee, R. (1994). Second-order Pitman closeness and Pitman admissibility. *Ann. Statist. 22*, 1133-1141.

Ghosh, M. (1972). On the representation of linear functions of order statistics. *Sankhyā Ser. A 34*, 349-356.

Ghosh, M., Mukhopadhyay, N., and Sen, P. K. (1996). *Sequential Estimation.* Wiley, New York.

Ghosh, M., Nickerson, D. M., and Sen, P. K. (1987). Sequential shrinkage estimation. *Ann. Statist. 15*, 817-829.

Ghosh, M., and Sen, P. K. (1970). On the almost sure convergence of von Mises' differentiable statistical functions. *Calcutta Statist. Assoc. Bull. 19*, 41-44.

Ghosh, M., and Sen, P. K. (1971a). Sequential confidence intervals for the regression coefficients based on Kendall's tau. *Calcutta Statist. Assoc. Bull. 20*, 23-36.

Ghosh, M., and Sen, P. K. (1971b). A class of rank order tests for regression with partially informed stochastic predictors. *Ann. Math. Statist. 42*, 650-661.

Ghosh, M., and Sen, P. K. (1972). On bounded width confidence intervals for the regression coefficient based on a class of rank statistics. *Sankhyā Ser. A 34*, 33-52.

Ghosh, M., and Sen, P. K. (1973). On some sequential simultaneous confidence intervals procedures. *Ann. Inst. Statist. Math. 25*, 123-135.

Ghosh, M., and Sen, P. K. (1977). Sequential rank tests for regression. *Sankhyā Ser. A 39*, 45-62.

Ghosh, M., and Sen, P. K. (1989). Median unbiasedness and Pitman closeness. *J. Amer. Statist. Assoc. 84*, 1089-1091.

Ghosh, M., and Sukhatme, S. (1981). On Bahadur representation of quantiles in nonregular cases. *Comm. Statist. Theor. Meth. A10*, 169-182.

Gill, R. D. (1989). Non- and semi-parametric maximum likelihood estimators and the von Mises method. Part 1. *Scand. J. Statist. 16*, 97-128.

Gutenbrunner, C. (1986). *Zur Asymptotik von Regression Quantil Prozesen und daraus abgeleiten Statistiken.* Ph. D. dissertation, Universitat Freiburg, Freiburg.

Gutenbrunner, C., and Jurečková, J. (1992). Regression rank scores and regression quantiles. *Ann. Statist. 20*, 305-330.

Gutenbrunner, C., and Jurečková, J., Koenker, R., and Portnoy, S. (1993). Tests of linear hypotheses based on regression rank scores. *J. Nonpar. Statist. 2*, 307-331.

Hájek, J. (1961). Some extensions of the Wald-Wolfowitz-Noether theorem. *Ann. Math. Statist. 32*, 506-523.

Hájek, J. (1962). Asymptotically most powerful rank order tests. *Ann. Math. Statist. 33*, 1124-1147.

Hájek, J. (1965). Extensions of the Kolmogorov-Smironov tests to regression alternatives. *Bernoull-Bayes-Laplace Seminar*, (ed. L. LeCam), Univ. Calif. Press, Los Angeles, Calif., pp, 45-60.

Hájek, J. (1968). Asymptotic normality of simple linear rank statistics under alternatives. *Ann. Math. Statist. 39*, 325-346.

Hájek, J., and Šidák, Z. (1967). *Theory of Rank Tests.* Academia, Prague.

Halmos, P. R. (1946). The theory of unbiased estimation. *Ann. Math. Statist. 17*, 34-43.

Hampel, F. R. (1968) *Contributions to the theory of robust estimators.* Ph.D. dissertation, Univ. California, Berkeley.

Hampel, F. R. (1971). A general qualitative definition of robustness. *Ann. Math. Statist. 42*, 1887-1896.

Hampel, F. R. (1974). The influence curve and its role in robust estimation. *J. Amer. Statist. Assoc. 62*, 1179-1186.

Hampel, F. R., Rousseeuw, P. J., Ronchetti, E., and Stahel, W. (1986). *Robust Statistics - The Approach Based on Influence Functions.* Wiley, New York.

Hanousek, J. (1988). Asymptotic relations of M- and P-estimators of location. *Comp. Statist. Data Anal. 6*, 277-284.

Hanousek, J. (1990). Robust Bayesian type estimators and their asymptotic representation. *Statist. Dec. 8*, 61-69.

Harrell, F. and Davis, C. (1982). A new distribution-free quantile estimator. *Biometrika 69*, 635-640.

He, X., Jurečková, J., Koenker, R., and Portnoy, S. (1990). Tail behavior of regression estimators and their breakdown points. *Econometrica 58*, 1195-1214.

Heiler, S. (1992). Bounded influence and high breakdown point regression with linear combinations of order statistics and rank statistics. L_1-*Statistical Analysis and Related Methods* (ed. Y. Dodge), North Holland, Amsterdam, pp. 201-215.

Heiler, S., and Willers, R. (1988). Asymptotic normality of R-estimates in the linear model. *Statistics 19*, 173-184.

Helmers, R. (1977). The order of normal approximation for linear combinations of order statistics with smooth weight function. *Ann. Probab. 5*, 940-953.

Helmers, R. (1980). Edgeworth expansions for linear combinations of order statistics. *Math. Centre Tract 105*, Amsterdam.

Helmers, R. (1981). A Berry-Esseen theorem for linear combination of order statistics. *Ann. Probab. 9*, 342-347.

Hodges, J. L. Jr. (1967). Efficiency in normal samples and tolerance of extreme values for some estimate of location. *Proc. 5th Berkeley Symp. Math. Statist. Probab.*, Univ. Calif. Press, Los Angeles, vol.1, pp. 163-186.

Hodges, J. L., and Lehmann, E. L. (1963). Estimation of location based on rank tests. *Ann. Math. Statist. 34*, 598-611.

Hoeffding, W. (1948). A class of statistics with asymptotically normal distribution. *Ann. Math. Statist. 19*, 293-325.

Hoeffding, W. (1961). The strong law of large numbers for U-statistics. *Univ. North Carolina, Institute of Statistics Mimeo Series*, No 302, Chapel Hill.

Hoeffding, W. (1963). Probability inequalities for sums of bounded random variables. *J. Amer. Statist. Assoc. 58*, 13-30.

Hoeffding, W. (1965a). Asymptotically optimal tests for multinomial distributions (with discussion). *Ann. Math. Statist. 36*, 369-408.

Hoeffding, W. (1965b). On probabilities of large deviations. *Proc. 5th Berkeley Symp. Math. Stat. Probab.*, Univ. Calif. Press, Los Angeles, vol 1, pp. 203-219.

Hössjer, O. (1994). Rank-based estimates in the linear model with high breakdown point. *J. Amer. Statist. Assoc. 89*, 149-158.

Huber, P. J. (1964). Robust estimator of a location parameter. *Ann. Math. Statist. 35*, 73-101.

Huber, P. J. (1965). A robust version of the probability ratio test. *Ann. Math. Statist. 36*, 1753-1758.

Huber, P. J. (1968). Robust confidence limits. *Zeit. Wahrsch. verw. Geb. 10*, 269-278.

Huber, P. J. (1969). Théorie de l'inférence statistique robuste. *Seminar de Math. Superieurs*, Univ. Montréal.

Huber, P. J. (1973). Robust regression: Asymptotics, conjectures and Monte Carlo. *Ann. Statist. 1*, 799-821.

Huber, P. J. (1981). *Robust Statistics.* Wiley, New York.

Huber, P. J. (1984). Finite sample breakdown of M- and P-estimators. *Ann. Statist. 12*, 119-126.

Huber, P. J., and Dutter, R. (1974). Numerical solutions of robust regression problems. *COMPSTAT 1974* (eds. G.Bruckmann et al.). Physica Verlag, Vienna, pp. 165-172.

Huber, P. J., and Strassen, V. (1973). Minimax tests and the Neyman-Pearson lemma for capacities. *Ann. Statist. 1*, 251-263.

Huber-Carol, C. (1970). Etude asymptotique de tests robustes. *Ph. D. dissertation,* ETH Zűrich.

Humak, K.M.S. (1983). *Statistische Methoden der Modellbildung. Band II, Nichtlineare Regression, Robuste Verfahren in linearen Modellen, Modelle mit Fehlern in den Variablen.* Akademie Verlag, Berlin. [English Translation (1989): *Statistical Methods of Model Building, Vol. II: Nonlinear Regression, Functional Relations and Robust methods.* (eds. H. Bunke and O. Bunke). Wiley, New York.]

Hušková, M. (1979). The Berry-Esseen theorem for rank statistics. *Comment. Math. Univ. Caroliae 20,* 399-415.

Hušková, M. (1981). On bounded length sequential confidence interval for parameter in regression model based on ranks. *Coll. Nonpar. Infer. Janos Bolyai Math. Soc.* (ed. B.V.Gnedenko et al.), North Holland, Amsterdam, pp. 435-463.

Hušková, M. (1991). Sequentially adaptive nonparametric procedures. *Handbook of Sequential Analysis* (eds. B. K. Ghosh and P. K. Sen), Dekker, New York, pp.459-474.

Hušková, M., and Janssen, P. (1993). Consistency of generalized bootstrap for degenerate U-statistics. *Ann. Statist. 21,* 1811-1823.

Hušková, M., and Jurečková, J. (1981). Second order asymptotic relations of M-estimators and R-estimators in two-sample location model. *J. Statist. Planning Infer. 5,* 309-328.

Hušková, M., and Jurečková, J. (1985). Asymptotic representation of R-estimators of location. *Proc. 4th Pannonian Symp.*, 145-165.

Hušková, M., and Sen, P. K. (1985). On sequentially adaptive asymptotically efficient rank statistics. *Sequen. Anal. 4,* 225-251.

Hušková, M., and Sen, P. K. (1986). On sequentially adaptive signed rank statistics. *Sequen. Anal. 5,* 237-251.

Ibragimov, I. A., and Hasminski, R. Z. (1970). On the asymptotic behavior of generalized Bayes estimators. *Dokl. Akad. Nauk SSSR 194,* 257-260.

Ibragimov, I. A., and Hasminski, R. Z. (1971). On the limiting behavior of maximum likelihood and Bayesian estimators. *Dokl. Akad. Nauk SSSR 198,* 520-523.

Ibragimov, I. A., and Hasminski, R. Z. (1972). Asymptotic behavior of statistical estimators in the smooth case I. *Theor. Probab. Appl. 17,* 445-462.

Ibragimov, I. A., and Hasminski, R. Z. (1981). *Statistical Estimation: Asymptotic Theory.* Springer-Verlag, New York.

Inagaki, N. (1970). On the limiting distribution of a sequence of estimators with uniformity property. *Ann. Inst. Statist. Math. 22,* 1-13.

Jaeckel, L. A. (1971). Robust estimation of location: Symmetry and asymmetric contamination. *Ann. Math. Statist. 42,* 1020-1034.

Jaeckel, L. A. (1972). Estimating regression coefficients by minimizing the dispersion of the residuals. *Ann. Math. Statist. 43,* 1449-1458.

James, W., and Stein, C. (1961). Estimation with quadratic loss. *Proc. 4th Berkeley Symp. Math. Statist. Probab.* (ed. J. Neyman), Univ. Calif. Press, Los Angeles, vol. 1, pp. 361-380.

Janssen, P., Jurečková, J., and Veraverbeke, N. (1985). Rate of convergence of one- and two-step M-estimators with applications to maximum likelihood and Pitman estimators. *Ann. Statist. 13*, 1222-1229.

Janssen, P., and Veraverbeke, N. (1987). On nonparametric regression estimators based on regression quantiles. *Comm. Statist. Theor. Meth. A16*, 383-396.

Johns, M. V. (1979). Robust Pitman-like estimators. *Robustness in Statistics* (eds: R.L. Launer and G.N. Wilkinson). Academic Press, New York.

Jung, J. (1955). On linear estimates defined by a continuous weight function. *Arkiv fűr Mathematik 3*, 199-209.

Jung, J. (1962). Approximation to the best linear estimates. *Contributions to Order Statistics* (eds. A. E. Sarhan and B. G. Greenberg). Wiley, New York, 28-33.

Jurečková, J. (1969). Asymptotic linearity of a rank statistic in regression parameter. *Ann. Math. Statist. 40*, 1889-1900.

Jurečková, J. (1971a). Nonparametric estimate of regression coefficients. *Ann. Math. Statist. 42*, 1328-1338.

Jurečková, J. (1971b). Asymptotic independence of rank test statistic for testing symmetry on regression. *Sankhyā, Seer. A 33*, 1-18.

Jurečková, J. (1973a). Almost sure uniform asymptotic linearity of rank statistics in regression parameter. *Trans. 6th Prague Conf. on Inform. Theor., Random Proc. Statist. Dec. Funct.*, pp.305-313.

Jurečková, J. (1973b). Central limit theorem for Wilcoxon rank statistics process. *Ann. Statist. 1*, 1046-1060.

Jurečková, J. (1977a). Asymptotic relations of least squares estimate and of two robust estimates of regression parameter vector. *Trans. 7th Prague Conf. and 1974 European meeting of Statisticians*, Academia, Prague, pp. 231-237.

Jurečková, J. (1977b). Asymptotic relations of M-estimates and R-estimates in linear regression model. *Ann. Statist. 5*, 464-472.

Jurečková, J. (1978). Bounded-length sequential confidence intervals for regression and location parameters. *Proc. 2nd Prague Symp. Asympt. Statist.*, Academia, Prague, pp. 239-250.

Jurečková, J. (1979). Finite-sample comparison of L-estimators of location. *Comment. Math. Univ. Carolinae 20*, 507-518.

Jurečková, J. (1981a). Tail behavior of location estimators. *Ann. Statist. 9*, 578-585.

Jurečková, J. (1981b). Tail behavior of location estimators in nonregular cases. *Comment. Math. Univ. Carolinae 22*, 365-375.

Jurečková, J. (1982). Tests of location and criterion of tails. *Coll. Math. Soc. J.Bolai 32*, 469-478.

Jurečková, J. (1983a). Robust estimators of location and regression parameters and their second order asymptotic relations. *Trans. 9th Prague Conf. on Inform. Theor., Rand. Proc. and Statist. Dec. Func.* (ed. J. A. Višek), Reidel, Dordrecht, 19-32.

Jurečková, J. (1983b). Winsorized least-squares estimator and its M-estimator counterpart. *Contributions to Statistics: Essays in Honour of Norman L. Johnson* (ed. P. K. Sen), North Holland, Amsterdam, pp. 237-245.

Jurečková, J. (1983c). Trimmed polynomial regression. *Comment. Math. Univ. Carolinae 24*, 597-607.

Jurečková, J. (1983d). Asymptotic behavior of M-estimators of location in nonregular cases. *Statist. Dec. 1*, 323-340.

Jurečková, J. (1984a). Rates of consistency of classical one-sided tests. *Robustness of Statistical Methods and Nonparametric Statistics.*(eds. D. Rasch and M. L. Tiku), Deutscher-Verlag, Berlin, pp. 60-62.

Jurečková, J. (1984b). Regression quantiles and trimmed least squares estimator under a general design. *Kybernetika 20*, 345-357.

Jurečková, J. (1985). Robust estimators of location and their second-order asymptotic relations. *A Celebration of Statistics*, (eds. A. C. Atkinson and S. E. Fienberg), Springer-Verlag, New York, pp. 377-392.

Jurečková, J. (1986). Asymptotic representations of L-estimators and their relations to M-estimators. *Sequen. Anal. 5*, 317-338.

Jurečková, J. (1989). Robust statistical inference in linear models. *Nonlinear Regression, Functional Relations and Robust methods: Statistical Methods of Model Building* (eds. Bunke, H. and Bunke, O.), Wiley, New York. vol. 2, pp. 134-208.

Jurečková, J. (1992a). Uniform asymptotic linearity of regression rank scores process. *Nonparametric Statistics and Related Topics.* (ed. A. K. Md. E. Saleh), North Holland, Amsterdam, pp. 217-228.

Jurečková, J. (1992b). Estimation in a linear model based on regression rank scores. *J. Nonpar. Statist. 1*, 197-203.

Jurečková, J. (1995a). Jaroslav Hájek and asymptotic theory of rank tests. *Kybernetika 31,(2)*, 239-250.

Jurečková, J. (1995b). Affine- and scale-equivariant M-estimators in linear model. *Probab. Math. Statist. 15*, 397-407.

Jurečková, J. (1995c). Trimmed mean and Huber's estimator: Their difference as a goodness of fit criterion. *J. Statist. Res. 29(2)*, in press.

Jurečková, J. (1995d). Regression rank scores: Asymptotic linearity and RR-estimators. *Proc. MODA'4*(eds. C. P. Kitsos and W. G. Müller), Physica-Verlag, Heidelberg, pp.193-203.

Jurečková, J., Koenker, R., and Welsh, A. H. (1994). Adaptive choice of trimming. *Ann. Inst. Statist. Math. 46*, 737-755.

Jurečková, J., and Malý, M. (1995). The asymptotics for studentized k-step M-estimators of location. *Sequen. Anal. 14*, 229-245.

Jurečková, J., and Portnoy, S. (1987). Asymptotics for one-step M-estimators in regression with application to combining efficiency and high breakdown point. *Commun. Statist. Theor. Meth. A16*, 2187-2199.

Jurečková, J., and Puri, M. L. (1975). Order of normal approximation of rank statistics distribution. *Ann. Probab. 3*, 526-533.

Jurečková, J., Saleh, A. K. Md. E., and Sen, P. K. (1989). Regression quantiles and improved L-estimation in linear models. *Probability, Statistics and Design of Experiment: R.C.Bose Memorial Conference* (ed. R. R. Bahadur),

Wiley Eastern, New Delhi, pp. 405-418.

Jurečková, J., and Sen, P. K. (1981a). Sequential procedures based on M-estimators with discontinuous score functions. *J. Statist. Plan. Infer. 5*, 253-266.

Jurečková, J., and Sen, P. K. (1981b). Invariance principles for some stochastic processes related to M-estimators and their role in sequential statistical inference. *Sankhyā, Ser. A 43*, 190-210.

Jurečková, J., and Sen, P. K. (1982a). Simultaneous M-estimator of the common location and of the scale-ratio in the two sample problem. *Math. Operations. und Statistik, Ser. STATISTICS 13*, 163-169.

Jurečková, J., and Sen, P. K. (1982b). M-estimators and L-estimators of location: Uniform integrability and asymptotically risk-efficient sequential versions. *Sequen. Anal. 1*, 27-56.

Jurečková, J., and Sen, P. K. (1984). On adaptive scale-equivariant M-estimators in linear models. *Statist. Dec. Suppl. 1*, 31-46.

Jurečková, J., and Sen, P. K. (1989). Uniform second order asymptotic linearity of M-statistics in linear models. *Statist. Dec. 7*, 263-276.

Jurečková, J., and Sen, P. K. (1990). Effect of the initial estimator on the asymptotic behavior of one-step M-estimator. *Ann. Inst. Statist. Math. 42*, 345-357.

Jurečková, J., and Sen, P. K. (1993). Asymptotic equivalence of regression rank scores estimators and R-estimators in linear models. *Statistics and Probability: A R. R. Bahadur Festschrift.* (eds. J. K. Ghosh et al.). Wiley Eastern, New Delhi, pp. 279-292.

Jurečková, J., and Sen, P. K. (1994). Regression rank scores statistics and studentization in the linear model. *Proc. 5th Prague Conf. Asympt. Statist.* (eds. M. Hušková and P. Mandl), Physica-Verlag, Vienna, pp. 111-121.

Jurečková, J., and Víšek, J. A. (1984). Sensitivity of Chow-Robbins procedure to the contamination. *Sequen. Anal. 3*, 175-190.

Jurečková, J., and Welsh, A. H. (1990). Asymptotic relations between L- and M-estimators in the linear model. *Ann. Inst. Statist. Math. 42*, 671-698.

Kagan, A. M. (1970). On ϵ-admissibility of the sample mean as an estimator of location parameter. *Sankhyā, Ser. A 32*, 37-40.

Kagan, A. M., Linnik, J. V., and Rao, C. R. (1967). On a characterization of the normal law based on a property of the sample average. *Sankhyā, Ser A 27*, 405-406.

Kagan, A. M, Linnik, J. V., and Rao, C. R. (1973). *Characterization Problems in Mathematical Statistics.* Wiley, New York.

Kaigh, W. D., and Lachebruch, P. A. (1982). A generalized quantile estimator. *Comm. Statist., Theor. Meth. A11*, 2217-2238.

Kallianpur, G., and Rao, C. R. (1955). On Fisher's lower bound to asymptotic variance of a consistent estimate. *Sankhyā, Ser A 16*, 331-342.

Karmous, A. R., and Sen, P. K. (1988). Isotonic M-estimation of location: Union-intersection principle and preliminary test versions. *J. Multivar. Anal. 27*, 300-318.

Keating, J. P., Mason, R. L., and Sen, P. K. (1993). *Pitman's Measure of Closeness : A Comparison of Statistical Estimators.* SIAM, Philadelphia.

Kendall, M. G. (1938). A new measure of rank correlation. *Biometrika 30*, 81-93.

Kiefer, J. (1961). On large deviations of the empirical D.F. of the vector chance variable and a law of iterated logarithm. *Pacific J. Math. 11*, 649-660.

Kiefer, J. (1967). On Bahadur's representation of sample quantiles. *Ann. Math. Statist. 38*, 1323-1342.

Kiefer, J. (1970). Deviations between the sample quantile process and the sample cdf. *Nonparametric Techniques in Statistical Inference* (ed. M. L. Puri), Cambridge Univ. Press, 299-319.

Kiefer, J. (1972). Skorokhod embedding of multivariate rv's and the sample DF. *Zeit. Wahrsch. verw. Geb. 24*, 1-35.

Koenker, R., and Bassett, G. (1978). Regression quantiles. *Econometrika 46*, 33-50.

Koenker, R., and d'Orey, V. (1993). A remark on computing regression quantiles. *Appl. Statist. 36*, 383-393.

Koenker, R., and Portnoy, S. (1987). L-estimation for linear models. *J. Amer. Statist. Assoc. 82*, 851-857.

Koenker, R., and Portnoy, S. (1990) M-estimate of multivariate regression. *J. Amer. Statist. Assoc. 85*, 1060-1068.

Komlós, J., Májor, P., and Tusnády, G. (1975). An approximation of partial sums of independent R.V.'s and the DF.I. *Zeit. Wahrsch. verw. Geb. 32*, 111-131.

Koul, H.L. (1969). Asymptotic behavior of Wilcoxon type confidence regions in multiple linear regression. *Ann. Math. Statist. 40*, 1950 – 1979.

Koul, H.L. (1971). Asymptotic behavior of a class of confidence regions in multiple linear regression. *Ann. Math. Statist. 42*, 42 –57.

Koul, H.L. (1992). *Weighted Empiricals and Linear Models.* Inst. Math. Statist. Lect. Notes Monographs, vol 21, Hayward, Calif.

Koul, H. L., and Mukherjee, K. (1994). Regression quantiles and related processes under long range dependent errors. *J. Multivar. Anal. 51*, 318-337.

Koul, H. L., and Saleh, A. K. M. E. (1995). Autoregression quantiles and related rank scores processes. *Ann. Statist. 23*,670-689.

Koul, H. L., and Sen, P. K. (1985). On a Kolmogorov-Smirnov type aligned test in linear regression. *Statist. Probab. Letters 3*, 111-115.

Koul, H. L., and Sen, P. K. (1991). Weak convergence of a weighted residual empirical process in autoregression. *Statist. Dec. 9*, 235-261.

Kraft, C. and van Eeden, C. (1972). Linearized rank estimates and signed rank estimates for the general linear hypothesis. *Ann. Math. Statist. 43*, 42-57.

Krasker, W. S. (1980). Estimation in linear regression models with disparate data points. *Econometrica 48*, 1333-1346.

Krasker, W. S., and Welsch, R. E. (1982). Efficient bounded-influence regression estimation. *J. Amer. Statist. Assoc. 77*, 595-604.

LeCam, L. (1953). On some asymptotic properties of maximum likelihood estimates and related Bayes' estimates. *Univ. Calif. Publ. Statist. 1*, 277-330.

LeCam, L. (1960). Locally asymptotically normal families of distributions. *Univ. Calif. Publ. Statist. 3*, 37-98.

LeCam, L. (1986). *Asymptotic Methods in Statistical Decision Theory.* Springer-Verlag, New York.

LeCam, L., and Yang, G. L. (1990). *Asymptotics in Statistics - Some Basic Concepts.* Springer-Verlag, New York.

Lecoutre, J. P., and Tassi, P. (1987). *Statistique non parametrique et robustesse.* Economica, Paris.

Lehmann, E. L. (1963a). Asymptotically nonparametric inference: an alternative approach to linear models. *Ann. Math. Statist. 34*, 1494-1506.

Lehmann, E. L. (1963b). Nonparametric confidence interval for a shift parameter. *Ann. Math. Statist. 34*, 1507-1512.

Lehmann, E. L. (1975). *Nonparametrics.* Holden-Day, San Francisco.

Lehmann, E. L. (1983). *Theory of Point Estimation.* Wiley, New York.

Levit, B. Ya. (1975). On the efficiency of a class of nonparametric estimates. *Theor. Prob. Appl. 20*, 723-740.

Lindwall, T. (1973). Weak convergence of probability measures and random functions in the function space $D[0,\infty]$. *J. Appl. Probab. 10*, 109-121.

Lloyd, E. H. (1952). Least squares estimation of location and scale parameters using order statistics. *Biometrika 34*, 41-67.

Malý, M. (1991). Studentized M-estimators and their k-step versions. Ph.D. Dissertation (in Czech). Charles Univ., Prague.

Marazzi, A. (1980). ROBETH, a subroutine library for robust statistical procedures. *COMPSTAT 1980*, Physica Verlag, Vienna.

Marazzi, A. (1986). On the numerical solutions of bounded influence regression problems. *COMPSTAT 1986*, Physica Verlag, Heidelberg, pp.114-119.

Marazzi, A. (1987). Solving bounded influence regression problems with ROBSYS. *Statistical data Analysis Based on L_1-norm and Related Methods*, (ed. Y. Dodge), North Holland, Amsterdam, pp. 145-161.

Maronna, R. A., Bustos, O. H., and Yohai, V. J. (1979). Bias- and efficiency-robustness of general M-estimators for regression with random carriers. *Smoothing Techniques for Curve Estimation, Lecture Notes in Mathematics*, 757, Springer-Verlag, Berlin pp. 91 – 116;

Maronna, R. A., and Yohai, V. J. (1981). Asymptotic behavior of general M-estimators for regression and scale with random carriers. *Zeit. Wahrsch. verw. Geb. 58*, 7-20.

Maronna, R., and Yohai, V. J. (1991). The breakdown point of simulataneous general M-estimates of regression and scale. *J. Amer. Statist. Assoc. 86*, 699-716.

Martin, R. D., Yohai, V.J., and Zamar, R. H. (1989). Min-max bias robust regression. *Ann. Statist. 17*, 1608-1630.

Martin, R. D., and Zamar, R. H. (1988). High breakdown-point estimates of regression by means of the minimization of the efficient scale. *J. Amer. Statist. Assoc. 83*, 406-413.

McKean, J. W., and Hettmansperger, T. P. (1976). Tests of hypotheses based on ranks in general linear model. *Comm. Statist. A5*, 693-709.

McKean, J. W., and Hettmansperger, T. P. (1978). A robust analysis of the general linear model based on one-step R-estimates. *Biometrika 65*, 571-579.

McKean, J. W., and Schrader, R. M. (1980). The geometry of robust procedures in linear models. *J. Royal Statist. Soc. B 42*, 366-371.

Millar, P. W. (1981). Robust estimation via minimum distance methods. *Zeit. Wahrsch. verw. Geb. 55*, 73-89.

Millar, P. W. (1985). Nonparametric applications of an infinite dimensional convolution theorem. *Zeit. Wahrsch. verw. Geb. 68*, 545-556.

Miller, R. G. Jr., and Sen, P. K. (1972). Weak convergence of U-statistics and von Mises differentiable statistical functionals. *Ann. Math. Statist. 43*, 31-41.

Moses, L. E. (1965). Confidence limits from rank tests. *Technomet. 3*, 257-260.

Müller, Ch. H. (1994). Asymptotic behaviour of one-step M-estimators in contaminated nonlinear models. *Asymptotic Statistics; Proc. 5th Prague Confer.* (eds. M. Hušková and P. Mandl), Physica-Verlag, Heidelberg, pp. 394-404.

Niemiro, W. (1992). Asymptotics for M-estimators defined by convex minimization. *Ann. Statist. 20*, 1514-1533.

Ortega, J. M., and Rheinboldt, W. C. (1970). *Iterative Solution of Nonlinear Equations in Several Variables.* Academic Press, New York.

Parr, W. C. (1981). Minimum distance estimation: A bibliography. *Commun. Statist. Theor. Meth. A10*, 1205-1224.

Parr, W. C., and Schucany, W. R. (1982). Minimum distance estimation and components of goodness-of-fit statistics. *J. Roy. Statist. Soc. B*, 178-189.

Parthasarathy, K. R. (1967). *Probability Measures on Metric Spaces.* Academic Press, New York.

Parzen, E. (1979). Nonparametric statistical data modelling. *J. Amer. Statist. Assoc. 74*, 105-131.

Pearson, E. S. (1931). The analysis of variance in cases of nonnormal variation. *Biometrika 23*, 114-133.

Pfanzagl, J., and Wefelmeyer, W. (1982). *Contributions to a General Asymptotic Statistical Theory, Lecture Notes in Statistics, 13*, Springer-Verlag, New York.

Pitman, E. J. G. (1937). The closest estimates of statistical parameters. *Proc. Cambridge Phil. Soc. 33*, 212-222.

Pitman, E. J. G. (1939). Tests of hypotheses concerning location and scale parameters. *Biometrika 31*, 200-215.

Pollard, D. (1984). *Convergence of Stochastic Processes.* Wiley, New York.

Pollard, D. (1990). *Empirical Processes - Theory and Applications.* NSF-CBMS reg. Cof. Series in Probab. and Statist., vol.2. IMS, Hayward, California.

Pollard, D. (1991). Asymptotics for least absolute deviation regression estimators. *Econometric Theory 7,* 186-199.

Portnoy, S. (1984). Tightness of the sequence of c.d.f. processes defined from regression fractiles. *Robust and Nonlinear Time Series Analysis* (eds. J. Franke et al.), Springer Verlag, New York, pp. 231-246.

Portnoy, S., and Koenker, R. (1989). Adaptive L-estimation for linear models. *Ann. Statist. 17*, 362-381.

Powell, J. L. (1984). Least absolute deviations estimation for the censored regression model. *J. Econometr. 25(3)*, 303-325.

Puri, M. L., and Sen, P. K. (1985). *Nonparametric Methods in General Linear Models.* Wiley, New York.

Rao, C. R. (1945). Information and accuracy attainable in the estimation of statistical parameters. *Bull. Calcutta Math. Soc. 37*, 81-91.

Rao, C. R. (1959). Sur une caracterisation de la distribution normal étable d'apres une propriéte optimum des estimations linéaries. *Colloq. Intern. C.N.R.S. France 87*, 165-171.

Rao, P. V., Schuster, E. F., and Littell, R. C. (1975). Estimation of shift and center of symmetry based on Kolmogorov-Smirnov statistics. *Ann. Statist. 3*, 862-873.

Reeds, J. A. (1976). On the definitions of von Mises functions. Ph. D. dissertation, Harvard Univ., Cambridge, Mass.

Ren, J. J. (1994). Hadamard differentiability and its applications to R-estimation in linear models. *Statist. Dec. 12*, 1-22.

Ren, J. J., and Sen, P. K. (1991). Hadamard differentiability of extended statistical functionals. *J. Multivar. Anal. 39*, 30-43.

Ren, J. J., and Sen, P. K. (1992). Consistency of M-estimators: Hadamard differentiability approaches. *Proc. 11th Prague conf. on Inf. Th. Stat. Dec. Funct. Random Proc.*, 198-211.

Ren, J. J., and Sen, P. K. (1994). Asymptotic normality of regression M-estimators: Hadamard differentiability approaches. *Asymptotic Statistics, Proc. 5th Prague Conf.* (eds. M. Hušková and P. Mandl), Physica Verlag, Heidelber, pp. 131-147.

Ren, J. J., and Sen, P. K. (1995a). Second order Hadamard differentiability with applications. (to appear).

Ren, J. J., and Sen, P. K. (1995b). Hadamard differentiability of functionals on $D[0,1]^p$. *J. Multivar. Anal. 55*, 14-28.

Rey, W. J. J. (1983). *Introduction to Robust and Quasi-Robust Statistical Methods.* Springer-Verlag, New York.

Rieder, H. (1978). A robust asymptotic testing model. *Ann. Statist. 6*, 1080-1094.

Rieder, H. (1980). Estimation derived from robust tests. *Ann. Statist. 8*, 106-115.

Rieder, H. (1981). On local asymptotic minimaxity and admissibility in robust estimation. *Ann. Statist. 9*, 266-277.

Rieder, H. (1987). Robust regression estimators and their least favorable contamination curves. *Statist. Dec. 5*, 307-336.

Rieder, H. (1994). *Robust Asymptotic Statistics*, Springer-Verlag, New York.

Rivest, L. P. (1982). Some asymptotic distributions in the location-scale model. *Ann. Inst. Statist. Math. A 34*, 225-239.

Rockafellar, R. T. (1970). *Convex Analysis.* Princeton Univ. Press. Princeton.

Rousseeuw, P. J. (1984). Least median of squares regression. *J. Amer. Statist. Assoc. 79*, 871-880.

Rousseeuw, P. J., and Bassett, G. W., Jr. (1990). The Remedian: A robust averaging method for large data sets. *J. Amer. Statist. Assoc. 85*, 97-104.

Rousseeuw, P. J., and Croux, C. (1993). Alternatives to the median absolute deviation. *J. Amer. Statist. Assoc. 88*, 1273-1283.

Rousseeuw, P. J., and Croux, C. (1994). The bias of k-step M-estimators. *Statist. Probab. Letters 20*, 411-420.

Rousseeuw, P. J., and Leroy, A. M. (1987). *Robust Regression and Outlier Detection.* Wiley, New York.

Rousseeuw, P. J., and Yohai, V. (1984). Robust regression by means of S-estimates. *Robust and Nonlinear Time Series Analysis, Lecture Notes in Statistics, 26* (eds. J. Franke et al.), Springer-Verlag, Berlin, pp. 256 – 272.

Roy, S. N. (1953). On a heuristic method of test construction and its use in multivariate analysis. *Ann. Math. Statist. 24*, 220-238.

Ruppert, D., and Carroll, R. J. (1980). Trimmed least squares estimation in the linear model. *J. Amer. Statist. Assoc. 75*, 828-838.

Ruschendorf, L. (1988). *Asymptotische Statistik.* B.G. Teubner, Stuttgart.

Sacks, J., and Ylvisacker, D. (1978). Linear estimation for approximately linear models. *Ann. Statist. 6*, 1122-1137.

Sarhan, A. E., and Greenberg, B. G. (eds.) (1962). *Contributions to Order Statistics.* Wiley, New York.

Schrader, R. M., and Hettmansperger, T. P. (1980). Robust analysis of variance based upon a likelihood ratio criterion. *Biometrika 67*, 93-101.

Schrader, R. M., and McKean, J. W. (1977). Robust analysis of variance. *Comm. Statist., Theor. Meth. 46*, 879-894.

Schuster, E. F., and Narvarte, J. A. (1973). A new nonparametric estimator of the center of a symmetric distribution. *Ann. Statist. 1*, 1096-1104.

Scott, D. J. (1973). Central limit theorems for martingales and processes with stationary increments using a Skorokhod representation approach. *Adv. Appl. Probab. 5*, 119-137.

Sen, P. K. (1959). On the moments of sample quantiles. *Calcutta Statist. Assoc. Bull. 9*, 1-20.

Sen, P. K. (1960). On some convergence properties U-statistics. *Calcutta Statist. Assoc. Bull. 10*, 1-18.

Sen, P. K. (1963). On the estimation of relative potency in dilution (-direct) essays by distribution-free methods. *Biometrics 19*, 532-552.

Sen, P. K. (1964). On some properties of the rank-weighted means. *J. Indian Soc. Agricul. Statist. 16*, 51-61.

Sen, P. K. (1966a). On a distribution-free method of estimating asymptotic efficiency of a class of nonparametric tests. *Ann. Math. Statist. 37*, 1759-1770.

Sen, P. K. (1966b). On nonparametric simultaneous confidence regions and tests in the one-criterion analysis of variance problem. *Ann. Inst. Statist. Math. 18* ,319-366.

Sen, P. K. (1967). U-statistics and combination of independent estimates of regular functionals. *Calcutta Statist. Assoc. Bull. 16* , 1-16.

Sen, P. K. (1968a). Estimates of regression coefficients based on Kendall's tau. *J. Amer. Statist. Assoc. 63*, 1379-1389.

Sen, P. K. (1968b). On a further robustness property of the test and estimator based on Wilcoxon's signed rank statistic. *Ann. Math. Statist. 39*, 282-285.

Sen, P. K. (1968c). Robustness of some nonparametric procedures in linear models. *Ann. Math. Statist. 39*, 1913-1922.

Sen, P. K. (1969). On a class of rank order tests for the parallelism of several regression lines. *Ann. Math. Statist. 40*, 1668-1683.

Sen, P. K. (1970a). A note on order statistics from heterogeneous distributions. *Ann. Math. Statist. 41*, 2137-2139.

Sen, P. K. (1970b). On some convergence properties of one-sample rank order statistics. *Ann. Math. Statist. 41*, 2140-2143.

Sen, P. K. (1970c). The Hăjek-Renyi inequality for sampling from a finite population. *Sankhyā, Ser. A 32*, 181-188.

Sen, P. K. (1971a). Robust statistical procedures in problems of linear regression with special reference to quantitative bioassays, I. *Internat. Statist. Rev. 39* ,21-38.

Sen, P. K. (1971b). A note on weak convergence of empirical processes for sequences of ϕ-mixing random variables. . *Ann. Math. Statist. 42*, 2132-2133.

Sen, P. K. (1972a). On the Bahadur representation of sample quantiles for sequences of mixing random vbariables. *J. Multivar. Anal. 2*, 77-95.

Sen, P. K. (1972b). Weak convergence and relative compactness of martingale processes with applications to nonparametric statistics. *J. Multivar. Anal. 2*, 345-361.

Sen, P. K. (1972c). Robust statistical procedures in problems of linear regression with special reference to quantitative bioassays, II. *Internat. Statist. Rev. 40*, 161-172.

Sen, P. K. (1973a). On weak convergence of empirical processes for random number of independent stochastic vectors. *Proc. Cambridge Phil. Soc. 73*, 135-140.

Sen, P. K. (1973b). An almost sure invariance principle for multivariate Kolmogorov-Smirnov statistics. *Ann. Probab. 1*,488-496.

Sen, P. K. (1974a). Almost sure behavior of U-statistics and von Mises' differentiable statistical functions. *Ann. Statist. 2*, 387-395.

Sen, P. K. (1974b). Weak convergence of generalized U-statistics. *Ann. Probab. 2*, 90-102.

Sen, P. K. (1974c). On L_p-convergence of U-statistics. *Ann. Inst. Statist. Math. 26*, 55-60.

Sen, P. K. (1975). Rank statistics, martingales and limit theorems. In *Statistical Inference and Related Topics* (ed. M. L. Puri), Academic Press, New York, pp. 129-158.

Sen, P. K. (1976a). Weak convergence of a tail sequence of martingales. *Sankhyā, Ser.A 38*,190-193.

Sen, P. K. (1976b). A two-dimensional functional permutational central limit theorem for linear rank statistics. *Ann. Probab. 4*, 13-26.

Sen, P. K. (1976c). A note on invariance principles for induced order styatistics. *Ann. Probab. 4*, 474-479.

Sen, P. K. (1977a). Some invariance principles relating to jackknifing and their role in sequential analysis. *Ann. Statist. 5*, 315-329.

Sen, P. K. (1977b). On Wiener process embedding for linear combinations of order statistics. *Sankhyā, Ser. A5*, 1107- 1123.

Sen, P. K. (1977c). Tied down Wiener process approximations for aligned rank order statistics and some applications. *Ann. Statist. 5*, 1107-1123.

Sen, P. K. (1978a). An invariance principle for linear combinations of order statistics. *Zeit. Wahrsch. verw. Geb. 42*, 327-340.

Sen, P. K. (1978b). Invariance principles for linear rank statistics revisited. *Sankhyā, Ser.A 40*,215-236.

Sen, P. K. (1979a). Weak convergence of some quantile processes arising in progressively censored tests. *Ann. Statist. 7*, 414-431.

Sen, P. K. (1979b). Asymptotic properties of maximum likelihood estimators based on conditional specifications. *Ann. Statist. 7*, 1019-1033.

Sen, P. K. (1980a). On almost sure linearity theorems for signed rank order statistics. *Ann. Statist. 8*, 313-321.

Sen, P. K. (1980b). On nonparametric sequential point estimation of location based on general rank order statistics. *Sankhyā, Ser. A 42*, 223-240.

Sen, P. K. (1981a). *Sequential Nonparametrics: Invariance Principles and Statistical Inference.* Wiley, New York.

Sen, P. K. (1981b). Some invariasnce principles for mixed-rank statistics and induced order statistics and some applications. *Commun. Statist. Theor. Meth. A 10*, 1691-1718.

Sen, P. K. (1981c). The Cox regression model, invariance principles for some induced quantile processes and some repeated significance tests. *Ann. Statist. 9*, 109-121.

Sen, P.K. (1982a). On M-tests in linear models. *Biometrika 69*, 245-248.

Sen, P. K. (1982b). Invariance principles for recursive residuals. *Ann. Statist. 10*, 307-312.

Sen, P. K. (1983a). On the limiting behavior of the empirical kernel distribution function. *Calcutta Statist. Assoc. Bull. 32*, 1-8.

Sen, P. K. (1983b). On permutational central limit theorems for general multivariate linear rank statistics. *Sankhyā, Ser A 45*, 141-49.

Sen, P. K. (1984a). Jackknifing L-estimators: Affine structure and asymptotics. *Sankhyā, Ser. A 46*, 207-218.

Sen, P. K. (1984b). Nonparametric procedures for some miscellaneous problems. *Handbook of Statistics, vol. 4 : Nonparametric Methods.* (eds. P. R. Krishnaiah and P. K. Sen), North Holland, Amsterdam, pp. 699-739.

Sen, P. K. (1984c). Invariance principles for U-statistics and von Mises functionals in the non-i.d. case. *Sankhyā, Ser A 46*, 416-425.

Sen, P. K. (1984d). On a Kolmogorov-Smirnov type aligned test. *Statist. Probab. Letters 2* , 193-196.

Sen, P. K. (1984e). A James-Stein detour of U-statistics. *Comm. Statist. Theor. Metd. A 13*, 2725-2747.

Sen, P. K. (1985). *Theory and Applications of Sequential Nonparametrics.* SIAM, Philadelphia.

Sen, P. K. (1986a). Are BAN estimators the Pitman-closest ones too? *Sankhyā, Ser A 48*, 51-58.

Sen, P. K. (1986b). On the asymptotic distributional risks of shrinkage and preliminary test versions of maximum likelihood estimators. *Sankhyā, Ser A 48*, 354-371.

Sen, P. K. (1986c). Whither jackknifing in Stein-rule estimation? *Comm. Statist. Theor. Meth. A 15*, 2245-2266.

Sen, P.K. (1988a). Functional jackknifing: Rationality and general asymptotics. *Ann. Statist. 16*, 450-469.

Sen, P.K. (1988b). Functional approaches in resampling plans: A review of some recent developments. *Sankhyā, Ser. A 40*, 394-435.

Sen, P. K. (1988c). Asymptotics in finite population sampling. In *Handbook of Statistics, vol. 6: Sampling.* (eds. P. R. Krishnaiah and C. R. Rao), North Holland, Amsterdam, pp. 291-331.

Sen, P. K. (1989a). Whither delete-k jackknifing for smooth statistical functionals. *Statistical Data Analysis and Inference* (ed. Y.Dodge), North Holland, Amsterdam, pp. 269-279.

Sen, P. K. (1989b). Asymptotic theory of sequential shrunken estimation of statistical functionals. *Proc. 4th Prague Confer. Asympt. Statist.* (eds. M. Hušková and P. Mandl), Academia, Prague, pp. 83-100.

Sen, P. K. (1990a). Optimality of BLUE and ABLUE in the light of the Pitman closeness of statistical estimators. *Coll. Math. Soc. Janos Bolyai, 57: Limit Theorems in Probability and Statistics.*, pp. 459-476.

Sen, P. K. (1990b). Statistical functionals, stopping times and asymptotic minimum risk property. *Probability Theory and Math. Statist.* (eds. B. Gregelionis et al.), World Sci. Press, Singapore, vol. 2, pp. 411-423.

Sen, P. K. (1991a). Nonparametrics: Retrospectives and perspectives (with discussion). *J. Nonparamet. Statist. 1*, 1-68.

Sen, P. K. (1991b). Nonparametric methods in sequential analysis. *Handbook of Sequential Analysis* (eds. B. K. Ghosh and P. K. Sen), Dekker, New York, pp. 331-360.

Sen, P. K. (1991c). Asymptotics via sub-sampling Bahadur type representations. *Probability, Statistics and Design of Experiment: R. C. Bose Memorial Conference*(ed. R. R. Bahadur), Wiley Eastern, New Delhi, pp. 653-666.

Sen, P. K. (1992). Some informative aspects of jackknifing and bootstrapping. In *Order statistics and Nonparametrics: Theory and Applications*(eds: P. K. Sen and I. A. Salama). North Holland, Amsterdam, pp.25-44.

Sen, P.K. (1993a). Multivariate L_1-norm estimation and the vulnerable bootstrap. *Statistical Theory and Data Analysis -III* (eds. K. Matusita et al.), North Holland, Amsterdam, pp. 441-450.

Sen, P. K. (1993b). Perspectives in multivariate nonparametrics: conditional functionals and Anocova models. *Sankhyā, Ser. A 55*, 516-532.

Sen, P. K. (1994a). Isomorphism of quadratic norm and PC ordering of estimators admitting first order AN representation. *Sankhyā, Ser.B 56*,

Sen, P. K. (1994b). Regression quantiles in nonparametric regression. *J. Nonparamet. Statist. 3*, 237-253.

Sen, P. K. (1994c). The impact of Wassily Hoeffding's research on nonparametrics. *The Collected Works of Wassily Hoeffding* (eds. N. I. Fisher and P. K. Sen), Springer-Verlag, New York, pp. 29-55.

Sen, P. K. (1995a). Robust and Nonparametric Methods in linear models with mixed-effects. *Tetra Mount. Math. J., Sp. Issue on Probastat'94 Confer.* Bratislava, Slovak., in press.

Sen, P. K. (1995b). Regression rank scores estimation in ANOCOVA. *Ann. Statist.* (to appear).

Sen, P. K. (1995c). The Hájek asymptotics for finite population sampling and their ramifications. *Kybernetrika 31*, 251-268.

Sen, P. K. (1995d). Censoring in Theory and Practice: Statistical Perspectives and Controversies. *Analysis of Censored Data, IMS Lecture Notes Mon. Ser. 27* (eds. H. Koul and J. Deshpande), Hayward, Calif. pp. 177-192.

Sen, P. K. (1995e). Bose-Einstein statistics, generalized occupancy problems and allied asymptotics. *Rev. Bull. Calcutta Math. Soc. 2*, 1-12.

Sen, P. K. (1995f). Statistical functionals, Hadamard differentiability and Martingales. *A Festschrift for Professor J. Medhi*, Wiley Eastern, New Delhi, in press.

Sen, P. K., and Bhattacharyya, B. B. (1976). Asymptotic normality of the extrema of certain sample functions. *Zeit. Wahrsch. verw Geb. 34*, 113-118.

Sen, P. K., Bhattacharyya, B. B., and Suh, M. W. (1973). Limiting behavior of the extrema of certain sample functions. *Ann. Statist. 1*, 297-311.

Sen, P. K., and Ghosh, M. (1971). On bounded length sequential confidence intervals based on one-sample rank order statistics. *Ann. Math. Statist. 42*, 189-203.

Sen, P. K., and Ghosh, M. (1972). On strong convergence of regression rank statistics. *Sankhyā, Ser. A 34*, 335-348.

Sen, P. K., and Ghosh, M. (1973a). A Chernoff-Savage representation of rank order statistics for stationary ϕ-mixing processes. *Sankhyā, Ser. A 35*, 153-172.

Sen, P. K., and Ghosh, M. (1973b). A law of iterated logarithm for one sample rank order statistics and some applications. *Ann. Statist. 1*, 568-576.

Sen, P. K., and Ghosh, M. (1974). On sequential rank tests for location. *Ann. Statist. 2*, 540-552.

Sen, P. K., and Ghosh, M. (1981). Sequential point estimation of estimable parameters based on U-statistics. *Sankhyā, Ser. A 43*, 331-344.

Sen, P. K., Kubokawa, T., and Saleh, A. K. M. E. (1989). The Stein paradox in the sense of the Pitman measure of closeness. *Ann. Statist. 17*, 1375-1386.

Sen, P. K., and Puri, M. L. (1977). Asymptotically distribution-free aligned rank order tests for composite hypotheses for general multivariate linear models. *Zeit. Wahrsch. verw. Geb. 39*, 175-186.

Sen, P. K., and Saleh, A. K. M. E. (1985). On some shrinkage estimators of multivariate location. *Ann. Statist. 13*, 272-281.

Sen, P. K., and Saleh, A. K. M. E. (1987). On preliminary test and shrinkage M-estimation in linear models. *Ann. Statist. 15*, 1580-1592.

Sen, P. K., and Saleh, A. K. M. E. (1992). Pitman closeness of Pitman estimators. *Gujarat Statist. Rev., C. G. Khatri Mem. Vol.*, 198-213.

Sen, P. K. and Singer, J. M. (1985). M-methods in multivariate linear models. *J. Multivar. Anal. 17*, 168-184.

Sen, P. K. and Singer, J. M. (1993). *Large sample Methods in Statistics: An Introduction with Applications.* Chapman and Hall, New York.

Seneta, E. (1976). *Regularly Varying Functions, Lectures Notes Math. 508*, Springer-Verlag, New York.

Serfling, R. J. (1980). *Approximation Theorems of Mathematical Statistics.* Wiley, New York.

Serfling, R. J. (1984). Generalized L-, M-, and R-statistics. *Ann. Statist. 12*, 76-86.

Shorack, G. R., and Wellner, J. A. (1986). *Empirical Processes with Applications to Statistics.* Wiley, New York.

Siegel, A. F. (1982). Robust regression using repeated medians. *Biometrika 69*, 242-244.

Sievers, G. L. (1978). Weighted rank statistics for simple linear regression. *J. Amer. Statist. Assoc. 73*, 628-631.

Sievers, G. L. (1983). A weighted dispersion function for estimation in linear models. *Comm. Statist. A 12*, 1161-1179.

Silverman, B. W. (1983). Convergence of a class of empirical distribution functions of dependent random variables. *Ann. Probab. 11*, 745-751.

Silverman, B. (1986). *Density Estimation for Statistics and Data Analysis.* Chaman and Hall, New York.

Simpson, D. G., Ruppert, D., and Carroll, R. J. (1992). On one step GM estimates and stability of inferences in linear regression. *J. Amer. Statist. Assoc. 87*, 439 – 450.

Singer, J. M., and Sen, P. K. (1985). Asymptotic relative efficiency of multivariate M-estimators. *Comm. Statist. Sim. Comp. B 14*, 29-42.

Skorokhod, A. V. (1956). Limit theorems for stochastic processes. *Theo. Probab. Appl. 1*, 261-290.

Staudte, R. J., and Sheather, S. J. (1990). *Robust Estimation and Testing.* Wiley, New York.

Stefanski, L. A., Carroll, R. J., and Ruppert, D. (1986). Otimally bounded score functions for generalized linear models with applications to logistic regression. *Biometrika 73*, 413-424.

Stein, C. (1956). Inadmissibility of the usual estimator for the mean of a multivariate normal distribution. *Proc. 3rd Berkeley Symp. Math. Statist. Probab.*(ed. J. Neyman), Univ. Calif. Press, Los Angeles, vol 1, pp 187-195.

Steyn, H. S., and Geertsema, J. C. (1974). Nonparametric confidence sequence for the centre of a symmetric distribution. *South Afr. J. Statist. 8*, 24-34.

Stigler, S. M. (1969). Linear functions of order statistics. *Ann. Math. Statist. 40*, 770-784.

Stigler, S. M. (1973a). The asymptotic distribution of the trimmed mean. *Ann. Statist. 1*, 472-477.

Stigler, S. M. (1973b). Simon Newcomb, Percy Daniell, and the history of robust estimation 1885-1920. *J. Amer. Statist. Assoc. 68*, 872-879.

Stigler, S. M. (1974). Linear functions of order statistics with smooth weight function. *Ann. Statist. 2*, 676-693.

Stigler, S. M. (1980). Studies in the history of probability and statistics XXXVIII. R.H.Smith, a Victorian interested in robustness. *Biometrika 67*, 217-221.

Stone, C. (1974). Asymptotic properties of estimators of a location parameter. *Ann. Statist. 2* 1127-1137.

Stout, W. F. (1974). *Almost Sure Convergence.* Academic Press, New York.

Strassen, V. (1964). An invariance principle for the law of iterated logarithm. *Zeit. Wahrsch. verw. Geb. 3*, 211-226.

Strassen, V. (1967). Almost sure behavior of sums of independent random variables and martingales. *Proc. 5th Berkeley Symp. Math. Statist. Probab.* (eds. L. Lecam et al.), Univ. Calif. Press, Los Angeles, vol 2, pp. 315-343.

Tableman, M. (1990). Bounded influence rank regression - a one step estimator based on Wilcoxon scores. *J. Amer. Statist. Assoc. 85*, 508-513.

Tableman, M. (1994). The asymptotic of the least trimmed absolute deviations (LTAD) estimator. *Statist. Probab. Letters 14*, 387-398.

Tsiatis, A. A. (1990). Estimating regression parameters using linear rank tests for censored data. *Ann. Statist. 18*, 354-372.

Tukey, J. (1958). Bias and confidence in not quite large samples. (Abstract). *Ann. Math. Statist. 29*, 614.

Tukey, J. W. (1977). *Exploratory Data Analysis.* Addison-Wesley, Reading, Mass.

Vajda, I. (1984). Minimum weak divergence estimators of structural parameters. *Proc. 3rd Prague Symp. on Asympt. Statist.* (eds. M. Hušková and P. Mandl), Elsevier, Amsterdam, pp. 417-424.

van der Vaart, A. (1991). Efficiency and Hadamard differentiability. *Scand. J. Statist. 18*, 63-75.

van Eeden, C. (1972). Ananalogue for signed rank statistics, of Jurečková's asymptotic linearity theorem for rank statistics. *Ann. Math. Statist. 43*, 791-802.

van Eeden, C. (1983). On the relation between L-estimators and M-estimators and asymptotic efficiency relative to the Cramér-Rao lower bound. *Ann. Statist. 11*, 674-690.

van Zwet, W. R. (1980). A strong law for linear functions of order statistics. *Ann. Probab. 8*, 986-990.

van Zwet, W. R. (1984). A Berry-Esseen bound for symmetric statistics. *Zeit. Wahrsch. verw. Geb. 66*, 425-440.

van Zwet, W. R. (1985). Van de Hulstx and robust statistics: A historical note. *Statist. Neerlandica 32*, 81-95.

van Zwet, W. R. (1992). *Wald Memorial Lecture*, IMS, Boston, Mass.

von Mises, R. (1936). Les lois de probabilité pour des fonctions statistiques. *Ann. Inst. H.Poincaré 6*, 185-212.

von Mises, R. (1947). On the asymptotic distribution of differentiable statistical functions. *Ann. Math. Statist. 18*, 309-348.

Wald, A. (1943). Test of statistical hypothesis concerning several parameters when the number of observations is large. *Trans. Amer. Math. Soc. 54*, 426-482.

Wald, A. (1947). *Sequential Analysis*, Wiley, New York.

Walker, H. M. and Lev, J. (1953). *Statistical Inference.* Holt, New York.

Welsh, A. H. (1986). Bahadur representation for robust scale estimators based on regression residuals. *Ann. Statist. 14*, 1246 -1251.

Welsh, A. H. (1987a). One-step L-estimators for the linear model. *Ann. Statist. 15*, 626-641.

Welsh, A. H. (1987b). The trimmed mean in the linear model. *Ann. Statist. 15*, 626-641.

Welsh, A. H. (1989). On M-processes and M-estimation. *Ann. Statist. 17*, 337-361.

Whitt, W. (1970). Weak convergence of probability measures on the function space $C[0,\infty)$. *Ann. Math. Statist. 41*, 939-944.

Wiens, D. P. (1990). Minimax-variance L- and R-estimators of locations. *Canad. J. Statist. 18*, 47-58.

Wiens, D. P., and Zhou, J. (1994). Bounded-influence rank estimator in the linear model. *Canad. J. Statist. 22*, 233-245.

Wijsman, R. A. (1979). Constructing all smallest simultaneous confidence sets in a given class with applications to MANOVA. *Ann. Statist. 7*, 1003-1018.

Wolfowitz, J. (1957). The minimum distance method. *Ann. Math. Statist. 28*, 75-88.

Yanagawa, T. (1969). A small sample robust competitor of Hodges-Lehmann estimate. *Bull. Math. Statist. 13*, 1-14.

Yohai, V. J., and Maronna, R. A. (1976). Location estimators based on linear combinations of modified order statistics. *Comm. Statist. Theor. Meth. A5*, 481-486.

Yoshizawa, C. N., Davis, C. E., and Sen, P. K. (1986). Asymptotic equivalence of the Harrell-Davis estimator and the sample median. *Comm. Statist., Theor. Meth. A 14*, 2129-2136.

Author Index

Trécourt, P., 434
Tsiatis, A. A., 457
Tukey, J. W., 21, 372, 434, 457
Tusnády, G., 210, 447

Vajda, I., 115, 457

Subject Index

WILEY SERIES IN PROBABILITY AND STATISTICS

Probability and Statistics

ANDERSON · An Introduction to Multivariate Statistical Analysis, *Second Edition*
*ANDERSON · The Statistical Analysis of Time Series
ARNOLD, BALAKRISHNAN, and NAGARAJA · A First Course in Order Statistics
BACCELLI, COHEN, OLSDER, and QUADRAT · Synchronization and Linearity: An Algebra for Discrete Event Systems
BERNARDO and SMITH · Bayesian Statistical Concepts and Theory
BHATTACHARYYA and JOHNSON · Statistical Concepts and Methods
BILLINGSLEY · Convergence of Probability Measures
BILLINGSLEY · Probability and Measure, *Second Edition*
BOROVKOV · Asymptotic Methods in Queuing Theory
BRANDT, FRANKEN, and LISEK · Stationary Stochastic Models
CAINES · Linear Stochastic Systems
CAIROLI and DALANG · Sequential Stochastic Optimization
CHEN · Recursive Estimation and Control for Stochastic Systems
CONSTANTINE · Combinatorial Theory and Statistical Design
COOK and WEISBERG · An Introduction to Regression Graphics
COVER and THOMAS · Elements of Information Theory
*DOOB · Stochastic Processes
DUDEWICZ and MISHRA · Modern Mathematical Statistics
ETHIER and KURTZ · Markov Processes: Characterization and Convergence
FELLER · An Introduction to Probability Theory and Its Applications, Volume 1, *Third Edition,* Revised; Volume II, *Second Edition*
FREEMAN and SMITH · Aspects of Uncertainty: A Tribute to D. V. Lindley
FULLER · Introduction to Statistical Time Series, *Second Edition*
FULLER · Measurement Error Models
GIFI · Nonlinear Multivariate Analysis
GUTTORP · Statistical Inference for Branching Processes
HALD · A History of Probability and Statistics and Their Applications before 1750
HALL · Introduction to the Theory of Coverage Processes
HANNAN and DEISTLER · The Statistical Theory of Linear Systems
HEDAYAT and SINHA · Design and Inference in Finite Population Sampling
HOEL · Introduction to Mathematical Statistics, *Fifth Edition*
HUBER · Robust Statistics
IMAN and CONOVER · A Modern Approach to Statistics
JUREK and MASON · Operator-Limit Distributions in Probability Theory
KAUFMAN and ROUSSEEUW · Finding Groups in Data: An Introduction to Cluster Analysis
LARSON · Introduction to Probability Theory and Statistical Inference, *Third Edition*
LESSLER and KALSBEEK · Nonsampling Error in Surveys
LINDVALL · Lectures on the Coupling Method
MANTON, WOODBURY, and TOLLEY · Statistical Applications Using Fuzzy Sets
MARDIA · The Art of Statistical Science: A Tribute to G. S. Watson
MORGENTHALER and TUKEY · Configural Polysampling: A Route to Practical Robustness

*Now available in a lower priced paperback edition in the Wiley Classics Library.

Probability and Statistics (Continued)

MUIRHEAD · Aspects of Multivariate Statistical Theory
OLIVER and SMITH · Influence Diagrams, Belief Nets and Decision Analysis
*PARZEN · Modern Probability Theory and Its Applications
PRESS · Bayesian Statistics: Principles, Models, and Applications
PUKELSHEIM · Optimal Experimental Design
PURI and SEN · Nonparametric Methods in General Linear Models
PURI, VILAPLANA, and WERTZ · New Perspectives in Theoretical and Applied Statistics
RAO · Asymptotic Theory of Statistical Inference
RAO · Linear Statistical Inference and Its Applications, *Second Edition*
RAO and SHANBHAG · Choquet-Deng Type Functional Equations and Applications to Stochastic Models
RENCHER · Methods of Multivariate Analysis
ROBERTSON, WRIGHT, and DYKSTRA · Order Restricted Statistical Inference
ROGERS and WILLIAMS · Diffusions, Markov Processes, and Martingales, Volume I: Foundations, *Second Edition;* Volume II: Îto Calculus
ROHATGI · An Introduction to Probability Theory and Mathematical Statistics
ROSS · Stochastic Processes
RUBINSTEIN · Simulation and the Monte Carlo Method
RUBINSTEIN and SHAPIRO · Discrete Event Systems: Sensitivity Analysis and Stochastic Optimization by the Score Function Method
RUZSA and SZEKELY · Algebraic Probability Theory
SCHEFFE · The Analysis of Variance
SEBER · Linear Regression Analysis
SEBER · Multivariate Observations
SEBER and WILD · Nonlinear Regression
SERFLING · Approximation Theorems of Mathematical Statistics
SHORACK and WELLNER · Empirical Processes with Applications to Statistics
SMALL and McLEISH · Hilbert Space Methods in Probability and Statistical Inference
STAPLETON · Linear Statistical Models
STAUDTE and SHEATHER · Robust Estimation and Testing
STOYANOV · Counterexamples in Probability
STYAN · The Collected Papers of T. W. Anderson: 1943–1985
TANAKA · Time Series Analysis: Nonstationary and Noninvertible Distribution Theory
THOMPSON and SEBER · Adaptive Sampling
WHITTAKER · Graphical Models in Applied Multivariate Statistics
YANG · The Construction Theory of Denumerable Markov Processes

Applied Probability and Statistics

ABRAHAM and LEDOLTER · Statistical Methods for Forecasting
AGRESTI · Analysis of Ordinal Categorical Data
AGRESTI · Categorical Data Analysis
AGRESTI · An Introduction to Categorical Data Analysis
ANDERSON and LOYNES · The Teaching of Practical Statistics
ANDERSON, AUQUIER, HAUCK, OAKES, VANDAELE, and WEISBERG · Statistical Methods for Comparative Studies
*ARTHANARI and DODGE · Mathematical Programming in Statistics
ASMUSSEN · Applied Probability and Queues
*BAILEY · The Elements of Stochastic Processes with Applications to the Natural Sciences
BARNETT · Interpreting Multivariate Data
BARNETT and LEWIS · Outliers in Statistical Data, *Second Edition*
BARTHOLOMEW, FORBES, and McLEAN · Statistical Techniques for Manpower Planning, *Second Edition*

*Now available in a lower priced paperback edition in the Wiley Classics Library.

Applied Probability and Statistics (Continued)

BATES and WATTS · Nonlinear Regression Analysis and Its Applications
BECHHOFER, SANTNER, and GOLDSMAN · Design and Analysis of Experiments for Statistical Selection, Screening, and Multiple Comparisons
BELSLEY · Conditioning Diagnostics: Collinearity and Weak Data in Regression
BELSLEY, KUH, and WELSCH · Regression Diagnostics: Identifying Influential Data and Sources of Collinearity
BERRY, CHALONER, and GEWEKE · Bayesian Analysis in Statistics and Econometrics: Essays in Honor of Arnold Zellner
BHAT · Elements of Applied Stochastic Processes, *Second Edition*
BHATTACHARYA and WAYMIRE · Stochastic Processes with Applications
BIEMER, GROVES, LYBERG, MATHIOWETZ, and SUDMAN · Measurement Errors in Surveys
BIRKES and DODGE · Alternative Methods of Regression
BLOOMFIELD · Fourier Analysis of Time Series: An Introduction
BOLLEN · Structural Equations with Latent Variables
BOULEAU · Numerical Methods for Stochastic Processes
BOX · R. A. Fisher, the Life of a Scientist
BOX and DRAPER · Empirical Model-Building and Response Surfaces
BOX and DRAPER · Evolutionary Operation: A Statistical Method for Process Improvement
BOX, HUNTER, and HUNTER · Statistics for Experimenters: An Introduction to Design, Data Analysis, and Model Building
BROWN and HOLLANDER · Statistics: A Biomedical Introduction
BUCKLEW · Large Deviation Techniques in Decision, Simulation, and Estimation
BUNKE and BUNKE · Nonlinear Regression, Functional Relations and Robust Methods: Statistical Methods of Model Building
CHATTERJEE and HADI · Sensitivity Analysis in Linear Regression
CHATTERJEE and PRICE · Regression Analysis by Example, *Second Edition*
CLARKE and DISNEY · Probability and Random Processes: A First Course with Applications, *Second Edition*
COCHRAN · Sampling Techniques, *Third Edition*
*COCHRAN and COX · Experimental Designs, *Second Edition*
CONOVER · Practical Nonparametric Statistics, *Second Edition*
CONOVER and IMAN · Introduction to Modern Business Statistics
CORNELL · Experiments with Mixtures, Designs, Models, and the Analysis of Mixture Data, *Second Edition*
COX · A Handbook of Introductory Statistical Methods
*COX · Planning of Experiments
COX, BINDER, CHINNAPPA, CHRISTIANSON, COLLEDGE, and KOTT · Business Survey Methods
CRESSIE · Statistics for Spatial Data, *Revised Edition*
DANIEL · Applications of Statistics to Industrial Experimentation
DANIEL · Biostatistics: A Foundation for Analysis in the Health Sciences, *Sixth Edition*
DAVID · Order Statistics, *Second Edition*
*DEGROOT, FIENBERG, and KADANE · Statistics and the Law
*DEMING · Sample Design in Business Research
DILLON and GOLDSTEIN · Multivariate Analysis: Methods and Applications
DODGE and ROMIG · Sampling Inspection Tables, *Second Edition*
DOWDY and WEARDEN · Statistics for Research, *Second Edition*
DRAPER and SMITH · Applied Regression Analysis, *Second Edition*
DUNN · Basic Statistics: A Primer for the Biomedical Sciences, *Second Edition*
DUNN and CLARK · Applied Statistics: Analysis of Variance and Regression, *Second Edition*
ELANDT-JOHNSON and JOHNSON · Survival Models and Data Analysis

*Now available in a lower priced paperback edition in the Wiley Classics Library.

Applied Probability and Statistics (Continued)

EVANS, PEACOCK, and HASTINGS · Statistical Distributions, *Second Edition*
FISHER and VAN BELLE · Biostatistics: A Methodology for the Health Sciences
FLEISS · The Design and Analysis of Clinical Experiments
FLEISS · Statistical Methods for Rates and Proportions, *Second Edition*
FLEMING and HARRINGTON · Counting Processes and Survival Analysis
FLURY · Common Principal Components and Related Multivariate Models
GALLANT · Nonlinear Statistical Models
GLASSERMAN and YAO · Monotone Structure in Discrete-Event Systems
GROSS and HARRIS · Fundamentals of Queueing Theory, *Second Edition*
GROVES · Survey Errors and Survey Costs
GROVES, BIEMER, LYBERG, MASSEY, NICHOLLS, and WAKSBERG · Telephone Survey Methodology
HAHN and MEEKER · Statistical Intervals: A Guide for Practitioners
HAND · Discrimination and Classification
*HANSEN, HURWITZ, and MADOW · Sample Survey Methods and Theory, Volume 1: Methods and Applications
*HANSEN, HURWITZ, and MADOW · Sample Survey Methods and Theory, Volume II: Theory
HEIBERGER · Computation for the Analysis of Designed Experiments
HELLER · MACSYMA for Statisticians
HINKELMAN and KEMPTHORNE: · Design and Analysis of Experiments, Volume 1: Introduction to Experimental Design
HOAGLIN, MOSTELLER, and TUKEY · Exploratory Approach to Analysis of Variance
HOAGLIN, MOSTELLER, and TUKEY · Exploring Data Tables, Trends and Shapes
HOAGLIN, MOSTELLER, and TUKEY · Understanding Robust and Exploratory Data Analysis
HOCHBERG and TAMHANE · Multiple Comparison Procedures
HOCKING · Methods and Applications of Linear Models: Regression and the Analysis of Variance
HOEL · Elementary Statistics, *Fifth Edition*
HOGG and KLUGMAN · Loss Distributions
HOLLANDER and WOLFE · Nonparametric Statistical Methods
HOSMER and LEMESHOW · Applied Logistic Regression
HØYLAND and RAUSAND · System Reliability Theory: Models and Statistical Methods
HUBERTY · Applied Discriminant Analysis
IMAN and CONOVER · Modern Business Statistics
JACKSON · A User's Guide to Principle Components
JOHN · Statistical Methods in Engineering and Quality Assurance
JOHNSON · Multivariate Statistical Simulation
JOHNSON and KOTZ · Distributions in Statistics
Continuous Univariate Distributions—2
Continuous Multivariate Distributions
JOHNSON, KOTZ, and BALAKRISHNAN · Continuous Univariate Distributions, Volume 1, *Second Edition;* Volume 2, *Second Edition*
JOHNSON, KOTZ, and KEMP · Univariate Discrete Distributions, *Second Edition*
JUDGE, GRIFFITHS, HILL, LÜTKEPOHL, and LEE · The Theory and Practice of Econometrics, *Second Edition*
JUDGE, HILL, GRIFFITHS, LÜTKEPOHL, and LEE · Introduction to the Theory and Practice of Econometrics, *Second Edition*
JUREČKOVÁ and SEN · Robust Statistical Procedures: Aymptotics and Interrelations
KADANE · Bayesian Methods and Ethics in a Clinical Trial Design
KADANE and SCHUM · A Probabilistic Analysis of the Sacco and Vanzetti Evidence
KALBFLEISCH and PRENTICE · The Statistical Analysis of Failure Time Data
KASPRZYK, DUNCAN, KALTON, and SINGH · Panel Surveys

*Now available in a lower priced paperback edition in the Wiley Classics Library.

Applied Probability and Statistics (Continued)

KISH · Statistical Design for Research
*KISH · Survey Sampling
LANGE, RYAN, BILLARD, BRILLINGER, CONQUEST, and GREENHOUSE · Case Studies in Biometry
LAWLESS · Statistical Models and Methods for Lifetime Data
LEBART, MORINEAU., and WARWICK · Multivariate Descriptive Statistical Analysis: Correspondence Analysis and Related Techniques for Large Matrices
LEE · Statistical Methods for Survival Data Analysis, *Second Edition*
LEPAGE and BILLARD · Exploring the Limits of Bootstrap
LEVY and LEMESHOW · Sampling of Populations: Methods and Applications
LINHART and ZUCCHINI · Model Selection
LITTLE and RUBIN · Statistical Analysis with Missing Data
MAGNUS and NEUDECKER · Matrix Differential Calculus with Applications in Statistics and Econometrics
MAINDONALD · Statistical Computation
MALLOWS · Design, Data, and Analysis by Some Friends of Cuthbert Daniel
MANN, SCHAFER, and SINGPURWALLA · Methods for Statistical Analysis of Reliability and Life Data
MASON, GUNST, and HESS · Statistical Design and Analysis of Experiments with Applications to Engineering and Science
McLACHLAN · Discriminant Analysis and Statistical Pattern Recognition
MILLER · Survival Analysis
MONTGOMERY and MYERS · Response Surface Methodology: Process and Product in Optimization Using Designed Experiments
MONTGOMERY and PECK · Introduction to Linear Regression Analysis, *Second Edition*
NELSON · Accelerated Testing, Statistical Models, Test Plans, and Data Analyses
NELSON · Applied Life Data Analysis
OCHI · Applied Probability and Stochastic Processes in Engineering and Physical Sciences
OKABE, BOOTS, and SUGIHARA · Spatial Tesselations: Concepts and Applications of Voronoi Diagrams
OSBORNE · Finite Algorithms in Optimization and Data Analysis
PANKRATZ · Forecasting with Dynamic Regression Models
PANKRATZ · Forecasting with Univariate Box-Jenkins Models: Concepts and Cases
PORT · Theoretical Probability for Applications
PUTERMAN · Markov Decision Processes: Discrete Stochastic Dynamic Programming
RACHEV · Probability Metrics and the Stability of Stochastic Models
RÉNYI · A Diary on Information Theory
RIPLEY · Spatial Statistics
RIPLEY · Stochastic Simulation
ROSS · Introduction to Probability and Statistics for Engineers and Scientists
ROUSSEEUW and LEROY · Robust Regression and Outlier Detection
RUBIN · Multiple Imputation for Nonresponse in Surveys
RYAN · Statistical Methods for Quality Improvement
SCHUSS - Theory and Applications of Stochastic Differential Equations
SCOTT · Multivariate Density Estimation: Theory, Practice, and Visualization
SEARLE · Linear Models
SEARLE · Linear Models for Unbalanced Data
SEARLE · Matrix Algebra Useful for Statistics
SEARLE, CASELLA, and McCULLOCH · Variance Components
SKINNER, HOLT, and SMITH · Analysis of Complex Surveys
STOYAN, KENDALL, and MECKE · Stochastic Geometry and Its Applications, *Second Edition*
STOYAN and STOYAN · Fractals, Random Shapes and Point Fields: Methods of Geometrical Statistics

Applied Probability and Statistics (Continued)

THOMPSON · Empirical Model Building
THOMPSON · Sampling
TIERNEY · LISP-STAT: An Object-Oriented Environment for Statistical Computing and Dynamic Graphics
TIJMS · Stochastic Modeling and Analysis: A Computational Approach
TITTERINGTON, SMITH, and MAKOV · Statistical Analysis of Finite Mixture Distributions
UPTON and FINGLETON · Spatial Data Analysis by Example, Volume 1: Point Pattern and Quantitative Data
UPTON and FINGLETON · Spatial Data Analysis by Example, Volume II: Categorical and Directional Data
VAN RIJCKEVORSEL and DE LEEUW · Component and Correspondence Analysis
WEISBERG · Applied Linear Regression, *Second Edition*
WESTFALL and YOUNG · Resampling-Based Multiple Testing: Examples and Methods for *p*-Value Adjustment
WHITTLE · Optimization Over Time: Dynamic Programming and Stochastic Control, Volume I and Volume II
WHITTLE · Systems in Stochastic Equilibrium
WONNACOTT and WONNACOTT · Econometrics, *Second Edition*
WONNACOTT and WONNACOTT · Introductory Statistics, *Fifth Edition*
WONNACOTT and WONNACOTT · Introductory Statistics for Business and Economics, *Fourth Edition*
WOODING · Planning Pharmaceutical Clinical Trials: Basic Statistical Principles
WOOLSON · Statistical Methods for the Analysis of Biomedical Data

Tracts on Probability and Statistics

BILLINGSLEY · Convergence of Probability Measures
KELLY · Reversibility and Stochastic Networks
TOUTENBURG · Prior Information in Linear Models